SEMI-ARID CLIMATE CHANGE

SEMI-ARID CLIMATE CHANGE

Jianping Huang
College of Atmospheric Sciences, Lanzhou University, China

Xiaodan Guan
College of Atmospheric Sciences, Lanzhou University, China

Haipeng Yu
Northwest Institute of Eco-Environment and Resources, CAS, China

World Scientific

NEW JERSEY • LONDON • SINGAPORE • BEIJING • SHANGHAI • HONG KONG • TAIPEI • CHENNAI • TOKYO

Published by

World Scientific Publishing Co. Pte. Ltd.
5 Toh Tuck Link, Singapore 596224
USA office: 27 Warren Street, Suite 401-402, Hackensack, NJ 07601
UK office: 57 Shelton Street, Covent Garden, London WC2H 9HE

Library of Congress Control Number: 2023020371

British Library Cataloguing-in-Publication Data
A catalogue record for this book is available from the British Library.

SEMI-ARID CLIMATE CHANGE

ISBN 978-981-127-617-0 (hardcover)
ISBN 978-981-127-618-7 (ebook for institutions)
ISBN 978-981-127-619-4 (ebook for individuals)

For any available supplementary material, please visit
https://www.worldscientific.com/worldscibooks/10.1142/13407#t=suppl

Desk Editors: Aanand Jayaraman/Amanda Yun

Typeset by Stallion Press
Email: enquiries@stallionpress.com

Preface

In June 2004, nearly 20 years ago, the College of Atmospheric Sciences was established at Lanzhou University, China, and I was appointed as its first dean. For a new college, there were many tasks to be addressed, and the most pressing of these tasks was to improve teaching conditions and create a research platform. Compared with the superior research and social environment in Eastern China, the working and living conditions in Northwest China were difficult. However, with years of experience in atmospheric science research, I realized that there were unique scientific advantages in Northwest China. Gansu is located in a typical semi-arid region of Northwest China and is a great area for studying climate change. If we equipped ourselves with advanced observation devices, we could obtain considerable research results. My team and I have spent twenty years building an international-level observation site. Because of this site, the College of Atmospheric Science at Lanzhou University has become an important research center.

It was difficult for me to make the decision to establish a modern climate and environment observatory at the top of Cuiying Mountain on Lanzhou University's Yuzhong Campus, as much funding was required to support it. The observatory was initially constructed under highly challenging circumstances; without running water or a passable road. Teachers and students stored rainwater for domestic use and used rickshaws to transport observation devices. After arduous efforts, we built the "Semi-Arid Climate and Environment Observatory of Lanzhou University" (SACOL), which meets international standards and has been open to the public since 2005. Additionally, we developed a mobile monitoring system suitable for field experiments in Northwest China. After years of hard work in Northwest

China, my colleagues and I have obtained encouraging research results, which have been acknowledged and accepted by our peers both locally and internationally.

With firsthand, high-quality observations from the site, we have the opportunity to obtain breakthrough results. I have led my team to obtain outstanding results in dryland climate research. We have elaborated on the direct, indirect, and semi-direct effects of dust aerosols on the semi-arid climate in Northwest China, and highlighted the semi-direct effect of dust aerosols and their aridification mechanism. From the perspectives of surface energy balance and the ice cloud heating effect, we propose a thermodynamic mechanism of enhanced warming in semi-arid regions, the land–atmosphere feedback mechanism of expansion, and the important role of the carbon cycle in the aridification process. Our research extends the scope from arid and semi-arid regions in Northwest China to global drylands. Based on long-term observations and data collection, our basic observation and theoretical results enrich the scope of global research on semi-arid regions. We have gradually developed a theoretical system for semi-arid climate change. Our research results have attracted considerable attention both locally and internationally, and won the second prize in 2013 at the National Natural Science Awards.

This book provides a summary of our research. The two coauthors, Professors Xiaodan Guan and Haipeng Yu, have been working on semi-arid climate change since they were my students and both have become experts in this field. In addition to the two coauthors, this book is enriched by the outstanding work of Professors Qiang Fu, Lei Zhang, and members of our team, including Professors Zhongwei Huang, Yuzhi Liu, Tianhe Wang, and Siyu Chen; Drs. Yongkun Xie, Dongdong Li, Zhiyuan Hu, and Ruixia Guo; and Mr. Tiangang Yuan. During the preparation of this book, Ms. Li Zou organized and supported the work and assisted in translation and proofreading. Ms. Wenmin Wei, Ms. Xiaoyou Zhou, Ms. Yu Ren, Mr. Yihui Xu, Dr. Yun Wei, Mr. Xiaolu Fan, Ms. Li Fu, Mr. Ming Peng, Mr. Xiaoyue Liu, Mr. Lei Ding, and Drs. Guolong Zhang, Fan Yang, Peng Zhao, and Wei Pu assisted in the preparation of Chapters 2–9. They created some of the figures, typed the text, and proofread the chapters. Some of them also suggested additions or corrections to parts of the book. Ms. Xiaojuan Bai created numerous figures for the book. Ms. Minna Zhao designed the cover and figures. Ms. Xuwen Chi formatted the book. I would also like to express my sincere gratitude to World Scientific Publishing Group for their support.

This book is a specialized textbook for graduate students of Earth sciences in higher education institutions, but it is also suitable for undergraduate or graduate students in related majors, as well as those engaged in researching atmospheric science, climate change, and Earth sciences. Due to the wide scope and rich subject matter of climate change research, as well as the shortage of time and limitations of the authors' knowledge, errors and omissions are unavoidable, and readers are invited to suggest corrections and provide critiques for revision in the reprint.

About the Authors

Dr. Jianping Huang is a distinguished professor and founder of the College of Atmospheric Sciences, Lanzhou University, China. He received his Ph.D. degree from Lanzhou University in 1988. His research focuses on semi-arid climate change by combining observations, numerical modeling, and theoretical studies. He established the theoretical framework of a semi-arid climate change mechanism that emphasizes the effects of aerosol-cloud-precipitation, land–atmosphere and ocean–atmosphere interactions on climate change in semi-arid regions. Dr. Huang is one of the most cited Chinese researchers according to Elsevier and was among the most highly cited researchers according to Web of Science in 2021 and 2022. His research achievements won the Second National Prize for Natural Science (ranked first) in 2013 and the First National Innovation Award. He has also received eight provincial and ministerial awards.

Dr. Xiaodan Guan is a professor at the College of Atmospheric Sciences, Lanzhou University, China. She received her Ph.D. degree from Lanzhou University. Her research focuses on climate change over semi-arid regions, including the influence of the global ocean, regional land and anthropogenic activities. Dr. Guan teaches classes on physical climatology and has authored more than 80 peer-reviewed journal articles. She serves as an associated editor in the field of climate change for the journal *Land Degradation & Development* and is a member of the youth editorial board of *Plateau Meteorology*. Her scientific projects have been funded by the Excellent Young Scientists Fund, General Program, and Young Scientists Fund. She is also a major contributor of the "Fourth National Assessment Report on Climate Change".

Dr. Haipeng Yu is a professor at the Northwest Institute of Eco-Environmental and Resources, Chinese Academy of Sciences. He received his B.S. and Ph.D. degrees from Lanzhou University, China. He is currently engaged in research on dryland climate change, focusing on dryland dynamic and thermal processes and climate predictions. Dr. Yu has published more than 40 SCI papers which have accrued over 2000 SCI citations. He serves as a member of the youth editorial board of *Plateau Meteorology, Journal of Glaciology* and *Geocryology*, and *Journal of Desert Research.* He is also a major contributor of the "Fourth National Assessment Report on Climate Change".

Acknowledgments

This book is supported by the National Natural Science Foundation of China (Grant Nos. 41991231, 42041004, and 91937302).

Jianping Huang
January 25, 2023

Contents

List of Figures

List of Tables

List of Abbreviations

AI	Aridity Index
AMO	Atlantic Multi-decadal Oscillation
AO	Arctic Oscillation
AOD	Aerosol Optical Depth
BC	Black Carbon
CCN	Cloud Condensation Nuclei
CMIP5	The Fifth Coupled Model Intercomparison Project
CN	Condensation Nuclei
DCCs	Deep Convective Clouds
DIT	Dynamic Induced Temperature
DMO	Decadal Modulated Oscillation
DRE	Direct Radiative Effect
DRH	Dust-Forced Radiative Heating
DTR	Diurnal Temperature Difference
EASM	East Asian Summer Monsoon
EEMD	Ensemble Empirical Mode Decomposition
ENSO	El Niño–Southern Oscillation
EOF	Empirical Orthogonal Function
ESAW	Enhanced Semi-Arid Warming
ESI	Ecological Security Index
ET	Evapotranspiration
EWT	Equivalent Water Thickness
GDP	Gross Domestic Product
GDVI	Global Desertification Vulnerability Index
GHG(s)	Greenhouse Gas(es)

GMSW	Global Mean Surface Warming
GPH	Geopotential Height
GPP	Gross Primary Productivity
ICV	Internal Climate Variability
IMFs	Intrinsic Mode Functions
IN	Ice Nuclei
IOBM	Indian Ocean Basin Mode
IOD	Indian Ocean Dipole
IPCC	Intergovernmental Panel on Climate Change
IPO	Interdecadal Pacific Oscillation
IWP	Ice Water Path
LAI	Leaf Area Index
LH	Latent Heat Flux
LUCC	Land Use/Cover Change
LW	Longwave
LWP	Liquid Water Path
MLR	Multiple Linear Regression
MTF	Meridional Thermal Forcing
NAO	North Atlantic Oscillation
NH	Northern Hemisphere
NPP	Net Primary Production
OC	Organic Carbon
PBL	Planetary Boundary Layer
PDO	Pacific Decadal Oscillation
PDSI	Palmer Drought Severity Index
P-E	Precipitation Minus Evaporation
PET	Potential Evapotranspiration
PNA	Pacific–North America Pattern
RCP(s)	Representative Concentration Pathway(s)
RF	Radiative Forcing
RFT	Radiative Forced Temperature
SACOL	Semi-Arid Climate and Environment Observatory of Lanzhou University
SAT	Surface Air Temperature
SH	Sensible Heat Flux
SLP	Sea-Level Pressure
SOC	Soil Organic Carbon
SSA	Single-Scattering Albedo

SST(s)	Sea Surface Temperature(s)
SW	Shortwave
SWC	Supercooled Water Cloud
TCZ	Transitional Climate Zone
TD	Taklimakan Desert
TOA	Top of Atmosphere
TP	Tibetan Plateau
TWS	Terrestrial Water Storage

Chapter 1

Introduction

Semi-arid regions, the transitional zone between arid and humid regions, are ecological security barriers that prevent arid regions from developing into deserts. The ecological environment of semi-arid regions is vitally important to the global ecosystem due to its ability to defend against further desertification. In addition, the ecological environment is the foundation for human survival and development, and its changes directly affect the rise and fall of human civilization. The development of the ecological environment is closely tied to the future of humankind, and by protecting the ecological environment, it is possible to experience "pains for the moment, gains for the millennia." It is necessary for people to eliminate environmental deterioration at the source and contribute to global ecological security. The protection of ecosystems should begin with protecting ecological security barriers, preventing the expansion of desertification, and studying semi-arid climate change. More than 38% of the world's population lives in semi-arid regions, and 90% of them reside in developing countries (Huang *et al.*, 2017c). Semi-arid regions comprise approximately 20.3% of the world's terrestrial land surface (Huang *et al.*, 2016b; Armah *et al.*, 2010). The low fertility of semi-arid soils makes these regions particularly vulnerable to degradation caused by climate change and human activities (Zhou *et al.*, 2016). The economy and society, particularly in developing countries, are significantly impacted by climate change in semi-arid regions (Huang *et al.*, 2017a). Thus, the top priority for climate research is to improve our knowledge and prediction of semi-arid climate change (Qi *et al.*, 2012).

The climate system of semi-arid regions is a complicated system involving the Earth's atmosphere, hydrosphere, biosphere, and anthroposphere. It is characterized by high temperatures, low precipitation, high occurrence of drought, relatively longer summers, low soil moisture and terrestrial water storage, and stronger potential evapotranspiration. These characteristics suggest that semi-arid regions are relatively more sensitive to global climate change and have undergone obvious changes in past decades. Moreover, interactions between different spheres of the Earth's system are significant over semi-arid regions, especially the anthroposphere effect, which shows violent changes in climate variables, and is more pronounced than that in humid regions. Both observations (Huang *et al.*, 2016a; Trenberth *et al.*, 2014a; Dai and Zhao, 2017) and model simulations (Fu and Feng, 2014; Zhao and Dai, 2017) suggest that global aridity has increased since approximately 1950, and this drying trend may continue in the 21st century due to global warming, particularly over arid and semi-arid regions. A significant portion of arid and semi-arid regions are expected to experience a relatively drier climate, and the expansion of arid and semi-arid regions is projected by 2100, primarily because of the significant potential evapotranspiration (PET) increase associated with rising surface temperature (SAT) (Fu *et al.*, 2016; Huang *et al.*, 2017a). In arid and semi-arid regions, especially in semi-arid regions, variations in temperature and precipitation may affect terrestrial ecosystems (Klein *et al.*, 2017; Wu *et al.*, 2016), hydrology (Angeler *et al.*, 2014; Hu *et al.*, 2015), and agricultural production (Valizadeh *et al.*, 2014; Hadgu *et al.*, 2015). Owing to the acute climate change and fragile ecosystems of semi-arid regions, more research on semi-arid climate change and its impact is necessary for policymaking.

On several temporal and spatial scales, the causes of semi-arid climate change have been examined (Huang *et al.*, 2016a). Focusing on semi-arid climate change around the world, Huang *et al.* (2016a) found that the semi-arid regions over East Asia, the Sahel in West Africa, Southern Africa, and Eastern Australia were becoming drier from 1948 to 2008, while those over Central Asia, Northern Australia, North America, and South America were becoming wetter. The spatial distribution of the precipitation trend agrees with that of the aridity index (AI). According to Cai *et al.* (2012), semi-arid regions of the Southern Hemisphere have been experiencing a drying tendency in austral autumn since the late 1970s, primarily in April and May. These regions include Southern Coastal Chile, Southern Africa, and Southeast Australia.

The changes in regionally averaged precipitation, PET, and AI over semi-arid regions in East Asia, Central Asia, and North America were also examined by Huang *et al.* (2016a). Since newly formed semi-arid regions evolved from a different climate, they differ from semi-arid regions that have existed for a long time. Therefore, each region is classified into old and newly formed semi-arid regions. Over the newly developed and existing semi-arid regions, annual precipitation, PET, and AI showed continuous increasing and declining patterns from 1948 to 2008. Precipitation fell in East Asia but rose in North America. Conversely, PET rose in East Asia but fell in North America. Therefore, the results for AI and precipitation agree, indicating that the climate is getting drier over East Asia and wetter over North America. These trends are highly correlated with the weaker East Asian summer monsoon (EASM) and strengthened westerlies. The multi-decadal sea surface temperature (SST) oscillations in the Pacific have been responsible for the wetting tendency over North America since the 1950s (Dai, 2013a).

Both natural decadal modulation oscillation (DMO) (Huang *et al.*, 2017a) and influences caused by humans, such as elevated greenhouse gas (GHG) emissions (Rotenberg and Yakir, 2010), have been explored in the cited studies. Oceans exert a considerable influence on regulating the climates of arid and semi-arid regions in the Sahel (Hoerling *et al.*, 2006; Lu, 2009), Asia (Fu *et al.*, 2008; Zhao and Zhang, 2015), and North America (H. Li *et al.*, 2010), according to atmospheric modeling studies with specified SSTs. As a result of climate change and human mismanagement, some semi-arid regions are at risk for desertification (long-term land degradation) and further expansion (Huang *et al.*, 2017b). Moreover, many other human activities, such as land use, may cause climate change (Li and Zhao, 2012). In recent decades, the world's ecosystems in arid and semi-arid regions have faced additional challenges due to population growth and migration, as well as changes in land-use patterns (Hoover *et al.*, 2015). Land degradation, caused by surface and soil water deficits, poses a severe threat to semi-arid regions (Huang *et al.*, 2017b). In these regions, semi-arid climate change is also threatening food security and water availability (Huang *et al.*, 2016a, 2016b, 2017a, 2017b).

This book aims to build a theoretical framework for research on semi-arid climate change. It provides an overview of recent progress in semi-arid climate change research, and systematically summarizes the temporal and spatial characteristics of climate change over semi-arid regions. This book focuses on the following areas: (1) the enhanced warming of semi-arid regions, (2) the impacts of dust–cloud–precipitation interactions, land–atmosphere interactions, ocean–atmosphere interactions, and human activity on semi-arid regions, (3) the dynamics of the semi-arid climate, (4) desertification in semi-arid regions and the responses of semi-arid ecosystems to the changing climate, and (5) future projections of semi-arid climate change.

This book consists of 10 chapters. Chapter 1 provides an introduction to the whole book, with a review of the progress and challenges in semi-arid climate change research. In the final chapter, a summary and some perspectives on future research on semi-arid climate change are proposed. The contents of other chapters will be introduced as follows.

Chapter 2 reviews climate changes in semi-arid regions from the perspectives of semi-arid regions, the semi-arid climate system, and aridity changes. Semi-arid regions are the most sensitive climate regions in the world. At present, the main climate classification indexes are the Köppen–Geiger climate classification index, Palmer drought index, and AI, and semi-arid regions are mainly concentrated at mid-latitudes in the Northern and Southern Hemispheres. These semi-arid regions exhibit a noticeable expansion trend under the effect of global warming, among which the expansion of Asian arid semi-arid regions is the most significant. The climate system of semi-arid regions mainly includes the atmosphere, hydrosphere, biosphere, and anthroposphere. As a result of interactions between different spheres of the Earth system, changes in the climate system of semi-arid regions have become more significant and sensitive, which have been clearly reflected in precipitation, AI, potential evaporation, soil moisture, lake temperature, terrestrial water storage, and seasonal changes during the process of aridity changes.

Chapter 3 describes the enhanced warming of semi-arid regions. Arid and semi-arid regions account for approximately one-third of the global area and are mainly concentrated at mid-latitudes. These areas are in the sinking branch of the three-cell circulation model. In the past few decades, arid and semi-arid regions have shown a strong expansion trend, which is reflected in the transition from subhumid regions to semi-arid regions. These regions are characterized by obvious climate change, among which the most significant factor is enhanced warming. In different arid and semi-arid regions, the degree of enhanced warming varies, and the largest amount of warming was observed over the semi-arid regions of Asia. The two key contributors to enhanced warming are dynamic temperature and thermal temperature. Dynamic temperature is closely related to large-scale circulation and ocean activities, while thermal temperature is significantly influenced by local factors. Thermal temperature is most pronounced in Asia, where human activities play an important role in the thermal warming process. In arid and semi-arid regions, enhanced warming occurs, and a series of variables related to energy and the water cycle are noticeably changed, such as daily maximum temperature, daily minimum temperature, daily range of temperature, precipitation, evapotranspiration, drought index, soil moisture, and groundwater. Among them, the variables that are closely related to energy show an apparent increasing trend, and the variables related to the water cycle show an apparent decreasing trend. These changes will further intensify the local energy cycle, weaken the water cycle, and reduce local water resources.

Chapter 4 discusses the impact of dust–cloud–precipitation interactions on semi-arid regions. Mineral dust particles are widely dispersed from arid and semi-arid regions and have significant direct, indirect, and semi-direct effects on the local energy balance (Huang *et al.*, 2006a, 2009, 2010b; Twomey, 1977; Albrecht, 1989; DeMott *et al.*, 2003). In fact, these particles alter the hydrological cycle through radiative forcing (Huang *et al.*, 2014) and have an impact on the microphysical characteristics of clouds (Huang *et al.*, 2006a, 2006b, 2010b). In addition, dust aerosols can serve as a means of long-distance nutrient delivery to oceans and terrestrial land surfaces (Okin *et al.*, 2011; Das *et al.*, 2013). Dust aerosols over arid and semi-arid regions in Asia have a warming effect within the atmosphere and a cooling effect at the surface, similar to the Sahel and Sahara. Due to higher surface albedos, Asian arid and semi-arid regions usually experience warming (Huang *et al.*, 2009; Ge *et al.*, 2010b), which is in contrast to dark surfaces. Asian dust plays a significant role in the global biogeochemical cycle by serving as a potential source of iron to marine ecosystems in the North Pacific (Mahowald *et al.*, 2010).

Chapter 5 presents the impacts of land–atmosphere interactions on semi-arid climate. Through turbulence at the planetary boundary layer, land-atmosphere interactions influence the energy balance, water cycle, and carbon cycle. They are crucial in climate change, particularly in arid and semi-arid regions. The driving role of oceanic warming and related atmospheric processes is highlighted in many mechanisms proposed to affect the climate changes occurring in arid and semi-arid regions; however, notably, the AI response is significantly increased by atmosphere–land feedbacks that respond to climate changes (Berg *et al.*, 2016). Drought-induced land surface changes in arid, semi-arid and subhumid regions can result in positive feedback that strengthens and prolongs droughts. Obviously, understanding how the energy, water, and carbon cycles respond to global warming is essential for understanding and forecasting future changes in semi-arid climate. Through the responses of the energy, water, and carbon cycles under climate change, enhanced semi-arid warming and decreased AI were observed (Huang *et al.*, 2012; Berg *et al.*, 2016). These effects were amplified by land–atmosphere interactions. The ability of the ecosystem to store carbon is dramatically impacted by changes in temperature and soil moisture, since desertification reduces soil water content and raises soil temperature (Sjögersten and Wookey, 2016). As a result, enhanced semi-arid warming and land–atmosphere interactions create a positive feedback loop that exacerbates global warming (Huang *et al.*, 2017a).

Chapter 6 discusses the impacts of ocean-atmosphere interactions on semi-arid climate. As the dominant factor in global climate change, oceanic activities are mainly represented by the El Niño–Southern Oscillation (ENSO), North Atlantic Oscillation (NAO), Pacific Decadal Oscillation (PDO), and Atlantic Multidecadal Oscillation (AMO). These oceanic activities affect the climate of arid and semi-arid regions on different temporal and spatial scales, with the effects of synergistic decadal oscillation being the most prominent. The effects of different oceanic factors can cause significant spatial differences in temperature, precipitation, and wet and dry distributions. Moreover, different oceanic factors interact with each other, resulting in compound spatial differences in impacts. In addition to oceanic oscillations, ocean–land thermal differences are crucial factors that affect climate change in semi-arid regions on a large scale. With global warming, the ocean–land thermal difference shows obvious changes. The change in ocean–land thermal difference, a key factor in large-scale circulation changes, further increases the meridional thermal difference, encourages the development of winter high-pressure blocking, and promotes more cold air events. Due to the differences in warming at different latitudes, the dimensional thermal difference continues to narrow, which is beneficial for increasing the frequency and intensity of cold air in winter.

Chapter 7 introduces the impacts of human activities on semi-arid climate. Semi-arid climate change is significantly influenced by anthropogenic activities, and anthropogenic activities are partly to blame for environmental degradation (Huang *et al.*, 2016a, 2017a). The rapid growth of the economy has increased the consumption of fossil fuels, which has increased energy output and GHG emissions (Barnett and O'Neill, 2010). Anthropogenic activities, especially in higher-density GHG scenarios, are responsible for the enhanced warming in semi-arid regions (Huang *et al.*, 2017a). Accompanied by the increase in GHG emissions, additional warming will bring about a series of terrible disasters to people living in arid and semi-arid regions, including worsening drought, reduced agricultural production, and increased disease. In addition, the rapid growth of GHG emissions will accelerate the expansion of arid and semi-arid regions and increase occurrences of extreme weather. Along with the increase in GHG emissions, irrational land use by humans has intensified over the past century, causing serious damage to the oxygen balance of the Earth's ecosystem and raising the risk of hypoxia. The water reserves of the Earth system are also at risks from direct human activities, such as excessive groundwater exploitation, which exacerbates water shortages in arid regions.

Chapter 8 illustrates the dynamics of the semi-arid climate system. Semi-arid climate change is a coupling of long-term trends caused by external forcing and interdecadal oscillations modulated by oceanic processes, the combination of which results in a long-term drought trend and a regional wet and dry oscillation. From the global point of view, the water cycle is accelerated under climate warming, which has expedited water dissipation in arid and semi-arid regions. This has resulted in a notable decrease in terrestrial evapotranspiration and water storage, as well as an overall trend of aridity intensification and expansion in arid and semi-arid regions. At the regional scale, oceanic processes perform a modulated function for regional wet and dry oscillations. Asymmetric meridional thermal forcing (MTF) and zonal thermal forcing (ZTF) can be altered by internal oceanic climate variability modes, which can also excite a global decadal modulated oscillation (DMO) (Huang *et al.*, 2017a). The DMO's modulation effect on SST was strongest when both weaker (stronger) MTF and stronger (weaker) ZTF occurred simultaneously. The combination of different phases of SST resulted in atmospheric circulation anomalies, causing interdecadal oscillations and significant regional differences in dry and wet changes in arid and semi-arid regions. In this chapter, we explore the dynamic mechanisms of semi-arid climate change theoretically.

Chapter 9 discusses the degradation of semi-arid regions. Arid and semi-arid regions dominate the interannual and long-term trends of the global carbon cycle in terrestrial ecosystems, with sparse vegetation, drastic climate change, and high ecosystem vulnerability. Desertification is a phenomenon of land degradation in arid and semi-arid regions caused by climate change and human activities. It not only has an important impact on the global ecological environment but also affects the global economy and sustainable development of human society. The construction of the global desertification vulnerability index (GDVI) is essential for risk assessment and advance policymaking, which can prevent damage from occurring and reduce the damage caused by desertification disasters. The GDVI that applies to the global scale considering climate change and human activities is helpful to visually reflect the global desertification risk distribution, compare the desertification risk of different regions under the same evaluation system, and provide a scientific basis for suppressing desertification expansion. When vegetation degrades, its capacity for photosynthesis, oxygen production, and water storage decreases, and it will spread to nearby areas. This phenomenon is known as "land cancer" and poses a severe danger to terrestrial ecological security. Reducing the amount of CO_2 produced by human activities is crucial to mitigating climate change. Desert ecosystems, which are ecosystems mainly concentrated in arid and semi-arid regions, cannot be ignored in the global carbon cycle, although their carbon sequestration capacity is relatively weak compared to that of ecosystems, such as forests and grasslands. Desertification has become a key environmental issue worldwide, and many countries are actively engaged in combating desertification. Due to the vast area of arid and semi-arid regions, artificially planted trees in these regions are also considered to have an important role in absorbing CO_2 from human activities, but their potential needs to be further evaluated.

Presently, climate change and its impact on economic and social development are a major concern for governments and scientific communities around the world. Arid and semi-arid regions are home to some of the most impoverished and underdeveloped countries with fragile social stability, lagging economic development and high environmental vulnerability. Furthermore, they are also key regions for achieving sustainable development, and five of the seventeen Sustainable Development Goals are directly related to the climate, environment, and ecosystem in arid and semi-arid regions. However, our current knowledge is insufficient to address all the scientific issues related to climate change in arid and semi-arid regions, including the significance of human forcing factors and natural factors. With climate change emerging in a seemingly endless stream, it is only by bolstering the range and depth of climate change research and continuously advancing our knowledge that we can fundamentally understand the laws of climate change and capture its essence from complex phenomena. To meet the demand for training talent in this field, there is an urgent need for a textbook on the basic principles and methods of semi-arid climatology. This book is suitable for undergraduates and graduate students in related disciplines, as well as those working in atmospheric science, climate change, and Earth sciences.

Chapter 2

Semi-Arid Climate Changes

2.1 Introduction

Around 15% of the Earth's land surface is covered by semi-arid regions, which in 2000 were home to 14.4% of the world's population (Hassan *et al.*, 2005). Most of the semi-arid regions can be found in Central and Western United States, the majority of Mexico, the west coast and northeastern corner of South America, the Sahel region and the southern part of North Africa, Northern Kazakhstan and Mongolia, Western India, some areas of Northern China, and most of the desert region outside of Central Australia. The surface vegetation types in the semi-arid zone are numerous, but grassland predominates (making up around 41.6% of the area of the semi-arid region), followed by open shrubland and farmland. The climate types in the semi-arid zone are mostly arid and semi-arid grassland climates.

Semi-arid regions have three different vegetation types, which are classified as tropical semi-arid, temperate semi-arid, and cold semi-arid. The primary features of the vegetation in these regions are that it is often sparse and composed primarily of annual plants with short growth cycles and tolerance to drought. Most of the semi-arid regions in the tropics are in the savannah zone, where grasses and shrubs predominate, the root structure is typically deeper, and the leaves are smaller (reducing evaporation). The semi-arid area in the temperate region mostly belongs to the temperate grassland zone, and pasture is the dominant plant type. For instance, the Inner Mongolia Plateau in China is a typical semi-arid temperate grassland region. The semi-arid area in the cold areas is mostly part of the boreal tundra zone, where the predominant plant is moss, and the vegetation and soil are extremely sterile.

The average annual precipitation over past 109 years has varied spatially, ranging from less than 10 mm yr^{-1} to a maximum of more than 1,200 mm yr^{-1}. The long-term average, frequency, and extremes of a number of meteorological factors, most notably temperature and precipitation, form a region's climate (Giddings *et al.*, 2005). The long-term mean precipitation is the most straightforward metric for categorizing climatic areas, despite the fact that precipitation and surface temperature are connected. The areas with precipitation ranging from 200 to 600 mm yr^{-1} are referred to as semi-arid zones. Due to the inability of moisture-bearing winds to penetrate and cool such areas, semi-arid regions operate as transition zones between arid and subhumid belts. These regions are characterized by low and restricted precipitation. Areas with precipitation less than the potential evaporation are sometimes referred to as semi-arid regions. Large portions of Western North America and Central Asia are covered by mid-latitude semi-arid climatic zones. The majority of North America's semi-arid area is found in the west. The majority of Central Asia's semi-arid region is situated in an area that is in between desert and semi-wet climates. These semi-arid regions often have temperatures that are similar to those of dry or desert mid-latitudes. Mid-latitude arid regions get somewhat less precipitation than semi-arid mid-latitude regions.

Due to the inherent characteristics of the global atmospheric circulation, the inter-annual variability of rainfall in semi-arid regions is unusually large. Most semi-arid regions are experiencing warming at rates above the global mean rate over land, rising evaporative demand, and less consistent rainfall, according to observed and anticipated climatic changes. Due to increased evaporation and decreased precipitation, both observations and modeling studies suggest that in locations already prone to dryness, droughts may become more severe and last longer in a warmer climate (Dai, 2013a). As a result of global warming, drying trends may be most pronounced in semi-arid and arid regions, intensifying the hydrological cycle and making climatologically wet regions wetter and dry regions drier (Chou *et al.*, 2009). By 2100, the area of drylands worldwide is predicted to increase by around 10%, except for India and Northern Tropical Africa, which would experience a wetter climate (Feng and Fu, 2013). These areas have also seen decadal-scale climate anomalies and land surface degradation, with the risk of land degradation being highest in the middle of the aridity gradient, which is primarily comprised of drylands (Hassan *et al.*, 2005).

Such locations have delicate ecosystems that are highly susceptible to interactions between human activity and climate change (Charney, 1975;

Huang *et al.*, 2010b). Most human impacts on semi-arid climates are caused by the use of land for agriculture. Several economic or societal considerations may be responsible for the decision to introduce agriculture into these areas, whose natural state is already marginal for agriculture (Liu *et al.*, 2008a). These include the necessity to feed a growing population, the declining productivity of the land now used for agriculture, and the desire for financial gain. Many semi-arid communities depend on rain-fed agriculture, and changes in precipitation have a significant impact on the availability of surface water resources needed to support agricultural irrigation. For instance, a small change in temperature or rainfall patterns could have severe effects. A large portion of the arid and semi-arid area has also undergone significant anthropogenic modification as a result of urbanization and changes in land use (Gao *et al.*, 2010).

Arid and semi-arid regions are at the edge of sustainability and are frequently threatened by severe water stress since they are some of the most environmentally delicate and economically vulnerable places (Huang *et al.*, 2016b). Recent research has revealed that most of the world's diminishing hotspots of terrestrial water storage (TWS), including surface waters, glaciers, canopy water, soil moisture, and groundwater, are in arid and semi-arid regions (Chang *et al.*, 2020; Rodell *et al.*, 2018; Wang *et al.*, 2018). In arid and semi-arid regions, the lack of water has several negative consequences for the health of the ecosystem, but it also has a negative impact on residents' socioeconomic growth. Therefore, it is crucial to promote stability and sustainability on a global scale to have a better understanding of the factors that contribute to TWS changes in arid and semi-arid regions, particularly in the setting of humid regions. The impacts of the three mechanistic drivers are frequently entangled, and the relative relevance of each driver varies among locations, which presents a significant barrier for correct attribution of TWS changes. As a result, there is still a lack of global-scale analysis for the mechanical causes of TWS shifts across arid and semi-arid regions.

The arrangement of this chapter is as follows. Section 2.1 is the introduction to this chapter. Section 2.2 illustrates the definition of semi-arid regions. Section 2.3 discusses the semi-arid climate system, including the atmosphere, hydrosphere, biosphere, and the Anthropocene. Section 2.4 depicts the characteristics of aridity changes in semi-arid regions. Section 2.5 provides a comprehensive analysis of the TWS changes in both semi-arid regions and humid regions around the world. Conclusions and discussion are summarized in Section 2.6.

2.2 Definition of Semi-Arid Regions

Large-scale topography and atmospheric general circulation play a major role in determining the locations of arid and semi-arid regions and arid climates, specifically the subsidence of the Hadley circulation. Since they are linked to the primary elements supporting aridity, the subtropical highs are crucial in the formation of arid and semi-arid region climates in the mid-low latitudes (e.g., subsidence, divergent airflow, and thermal stability). The subtropical highs and the descending branch of the Hadley cells are strongly related to the arid and semi-arid regions across Australia, Northern and Southern Africa, the Middle East, and Southwestern North America (Shin *et al.*, 2012). Furthermore, because high terrain creates lee rain shadows and obstructs the movement of rain-bearing disturbances, topography has a significant impact on the aridity of various arid and semi-arid region areas in Asia, South America, the Western United States, and the Horn of Africa. For instance, as a result of the fast elevation of the Tibetan Plateau, enormous arid and semi-arid regions were created in Asia (An *et al.*, 2001). The fact that Central Asia and the Sahara are far away from significant sources of rainfall also contributes to their arid climates. Additionally, Rossby wave responses to the west caused by diabatic heating in monsoon regions can interact with mid-latitude westerlies to create an area of adiabatic descent. According to this process, in addition to local effects, distance alterations in monsoon regions can also have an impact on desertification. In conclusion, synoptic (such as cyclones, anticyclones, and tropical storms) and mesoscale systems (such as sea breezes and mountain winds) interact with large-scale atmospheric circulations and regional topography influences to govern the crucial elements of arid and semi-arid region climates. This section discusses the distribution of several types of arid and semi-arid regions around the world, particularly in China, and presents three markers for classifying arid and semi-arid regions.

2.2.1 *Köppen–Geiger climate classification*

The Köppen–Geiger climatic classification was created at the end of the 19th century and was initially proposed by Köppen (2011). Local temperature and precipitation — two essential climatic variables — are dependent on vegetation so forests are more prevalent in locations with higher temperatures and more rainfall, while sparsely vegetated or desert areas are more common in areas with less rainfall. The approach typically uses temperature thresholds for distinct vegetation development to divide the world into five climate zones. It uses the 0.5° resolution Köppen–Geiger climate classification map for the years 1951–2000 that was created by Kottek *et al.* (2006). At http://koeppen-geiger.vu-wien.ac.at/present.htm, both the map and the underlying digital data are accessible online. Many climate zones around the world were depicted on Köppen's map using various hues and tones. Most of the zones are arranged according to local temperature. The zones are as follows: tropical or equatorial zone (most maps show this zone in blue), arid or dry zone (most maps show this zone in red, pink, and orange), warm/mild temperate zone (most maps show this zone in green), continental zone (most maps show this zone in purple, violet, and light blue), and polar zone (most maps show this zone in gray). These climate zones are dispersed globally in intricate patterns. However, there are certain broad generalizations that one can make about their location, such as deserts in subtropical areas, tundra and polar climates close to the poles, and tropical and equatorial savannah climates. The five main categories of the Köppen–Geiger climate classification are mentioned, along with their distinctive distribution features. Low latitudes, primarily between 15°N and 15°S, are where Köppen's A climates can be found in an almost continuous strip around the globe. A quarter of the Earth's land surface is covered by Type B climates (arid and semi-arid climates), which are primarily located between 50°N and 50°S, but primarily in the 15°–30° latitude belt in both hemispheres. The Köppen climatic classification system classifies most of the middle and high latitudes (mainly from 25°N to 70°N and S) as C and D types. The polar and arctic air masses at high latitudes (60°N and 60°S and higher) govern Köppen's type E climates.

Both advantages and disadvantages of the Köppen Climate Classification System exist. Along with its division, the Köppen climatic structure is quantitative in nature. Both temperature and precipitation are important in relation to the Köppen climate classification system. The Köppen climate classification system is so comprehensive that it truly encompasses all the climate conditions and patterns that can be found on Earth. The classification structure is uncomplicated and obvious. Any region's vegetation pattern matches the Köppen climate classification system for climatic zones. High consistency, significant biological association, and greater applicability can be seen in the climatic and vegetational boundaries, which are more prevalent at low latitudes. The Köppen climate classification has several significant advantages over other classification systems, including strict standards, clear boundaries, easy application, and better suitability for landscape belts.

The primary climate groups of the Köppen climate classification are based on what kinds of flora thrive in a certain climate classification zone. Köppen developed the system based on his expertise as a botanist. The system can be used to categorize climates, examine ecosystem conditions, and determine the primary forms of flora within climates. However, the Köppen climatic classification idea is broad and contains many classifications, making it sometimes difficult to remember. It relies heavily on average values because in order to use it, average temperature and precipitation values are required. The Köppen climate classification system ignores many factors, such as the amount of rain, the frequency of wet days, cloud cover, and changes in temperature or precipitation. The air masses and their importance in the creation of the climate, which play a significant role in climatology, were also disregarded by Köppen in this system of climate classification.

The Köppen climate categorization is not good for extreme events. It has been noted that extreme events, like periodic droughts or extremely cold temperatures, are as crucial in determining the spread of vegetation as the average conditions on which Köppen climate classification is based. Since natural vegetation reacts slowly to climatic changes, the distribution of vegetation that is currently visible is not adapted to the current climate but rather to the previous one. The distribution of plants found in the Köppen climate classification and those found in many regions of the Earth today clearly diverge from one another. The restricted Köppen climate classification ignores the significant annual temperature variation in the eastern section of the Eurasian continent, which makes the dry climate requirements artificial to some extent. The dry climate is not suitable to be divided into an independent climatic zone, as it is caused by a variety of variables and occurs in zones A, C, and D. Moreover, the Köppen climate classification overlooks the vertical distribution of temperature and precipitation at high altitudes and does not consider how altitude affects climate. Therefore, the high-latitude low-precipitation area with annual precipitation less than 500 mm has not been classified as a semi-arid region.

2.2.2 *Palmer drought severity index*

An analysis of soil moisture supply and demand is the foundation of the Palmer Drought Severity Index (PDSI). Demand is more difficult to calculate than supply because it depends on numerous variables, including evapotranspiration and recharge rates in addition to temperature and soil moisture content, both of which are difficult to measure. Meteorologist

Wayne Palmer tried to get around these problems by creating an algorithm that roughly approximated the variables using the two most accessible variables, precipitation and temperature. Using a straightforward water balance model, the PDSI equation always includes temperature and precipitation data to investigate the supply and demand for moisture. Palmer initially published his technique in the 1965 publication Meteorological Drought for the U.S. Weather Bureau's Office of Climatology. Different from the Köppen–Geiger climate classification, the PDSI is good at reflecting extreme events. The indicator has shown to be most accurate in predicting long-term drought, which lasts several months, but it is less accurate in predicting conditions over a few weeks. Drought is depicted using negative integers; for instance, negative 2 represents a moderate drought, negative 3 a severe drought, and negative 4 an extreme drought, and it uses 0 as the reference point. Palmer also created a technique that uses the local temperature and precipitation variability to standardize drought calculations for each specific area. The Palmer index can therefore be used at any location where adequate temperature and precipitation data are available.

The PDSI change exhibits a clear geographic difference, for example, Northern and Southeastern North America, Argentina, the middle of Africa, Northern Europe, and Northern and Central Asia, as well as Northern and Southern Africa, the Middle East, Southeast Asia, Eastern Australia, Brazil, and the Mediterranean region are just a few examples. The rest of the world has not undergone any significant change. However, these originally dry lands have changed into drier, humid lands that have become wetter. As a result, the difference between dry lands and humid regions would widen even more, having a significant impact on ecosystems, crop yields, and water resources. According to Dai *et al.* (1997), much of Eurasia, Southern North America, and Argentina have the highest increases in precipitation. Additionally, Dai *et al.* (2004) concluded that the 20th century has seen an increase in precipitation when they compared the temperature and precipitation in the first and second halves of the 20th century. This also suggests that the primary cause of the aforementioned three regions' increased wetness was higher rainfall. But because these areas are more vulnerable to climate change than others, such as Eurasia, Africa, and Eastern Australia, they warm up more quickly than other areas. In addition, surface warming has significantly increased aridity, making the arid and semi-arid regions drier, particularly in the Sahel and Central Asia. When the temperature is kept constant, the PDSI value will rise by 0.5–1.5.

PDSI has been utilized in numerous studies to distinguish between arid and semi-arid regions and humid lands and to identify the dry/wet types of land throughout the world. However, because the derived data for the original PDSI comes from nine different climatic areas within the United States, it can only be used to assess drought in that country. New PDSI versions, such as the self-calibrating PDSI (sc_PDSI), have emerged in order to use the indices to evaluate drought broadly around the world. According to the sc_PDSI, Dai (2011) divided the world's land regions into nine arid and semi-arid regions, including the Central United States, Central Eurasia, Southeast Australia, Southwest United States, South India, the Amazon, West Europe, Indonesia, and the Sahel region. The remaining humid regions of the world were classified as the rest of the world. Additionally, he examined the differences between sc_PDSI and PDSI using revised formulas for possible evapotranspiration. The outcome of this calculation is expected to be solid. In the meantime, this arid and semi-arid region distribution resembles distributions with other aridity indices. Additionally, research on regional scale application of PDSI is also common.

2.2.3 *Aridity index (AI)*

The precipitation (P) to potential evapotranspiration (PET) ratio is a typical way to describe aridity (Middleton and Thomas, 1997). Increased aridity affects landscapes, ecosystems, and human habitation by causing severe soil moisture deficits, lowering carbon uptake, and expanding the world's arid and semi-arid regions. Drylands are designated as areas with $P/PET < 0.65$, which are sensitive and vulnerable to climate change. The NH's mid-latitudes contain significant arid and semi-arid region areas that are home to a sizable population, with the biggest densities of people found in South and East Asia, North Africa, and the Middle East.

The aridity index (AI) is denoted as follows:

$$\mathrm{AI} = \frac{P}{\mathrm{PET}} \tag{2.1}$$

Feng and Fu (2013) use the Precipitation Reconstruction over Land (PREC/L) dataset, which has a 0.5° resolution and covers the years 1948 to 2008. It was created at the Climatic Prediction Center (CPC) of the National Oceanic and Atmospheric Administration (NOAA). The PET data come from the Climatic Research Unit (CRU) TS 4.06 dataset, which has a 0.5° resolution and spans the years 1901–2021. Feng and Fu (2013) offer the PET data and the observed precipitation.

AI is simple to calculate and useful for historical and contemporaneous climate drought evaluation as well as real-time climate drought monitoring. In order to avoid the one-sidedness that comes from looking at one of the numbers separately, AI highlights the integrated link between precipitation and evapotranspiration. Because of its extremely low temperatures and yearly evapotranspiration capacity, which is occasionally even lower than the annual precipitation, AI can be compared to the Qinghai-Tibet Plateau even if its precipitation is quite minimal. The Qinghai-Tibet Plateau has numerous lakes, as is evident, and these lakes are the result of the last of the precipitation's accumulation. The annual precipitation in Northeast China is even lower than that in the Central Plains; however, some locations can be described as humid regions. On the other hand, the Central Plains can still be classified as arid and semi-arid regions while having annual precipitation that may even surpass 700 mm.

However, AI has a number of drawbacks. The first is that evapotranspiration capacity is an estimated value following observation rather than an evapotranspiration amount or an observed value. The relative humidity of the air in the observation area, the observation value, and the amount of evapotranspiration are all different. Since evapotranspiration is constrained by a variety of factors, the observation values of evapotranspiration in the same region vary depending on factors such as the relative humidity of the surrounding air, whether the slope is sunny or shaded, windy or calm, or whether it is sunny or cloudy. Of course, it is undeniable that the average value of observation data from multiple weather stations in extreme drought areas should be more scientific. This is because there is no liquid water available for evapotranspiration, particularly when all the precipitation evaporates, so the precise amount of evapotranspiration is determined by artificial observation and estimation.

Because of this, the precipitation indicator can only detect vertical precipitation and is unable to detect or account for horizontal precipitation. As long as there are trees, there is constant perennial runoff coming down from the mountains in the Eastern Northwest and Central Plains, but there is almost no perennial runoff in the arid alpine regions. Take the Qinling Mountains as an example. The north slope of the mountains is very steep, the mountains are essentially above 1,200 m in elevation, there is a lot of forest cover and depression, and small rivers flow out as the mountain's runoff heavily and continuously throughout the year, even in the driest month of the year, i.e., July. The difference is very noticeable; several rivers in the Qianshan region, which is barely 30 km away, have dried up.

Another example is the Yan'an area, which was once bare in the desolate mountains with little drainage; but in recent years, the forest cover has substantially increased and many valleys have runoff again. This might be a result of the mountains' high altitude and dense forest cover, which favor the development of horizontal precipitation. However, incoming water weather stations rarely record horizontal precipitation. Horizontal precipitation has the characteristic that it cannot create precipitation but can form runoff, and the runoff that results is clear, small, and persistent.

The exact quantity of the water deficit and the exact amount of stored water cannot be determined by AI. The gap between its demand and availability cannot be precisely determined, which is regrettable because the absolute value is crucial for biological growth, human society, and climatic regulation. The runoff should be subtracted from the overall amount of water, and the demand for agriculture, industry, residential use, forest and wildlife, and groundwater recharge should then be considered. In addition, AI cannot adequately depict how much precipitation falls during each season and month of the year. The year may be wet but a particular month may be dry when there is a lot of annual precipitation, which does not always imply a lot of precipitation in that month. Since the drought index is unable to accurately reflect the volume and severity of each rainfall, it cannot be used to forecast floods or droughts. The water demand and organism shortage are not accurately reflected by the drought index. This indicator is not very useful for directing agricultural production, because the water deficit experienced by organisms occurs primarily during the growing season, which in the north is from April to October.

2.2.4 *Distribution of semi-arid regions*

Figures 2.1(a–c) displays the spatial distributions of drylands (as determined by AI, annual precipitation from 1961 to 1990, or the Köppen–Geiger classification of climate for 1951–2000). Savanna, grassland, and desert are the three main biomes of drylands, although there are no set or exact boundaries (Hassan *et al.*, 2005). The main types of dryland surface vegetation from MODIS are displayed in Fig. 2.1(d). As can be seen in Fig. 2.1(a), the worldwide drylands that AI has described are largely found in the middle and low latitudes and roughly resemble the map from UNEP (Middleton and Thomas, 1997). Major hyper-arid regions include the Atacama Desert, the Gobi Desert, the Taklimakan Desert, the Central and Northern Sahara Desert, the Rub al Khali Desert, the Eastern Arabian

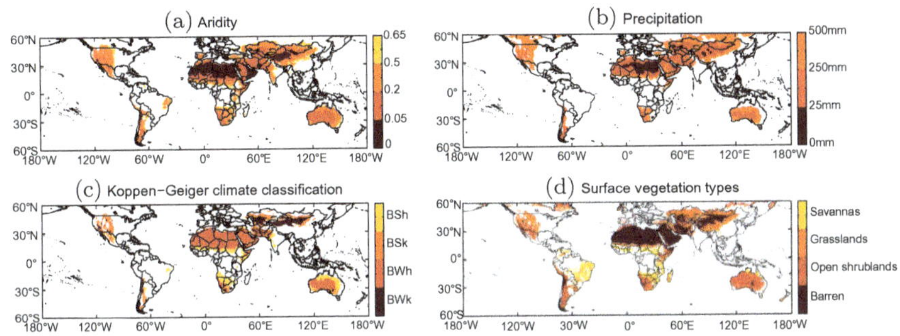

Fig. 2.1. The global distribution of drylands over 1961–1990 according to three definitions: aridity (a), precipitation (b), and Köppen–Geiger climate classification (c); surface vegetation types from the moderate-resolution imaging spectroradiometer (MODIS) (ISLSCP II MODIS (Collection 4) IGBP Land Cover) (d), available online at http://daac.ornl.gov/cgi-bin/dsviewer.pl?ds_id=968 (Huang *et al.*, 2016a).

Peninsula's Rub al Khali Desert, and the Iran/Afghanistan/Pakistan border area (Fig. 2.1(a)). These regions have barren MODIS land (Fig. 2.1(d)). Arid regions are found in the Southern Sahara Desert, Southern Africa, the Western Arabian Peninsula, Central Asia, parts of Mongolia and Northern China, and much of Australia (Fig. 2.1(a)), where the land cover types are barren or sparsely vegetated, i.e., shrubland and grassland (Fig. 2.1(d)). Outside of the Central desert regions, where the predominant land cover type is grassland, semi-arid and dry subhumid regions can be found in the Western United States, the West Coast of South America, Central and East Asia, and a significant part of Australia (Fig. 2.1(d)).

Compared with Fig. 2.1(a), apart from the latitudes between 50°N and 60°N, the dryland area described by annual precipitation (Fig. 2.1(b)) is substantially smaller than that defined by AI. At about 60°N, such as in Siberia, the dryland area defined by precipitation is larger in Asia and North America, although the AI considers these regions as humid because of the low PET. The Köppen–Geiger hot desert climate subtype is found in the subtropical latitudes of 30°S and 30°N in Fig. 2.1(c). The entirety of the Sahara Desert, the Arabian Peninsula, sizable portions of Iran, Southern Pakistan, Southern Afghanistan, Southern Africa, and a huge portion of Australia are examples of such places. Cool desert climate areas are primarily found in East and Central Asia, with a few minor patches also present in Northern Africa, Southern Australia, and the Great Basin Desert in the Western United States. The tropics and subtropics, which are typically found on the periphery of subtropical deserts, are where the hot steppe climate regions most frequently exist. These areas are mostly found in places like the

Sahel in Northern Africa, Southern Africa, minor parts of India and Mexico, and a sizable portion of Australia outside of the Central desert regions. The deserts of Northern Africa, which normally have a Mediterranean climate, also have this type of terrain on their poleward side. The bulk of these regions are in East Asia, Central Asia, and North America in the Northern Hemisphere, and in parts of South America and interior Southern Australia in the Southern Hemisphere. The cool steppe climate subtype is most found in temperate zones.

The AI definition of drylands is the most rational and trustworthy of the three, and the arid and semi-arid region coverage is in good agreement with the surface vegetation types. Precipitation alone obviously is not enough to establish climatic limits (Hassan *et al.*, 2005). In Australia, for instance, the acknowledged threshold for an arid zone is 250 mm, yet several regions of the tropical Northwest are blatantly desert despite having mean annual rainfall levels that are higher than 500 mm. In addition, Siberia receives less than 500 mm of precipitation annually while having a predominantly forested land cover and a humid climate. The Köppen–Geiger classification of climate may inadvertently lump together locations with noticeably dissimilar weather patterns and vegetation, as investigated in earlier studies (Trewartha and Horn, 1980). For instance, even though Southern California (in Southwest US) and Western Washington and Oregon (in Northwest US) have noticeably different weather patterns and flora types, the two locations are categorized as having the same climate type. Additionally, specific drylands can be categorized as other climate types. For example, the Köppen–Geiger climate classification classifies Mediterranean semi-arid regions as Csb or Csa, where letter "s" stands for "summer dry" and letters "b" and "a" stand for "warm summer" and "hot summer," respectively. Because there are several climate subtypes classified within each major climate type based on seasonal variations in temperature and precipitation, the arid climate (type B) as defined by the Köppen–Geiger climate classification does not represent the entirety of the drylands and is considerably smaller than the global drylands. The Köppen–Geiger drylands are substantially smaller in most regions, including East Asia, Central Asia, and North America, when compared to Fig. 2.1(a). According to Kottek *et al.* (2006), Köppen–Geiger arid climates cover a total of 29.1% of the world's land area, whereas the area of all drylands according to the AI definition is 6.1×10^7 km^2, or 41% of the world's land area (Hassan *et al.*, 2005). As a result, the concept of AI is the foundation for the drylands in the following sections.

The climatological distribution of Chinese semi-arid regions for the years 1961–1990 was displayed by Li *et al.* (2015a). The semi-arid regions of China are primarily distributed in a zonal band from the west to the Northeast, with arid and hyper-arid regions to the Northwest and dry subhumid regions to the Southeast. The areas of semi-arid and dry subhumid regions defined by the AI are smaller than those defined by P. The AI-based geographical distribution of semi-arid areas is likewise in good agreement with measurements of China's grasslands and other surface vegetation types from the Moderate Resolution Imaging Spectroradiometer. Semi-arid regions generally cover 2.14×10^6 km^2, whereas the areas of hyper-arid, arid, and dry subhumid regions are 1.06×10^6, 1.42×10^6, and 1.00×10^6 km^2, respectively. This suggests that semi-arid regions predominate in the covering of drylands. The semi-arid region of Northern China is a transitional zone of instability between the humid monsoon regions of Southeast China and the inland arid regions of Northwest China, and vegetation growth in these regions is relatively unstable. This zone is related to strong climate gradients and biological variables. Furthermore, due to interactions between the East Asian summer monsoon (Huang *et al.*, 2016a) and westerlies, and mixed agricultural and pastoral activities, these regions are extremely vulnerable to natural disasters, climate change, and human activities. For instance, the Transitional Climate Zone (TCZ) front and back edges historically show large year-to-year fluctuations, and the interannual variability of the TCZ is mostly influenced by P fluctuation while the long-term trend of the TCZ is controlled by potential evaporation. In the meantime, semi-arid regions have experienced the TCZ's greatest expansion and heightened warming during the last century (Huang *et al.*, 2012; Ji *et al.*, 2014; Guan *et al.*, 2015b). The vegetation cover change in the eastern semi-arid regions is sensitive to the fluctuation in regional P.

2.2.5 *Variation of semi-arid regions*

The drylands subcategories of hyper-arid, arid, semi-arid, and dry subhumid, as described by AI, cover 1.1×10^7 km^2, 1.9×10^7 km^2, 2.3×10^7 km^2, and 0.9×10^7 km^2, accounting for 7, 13, 15, and 6% of the land surface, respectively. The semi-arid region, which makes up more than one-third of the total drylands, is the largest of all these subtypes of global drylands. From 1948 to 2008, the area of total drylands and its subcategories is changing, as shown in Fig. 2.2. The area of the world's drylands has significantly increased over the past 15 years (1990–2004), growing by 2.6×10^6 km^2

(or 4%), compared to the period between 1948 and 1962 (Fig. 2.2(a)). By examining the observations from 1948 to 2008, Feng and Fu (2013) likewise came to similar conclusions. The dryland subtype expansions for the hyper-arid, arid, semi-arid, and dry subhumid categories are 0.6×10^6, -0.1×10^6, 1.6×10^6, and 0.5×10^6 km^2, respectively (Figs. 2.2(b)–2.2(e)). Semi-arid regions, where the area has grown since the early 1960s, have had the biggest arid and semi-arid region expansion, accounting for more than half of the overall dryland increases (Fig. 2.2(c)). Prior to the 1980s, there were few changes in the extent of dry subhumid regions; they shrank in the early 1980s and considerably increased after that decade. Strong decadal oscillations can be seen in the area changes in arid regions, with enormous areas before the 1970s, shrinking areas in the 1970s, and expansion again in the early 1980s; the area for the past 15 years is comparable to that present before the 1970s. Due to significant positive precipitation anomalies in the Southern Hemisphere in the 1970s, which caused a change from arid to semi-arid regions, the arid regions had the smallest areas (Feng and Fu, 2013).

According to Huang *et al.* (2012), a semi-arid region has an AI of 0.2–0.5. If the AI is outside the range of 0.2–0.5, transitions from semi-arid regions to other dryland subtypes will take place. The first 15 years of the monitoring period, from 1948 to 1962, and the most recent 15 years, from 1990 to 2004, are selected in order to better comprehend the changes that have occurred in semi-arid regions during the past 61 years. All transitions in the semi-arid regions, including those from arid to semi-arid, subhumid/humid to semi-arid, semi-arid to arid, and semi-arid to subhumid/humid regions, can be seen in Table 2.1. There have been two transitions: from arid to semi-arid regions (1.1×10^6 km^2) and from subhumid/humid to semi-arid regions (3.0×10^6 km^2), totaling 4.1×10^6 km^2 of newly produced semi-arid regions that have transitioned from other subtypes. In addition to two transitions from semi-arid to arid areas (1.6×10^6 km^2) and from semi-arid to subhumid/humid regions (0.9×10^6 km^2), the area from semi-arid regions to other subtypes decreased by 2.5×10^6 km^2. Therefore, after deducting the reduced areas from the newly generated areas, the semi-arid region has seen a net change of 1.6×10^6 km^2. The semi-arid area that shifted from subhumid/humid regions is about 3 times bigger than that from arid regions, when comparing the two types of movements. This finding suggests that the newly produced semi-arid regions, which make up 73% of the area, mostly originated from subhumid and humid regions. When

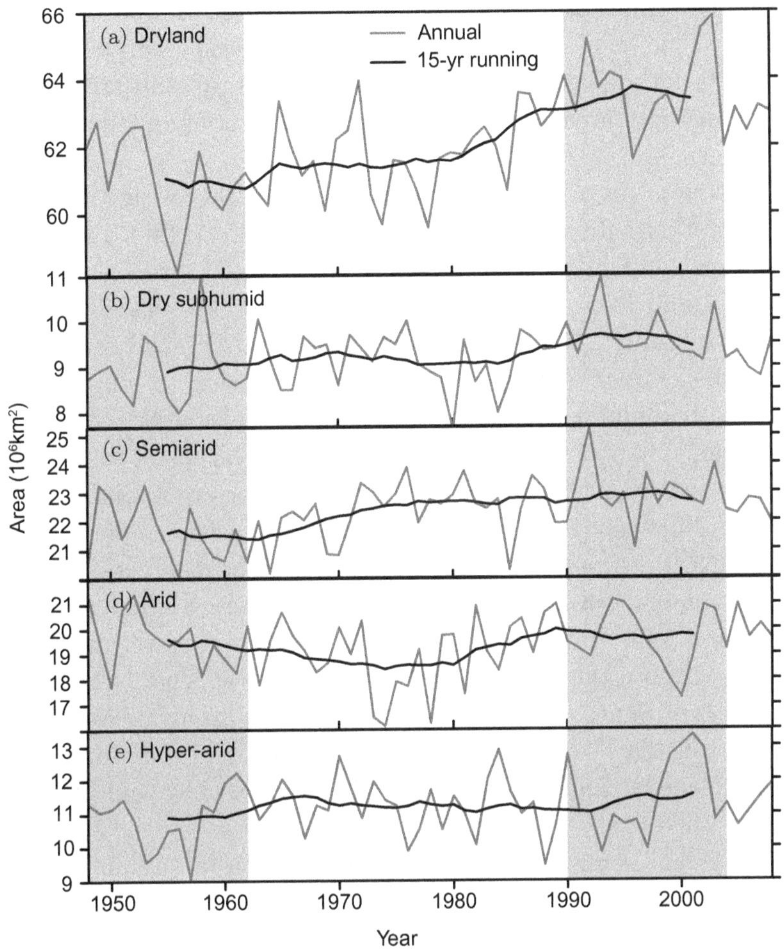

Fig. 2.2. Temporal variations in the area (units: 10^6 km^2) for total drylands (a), dry subhumid regions (b), semi-arid regions (c), arid regions (d), and hyper-arid regions for the period of 1948–2008 (e). The 15-year running mean (thick black curves) is applied to emphasize climate change (Huang *et al.*, 2016a).

Table 2.1. The area transitions (10^6 km^2) between semi-arid regions and other classes for 1990–2004 relative to 1948–1962 (Huang *et al.*, 2016a).

Areas (10^6 km^2)		
Semi-arid to arid	Subhumid to semi-arid	Humid to semi-arid
1.6	2.84	0.11
Arid to semi-arid	Semi-arid to subhumid	Semi-arid to humid
1.1	0.87	0.01

comparing the transitions from semi-arid (subhumid/humid) to arid (semi-arid) regions, the semi-arid (subhumid/humid) to arid (semi-arid) transition area is significantly greater than the arid (semi-arid) to semi-arid (subhumid/humid) transition area, and more semi-arid (subhumid/humid) regions became drier.

The distribution of semi-arid regions around the world, and how they change into other subtypes, is shown in Fig. 2.3. East Asia, East Australia, and Northern and Southern Africa are the main regions where the shift from subhumid/humid (semi-arid) to semi-arid (arid) takes place, as illustrated in Fig. 2.3. In Central/West Australia, the Central US, and Southern South America, the climate changes from semi-arid (arid) to subhumid/humid (semi-arid). The findings show that, except for Central/Western Australia, the transitions to drier types predominantly took place in the Eastern Hemisphere, whereas the transitions to wetter types took place in the mid-latitudes of North and South America. Notably, two temperate semi-arid regions in North America and East Asia exhibit divergent trends, with semi-arid and dry subhumid regions in East Asia becoming drier while more arid and semi-arid regions in North America grew wetter. Eight semi-arid regions around the world — East Asia, Central Asia, Northern Africa, North America, South America, Southern Africa, Central/West Australia, and East Australia — have been chosen to study regional climate changes. In Table 2.2, the latitude/longitude details are presented, and Fig. 2.3 labels them.

Eight typical semi-arid regions' transitions from subhumid/humid to semi-arid, semi-arid to arid, semi-arid to subhumid/humid, and arid to semi-arid are depicted in Fig. 2.4. The findings indicate that semi-arid areas in the most recent 15 years (1990–2004) are larger than those during 1948–1962, apart from Central Asia and Northern Africa. East Asia, which makes up about 50% of the world's semi-arid expansion, experienced the largest semi-arid expansion. The area that switched to a drier class is significantly bigger than the area that shifted to a wetter class when comparing the four semi-arid regions of the Northern Hemisphere, except for North America. This is especially true in East Asia and Northern Africa (Figs. 2.4(a) and 2.4(b)). In East Asia, for instance, the area that moved to the drier classes is four times larger than that shifted to wetter types. The area where the climate predominantly changed to drier categories in larger than the area where the climate changes to the wetter classes. However, the region in North America that switched to wetter types is almost ten times larger than that in Southern Africa and East Australia, while the Central/West Australia

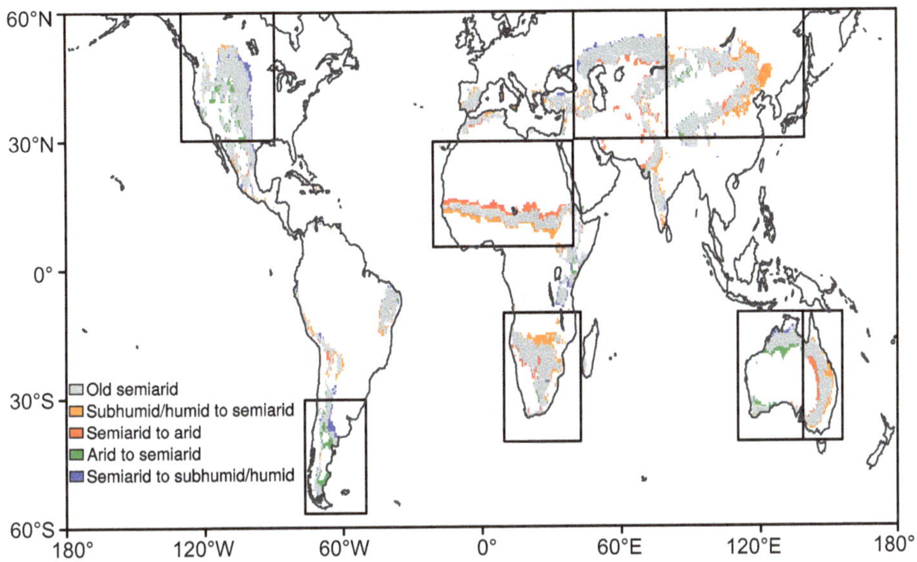

Fig. 2.3. Global distribution of semi-arid regions and their transitions to other climate classifications, including old semi-arid regions (grey), subhumid/humid to semi-arid regions (orange), semi-arid to arid regions (dark red), arid to semi-arid regions (dark green), and semi-arid to subhumid/humid regions (blue) for 1990–2004, relative to 1948–1962 (Huang *et al.*, 2016a).

Table 2.2. Latitude/longitude information for eight regions (Huang *et al.*, 2016a).

	Region	Longitude (E)	Latitude (N)
1	East Asia	80° to 140°	30° to 60°
2	Central Asia	40° to 80°	30° to 60°
3	Northern Africa	−20° to 40°	5° to 30°
4	North America	−130° to −89°	30° to 60°
5	South America	−77° to −50°	−60° to −30°
6	Southern Africa	9° to 43°	−40° to −10°
7	Central/West Australia	110° to 139°	−40° to −10°
8	East Australia	139° to 156°	−40° to −10°

and South America in the Southern Hemisphere are also changing to wetter categories (Figs. 2.4(c) and 2.4(d)). Two of the three places that changed to wetter types throughout the eight semi-arid regions analyzed are found on the American continent.

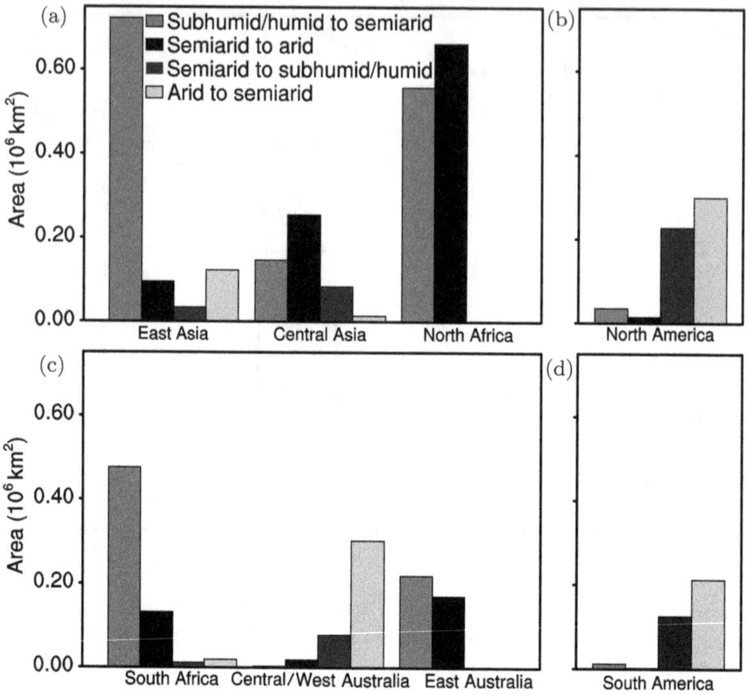

Fig. 2.4. The transitions from subhumid/humid to semi-arid, semi-arid to arid, semi-arid to subhumid/humid, and arid to semi-arid regions over the eight semi-arid regions of East Asia, Central Asia, and Northern Africa (in the Eastern Hemisphere) (a); North America (in the Northern Hemisphere) (b); Southern Africa, Central/West Australia, and East Australia (in the Eastern Hemisphere) (c); and South America (in the Southern Hemisphere) (d) (Huang *et al.*, 2016a).

Table 2.3. The areal changes (10^6 km^2) in semi-arid regions over the American continents and the continents in the Eastern Hemisphere (Huang *et al.*, 2016a).

Change	Eastern Hemisphere	America
Subhumid/humid to semi-arid	2.6	0.4
Semi-arid to arid	1.5	0.1
Semi-arid to subhumid/humid	0.4	0.5
Arid to semi-arid	0.5	0.6

Table 2.3 displays the changes in the semi-arid areas between the American continents and the Eastern Hemisphere. Within the American continents and the Eastern Hemisphere, the semi-arid regions in the most recent 15 years (1990–2004) are 0.4×10^6 km^2 and 1.2×10^6 km^2 bigger,

respectively, than those during 1948–1962. Both regions see the expansion of semi-arid regions, but the patterns of this development are evolving in distinct ways. Within the American continents, the climate changes from desert to semi-arid regions, where it gets more humid. In contrast, semi-arid zones are replacing subhumid/humid regions on the continents of the Eastern Hemisphere, where the climate is becoming drier. For instance, in the continental Eastern Hemisphere, the change from subhumid/humid to semi-arid classes occurred over an area of 2.6×10^6 km^2, which is five times larger than the area that went from arid to semi-arid (0.5×10^6 km^2). The area that changed from subhumid/humid to semi-arid conditions inside the American continent is smaller (0.4×10^6 km^2) than the region that changed from arid to semi-arid conditions (0.6×10^6 km^2). The findings show that the growth of semi-arid areas in the American continents is significantly influenced by the wetting of arid regions as opposed to the drying climate trend in the Eastern Hemisphere's continents.

Numerous studies conducted over the past 60 years have revealed considerable aridity trends and continuing expansion in China's semi-arid regions (Feng and Fu, 2013; Huang *et al.*, 2016a, 2017a). While the areas of arid and dry subhumid regions have increased by 0.013×10^6 km^2 (10 yr)$^{-1}$ and 0.017×10^6 km^2 (10 yr)$^{-1}$, respectively, the rate of expansion for semi-arid regions between 1948 and 2008 was the fastest [0.111×10^6 km^2 (10 yr)$^{-1}$]. This suggests that the semi-arid regions have experienced the largest expansion in China over the past 61 years. In arid and semi-arid regions, where the area rise of semi-arid areas is more than 10 times greater than that in arid and dry subhumid regions, the expansion of semi-arid regions, with relative increases of 33%, is the most severe relative to the 1948–1962 period. As a result, the area increase in the total drylands has been dominated by the rapid expansion of semi-arid regions; in contrast, the area of hyper-arid regions has declined by 6%, but the reduced amplitude is much smaller than that of semi-arid expansion. Furthermore, according to Huang *et al.* (2016a), 50% of the world's semi-arid expansion has occurred across East Asia, making it the largest semi-arid expansion in a single region. According to Feng and Fu (2013), the population of areas impacted by water scarcity and land degradation has increased, and desertification is a result of the growth of semi-arid regions as a result of climate change. Li *et al.* (2015a) and Huang *et al.* (2016a) compared the spatial shifts in semi-arid regions toward drier/wetter types for the last 15 years relative to the 1961–1990 climatology, considering that the newly formed semi-arid regions have different features compared with the original type of region. Northeast China was primarily affected by the newly formed semi-arid regions, which evolved from subhumid/humid regions. These regions spanned from the Western Heilongjiang and Jilin Provinces and Eastern Inner Mongolia, through a majority of Hebei and Shanxi Provinces, and finally toward the Southern Gansu Province. In Northeastern China, semi-arid regions have recently formed next to newly formed dry subhumid regions, but the expansion in the former is smaller than that in the latter. The semi-arid regions have mostly contributed to the increasing of arid and hyper-arid regions, since the transformation of arid and hyper-arid regions from other types of regions have been small. The retreat of arid to semi-arid zones, in contrast to the movements toward drier types, was primarily found in small patches of the Central TP and along the margin of the Junggar Basin. The areal change in newly formed semi-arid areas (from other types to semi-arid regions, 0.73×10^6 km^2) is six times larger than that in semi-arid decreasing areas (from semi-arid regions to other types, 0.105×10^6 km^2), according to the contributions of different types of transitions to the area change in semi-arid regions in China. The expansion of semi-arid regions is most severe in Northern China, and the newly formed semi-arid regions primarily originated from subhumid and humid regions, so more semi-arid regions have become drier, and the aridity has intensified dramatically, especially in the newly formed semi-arid regions. In the meantime, the areas of semi-arid regions with intensified aridity (drying) are much larger than those with decreased aridity (wetting). According to Huang *et al.* (2016a), the semi-arid expansion in East Asia is the largest when compared to other regions of the world when it comes to transformation types. Numerous studies have claimed that global warming is to blame for the spread of semi-arid regions along with an increase in aridity (Feng and Fu, 2013; Huang *et al.*, 2017c).

2.3 Semi-Arid Climate System

Both regional and global-scale factors via physical processes and dynamic mechanisms, such as dust–cloud–precipitation interactions, land–atmosphere interactions, ocean–atmosphere interactions, and human activities, can contribute to the temporal and spatial changes in the arid and semi-arid region climate (Fig. 2.5). The strength of the response to climate change varies across different regions based on interactions between the atmosphere and the land. In arid, semi-arid, and subhumid locations, drought-related changes to the land surface appear to trigger a positive feedback loop that reinforces such droughts, lengthening and deepening them. On a global scale, the decadal and interdecadal fluctuations in the arid and semi-arid region

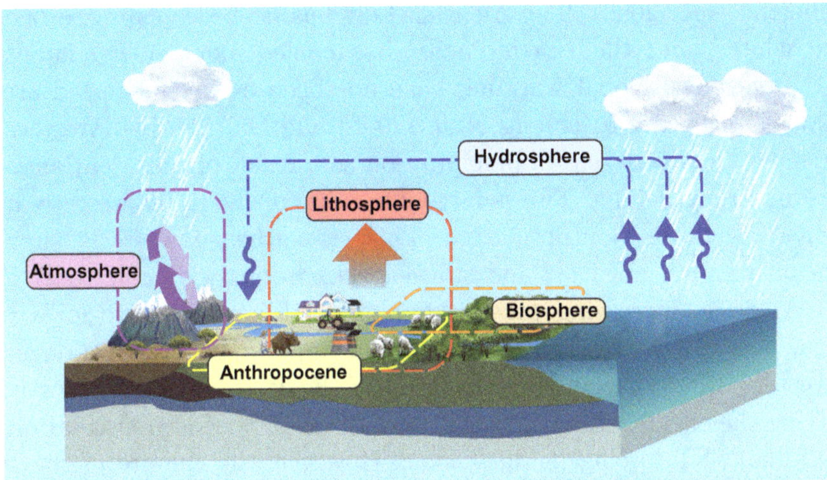

Fig. 2.5. Climate system over arid and semi-arid regions. The semi-arid climate system is a coupling system of atmosphere, hydrosphere, biosphere, lithosphere, and Anthropocene.

climate are primarily driven by ocean–atmosphere interactions. Through teleconnection, the decadal to interdecadal variability of SSTs can change the overlying air circulation and have an impact on arid and semi-arid regions. Through activities like deforestation, agriculture, and the combustion of fossil fuels, humans have significantly changed the global land surface and atmospheric composition. The long-term trends in climate change, notably those of temperature and precipitation, will be determined by human activity. The following four paragraphs provide a summary of the progress made in understanding how these key mechanisms impact arid and semi-arid region climate change.

2.3.1 *Atmosphere*

Changes in the semi-arid climate could be brought on by both global- and regional-scale factors. The planetary-scale internal dynamical processes, SST anomalies, well-mixed greenhouse effects, and external forcing are among the global-scale factors. Regional differences may exist in how these factors affect climate change in semi-arid regions. This section explores the impact of planetary-scale internal dynamical processes and ocean–atmosphere interactions on the causes of the various behaviors in semi-arid climatic changes over the eight typical semi-arid regions. When comparing effects of the westerly circulation and monsoon circulation on semi-arid regions in North America and East Asia, the westerly circulation has a stronger influence in North America, while the monsoon circulation has a stronger impact in East Asia

(Trenberth *et al.*, 2000). The Aleutian Low has been deepening, especially in the winter, and the North Pacific has cooled along 40°N, indicating that the effects of global warming have not been evenly dispersed around the globe. According to Huang *et al.* (2005) and Dai (2013b), Western US experiences more cool-season precipitation as a result of the Aleutian Low's long-term strengthening. This is largely because of the faster westerly along the Aleutian Low's Southern edge. This strengthened westerly supports significantly more moist air and storm tracks from the Pacific to Western North America, increasing precipitation and adding to the region's rainy pattern. Warm Atlantic SSTs also have an impact on precipitation across the Western US, especially in the summer (Feng *et al.*, 2011), mostly via modifying the effect of the relatively weak El Niño–Southern Oscillation-like (ENSO-like) SST forcing in the Pacific. However, the climate of the semi-arid region of East Asia is heavily influenced by the Asian monsoon system, especially the summer monsoon. The rainfall advected by the monsoon cannot reach the margins of semi-arid areas because the East Asia Summer Monsoon (EASM) is diminishing; as a result, drying is induced across these areas. Overall, the westerlies and the EASM, two separate circulation patterns, play significant roles in the wetness of the semi-arid climate in North America and the drying of the climate in East Asia, respectively.

Westerly winds dominate atmospheric circulation in arid Central Asia, in contrast to East Asia. The anomalous high-pressure center causes the westerlies, which cause wetting in the Northernmost portions of Central Asia and drying across most of the southern region to become stronger in the north and weaker in the south. The West African Monsoon (WAM) system, which blows across the Sahel from the tropical Atlantic and brings rain to this region, has an impact on the climate of the semi-arid region of North Africa (Nicholson and Kim, 1997). Numerous studies have demonstrated that the significant warming of the tropical Indian sea surface temperature (SST) facilitates the formation of deep convection over the ocean, weakening the continental convergence associated with the monsoon and being a major factor in the Sahel's drying trend. The Indian Ocean also has an impact on Southern Africa's semi-arid regions. A rise in pressure in Southern Africa coincides with the increase in SST. The semi-arid regions of Southern Africa are seeing a drying trend as a result of these changes, which limit convection. Additionally, ENSO has an impact on rainfall in Southern Africa through changes in SST in the Western Indian Ocean, with El Niño and La Niña bringing drier-than-average conditions (Nicholson and Kim, 1997; Kane, 2009).

Due to an atmospheric anomalous ascending motion west of 140°E and anomalous subsidence east of 140°E, precipitation has declined in Eastern Australia during the past few decades while increasing in Central/Western Australia. Furthermore, an anomalous anticyclonic circulation at low altitudes in the east reduces the easterlies that deliver water vapor from the tropical Western Pacific and dry up Eastern Australia. As a result of the unusual cyclonic circulation across Central and Western Australia, more precipitation occurs here. The normal westerly flow has a significant impact on the semi-arid region of South America, which is situated leeward of the Andes Mountains and is oriented North–South (Garreaud *et al.*, 2013). The Andes obstruct most of the humid air masses carried by the westerlies from the Pacific, leaving the region east of the Andes dry. As a result, the 850-hPa zonal wind and precipitation are substantially and adversely affected across the semi-arid region east of the Andes (Garreaud *et al.*, 2013).

2.3.2 *Lithosphere*

Land systems include all processes and activities connected to human land use, including investments and organizational arrangements in the fields of socioeconomics and technology. The local environment and human well-being are significantly impacted by changes in land systems, which are also widespread contributors to global climate change. Food, fuel, fibers, and a variety of other ecosystem services that support production activities, control the danger of natural disasters, or provide cultural and spiritual services are all provided by the land as essential resources to society. Society adjusts and modifies the quantity and quality of these services by exploiting the land. Changes in the land system are the product of human decision-making at various stages, from local landowner decisions to large-scale national land-use planning and international trade agreements. The combined effect of numerous small land system changes has far-reaching effects on ecological services, human well-being, and decision-making within the Earth system. As a result, changes in the land system have both causes and effects on socio-ecological processes.

Numerous large-scale, comprehensive observation experiments have also been conducted in arid and semi-arid regions of Asia. Many large-scale experiments on land surface processes have been conducted in various climatic and biological zones throughout the world. The “Atmosphere-Land Surface Processes Experiment in Heihe River Basin, Western China” (HEIFE), a comprehensive observation study of land–atmosphere interaction centered on water and heat exchange, was collaboratively carried out by China and Japan in Heihe River Basin, Gansu Province, China. In order to provide an observational basis for the parameterization scheme of water and energy budgets in arid and semi-arid regions of the mid-latitude, this study focuses on the land surface physical process of the production and modification of arid climate. Moreover, the following studies were also conducted: the "Inner Mongolia Semi-Arid Grassland Soil-Vegetation-Atmosphere Interaction" (IMGRASS) from 1997 to 2001, the "Land-Atmosphere Interaction Experiment in Arid Region of Northwest China" (NWC-ALIEX) in the Gobi Desert of Dunhuang, Gansu Province, the field observation experiment of "Study on Interaction between Underlying Surface with Sparse Vegetation and Atmosphere" in the climate transition zone and eco-geographic transition zone of arid, semi-arid, and humid areas in Ordos, and the "Experimental Study of Land Surface Processes in Chinese Loess Plateau" (LOPEX).

Numerous important scientific findings were obtained from the systematic monitoring in these studies of the energy and water exchange and circulation between the land and the atmosphere in China's arid and semi-arid regions. A few long-term observatories have been conducted in arid and semi-arid locations over the past few years, aside from these brief field observation studies. These long-term observational studies not only contribute to a better understanding of arid and semi-arid climates from the perspective of the interaction between the land and the atmosphere but they also lay a strong foundation for future research into the mechanism of aridification, which will help manage human activities. Model simulations also consider interactions between the land and atmosphere in addition to empirical investigations. An important criterion for assessing the performance of hydrologic and atmospheric models is the extent to which land surface–atmosphere interactions, or anomalies in the land surface state (i.e., soil wetness, soil texture, and surface roughness), can influence the planetary boundary layer (PBL) and, in extreme cases, rainfall generation (Ferguson, 2012). The Global Land–Atmosphere Coupling Experiment (GLACE) is a model intercomparison study focusing on the land–atmosphere coupling strength, or the extent to which anomalies in land surface states (e.g., soil moisture) can affect rainfall generation and other atmospheric processes. This is a crucial but frequently overlooked aspect of numerical weather and climate modeling.

2.3.3 *Hydrosphere*

The hydrosphere is a term used to describe an almost continuous sphere of liquid, gaseous, and solid water on the surface of the globe. Water bodies refer to a variety of aggregations of natural or manufactured water on the earth's surface. The hydrosphere is the collective name for the fluid that surrounds the earth's surface on occasion. The hydrosphere is made up of all the water in the seas, rivers, lakes, and the crust of the globe. This water is continually moving and interacting with one another to create the hydrosphere. The hydrosphere is the most active sphere on earth when compared to other spheres. The evolution of the system that influences human activities is intimately correlated with the interactions between the hydrosphere and the atmosphere, the biosphere, and the inner ring of the planet. About 1.386 billion cubic kilometers of water makes up the hydrosphere overall, with fresh water making up 2.53% and salt water making up 97.47%. The hydrosphere's largest body of water is the ocean.

The ocean, often known as the world ocean, is a body of salt water that makes up 97% of the planet's water and covers around 70.8% of its surface. It includes the Arctic, Pacific, Atlantic, Indian, and Southern oceans. Approximately 3.61×10^8 km^2 of the planet is covered in water. The main element of the hydrosphere on Earth is the ocean. It has had a significant impact on the climate and weather patterns, the carbon cycle, and the water cycle. The weather, climate, and climate change all have a significant impact on the ocean. The water and the atmosphere are constantly exchanging heat, momentum, and material. Typhoons, El Niño events, La Niña events, and other extreme weather phenomena can be brought on by ocean–air interactions, which can lead to local or global climate calamities. The seas are where most solar radiation that enters the Earth's atmosphere lands and is absorbed. The seas are a massive source of energy because of their high bulk and specific heat. Sea surface temperature variation is substantially smaller than land surface temperature variation due to the enormous thermal inertia of the seas, which also serves as a buffer and regulator for atmospheric temperature change. Due to higher levels of direct solar radiation in the tropics, there is an excess of energy in the ocean, and the water temperature near the equator is dramatically higher than the high-latitude sea areas, which causes warm currents to flow from the equator to high latitudes and cold currents to flow from high latitudes to the equator in large-scale circulation. This redistribution of energy keeps climate of the equatorial region and the poles from being too disparate. In addition to moving a portion of the energy from the tropics to the colder middle and high latitudes, the currents also release energy into the atmosphere in the form of sensible and latent heat and supply a lot of water vapor to the atmosphere. The oceans are a part of the water cycle on the surface of the earth; the water vapor from the sea surface condenses into rain and snow and falls on the surface of all continents as a source of fresh water. As a result, the ocean contributes to the processes of surface material and energy balance, which makes it a crucial component of the entire climate system. The ocean also serves as a significant source of heat and water vapor for the atmosphere. Additionally, currents transport a great deal of heat and water around the world.

Ocean circulation has a considerable impact on the world's climate over comparatively longer decadal and multi-decadal timescales. In the realm of seasonal forecasting, the ocean effect is also heavily considered. On the other hand, the ocean is also greatly impacted by climate change. Sea level and seawater temperature will rise together as a result of rising temperatures. The probability of maritime disasters is rising as a result of climate change's effects on ocean climate patterns. In addition, contemporary research not only focuses on the ocean's role in climate change but also examines how pollutants like carbon dioxide are taken up and redistributed by the deep ocean. The oceans' biological structure may be harmed by acidification caused by an excessive absorption of carbon dioxide by the seas. As a result, studying how the water and atmosphere interact is crucial to understanding the problem of climate change. Because oceans play a significant role in the climate system of the planet, it is crucial to understand how oceans affect climate change in arid and semi-arid regions. Previous research has demonstrated that oceanic activities, particularly sea surface temperature (SST) fluctuations in the Pacific, Atlantic, and Indian Ocean Basins, can have a variety of timescale effects on precipitation, temperature, and dry and wet shifts in the world's arid and semi-arid regions. Global dry and wet fluctuations on an interdecadal scale are determined by the interplay between the ocean and the atmosphere (Guan *et al.*, 2017). The influence of oceans on climate change in arid and semi-arid regions has also been extensively researched from the perspectives of the combination of oceanic oscillation components and the thermal differential between the land and the sea (Huang *et al.*, 2017b, 2017c).

2.3.4 *Biosphere*

The biosphere, also known as the top-down environment of microorganisms, is the area of the Earth where life activities take place, the place where humans live and give birth, and the planet's greatest ecosystem. The first point to note with respect to the biosphere and climate change is the connection between vegetation and climate. The biosphere's main source of biomass is vegetation, and many creatures are dependent on plants. As a fundamental component of terrestrial ecosystems, vegetation contributes considerably to the regulation of the global carbon cycle, the slowing of greenhouse gas emissions like CO_2 in the atmosphere, and the stability of the global climate. For the sake of the ongoing research on climate change, it is crucial to examine the traits and patterns of vegetation changes as well as the relationship between vegetation activities and climate change. Changes in vegetation ecosystems, including photosynthesis, respiration, and growing seasons, as well as changes in phenology, which will affect the global carbon balance, will result from rising temperatures, increasing precipitation, and increasing CO_2 concentration. According to studies, average precipitation globally drives 40.29% of the growth of arid and semi-arid region plants. Second, arid and semi-arid region vegetation benefits from both intense precipitation and intense heat. Vegetation and precipitation have a more pronounced phase relationship. The evaporation of water vapor from the ocean and rivers is augmented by the transpiration of vegetation, which also mixes with groundwater created by underground seepage. This additional evaporation of water vapor provides continuous water vapor transport for a subsequent round of precipitation, resulting in an increase in precipitation. It is important to note that when vegetation and ground cover increase, the earth's specific heat capacity decreases, the diurnal temperature difference decreases, and the temperature tends to stabilize. The temperature difference primarily reflects how vegetation affects temperature.

The global carbon cycle has become the most important part in the field of climate change. The exchange of carbon in the biosphere, lithosphere, hydrosphere, and atmosphere maintained classified into natural cycle and anthropogenic effect changes. The primary mechanism of the natural carbon cycle is the absorption of atmospheric CO_2 by land and marine plants. However, the anthropogenic contribution majorly has an effect on the process of releasing CO_2 as a result of biological or geological processes. In the typical arid ecosystem, such as deserts and dry shrublands ecosystems, which can reach almost 40% of the world's net primary productivity (Wang *et al.*, 2012). Semi-arid regions are the largest contributors to the trend and interannual variability of the terrestrial carbon sink as well as the recipients of the most amplified climate changes brought on by rising CO_2 (Huang *et al.*, 2017b). With annual precipitation ranging from 200 mm to 600 mm, semi-arid forests were able to retain high levels of carbon sequestration by moving the peak time for photosynthetic activity from summer to early spring. If Savannas were protected from fire and grazing such that carbon accumulation could occur, estimates based on data for Savannas demonstrate the enormous potential for harnessing these ecosystems for carbon sequestration, an increasingly significant factor of global carbon interannual variability. The quantitative significance of terrestrial carbon cycles is now acknowledged to be equal to that of ocean–atmosphere interactions in global climate models. Even though the combined land and ocean sinks remove around 50% of the world's CO_2 emissions, coupled climate–carbon feedback models have forecast decreasing carbon absorption by terrestrial sinks, which is caused by declining productivity and increased respiration brought on by high temperatures. Recent research demonstrates that the evolution of the terrestrial carbon sink has taken an increasing trend over the past 30 years. This trend may be caused by a variety of factors, including changes in the length of the growing season and land management, as well as the fertilization effect of increased CO_2 and atmospheric nitrogen deposition on plant growth. Increased tree cover and leaf area index (LAI) were linked to an increase in net primary production that was driven by CO_2 and the climate, according to Wu *et al.* (2016). But given that the vegetation greening brought on by CO_2 fertilization in semi-arid places may cause a rise in evapotranspiration and a fall in soil moisture, this phenomenon is probably going to disappear soon. Due to desertification, gross primary output will abruptly decline as soil moisture continues to drop and it becomes incapable of supporting the present ecosystem.

2.3.5 *Anthropocene*

Current climate change encompasses the effects of global warming on the planet's weather patterns. Although there have been periods of climate change in the past, the current rise in the average global temperature is more pronounced and is largely due to humans. Anthropogenic causes, which include deforestation, blind land reclamation, overgrazing, irrational land use, and the rise in atmospheric concentrations of carbon dioxide and other trace gases as a result of industrialization, are actions taken by humans that have had an impact on climate change. There are three ways that human activity can affect climate change: anthropogenic changes in the nature of the subsurface, like the removal of native vegetation to alter the albedo and roughness; anthropogenic changes in some atmospheric components, like the increase in carbon dioxide concentration; and anthropogenic heat release, from various energies consumed by humans, like a furnace to warm the atmosphere. Carbon dioxide (CO_2) and methane are the two main greenhouse gases that are released into the atmosphere when fossil fuels are burned. In the atmosphere, there exist substances known as greenhouse gases that absorb and release infrared radiation emitted by the Earth's surface, atmosphere, and clouds. The most significant greenhouse gases in the Earth's atmosphere are ozone (O_3), methane (CH_4), nitrous oxide (N_2O), water vapor (H_2O), and carbon dioxide (CO_2). Agriculture, industrial operations, and forest loss have smaller impacts. By absorbing the heat that the Earth radiates and keeping it close to the surface, greenhouse gases warm the air. This impact is amplified by greenhouse gas emissions, which result in the Earth absorbing more solar energy than it can emit back into space.

According to studies, many nations see a 1% rise in greenhouse gas emissions for every 1% increase in population. It is obvious that human activity is a major contributor to climate change. Deforestation and particular agricultural practices, among others, have contributed to rising atmospheric concentrations of greenhouse gases over the years of industrialization. The overall amount of greenhouse gas emissions has increased along with population growth, economic expansion, and a rise in living standards. Since the Industrial Revolution, humans have used fossil fuels like coal, oil, and natural gas, which have increased the amount of greenhouse gases in the atmosphere. Increased human activity and population impact local ecological conditions as well as the regional and temporal distribution of water resources, which are sensitive and fragile ecosystems. Numerous studies have also shown that high-emission scenarios greatly increase climate factors like precipitation. Additionally, the natural environment, ecosystems, and biodiversity are significantly impacted by urbanization and changes in land use. Globally, urbanization is undergoing a substantial process that has changed people's productivity and lifestyles, as well as the emissions of greenhouse gases and other pollutants as a result of urban development and extensive population clustering. Urban flooding, the heat island effect, excessive heat, and other aspects of climate change are all becoming more serious as the process of global urbanization advances.

Arid and semi-arid region climate change is significantly influenced by anthropogenic activity. Anthropogenic activities contribute to salinization, desertification, the loss of vegetative cover, biodiversity loss, and other types of environmental deterioration. More fossil fuels have been consumed as economic development has accelerated, resulting in significant GHG emissions and energy production. The spatial distribution of temperature has been significantly impacted by the released GHGs and energy in recent years (Zhang *et al.*, 2013), particularly in developing nations where economic policies result in the inefficient use of resources and, consequently, increased energy waste. The effects of human activity, particularly in high-density GHG scenarios, are closely related to the heightened warming that has occurred in arid and semi-arid regions (Huang *et al.*, 2012).

2.4 Aridity Changes

Research has demonstrated that over the past 100 years, the arid and semi-arid regions have become increasingly drier. There are notable geographical disparities in the precipitation changes in arid and semi-arid regions as a result of global warming. The understanding of the variations in global and regional precipitation has advanced significantly in previous studies (Huang *et al.*, 2011). At the same time, aging-scale climate change has been mostly responsible for the significant drought occurrences in the arid and semi-arid regions over the past century. Numerous studies on climate changes at the chronological scale have been conducted (Huang *et al.*, 2006c; Fu *et al.*, 2008). For the study of decadal climate in the arid and semi-arid regions, precipitation has been the main contributor to the development of drought on a decadal scale. Aridity and the related water shortage are long-term hydrologic and climatic conditions that have a significant impact on the dynamics of both terrestrial ecosystems and human society. Globally, aridity, which is determined by the supply of water in the atmosphere (precipitation) and the demand for water (potential evapotranspiration, or PET), has grown recently. Aridity changes are closely related to the vast terrain, natural processes, and human-made mechanisms that affect such aridity. Through atmosphere–ocean feedbacks and atmospheric teleconnections, natural variability (such as the Pacific Decadal Oscillation and El Niño–Southern Oscillation) has been found to modulate large-scale aridity. However, natural variability cannot account for the increasing aridification over long timescales. Overall, anthropogenic greenhouse gas emissions have the potential to increase worldwide precipitation, alter stomatal conductance, and change how much water plants consume. However, the widespread rise in PET caused by greenhouse gas-induced warming may make the aridity problem worse. As a result, both natural and anthropogenic effects are demonstrated in a variety of variables, including precipitation, aridity index, terrestrial water storage (TWS), soil moisture, evaporation and potential evaporation, and lake temperature. These variations illustrate various changes in the typical semi-arid region of the world.

2.4.1 *Precipitation (P) and aridity index (AI)*

Changes in precipitation and PET brought on by climate change would alter AI, which would alter the overall extent of drylands and its subtypes. Fig. 2.6 displays the linear trends in AI, precipitation, and PET in semi-arid regions from 1948 to 2008. The AI declines, indicating drier trends over East Asia over the past 61 years, including Northeastern China, Northern Mongolia, East Australia, the Sahel in Northern Africa, Southern Africa, and the Mediterranean regions, while wetter trends are evident in Central/West Australia, South America, North America, and other small areas, including Northern Central Asia and Western China (Fig. 2.6(a)). The Sahel, Southern Africa, East Australia, and East Asia see a drop in precipitation, whereas Central/West Australia and the American continents see an increase (Fig. 2.6(b)). Only certain regional differences separate the precipitation patterns from those for AI. This finding indicates that precipitation is a significant factor influencing AI. Over the American continent, the PET drops while rising over the other continents in the Northern Hemisphere. However, in the Southern Hemisphere, this tendency is reversed, and the PET rises in South America while falling in Southern Africa and Northern Australia (Fig. 2.6(c)). The surface air temperature (SAT) warms in most semi-arid areas throughout the same time period; this trend is most prominent in the Eastern Sahel and higher latitudes across the Northern Hemisphere, i.e., East and Central Asia (Fig. 2.6(d)). The different patterns of the PET and temperature in specific areas, such as North America and Southern Africa, suggest that temperature is not always the main factor regulating the PET (Hobbins *et al.*, 2012), as it is also influenced by other factors, like winds, net radiation, and humidity.

The differences in the regionally averaged precipitation, PET, and AI in three mid-latitude semi-arid regions are over East Asia, Central Asia, and North America in the Northern Hemisphere. Each semi-arid region is separated into old and newly formed semi-arid regions because the newly formed semi-arid regions transitioned from other climate types and are different from the semi-arid regions of the same domain already in place. In East Asia, precipitation declines in both zones, but the newly developed semi-arid regions have a precipitation decline of 57 mm, which is three times

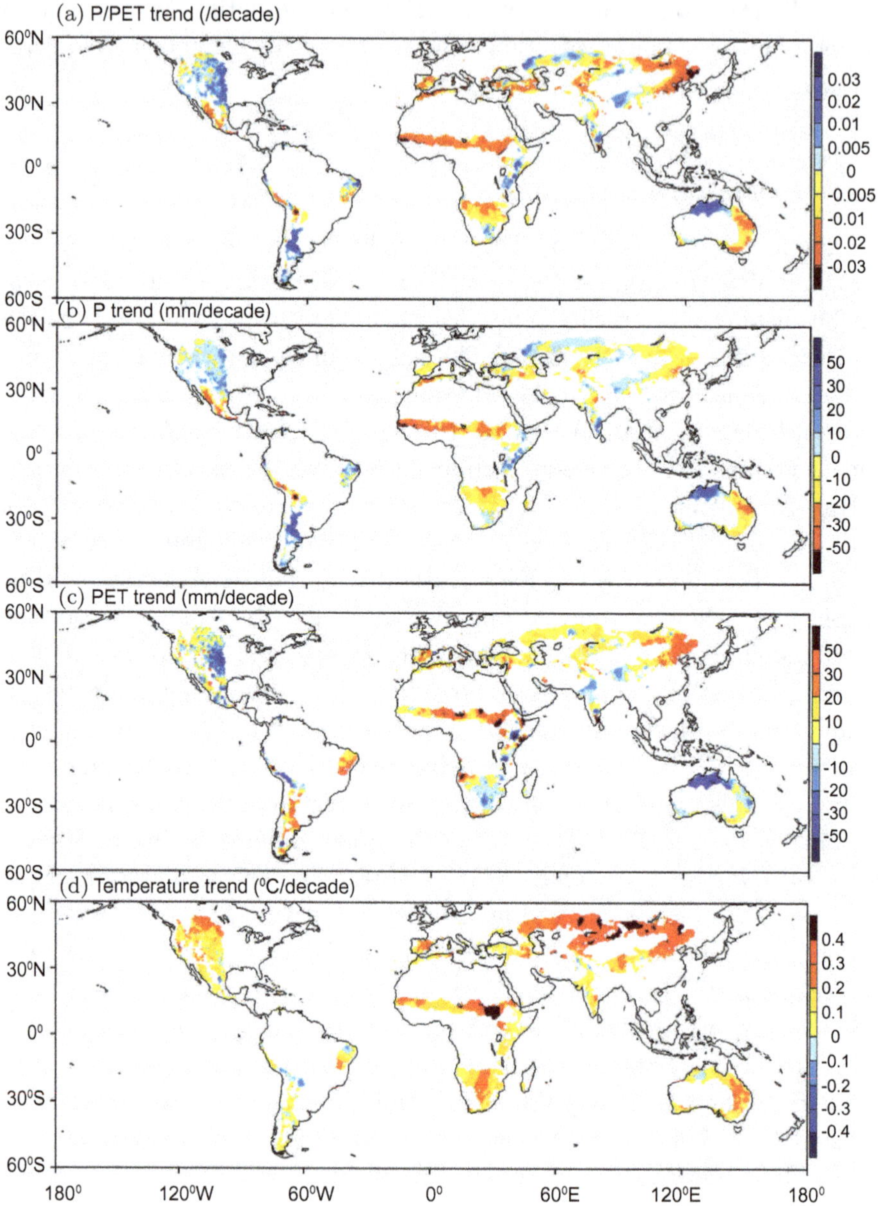

Fig. 2.6. The distributions of the linear trends in the AI, precipitation, PET, and SAT in semi-arid regions globally from 1948 to 2008. P/PET trend (/decade) (a), P trend (mm/decade) (b), PET trend (mm/decade) (c), and temperature trend (°C/decade) (d) (Huang *et al.*, 2016a).

greater than that of the older semi-arid regions. In the recently formed semi-arid zones, the PET rises by 132 mm; this value is twice as high as that in the earlier semi-arid regions, indicating that the two types of semi-arid regions got drier over the last 61 years, particularly the newly developed semi-arid regions, and the decrease in precipitation and increase in PET led to a large fall in AI. In both types of semi-arid regions in Central Asia, the PET rises dramatically. The old and newly developed semi-arid zones experience precipitation changes at notably varied rates. In the newly established semi-arid regions, the regionally averaged precipitation shows a dramatic decline of 81 mm; this value is about 9 times larger than that in the old semi-arid regions. As a result, the decline in AI is 4 times greater in the newly developed semi-arid regions than it is in the older semi-arid regions. In North America, there have been minimal changes to the time series of regionally averaged precipitation, PET, and AI in both the old and newly developed semi-arid zones. Strong decadal changes may be seen in the PET, with the maximum value occurring in the 1950s, documented declines in the mid-1960s, and minimal values in the mid-1980s before increasing again in the early 1990s. In both the older and more recently established semi-arid regions, the rising trend in precipitation and the decreasing trend in PET led to an increasing trend in the AI, indicating that the semi-arid regions have become wetter during the past 61 years. The annual variability of the AI is perfectly compatible with that of precipitation, suggesting that precipitation is the dominant contributor to North America's wet trend. The greater AI and precipitation in the past 20 years provide evidence that the current climate is really becoming wetter compared to the climate of the 1950s in North America, even if historical records show that droughts occurred in the late 1980s and at the start of the 21st century.

The regionally averaged precipitation, PET, and AI indicate consistent rising or decreasing trends between the newly formed and existing semi-arid zones when these regions are compared. The PET indicates an increasing tendency in East Asia and Central Asia, in contrast to the declining trend in North America, while the amount of precipitation declines in East Asia and Central Asia but increases in North America. As a result, the AI is consistent with precipitation, showing that the climate changes from wet to dry in North America but from dry to wet in East Asia and Central Asia. Notably, in the drying semi-arid regions over East Asia and Central Asia, the changes in precipitation, PET, and AI in the newly developed semi-arid regions are substantially larger than those in the previous semi-arid regions. In contrast, there is no difference between the newly developed and older semi-arid zones in the wetting region over North America as seen by fluctuations in precipitation, PET, and AI.

Another significant semi-arid region is Northern Africa, which has drawn increasing attention due to the region's lack of rainfall and the prolonged droughts that occurred during the 1970s and 1980s. The PET rises in both the preexisting and newly developed semi-arid zones, and it has been stable since the 1980s. In both the old and the newly developed semi-arid zones, precipitation and AI both decline dramatically, and the time series shows high decadal variability. The peak values are in the early 1950s; following that, with less precipitation and AI, the values decline and hold stable until the mid-1980s. Since the driest time in the middle of the 1980s, these numbers have partially recovered. In earlier investigations, the same decadal variance in precipitation was also shown (Zeng *et al.*, 1999).

The decrease in precipitation in the old semi-arid regions of Northern Africa is 8 times greater than the changes in East Asia for two severely drying semi-arid regions in the Northern Hemisphere. The rate of change in the decreasing PET also shows a rate twice as large as that in East Asia. As a result, the former semi-arid regions of Northern Africa are experiencing a more severe drying trend than East Asia. Contrarily, during the same time period, the precipitation drops in the newly developed semi-arid zones over Northern Africa by 173 mm, which is three times greater than that in East Asia (57 mm); just a tiny difference is detected in the PET variability between East Asia and Northern Africa. Although the AI changes in the newly formed semi-arid regions of East Asia and Northern Africa are comparable, the drying trend in the newly formed semi-arid regions of East Asia is more sensitive to the decline in precipitation than the trend in Northern Africa because the climatological PET of the newly formed semi-arid regions in Northern Africa (1785 mm) is much greater than that in East Asia (902 mm).

The Southern Hemisphere's low latitudes are represented by the time series of regionally averaged precipitation, PET, and AI in semi-arid regions over Southern Africa, Central/West Australia, and East Australia. The semi-arid regions become drier, as seen by the declining trends in the AI in Southern Africa and Eastern Australia. Furthermore, compared to the changes in the previous semi-arid zones, the rates of change in the AI in the newly established areas are substantially higher. In Central/West Australia, both the old and new semi-arid zones exhibit the same yearly variability and linear trends in regionally averaged precipitation, PET, and AI. The rising trend in the AI suggests that the climate in Central/West Australia became wetter, even though the wetting trend is not statistically significant. This is due to the growing trend in precipitation and the considerable falling trend in the PET. The existing and newly developed semi-arid zones over South America have significantly increased, according to similar results for the regionally averaged precipitation and the PET, with AI revealing increasing trends of 0.06 and 0.04, respectively, which demonstrate that South America's semi-arid zones have gotten wetter over the past 61 years.

2.4.2 *Evapotranspiration (ET) and potential evapotranspiration (PET)*

Over semi-arid regions, as the basic variable, the PET and Evapotranspiration (ET) have unique regional changes. The typical aridity index is calculated from the ratio of precipitation (P) to potential evapotranspiration (PET) in an attempt to capture changes in background status rather than transient anomalies like droughts as a different method of approaching the problem. As shown in Fig. 2.7, aridity increases in warmer climates lead to the expansion of dry climate zones. Evaporation and precipitation increase modestly, while on land, evaporative demand increases faster than precipitation. The primary method of surface water consumption, evaporation is most directly linked to lower surface characteristics and climatic change in all water cycles. Evapotranspiration (ET), which regulates the size of latent heat over the soil surface, also includes the plant transpiration. Additionally, an increase in ET can raise the relative humidity of the air, which has a negative feedback effect on the climate by adjusting the maximum or minimum temperature. Changes in evaporation's temporal and spatial patterns not only have a direct impact on the availability of water resources and the effectiveness of water usage but also have a significant impact on the growth of crops and ecological vegetation. Nearly half of the auxiliary energy from the sun that is absorbed by the earth is used for evaporation consumption, and global warming will inevitably lead to changes in the spatial and temporal distribution of surface evaporation. Surface water is vaporized under the action of heat energy and transported to the air through atmospheric movement. In addition to climatic factors like surface thermal energy conditions and atmospheric motion state, the actual evaporation of the surface is also influenced by soil moisture stress, or drought stress, particularly in arid and semi-arid regions. As a result of the interference of soil moisture stress, the actual evaporation of the surface may not clearly demonstrate the response characteristics to climate warming.

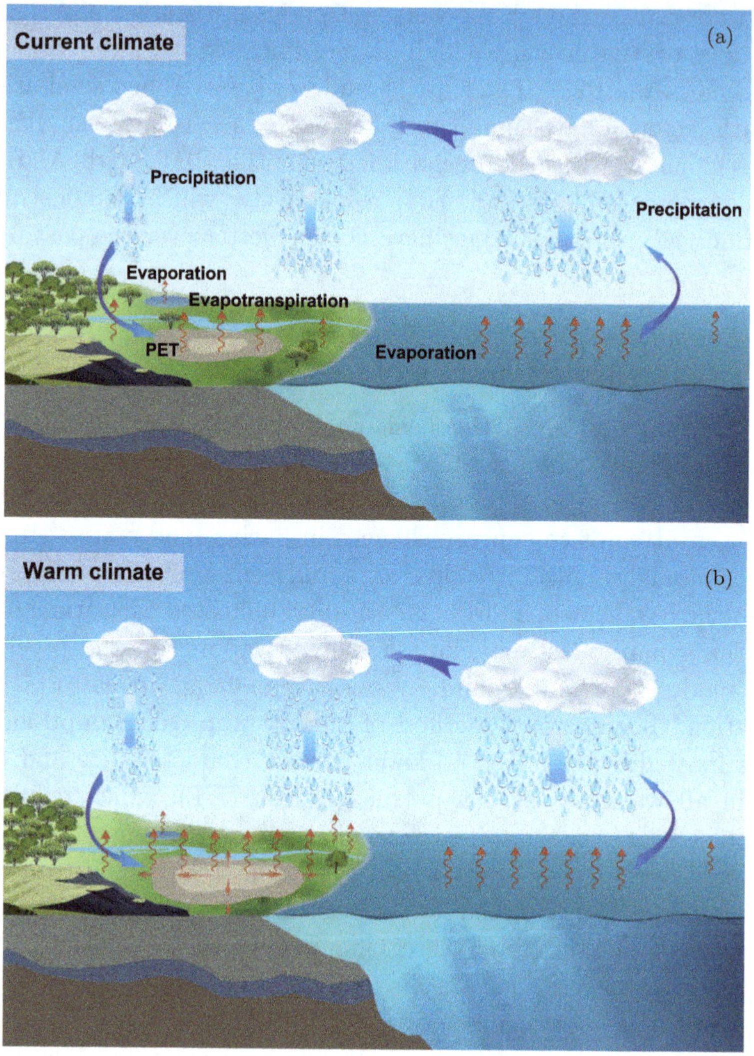

Fig. 2.7. Aridity increases in warmer climates, leading to the expansion of dry climate zones.

The ET calculated using Penman's method is the ET of the land surface free from soil moisture constraints. This estimate reflects the capability of the ET process as determined by weather and climate conditions and represents the theoretical upper limit of the actual ET, which includes the PET amount and the ET of the reference crop. It is more in line with the real ET on the land surface in humid locations, but it varies substantially in arid regions. The quantity of evaporation one would expect from a completely wet surface with the given air characteristics is used to compute the PET, or the evaporative demand of the atmosphere. On land, the true evaporation will be lower than the PET until the soil is completely saturated with water. Over a body of water, PET equals true evaporation. Using evaporation dishes to estimate PET is a widely used technique. Evaporation dishes are a straightforward way to measure complex meteorological interactions. The gap between water surface evaporation and plant surface evaporation is significant, although evaporation dishes roughly correspond to crop transpiration under the same climatic conditions. While most crop transpiration only happens during the day, the heat stored in the evaporation dish should be considered since this heat may have a substantial impact on the evaporation of the evaporation dish at night. Due to variations in air temperature, humidity, and vorticity, there are many differences between them. The energy balance is upset when the heat is dispersed outside through the evaporation dish's edge. The quantity of heat energy available for evaporation consumption on the surface will rise as the climate heats, and the water cycle of the climate system will speed up, causing more water to evaporate. The "evaporation fallacy" or "evaporation paradox" is a term used in academia to describe the finding that, against the backdrop of global warming, the evaporation of evaporation dishes shows a considerable downward trend in most places of the world. In a desert, the P/PET ratio may be close to zero, while in humid regions, it may be higher. When the P/PET ratio decreases, the weather is getting drier; when it increases, the weather is getting wetter. Recent observational studies have demonstrated that P/PET typically decreases as the globe warms. According to simulations of climate models, packing restrictions on cells moving from a mesenchymal state to a polarized epithelium will occur. However, additional long-range signal transfer methods, including mechanical stress, cell projection, and diffusion, may also play a role in establishing the regular size. P/PET will decrease significantly in most tropical and mid-latitude land regions by 2100 under a high-emissions scenario, when the temperature is anticipated to be several degrees warmer than it is now. Such decreases can cause a place to move into the next drier climate group, such as subhumid, arid, or hyper-arid (the latter four together are denoted as arid and semi-arid regions). According to one simulation, the worldwide arid and semi-arid region area will increase by about 10% by the year 2100. Nearly all other geographical areas are anticipated to become drier, except for India and Northern Tropical Africa. Most likely, by the 22nd century, the drying would become even more pronounced. On climate timescales, the global averages of precipitation and evaporation must remain constant. According to the observed and anticipated drying trend in P/PET over land, the amount of PET is therefore expected to increase more quickly than worldwide evaporation (noting that precipitation changes similarly on land and oceans). PET worldwide could not rise faster than P; they would always be equal if there were no land on Earth. The rise in P/PET must thus be exclusive to terrestrial surfaces.

2.4.3 *Soil moisture*

The annual mean soil moisture patterns were measured across the world from 1948 to 2010. High-latitude regions (above 60°S or 70°N) were disregarded because permafrost predominated there (similarly hereinafter). For 40.6% of the global area, significant trends ($p = 0.05$) were seen; 39.8% were positive and 60.2% were negative. Eastern Australia, Northern Africa, East Asia, and Southern Europe all experienced significant drying trends. Numerous areas with low average soil moisture values had substantial drying trends. For the Central United States, South America, and Western Australia, there was a slight wetness trend. Given that there has been a considerable rise in precipitation in these areas over the previous few decades, this may have been precipitation induced. The soil moisture trend was largely consistent with many drought indices, including the PDSI (Wang *et al.*, 2014) and the AI (Feng and Fu, 2013), indicating that the broad patterns displayed by soil moisture data are probably trustworthy.

The continents were split into nine subregions in order to account for the spatial variety of soil moisture trends. The regions (Table 2.4 for details) are categorized using Huang's definition (Huang *et al.*, 2012). Table 2.4 shows the geographical percentage of acreage and the regional contributions of soil moisture drying to global drying. Comparing PTA (percentage of total acreage) and PDA (percentage of acreage with soil moisture drying) yields the relative drying acreage; PDA > PTA indicates that the area is dominated by soil moisture drying. Four of the nine subregions exhibited PDA > PTA: Europe (EU), East Asia (EAS), North Africa (NAF), and South Africa (SAF). They were dominated by soil moisture drying at the respective proportions of 63.7%, 78.1%, 84.9%, and 78.9%. Comparing contributions to global drying and PDA yields the other characteristic of drying — severity; contributions > PDA indicates that the drying in the area is extremely severe. Of the EU, EAS, NAF, and SAF regions, only EAS and NAF exhibited contributions > PDA. The PTA values for EAS and NAF were 12.3% and 16.6%, respectively, but the contributions of soil moisture drying to global drying were 28.3% and 31.8%. Despite drying in a sizable

Table 2.4. PTA (percentage of total acreage, the ratio between acreage of subregions and global acreage, units: %), PDA (percentage of acreage with soil moisture drying, the ratio between drying acreage of subregions and global drying acreage, units: %), and contributions of soil moisture drying to global drying for different subregions. Note that the ranges of subregions may differ from the geographic position, and the names may not be entirely accurate (Cheng and Huang, 2016).

Regions	Latitude/Longitude	PTA	PDA	Contributions
North America (NAM)	15°N–70°N, 180°W–25°W	16.3	8.7	7.4
South America (SAM)	60°S–15°N, 180°W–25°W	13.9	7.0	3.4
Europe (EU)	30°N–70°N, 25°W–60°E	11.7	13.3	10.2
West Asia (WAS)	30°N–70°N, 60°E–90°E	7.2	6.0	4.4
East Asia (EAS)	30°N–70°N, 90°E–180°E	12.3	17.1	28.3
South Asia (SAS)	12°S–30°N, 60°E–180°E	8.2	7.4	2.7
North Africa (NAF)	0°N–30°N, 25°W–60°E	16.6	25.1	31.8
South Africa (SAF)	60°S–0°N, 25°W–60°E	7.6	10.6	7.2
Australia (AUS)	60°S–12°S, 60°E–180°E	6.1	4.8	4.6

percentage of the EU and SAF regions, the intensity of drying was modest when PTA, PDA, and contributions were considered. The EAS and NAF regions had the most notable drying.

Since the climatic region greatly influences the soil moisture variability, Fig. 2.8 shows the regionally averaged soil moisture trend as a function of the climatological mean AI. The annual mean AI for the years 1961–1990 that depicts the climatological state is known as the climatological mean AI. Two minimal values were found, one for the dry regions between dry and wet climates with an AI range from 0.4 to 0.8, and the other for the hyper-arid regions with an AI of less than 0.1. The significant drying over the Sahara with an AI of less than 0.02 was mostly to blame for the pronounced drying trend in the hyper-arid regions with an AI of less than 0.1. The trend decreased to 1.2×10^{-4} $m^3/10$ yr when this area was masked out, which is very little. We disregarded the obvious drying trend over this area due to the region's relatively small size and the lack of relevance of studying drying over hyper-arid zones. A drying peak with an AI ranging from 0.4 to 0.8 in the transitional regions between dry and wet climates was of special interest. In this book, it refers to the areas within the AI range between 0.4 and 0.8 being transitional regions. The highest variation, which occurred at intermediate moisture contents, is consistent with observations of short-term and small-scale soil moisture variation in temperate regions. The percentage of drying areas for the various climate regions is also shown, and it demonstrates the extraordinary contributions made by the transitional regions. More than 60%

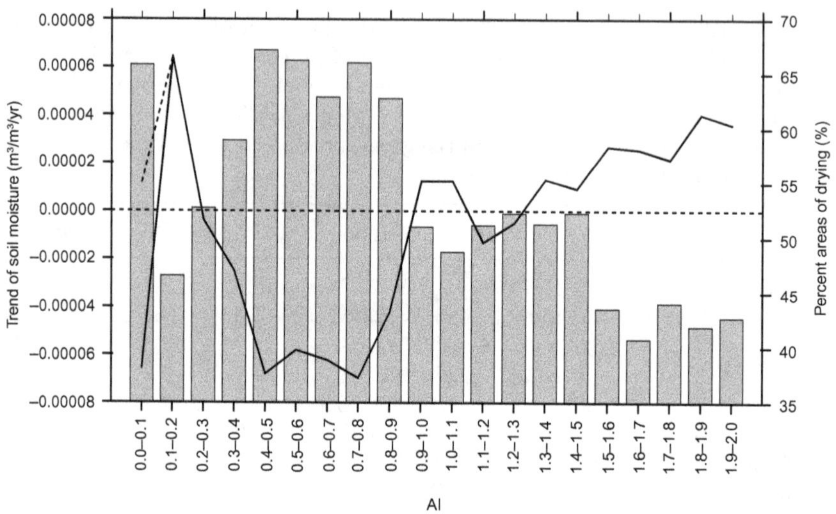

Fig. 2.8. The linear trends of annual mean soil moisture (black line) and percent areas of drying (bars) as a function of the climatological mean AI. The climatological mean AI on the x-axis is the annual mean AI for the period from 1961 to 1990 (Cheng and Huang, 2016).

of the areas in the transitional regions had negative trends in soil moisture. This suggests that, both in terms of area and intensity, the transitional zones dominate the soil moisture drying process.

2.4.4 *Lake temperature*

Based on the GloboLakes and Global Lake Temperature Collaboration (GLTC) datasets, the time series and global distribution of summer Lake Surface Water Temperatures (LSWT) anomalies were estimated and compared to the climatology values over the overlapping period of 1995–2005. The findings indicate that for the GLTC and GloboLakes datasets, the mean warming trends of the summer LSWT are 0.39° decade^{-1} (95% CI: 0.35–0.43) and 0.25° decade^{-1} (95% CI: 0.23–0.26), respectively. The summer LSWT exhibits interannual fluctuation, but lakes are warming across the board, which is in line with previous studies. The summer LSWT trends for specific lakes, however, range from −0.19 to 0.75° decade^{-1} (GloboLakes) and from −0.12 to 0.87° decade^{-1} (GLTC) (the 5th and 95th percentiles). Lakes in Western and Central North America and Europe are warming much more quickly than the average worldwide. On the other hand, lakes at low latitudes (<23.5°) often exhibit a negative LSWT trend. Furthermore, even within the same region, the warming rates in these lakes vary widely.

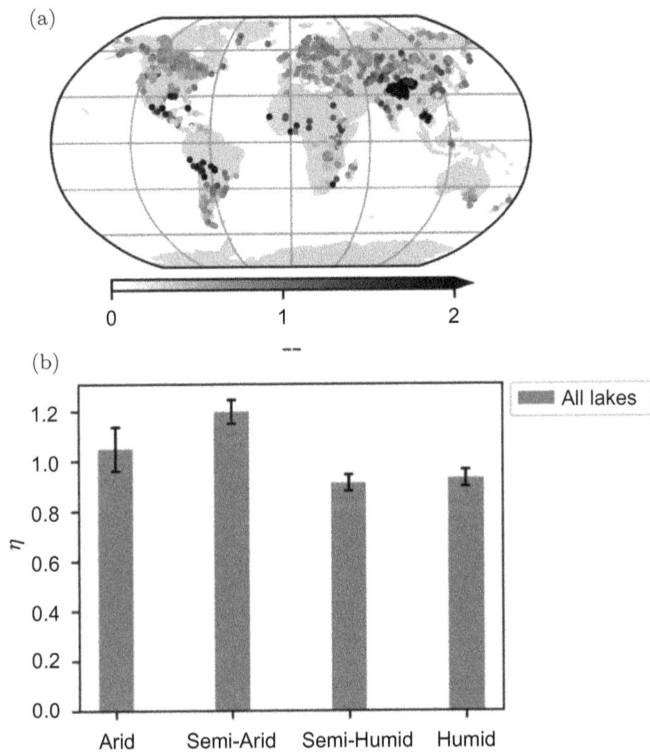

Fig. 2.9. Global spatial distribution of summer warming efficiency η of lakes during 1995–2016 (a). The mean warming efficiency of lakes in different climate zones (b). The error bars represent 1σ uncertainty (Wang *et al.*, 2022a).

Between 1995 and 2016, the average air temperature rose, except for a few spots in Africa and Asia, the center of South America, and the Northeastern part of North America that had weakly negative trends. It was discovered that these lakes are typically dispersed in places where the air temperature increased after taking the position of the lakes into consideration (hence, only the GloboLakes dataset is analyzed for consistency's sake). The number of lakes and the total area occupied within each climate region clearly distinguish the differences between them. The most number of lakes are found in semi-arid regions, followed by semi-humid and humid regions, and the least lakes are found in arid regions. The humid and semi-humid regions have the biggest cumulative area. Therefore, assuming homogeneous coverage of the GloboLakes dataset, 49.6% of all global lakes are found in arid and semi-arid regions. The humid and semi-humid areas have the highest lake coverage, which suggests that the lakes there have an averagely greater individual surface area. Last but not least, this result contrasts the

LSWT's warming trends with those of air temperature across various climate zones. Lakes in arid and semi-arid regions are generally warming at a pace of 0.28° decade^{-1} in summer, which is almost 50% quicker than those in semi-humid and humid regions (0.19° decade^{-1}). This is evident when comparing the warming of lakes in different climate zones. In addition, it was found that while lakes in semi-arid regions showed the largest LSWT rising in summer, air temperatures warmed faster in semi-humid regions than in other climate regions. More significantly, lakes in arid and semi-arid regions have the maximum sensitivity to air temperature changes when the summer warming occurs. This efficiency links LSWT warming to the concurrent air temperature changes (Fig. 2.9). The average (95% CI: 1.10–1.27) value of η in arid and semi-arid regions is 1.18, which is higher than the values of 0.91 (95% CI: 0.85–0.98) for semi-humid lakes and 0.93 (95% CI: 0.87–1.00) for humid lakes. It is interesting to notice that lakes across the Qinghai–Tibet Plateau have a significantly higher warming efficiency than lakes in other parts of the world (Fig. 2.9). Overall, these findings show that the summer LSWT in arid and semi-arid regions warms more quickly than in semi-humid and humid places, and they imply that arid and semi-arid region lakes may be more vulnerable to threats from a warming climate than other aquatic ecosystems.

2.4.5 *Seasons in semi-arid regions*

Seasonal durations have changed dramatically from 1985 to 2014 compared with the durations from 1951 to 1980: Summers are increasing longer, while the other three seasons are getting shorter, apart from specific regions of North America where alterations were incongruous with the overall trend. The high influence of internal climate variability is probably responsible for the reversal of changes (Huang *et al.*, 2017b; Guan *et al.*, 2019). The summer and winter seasons saw the biggest alterations, with summer lengthening by more than 10 days and winter lengthening by more than 8 days in most places. The strongest seasonality has been observed around the Mediterranean since the 1950s, which is compatible with the fact that seasonality is weakening in the Northern Hemisphere's mid-high latitudes. The Mediterranean region is close to where the changes in seasonal durations are most noticeable.

While the other three seasons exhibited an overall drop from 1951 to 2000, the variance trend of summer duration in each decade increasingly grew with time. In contrast, all seasons' trends from 2001 to 2014 markedly diverged from those of the preceding decades. The shortening trends in spring

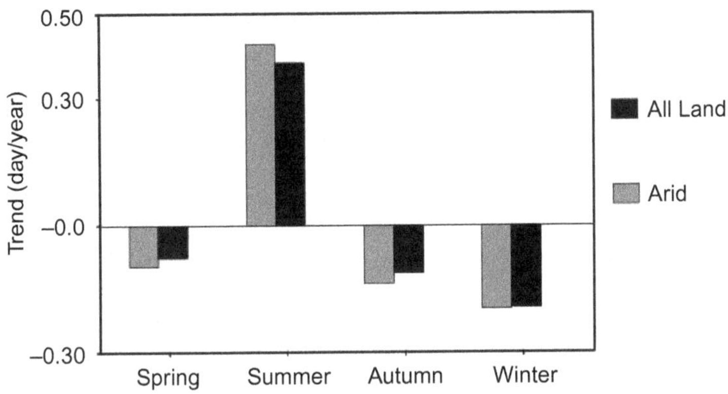

Fig. 2.10. The averaged seasonal duration trend for arid region and all land over the middle latitudes of the Northern Hemisphere.

and autumn also slowed down, while the shortening trend in winter changed to an extending trend. The extending trend in summer duration drastically fell. These indicate that the increase (decrease) in summer (spring/autumn) duration slowed down, and winter changed from shortening to lengthening. The changes might be related to the warming pause between 1998 and 2013 (Huang *et al.*, 2017b).

The averaged season duration linear trend and all grids' season lengths take a function of the climatological mean AI. It shows that, aside from winter, season lengths do not vary uniformly around the globe, but that trends in arid regions are clearly bigger than those in other regions, and the drier the land, the greater the changes in season lengths. Arid regions are more vulnerable to changes in the length of the seasons than other regions, so rising temperatures may result in more significant changes in the length of the seasons and a more serious threat. Comparing regionally averaged season duration patterns for the arid region and entire land areas over the middle latitude of Northern Hemisphere (MLNH) is shown in Fig. 2.10. Visual observation reveals that the lengths of the seasons in arid regions are shifting more quickly than in the entire study area, particularly for spring, summer, and autumn. Additionally, arid areas contribute even more to changes in season lengths across the MLNH. Less than half of the MLNH is made up of arid regions, which are responsible for 65.21%, 57.05%, 61.50%, and 52.08% of the variation in spring, summer, autumn, and winter lengths, respectively, across the MLNH territory.

2.5 Terrestrial Water Storage (TWS) Decline

Terrestrial water storage (TWS) losses are becoming more noticeable on both a global and regional scale, especially in arid and semi-arid regions that are more sensitive. In arid and semi-arid regions, the lack of water has numerous negative consequences for the health of the ecosystem, but it also has a negative impact on residents' socioeconomic growth. The hydrologic cycle, interactions between the surface and the atmosphere (such as the carbon and dust cycles), virtual water trades, and population migration can all have an impact on larger geographic areas. Therefore, for the sustainability and conservation of water resources, a correct attribution of TWS changes is crucial. This section provides a thorough analysis of recent TWS trends in each climate zone as measured by satellites, along with quantitative attributions to natural and anthropogenic sources. Although direct human activities and climate change are the main causes of zonal TWS alterations, the effects on arid and semi-arid regions and humid regions are very different.

2.5.1 *TWS dataset*

The monthly anomalies of equivalent water thickness (EWT) detected by GRACE, which fully capture the dynamic of vertically integrated water mass on land, were used to compute the trends in terrestrial water storage (TWS). Six solutions make up the most recent GRACE products, including 3 mascon solutions from the Jet Propulsion Laboratory (JPL-M), the Center for Space Research (CSR-M), and the Goddard Space Flight Center, and 3 spherical-harmonic solutions from JPL (JPL-SH), CSR (CSR-SH), and the German Research Centre for Geosciences (GFZ-SH).

Table 2.5 provides information and access sources for these solutions. The spatial variance of TWS trends across various GRACE products revealed that glacierized regions (such as Alaska, Arctic Canada North, Arctic Canada South, Iceland, the Southern Andes, and the Tibetan Plateau), where the total variance accounts for about 70% of the global variance, are where the discrepancy in the 6 GRACE solutions primarily originates. According to the results, JPL-M is the most accurate in detecting mass trends in glaciated areas. In order to compute the TWS trend in each climate zone, we chose to use JPL-M (RL06 v02), while all solutions were used to determine the uncertainty. Additionally, the JPL-M solution provided a set of 0.5-degree scale factors that the Community Land Model could replicate for amplitude recovery and improved spatial resolution. The CLM does not, however, consider all water components (such as mountain glaciers or ice sheets),

Table 2.5. List of the 6 GRACE products applied in this book, including 3 mascon solutions (JPL-M, GSFC-M, and CSR-M) and 3 spherical harmonics solutions (JPL-SH, CSR-SH, and GFZ-SH) (An *et al.*, 2021).

Name	Solutions	Time period	Spatial resolution	Source
JPL Mascons RL06 v02 with CRI	Mascons	200204–202007	3° equal-area mascons (sampled on a 0.5° grid)	https://podaac.jpl.nasa.gov/dataset/TELLUS_GRAC-GRFO_MASCON_CRI_GRID_RL06_V2 (obtained on 8 October 2020)
GSFC Mascons v02.4, ICE6G	Mascons	200301–201607	1° equal-area mascons (sampled on a 0.5° grid)	https://earth.gsfc.nasa.gov/geo/data/grace-mascons (obtained on 8 October 2020)
CSR Mascons RL06 all-corrections v02	Mascons	200204–202007	1° equal-area mascons (sampled on a 0.25° grid)	http://www2.csr.utexas.edu/grace/RL06_mascons.html (obtained on 8 October 2020)
GRACE and GRACE-FO Tellus Land Mass JPL RL06 v03	Spherical harmonics	200204–201706; 201806–202007	1° grid	https://podaac.jpl.nasa.gov/dataset/TELLUS_GRAC_L3_JPL_RL06_LND_v03; https://podaac.jpl.nasa.gov/dataset/TELLUS_GRFO_L3_JPL_RL06_LND_v03 (obtained on 8 October 2020)
GRACE and GRACE-FO Tellus Land Mass CSR RL06 v03	Spherical harmonics	200204–201706; 201806–202007	1° grid	https://podaac.jpl.nasa.gov/dataset/TELLUS_GRAC_L3_CSR_RL06_LND_v03; https://podaac.jpl.nasa.gov/dataset/TELLUS_GRFO_L3_CSR_RL06_LND_v03 (obtained on 8 October 2020)
GRACE and GRACE-FO Tellus Land Mass GFZ RL06 v03	Spherical harmonics	200204–201706; 201806–202006	1° grid	https://podaac.jpl.nasa.gov/dataset/TELLUS_GRAC_L3_GFZ_RL06_LND_v03; https://podaac.jpl.nasa.gov/dataset/TELLUS_GRFO_L3_GFZ_RL06_LND_v03 (obtained on 8 October 2020)

and the variations in terrestrial water storage that occur annually appear to dominate the scale factors. As a result, it is possible that the scale factors cannot be used to determine TWS trends.

2.5.2 *Unanimous TWS declines*

Using data on equivalent water thickness (EWT) from GRACE satellites, TWS trends in each of these climate zones were quantified (Fig. 2.11). The Jet Propulsion Laboratory mascon solution (RL06 v02) (An *et al.*, 2021) was chosen to be used for trend computation and uncertainty propagation after examining all the GRACE solutions to local observations. The TWS trends, which completely reflect the dynamic of vertically integrated water mass on land, are calculated using the monthly anomalies of EWT detected by GRACE (Tapley *et al.*, 2004). The most updated GRACE products consist of 6 solutions, including 3 mascon solutions from the Jet Propulsion Laboratory (JPL-M), the Center for Space Research (CSR-M), and the Goddard Space Flight Center, and 3 spherical-harmonic solutions from JPL (JPL-SH), CSR (CSR-SH), and the German Research Centre for Geosciences (GFZ-SH). Here, we analyzed the performance of each GRACE product in glacierized regions to increase accuracy. The discrepancy in the 6 GRACE solutions is primarily caused by glacierized regions, where the total variance accounts for about 70% of the global variance (e.g., Alaska, Arctic Canada North, Arctic Canada South, Iceland, the Southern Andes, and the Tibetan Plateau). This was revealed by the spatial variance of TWS trends across different GRACE products. The results indicate that JPL-M is the most accurate in detecting mass trends in glaciated areas. As a result, we chose to use JPL-M (RL06 v02) to determine the TWS trend for each climate zone while using all available solutions to determine the uncertainty. The JPL-M solution also provided a set of 0.5-degree scale factors for amplitude recovery and improved spatial resolution that were simulated by the Community Land Model. As a result, it is possible that the scale factors cannot be used to determine TWS trends. We also discovered that the introduction of scale factors had a minimal impact on the zonal TWS trends when comparing trends estimated with and without scale factors. As a result, we did not use scale factors to calculate the global and zonal TWS trends; instead, we cited Wang *et al.* (2018) to account for the rescaling errors that were introduced during the transition from mascons to the study region.

The monthly deseasonalized TWS anomaly dataset's best-fit linear regression was used to determine the TWS trends seen by GRACE. First,

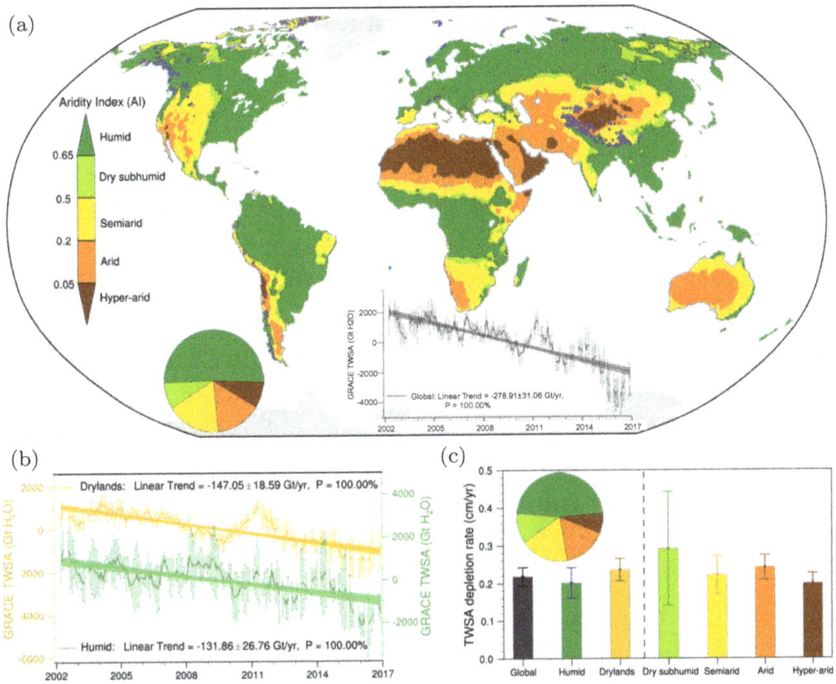

Fig. 2.11. TWS trends over different climate zones from GRACE observations, April 2002–December 2016. Map of climate zones based on the long-term aridity index (1961–1990). Blue dots show glacierized regions. The inset pie chart illustrates the proportion of global land area for each climate zone. The inset time series illustrates monthly deseasonalized TWS anomalies in global landmasses (black), with error bars and shading denoting 95% confidence intervals of monthly anomalies and the best-fit linear trend, respectively (a). Same as the inset time series in a, but for monthly deseasonalized TWS anomalies in arid and semi-arid regions (yellow) and humid (green) regions (b). TWS decline rates (in cm yr^{-1}) for each climate zone (c). Error bars show 95% confidence intervals. The inset pie chart illustrates the relative contributions of each climate zone to the global TWS change (in Gt per year) (An *et al.*, 2021).

area-weighted scaling was used to determine the monthly anomalies of EWT in each climate zone and on the entire planet (i.e., $\mathrm{A} = \sum r_i a_i / \sum r_i$, where A stands for the monthly anomaly for the study region, a_i for each mascon i that intersects this region, and r_i for the intersection area). We used a linear interpolation to fill in the missing months in order to generate deseasonalized series, which was necessary to calculate the mean monthly seasonal cycle (Rodell *et al.*, 2018). The climatological mean of the corresponding month was then subtracted from the initial time series.

The systematic errors in JPL-M (es), the discrepancy between the various GRACE solutions (em), and the rescaling errors (er) from the GRACE masons to the climate zone all contributed to the uncertainty of the monthly TWS anomaly. The original datasets provided *es*, which stands for the intrinsic errors associated with the initial 3° equal-area mascons; *em*, which stands for the standard deviation between the estimates from the 6 GRACE solutions; and *er*, which was computed in accordance with Wang *et al.* (2018) and was caused by signal leakage in fringe mascons. In short, it calculated the ratios (RA) of each mascon's intersecting area with the research region to the overall mascon area initially. The average EWT anomalies in the area bounded by the mascons were then determined for each month using RA $\geq t$ (where t grew from 0 to 1 with a 0.05 interval). The standard deviation of average EWT anomalies with various RA ranges was then used to determine monthly *er*. To obtain the monthly TWS anomaly series' uncertainty (E), *es*, *em*, and *er* were propagated. Then, using a Monte Carlo approach that takes E and the linear fitting residuals into account, we determined the TWS trend's confidence interval.

The contributions of first-order controls or outcomes of the three mechanistic drivers — precipitation for climatic variability, glacier mass changes for climate change, and water use and regulation for direct human activities — were then broken down into the GRACE-observed TWS trends. We believe that the precipitation fluctuation during the analyzed 15-year GRACE era was mostly a response to interannual and intraannual climate variability, even though long-term precipitation trends could be a result of climate change. The ensemble of the GRACE-REC datasets (GRACEREC-EM), which were statistically trained with GRACE observations and multi-source meteorological forcing, was used to calculate the precipitation-induced TWS trends.

The GRACE-REC datasets were used to evaluate the contributions of precipitation-driven climate variability to TWS trends. These datasets, which were produced using a statistical model that was forced by meteorological datasets and calibrated by GRACE measurements, allowed for the reconstruction of climate-driven TWS variations. The reconstruction excluded all long-term trends except for precipitation from the input data; therefore, the residual trends in the GRACE-REC TWS anomalies were essentially entirely caused by variations in precipitation. Since the statistical model was calibrated with two GRACE observations (JPL mascons and GSFC mascons) and forced by the climatic variables (such as precipitation and temperature) from three meteorological datasets (ERA5, GSWP3, and MSWEP), it created six separate GRACE-REC datasets. It is accessible at https://doi.org/10.6084/m9.figshare.7670849, and has 100 ensemble members for each dataset. The six GRACE-REC datasets include three different time periods because the meteorological datasets have a limited time span: 1979–2016, 1979–present, and 1901–2014. We constructed an ensemble (GRACE-REC-EM) to limit the uncertainties brought on by a particular dataset and selected the 4 separate GRACE-REC datasets (i.e., GSFC-ERA5, GSFC-MSWEP, JPL-ERA5, and JPL-MSWEP), whose time periods approached December 2016. Furthermore, the individual errors and ensemble mean errors of the four applicable GRACE-REC datasets were used to propagate the final uncertainties of precipitation-induced TWS trends. The ensemble mean errors were estimated as the standard deviations of the four GRACE-REC datasets, while the individual errors were created from the 100 members of each GRACE-REC dataset.

We evaluated the contemporaneous precipitation from direct measurements in order to verify the zonal precipitation contributions from GRACE-REC. Changes in TWS in a location indicate the difference between water ingress and outflow, according to the water balance equation. Precipitation (P) dominates the input, while evapotranspiration (ET), runoff (R), and human abstraction make up the outflow (HA). In this way, the availability of TWS is directly impacted by the amount of precipitation. We compared the yearly mean precipitation during our study period (2002–2016) with the average precipitation across the lengthy period from 1979 to 2016, using Rodell *et al.*'s (2018) research as a guide, considering that TWS maintains a dynamic equilibrium under a normal precipitation state. As a result, precipitation that is more (or less) than typical may indicate a positive (or negative) impact on TWS. The Global Precipitation Climatology Center (GPCC), the Global Precipitation Climatology Project (GPCP), the Precipitation Reconstruction over Land Precipitation (PREC/L), and the Climate Prediction Center Merged Analysis of Precipitation were used to obtain the annual means of precipitation to account for the uncertainties in precipitation datasets (CMAP).

The world's glaciers were all covered by fine-spatiotemporal-resolution surface elevation time series that were reconstructed from stereo imagery and confirmed against highly accurate measurements to determine glacier mass changes (Hugonnet *et al.*, 2021]. By integrating the individual mass changes of all the glaciers in each climate zone, we were able to determine the zonal glacier mass changes (Hugonnet *et al.*, 2021). These estimations are based on volume change time series derived mostly from the stereo imagery library of NASA's Advanced Spaceborne Thermal Emission and Reflection Radiometer and calculated from repeated digital elevation models

(ASTER). All glaciers whose outline centroids were within the range of our climate zones were examined. This entire glacier area, excluding Greenland and Antarctica, is 433,301 km^2, or nearly 90% of the total glacier area on Earth. Within each climate zone, we added up the contribution of each glacier to the volume change. The Randolph Glacier Inventory 6.0 first-order areas intersect with the clusters of glaciers that can be found in each climate zone. Based on multi-range spatial correlations, the uncertainty of individual glacier volume changes was propagated at the regional scale for the clusters in each climate zone (Hugonnet *et al.*, 2021). When converting glacier volume changes to mass changes, a density factor of 850 kg m^{-3} and a 1-sigma uncertainty of 60 kg m^{-3} were used (Huss, 2013). The final step was to assume independence for the mass change uncertainty for each crossing RGI cluster and then add them quadratically. The geographic correlations for volume changes considered potential temporal interpolation problems and were based on comparisons with highly accurate ICES at elevation readings. By assuming a negligible temporal autocorrelation of density conversion factors, we presented the annual mass change uncertainty at wide scales (i.e., the entire planet's landmass, wet areas, and arid and semi-arid regions).

We used the most recent iteration of the cutting-edge hydrological model WaterGAP2.2d to calculate the effect of direct human activities on TWS trends. This model fully represents human interference with the water cycle, including irrigation, reservoir regulation, groundwater abstraction, and multi-sectoral water use (Müller Schmied *et al.*, 2021). Through sensitivity testing, TWS trends caused directly by human activity were extracted from the Water Global Assessment and Prognosis (WaterGAP) hydrological model. WaterGAP, which mimics global water flows, storage, withdrawals, and consumption uses, is frequently used in both large-scale and worldwide investigations. Here, we used WaterGAP2.2d (Müller Schmied *et al.*, 2021), the most recent version, which has demonstrated acceptable model performance in the assessments of important output variables, such as total water storage, withdrawal water uses, and streamflow (Müller Schmied *et al.*, 2021). Specifically, for the course of our study period, WaterGAP2.2d was compelled to employ WFDEI-GPCC meteorological datasets to simulate real evapotranspiration, sectoral water usage, and TWS anomalies. TWS incorporates water storage in canopies, lakes, wetlands, groundwater, manmade reservoirs, rivers, soil, and snowpacks in non-glacierized regions but leaves out glaciers. The prospective and actual water withdrawal and consumption for domestic, industrial (i.e., the total of manufacturing and thermal power), and livestock sectors are all included in sectoral water usage.

There are specifically two WaterGAP simulation scenarios: WaterGAP-histsoc and WaterGAP-nosoc. While WaterGAP-nosoc is the naturalized mode, which solely accounts for climate-induced variabilities without taking reservoirs and human water abstraction into consideration, WaterGAP-histsoc is the standard mode that accounts for both climate-induced and human-induced variabilities (i.e., direct human activities). The contributions of each land water storage component and the human-induced AET were determined by comparing the outputs from the standard and naturalized modes.

Through sensitivity studies with WaterGAP2.2d that contrasted the simulations from a "normal" run considering both climatic and human influences with a "naturalized" run considering no direct human interventions, direct human impacts on TWS and each water component were identified (although land use is included). In order to further validate our calculations from GRACE-REC-EM and WaterGAP2.2d and to simplify result interpretation, we also used the simulated climate-driven TWS changes from WaterGAP2.2d, human water-use data from the Inter-Sectoral Impact Model Intercomparison Project (ISIMIP; https://www.isimip.org), and auxiliary irrigation data.

All examined temperate zones have had evident TWS losses in recent decades, although arid and semi-arid regions in particular show a more pronounced and quicker drop in TWS than humid regions (Fig. 2.11). Evident TWS losses have been observed in all the climate zones that have been investigated between April 2002 and December 2016. The net TWS change in the world's humid areas was -131.9 ± 26.8 Gt yr^{-1} (uncertainties are 95% confidence intervals), which accounts for almost 47% of the concurrent total change in the landmass (-278.9 ± 31.1 Gt yr^{-1}) (Fig. 2.11). Despite having a restricted supply of freshwater resources, arid and semi-arid regions were responsible for the remaining more than 50% of the global net loss (147.1 ± 18.6 Gt yr^{-1}) (Chang *et al.*, 2020). In arid and semi-arid regions, the area of significant TWS loss (3.1×10^7 km^2) is greater than that of significant TWS gain (2.2×10^7 km^2) by more than 40%. In worldwide humid regions, the areas that experienced significant TWS loss and gain (with a 0.05 significant level) are generally comparable. All four dryland climate zones have net TWS losses, with an area of considerable water loss in each exceeding 50%. These findings imply that although there was a net TWS drop in both humid regions and drylands, the former TWS decline was more pronounced and extensive.

In addition to losing more water overall, global arid and semi-arid regions have higher TWS depletion per unit area (or EWT) than humid regions (Fig. 2.11(c)). Although humid regions are home to a majority of the world's glaciers, TWS in arid and semi-arid regions was depleted at a pace of 2.4 ± 0.3 mm yr^{-1}, which is higher than the rate of 2.0 ± 0.4 mm yr^{-1} in those regions. The hyper-arid zone, which includes most of the interior of the Sahara and Taklamakan deserts, exhibits a decreasing rate of 2.0 ± 0.3 mm yr^{-1} or about 7% (20.5 ± 2.9 Gt yr^{-1}) of the worldwide net loss. It is the fastest declining of the four arid and semi-arid region subregions. Given that the hyper-arid zone has highly restricted water supplies but saw a water loss per area equivalent to that in globally humid regions, this dropping pace is startling.

The rates of TWS decline per unit area in the remaining arid and semi-arid region subregions are all higher than the humid regions. The arid and semi-arid regions, which are distributed in the periphery of hyper-arid regions, suffer from a net TWS loss of 2.4 ± 0.3 mm yr^{-1} (45.6 ± 6.2 Gt yr^{-1}) and 2.2 ± 0.5 mm yr^{-1} (50.0 ± 11.4 Gt yr^{-1}), respectively. These two subregions alone account for more than one-third of the net worldwide loss due to TWS depletion. The outermost dry subhumid zone has had the most significant TWS depletion per unit area among the four subregions (2.9 ± 1.5 mm yr^{-1}), accounting for about 8% of the world land surface while suffering from about 11% of the global TWS loss (31.0 ± 16.1 Gt yr^{-1}). These widespread zonal TWS losses not only highlight the extreme water scarcity in arid and semi-arid regions but also point to significant climatic and direct human activity-related consequences. Additionally, the rate of TWS drop in arid and semi-arid regions is more monotonic with lower uncertainty (Fig. 2.11(b)), indicating a longer-lasting water deficit. These widespread zonal TWS losses demonstrate the extreme water scarcity in drylands and indicate the urgent need to investigate the cause.

2.5.3 *Limited impacts of climate variability*

Over the 15 years we analyzed, climate variability had only minor effects on TWS trends in each of the climate zones (Figs. 2.12(a), 2.12(b)). Changes in precipitation have produced spatially varied TWS trends at regional levels as the main indicator of inter-annual climatic variability, as shown by the Okavango Delta in Africa and the moist Northern Great Plains of North America, as well as the dry Eastern Brazil and Northwestern Australia (Rodell *et al.*, 2018). But when the regional trends are combined over the non-glacierized continents of the world, the net impact of precipitation is only approximately 16.7 ± 35.3 Gt yr^{-1}, or 6% of the reported global TWS drop (Figs. 2.12(a) and 2.12(b)). With a contribution of little more than 15 Gt yr^{-1} or −20%, precipitation likewise provides only a negligible explanation for the observed TWS drop in each climate zone. The direct observations of precipitation support these minimal contributions, and they are also consistent with the ENSO trend between 2002 and 2016, during which the effects of the major La Niño events in 2008–2014 were essentially offset by those of the subsequent El Niño (Wang *et al.*, 2018).

Although other types of climate forcing, such as temperature, also had an impact on the observed TWS trends, the attribution to climate variability focused primarily on precipitation. We used the hydrological model WaterGAP2.2d (Müller Schmied *et al.*, 2021) with full climate forcing (including both precipitation and temperature) to simulate climate-driven TWS variations during the examined 15 years in order to get a basic sense of the impact of other climatic variables. The modeled TWS trends, with their deviations within the uncertainty intervals, are comparable to those of GRACE-REC-EM. This comparison strengthened the case for why we chose to use precipitation as a primary indicator of the magnitude of the influence of climate variability on zonal TWS trends.

2.5.4 *Net TWS decline in humid regions*

The study reveals that the net TWS loss across the world's humid regions is mostly caused by changes in glacier mass as a result of a warming environment (Figs. 2.12(c) and 2.12(d)). Four percent (or about 210,000 km^2) of the world's wet regions are covered by glaciers. The near-surface temperature over glacierized regions rose by 0.6°C from 2002 to 2016, which caused a startling glacier mass loss of 136.4 ± 23.8 Gt yr^{-1} in humid areas. The total glacier mass loss in humid areas matches its net TWS drop (131.9 ± 26.8 Gt yr^{-1}) despite limited spatial coverage. Climate change and human activity appear to balance each other out in the remaining 96% of the humid regions that are not glacierized, resulting in a net TWS equilibrium throughout the course of the study period. The overall area of non-glacierized humid regions with significant drying tendencies is equal to that of humid regions with wetting trends, according to GRACE measurements, which also recorded this occurrence.

Glacier retreat only accounts for around one-third of the TWS loss in the world's arid and semi-arid regions, in contrast to the dominant role that

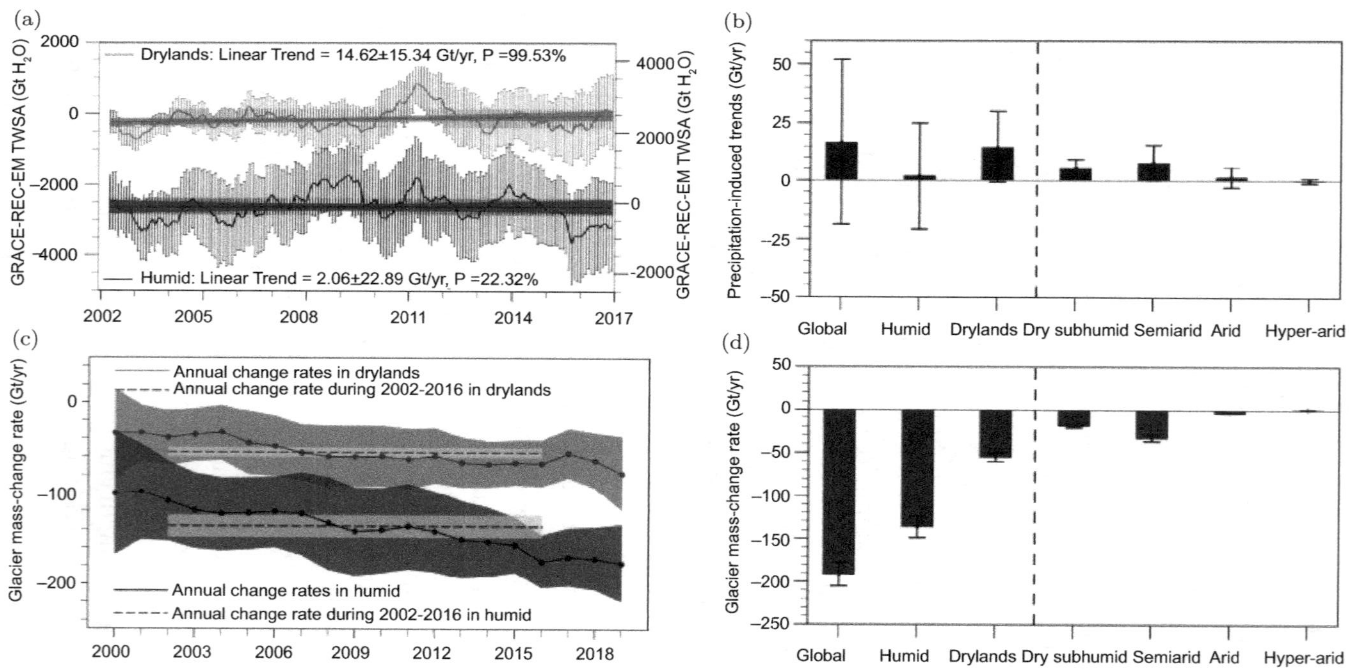

Fig. 2.12. Contributions of precipitation and glaciers to terrestrial water storage (TWS) trends in studied climate zones. Monthly time series of precipitation-induced TWS anomalies in arid and semi-arid regions and humid regions from April 2002 to December 2016. Error bars and shading illustrate 95% confidence intervals of monthly anomalies and best-fit linear trends, respectively (a). Precipitation-driven TWS trends in each of the climate zones (b). Annual time series of glacier mass change rates in arid and semi-arid regions and humid regions. The shading indicates 95% confidence intervals. Dotted horizontal lines denote the 15-year averages (during 2002–2016) (c). Glacier mass change rates in each climate zone (d) (An *et al.*, 2021).

climate change plays in humid regions. From 2002 to 2016, we estimated a cumulative glacier mass loss of 54.9±5.1 Gt yr^{-1}. The semi-arid region, which includes the glaciers in Arctic Canada and a portion of those in the HMA, has the biggest glacier extent (about 120,000 km^2) and the highest mass loss rate (32.9 ± 3.7 Gt yr^{-1}) of the four dryland zones. The rate of glacier mass loss in the dry subhumid zone is 19.4 ± 1.6 Gt yr^{-1} (Fig. 2.12(d)). Most of the glaciers in Central HMA are in the remaining two zones, which are the hyper-arid and arid regions. These two regions, which are the only ones in the world outside of the ice sheet peripheries to have seen a net glacier accumulation over the past two decades, were impacted by the slight mass gain in the Karakoram and West Kunlun glaciers (Hugonnet *et al.*, 2021). Therefore, the TWS decreases in these two zones were only little impacted by the glacier mass changes in these two regions (−3.4 ± 0.9 Gt yr^{-1} in the arid zone and 0.7 ± 0.4 Gt yr^{-1} in the hyper-arid zone). Though not yet the primary factor in the TWS declines in global drylands, the increased glacier mass losses caused by rising temperatures in all dryland climate zones show that climate change is a significant, if not yet dominating, contributor to those declines.

2.5.5 *TWS declines in arid and semi-arid regions*

While the recent net TWS drop in humid regions had been caused by climatic factors, the widespread water loss across the world's arid and semi-arid regions was primarily due to direct human activities (Fig. 2.13). In the non-glacierized world, direct human activities such as human water withdrawals and reservoir management caused a total TWS loss of 104.3±5.3 Gt yr^{-1}, of which 94.3 ± 3.0 Gt yr^{-1} (90%) originated from arid and semi-arid regions. Regional hotspots where the rate of water consumption exceeds 10 mm per year are where the majority of human-induced TWS loss in arid and semi-arid regions is concentrated (Fig. 2.13(a)). Even though these consumption hotspots seem sporadic and only make up about 4% of global dryland area, they are widely distributed, especially in the Northern Hemisphere, where they can be found in places like the Arabian Peninsula, Western India, the Northern Sahara, the Western US, the Southern Great Plains, and the North China Plain. In total, drylands made up around 86% of the world's hotspots for human consumption. These discrepancies show a disproportionate human influence and dependency on the already fragile ecology of arid and semi-arid regions (Figs. 2.13(a) and 2.13(b)). The substantial TWS decline in arid and semi-arid regions was caused by unsustainable human water abstraction

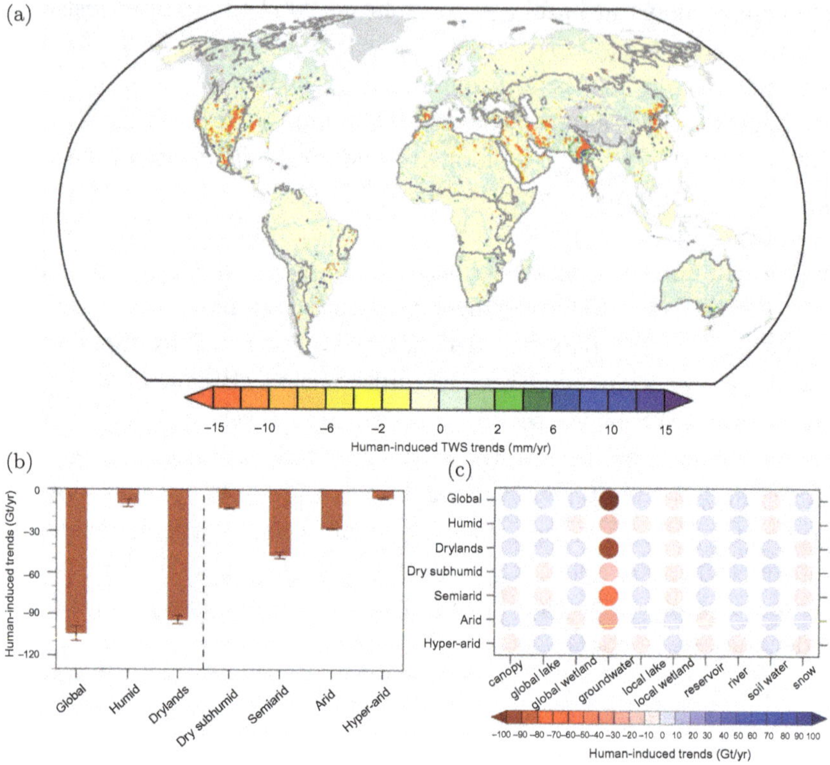

Fig. 2.13. Terrestrial water storage (TWS) trends induced by direct human activities. Global map of human-induced TWS trends simulated by the WaterGAP2.2d hydrological model during April 2002 through December 2016. Lines delineate the extent of global drylands (a). Human-induced TWS trends in each climate zone. Error bars show 95% confidence intervals (b). Human-induced TWS trends for different water components in each climate zone (c). Glacierized regions are excluded in this result (An *et al.*, 2021).

to meet the rising demands from the livestock, irrigation, domestic, and industrial sectors. This increased regional water outflow through increased evapotranspiration ultimately resulted in the TWS declines of arid and semi-arid regions.

We found that widespread irrigation techniques frequently coexist with larger human-induced TWS decreases. The semi-arid region saw the most human-induced TWS loss (47.3 ± 2.4 Gt yr^{-1}) and has the largest irrigation area and population of the four dryland climate zones. In order of decreasing irrigation area, these are followed by the arid zone (27.9 ± 0.6 Gt yr^{-1}), the dry subhumid zone (13.3 ± 0.6 Gt yr^{-1}), and the hyper-arid zone (5.7 ± 0.7 Gt yr^{-1}) (Fig. 2.13(b)). Irrigation typically uses more water than other water-use sectors, and a large portion of the total water extraction is used to meet crop biophysiological needs like transpiration. This is especially true in drylands, where there is a greater need for irrigation both in terms of area and percentage due to the low yearly precipitation and high atmospheric water demands. Due to these factors, global drylands dominate the world's irrigation water consumption, which significantly outweighs the total consumption in the other sectors (e.g., by nearly seven times globally). This is true even though humid regions account for most water consumption in the industrial and domestic sectors. These findings imply that the excessive TWS reduction is probably caused by widespread irrigation throughout drylands.

Direct human activities had an impact on the budgets of all water components, but a thorough division using hydrological modeling showed that groundwater depletion was a major factor in human-caused TWS decreases (Fig. 2.13(c)). The global anthropogenic groundwater depletion reached 111.6 ± 1.0 Gt yr^{-1} from 2002 to 2016, accounting for 40% of the contemporaneous global net TWS loss (278.9 ± 31.1 Gt yr^{-1}). However, water controls in large reservoirs only partially alleviated the global TWS decline (6.9 ± 4.9 Gt yr^{-1}). According to Fig. 2.13(c), drylands account for around 88% of the global groundwater depletion, which means that within drylands, water loss from aquifers caused by humans (98.6 ± 0.9 Gt yr^{-1}) accounts for close to 70% of the net TWS drop (147.1 ± 18.6 Gt yr^{-1}). The overall trend of this excessive groundwater depletion in arid climates is consistent with other findings (e.g., Wang *et al.*, 2018).

The severity of irresponsible water abstraction in arid areas is highlighted by the startling dominance of groundwater depletion over dryland TWS decreases. Due to the scarcity of surface water in drylands and the distinct seasonality of their precipitation, society has come to rely more and more on groundwater, which in some extremely arid areas is the only reliable water source. This is demonstrated by the fact that in each dryland zone, groundwater-fed irrigation accounts for more land area (between 3.0% and 6.4%) than humid regions (2.0%). Aridity, meanwhile, causes incredibly poor rates of groundwater recharge. Aquifers in arid and semi-arid regions also frequently have deeper water tables and can only be refilled by concentrated stream leakage or by severe rains (Bierkens and Wada, 2019). Due to a chronic imbalance between water abstraction and recharge caused by all of these circumstances, many aquifers in arid and semi-arid regions have insufficient budgets (Wang *et al.*, 2018). Several of the depletion hotspots shown in Fig. 2.13, including those in North China, the Indus Basin, and the Arabian Aquifer, serve as examples. In light of this, we come to the conclusion that, despite significant climate impacts on regional water budgets (Chang *et al.*, 2020; Rodell *et al.*, 2018; Wang *et al.*, 2018), the net TWS decline in global drylands is primarily attributed to direct human water-use activities, particularly unsustainable groundwater abstraction.

The attribution of the TWS losses focused on the effects of primary controls for each of the three mechanistic causes and included satellite data and model simulations (i.e., precipitation, glacier mass changes, and human water consumption and regulation). These techniques inevitably entailed some simplification and could not completely close the GRACE-observed TWS budgets. The magnitude of the effects of other factors, such as temperature trends over non-glacierized regions, new reservoir impoundments, and changes in land use and land cover, may be suggested by the residuals, but they do not appear to influence our main conclusions. Satellite data, model simulations, and attribution process assumptions are some of the sources of the residuals. In particular, measurement and leakage errors are included in GRACE products, which determine the TWS anomalies from gravity fields recorded by satellites. Aside from the inaccuracies brought on by diverse input data, certain mistakes originated from the model simulation or rebuilt process itself, which are unavoidable and challenging to quantify for the WaterGAP2.2d model outputs and GRACE-REC datasets. For instance, global hydrological models (such as WaterGAP and PCR-GLOBWB) tend to underestimate large water storage trends in comparison to GRACE trends at regional scales. This may be because the models have a low storage capacity, the climate forcing is uncertain, and human activity cannot be accurately modeled. But when used at large scales, these model outputs, such as GRACE-REC observations and WaterGAP2.2d model simulations, generally perform reasonably well (Müller Schmied *et al.*, 2021). To reduce the uncertainty, we decided to investigate the causes of TWS patterns in vast climate zones. Additionally, we highlighted the role of climatic variability brought on by precipitation, and residuals may also be influenced by other climate forcings. Theoretically, the glacier mass losses in endorheic basins may not fully convert into zonal TWS losses since the glacier mass losses may be transferred from the ocean to terminal water bodies. Although Wang *et al.* (2018) estimated that the total glacier mass change in endorheic basins was minimal (-4.9 Gt yr^{-1} between 2000 and 2016), the associated water relocation may have had an impact on the residuals in semi-arid, arid, and hyper-arid regions, which are typically found in endorheic basins.

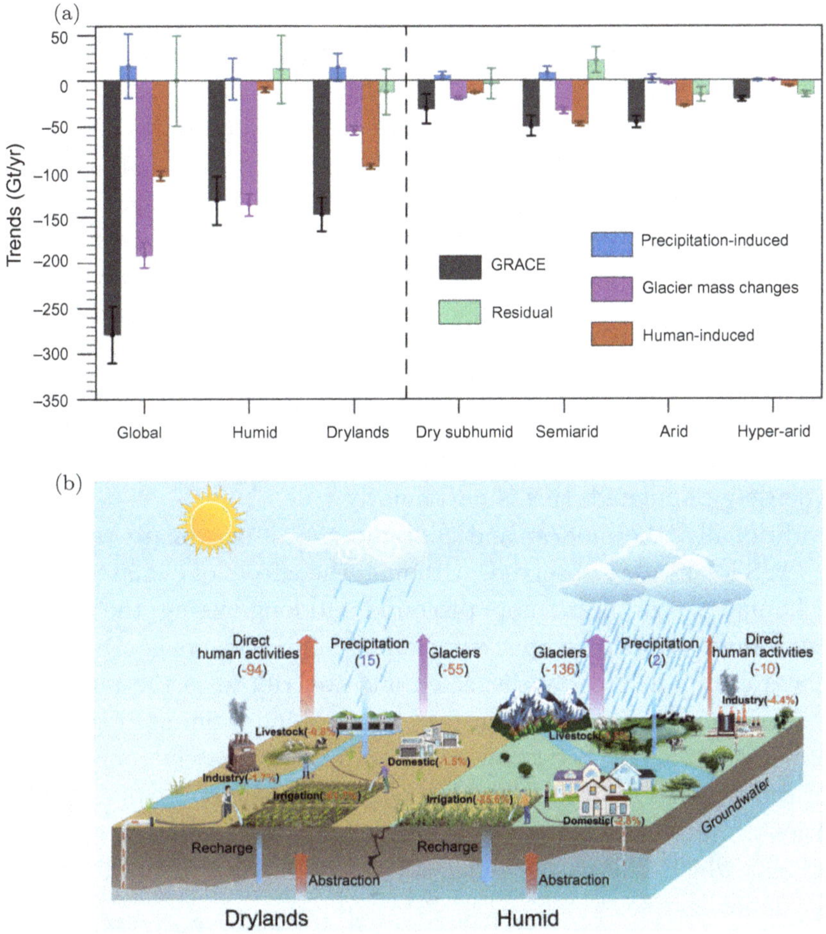

Fig. 2.14. Quantitative attribution of terrestrial water storage (TWS) trends in each climate zone during April 2002 through December 2016. Contributions of each driver to zonal TWS trends (a). Schematic diagram summarizing the contributions of major drivers to the net TWS trends in global arid and semi-arid regions and humid regions (b). Contributions of direct human activities, precipitation, and glacier mass changes are in Gt yr^{-1}, whereas sectoral water consumptions are in percentage (An *et al.*, 2021).

On the one hand, the residuals for humid and arid regions are both within their respective uncertainty ranges. The overall residual for drylands, on the other hand, is -12.5 Gt yr^{-1}, which somewhat undershoots the GRACE-observed trend. Since direct human activities constitute the main cause of the global dryland water loss, a more thorough attribution of this residual may help to further close the gap in the dryland TWS drop. However, it is unlikely that this attribution will change the overall conclusion.

It is also important to note that the spatial variations in the accuracy of applied data and model simulation contribute to a general reduction in the capacity to attribute TWS variation in smaller areas. Because of this, only the major climate zones were included in the attribution, and caution should be used when applying the findings to smaller spatial scales. At continental and subcontinental scales, we repeated the attribution for six regional drylands as part of an experiment. The outcomes show regional variation as expected. There is no discernible net TWS decline in drylands of Southern Africa and Australia. Declining TWS signals coexisted with generally smaller water gains in other dryland regions (such as the majority of sub-Saharan Africa) (MacDonald *et al.*, 2021). However, most dryland areas that saw a net TWS drop continue to be dominated by direct human activity. In the Northern Hemisphere, where drylands are both larger and more densely populated, this is particularly true.

Additionally, the modest and sporadic effects of climate variability on zonal TWS losses were observed. Although the effects of climate change and direct human activity were more profound and long-lasting, the distribution of their dominance differs between humid and arid areas. Direct human influences are more broadly dispersed and contribute to the main cause of the water loss in drylands, in contrast to the trend in humid regions, where climate-dominated TWS decreases are disproportionately concentrated on glacierized regions. Additionally, we show that the influence of direct human activities on dryland TWS was primarily due to excessive groundwater depletions, which led to declining water tables and jeopardized shallow aquifer-dependent ecosystems (Fig. 2.14).

2.6 Conclusions and Discussion

Semi-arid areas make up more than one-third of the world's arid and semi-arid region area, making them the largest subtype of arid and semi-arid regions overall. Half of the entire expansion of arid and semi-arid regions occurs in semi-arid areas, where arid and semi-arid regions also grow to their maximum extent. Semi-arid areas have grown significantly over the past 61 years (1948–2008), with the area between 1990 and 2004 being 1.6×10^6 km^2 (or 7%) bigger than that between 1948 and 1962 on a global scale. Semi-arid zones in the Eastern Hemisphere have spread across the continents at a rate of 75% compared to 25% across the American continents, totaling 1.6×10^6 km^2. Although semi-arid zones are expanding in both regions, the patterns of expansion are changing in various ways. Over the American continents, the newly established semi-arid zones have replaced the dry regions where the climate grows wetter. In contrast, semi-arid zones are replacing subhumid/humid regions on the continents of the Eastern Hemisphere, where the climate grows drier. Five of the eight semi-arid regions under consideration see a dryness increase, whereas three of them — North America, South America, and Central/West Australia — experience a wetness increase. In the drying semi-arid regions, the rates of change in precipitation, PET, and AI are substantially higher than those in the older semi-arid regions. However, there are only very minor differences between the newly formed and old areas in the wet semi-arid regions that changed from drier climate types, and the precipitation, PET, and AI exhibit similar trends.

The effects could be a result of multiple forced temperature shifts. On a global and regional scale, effects can be primarily separated into two circulations: that of the ocean and atmosphere and that of the land and atmosphere (Fig. 2.15). We received various combinations of phases of ocean oscillation and atmospheric activity from large-scale circulation. Air circulation, sea surface temperature anomalies brought on by greenhouse gas forcing, interactions between the land and the atmosphere, changes in the snow and frozen soil cover, and local human activities may all have a major role in the arid and semi-arid region warming (Huang *et al.*, 2008a;

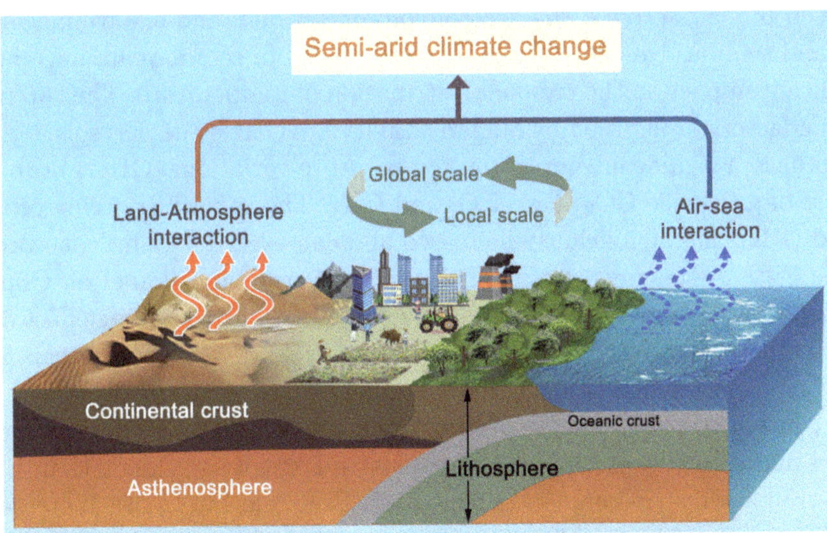

Fig. 2.15. Schematic illustration of the semi-arid climate change.

Guan *et al.*, 2009). The various reactions of semi-arid regions to the decadal variability in the westerlies and monsoon systems under global warming are revealed by the distinct behaviors of the ancient and newly developed semi-arid zones in North America and East Asia. While the wetting semi-arid regions over North America are primarily controlled by enhanced westerlies, which are connected to the decreased ocean–land temperature gradient under global warming, the climate change in the drying semi-arid region over East Asia is primarily driven by a weaker East Asian summer monsoon.

Furthermore, by acting as boundary forces, starting and maintaining specific circulation anomalies in the atmosphere at the appropriate time scales, and exhibiting decadal variations and long-term trends, the world oceans' sea surface temperatures (SSTs) also play a significant role in the widespread drying of semi-arid regions. There is evidence that ocean-atmosphere interactions also govern the widespread drying that has taken place over the years in semi-arid places (Huang *et al.*, 1998). Studies employing atmospheric models pushed by long-term SSTs provide evidence in favor of the idea that the oceans regulate the climate of the Sahel and other extra-tropical semi-arid regions. The present recovery can be related to Atlantic Ocean variability; warming of the tropical oceans is a non-negligible component of historical Sahelian droughts. Warming in the Tropical Pacific could diminish the summer monsoons, leading to drier weather in Eastern China.

In the meantime, Guan *et al.* (2015b) discovered a close association between human activity, the aerosol column burden, and a warming in the arid and semi-arid regions. The GHGs produced in recent years have had a significant impact on the regional distribution of temperature. The two main driving factors brought on by human activity are GHGs and aerosols, both of which have a significant impact on temperature variations. GHGs primarily consist of gases like CO_2, O_3, N_2O, and CH_4. The radiative forcing of well-mixed GHGs has significantly increased and can exceed 3 W m^{-2}, according to the Fifth Assessment report of the Intergovernmental Panel on Climate Change (IPCC). Increased CO_2 has the biggest radiative forcing effect of all the components, with a radiative forcing of 1.68 W m^{-2}, which can grow to 1.82 W m^{-2}, when other carbon-containing gases that can increase the CO_2 concentration are taken into account. Additionally, model simulation results indicate that rising CO_2 levels would result in a decline in mean terrestrial aridity, making land areas drier (Feng and Fu, 2013). Increases in GHGs, such as CO_2, may further affect the vegetation of arid and semi-arid regions due to the effects of GHGs on global precipitation patterns and the fertilization effect (Ukkola *et al.*, 2015).

The interactions between the atmosphere, land surface, snow/ice and frozen ground cover change, and local human activity have become more active in the past decades. In the last 30 years, satellite measurements of the Northern Hemisphere show that snow cover has significantly decreased, especially from early spring through summer (Zhai and Zhou, 1997). In the cold season, a decrease in snow cover could result in a rise in surface temperature. Due to winter warming, the thickness of the seasonally frozen ground has diminished. However, when deposited on snow, both dust and biomass burning aerosols may have an impact on the surface albedo; soot has significant effects on sunlight absorption (Qian *et al.*, 2011). A variety of models and observations were used to examine the effects of these processes on springtime climate. Qian *et al.* discovered that equilibrium climate experiments indicate that fossil fuel and biofuel emissions of black carbon plus organic matter (BC + OM) cause 95% as much loss in springtime snow cover over Eurasia as anthropogenic carbon dioxide, as a result of strong snow feedback and significant BC + OM emissions from Asia. BC combined with snow caused a 0.5–3% disturbance in the snow albedo and elevated the temperature of the snow skin (2-m air) over most snow-covered locations in the Western US from late winter to early spring by 0.2°C–1.4°C (0.1°C–1.0°C). In semi-arid regions, previous findings mostly identified a single thermodynamic factor, the enhanced altered variables. However, it is difficult to discern the linked or even triple combined effect of many thermodynamic parameters. To better understand how these elements contribute, more research is required.

Human water use, which we measured to be the primary cause of the loss in arid and semi-arid region TWS, is anticipated to rise toward the end of this century as population and living standards continue to rise. Future climate change may lessen or even reverse the impact of water depletion in some regions, but as human activity increases and water use is improperly managed, this benefit may be partially offset and water stress may worsen in other areas. This likelihood is particularly high in drylands, where both population increase and environmental fragility are more pronounced. In the meantime, regional warming in arid and semi-arid regions may amplify water scarcity (Huang *et al.*, 2017c) and hasten the deterioration of nearby and neighboring arid and semi-arid regions through water and dust cycles. Once the land has been degraded, human ecological restoration will need more water, which will lead to more water depletion. The altered TWS balance in the sensitive global drylands calls for immediate adaptation measures, such as the establishment of thorough water resource observation systems and the development of more efficient water management policies.

These processes in arid and semi-arid regions will further alter the regional biogeochemistry in the context of an increase in warming, drought frequency and severity, and evaporative demand in arid and semi-arid regions at faster rates than global means. And there are two key challenges faced by us for evaluating biogeochemical responses to climate change in these regions and for improving our capacity to predict ecosystem changes, which include high levels of spatial and temporal heterogeneity in environmental conditions and generalized resource scarcity. As the Earth's largest terrestrial biome, arid and semi-arid regions are distributed across all continents and encompass enormous variability in soil-forming factors (i.e., parent material, topography, soil age, climate, and organisms), which cause fundamental differences in carbon and nutrient cycling dynamics from local to continental scales. Some studies have shown that climate and local-scale topography can interact to strongly mediate soil respiration responses to rainfall pulses. Besides, biogeochemical cycling in arid and semi-arid regions is sensitive to interannual and intraannual fluctuations in precipitation and temperature. These fluctuations can result in pulse–dynamic biogeochemical cycling, generating complex temporal patterns in ecosystem processes. The understanding of biogeochemical dynamics across space and through time in drylands is quite crucial for improving our ability to forecast climate change effects on ecosystem structure and function.

Chapter 3

Enhanced Semi-Arid Warming

3.1 Introduction

Since the 1970s, increasing scientific evidence has indicated that global warming is occurring. Numerous independent climate indicators have shown that the air temperature is rising, including changes in land cover, atmospheric and oceanic temperatures, glaciers, snow cover, sea ice, sea levels, and atmospheric water vapor. However, the temperature change was not consistent, showing distinct seasonal and regional variances. The temperature over land has risen more quickly than that over the ocean, and the ratio of land to sea warming has fluctuated between 1.36 and 1.84 (Sutton *et al.*, 2007). Additionally, various warming tendencies were seen throughout the various climatic regions during both the cold and warm seasons. According to earlier research, the semi-arid regions in the Northern Hemisphere (NH) spanning mid to high latitudes have witnessed significantly heightened warming during the boreal cold seasons (Huang *et al.*, 2012). Cumulative warming in cold seasons from 1901 to 2009 has been examined (Ji *et al.*, 2014) using the ensemble empirical mode decomposition (EEMD) approach. It was discovered that the strongest warming over the past century occurred in the semi-arid areas of NH's mid-latitude.

As the strongest climate change is witnessed in semi-arid regions, different locations may be vulnerable to global warming. The behavior of semi-arid regions on potential evaporation is greater than potential precipitation on a yearly average basis, and the landscapes are characterized by dry climates, scant vegetation cover, scant nutrient content, and scant soil water retention capacity. And both observational and modeling studies have shown that an aridity trend is present and will be most pronounced in the semi-arid regions in terms of precipitation, soil moisture, and drought frequency. For instance, two distinct mechanisms have been identified to analyze vertical stability and net energy in the semi-arid African Sahel: One is the indirect effect of anthropogenic warming on continental climates and the other is the direct impact of an increase in anthropogenic greenhouse gases (GHGs) on continental climate change. Concerns about whether regional patterns may have skewed estimates of large-scale temporal trends have been raised in light of significant regional disparities in surface air temperature trends (Wu *et al.*, 2011). Several factors, such as variations in air circulations and sea surface temperatures, interactions between the land and the atmosphere, and feedback from snow, can contribute to regional warming (Huang *et al.*, 2008a). Wallace *et al.* (1995, 1996) pointed out that dynamic reasons, in addition to GHGs, had significant impact on temperature fluctuations. Geographical patterns of industrial and socioeconomic growth are closely connected to geographic patterns of warming trends over land.

Studying this warming is essential to comprehending the non-uniformity of global warming since enhanced semi-arid warming (ESAW) is a clear signal in the mid to high latitudes of the NH. Several variables, including continental and local influences, contribute to ESAW (Huang *et al.*, 2012). The Atlantic multi-decadal oscillation (AMO) (Mann *et al.*, 2014), the North Atlantic oscillation (NAO) (Huang *et al.*, 1998), and the Pacific decadal oscillation (PDO) (Trenberth and Hurrell, 1994), which are all components of the climate system, are thought to dominate the dynamically driven component of this warming. In recent decades, a great deal of research has been done on how these events affect rainfall and continental temperatures (Wallace *et al.*, 2012; Smoliak *et al.*, 2015). In particular, Wallace *et al.* (2012) proposed a dynamic adjustment approach. Their findings revealed that during the substantial warming period of 1965–2000, over land areas poleward of 40°N, changes in air circulation accounted for about 41% of the warming.

Thermodynamic temperature is closely related to thermodynamic factors like GHGs, land use/cover change (LUCC), and anthropogenic aerosol emissions, many of which are local factors. The thermodynamic temperature is equal to the raw temperature minus the dynamic temperature. A single thermodynamic factor's impact has been the subject of numerous investigations. Qian *et al.* (2006) discovered that urban growth causes significant temperature increases. Model simulations have shown that LUCC can affect regional climate extremes and their variability. Additionally, the burning of fossil fuels and agricultural fires result in black carbon particles that have an impact on warming in winter by considerably lowering the snow's albedo and increasing the absorption of solar radiation (Huang *et al.*, 2011). Asia has seen significant anthropogenic black carbon, organic matter emissions, and substantial snow feedback.

Recent studies (Ma and Fu, 2006, 2007; Fu *et al.*, 2008) have shown that drying trends over arid and semi-arid regions under global warming have been detected from both observations and model simulations. Increases in the scope and intensity of desertification or degradation are related to climate change and human activities (Wang *et al.*, 2006). Desertification, a vital issue for sustainable development, is a significant result of the growth of arid and semi-arid regions. The loss of vegetation cover may also exacerbate the effects of drought. Increased aridity will encourage dust mobilization, resulting in more frequent dust storms from dust source regions. It will be easier to see how dust aerosols affect the climate across arid and semi-arid regions (Huang *et al.*, 2010b). As a result, studies on the effect of dust aerosols on the climate in Northern China have been conducted by using satellite observations (Huang *et al.*, 2007a, 2010b), ground-based measurements (Huang *et al.*, 2008a), and model simulations.

The arrangement of this chapter is as follows. Section 3.1 is the introduction to this chapter. Section 3.2 illustrates the data and methods used in this chapter, including reanalysis data, simulation data, satellite data, statistical methods, dynamical adjustment method, EEMD, and Estimation of PET. Section 3.3 introduces enhanced semi-arid warming (ESAW), and Section 3.4 discusses regional enhanced warming. Section 3.5 is the summary of this chapter.

3.2 Data and Methods

3.2.1 *Reanalysis data*

3.2.1.1 *Precipitation data*

The National Centers for Environmental Prediction (NCEP) and Climate Prediction Center (CPC) provided the precipitation dataset used on the arid and semi-arid regions. It is an observation station-based dataset produced from gauge observations made at more than 17,000 stations. It has a globally covered $0.5° \times 0.5°$ latitude–longitude resolution for the period spanning from 1948 to the present. Additionally, the data were gathered from two sizable individual datasets: the Climate Anomaly Monitoring System (CAMS) and the Global Historical Climatology Network version 2 (GHCN2).

The Global Precipitation Climatology Centre (GPCC) provided the gauge-based gridded precipitation dataset for the semi-arid land surface. The dataset spans the years 1891–2016. It has a spatial resolution of $0.5° \times 0.5°$ (https://opendata.dwd.de/climate_environment/GPCC/html/fulldata-monthly_v2018_doi_download.html). The substantially denser station network of GPCC's revised global precipitation climatology V.2018, which is based on data from more than 79000 stations, it is adequate for predicting inland precipitation (covering a majority of the semi-arid regions). The results from GPCC are validated using the precipitation in CRU TS v. 4.05 (Climatic Research Unit gridded Time Series Version 4.05). It is a commonly used dataset that spans from the year 1901 to 2018 on a 0.5° latitude by 0.5° longitude grid, which satisfies the requirements for the long timescale.

3.2.1.2 *Temperature data*

The GHCN_CAMS Gridded 2 m temperature dataset, which corresponds to CPC's surface air temperature (SAT) dataset, is utilized in this chapter. The PREC/L and this dataset both have the same temporal coverage and spatial resolution. The GHCN_CAMS land SAT analysis has a respectable level of quality and is capable of capturing a majority of the typical temporal–spatial patterns in observed climatology as well as the anomaly fields throughout both the regional and global domains.

The Met Office Hadley Centre provides the HadISST 1.1 monthly average sea surface temperature (SST) data (https://www.metoffice.gov.uk/hadobs/hadisst/data/download.html), which have a spatial resolution of $1° \times 1°$ and cover the period from 1870 to the present.

The GISTEMP dataset from the Goddard Institute for Space Studies (GISS) of the National Aeronautics and Space Administration has a resolution of $2° \times 2°$. It spans from the year 1880 to the present. Compared to GISTEMP, the MLOST and HadCRUT4 datasets have more missing data over high latitudes. As a result, the study only published the GISTEMP dataset's findings. The Extended Reconstructed Sea Surface Temperature (ERSST) dataset is provided by the National Oceanic and Atmospheric Administration and available at the website of https://psl.noaa.gov/data/gridded/data.noaa.ersst.v4.html. Version 4 of this dataset covers the period from January 1854 to the present. Using the ERSST data, the Nino3.4, PDO, and AMO indices were constructed.

3.2.1.3 *Wind speed and geopotential height data*

NCEP/NCAR Reanalysis I, NCEP-DOE Reanalysis II, and ERA-interim reanalysis product sets were used to obtain the wind speed and geopotential height data. The resolution of each of these reanalysis products is $2.5° \times 2.5°$. The two additional products, the NCEP/NCAR Reanalysis I and II, covered the period from 1979 to the present. The ensemble mean of these three products was used to show all the data linked with them in order to access some of the uncertainties in the results based on these reanalysis products. Additionally, the atmospheric dynamics during the last warming pause were examined using the wind speed and geopotential height data from the NOAA-CIRES 20th Century Reanalysis V2c and the ERA-20C reanalysis product sets. This chapter describes how the AMO controls the precipitation in semi-arid locations using the sea-level pressure (SLP), geopotential height, omega, and wind fields. These fields are taken from the 20th Century Reanalysis V3 dataset, which spans 1836–2015 with a spatial resolution of $1° \times 1°$ (https://psl.noaa.gov/data/gridded/data.20thC_ReanV3.html).

3.2.1.4 *Aerosol data*

The Goddard Earth Observing System (GEOS) atmospheric model and a cutting-edge data assimilation method, the Gridpoint Statistical Interpolation analysis, were used to obtain the most recent MERRA-2 data. To accurately simulate aerosol-related atmospheric processes, the GEOS model is radiatively coupled with the Goddard Chemistry Aerosol Radiation and Transport (GOCART) model. In order to comprehend the evolution of dust storms and their transport mechanism across the TP, this chapter uses the meteorological fields and the dust aerosol mixing ratio with six distinct effective radii of MERRA-2 at $0.625° \times 0.5°$ spatial resolution.

3.2.1.5 *Other meteorological data*

This chapter also uses the following kinds of meteorological data with numerous variables. Using a horizontal resolution of $1° \times 1°$, the monthly GLDAS Version 2 product (GLDAS-2) from January 1948 to December 2010 provided the surface air temperature, precipitation, and soil moisture data. The soil moisture data are organized as per four vertical layers from the surface to the bottom at 200 cm; precipitation is the sum of rainfall and snowfall. The four soil layers vary in depth from 0 to 10 cm, 10 to 40 cm, 40 to 100 cm, and 100 to 200 cm, and volumetric (m^3/m^3) units of soil moisture have been adopted. The mean soil moisture between 0 and 200 cm is chosen because different strata can be compared to one another. In accordance with the available LSWT observations, monthly mean data for precipitation, air temperature, evaporation, net shortwave, net longwave, wind, and cloudiness were derived from the ERA5 reanalysis datasets for the years 1995–2016. These data were utilized to categorize climate zones and investigate the key variables influencing LSWT in the lakes of each zone. The information from the University of East Anglia's Climate Research Unit (monthly precipitation and monthly mean surface air temperature), version TS3.1, is examined in this chapter. The data had a precision of $0.5° \times 0.5°$ and covered the years 1901–2009. This gridded dataset was created based on an archive of monthly mean temperatures provided by more than 4000 meteorological stations dispersed over the world.

3.2.2 *Simulation data*

3.2.2.1 *CMIP5 models*

In order to evaluate the correlation with the observational dataset, which is based on medium-to-low-scenario simulations in CMIP5 for the RCP 2005–2008 simulation, this chapter uses the precipitation and PET datasets from the historical simulations of 20 CMIP5 climate models (Table 3.1). The National Oceanic and Atmospheric Administration and the Cooperative Institute for Research in Environmental Sciences 20th Century Reanalysis package version 2, which has a geographic resolution of $2° \times 2°$ and covers the period 1901–2011, provided the observed sea-level pressure (SLP) data. From the Climate Explorer (http://climexp.knmi.nl/), the North Atlantic Oscillation (NAO), Pacific Decadal Oscillation (PDO), and Atlantic Multi-decadal Oscillation (AMO) indexes were downloaded.

BC and SO_2 emissions are calculated using the standard inventory used by CMIP5 models to support the IPCC AR5. Both BC and SO_2 emissions are mostly anthropogenic emissions, which include emissions from burning biomass, shipping, and aircraft, as well as emissions from burning agricultural waste, the domestic sector, the transportation sector, the industrial sector, and the waste treatment sector (BC only). The spatial resolution of the inventory is $0.5° \times 0.5°$, with monthly intervals every 10 years, covering the period from 1850 to 2000. In order to prepare the IPCC AR5, the atmospheric CO_2 concentrations were those utilized in the CMIP5 historical and Representative Concentration Pathway (RCP) 8.5 simulations. This chapter used the anticipated data for the RCP8.5

Table 3.1. CMIP5 models examined in this chapter (Guan *et al.*, 2017).

Model name	Modeling center
BCC-CSM1.1	Beijing Climate Center, China
CanESM2	Canadian Centre for Climate, Canada
CCSM4	National Center for Atmospheric Research, USA
CNRM-CMS	Centre National de Recherches Météorologiques, France
CSIRO-Mk3.6.0	Commonwealth Scientific and Industrial Research, Australia
GFDL-CM3	Geophysical Fluid Dynamics Laboratory, USA
GFDL-ESM2G	Geophysical Fluid Dynamics Laboratory, USA
GFDL-ESM2M	Geophysical Fluid Dynamics Laboratory, USA
GISS-E2-R	NASA Goddard Institute for Space Studies, USA
HadGEM2-CC	Met Office Hadley Centre, UK
HadGEM2-ES	Met Office Hadley Centre, UK
INM-CM4	Institute for Numerical Mathematics, Russia
IPSL-CMSA-LR	Institute Pierre Simon Laplace, France
IPSL-CMSA-MR	Institute Pierre Simon Laplace, France
MIROC- ESM	Japan Agency for Marine Earth Science and Technology, Japan
MIROC-ESM-CHEM	Japan Agency for Marine Earth Science and Technology, Japan
MIROCS	Atmosphere and Ocean Research Institute, Japan
MPI-ESM-LR	Max Planck Institute for Meteorology, Germany
MRI-CGCM3	Meteorological Research Institute, Japan
NorESM1-M	Norwegian Climate Centre, Norway

simulation for the period after 2005 because the other RCPs did not reveal significant variations. After all, this chapter only used the CO_2 concentration from 1902 to 2011. The annual CO_2 concentration data show a consistent global spatial distribution.

3.2.2.2 *CMIP6 models*

The CESM experimental data repository (https://www.earthsystemgrid.org/search.html) was used to get the monthly results of the historical simulation (1850–2005) and future predictions (2006–2100) for the two RCP scenarios (i.e., RCP45 and RCP85). Additionally, the results were obtained from the CESM low warming runs under the LW1.5deg and LW2deg scenarios, which are freely available at http://www.cesm.ucar.edu/experiments/1.5-2.0-targets.html and designed so that the expected multi-year global mean temperatures never exceed 1.5°C and 2°C, respectively, above the pre-industrial levels (where the pre-industrial level is taken as 1850–1920 mean). Historical, RCP45, RCP85, LW1.5deg, and LW2deg experiments from the CESM archive have 30, 15, 30, 9, and 10 ensemble members, respectively,. In order to eliminate the internal climate variability, this chapter studied the ensemble mean. Additionally, the model simulations are modified to have the same climatology for 1961–1990 as the observations to concentrate on temporal variation and long-term climate change. It should be highlighted that eliminating the mean bias can only lower the systematic departure of the model from the data; it cannot, however, enhance simulation performance or reduce model sensitivity. Future air temperature changes were predicted using the daily air temperature data from nine models used in the sixth phase of Coupled Model Intercomparison Project (CMIP6). Based on the data availability, the nine models were chosen. A rectangular combination of the various shared socioeconomic pathways (SSPs) and RCP, which respectively, reflect the future global growth and climatic radiative forcing level, makes up the CMIP6 scenario experiment. For future projections from 2015 to 2100, low-force and high-force scenarios are examined using two scenarios (SSP1-2.6 and SSP5-8.5). x and y for a given SSPx-y scenario represent the exact SSP and RCP forcing routes, respectively. SSP1-2.6 and SSP5-8.5 represent the low and high edges of the future forcing pathway spectrum, respectively. RCP2.6 and RCP8.5 are roughly equivalent to SSP1-2.6 and SSP5-8.5, respectively. Although the future is expected to be well above the SSP1-2.6 scenario at present emission rates, this chapter chooses the lowest and highest scenarios to maximize the range of future LSWT variability. The nearest grid point to each lake's center was used to derive the climate variables for the investigated lakes for the ERA5 and CMIP6 datasets. The lapse rate (Γ) was used to adjust the surface air temperature to the over-lake value when the lake's elevation and the model grid were not equal.

3.2.3 *Satellite data*

3.2.3.1 *LSWT data*

The Global Lake Temperature Collaboration (GLTC) and GloboLakes LSWT v4.0 provided satellite observations to determine the Lake Surface Water Temperatures (LSWTs). The GloboLakes dataset was obtained from the AVHRR on MetOpA, the AATSR on Envisat, and the ATSR-2 on ERS-2 and is available at http://www.laketemp.net. This chapter determined the lake-averaged LSWT for the 979 lakes in the GloboLakes dataset using daily time series of satellite-derived LSWT for the years 1995–2016 with a spatial resolution of 0.05°. The quality level index for each LSWT pixel ranges from low (0) to best quality (5). This chapter computed the lake-averaged LSWT only using pixels with a quality index bigger than 4, indicating the acceptable and highest quality, in order to eliminate data of poor quality.

The GLTC dataset comprises mean summer lake-averaged LSWTs from 291 lakes throughout the world between 1985 and 2009. Summer was defined as July–September (JAS) for lakes in the NH and January–March (JFM) for lakes in the Southern Hemisphere (SH). The JFM mean was used between the equator and 23.5°N and the JAS mean between the equator and 23.5°S to avoid the cloudy wet season of tropical climates and to account for the fact that the air temperature range in tropical regions is very constrained (absence of the traditional four thermal seasons). This is in line with the extremely narrow range of air temperatures found in tropical climes. The mean summer LSWT for the GloboLakes dataset was calculated using the same summer season convention.

3.2.3.2 *MODIS data*

The Terra and Aqua satellites launched the Moderate Resolution Imaging Spectroradiometer (MODIS) instrument in 2000 and 2002, respectively. The MODIS instruments from Terra and Aqua are crucial for studying planetary environmental and climatic changes. Since March 2000 for Terra and July 2002 for Aqua, the satellites have gathered an enormous amount of data in 36 spectral bands covering the land, atmosphere, ocean, and cryosphere at wavelengths between 0.4 and 14.4 μm. By integrating two algorithms — enhanced deep blue over bright surfaces and dark target primarily over vegetated or water surfaces — the collection of MODIS 6.1 aerosol products constitutes a new "merged" dataset over land and ocean. The Level 3 daily aerosol optical depth at 550 nm (MOD08_D3 and MYD08_D3) with a spatial resolution of $1^\circ \times 1^\circ$ was used in this chapter.

3.2.4 *Statistical methods*

3.2.4.1 *Regional contributions*

- The following formula was used to compute the contribution rate in order to examine the contributions of various precipitation regions to trends in the earth's surface temperature: To calculate the regional averaged temperature $\bar{T}_k$ of region k,

$$\bar{T}_k = \frac{\sum_{i=1}^{N_k} W_{ki} \cdot T_{ki}}{\sum_{i=1}^{N_k} W_{ki}}, \tag{3.1}$$

  where $W_{ki} = \cos(\theta_i \cdot \pi/180.0)$, where θ_i is the latitude of the grid I, N_k is the number of grids in region k, T_{ki} is the temperature of grid i in region k, and to determine the average temperature for a given region, apply Eq. (3.1). For each type of precipitation region, regional averaged temperature values will be computed for the yearly averaged temperature series.

- To calculate the contribution rate (CR_k) of region k,

$$CR_k = \frac{a_k \cdot \sum_{i=1}^{N_k} W_{ki}}{A_g \cdot \sum_{i=1}^{N_g} W_i}, \tag{3.2}$$

  where N_g is the total number of grids over the entire planet, A_g is the trend of the global land surface mean temperature, and a_k is the trend of $\bar{T}_k$ for region k. Contribution is one of the crucial elements used to support the degree of heat caused by global warming. The contribution of several climatic types to a location can show how the warming of regional temperatures has contributed to global warming.

3.2.4.2 *Variance*

Climate change has been receiving more attention, and there is increasing damage associated with extreme weather and climate. Since the 21st century, with improved computing power, numerical models have been frequently used in meteorological analysis and forecasting. Although the atmospheric model will never be able to perfectly capture every physical process that governs atmospheric activities, particularly small-scale disruption, it can come incredibly close to simulating actual atmospheric activity. Statistical methods are therefore essential for weather forecasting. Numerous techniques are needed to study historical climate change and predict the weather in the future. The most fundamental statistic among them is the mean. The quantity that describes the average level of a certain meteorological element is called the mean value. It can be determined using the following method:

$$\bar{x} = \frac{1}{n}\sum_{t=1}^{n} x_t, \tag{3.3}$$

where n stands for various integers according to different timescales. The idea of mean value is used in meteorological research to describe various data with multiple resolutions, including daily precipitation data, monthly precipitation data, and annual precipitation data. The deviation is a variable that is connected to the mean value and expresses how far the data deviate from the average value, or how aberrant the data are. As a result, the mean value is very important when analyzing climate change. For instance, given monthly temperature data, the global mean temperature must be determined on spatial scales in 2020. To determine the mean temperature in 2020, n (=12) can be divided by the sum of the temperatures measured at each grid from January to December (3.3). Furthermore, before using monthly data to compute the year average temperature using meteorological station data, daily data must be transformed into monthly data first (n denotes the number of days in each month).

To observe the reality of global warming, the global mean temperature can be calculated since the industrial revolution on timescales. Given that the world is spherical, it is essential to remember that the grids this chapter currently uses were created by map projection. Therefore, when computing the regional average using the following formula, the weight of latitude must be considered:

$$\bar{x}_t = \sum_{i=1}^{n} x_{t,i} \times \cos(\text{lat}_i \times \text{rad}) \Big/ \sum_{i=1}^{n} \cos(\text{lat}_i \times \text{rad}) \quad (t = 1, 2, \ldots), \tag{3.4}$$

where rad is radian ($\pi/180$) and lat is latitude. To determine the typical temperature for a given area, apply Eq. (3.4).

3.2.4.3 *Standard deviation*

In addition to mean, variance is a crucial fundamental statistic (standard deviation). It can be estimated by using the following process and suggests the size of the fluctuations for meteorological elements:

$$s^2 = \frac{1}{n}\sum_{t=1}^{n}(x_t - \bar{x})^2 \tag{3.5}$$

$$s = \sqrt{\frac{1}{n}\sum_{t=1}^{n}(x_t - \bar{x})^2} \tag{3.6}$$

The variance and standard deviation (std) are represented, respectively, by Eqs. (3.5) and (3.6). The standard deviation is consistent with the average value, even though both statistics show the degree of dispersion of meteorological components. As a result, standard deviation is frequently employed to show the variability of the climate. If the standard deviation of the temperature for the same place in 2000 is higher than in 1961, the temperature in 2000 is considered to be changed significantly over time. If the average temperature in the Tropics is lower than that in the NH's mid-latitudes for a given year (such as 2000), the temperature difference in the boreal mid-latitudes is more significant. Inland temperatures vary more daily than coastal temperatures, and this comparison of the size of the daily variance typically is related to the size of the standard deviation. Similar to this, while computing std in a certain place, latitude must be considered. First, using Eq. (3.4), this chapter determines the weighted average. The following formula can be used to determine the regional standard:

$$s_t = \sqrt{\sum_{i=1}^{n}(x_{t,i} - \bar{x_t})^2 \times \cos(\text{lat}_i \times \text{rad}) \Big/ \sum_{i=1}^{n} \cos(\text{lat}_i \times \text{rad}))}$$
$$(t = 1, 2, \ldots), \tag{3.7}$$

where lat represents latitude and rad is radian ($\pi/180$).

3.2.5 *Dynamical adjustment methodology*

Wallace *et al.* (2012) suggested the dynamical adjustment approach, which they utilized to examine the non-uniformity of spatial warming over the NH. The dynamical adjustment method divides the surface air temperature (SAT), or the raw temperature data, into two parts: dynamic induced temperature (DIT) and radiative forced temperature (RFT). Usually, when talking about radiative forcing, one refers to changes in net radiation in the vertical direction at the top of the troposphere brought on by internal climate system changes like variations in CO_2 concentration or external forcing like variations in solar radiation. Radiative forcing is both an index that shows the significance of a factor in probable climate change mechanisms and a measure of the extent to which a factor affects the equilibrium of incoming and outgoing energy from the Earth-atmosphere system. Dynamic forcing, on the other hand, typically refers to elements like SLP that are connected to modifications in air circulation patterns. In the cold season (November–April) over land areas poleward of 20°N, the dynamical adjustment approach could eliminate the dynamic component of the SAT caused by the atmospheric circulation pattern from the raw SAT.

Smoliak *et al.* (2015) have refined the dynamical adjustment methods employed. The NH land surface temperature field is precisely adjusted monthly using the partial least squares (PLS) method, where each grid point's temperature time series serves as the predictand. Before performing the subsequent dynamic adjustment procedures, the SLP data are standardized, and the temperature time series are high-pass filtered by subtracting the cold season mean from the mean for each month within a specific cold season. (1) To create a one-point cross-correlation map, correlate the grid point temperature time series with the SLP field. (2) To obtain the first PLS predictor time series Z1, project the monthly SLP field onto the correlation pattern and weigh each grid point by the cosine of its latitude. (3) Using traditional least squares fitting techniques, regress this PLS predictor Z1 out of each grid point temperature time series and its SLP predictor field to generate a residual temperature time series and residual SLP field. Repeat similar procedures on the residual temperature time series and residual SLP field to get the corresponding, mutually orthogonal PLS predictors Z2 and $Z3, \ldots, Zn$. Take care to high-pass filter each subsequent residual temperature time series before computing corrective patterns. The temperature dataset has been subjected to the enhanced dynamical adjustment methodology in this chapter, and three predictors that have undergone cross-validation are kept.

The components linked to changes in atmospheric circulation patterns that are expressed in terms of SLP are partitioned and referred to as DIT variability after the procedure described above. The RFT, or residual portion, is connected to radiatively forced components. The accumulation of GHGs, stratospheric ozone depletion, volcanic eruption, aerosol emission, and local anthropogenic forcing is thought to be a contributing factor to the RFT. RFT also includes non-radiative effects originating from thermodynamic processes. The dynamical adjustment approach may be used to pinpoint DIT and RFT's contributions to enhanced warming.

3.2.6 *Ensemble empirical mode decomposition (EEMD)*

It is common practice to deal with nonlinear and non-stationary time series, such as climate variables, using EEMD. With enough trials, this method sorts a group of white-noise-added signals, and it treats the mean as the final, accurate result. The remaining signal components are real and physically significant, and the white noise will be averaged out during the process. In this chapter, the EEMD separates the regional precipitation indices and the decadal to multi-decadal components of the AMO from the filtered data. To create the signals this chapter evaluates, the intrinsic mode functions (IMFs) that satisfy the timescale's requirements are chosen. The additional white noise in the EEMD has a standard deviation of 0.2 and an ensemble size of 400. EEMD is an advancement of the empirical mode decomposition (EMD) technique, which is a powerful non-stationary data processing tool. Huang *et al.* (1998) initially proposed the EMD method. Both non-stationary and nonlinear time series can be broken down into a group of IMFs using the

EMD approach. The decomposition is based on three presumptions: (1) that the time series have at least two extrema, a maximum and a minimum; (2) that the characteristic timescale is determined by the interval between the extrema; and (3) that even if the data were entirely devoid of extrema but contained only inflection points, the extrema could still be discerned by differentiating the data once or more times.

The IMFs do not have to be a narrow band signal; they represent the oscillation mode embedded in the data. Each cycle's IMF, which is determined by zero crossings, only involves one mode of oscillation; complicated riding waves are not permitted. Additionally, it can be both amplitude and frequency modulated, and it is not limited to a narrow band signal. A function that meets both requirements is an IMF. The first, which is obviously identical to the conventional narrow band requirements for a stationary Gaussian process, is that the number of extrema and the number of zero crossings in the entire dataset must either equal or differ by no more than one. The second requirement is that the mean value of the envelope defined by the local maxima and the envelope defined by the local minima is zero at any location where data are present. This condition is a novel concept that converts a traditional global requirement into a local requirement. It is required to prevent asymmetric wave patterns from causing undesirable fluctuations in the instantaneous frequency. These IMFs can serve as the foundation for that expansion, which can be linear or nonlinear, depending on the data. They are based on and derived from the data. In addition to being full and orthogonal, the technique is also adaptable and this is crucial. The following is a review of a few characteristics of the original EMD:

- Because the EMD is an adaptive data analysis technique based on local data properties, it more efficiently detects nonlinear, non-stationary oscillations;
- For any white-noise-only (or fractional Gaussian) series, the EMD is a dyadic filter bank;
- The dyadic property of the original EMD is frequently harmed when the data are intermittent;
- Adding noise to the data could offer a reference scale with a uniform distribution, allowing EMD to restore the dyadic quality that has been degraded;
- The corresponding IMFs of various noise series do not correlate with one another. As a result, the means of the related IMFs of various white noise series are likely to cancel one another.

Even though EMD has proven to be helpful, there are still some bothersome issues that need to be fixed. Huang *et al.* (1998) attempted to stop the low-frequency mode from enlarging into the quiescent area by introducing noise of microscopic magnitude to earthquake data. However, they were not completely aware of the effects of the additional noise in the EMD approach. The predicted data turned out to be a combination of sporadic high-frequency oscillations riding on a steady low-frequency sinusoidal signal. The intermittent nature of the signal could not only result in significant aliasing in the time–frequency distribution but also obscure the physical significance of each IMF. Any IMF that consists of oscillations on significantly different scales is referred to as "mode mixing," and the intermittent nature of the driving forces frequently brings it on. An IMF might lose its physical significance when mode mixing takes place, giving a false impression that a mode may reflect various physical processes. Because the EMD is entirely dependent on the existence of extrema, the approach fails if the data lack the required extrema. Then, the EEMD approach, a new noise-assisted data analysis technique, is suggested to solve this issue without using a subjective intermittence test. According to the EEMD approach, the genuine IMF components are the average of an ensemble of trials, each of which included the signal and a small amount of white noise. The following is how the proposed EEMD is created:

- add a white noise series to the targeted data;
- decompose the data with added white noise into IMFs;
- perform steps 1 and 2 repeatedly, using a different white noise series each time;
- as for the outcome, obtain the (ensemble) means of corresponding IMFs of the decompositions.

The added white noise series cancel each other out in the final mean of the corresponding IMFs due to the decomposition using the EEMD. The mean IMFs also remain within the natural dyadic filter windows, significantly lowering the likelihood of mode mixing and preserving the dyadic property.

EMD is now the preferred tool in many scientific disciplines. Numerous people have created their own EMD applications. Programming in Matlab can be used to carry out the EMD/EEMD computation process. An accessible replacement package is offered for the Matlab EMD/EEMD program. The program incorporates the majority of the EMD's advancements and has the following parts:

- The foundational EMD/ERMD program;
- The statistical significance test of IMFs;
- An EMD-based approach for calculating instantaneous frequency.

Visit https://github.com/psharma15/EEMD-Huang to get the program and its instructions.

3.2.7 *Estimation of potential evaporation temperature (PET)*

Temperature-based methods, radiation-based methods, and combination methods are the three main categories of techniques used to estimate PET in climate studies. Temperature-based PET methods seem to work rather well in climatological applications because air temperature and net radiation are connected. However, recent research suggests that simple temperature-based PET estimates are insufficient to predict trends like long-term evaporative demands and drought projections since temperature is not necessarily the main factor influencing PET. In fact, PET depends on radiative and aerodynamic factors in addition to air temperature.

The Priestley and Taylor approach is a condensed version of the Penman–Monteith method for radiation-based PET. The former clearly disregards the effects of the aerodynamic control term, which are the effects of the water vapor pressure deficit (VPD), wind speed, and canopy resistance, and only keeps the radiation term. It has been noticed that some drought indices may mistakenly show a tendency toward greater droughts as a result of global warming if changes in available energy, air humidity, and wind speed are not specifically taken into consideration. The Penman–Monteith method is preferable to empirically based formulations, which often just take the influence of temperature and radiation into account, because it is developed from physical principles. The Penman–Monteith algorithm employed in this chapter considers a wide range of meteorological factors linked to the evapotranspiration process, including data on vegetation features, net radiation, relative humidity, wind speed, and temperature.

$$\mathrm{PET} = \frac{0.408\Delta(R_n - G) + \gamma \frac{900}{T_a + 273} u_2(e_s - e_a)}{\Delta + \gamma(1 + 0.34u_2)}, \tag{3.8}$$

in which R_n is the net radiation at the crop surface (MJ $\mathrm{m}^{-2}\mathrm{d}^{-1}$), G is the soil heat flux density at the soil surface (MJ m^{-2} d^{-1}), Ta is the mean daily air temperature at a 2 m height (°C), u_2 is the wind speed at a 2 m height (m/s), e_s is the saturation vapor pressure at a 2 m height (kPa),

e_a is the actual vapor pressure at a 2 m height (kPa), Δ is the slope of the vapor pressure–temperature curve (kPa/°C), and γ is the psychometric constant (kPa/°C). The employed data of net surface radiation (including shortwave and longwave net radiation), specific humidity, wind speed, and soil heat flux density are from the data products of the Global Land Data Assimilation Systems (GLDAS) at a 0.5° resolution. Based on two global radiation products, the solar radiation values for 1983–2000 are revised. According to the observationally based climatological methodologies created by the CRU at the University of East Anglia, specific humidity and wind speed data are corrected.

3.3 Enhanced Semi-Arid Warming (ESAW)

Global warming and regional aridification research have historically grown to be significant in the subject of global change. The troposphere near the 30° latitude band in the northern and SHs was the fastest warming region globally, which caused the tropical circulation to widen, tropospheric rapids to move to the poles, the expansion of arid regions, and an increase in the number of drought problems worldwide. The IPCC Sixth Assessment Report also noted that the increase in GHGs and the intensification of human activities have increased tropospheric temperatures and decreased stratospheric temperatures, resulting in a greater top tropospheric height, weakening the equatorial circulation, shifting the mid-latitude storm tracks and subtropical jet stream to the poles, and expanding the subtropical arid regions to the north and south. There were distinct regional differences in global dry and wet trends in the second half of the 20th century against the backdrop of warming, with the African, Eurasian, Australian, and South American continents being dominated by aridification trends and the North American continent showing a wetting trend. They compared the characteristics of dry and wet changes in each continent and their correlation within the large-scale context affecting climate change; simultaneously, it was discovered that warming had altered the distribution of global environmental dry and wet changes, and the continents of South America and Australia continue to exhibit aridification trends despite an increase in precipitation. However, the temperature increase is not insignificant. All arid regions have a much higher proportion of dry years in warm than in cold times against the backdrop of global warming, and the continuation of global warming may result in an increase in aridification in arid regions. This rapid warming will accelerate the regional aridification. The trend of temperature increase is clearly different in different parts of the world, and the warming in the middle and high latitudes of the NH's arid and semi-arid regions is most noticeable, particularly in winter. This rapid warming will further intensify the region's aridification.

3.3.1 *Global mean enhanced semi-arid warming*

Since the 1970s, more and more scientific data have indicated that global warming is occurring. The air temperature is rising, as shown by many independent climatic indicators, including changes in land cover, ocean and atmospheric temperatures, glaciers, snow cover, sea ice, sea level, and atmospheric water vapor. However, the temperature change has not been consistent, showing distinct seasonal and regional variances. The land temperature has risen more quickly than the ocean temperature, and the ratio of land to sea warming has ranged between 1.36 and 1.84 (Sutton *et al.*, 2007). Additionally, various warming tendencies were seen throughout various climatic regions during both the cold and warm seasons. The warming trend in the cold season is more than twice as strong as the trend in the warm season, according to the regionally averaged land surface linear temperature trend as a function of climatological mean precipitation for the annual, the cold season, and the warm season (Fig. 3.1). There is an apparent rise in temperature in semi-arid regions, with a warming

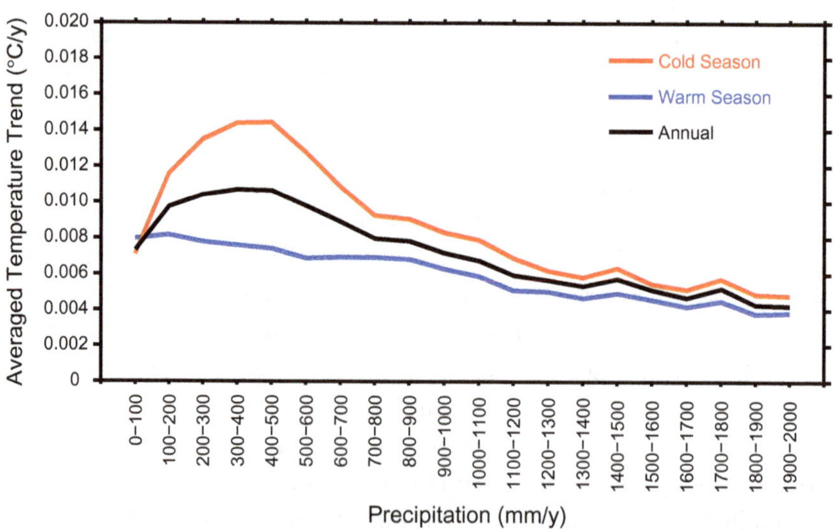

Fig. 3.1. Comparison of the regional averaged land surface linear temperature trend as a function of climatological mean precipitation for annual, cold season, and warm season of the world. The precipitation interval is 100 mm yr^{-1} (Huang *et al.*, 2012).

peak around the rainfall range of 300–400 mm yr^{-1}. The inference is that the temperature fluctuation in semi-arid regions during the cold season is more responsive to climate change than the temperature variability in other locations. This suggests that semi-arid warming is amplified or accelerated in the cold season. The warming trend for wet locations (rainfall >1000 mm yr^{-1}) does not change much between the cold and warm seasons. The annual climatological mean precipitation trend increases monotonically as the temperature trend declines. This suggests that drier land has warmed up. A temperature rise would cause greater evapotranspiration, making dry terrain drier and worsening the ecological processes there. Semi-arid zones, where warmth has increased quicker than in any other region, are notably intensified by it.

The global distribution of temperature trends for the annual, cold season, and warm season reveals that the temperature trend is positive in most of the mid-high-latitude region, though there are a few small dark blue patches with small negative trends in Southern Australia, Western South America, Southeastern North America, and a few other small domains in Asia and Africa. Generally speaking, the temperature trend across continents is favorable, most obvious over Asia and North America. The main rise areas are in the mid-high-latitude zones, including semi-arid parts of North America and Central Asia. The cold season's pattern is significantly dissimilar from the warm season. The NH's mid-high latitudes experience the most warming throughout the winter. The fact that some regions do not experience enhanced strong warming during the warm season shows that surface warming is not constant throughout the year. Particularly in Asia, the pace of temperature increase during the warm season is weaker than the rate during the cold season, and the degree of temperature variation is lower than in North America. Compared to other parts of the world, the warming trend in Central Asia during the cold season is significantly more pronounced. In addition to being distinct from the trend during the cold season, the warming in Central Asia during the warm season is also weaker in terms of scope and intensity. The accelerated warming is over arid and semi-arid areas during the cold season. The rate of temperature increase in the cold season is significantly faster than in the warm season. For instance, the warming rate (0.0194°C yr^{-1}) in areas with 300–400 mm of annual precipitation is 2.4 times higher than that during the warm season. Such a warming trend always occurs across semi-arid regions between 20°N and 60°N. The temperature warming amplitudes for various semi-arid regions and seasons demonstrate that warming has been most pronounced in semi-arid regions.

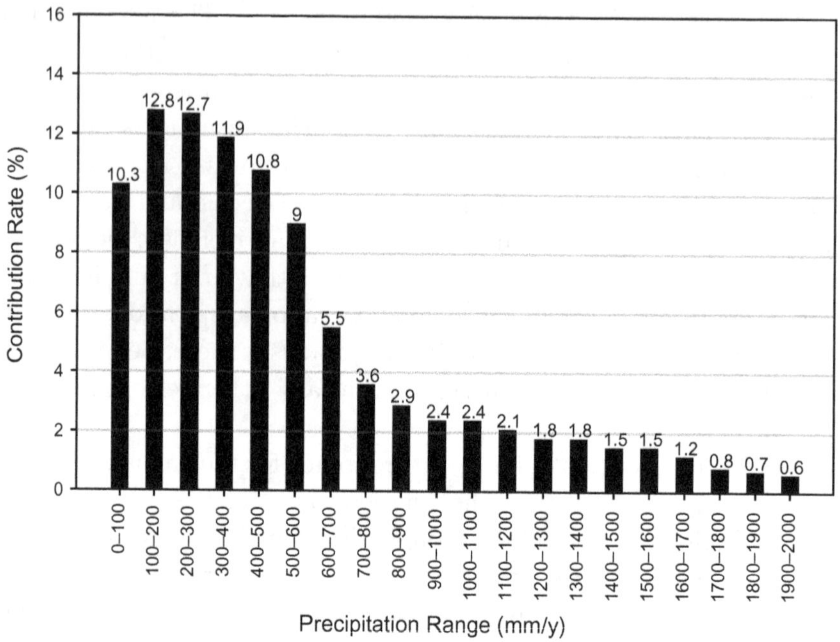

Fig. 3.2. The regional contribution of land surface linear air temperature trend to global as function of climatological mean precipitation. The precipitation interval is 100 mm yr^{-1}. The result of the regions where climatological mean precipitation is larger than 2000 mm is omitted due to smaller contribution rate (Huang *et al.*, 2012).

Over the mid-high latitudes of the NH, annual, cold season, and warm season temperatures increased by 1.33°C, 1.89°C, and 0.85°C, respectively, from 1901 to 2009. Mid-latitude semi-arid regions in Europe, Asia, and North America all had an increase in cold season temperatures during the same time period of 1.41°C, 2.42°C, and 1.50°C, respectively, with semi-arid Asia displaying the largest warming. Mid-latitude semi-arid regions in Europe, Asia, and North America warmed throughout the research period by 0.95°C, 0.68°C, and 1.05°C, respectively. While warming in the warm season is more noticeable in North America than in Asia, it is more pronounced in semi-arid regions of Asia during the cold season.

Figures 3.2 and 3.3 show the contributions of surface temperature trends in various precipitation regions to the overall trend for the annual and cold season means. The graph clearly shows how significantly diverse rainfall regions contribute to the overall mean trend. Nearly half of continental warming occurs in semi-arid regions, contributing significantly to global warming. Semi-wet and wet regions have far weaker temperature changes. In the cold season, semi-arid regions contribute even more to global warming

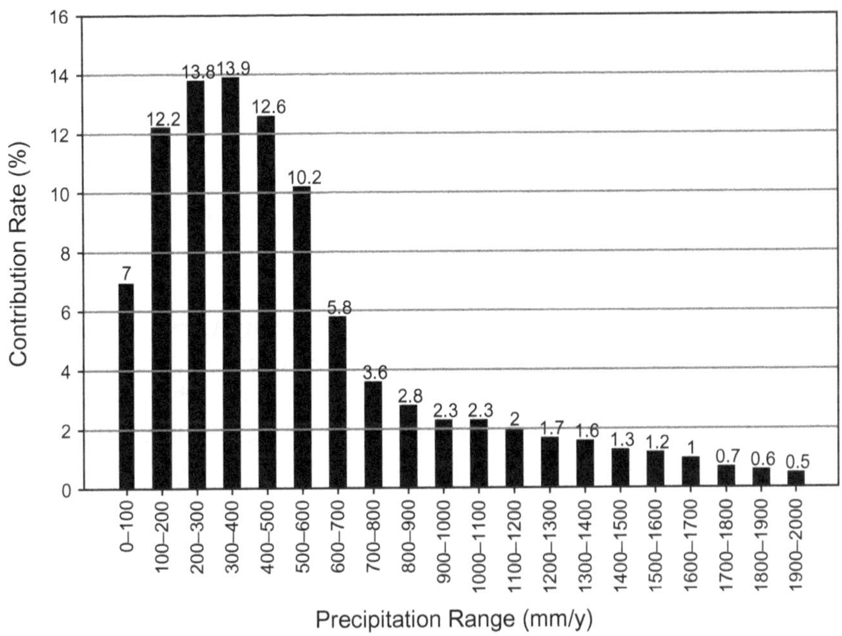

Fig. 3.3. Same as Fig. 3.2 but for cold season (November–March) (Huang *et al.*, 2012).

Table 3.2. Contribution of different mid-latitude regions of land surface temperature trends to global trend, where Europe is defined as the region of 20°N–60°N, −17 W–60°E, Asia is the region of 20°N–60°N, 60°E–180°E, and North America is the region of 20°N–60°N, −180∼ −17°W (unit: %) (Huang *et al.*, 2012).

	Europe	Asia	North America
Arid	8.76	5.65	0.64
Semi-arid	6.29	13.81	6.85
Semi-wet	3.23	2.48	3.54
Wet	0.73	3.11	2.20

than semi-wet regions. The contribution numbers for the cold season further highlight the distinct differences between semi-arid and semi-wet locations. The semi-arid region of Asia is shown to have contributed the most to global warming in Table 3.2, which also shows these results. Semi-arid areas in Europe, Asia, and North America correspondingly contributed 6.29%, 13.81%, and 6.85% of the earth's surface temperature rise. There were fewer contributions from arid, semi-arid, and wet regions.

3.3.2 *Evolution of enhanced warming*

The spatial evolution of the accumulated warming by a given time is shown in Fig. 3.4. Before 1950, noticeable accumulated warming (>0.5 K) or cooling (<−0.5 K) seemed to be sporadically distributed in space, mainly along subtropical bands around 30°S and 30°N and in the subpolar region around 60°N. The subtropical bands coincide with the downward branches of Hadley cells. The other noticeable warming region appeared in Western Africa near the southern edge of the Sahara Desert. These are all arid or semi-arid areas. Since 1950, the earlier, erratic warming has become more pronounced. By 1990, nearly all the northern mid-latitudes were covered by warming areas. Instead of corresponding to the original two bands near 30°N or 60°N, the NH's warmest areas are in between, that is, in mid-latitude semi-arid zones. The statistically significant EEMD trend from 1901 to 2009 differs slightly from the linear trend over the same time period in terms of amplitude and spatial patterns. However, the linear trend cannot tell how the warming pattern is evolving (Fig. 3.4).

Because the trends obtained are time varying, the warming/cooling rates that correspond to them are both temporally and spatially local entities. These rates can be calculated by computing the temporal derivatives of trends. In Fig. 3.4, the warming and cooling rates are depicted. Both mild cooling and moderate warming regions existed before 1950. Over the following three decades, the cooling regions dwindled, and most of them changed from cooling to warming. By 1980, nearly all the world's land had been warming, except for some minor cooling in Greenland's Northernmost tip and areas close to the Andes. There has been little change in global land warming rates since then. The northern mid-latitudes experienced the most intense warming. The spatial pattern of the warming rate in later decades is similar to that produced by fitting a straight line to the entire temporal domain (Fig. 3.4). The latter warming, however, is significantly more intense than that identified by the straight-line fitting method.

The essential characteristics of the spatial–temporal history of the warming are more clearly visible when the zonally averaged trend (across only the land area) is presented (Fig. 3.5). To remove the noisy pattern caused by spatially erratic warming, one can use a running mean across a 5° band in the meridional direction. The zonally averaged warming was plotted (Fig. 3.5). The subtropical and subpolar parts of the NH experienced the first significant zonally averaged warming (>0.2 K since 1900), followed by subtropical warming in the SH. Around 1920, the northern subtropical

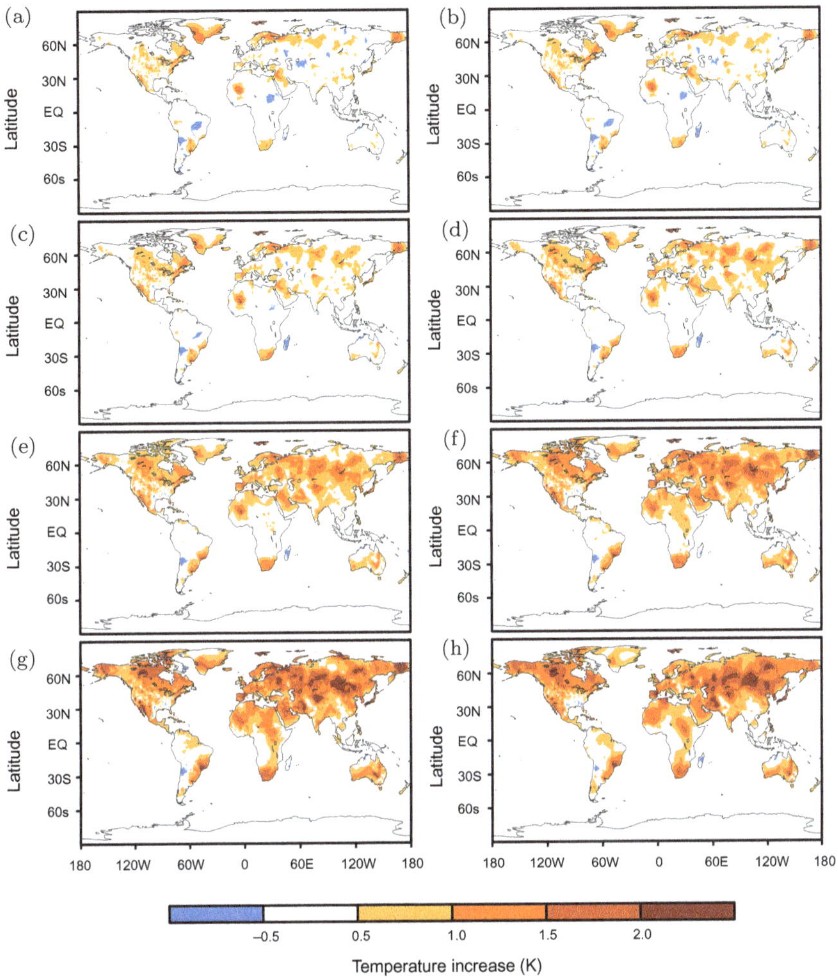

Fig. 3.4. Spatial evolution of the EEMD trend of global land surface air temperature. Ensemble EMD trends ending in 1950, 1960, 1970, 1980, 1990, 2000 and 2009, respectively (a–g). The spatial structure of temperature increase based on time-unvarying linear trend over the whole data domain from 1901 to 2009 (h) (Ji *et al.*, 2014).

and subpolar parts of the NH began to noticeably warm up. The latitudinal scope of warming expanded toward mid-latitude over the following three decades, and the 0.2 K lines of these two bands eventually joined around 1955. The amplitude of warming increased slowly for both bands. At all latitudes, the warming picked up speed after 1955. The original high-latitude and subtropical bands were the areas of the NH that experienced the most warming in recent decades. However, the tongue-like expansions of the

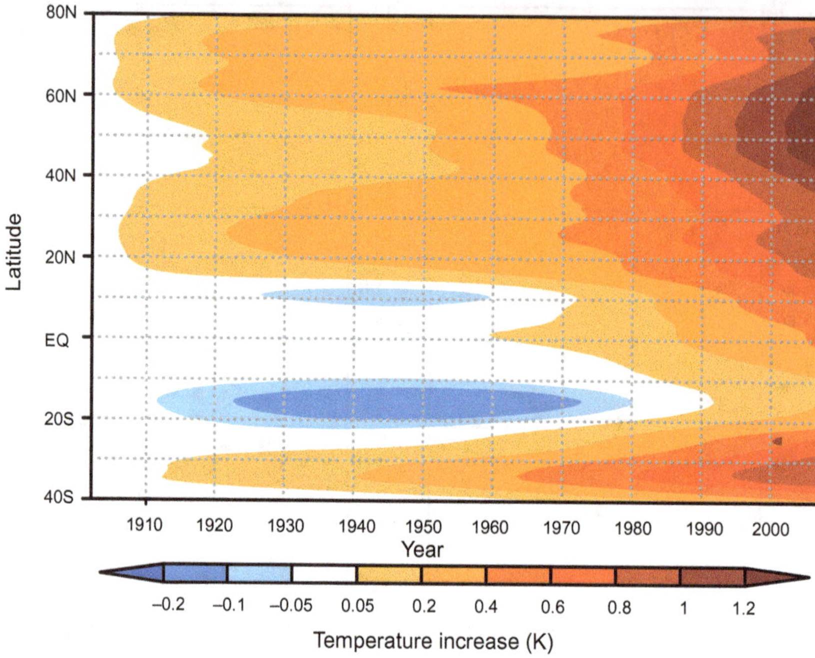

Fig. 3.5. Evolution of the zonally averaged trend of surface air temperature. Note that color intervals are uneven (Ji *et al.*, 2014).

contours toward the left at around 26°N, 62°N, and 75°N can still be used to distinguish these two bands.

The zonally averaged warming of more than 0.2 K in the subtropical belt in the SH lagged behind that in the NH and was relatively meridionally thin. The SH's poleward warming expansion is not as strong as that of the NH, a phenomenon that may be related to the SH's mid-latitude lower land covering. It should be noted that analyzing only later data, such as data from 1950 onward, will not reveal the evolution mentioned above, i.e., characteristics of the land surface air temperature trend over centennial and longer timescales because multi-decadal variability cannot be effectively separated from the trend of later (shorter) data using any analysis method. Additionally, there is insufficient evidence claiming that the extracted changing trend fully captures the impact of human activity, as it has been shown that multi-decadal variability of land surface air temperature can be driven by natural or human-induced forcing on a variety of timescales. However, the modest variations of trends are compatible with the atmosphere's slowly rising carbon dioxide levels. According to recent studies on the relationship between global warming and changes in Hadley cells, comparatively earlier warming in the subtropical regions shows that the warming may be linked to changes in atmospheric circulations (Hu and Fu, 2007). However, the NH's mid-latitude arid and semi-arid regions show that the greatest warming was thus far associated with either linear or EEMD trends, suggesting that the regional limited heat capacity may have been a factor. It is still unknown what significant physical factors caused the warming trend to change in this manner.

3.3.3 *Dynamic and thermodynamic warming*

ESAW has been discussed for a while, but little research has been done to clarify the effects of different elements on the ESAW processes. A dynamical adjustment method to the temperature field was used, which was split into dynamic and thermodynamic components (Wallace *et al.*, 2012; Smoliak *et al.*, 2015). The primary cause of ESAW should be revealed by the varied behaviors of the dynamic and thermodynamic temperatures, particularly in the three typical semi-arid regions of the mid to high latitudes in Asia, Europe, and North America. Additionally, the dynamic and thermodynamic temperatures were compared with the Coupled Model Intercomparison Project Phase 5 (CMIP5) models in order to understand the warming effect of thermodynamic forcing better.

The raw, dynamically generated, and thermodynamically driven temperatures over land poleward of 20°N are shown in Fig. 3.6 as temporal fluctuations of temperature anomalies during the boreal warm seasons (May–September) and the boreal cold seasons (November–March) (Fig. 3.6(b)). The raw temperature increase trend from 1902 to 2011 during the cold seasons is visually noticeable. Although its interannual volatility was significantly decreased, the thermodynamically forced temperature in the cold seasons displays a curve that is comparable to that of the raw temperature. The interannual variability in the raw temperature during the cold seasons is mostly explained by the dynamically induced temperature, which has a significant year-to-year change. However, compared with the cold seasons, the dynamic and thermodynamic temperatures fluctuated less during the hot seasons. While the thermodynamic temperature predominantly contributes to the long-term warming trend, the dynamic temperature during the cold season accounts for most of the raw temperature's interannual and decadal fluctuation.

The spatial distribution of land surface air temperature trends is shown in Fig. 3.7 for the raw, thermodynamically forced, and dynamically generated temperatures during the cold seasons of 1902–2011. Except for a few

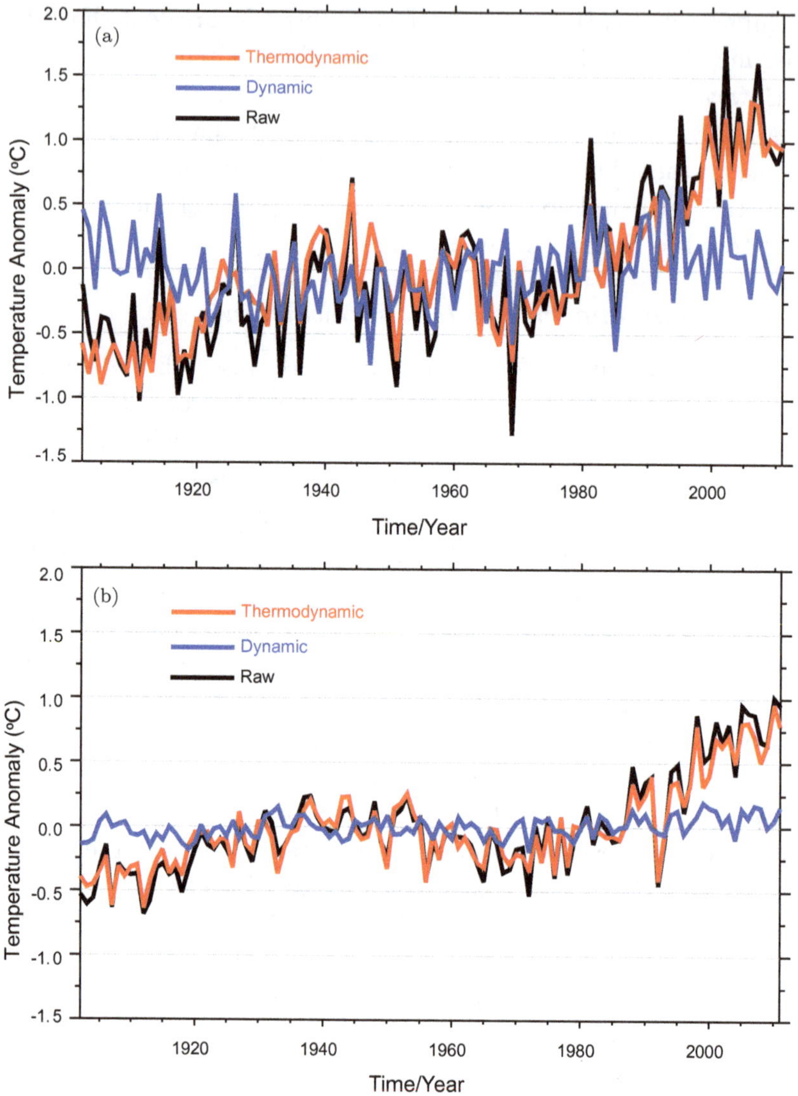

Fig. 3.6. Mean time series of surface air temperature anomalies from 1902 to 2011 over the NH poleward of 20°N during the cold season (November–March) (a) and the warm season (May–September) (b) for the raw, dynamically induced and thermodynamically forced temperatures (Guo *et al.*, 2018).

minor domains, the raw temperature trend was positive across many NH areas poleward of 20°N (Fig. 3.7(a)). This trend, nevertheless, was quite unevenly distributed. The warming rate exceeded 0.024°C/yr in the mid to high latitudes across Eurasia and North America, where the temperature increase was most noticeable. Similar warming patterns may be seen in

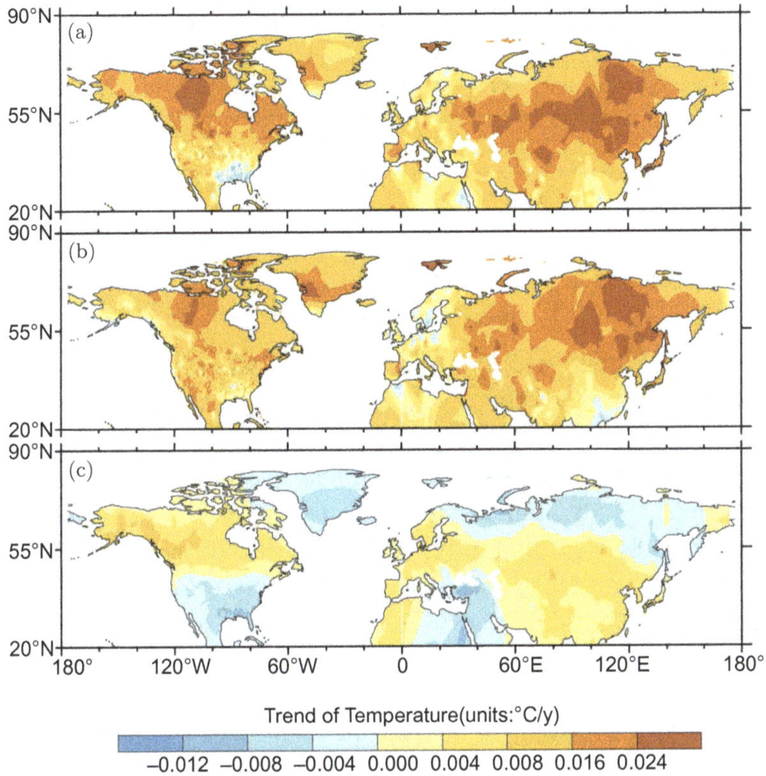

Fig. 3.7. Spatial distribution of the linear surface air temperature trends during the cold season in 1902–2011 for the raw (a), thermodynamically forced (b) and dynamically induced temperatures (c) (Guo *et al.*, 2018).

thermodynamically forced temperature (Fig. 3.7(b)), which displays stronger warming than the raw temperature, particularly over Eurasia and North America. In contrast to the thermodynamic temperature, the dynamic temperature change distribution is displayed in (Fig. 3.7(c)). In contrast to a majority of Canada and Central and Southern Eurasia, a cooling effect appears poleward of 55°N over Eurasia, the US, and Eastern Africa. The distinct ways in which the dynamic and thermodynamic temperatures behave suggest that the large-scale rise in the thermodynamically forced temperature is a key factor in the global warming process, particularly in the accelerated warming over the mid to high latitudes of the NH. Three typical semi-arid regions to research regional warming are chosen to examine the apparent regional differences in raw and thermodynamic temperature trends: Asia (20°N–60°N, 60°E–180°E), North America (20°N–60°N, −180°W–17°W), and Europe (20°N–60°N, −17°W–60°E).

Table 3.3. Cold season warming amplitudes during 1902–2011 for the raw, thermodynamically forced and dynamically induced temperatures over regions poleward of 20°N in the NH, Asia, Europe, and North America (unit: °C) (Guo *et al.*, 2018).

	20°N–90°N	Asia	Europe	North America
Raw	1.49	1.85	1.33	1.41
Thermodynamic	1.36	1.60	1.19	1.32
Dynamic	0.14	0.26	0.14	0.09

Given the regional variation in the extent of warming, Table 3.3 lists the specific warming rates for the area-averaged raw, dynamically induced, and thermodynamically forced temperatures. Table 3.3 shows that from 1902 to 2011, the thermodynamically forced temperature showed a greater warming amplitude than the dynamically induced temperature. Area-averaged raw, thermodynamically forced, and dynamically induced temperatures increased by 1.49°C, 1.36°C, and 0.14°C, respectively, in the regions poleward of 20°N. The warming amplitude of the thermodynamically forced temperature was around ten times greater than that of the dynamically induced temperature. Both the thermodynamically forced temperature (1.60°C) and the raw temperature (1.85°C) were the highest in Asia. North America, the second strongest warming region, showed warming amplitudes of 1.41°C and 1.32°C for the raw and thermodynamically induced temperatures, respectively. The raw and thermodynamically induced temperatures of Europe showed warming amplitudes of 1.33°C and 1.19°C, respectively.

The variability of the dynamically induced and thermodynamically forced temperatures is determined as a function of the annual precipitation, as shown in Fig. 3.8, because semi-arid locations between 20°N and 60°N are the most sensitive to warming (Huang *et al.*, 2012). The thermodynamic and dynamic temperature curves for the regionally averaged trends were different from the yearly precipitation variability (Fig. 3.8). In arid locations, the thermodynamically forced temperature trend rose as precipitation rose, peaked at about 300–400 mm/yr in semi-arid regions, and then monotonically fell as precipitation rose. Over all levels of precipitation, the dynamic temperature exhibited a relatively flat curve, nonetheless. The faster rate of thermodynamic temperature increase over semi-arid areas may indicate that semi-arid areas are more vulnerable to thermodynamic warming, which may have caused ESAW in the mid-latitudes of the NH.

Figure 3.9 contrasts the dynamic and thermodynamic temperatures for Asia, Europe, and North America. The warming trend bars in these three

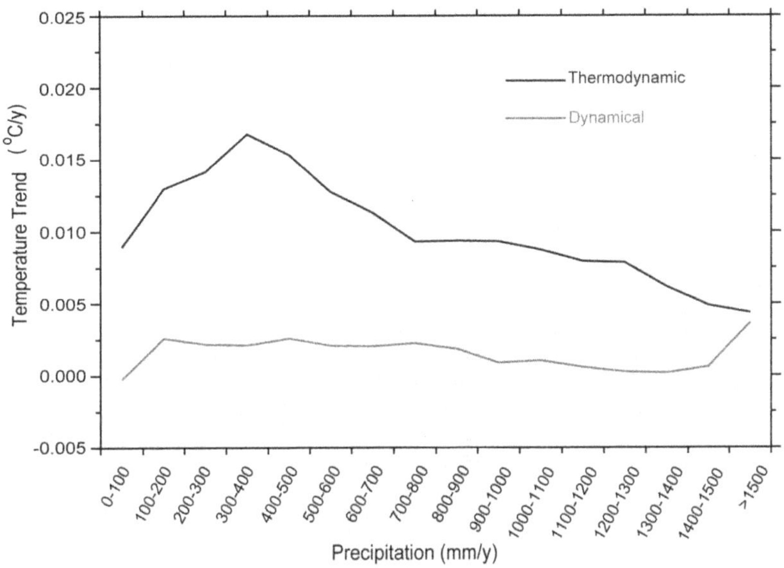

Fig. 3.8. The regionally averaged surface air temperature trend during the cold season in 1902–2011 as a function of the climatological annual precipitation for the thermodynamically forced and dynamically induced temperatures over the mid-latitudes of the NH (20°N–60°N) (Guo *et al.*, 2018).

regions vary depending on the amount of precipitation. This graph demonstrates that peaks in the thermodynamic temperature bars appear at 400–500 mm yr^{-1} of annual precipitation, that the warming difference observed in the thermodynamic temperature between semi-arid and wet regions is more obvious in Asia than in Europe, and that the thermodynamic temperatures of Asia and Europe have similar relationships to annual precipitation in the 20°–60° range of the NH (Fig. 3.8). In North America, there was just a slender correlation between annual precipitation and thermodynamic temperature. While North America shows a peak in semi-arid regions with a significantly smaller value than that reported for thermodynamic warming, the dynamic temperature variability shows a relatively flat variation over the various regions of Asia and Europe. The various thermodynamic warming patterns over Asia, Europe, and North America suggest that there are apparent regional differences in the variability of the thermodynamic and dynamic temperatures. Asia had more significant thermodynamic warming in its semi-arid zones than Europe and North America, indicating a pivotal role for Asia in ESAW.

Table 3.4 provides extensive thermodynamically induced temperature data for semi-arid and wet regions in the mid-latitudes of the NH in Asia,

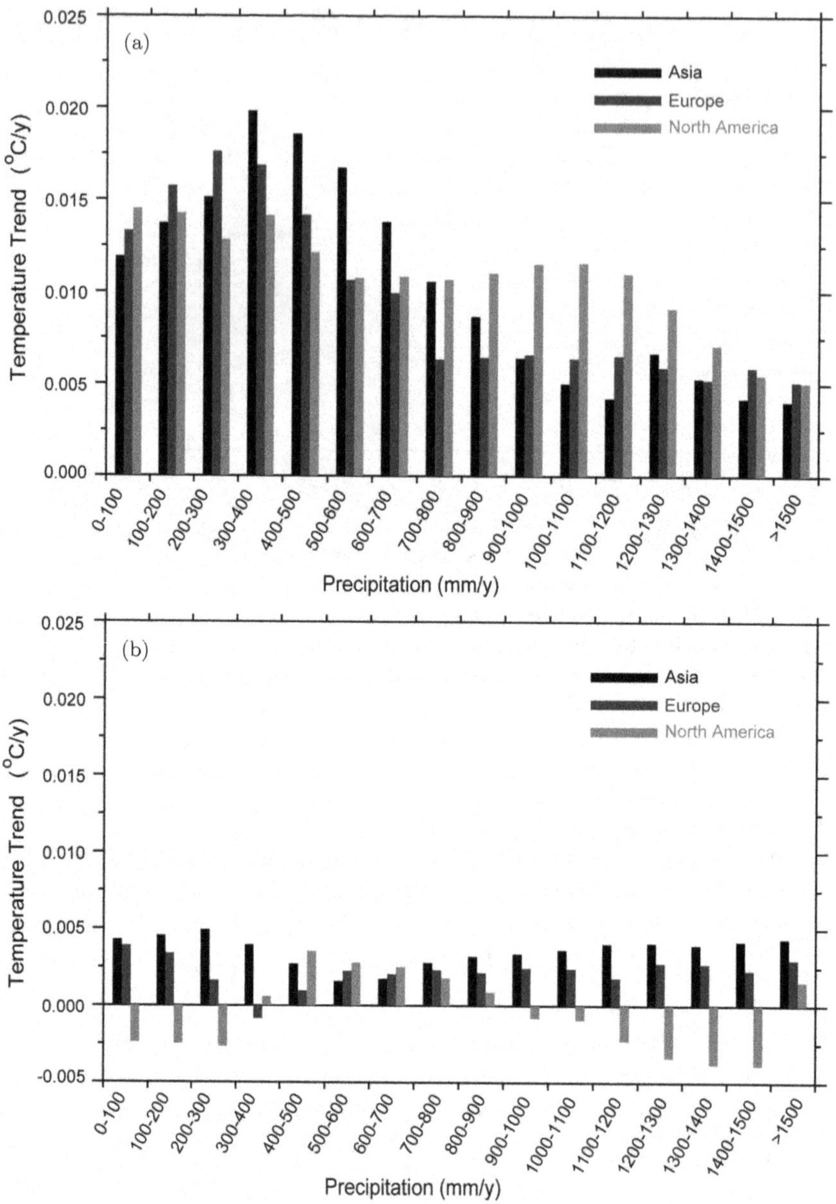

Fig. 3.9. The regionally averaged temperature trend during the cold season in 1902–2011 as a function of the climatological annual precipitation for the thermodynamically forced (a) and dynamically induced (b) temperatures in Asia, Europe and North America. Asia: 20°N–60°N, 60°E–180°E; Europe: 36°N–60°N, 17°W–60°E; and North America: 20°N–60°N, 180°W–17°W.

Table 3.4. Cold season mean thermodynamic temperature warming amplitudes during 1902–2011 over the mid-latitudes (20°N–60°N) of the NH, Asia, Europe, and North America (unit: °C/109 years) in semi-arid and wet regions (Guo *et al.*, 2018).

	20°N–60°N	Asia	Europe	North America
Semi-arid regions	1.63	1.95	1.43	1.39
Wet regions	0.72	0.52	0.65	0.97

Europe, and North America due to the crucial role that thermodynamic temperature plays in ESAW. As shown in the table, in the range of 20°N–60°N, the thermodynamically induced temperature increased by 1.63°C in the semi-arid region, which was 2.3 times more than in wet areas. A comparison of the semi-arid warming trends over Asia, Europe, and North America showed that Asia had the highest thermodynamic warming, with temperature increases as high as 1.95°C, or almost 3.7 times that of wet regions. The warming amplitudes over semi-arid regions in Europe and North America were similar, increasing by 1.43°C and 1.39°C, respectively, and were higher than those over wet regions (0.65°C for Europe and 0.97°C for North America). As a result, Asia had the strongest warming trend in semi-arid regions, which was around 1.4 times greater than Europe and North America. This stark contrast draws attention to Asia's thermodynamic warming tendency, which may be closely linked to regional thermodynamic issues (Guan *et al.*, 2015a).

The quantitative warming contributions of the thermodynamically forced and dynamically generated temperatures in each climate zone to the raw temperature mean warming in the 20°N–60°N region are evaluated to clarify their precise roles in the warming process. It shows how the contributions of thermodynamic and dynamic temperatures differ noticeably: The raw temperature increase over the 20°N–60°N region was warmed by 87.2% due to the thermodynamically forced temperature, with semi-arid regions accounting for 45.3% of the warming. In comparison to moist areas, this contribution was more significant. In contrast, the dynamically driven temperature in all climate regions showed a far lesser influence than the thermodynamic temperature. The clear difference between the contributions of the dynamic and thermodynamic temperatures leads one to believe that the mid-latitudes of the NH experienced thermodynamic warming as the main cause of ESAW.

The contributions made by each climate region in Asia, Europe, and North America to the mean raw temperature warming over the 20°N–60°N region by thermodynamically and dynamically induced temperature are different. In all the chosen areas, particularly Asia, the contribution of the thermodynamic temperature was more significant than that of the dynamic temperature. In Asia, North America, and Europe, the thermodynamic temperature in the semi-arid regions contributed to warming by 24.7%, 10.9% and 8.2%, respectively. In these regions, the dynamically generated temperature's contributions were significantly less than the thermodynamically forced temperature. This disparity between the contributions from North America, Europe, and Asia points to the importance of local impacts in ESAW, particularly in Asia, where frequent anthropogenic activities have significantly impacted the observed ESAW. According to the previous explanation and discussion, the thermodynamic temperature dominates in ESAW. The fact that the change in the CMIP5 ensemble mean matched the change in the thermodynamically induced temperature shows that the model did a good job simulating thermodynamic warming. In contrast to the thermodynamically forced temperature, the regional distribution of the CMIP5 ensemble mean temperature trend (Fig. 3.10(b)) was very homogenous (Fig. 3.10(b)). The main zonal feature of the CMIP5 ensemble mean temperature trend, which accentuates the warming along the Arctic coast, is the increasing warming rate from low to high latitudes.

The more significant warming of the thermodynamic temperature that occurred in semi-arid areas did not emerge in the CMIP5 ensemble mean, according to the temperature anomalies of the simulated and thermodynamic temperatures as a function of yearly precipitation. The amplitude of the thermodynamic warming that was seen in semi-arid regions was almost 1.55 times larger than that predicted by CMIP5 simulations. This significant difference between the thermodynamically forced temperature and the temperature predicted by CMIP5 in semi-arid regions shows that well-mixed GHGs alone cannot fully explain the ESAW. Therefore, the impacts of strong local thermodynamic forcing cannot be disregarded in the simulation process. The climatological distribution of precipitation from CMIP5 indicated that there were discrepancies between the simulations and the data. The values of CMIP5 in semi-arid regions across Asia and North America, particularly in the western regions of North America, were lower when compared with the observations. As a result, all of Asia, Europe, and North America showed relatively flat curves in the simulated temperature trends as a function of yearly precipitation. Additionally, not all three

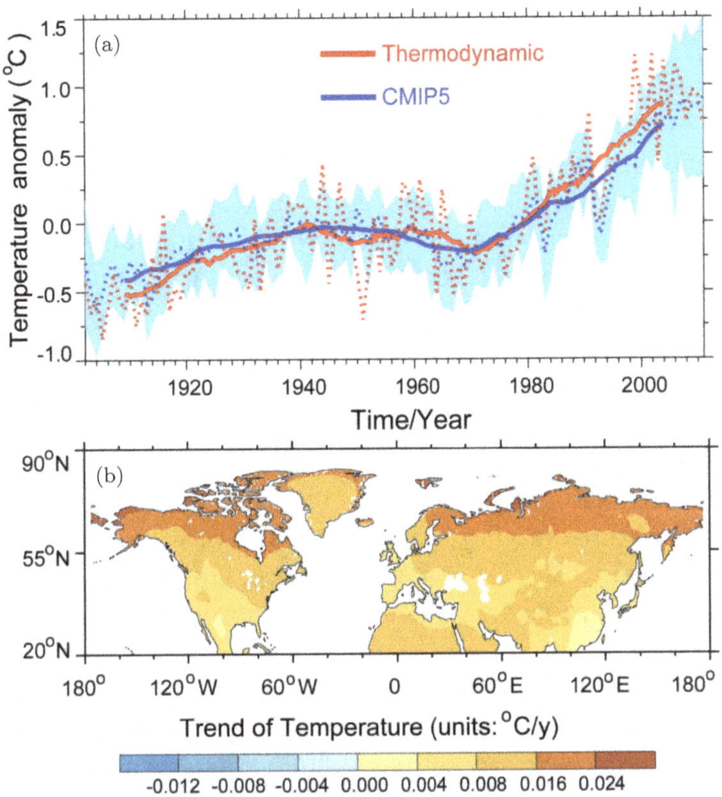

Fig. 3.10. Time series of mean temperature anomalies in the cold season over the mid-latitudes (20°N–60°N) of the NH during 1902–2011 for the thermodynamically forced temperature (red dashed) and the ensemble mean of 20 CMIP5 models (blue dashed). Solid lines indicate the 15-year running means of the dashed lines, and blue shading denotes the standard deviation of the temperature simulated by the CMIP5 models. The model data are masked so that they have the same coverage as observations (a). Spatial distribution of the temperature trend in the cold season of the ensemble means of 20 CMIP5 models during 1902–2011 (b) (Guo *et al.*, 2018).

selected locations had the reported temperature maxima over semi-arid regions, highlighting the significance of integrating local thermodynamic forcing in simulations.

3.3.4 *Daily max, min temperature, and diurnal temperature range*

Over the past century, there has been an increase in the average global surface air temperature, with a quicker pace of warming since 1950. Both the daily lowest and maximum air temperatures have grown dramatically over the past several decades, but due to their asymmetrical increases ($T_{\min}$ increased more quickly than $T_{\max}$), the diurnal temperature range (DTR) has clearly decreased on a worldwide scale. DTR is not merely a temperature-related variable; its variation can also impact ecological systems, the growth of plants, and even the carbon economy, demonstrating that it warrants more studies. According to Huang *et al.* (2012), the semi-arid region of the NH has the highest growing trend of $T_{\min}$ and $T_{\max}$ (Fig. 3.11(a)), making it the region most sensitive to the warming change. A majority of the world's regions, except for arid and semi-arid regions, have a negative trend in DTR as measured by $T_{\min}$ and $T_{\max}$ (Fig. 3.11(c), black line). In contrast to temperature variables, local factors can more easily affect DTR change. To compare the DTR change between the opposing regional processes of the drying Eastern North Hemisphere and wetting Western North Hemisphere, it also examines the impacts of CO_2, black carbon (BC), sulfur dioxide (SO_4), and the three driving factors taken together on the regional DTR trend. Additionally, this chapter will contrast how the semi-arid region's DTR changes in response to various CO_2 emission scenarios, from low to high.

The highest daily $T_{\min}$ and $T_{\max}$ peaks can be seen in semi-arid locations, according to Fig. 3.11(a). Daily $T_{\min}$ has a larger amplitude than daily $T_{\max}$, and the DTR is inconsistent across all areas. The raw $T_{\max}$ and $T_{\min}$ are first divided into dynamic and radiative components, respectively, by using the dynamic adjustment method to investigate the mechanisms of DTR fluctuation over various locations. In semi-arid areas, the radiative-forced daily $T_{\max}$ clearly peaks and exhibits similar variability to the raw $T_{\max}$ (Fig. 3.11(b)). This is especially true in semi-arid areas. With a radiative daily $T_{\max}$ of 0.012°C/yr, the noteworthy peak of the raw daily $T_{\max}$ in semi-arid environments approximately reaches 0.014°C/yr. However, with a value of 0.002°C/yr, the dynamic $T_{\max}$ shows a very flat variation across all locations.

The raw, dynamic, and radiative $T_{\min}$ curves in Fig. 3.11(c) resemble the $T_{\max}$ curves in Fig. 3.11(b) but have a higher amplitude. With a radiative $T_{\min}$ of 0.014°C/yr and a dynamic $T_{\min}$ of 0.002°C/yr, the peak daily $T_{\min}$ in semi-arid environments approaches 0.016°C/yr. When the differences in $T_{\max}$ or $T_{\min}$ between dynamic and radiative systems are compared, it can be seen that the radiative effect dominates the regions' apparent daily increases in $T_{\max}$ and $T_{\min}$, and it results in the obvious changes of radiative DTR trend (Fig. 3.11(d)).

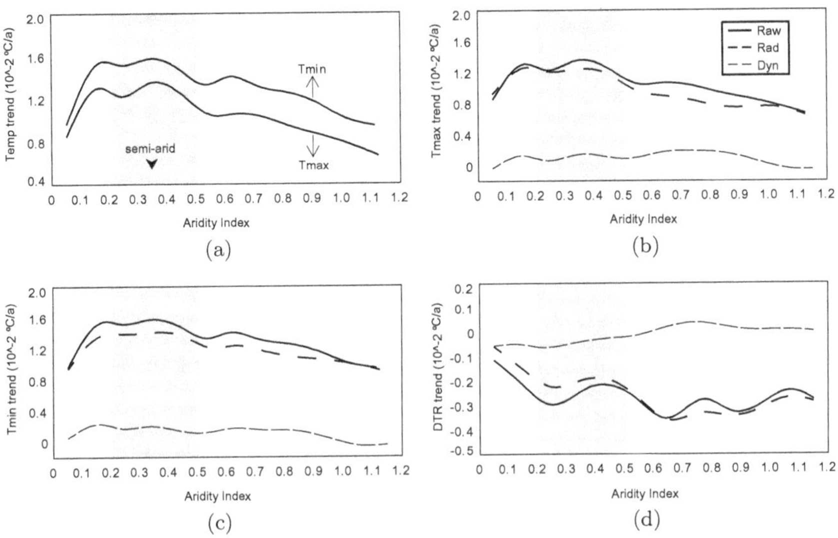

Fig. 3.11. Regionally averaged observed temperature trends in the cold season (November–March) as a function of the annual aridity index for daily maximum temperature and daily minimum temperature from 1902 to 2011 over 20°N–50°N (a). Regionally averaged observed temperature trends in the cold season (November–March) as a function of the annual aridity index for raw (solid black line), dynamically induced (gray dotted line), and radiatively forced (black dotted line) temperatures from 1902 to 2011 over 20°N–50°N for daily maximum temperature (b). (c) same as (b) but for daily minimum temperature. (d) same as (b) but for the diurnal temperature range. Shaded areas represent 99% confidence intervals (Guan *et al.*, 2022a).

According to Huang *et al.* (2016a), the Eastern NH and Western NH in the mid to high latitudes of the NH experience opposite regional processes of drying and wetting. Figure 3.12 contrasts the trends in DTR distributions across various locations between the Eastern and Western NHs. The lowest peak of DTR appears in the semi-arid zone at a declining rate of 0.04°C/decade in the Eastern NH, where DTR exhibits fluctuation as a function of the AI. The DTR has a modestly negative trend or even a positive trend in semi-arid regions with an AI of 0.3–0.5; however, the lowest value of the DTR trend in the Western NH is in the humid region with an AI greater than 0.6. Different DTR trend patterns in the eastern and western parts of North Hemisphere imply that DTR change is closely related to regional causes.

Previous findings show that CO_2 (Huang *et al.*, 2016a), BC, and other radiative factors are invariably linked to the local process. Model output over the last 60 years in the mid-to-high latitudes of Eastern NH shows that

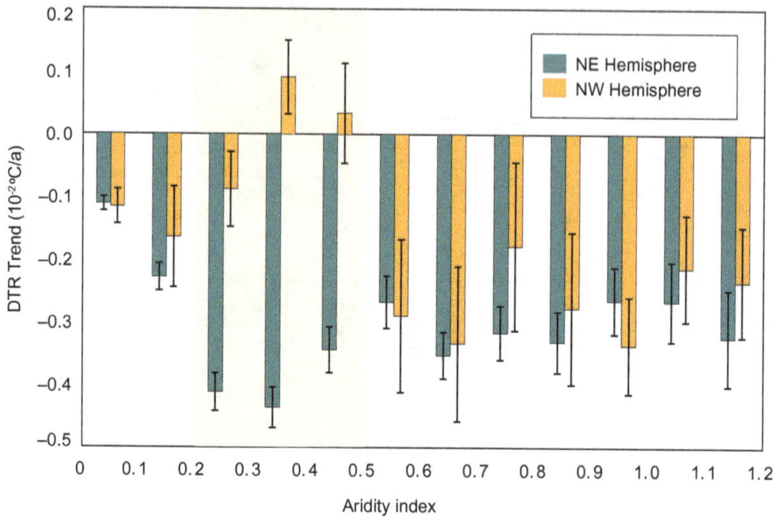

Fig. 3.12. The trend of diurnal temperature range as a function of the aridity index over the Eastern NH (green) and Western NH (yellow) from 1902 to 2011 (Guan *et al.*, 2022a).

average temperature, $T_{\min}$, $T_{\max}$, and DTR function of AI all take peaks in semi-arid regions. According to Fig. 3.13(a), in arid and semi-arid regions, CO_2, combined forcings, and BC all contributed to a noticeable warming of the daily mean temperature in comparison to the control trial. Due to the ability of SO_4 to counterbalance warming, compared to the control, the combined forcing experiment exhibits warming with a smaller amplitude than the CO_2 warming experiment. $T_{\max}$ and $T_{\min}$ distributions are similar (Figs. 3.13(b) and 3.13(c)).

Particularly in semi-arid regions, the warming brought on by CO_2 in the sensitivity experiment results in higher values of $T_{\max}$ and $T_{\min}$ than in the control experiment. In a warmer environment, the same amount of heating causes arid and semi-arid regions to warm more quickly than humid areas. $T_{\min}$ increases more rapidly in arid and semi-arid regions than $T_{\max}$ because there is more high upward net longwave radiation.

DTR variations between the sensitivity and control tests are depicted in Fig. 3.13(d). The most significant differences are evident in the combined forcing experiment, followed by those in the CO_2 and BC studies. The semi-arid regions show the most obvious variation in DTR. Regarding these radiative factors, CO_2 contributes more in a single experiment than BC and SO_4 and has a more significant impact on DTR decline in semi-arid locations. Compared to the low-emission scenario, the high-emission scenario displays a more pronounced declining trend in the semi-arid parts of Eastern NH for

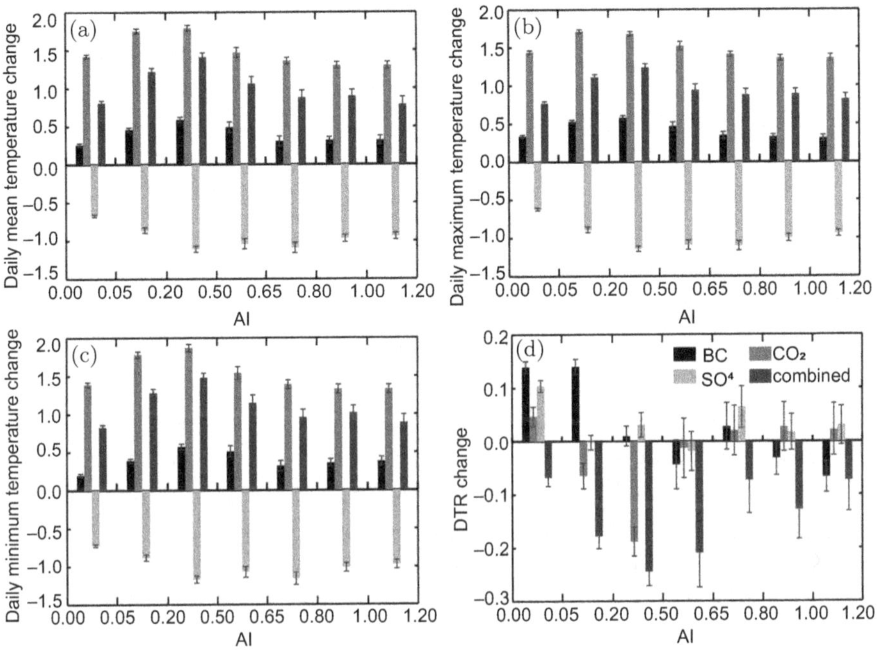

Fig. 3.13. Differences between the sensitivity and control experiments concerning black carbon, CO_2, SO_4, and combined forcings as a function of the aridity index over 20°N–50°N in the Eastern NH in the mean daily temperature (a), daily maximum temperature (b), daily minimum temperature (c), and diurnal temperature range (d) in winter for 60 years (Guan *et al.*, 2022a).

the dominant CO_2 in the DTR trend. All the regions in SSP 126 exhibit a marginally favorable DTR trend. The DTR reveals the strongest decreasing tendency in semi-arid regions for the highest emission scenario of SSP 585, which is substantially faster than that of SSP 126. This trend is roughly 0.094°C/decade. The stark difference in DTR change between SSPs 126 and 585 demonstrates the crucial contribution of the CO_2 effect to the improved DTR trend in the semi-arid area of Eastern NH.

3.3.5 *Contribution of radiative factors*

The global water cycle is predicted to be affected by the rise in global mean surface temperature, with climatologically wet regions likely to become wetter, arid and semi-arid regions expected to become dryer, and rainfall events expected to become more frequent. It shows that arid and semi-arid regions are warming up more quickly than semi-arid regions, especially during the winter. The annual, cold, and warm seasons' mean temperatures climbed by 1.13°C, 1.53°C, and 0.79°C, respectively, throughout the period from 1901 to 2009 in semi-arid locations (annual rainfall reaching 200–600 mm yr^{-1}), where there was a particularly strong warming trend. In semi-arid regions of mid-latitude Asia, Europe, and North America, the cold season temperatures rose by 1.41°C, 2.42°C, and 1.50°C, respectively, during the same time period. Semi-arid regions contribute 44.46% to worldwide land surface air temperature trends. Numerous factors, such as the dynamical adjustment of the circulation and SST anomalies, land surface processes, snow/frozen ground, and local human activities, may contribute to this ESAW (Fig. 3.14). However, the leading causes of the increased warmth in semi-arid environments can vary depending on the locality.

According to Fu *et al.* (2006), the tropical belt extended between 1979 and 2005, expanding by an estimated 2° of latitude. A broadening of the tropics like this would change circulation patterns, temperatures, and precipitation patterns. Recent developments in modeling have demonstrated that a poleward expansion of the Hadley cells and a shifting of the precipitation pattern minus evaporation ($P-E$) and storm tracks are robust characteristics of anthropogenic climate change. This results in a drying pattern over much of the already dry subtropics in a warmer world (15°–40° latitude in each hemisphere). Gong *et al.* (2001) attributed the increased surface air warming in the boreal cold season across the NH mid-high-latitude land to changes in atmospheric circulation based on observational data studies. However, dynamic adjustment cannot account for the striking

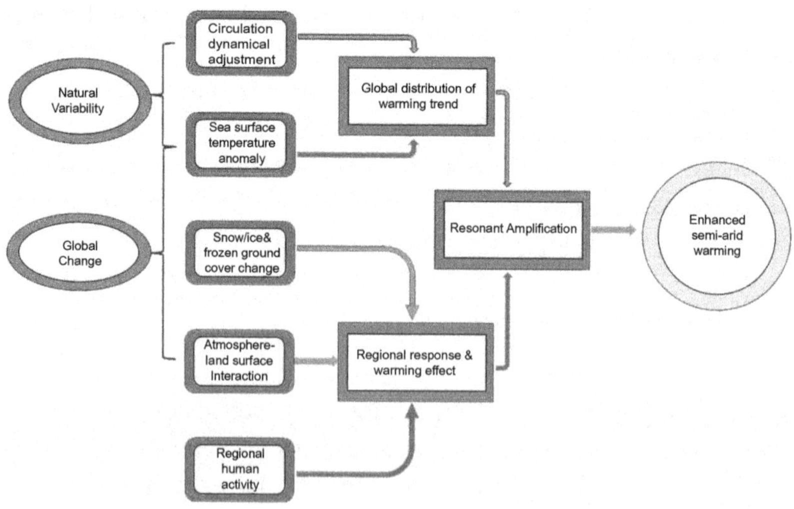

Fig. 3.14. Schematic diagram of ESAW (Huang *et al.*, 2012).

variation in warming trends found in semi-arid regions of Europe, Asia, and North America. The feedback system of surface albedo, desertification, and biogeophysics may have partially driven the considerable warming over semi-arid regions. Higher temperatures would dry out rocky or sandy soil, increasing the surface albedo and reflecting more solar energy back into space. When compared to their surroundings, semi-arid environments act as heat radiation sinks. According to some observational analyses, during the past few decades, desertification may have moderated warming trends by providing negative forcing at the Earth's surface that is roughly equal to $\sim$20% of the global anthropogenic CO_2 effect. According to Shuman *et al.* (2011), changes in the dominant flora and related regional surface albedo may significantly impact the Earth's temperature in the NH boreal forests, notably those in Siberia. Eventually, Siberia's deciduous larch forests will probably turn into evergreen conifer forests, lowering the region's surface albedo. In low-diversity areas of Southern Siberia, Shuman *et al.* (2011) discovered that the conversion of larch to evergreen would result in a local positive radiative forcing of 2.6 W m^{-2}. This radiative heating would amplify the anticipated warmth in the area affected by climate change.

3.4 Regional Enhanced Warming

However, when viewed globally from various climate zones, the warming of semi-arid regions is significant. The typical semi-arid regions of the world vary in their geographical distribution, climatic traits, substratum vegetation, and human activities, which has led to variations in the rate of warming in these typical semi-arid regions and the causes of these variations. In general, the annual mean surface temperature in the NH has been increasing significantly over the past 100 years, especially in the middle and high latitudes. The global average winter surface temperature is distributed spatially like that of the annual average temperature trends, but the NH is experiencing more noticeable warming in the middle and high latitudes, particularly in the semi-arid regions of East Asia, where the warming is significantly more intense than it in other parts of the world. Both the spatial extent and the intensity of the warming have also increased significantly. East Asia is a complex system that involves many different forms of human activities, in addition to the influence of circulation and human activities. It is one of the places with the most intense human activity. As a result, the semi-arid region of East Asia is much warmer than other areas, and the causes are radiation and power-related influences.

3.4.1 *Enhanced warming over East Asia*

Given that it makes up about 39% of the planet's surface area (Huang *et al.*, 2013), contains sizable arid and semi-arid regions (Huang *et al.*, 2014), and is home to 67% of the world's population, Asia is perhaps the region most vulnerable to climate change. During the boreal cold season, a significant portion of Asia's arid and semi-arid regions experience dramatically increased surface atmospheric warming (Huang *et al.*, 2012), making the regions more parched and hastening arid and semi-arid region development (Huang *et al.*, 2016b). Additionally, global warming in arid and semi-arid regions has been observed to be intensified (Huang *et al.*, 2012). Most arid and semi-arid regions are found in the Northwest and the humid regions are found in Southeast China, according to the distribution of the 60-year averaged AI in East Asia from 1948 to 2008. The majority of East Asia is made up of arid and semi-arid regions, and the mean AI has a general trend of increasing values from the Northwest to the Southeast. This broad distribution agrees with the distribution of yearly mean precipitation shown by Guan *et al.* (2015b).

According to Huang *et al.* (2012), the warming trend in East Asia was more pronounced across semi-arid regions during the boreal cold season (November–March). This tendency was much more pronounced in semi-arid areas of Asia, where the average yearly temperature rose by 1.13°C over land while it rose by 2.42°C during the cold season. It was also successfully demonstrated that RFT plays a dominant role in the enhanced warming in East Asia using the dynamical adjustment methodology (Wallace *et al.*, 2012; Guan *et al.*, 2015b); however, only a statistical approach was used to analyze the data, which did not qualify the warming effects of particular radiative factors. The raw, DIT, and RFT trends across East Asia in winter between 1950 and 2011 are distributed spatially in Fig. 3.15. Figure 3.15(a) depicts the substantial warming trend over Northern East Asia, particularly in the Junggar Basin and Northeast China, as well as the gradually increasing RFT distribution, which induced a generally warming over East Asia.

The long-term trends of the raw temperature, RFT, and DIT in the winter are depicted in Fig. 3.16 as a function of the annual mean AI. It is consistent with the heightened warmth previously noted in arid and semi-arid regions that appears on both the raw temperature and the RFT. The raw temperature was distributed similar to how the RFT trend was distributed (Fig. 3.15(c)), and apparent warming was evident over

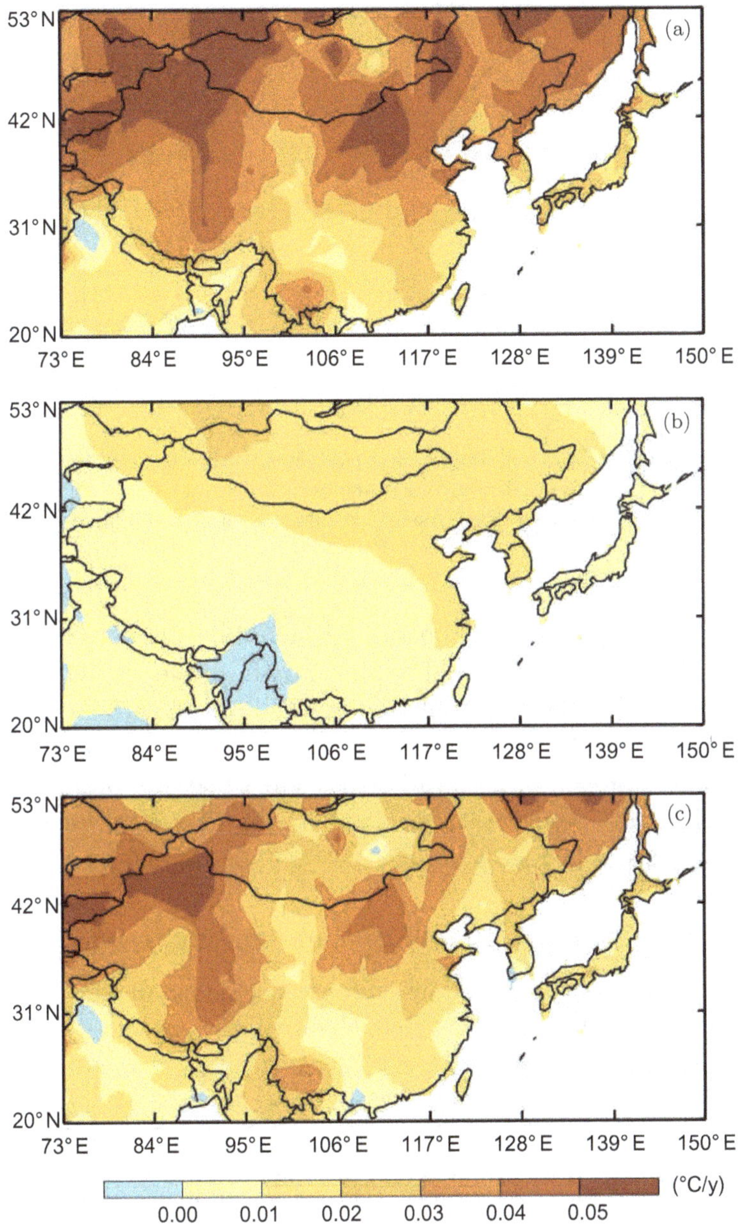

Fig. 3.15. Spatial distribution of trend of raw (a), dynamically induced (b), and radiatively forced temperature (c) (unit: °C/yr) in winter from 1950 to 2011 over East Asia (Zhang *et al.*, 2017).

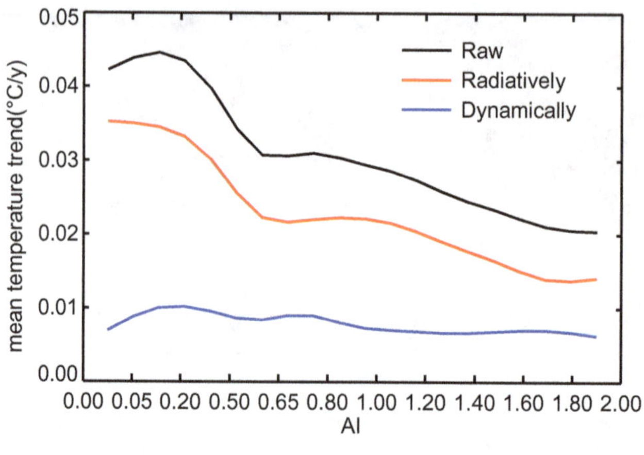

Fig. 3.16. Regionally averaged trend of the raw temperature (black), radiatively forced temperature (red), and dynamically forced temperature (blue) (unit: °C/yr) change in winter from 1950 to 2011 over East Asia as a function of the annual mean AI (Zhang *et al.*, 2017).

Northern East Asia, reaching 0.05°C yr^{-1} in certain places. The DIT trend distribution (Fig. 3.15(b)) differed significantly from the raw temperature by 0–0.2. The lack of a clear difference in the DIT trend across climate zones suggests that the RFT significantly contributed to the regional temperature variation. The trend shows that radiative factors played a prominent role in the elevated warmth seen over the arid and semi-arid regions when compared to the results shown in Fig. 3.15.

3.4.2 *Effect of CO_2, BC, and SO_4 on regional warming*

GHGs, aerosols, changing land use cover, declining albedo, clouds, and local human activity are the main contributors to RFT variability (Huang *et al.*, 2011; Guan *et al.*, 2016). A significant amount of anthropogenic GHGs, in particular CO_2 and aerosols, have been created and released into the atmosphere as a result of East Asia's rapid economic expansion and industrialization, and they have eventually had an impact on regional climate. Investigating how the surface temperature reacts to these radiative factors is therefore required. CO_2, BC, and SO_4 are the three typical components used in this study to examine the impact of surface temperature.

The stippling denotes the 95% confidence level, and the graph displays the distribution of the association between detrend CO_2 and detrend radiatively forced temperature. Because the CO_2 dataset utilized here is a time series dataset without a spatial distribution, it is assumed that the amount of CO_2 in the atmosphere is constant across all locations. As a result, the non-uniform distribution of the correlation coefficients can reveal how sensitive various regions are to CO_2. RFT and CO_2 concentrations are shown to be strongly correlated. In most of East Asia, the correlation coefficients show confidence levels above 95%, predominantly concentrated in two regions: Central China and Western China. With a correlation coefficient of 0.77, the area is more susceptible to CO_2 in that direction. The modifications in the regionally averaged correlation coefficients are caused by AI. In highly arid areas, the correlation coefficient is the highest. It is possible that increased arid and semi-arid region warming is closely related to CO_2 because the correlation coefficient over northern arid and semi-arid regions is typically higher than that over humid regions. This indicates that arid and semi-arid regions are more susceptible to CO_2 than humid parts. The following section will investigate CO_2's role in accelerating arid and semi-arid region warming in more detail.

The regionally averaged BC emission changes as a function of the AI. The BC emissions in arid and semi-arid regions are low and mainly distributed in humid regions. Most BC emissions in arid and semi-arid regions (with AI values less than 0.5) are less than 1×10^{-5} kg m^{-2}. The BC emissions in humid regions are much higher than in arid and semi-arid regions. The largest regionally averaged BC emissions are observed in the region with AI values between 1.4 and 1.6, where the BC reaches 7.6×10^{-5} kg m^{-2}, which is almost ten times greater than the regionally averaged BC emissions of arid and semi-arid regions. According to Xu (2014), China's BC emissions, which primarily come from the industrial and residential sectors, cause this disparity. In terms of industry and people, Eastern China is more developed than the dry Western provinces. The fact that the mean temperature changes as a function of BC emissions suggests that BC has a warming effect on temperature. The with low BC emissions (less than 1×10^{-5} kg m^{-2}) exhibit a more significant temperature response than regions with higher BC emissions. This shows that the temperature in arid and semi-arid regions is more susceptible to BC emissions than that in humid regions since the areas with BC emissions less than 1×10^{-5} kg m^{-2} are primarily dispersed in these regions.

The variations in the distribution of SO_2 (the precursor to SO_4) concentrations are a function of the AI and the variations in the mean temperature response to SO_4 are a function of SO_2 emissions. Despite having comparable distributions, SO_2 emissions are larger than BC emissions. The

SO_2 emissions are minimal in arid locations with AI values less than 0.5, while they are substantially higher in humid places. This disparity may have a similar cause as the higher BC emissions. Because the eastern part of East Asia is more developed and home to wet regions, anthropogenic aerosol emissions are higher there. The mean temperature response to SO_4 aerosols as a function of SO_2 emissions shows no significant changes. As a result, the temperature response is comparable even though SO_4 emissions in arid and semi-arid regions are far lower than those in humid regions.

The daily mean temperature differences between the control and perturbed experiments are computed to assess the relative contributions of CO_2, BC, and SO_4 to temperature changes. In the model simulation, the forcings immediately raised the precursor emissions of CO_2/BC/SO_4 to current levels. As a result, the model output does not seek to replicate 20th-century reality. In addition to the three representative variables in this analysis, numerous other GHGs and aerosols impact the climate system, such as ozone and land use. Thus, there are some disparities between the model results and the observations. Figures 3.17(a) and 3.17(d) demonstrate that the whole region has passed the 95% confidence level from a two-sided *t*-test for the daily mean temperature. Over East Asia, the mean temperature exhibits a warming trend in reaction to CO_2. A significant center is found in Inner Mongolia and Xinjiang Province, and the temperature increase is most significant in the arid regions of the Northwest. According to the results of model simulations, rising CO_2 levels would cause land areas to become drier, leading to the spread of arid and semi-arid regions. A higher temperature response would occur over arid and semi-arid regions than over humid regions (Huang *et al.*, 2016b) as a result of the expansion of arid and semi-arid regions, which would diminish carbon sequestration and increase regional warming (Huang *et al.*, 2012; Ji *et al.*, 2014). Nearly all of East Asia experiences an increase in temperature due to BC, with India and Burma experiencing a minor reduction. A larger center is seen over Tibet and Northeastern China, and the BC warming magnitude is also substantially lower than that of CO_2. This is because places with snow and ice exhibit a greater temperature sensitivity to BC. Due to the high concentration of sulfate aerosols, the cooling effect of SO_4 on surface temperatures was more significant in the Northeast and Southeast of China than in other areas. The Tibetan Plateau also has a sizable center, which is consistent with the findings of Kasoar *et al.* (2016). It demonstrates how several factors, including the cloud effect via SO_4 and atmospheric circulation, in addition to SO_4 concentration, influence the temperature response.

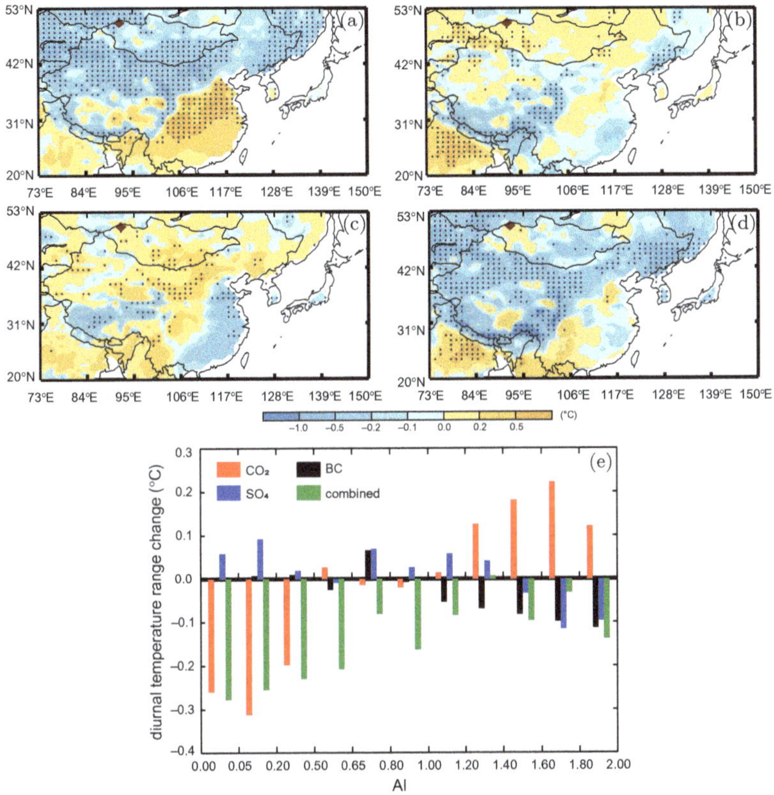

Fig. 3.17. Distribution of the daily temperature range (unit: °C) response to CO_2 (a), BC (b), SO_4 (c), and all three combined forcing (d) in winter; grid points are stippled for statistically significant changes based on a 95% confidence interval from a two-sided *t-test*. Regionally diurnal temperature range (unit: °C) change in winter to CO_2 (red), BC (black), SO_4 (blue), and the combined forcing (green) as a function of annual mean AI over East Asia (e) (Zhang *et al.*, 2017).

Figure 3.17 displays the distributions of the DTR response to CO_2 (a), BC (b), SO_4 (c), and all three forcings combined (d) in the winter. The DTR response to these radiative components is smaller than the daily mean, maximum, and minimum temperatures, but it still exhibits a clearly non-uniform distribution. With the boundary along the Taihang Mountains, the DTR response to CO_2 (Fig. 3.17(a)) increases in the Southeast wet regions and declines in the Northwest arid and semi-arid regions. Most of the area passed the 95% significant *t*-test. The DTR response to BC (Fig. 3.17(b)) is reduced in most areas, but an expanding mosaic is seen over Central and Western China as well as the majority of Inner Mongolia. However, the DTR's response to BC only partially passed the 95% significant *t*-test. As with CO_2, the DTR response to SO_4 (Fig. 3.17(c)) increases in the arid Northwest China region and diminishes in the humid Southeast China region. The three radiative components worked together to cause a DTR drop in most locations, as seen in Fig. 3.17(d). This finding is in line with that of earlier research on DTR based on observational data, which discovered that in the decades preceding, the DTR dropped globally as a result of the significant increase in the minimum diurnal temperature. However, few studies have investigated the underlying mechanism, with a majority concentrating on the effects of clouds, solar radiation, soil moisture, surface wind, and precipitation on the DTR based on observational data. According to research, the DTR in China decreased most noticeably in the winter. Therefore, research on how CO_2 and anthropogenic aerosols affect the DTR is sparse.

The depiction of the various regional mean temperature responses to CO_2, BC, and SO_4 in Fig. 3.17(e) revealed that the increases in temperature caused by BC, CO_2, SO_4, and the three forcings taken together are more significant in arid and semi-arid regions than they are in humid regions. Temperature shows the largest response to CO_2, with the next highest being the response to BC. In reaction to SO_4, the temperature drops, but the combined effect of these radiative variables raises the temperature. This result shows that CO_2 is the leading cause of the accelerated arid and semi-arid region warming, followed by BC. The increased warmth does not appear to be significantly attributable to SO_4.

3.4.3 *Snow/ice albedo feedback in enhanced warming*

The most reflective natural material on Earth's surface, snow, which covers more than 30% of the planet's land area, is another sensitive radiative component. One of the critical energy balance elements for snow-covered areas at medium and high latitudes is snow albedo feedback, which directly influences the surface radiation budget and exerts a cooling effect on the global temperature. However, light-absorbing particles (LAPs) like black carbon (BC), organic carbon (OC), dust, and algae that cause the snow's surface to darken can substantially change how reflective snow is (Hadley and Kirchstetter, 2012). LAPs reduce the snow's albedo by increasing solar radiation absorption when they are deposited on the snow surface. Because of this, the radiative forcing of LAPs in the snow is a significant factor in the global and regional loss of snow cover, which disturbs the climate system and affects hydrological cycles.

Over the last 10 years, researchers have become more interested in investigating the contents of LAPs in seasonal snow over Northern China's semi-arid regions. The first extensive field campaign was carried out in January and February 2010 in six provinces, i.e., Qinghai, Gansu, Inner Mongolia, Heilongjiang, Jilin, and Liaoning. At 46 sites, about 400 snow samples were gathered. Based on the various spectrum dependences of their light absorption, an enhanced two-sphere integration (TSI) technique was employed to estimate the concentrations of BC and non-BC constituents. The findings showed that the industrial areas of Northeastern China had the dirtiest snow, with a median BC content in surface snow of 1220 ng g^{-1}. In contrast, the distant Northeast region near the border with Siberia had the lowest BC contents, with a median concentration of 117 ng g^{-1}. More importantly, the field campaign's discovery of the sharp decline in BC content of snow with latitude in Northeastern China provided a groundbreaking indication that there is not much BC entering the Arctic from China (Huang *et al.*, 2011). The sources of LAPs in seasonal snow are then investigated using the LAPs and 14 other analytes as input to a positive matrix factorization (PMF) receptor model. According to the analysis, three elements/sources — soil dust, industrial pollution, and biomass and/or biofuel burning — were responsible for the measured light absorption of snow. The primary source, which accounted for 53% of LAPs on average, was soil dust. Continued snow readings were taken over Northern China after 2010. For instance, a significant field effort in seasonal snow was conducted in the Western Chinese Provinces of Xinjiang and Qinghai (Northeastern part of Tibetan Plateau) in January and February 2012 (Pu *et al.*, 2017). 284 snow samples were collected at 38 sites. The findings show that the purest snow, with a measured BC concentration of ~10 ng g^{-1}, was found in Northern Xinjiang along the Mongolian border. This concentration is comparable to that of the Arctic Ocean's shore. The dirtiest snow in Xinjiang had a BC concentration of ~450 mg g^{-1} close to industrial centers. Additionally, it was shown that height causes the BC content of snow to decrease. Additional vertical fluctuations in the snowpack LAPs suggested that the emission sources may have changed as winter progressed. For LAPs in snow in Northwest China, the source analysis results based on a PMF model identified three significant sources: industrial pollution, biomass burning, and soil dust. Between May 2013 and October 2015, a different field expedition collected 67 samples of ice and snow from seven glaciers on the Tibetan Plateau (Wang *et al.*, 2019a). They discovered that BC concentrations are significantly higher in the dry season than in the wet, which was explained by the wet season's scavenging effect. The BC concentrations were generally between 10 and 2000 ng g^{-1} over semi-arid regions of Northern China and the Tibetan Plateau.

Snow LAPs can substantially lower snow albedo and result in a positive radiative forcing. In January 2014, Wang *et al.* (2017) used a field spectroradiometer to determine the snow albedo throughout Northeastern China. Due to various LAP concentrations and snow characteristics, such as snow grain size, the measured snow albedos at 550 nm ranged greatly from 0.99 to 0.61. They further deduced that 100–500 ng g^{-1} of BC could reduce snow albedo by 0.014–0.039 compared to pure snow with a snow grain radius of 100 μm by using a newly created radiative transfer model (Spectral Albedo Model for Dirty Snow, or SAMDS). In January 2018, Shi *et al.* (2020) measured the snow albedo in Northwest China and discovered that at the visible wavelength band (400–750 nm), 40–1850 ng g^{-1} BC decreased the snow albedo by 0.01–0.20. The BC-induced radiative forcing was subsequently calculated to be 0.2–6.9 W m^{-2} by combining a Snow, Ice, and Aerosol Radiation (SNICAR) model with the surface-level downward shortwave flux from CERES, revealing a significant acceleration of snowmelt due to BC in Northwest China. By assuming a sphere-shaped snow grain with a radius of 200 μm at a solar zenith of 60°, Wang *et al.* (2020b) reviewed the snow albedo sensitivity to BC based on experimental datasets across the NH. They found that 10–2000 ng g^{-1} BC in Northern China can reduce snow albedo from 0.01 to 0.15. Along with BC, OC, and dust in Northern China can have a negative impact on snow albedo, especially over UV bands (Zhou *et al.*, 2019b, 2021b, 2022). Additionally, recent studies have gone into detail about how the shape of the snow grain, the mixing status of the snow and LAPs, and the solar zenith can all affect the snow albedo. Overall, the field observations and remote sensing retrievals showed that the daily radiative forcing by LAPs was in the range of ~1.5–15 W m^{-2} during the months of December–May for the years 2003–2018 over snow-covered parts of Northern China. The radiative forcing can further make a significant contribution to accelerate glacier ablation and cause the onset of snow melt earlier, thus influencing seasonal variability of snow and glacier melt runoff. For example, BC and dust have accelerated snowmelt over the Tibetan Plateau and Himalaya, and reduced the snow cover duration by several days to weeks (Sarangi *et al.*, 2019). The abundant water supply provided by snow and glacier meltwater over semi-arid regions of Northern China (Qin *et al.*, 2018) emphasizes the significant impact that LAPs in snow have on regional hydrologic cycles, societal sustainability, and downstream ecosystems, particularly over Xinjiang Province and the Hexi Corridor of Gansu Province.

3.4.4 *Variability of regional temperatures over different timescales*

Even though these radiative factors significantly contributed to the heightened warming, most of the warming did not have a phase, and the regional temperature series does not usually follow an unbroken upward trend. This demonstrated the importance of the dynamic forced effect on global warming. The EEMD method divided the SAT time series into various timescales. Three timescales were separated from the decomposed SAT time series, including interannual variability, decadal variability, and long-term trends. The first and second IMFs from the EEMD decomposition are added to provide the interannual variability. The third through fifth IMFs are included to determine decadal variability. The sixth IMF is the long-term trend.

The annual mean SAT anomalies in China are depicted in Fig. 3.18(a), along with their interannual variability, decadal variability, and long-term trends. The annual mean SAT in China, as shown in Fig. 3.18(a), clearly displayed considerable background long-term warming trends as well as large interannual and decadal variability. Decadal variability for the most recent warming pause period revealed a strong cooling trend that was consistent with the warming hiatus (Fig. 3.18(a)). Although the interannual variability indicated a slight cooling tendency during the current warming hiatus, the interannual variability has minimal impact on trends at the decadal scale (Fig. 3.18(a)). The mean SAT in China is depicted in Fig. 3.18(b), together with its interannual variability, decadal variability, and long-term trends for the warm and cold seasons. It is clear that atmospheric circulation variability has a stronger impact on SAT during the cold seasons because the amplitude of the interannual variability in the cold seasons is higher than in the warm seasons (Wallace *et al.*, 2012). Regarding decadal variability, during the current warming hiatus, the cold season showed a fairly significant cooling tendency, but the warm season underwent little change (Fig. 3.18(b)). The cold seasons had a longer-term warming tendency than the warm ones. Furthermore, during the preceding accelerated warming period, the decadal variability for the cold seasons similarly revealed a stronger warming trend than the warm seasons. Therefore, during both the present warming hiatus and the prior accelerated warming, the temperature in the cold seasons

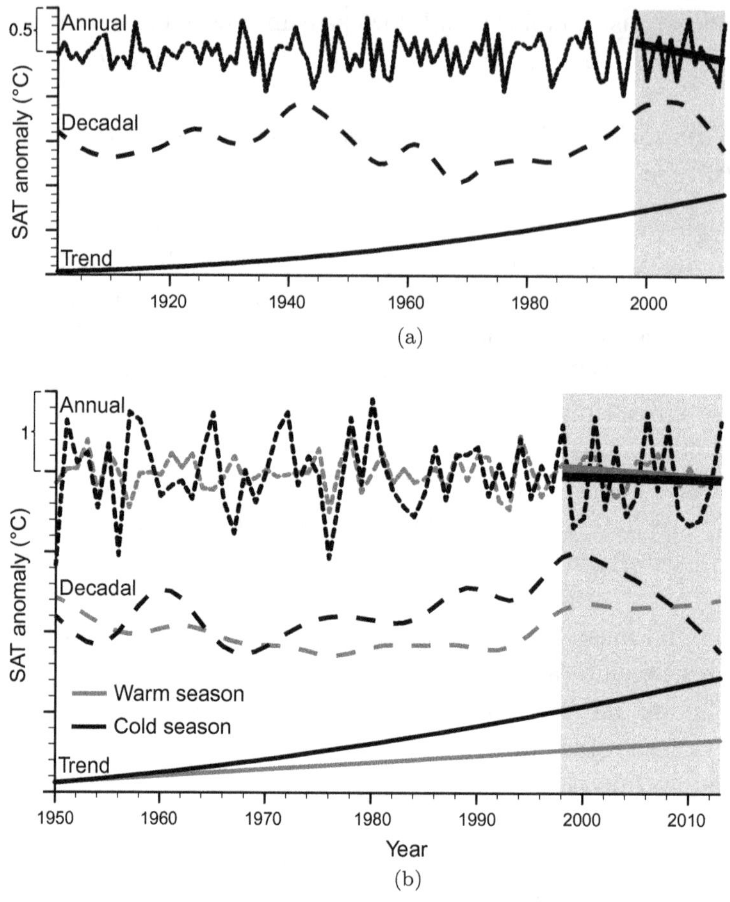

Fig. 3.18. Decomposition of the annual mean SAT anomaly time series averaged over China into different timescales using the ensemble empirical mode decomposition (EEMD) method. The dashed curve represents the interannual variability series, which is the sum of the first two IMFs from the EEMD (for details, see Section 2.4). The solid line is the trend line based on the LSE method for 1998–2013; the two solid curves represent the decadal variability and the background long-term warming trend, which were the sum of the third to fifth, and the sixth IMFs from the EEMD, respectively (a). (b) Same as (a), but for warm and cold seasons, respectively (Xie *et al.*, 2017).

underwent more significant decadal variations than the temperature in the warm seasons (Fig. 3.18(b)).

3.4.5 *Dynamics from accelerated warming to warming hiatus*

In the past decade, the typical semi-arid regions of East Asia showed an obvious warming hiatus, which is the result of the dynamic cooling effect. Therefore, the dynamic temperature effect over small regions cannot be ignored. From Fig. 3.18, it can be seen that both the recent warming pause and the earlier enhanced warming in China were mainly caused by the downward decadal temperature fluctuation during the cold seasons. Such an obvious abrupt change of temperature is a result of the dynamic effect. The question of which dynamic mechanisms account for China's decadal temperature variations is thus critical. The surface solar radiation in China transitioned from solar dimming to a leveling off around 1990, indicating that the surface solar radiation changed little during the recent warming hiatus period. As a result, this chapter concentrated on how changes in air masses and large-scale advection affected China's temperature. This started by looking at significant circulation changes during the cold seasons of both the recent warming hiatus and the prior accelerated warming period. Figure 3.19 displays the 500 hPa circulation trends around China for both 1980–1998 and 1998–2013, as well as the climatology of the turning period from 1994 to 2003. As depicted in Fig. 3.19(a), China was affected by the deep trough in East Asia and the backdrop westerlies circulation. For the previous accelerated warming period, a general figure is the opposite change between the high-latitude and the mid-latitude GPH. In this situation, the overall figure induced higher westerly winds and consequently stronger meridional GPH gradients over the northern parts of China (Fig. 3.19(b)). However, the opposite occurred during the recent warming pause, with weaker meridional GPH gradients and westerly winds across the north of China (Fig. 3.19(c)). The increasing westerly winds across the northern part of China can prevent Arctic cold air from invading, which led to the accelerated warming. The opposite is true for the recent warming hiatus.

Stronger westerly winds over Northern China can prevent the infiltration of cold air from high latitudes and the development of atmospheric blocking, and weaker westerly winds can prevent the opposite. Therefore, this chapter also looked at the variations in atmospheric blocking. The development of China's blocking frequency was determined using the D'Andrea *et al.* (1998) approach, the time series for the sector blocking frequency, and the local blocking frequency. The sector blocking frequency in China fell during the period of accelerated warming and increased during the recent warming hiatus. In response, the SAT in China rose with lower blocking but fell with higher blocking. Nearly all longitudes experienced a decrease in local blocking frequency in the 70°E–140°E regions, from a high to a low frequency, and an increase during the recent warming hiatus. The rise in the extreme and moderate cold weather events during the most recent warming hiatus can also be attributed to China's increased frequency of blockages.

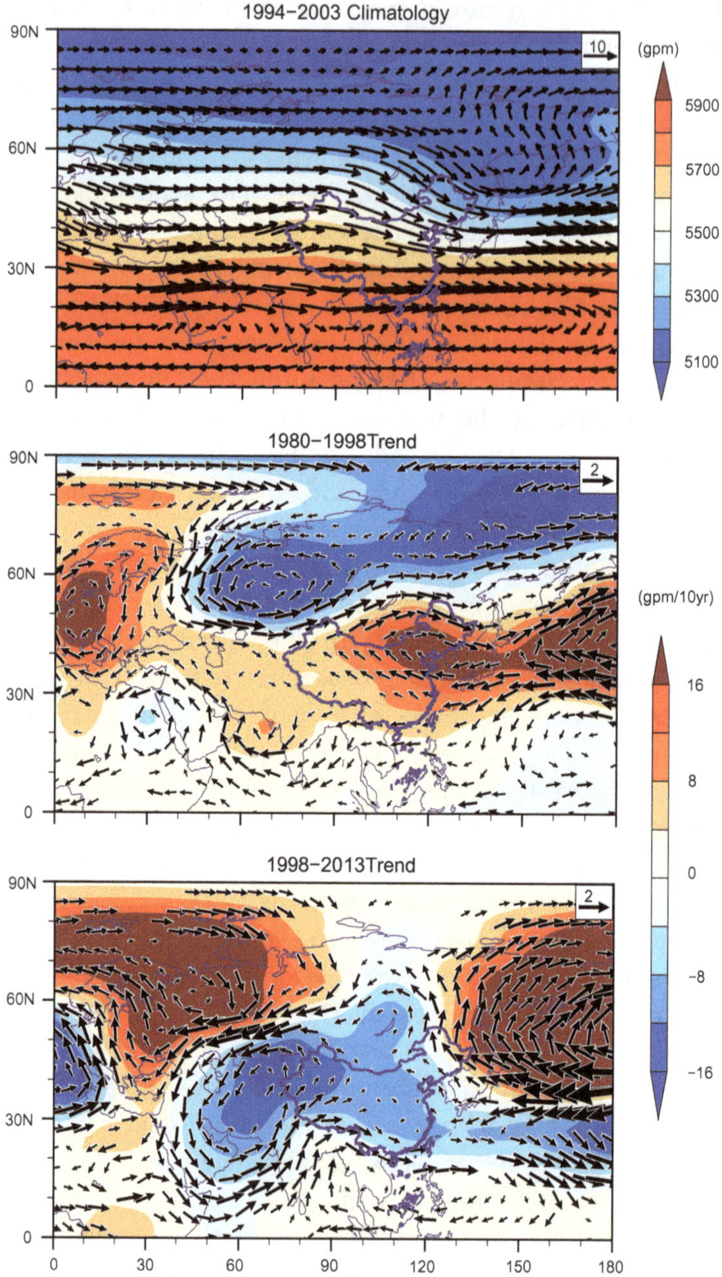

Fig. 3.19. Climatology of the 500 hPa circulation around China for 1994–2003 (a). The filled contours represent the GPH field. The vectors represent the corresponding wind field. Note that the results are the mean of the ERA-interim and NCEP II datasets. (b) Same as (a) but for trends for 1980–1998. The trends are calculated using the simple linear regression based on the LSE method. Only the wind vectors over the regions where the meridional and zonal wind speed trends from the ERA-interim and NCEP II datasets had the same signs are presented. (c) Same as (b) but for the period 1998–2013 (Xie *et al.*, 2017).

3.5 Conclusions and Discussion

As shown in Fig. 3.20, climate system refers to a physical system that has an impact on climate change at a certain timescale. It is an integration of multi-layer and multi-factor interactions. The intensified warming in semi-arid regions is also a result of multiple factors. According to recent research on semi-arid region warming, the inter-annual and decadal variability in raw temperature was primarily explained by the dynamic temperature over a long period. On the other hand, the large cold season warming in the NH from 1902 to 2011 was primarily driven by the thermodynamic temperature. The thermodynamic temperature showed uneven warming throughout the various climate zones, with semi-arid parts showing significantly elevated warming. In contrast, the dynamic temperature showed a trend that was essentially constant throughout all climate zones. By contrasting the warming contributions of the dynamic and thermodynamic temperatures, it is clear that the thermodynamic temperature dominates in ESAW. How the thermodynamic temperatures behaved in the three regions of Asia, Europe, and North America was examined. The findings indicated that Asia had the highest temperature increase, with semi-arid regions experiencing a more significant thermodynamic temperature increase than other regions.

These findings reveal that CO_2 and BC were primarily responsible for regional warming in East Asia since the temperature in the arid and semi-arid regions was more sensitive to both CO_2 and BC. Northwestern China is where the CO_2's overall impact is the most significant, and this region is where its effects on temperature are the most noticeable. Because BC exhibits an elevated impact on temperature increases in snow/ice regions, its influence on temperature is particularly pronounced over Tibet and Northeastern China (Xu *et al.*, 2016). Compared to CO_2 and BC, SO_4 has a negative effect on temperature, with Northeastern China being the region where this effect is the most pronounced.

While this is happening, the radiative factors also affect the daily maximum and minimum temperatures, in addition to the mean temperature. Climate-related regions have asymmetric reactions that cause changes in the DTR. In general, the three radiative components working together cause the DTR to decrease. However, the DTR response varies across different locations for CO_2 and SO_4. The changes in the DTR were computed as a function of the AI to further investigate the DTR response to CO_2, BC, and SO_4 variations over various climate zones. Nearly all dry and humid locations experienced a decline in DTR due to the three causes combined (green),

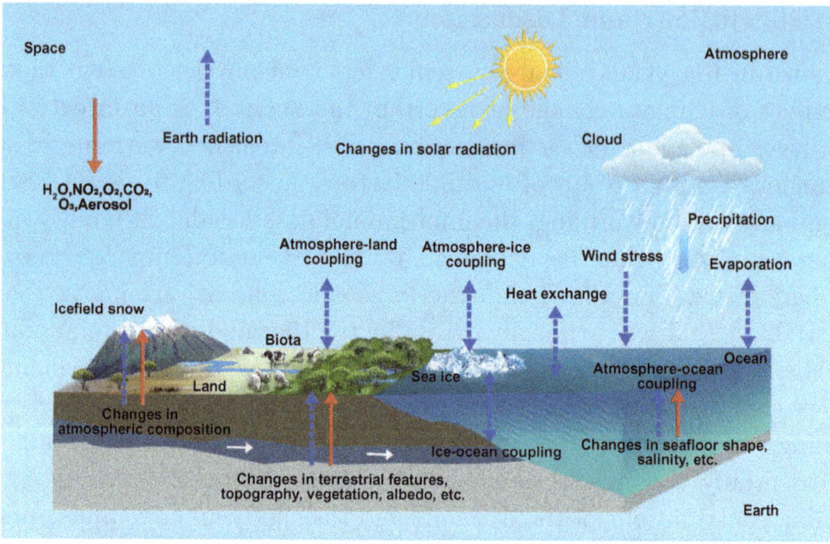

Fig. 3.20. Diagram of the climate system. Solid arrows are examples of processes external to climate change and dotted arrows are examples of processes internal to climate change.

with changes in arid and semi-arid regions being more pronounced than in humid areas. The most notable factor affecting the DTR was CO_2 (red). The DTR dropped in the drier areas where the AI values were lower. The DTR demonstrated the most significant drops in temperature in arid locations and may drop as low as $-0.31°C$, which was higher than the total of the three radiative forcings. The DTR (reaction to CO_2) rose in the wetter regions with higher AI values, and in the regions with AI values between 1.6 and 1.8, the enhanced amplitude was most significant. There was no noticeable effect of BC or SO_4 on the DTR. DTR is determined by subtracting the daily lowest temperature from the daily highest temperature. And although longwave radiation determines the daily minimum temperature, shortwave radiation mostly determines the daily maximum temperature. Because CO_2 primarily affects temperature through the absorption and emission of longwave radiation, both the daily maximum and minimum temperatures are warmed up more.

In comparison to the mean temperature change, the DTR can indicate local and regional temperature change characteristics that are significant from an ecological standpoint (Chen and Chen, 2007). This measure has been promoted as a sign of climate change (Lewis and Karoly, 2013) and is thought to be a valuable way to investigate the opposing effects of longwave and shortwave radiative forcing (Makowski *et al.*, 2008). Large changes in cloud cover, sunshine duration, soil moisture, snow cover, and precipitation variability have been observed in previous studies on reductions in the DTR in a warming climate (Cao *et al.*, 1992; Zhou *et al.*, 2009). According to Liu *et al.* (2016), these forcings can be broadly separated into two categories: natural forcings (volcanic aerosols and solar variability) and anthropogenic forcings (anthropogenic GHG and aerosols). According to observational analysis, increases in cloud cover and precipitation and increases in soil moisture are the leading causes of, for example, DTR reductions. It might only be relevant in some places. For instance, significant drops in cloud cover were seen over Eastern China, where huge drops in the DTR were also reported (Zhou *et al.*, 2009). According to several studies, e.g., (Lewis and Karoly, 2013), the observed global trend in DTR is predominantly the result of anthropogenic forcing. Studies using model simulations have shown that anthropogenic forcing, where GHG forcing is the primary component, dominates the downward trend in DTR over Asia. According to Liu *et al.* (2016), anthropogenic aerosol forcing also played a role in DTR's declining trend in China, particularly in Eastern China. Over arid and semi-arid regions and humid areas, however, DTR exhibits the opposite response to CO_2. Because of this, further research is needed to examine the mechanisms underlying the impacts of CO_2, anthropogenic aerosols, and their interactions on DTR, particularly in light of variations in DTR over regions with diverse climatic types.

In earlier research, the impact of a single thermodynamic component, such as black carbon, dust, or aerosol, was primarily calculated. Strong snow feedback and large anthropogenic black carbon and organic matter emissions have been found in Asia, which may have contributed to the temperature change. A thermodynamically induced warming effect results from the deposition of these particles on snow, which can also affect the surface albedo and radiation absorption (Qian *et al.*, 2011). According to Huang *et al.* (2010b) and Guan *et al.* (2016), agricultural operations in semi-arid areas boost dust emissions and the growth of agriculture, and two significant dust source regions (the Taklamakan and Gobi deserts) over East Asia produce a significant amount of dust aerosols (Huang *et al.*, 2010b). According to Huang *et al.* (2006b, 2006c, 2010b), these dust aerosols can decrease the particle size, optical depth, and liquid water path of cloud water. This can alter cloud growth and exacerbate dryness in the semi-arid regions of Northwest China (Huang *et al.*, 2010b). Additionally, an increase in anthropogenic aerosols may disrupt the hydrological cycle, which would directly affect ecosystems and droughts. According to Feng and Fu

(2013) and Li *et al.* (2015a), the semi-arid regions have grown during the past 50 years, and Huang *et al.* (2015) predicted that the arid and semi-arid regions would keep growing in the future. Land degradation and desertification will be brought on by increased aridity, increased warmth, and inappropriate land use in semi-arid regions during drought conditions. Surface albedo feedbacks, desertification, drought, and fragile ecosystems in semi-arid locations can all greatly impact how much the local temperature rises. Other thermodynamic mechanisms may also be responsible for the enhanced warming in semi-arid regions and the variations in warming across semi-arid regions. To better understand this problem, more research is required.

Aerosols also have a significant impact on climate change, particularly the clear warming effect in arid and semi-arid regions (Huang *et al.*, 2006b, 2008a; Jin *et al.*, 2015), which causes an obvious radiative effect in these regions. In addition, aerosols typically have a relatively large optical depth (Huang *et al.*, 2010a). Aerosols in arid and semi-arid regions are often separated into two categories: natural aerosols and anthropogenic aerosols. According to a recent study (Guan *et al.*, 2016), anthropogenic aerosols may significantly influence the regional heavy aerosolization in Asia, a region with increasing trends in both anthropogenic emissions and desert dust (Hsu *et al.*, 2012). Anthropogenic aerosol emissions have probably increased over many regions due to intense human activity and rapid economic development. These increases can enhance dust loading and impact radiative forcing (Huang *et al.*, 2015). According to earlier studies, research on climate change mainly focuses on changes in energy and water character caused by a single component that affects cloud, snow, land, sea ice, evaporation, and terrestrial features. The single effect focus of the existing works makes it difficult to discern the combined influence on these elements. Future work will be more difficult as methods and levels of calculation advance. Work that is built on multiple aspects will get greater attention.

Additionally, the CMIP5 ensemble mean temperature did a poor job of simulating the clear ESAW, in which considerable local thermodynamic forcing played a large role. Additionally, analysis of the simulated temperature changes from CMIP5 over Asia, North America, and Europe shows that the regional variations in semi-arid warming were not considered in the model simulations. Due to the fragile ecosystems in arid and semi-arid regions, which are vulnerable to strong interactions between human activities and climate change, arid and semi-arid regions may suffer from intense human activities that can induce strong local thermodynamic warming. (Charney, 1975; Huang *et al.*, 2008a, 2010b). In the past 50 years, the semi-arid regions have grown, and Huang *et al.* (2016b) predicted that the arid and semi-arid regions will keep growing in the future. Feedbacks from semi-arid regions' fragile ecosystems, drought, and desertification may, in turn, substantially impact the rise in local temperatures.

Chapter 4

Impacts of Dust–Cloud–Precipitation Interactions

4.1 Introduction

Due to their direct effects on solar and thermal radiation and their indirect/semidirect effects on cloud and precipitation processes, desert dust aerosols — one of the four main terrestrial sources of atmospheric aerosols (desert dust, biomass burning, and biogenic and anthropogenic air pollutions) — are considered to be a significant climate forcing agent (Sassen, 2002). Owing to the significant amounts of mineral dust particles released from dry arid and semi-arid regions around the world, dust aerosols also have a significant impact on local meteorology and climate (Huang *et al.*, 2010b). These areas, which make up about one-third of the continental surface of the planet, are where most dust storms originate. Each year, arid and semi-arid regions release between 1,000 and 3,000 trillion gram (Tg) of dust into the atmosphere. The North African (50%–70%) and Asian (10%–25%) deserts account for most of the global dust emissions. According to Uno *et al.* (2009), roughly 800 Tg/yr of dust is pumped into the atmosphere over Asia each year, of which about 30% is redeposited into deserts, 20% is carried on a regional scale, and the remaining about 50% is transported to the Pacific Ocean and beyond. As a result, Asian dust is crucial to the global biogeochemical cycle because it may provide iron to marine ecosystems in the North Pacific (Bishop *et al.*, 2002), which is vital for the energy budget and hydrological cycle.

As bigger soil particles crawl and salt over deserts and other arid/semi-arid surfaces, aggregates begin to disintegrate, resulting in dust particles. According to Twohy *et al.* (2009), these could contain iron oxides (e.g., hematite and goethite), carbonates (e.g., calcite and dolomite), quartz, and clays (e.g., kaolinite and montmorillonite). Increased desertification, which is fueled in turn by dust events that worsen drought conditions across semi-arid areas, is the cause of the rising number of dust storms (Huang *et al.*, 2010b). The Gobi Desert and Loess, the two main sources of Asian dust, are growing in Northwest China (Huang *et al.*, 2010). A significant source of the world's aerosol burden is mineral dust, which is released by deserts, including the Sahara, the Arabian, the Gobi, and the Taklamakan (Chen *et al.*, 2018). Strong frontal/postfrontal convection from significant desert source regions can lift dust aerosols into the upper troposphere, where they can then be carried to downwind regions by the westerlies (Hu *et al.*, 2019b), with some of the dust even traveling for one full circuit around the world (Uno *et al.*, 2009). During this transcontinental travel, dust has a considerable impact on the regional and global climate systems. For instance, earlier research indicated that dust could alter the heating of the atmosphere and have an impact on the patterns of atmospheric circulation in the middle and upper troposphere (Lau *et al.*, 2006). Dust may also trigger abnormal feedback in the water cycle, changing the pattern of rainfall over East Asia and South Asia. Additionally, it was discovered that dust particles alter the regional radiation budget by diffusing and absorbing solar radiation and altering the properties of clouds (Li *et al.*, 2017; Li *et al.*, 2018) by acting as cloud condensation nuclei (CCN) and ice nuclei (IN). Dust aerosols also worsen the quality of the air and reduce regional visibility. Dust can disrupt the hydrological cycle when it is deposited on the surface of snow/ice (Huang *et al.*, 2011). When it is deposited in rainforests and oceans, it can feed ecosystems and have an impact on the carbon cycle.

In dry arid and semi-arid regions, precipitation is one of the most crucial water resources. The surface radiation budget and the hydrological cycle can be affected by minor variations or changes in the amount, height, physical thickness, and/or microphysical features of clouds as a result of natural and human effects (Huang *et al.*, 2006b, 2006c, 2006d, 2010b). Any change in the quantity and/or composition of aerosols can have a range of effects on clouds because aerosol particles, particularly dust aerosol, are an essential component in the formation of cloud droplets. It is crucial to understand the interactions between aerosols, clouds, radiation, and precipitation over these arid and semi-arid regions, maybe more so than for any other place on Earth. Aerosols typically reduce precipitation as CCN in shallow clouds that have smaller cloud droplets and less effective coalescence (Li *et al.*, 2016), but they also increase precipitation as it relates to deep convective clouds (DCCs) through an effect known as the "invigoration effect" caused by aerosols (Rosenfeld *et al.*, 2008). According to Fan *et al.* (2018), the invigoration theory is also valid for shallow water clouds when ultrafine-mode aerosol pollution is present. However, knowledge about heterogeneous dust IN in DCC formation remains limited because of insufficient measurements. Based on a cloud model, Yin and Chen (2007) discovered that dust acting as IN can speed up the formation of ice and possibly even encourage the development of precipitation when a dust layer arises above a −5°C level. Zhang *et al.* (2015) discovered that dust aerosols can glaciate mixed-phase clouds by contributing a significant amount of effective IN across East Asia, using a combined dataset from the 4-year Cloud-Aerosol Lidar and Infrared Pathfinder Satellite Observation (CALIPSO) and CloudSat.

Numerous observational and modeling studies that concentrated on the direct, indirect, and semidirect effects of aerosols discovered that even small changes in the microphysical characteristics of clouds (such as droplet size, phase, and water path) could result in significant changes in the amount of precipitation and radiative forcing (RF) (IPCC, 2007). It is crucial to comprehend and quantify these aerosol effects, particularly how these forcing fields interact to change the hydrological cycle (modification of precipitation regimes), cloud properties, atmospheric dynamics, and surface budget (heat flux). Aerosols might be emitted from North Africa, Europe, and East Asia and hoisted into the middle and upper troposphere, where they could then be carried to North America by the strong westerlies over the North Pacific basin (Hu *et al.*, 2016). In downwind regions, such as the United States (US), the trans-Pacific aerosols can have an impact on the atmosphere's composition, stratospheric ozone depletion, surface air quality (Chin *et al.*, 2007), regional visibility, human health, regional climate (Huang *et al.*, 2009), and ecological integrity.

The arrangement of this chapter is as follows. Section 4.1 provides the introduction to this chapter. Section 4.2 introduces the models and methodology used in this chapter. Sources and transports of global dust are analyzed in Section 4.3. Section 4.4 reviews the climate effects of East Asian dust, including direct radiative, indirect radiative, and semidirect effects. Section 4.5 discusses dust–cloud–precipitation interactions over semi-arid regions. Conclusions and discussion are summarized in Section 4.6.

4.2 Data and Models

4.2.1 *CALIPSO/CERES data*

By providing 532 nm attenuated backscatter that separates dust from other types of aerosols based on its capability for depolarization measurements (Cesana *et al.*, 2016), the CALIPSO satellite, jointly developed by the US and France, has been designed to fill the observational gap between the global distribution and nature of aerosols and clouds. In order to comprehend dust aerosol features, such as frequency of occurrence (Liu *et al.*, 2008b), transport path, the dominant aerosol types (Wang *et al.*, 2020a), dust-forced radiative heating (DRH), and even global particulate mass concentration, it is crucial to have data from CALIPSO-based measurements of aerosol backscatter, extinction, and types. In this chapter, the CALIPSO version 4 level 2 dust and contaminated dust aerosol profiles, covering the 10-year period from 2007 to 2016, were used. These profiles have a nominal resolution of 5 km along the track and a vertical resolution of 60 m in the troposphere. In order to eliminate interference from sunlight during the day and ensure proper categorization of dust and contaminated dust aerosol layer features (Liu *et al.*, 2019), we exclusively used nighttime data for high-confidence data with a cloud–aerosol discrimination (CAD) score between −70 and −100. In order to eliminate troublesome retrievals for dust and contaminated dust aerosols, the values 0, 1, 2, 18, and 16 were chosen for the extinction quality control flag (Ext_QC).

The Tropical Rainfall Measuring Mission (TRMM) observatory is carrying five instruments, including the Clouds and the Earth's Radiant Energy System (CERES) sensor (Minnis *et al.*, 2011). The energy exchanged between the Sun, the Earth's atmosphere, surface, clouds, and space was studied using the CERES instrument's data. However, the data record that is now available is rather brief because it only ranges from January to August 1998 and March 2000. The NASA Langley Earth Radiation Budget Experiment, which employed three satellites to give measurements of the global energy budget from 1984 to 1993, served as the model for the CERES experiment. In addition to measuring energy levels in the atmosphere and at the Earth's surface, CERES also estimated energy levels at the top of atmosphere (TOA). Cloud parameters, such as quantity, height, thickness, and particle size, were also assessed by CERES. The improvement of climate prediction models and our understanding of the Earth's overall climate system depend on all of these measurements. Starting with TRMM and continuing with the Earth Observing System (EOS) AM satellite in 1998 and the EOSPM satellite in 2000, five CERES experiments have been flown on various spacecraft. For improved climate analysis, more CERES satellite missions are planned to produce a continuous 15-year history of very precise energy budget and cloud data.

4.2.2 *SBDART model*

The Santa Barbara DISORT Atmospheric Radiative Transfer (SBDART) (Ricchiazzi *et al.*, 1998) (https://Github.com/paulricchiazzi/SBDART) is a software tool used to solve the problem of Earth-plane parallel atmosphere radiative transfer in clear and cloudy conditions. This model's calculation incorporates a discrete coordinate radiation transmission module, a low-resolution atmospheric transmission mode, and the meter scattering theory of light radiation to water droplets and ice crystals in addition to the transmission processes for ultraviolet, visible, and infrared light. The radiation process is affected by all significant spectral bands, including the surface, the clear sky atmosphere, clouds, and aerosols. Cloud parameter sensitivity tests can be performed. This model lays the groundwork for the computation of satellite remote sensing and the atmospheric energy budget. It is useful for the study of various atmospheric radiation energy balances and remote sensing.

There are four presumptions in the model: (1) There are five intervals of altitude in the atmosphere. (2) Assume that the atmosphere can be divided into a number of thin, spherical layers between sea level and 100 km. (3) In order to calculate air transmittance, continuous absorption, molecular scattering, and aerosol weakening are taken into account. (4) When calculating the total radiation, the radiation from the atmosphere itself, the radiation emitted by the sun (moon), and the radiation reflected by the ground are all taken into account. In addition, there are seven patterns: (1) cloud model, (2) gas absorption model, (3) extraterrestrial source spectra, (4) extraterrestrial source spectra, (5) standard aerosol models, (6) radiative transfer equation solver, and (7) surface models.

SBDART is a practical, quick, and highly accurate method of transmitting air radiation. The basic information of a particular model can be accurately produced by employing satellite data, ground data, sounding data, and airborne data because the actual situation can be merged with specific difficulties, bringing the computed results closer to those from the actual atmosphere.

The primary input control parameters and the purpose of each subroutine make up the majority of the two structures that constitute the SBDART mode. Surface reflectance characteristic value (ISALB), filter function (ISAT), atmospheric profile type (IDATM), cloud height (ZCLOUD), optical thickness of cloud (TCLOUD), effective particle radius of cloud droplet

(NRE), and solar zenith angle (SZA) are the major input control parameters. The primary purposes of each subroutine are to perform the following tasks: computing the transmission gas absorption data (Absgas), computing the absorption of various gases for the whole gas layer (Absint), choosing various atmospheric profiles (Atms), computing the radiation transfer characteristics of clouds (Cloudpar), computing the phase function (Getmom), etc.

SBDART's installation procedure is simple, and the Unix system is given as an illustration here. On a Unix system, download the SBDART zip package and carry out the following steps:

```
unix > gunzip sbdart_2.4.tar.gz
unix > tar xvf sbdart_2.4.tar
unix > make
```

Similar steps can also be used to install SBDART in Cygwin. The model's operating results can be achieved by choosing various input settings, wavelength ranges, atmospheric profiles, gas concentrations, and surface types of the model, altering droplet size, height, and optical thickness, and choosing the type of vertical profile to be calculated.

In this chapter, we estimate the DRH at the shortwave (SW; 0.25–4.0 μm) and longwave (LW; 4–80 μm) wavelengths using the SBDART model. The SBDART model's interpolated atmospheric background characteristics are shown in Table 4.1 for $2^\circ \times 2^\circ$ grid cells. Additionally, the dust extinction coefficient (DEC) profiles at 532 nm were obtained directly from CALIPSO. By utilizing the benefits of polarization lidar, this method provides accurate aerosol types and vertical structure at a global scale.

The SBDART model's other crucial input parameters are the dust aerosol's single-scattering albedo (SSA) and asymmetry factor (g), which vary depending on the type of dust aerosol. We compared the instantaneous SW and LW radiative fluxes at the TOA between observations from the Cloud and the Earth's Energy Budget Scanner (CERES) and simulations driven by the CALIPSO orbit with four dust aerosol types from the Optical Properties of Aerosol and Cloud (OPAC) package, as shown in Table 4.2 (Hess *et al.*, 1998), in order to optimize these two parameters, as in the work of Huang *et al.* (2009). Two typical pure dust instances from the TD and ThD, respectively, were chosen for this chapter (refer to Fig. 4.1 for more details). The SSA and g of conveyed mode performed the best according to our research. The SW and LW flux at the TOA are compared between SBDART simulations and CERES measurements along the CALIPSO orbit

Table 4.1. Descriptions of the input parameters of SBDART (Wang *et al.*, 2020a).

Input	Datasets	Units
Surface spectral albedo	MODIS (BRDF/Albedo) Products (MCD43C3)	—
Temperature profile	ERA-Interim	K
Water vapor profile	MERRA-2 & ERA-Interim	g/m^3
Ozone density profile	OMO3PR	g/m^3
Column moisture content	ERA-Interim	g/cm^2
Column ozone content	OMO3PR	ATM-cm

Table 4.2. Single-scattering albedo (SSA) and asymmetry factor (g) for four dust aerosol types of the OPAC package (Wang *et al.*, 2020a).

	Nucleation mode	Accumulation mode	Transported mode	Coarse mode
SSA	0.9767	0.9203	0.89	0.7266
g	0.6471	0.7143	0.7460	0.8613

over the TD region (37°N–40°N) on April 19, 2012, and over the ThD region (22°N–26°N) on April 27, 2008, throughout the day. Only 28 and 12 W/m^2 on average separate the SW and LW wavelength differences. Therefore, in the subsequent simulations, SSA and g from the carried dust mode were used.

4.2.3 *Calculation of dust-forced radiative heating*

In this chapter, the total dust aerosol extinction coefficient profiles are calculated using the CALIPSO Version 4.2, level 2 all-day aerosol profiles covering the 12-year period from 2007 to 2018, with a nominal resolution of 5 km along the track and a vertical resolution of 60 m in the troposphere. To guarantee high-quality retrieval, the CAD score of between −70 and −100 (Liu *et al.*, 2019) and extinction quality control flag values of 0, 1, 2, 18, and 16 were first chosen. In order to isolate the backscatter coefficient profiles of dust, contaminated dust, and dirty marine from the overall one, the vertical feature mask (VFM) products were used. The total DEC (α_d) was then obtained from an assumed dust lidar ratio (S_d) and the total dust backscatter coefficient (β_d), which is separated from the total aerosol backscatter coefficient (β_a) according to the determined PDR for dust (δ_d)

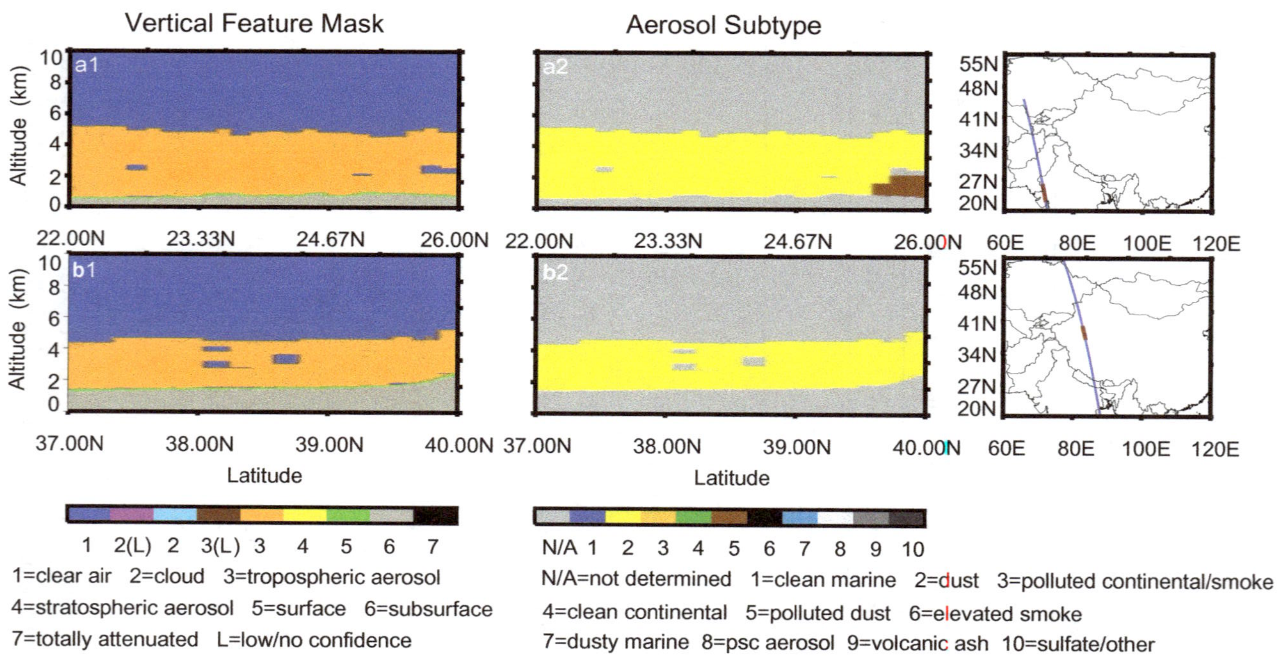

Fig. 4.1. The altitude–orbit cross-section of the vertical feature mask (left panels) and aerosol subtype (middle panels) for the red-shaded portion of each track (right panels) on April 27, 2008 (a1, a2) and April 19, 2012 (b1, b2) (Wang *et al.*, 2020a).

and non-dust (δ_{nd}) aerosols. This is similar to Wang *et al.*'s work (2020a):

$$\alpha_d = S_d\beta_d = S_d\beta_a \times \frac{(\delta - \delta_{\text{nd}})(1 + \delta_d)}{(1 + \delta)(\delta_d - \delta_{\text{nd}})}, \tag{4.1}$$

in which δ is the PDR observed by CALIOP and the values of δ_d and δ_{nd} in this chapter are set at 0.31 and 0.05, respectively, to avoid unnecessary exclusion or inclusion of dust events. The value of β_d is set at β_a or 0 for dust bins or non-dust bins, respectively. The S_d at 532 nm was set to 44 sr^{-1}, as in previous studies. The single aerosol species re-grid average approach, which assumes extinction to be 0.0 km^{-1} for all other species before computing the average dust extinction inside a grid, is then used to grid the entire DEC profile with a resolution of 2° at latitude and 2° at longitude.

The difference in SW and LW wavelengths between dust-laden and dust-free circumstances is typically used to calculate the DRH in the atmosphere. However, the SBDART model's simulation of the radiative heating rate along the CALIPSO orbit only depicts an instantaneous heating of the dust at 13:30 local time. In order to account for the diurnal variation of solar radiation, the daily mean DRH must be corrected from the instantaneous one, as shown in the following, by multiplying a normalized factor:

$$\text{DRH} = (H_{\text{SW}}^{\text{dust}} - H_{\text{SW}}^{\text{clear}}) \cdot f + (H_{\text{LW}}^{\text{dust}} - H_{\text{LW}}^{\text{clear}}), \tag{4.2}$$

$$f = \frac{h_o \sin(\varphi)\sin(\delta) + \cos(\varphi)\cos(\delta)\cos h_o}{\pi \sin(\varphi)\sin(\delta) + \cos(\varphi)\cos(\delta)\cos(h)}, \tag{4.3}$$

in which H_A^B is the instantaneous DRH of an A (SW or LW) wavelength under a B condition (dust or clear) and f is the normalization factor defined as the ratio of the daily mean flux density of solar radiation and the instantaneous one at some local solar time (Haynes *et al.*, 2013), which takes into account the diurnal variation of insolation at each grid. Equation (4.2) is a specific definition based on the combination of the inverse-square law, Lambert's cosine law, and the spherical law of cosines (Johansson *et al.*, 2015), where φ is the latitude, δ is the solar declination of each grid, and h and ho are the angles at 13:30 local solar time and the sunrise at each grid, respectively.

Given the geometry of CALIPSO lidar, it is challenging to extract the climatology of dust aerosol extinction and even heating rate on a daily basis because there are few high-frequency measurements within a research region. To show the climatology of the DRH with a unit of K/month over the TPS, this chapter was rescaled to the monthly DEC values at each grid, and the corresponding monthly cumulative DRH was calculated.

4.2.4 *Cloud phase identification*

In this chapter, the vertical distribution of ice, water, and oriented ice crystal clouds along the CALIPSO orbit at two periods is provided by the CALIPSO level 2 VFM data. Based on CALIPSO cloud products, Hu *et al.* (2009) created a method for recognizing supercooled water clouds (SWCs). They discovered that Eq. (4.4) can be used to parameterize the link between SWC fraction and cloud mid-layer temperature:

$$p(T_{\text{mid}}) = 5.3608 + 0.4025T_{\text{mid}} + 0.08387T_{\text{mid}}^2 + 0.007182T_{\text{mid}}^3 + 2.39 \times 10^{-4}T_{\text{mid}}^4 + 2.87 \times 10^{-6}T_{\text{mid}}^5, \tag{4.4}$$

$$f(T) = \frac{1}{1 + e^{-}p(T)}, \tag{4.5}$$

in which p(T) is the polynomial fitting function between the CALIPSO SWC fraction f(T) and the cloud mid-layer temperature. Equation (4.5) can be used to estimate the SWC fraction. Here, we followed Wang *et al.* (2019b), who replaced the original mid-layer cloud temperature in the work of Hu *et al.* (2009) with the CALIPSO cloud layer mid-temperature and considered the clouds with the SWC fraction greater than 80% as SWCs.

The cloud particle phase can be used in the CERES SYN1deg Ed4A product to differentiate between ice clouds and liquid clouds. We employ the technique created by Bodas-Salcedo *et al.* (2016), who explained that any liquid cloud with a cloud top temperature <0°C is known as an SWC. As a result, we chose the clouds from the CERES and MODIS cloud products that had LWP > 0 and temperature <0°C as SWC.

4.2.5 *WRF-Chem*

The WRF-Chem model, based on version 3.5.1, was upgraded by the University of Science and Technology of China (USTC), so as to include the capabilities of quasi-global simulation and tracer source tagging (Hu *et al.*, 2019b). For modeling the complex photolysis rates and aerosol radiative characteristics, it incorporates the model for simulation of aerosol interactions and chemistry (MOSAIC) aerosol scheme and the carbon bond mechanism Z (CBM-Z) photochemical mechanism. The MOSAIC aerosol scheme uses eight size bins to characterize the aerosol size distributions using the bin technique, which may more accurately depict the geographical distribution of the dust particle loading and size distribution of the dust than the three-mode and four-bin approaches. Meanwhile, the key aerosol components, including mineral dust, sulfate (SO_4^-2), ammonium (NH_4^+), nitrate (NO_3^-), organic matter (OM), black carbon (BC), and sea salt, are simulated in the model to depict the sophisticated chemistry during the long-range transport. Coagulation, condensation, nucleation, water uptake, and aqueous phase chemistry are among the aerosol physical and chemical processes in MOSAIC. The aerosol dry deposition method incorporates turbulent diffusion and gravitational settling. The grid box of each model's aerosol optical properties is a function depending on wavelength. With the exception of dust, whose SW refractive index is defined as $1.53 + 0.003i$, the LW and SW refractive indices of aerosols are based on the OPAC dataset. Aerosol radiative feedback for LW and SW radiation is included in WRF-Chem, which has been connected with the rapid radiative transfer model (RRTMG) for LW and SW radiation.

4.3 Sources and Transports of Global Dust

Mineral dust affects the radiation budget, the macro- and microphysics of clouds, precipitation, and snow albedo, which all have an impact on the climate. By modifying the surface albedo, dust aerosol, one of the primary light-absorbing contaminants in snow and ice sheets on the Tibetan Plateau (TP), can have a big impact on the speed and timing of snow melting and glacier retreat. Therefore, it is crucial to comprehend the possible origin and mode of transportation of the dust aerosol around the world.

4.3.1 *Sources of global dust*

The dust mass loading (DML) for the entire world was found to be 36.6 mg/m^2 for particles ranging from 0.1 to 10 mm, translating to a burden of 18.7 Tg (Tegen and Lacis, 1996). For the 0.1–10 mm size range, the dust load was calculated to be 13.8 Tg by Takemura *et al.* (2000), 31–40 Tg by Ginoux *et al.* (2001), and 18.1 Tg by Liao *et al.* (2004). Within the AeroCom project, Huneeus *et al.* (2011) analyzed 15 global aerosol models and discovered that there are significant discrepancies among them. The emission ranges for Asia are 27–873 Tg/yr. Large emission uncertainties as well as the various meteorological fields and deposition calculation methodologies used in this research are to blame for the significant discrepancies.

Around the world, dust emissions frequently happen over source areas where the terrain height is typically less than 3 km, and they can then be carried into the upper troposphere (Yu *et al.*, 2008). Xu *et al.* (2018) used the CALIPSO retrievals to study the transport of dust at various altitudes and discovered that the TP had a significant influence on the long-range global

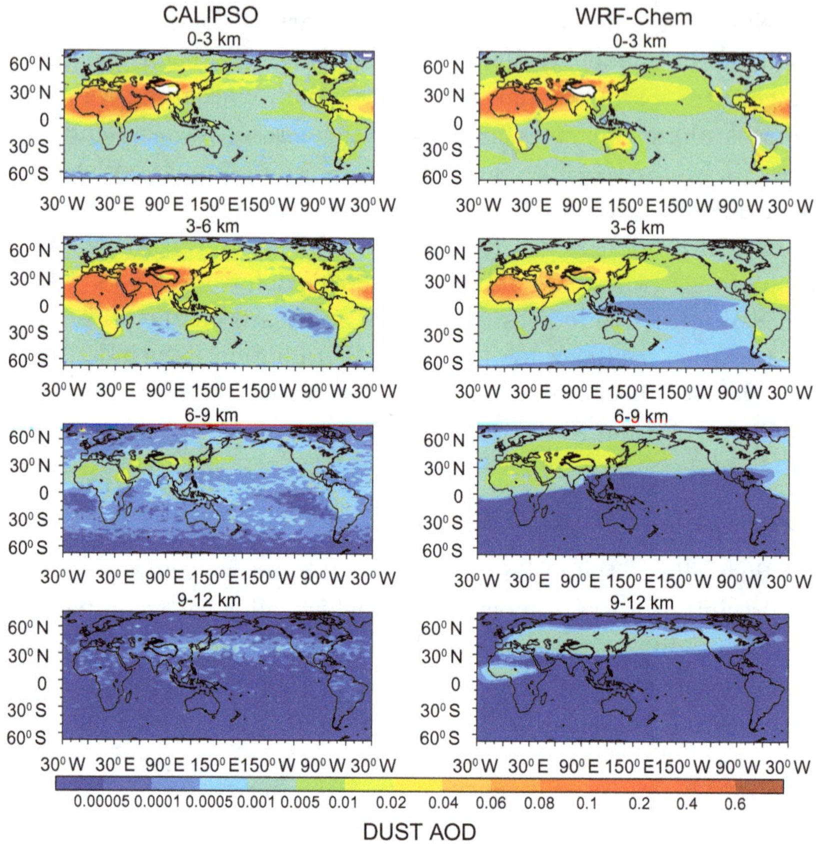

Fig. 4.2. Spatial distributions of annual mean dust AOD from CALIPSO (left) and WRF-Chem (right) in the ranges of 0–3, 3–6, 6–9, and 9–12 km for the period of 2010–2015 (Hu *et al.*, 2020).

transport of dust to the upper troposphere. We compare the geographical distribution of yearly dust aerosol optical depth (AOD) from WRF-Chem in the 0–3, 3–6, 6–9, and 9–12 km ranges with the CALIPSO results in order to assess the model's output (Fig. 4.2). The correlation coefficients of 0.71 (0–3 km), 0.86 (3–6 km), 0.49 (6–9 km), and 0.32 (9–12 km) demonstrate that the model simulations accurately capture the geographical distributions of dust AOD at various altitudes. The simulation overestimates the dust AOD across the North African deserts (more than 0.06) and underestimates it (less than −0.06) over the Middle East and India. The largest dust AOD emerges around the dust sources between 0 and 3 km, with a peak value of 0.6. The model's spatial distribution at 3–6 km is comparable to the CALIPSO results at 0–3 km. The positive difference is roughly 0.02 over the deserts of North Africa, whereas the negative difference is roughly −0.02 over the northern slope of the TP, the Southern Arabian Peninsula, and the Somali Peninsula. The predicted dust AODs are greater than CALIPSO between 6–9 and 9–12 km, with a positive difference of 0.001–0.01, which may be caused by model biases, such as those associated with circulation conditions.

The Taklimakan and Gobi deserts, which are located on the northern side, generate mineral dust particles that can be carried into the TP and elevated to very high altitudes, i.e., about 4–7 km (Huang *et al.*, 2007b) in most years and up to 12 km (Liu *et al.*, 2008) during some extreme conditions. The westerlies may also bring dust from the Thar and middle eastern deserts, which are on the western side of the planet, into the TP. According to Hu *et al.* (2019b), the dust from North Africa and the Middle East mixes before traveling to East Asia. The horizontal spatial distribution of the average annual dust mass load emitted from East Asia is used as an example in this chapter (Fig. 4.3). It is obvious that the East Asian dust dominates the TP's northern slope and even reaches the Southern TP. In the meantime, altitude causes a reduction in the DML. The source locations experience the highest DML at 0–3 km, with a value of more than 200 mg/m^2. Dust can rise to the TP and even the southern boundary of the TP at 3–6 km. At distances of 3–6 and 6–9 km, respectively, the DML over the northern slope is approximately 50 and 10 mg/m^2. The influence of the TP orography causes the dust (5 mg/m^2) to appear over the TP at a distance of 9–12 km. The East Asian desert supplies more than 90% at 0–3 km along the northern slope of the TP, compared to other sources. north of 35°N, the East Asian dust contribution dominates at 3–6 km, with a contribution of roughly 80%. It is important to note that the TP (Fig. 4.2) is covered in sizable dust plumes, particularly in the Northeast. This dust transport can be attributed to the TP's influence, which may create more ideal circumstances for dust uplift into the upper troposphere (such as cyclonic and dry convection).

African easterly waves, which are a significant worldwide long-range dust transport and have been studied extensively, can carry North African dust westward across the Atlantic Ocean (Liu *et al.*, 2008). Additionally, the Middle East and Europe could be affected by the North African dust (Hu *et al.*, 2019b). Hu *et al.* (2019b) found that the dust from North Africa could travel eastward and be conveyed into East Asia after mixing with the dust from the Middle East. The TP blocks eastward transport and has a significant impact on the vertical distribution and dust transfer over the TP.

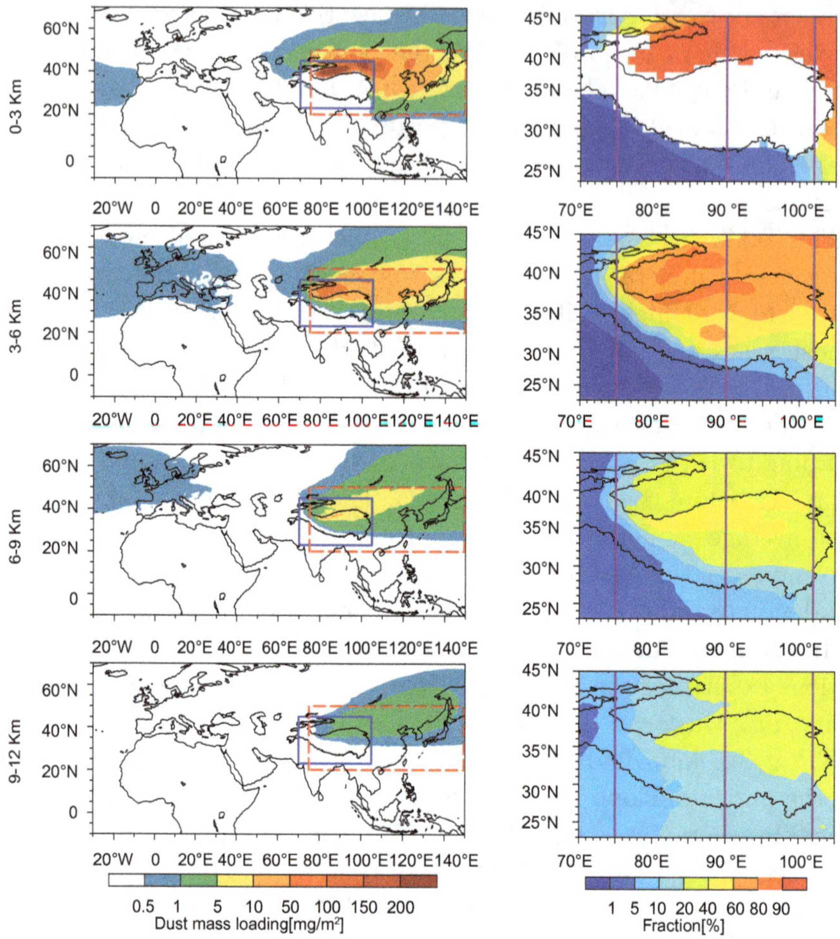

Fig. 4.3. Spatial distributions of annual mean DML (in milligrams per square meter) emitted from East Asia in the 0–3, 3–6, 6–9, and 9–12 km ranges (left) and the corresponding fractions compared with total DML over the TP (right). The box in red dotted lines shows the dust source regions, while the box in blue solid lines shows the TP regions considered in this chapter. The purple solid lines are at 75°E, 90°E and 102°E for the dust height–latitude cross-sections (Hu *et al.*, 2020).

The middle eastern dust can also be transferred into East Asia, much like how North African dust is moved eastward (Hu *et al.*, 2019b).

4.3.2 *Transportation of East Asian dust*

According to the definition given in Chapter 2, the arid and semi-arid regions are characterized by the aridity index, the ratio of precipitation to potential evapotranspiration, and the cutoff values for classifying the hyperarid, arid,

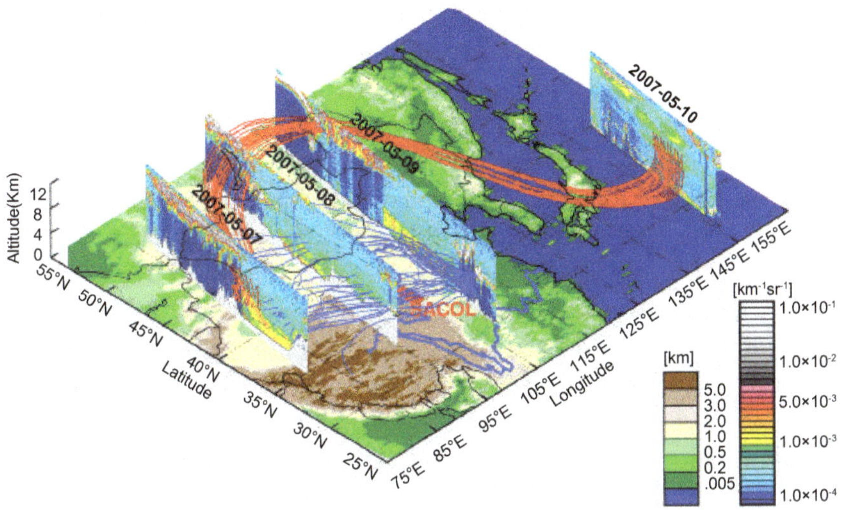

Fig. 4.4. Illustration of a dust event, which originated in the Taklamakan Desert on May 7, 2007, and transported the dust to the Pacific Ocean. Red lines represent back trajectories initialized over the Eastern Pacific. Blue lines represent back trajectories initialized over Central China. Total attenuated backscatter profiles at 532 nm from CALIPSO are shown in the vertical images. Color scales on the left represent the topographical elevation in km, while color scales on the right represent the 532 nm total attenuated backscatter (unit: km^{-1} sr^{-1}) (Huang *et al.*, 2008b).

and semi-arid regions. The two main sources of dust in East Asia are the Taklimakan Desert and the Inner Mongolian Gobi Desert. Compared to the Gobi Desert, dust storms happen more frequently in the Taklimakan Desert. While blowing dust aerosols are more common in the Gobi region, floating dust aerosols are more common in the Taklimakan Desert.

Huang *et al.* (2007b, 2008b, 2010b) provide a clear illustration of the vertical distribution and long-range transport pattern of typical dust particles in East Asia using a few ground-based measurements, satellite retrievals, and numerous field studies. Typically, dust aerosols are found in the boundary layer in and around the dust source regions. They can also travel to the upper troposphere via strong convective updrafts, where they can then form dust plumes. These plumes can travel from the continent to the open seas near Korea and Japan and even to Eastern North America, having an impact on the atmosphere's hydrological and radiative budgets (Huang *et al.*, 2007b, 2008b). High-altitude dust particles are guided by the jet stream, but topography and local weather patterns govern their paths in the lower atmosphere. A dust event that occurred on May 7, 2007, according to Huang *et al.* (2008b), is illustrated in Fig. 4.4

The dust event started that day (or perhaps somewhat earlier) over the Taklimakan Desert, combined with dust over the Gobi Desert, and was subsequently advected over the Pacific, according to images from the CALIPSO satellite and back trajectory analysis. The following are two main routes for dust aerosols to travel to downwind areas: (1) the route from the Qilian Mountain in Mongolia to the Pacific Ocean, which can take 3–4 days; and (2) the route from the Qaidam Basin through the provinces of Qinghai and Gansu, which impacts more areas than the first route but takes longer.

Furthermore, using the CALIPSO DEC profiles and the Dust Constraints from the joint Observational-Modeling-experimental analysis (DustCOMM) dataset, the climatology of Asian DML and its transport flux were approximated to provide fresh insights into the Asian dust cycle. The geography of Northwest Inland China controls the mass-weighted dust transport central axes (TCA) over East Asia; however, the Aleutian low and Western Pacific subtropical high cause alterations in isolated areas between 35°N and 50°N. As the altitude rises, the TCA tends toward a zonal straight line. In addition, the East Asian deserts contribute up to 214.28 Tg annually to transporting dust to mainland China and the oceans nearby, with the Gobi Desert being the primary source in Asia, contributing 145.88 Tg annually.

In addition to these two routes, dust from the Taklimakan Desert in Northwest China has the potential to cross the Loess Plateau and the Hexi Corridor on its way to Southeast China's Yangtze River Delta (Liu *et al.*, 2011a). Three primary varieties are distinguished based on the altitude of the transportation: the upper-level type (U type), the lower-level type (L type), and the descending type (D type). For the U- and D-type dust trajectories, dust particles are first propelled upward to the middle troposphere and then rapidly descended. For instance, during a dust outbreak in the Yangtze River Delta from March 14 to 17, a dense dust cover was seen above 5 km. In the Yangtze River Delta in spring, dust particles from Northwest China can dramatically increase seasonal mean aerosol extinction near the surface and at higher altitudes (Liu *et al.*, 2012).

Dust samples have been collected along dust paths, and observations have been performed at a few highly instrumented supersites and mobile facilities in order to provide direct evidence of the effects of dust aerosols across Asian arid and semi-arid regions. As an international long-term climate observatory, the Semi-Arid Climate and Environment Observatory of Lanzhou University (SACOL) was established in 2005 (Huang *et al.*, 2008a). It is situated in the loess inland and adjacent to the Gobi Desert, which is the main source of Asian dust. Additionally, two field experiments, called the East Asian Studies of Tropospheric Aerosols: An International Regional Experiment (EAST-AIRE) (Li *et al.*, 2007a) and the East Asian Studies of Tropospheric Aerosols and their Impact on Regional Climate (EAST-AIRC) (Li *et al.*, 2011), were carried out to study Asian dust and anthropogenic aerosols. The characteristics and climatic impacts of dust aerosols were studied in the EAST-AIRE over source regions in Northern and Western China (Huang *et al.*, 2010b) and over downstream regions as far as the Pacific. Based on these findings, numerous studies were conducted on the characteristics of dust, the adiabatic forcing of dust aerosols (Ge *et al.*, 2010a), and the direct and indirect impacts of dust and anthropogenic aerosol mixes on regional climate.

4.3.3 *Transportation of Southern Asian dust*

The dust column mass density (CMD) and total AOD derived from the MERRA-2 reanalysis, as shown in Figs. 4.5(a) and 4.5(b), respectively, were used to understand synoptic causes and a transport mechanism to the TP for a typical dust storm that occurred in the Thar Desert (ThD) in South Asia on May 1–4, 2018. On May 1, the ThD was where the majority of the many dust particles, with an AOD of 0.7 and a dust CMD of 0.8 g/m^2, were concentrated. The peak of AOD and dust CMD in the ThD rose above 1 and 1.8 g/m^2, respectively, along with the eruption and amplification of the dust storm on the following day (May 2) and constant emission from the ThD.

Additionally, the downstream and transport zones that are close to the ThD are now affected by AOD and dust CMD. On May 4, the dust storm's strength and impact area started to progressively decrease. The Indo-Gangetic Plain (IGP), Southern India, and Bangladesh were all affected by the dust storm, which is the most likely phenomenon of aerosol transport over South Asia during the pre-monsoon (March–May) period. This dust storm accumulated over the Himalayan foothills and moved eastward and southward (as shown in Fig. 4.6), which is the most likely direction of aerosol transport over the South Asian region during this time. Dust discharged from the ThD during the process of eastward transport was discovered to be superimposed on a dust cloud with an AOD of 0.5 and a CMD of 0.6 g/m^2 over the southern region of Afghanistan. Also interesting is the fact that the dust plume, which was traveling from southwest to northeast, was also visible on the southeast and northwest sides of the TP (e.g., Figs. 4.5(a), 4.5(b) and 4.5(d)).

Additionally, MERRA-2 data and the satellite data's description of this dust process are consistent. The regional distribution of the dust

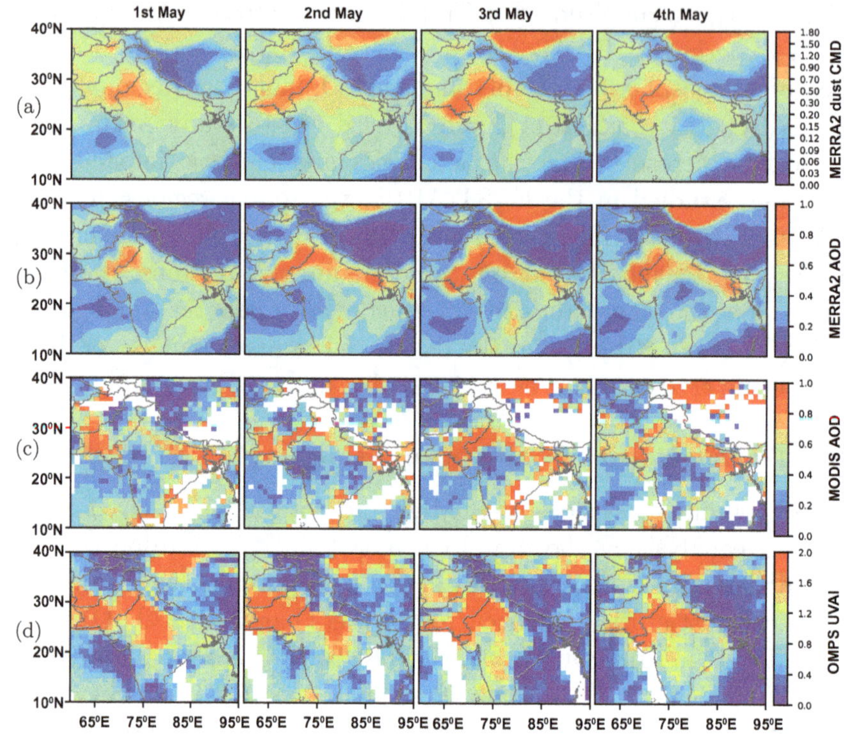

Fig. 4.5. Spatial distribution of the typical dust storm that occurred on May 1–4, 2018, for multi-source datasets: MERRA-2 dust CMD (unit: g/m^2) (a), MERRA-2 AOD (b), MODIS AOD (c), and OMPS UVAI (d) (Wang *et al.*, 2021).

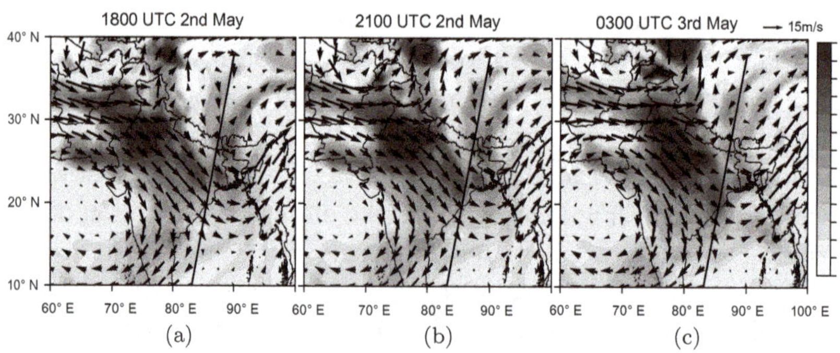

Fig. 4.6. Horizontal distribution of the column-integrated ADMR (unit: ppbm; shading) from 337.5 hPa to 525 hPa and total wind vectors (unit: m/s; arrows) at 412.5 hPa level at different times. The black line represents the CALIPSO track on May 2, 2018, over the night time (Wang *et al.*, 2021).

event as seen by the OMPS UVAI and the MODIS AOD is depicted in Figs. 4.5(c) and 4.5(d). Overall, except for spreading across a significantly greater area, the spatial distribution of light-absorbing aerosol, as shown by OMPS UVAI (>1.0), was similar to the dust CMD and AOD of MERRA-2. Even though MERRA-2 has a bigger edge over MODIS measurements in terms of detailing the dust event's evolution phase, some places still differ significantly. The MERRA-2 measurements revealed a greater regional dispersion of dust aerosols when compared to the MODIS and OMPS observations. The peak AOD values from MERRA-2 were overestimated in the TP, Southern Afghanistan, the Himalayan foothills, and the ThD, ranging from 0.1 to 0.4 on average, while underestimating dust emission from the ThD on May 1 and 4 and dust transport to Central India. The disparity between MODIS's observations of polar-orbiting satellites and the daily average of MERRA-2 reanalysis data may be the explanation. The outcomes additionally demonstrated that over the chosen six, the temporal variations of the MERRA-2 AOD and those of AERONET were in agreement (Wang *et al.*, 2021).

The air circulation and meteorological factors typically play a major role in the emission and movement of dust. The subtropical high located in the Arabian Sea had an impact on the Western Indian subcontinent, according to the geopotential height at 500 hPa level. Prior to the onset of this dust storm, a relatively robust southern branch trough (SBT) over the Bay of Bengal was discovered. Therefore, the Northwest airflows behind the SBT were what caused a cold advection over the ThD. The entire ThD region experienced a higher temperature based on the temperature at 850 hPa and sea levels. On May 2, in particular, there was a notable increase in the sea-level surface temperature across India. The most common source of such a vertical temperature profile in the atmosphere is the cooling of the upper air and the warming of the lower air, which results in a significant increase in convective effective potential energy and atmospheric instability. The temperature of the upper air over the ThD continued to drop as the low-pressure trough (LPT) at 500 hPa over Northern Afghanistan moved Southeastward by night on May 2. This led to the development of strong convective weather, which eventually generated momentum to lift the mineral dust in the ThD into the air and form a severe dust storm. A crucial dynamic element that moves the elevated dust aerosols to the TP is the southerly airflows at the upper level.

Additionally, the upper-level jet stream (ULJS), another crucial element of a macro-scale weather system, is accompanied by stronger dust storms. The characteristic dust storm across South Asia has always accompanied the ULJS between 25°N and 35°N. Under the impact of a non-geostrophic wind component, the axis of ULJS curled southwardly in the axial direction, creating a dynamic situation for the southward transport of dust aerosols (Fig. 4.6). Due to a dynamic instability state and ULJS-driven downward momentum transmission, the ThD dust storm's eruption was mostly caused by the proximity of the ULJS' departure area to the ThD. The updrafts on the South side of the TP were further increased by the lateral secondary circulation triggered by the ULJS, which led to more uplifted dust aerosols reaching higher altitudes. The southerly airflows in front of the upper trough created the ideal circumstances for the dust aerosols to be transported from South Asia to the TP through the magnificent Himalayas. A low-level jet stream (LLJS) at 600 hPa also appeared and intensified along with the downward transport of energy, which was located in the north portion of the ULJS axis and the South part of the TP. In addition, the wind speed of the ULJS axis at 200 hPa fell significantly from day to night. Additionally, the LLJS's higher inflow strength over the Indian subcontinent as a result of the disruption of the ULJS axis increased the dust storm's critical wind field.

Above all, the combination of numerous macro-scale weather systems was what caused the transit of dust aerosols from South Asia across the beautiful Himalayas to the TP. The schematic model in Fig. 4.7 shows a potential transport mechanism for dust particles over South Asia to the TP. The south–north secondary circulation and the LLJS saw a stronger upward movement as a result of the ULJS's downward energy transport, which further lifted the dust aerosols from the ThD to higher altitudes. Due to the LPT and SBT's continued existence, as well as the southerly wind blowing in front of the troughs, dust aerosols will eventually be transported to the TP via two different routes, as well as even further downstream, having a significant impact on the atmosphere and climate over these areas.

4.3.4 *Transportation of Saharan dust*

Previous research has demonstrated that WRF-Chem accurately captures the geographical and seasonal variations in AOD and aerosol size distribution (Hu *et al.*, 2016). In order to investigate how dust aerosol long-range transport evolution affects the regional climate and surface air quality in Western North America (Hu *et al.*, 2016), as well as how desert dust affects the summer monsoon precipitation over Southwestern North America, WRF-Chem quasi-global simulations were used. Figure 4.8 depicts the spatial

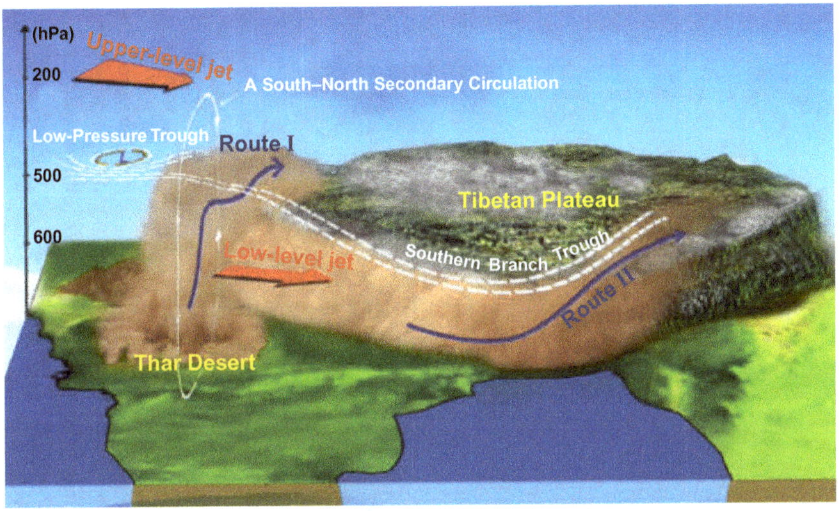

Fig. 4.7. A schematic diagram of the transport mechanism of dust aerosols over South Asia to the TP. The red arrows represent the ULJS at 200 hPa and the LLJS at 600 hPa. The white directional circle indicates the vertical secondary circulation triggered by the ULJS. The white dashed lines symbolize the atmospheric flow fields at 500 hPa level. The blue arrows represent the two transport routes to the TP (Wang *et al.*, 2021).

distribution of the seasonal mean (2010–2015) dust emission in the spring and the dust mass concentration at 700 and 500 hPa over the desert regions, whereas Figs. 4.9(a)–4.9(c) depict dust emission in other seasons. As the dust intercontinental transport over East Asia is the main subject of this study, we establish two meridional boundaries at 75°E and 145°E (red dotted lines in Fig. 4.8 to measure the amount of dust aerosols that are imported into and released from East Asia, respectively. We additionally establish 92°E as another meridian in order to distinguish the contributions of the Taklamakan Desert and the Gobi Desert to the total amount of dust particles in East Asia (black dotted line in Fig. 4.8). Over the East Asian (EAD), North African (NFD), and Middle Eastern (MED) deserts, the model predicts considerable volumes of dust emission (left), as reported by Huneeus *et al.* (2011). This amount, however, is almost 50% higher than the about 3,000 Tg/yr estimate made by Kim *et al.* (2013) for the period of 2000–2007 using the GOCART model. The impact of meteorological fields (such as surface winds) and land surface conditions (such as soil moisture) as simulated by various models, which results in variable soil erodibility and consequently different dust emissions, is one possible explanation for the disparity (Zhao *et al.*, 2012). Another factor might be the various simulation timeframes.

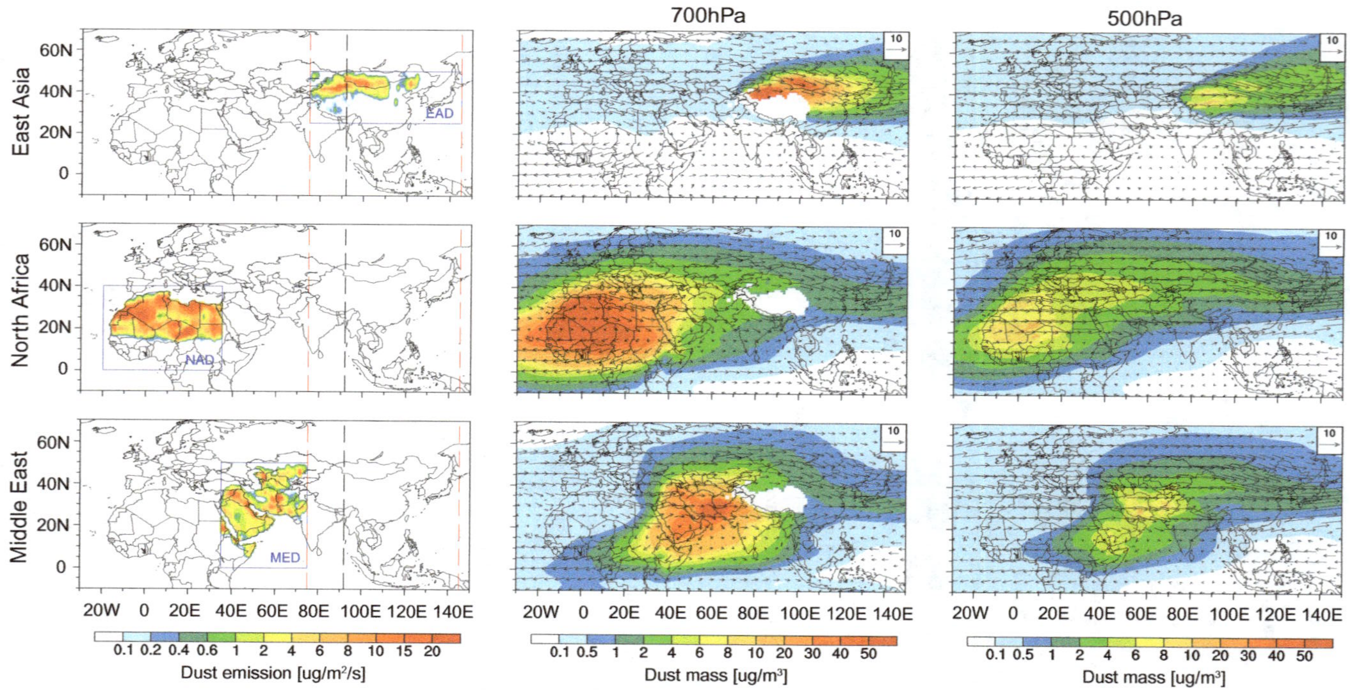

Fig. 4.8. Spatial distributions of seasonal mean dust emission rate (unit: $\mu g/m^2s$) over East Asian desert (EAD), North Africa desert (NFD), and Middle East desert (MED) (left), dust mass concentration (unit: $\mu g/m^3$), and winds (unit: m/s) at 700 hPa (middle) and 500 hPa (right) during MAM 2010–2015. In the left panels, the red dotted lines mark 75°E as the boundary of dust imported into East Asia and 145°E as the boundary of dust outflow from East Asia. The black dotted line marks 92°E, which is used as the dividing line between the Gobi and Taklamakan deserts (black dotted line) in East Asia (Hu *et al.*, 2019b).

Following this, the dominant westerlies can carry the enhanced dust thousands of miles downwind (Yu *et al.*, 2008; Hu *et al.*, 2016). Figure 4.8 (middle and left column panels) depicts a distinct pathway for the movement of dust from the NFD and MED source regions to East Asia. This dust then combines with EAD dust to form the "Asian outflow" dust in MAM, which is then transported to the North Pacific. At 700 and 500 hPa, the largest dust mass concentration is seen over desert areas, with values of more than 50 and 20 $\mu g/m^3$, respectively. Around the TP, the NFD and MED dust are split into two branches at 700 hPa. The concentration of NFD dust in the northern and southern branches is equal to about 6 $\mu g/m^3$. However, the south of the TP has larger amounts of MED dust than the north. The short distance from the source to the TP and the blocking effect of the high plateau in its southern region may be responsible for the south–north gradient in dust concentration (Xu *et al.*, 2018). The highest amount of outflowed dust at 500 hPa comes from the EAD with an 8 $\mu g/m^3$ concentration, followed by the NFD and MED dust with a 6 $\mu g/m^3$ concentration. Additionally, we observe that the EAD's high dust mass concentration is scattered in the northern 45°N, while NFD and MED's is distributed mainly south of 45°N.

The vertical cross-sections of meridional circulation and zonal dust mass flux over the source regions at 75°E (imported) and 145°E (outflowed) from 0°N to 60°N in MAM, which is the most significant scenario compared to other seasons, are shown in Fig. 4.9 to illustrate the transport characteristics of dust from various deserts. Clearly, the mass concentration of the desert dust is reduced from 40 to 1 $\mu g/m^3$, allowing it to be elevated to 300 hPa or higher. While the Gobi dust has a wider rising branch between 35°N and 55°N, the Taklamakan dust's rising branch is restricted to the area between 35°N and 45°N. The maximum mass concentration occurs between 40°N and 50°N at 900–500 hPa with a 6 $\mu g/m^3$ concentration when the dust from the two sources mixes and outflows from East Asia. The MED dust is imported into East Asia between a significantly larger range of 3°N–60°N and has a much higher concentration of 40 $\mu g/m^3$ than the NFD dust, which is imported between 25°N and 55°N. Although the MED dust is mostly transported between 0°N and 30°N, where wet scavenging caused by rainfall from the Indian summer monsoon is strongest, the NFD dust has a substantially greater dust concentration than the MED dust in the outflow zone (Yu *et al.*, 2008).

Clearly, there is greater spatial variability in the dry deposition fluxes from various dust sources. DML determines the dry deposition, whereas precipitation and DML control the wet deposition (Zhao *et al.*, 2013). Due to less precipitation over deserts and less DML over areas with

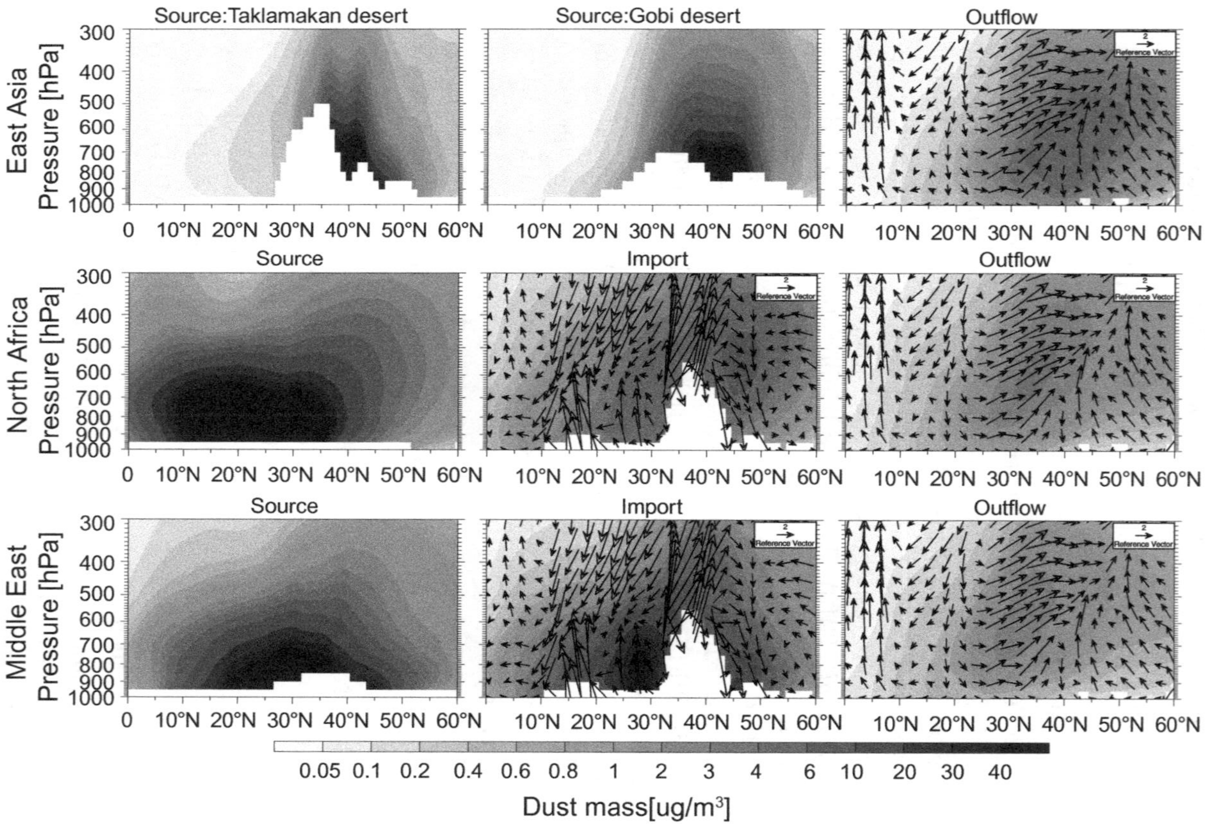

Fig. 4.9. Vertical cross-sections of zonal dust mass concentration (unit: $\mu g/m^3$) over the source regions, in which dust mass concentration (unit: $\mu g/m^3$) from the Taklamakan and Gobi deserts is averaged over 75°E–92°E and 92°E–145°E, respectively, and the sources in North Africa and the Middle East are averaged over 20°W–35°E and 35°E–75°E, respectively; Vertical cross-sections of meridional circulation (unit: m/s) and zonal dust mass concentration (unit: $\mu g/m^3$) at 75°E (imported) and 145°E (outflowed) from 0°N to 60°N during MAM 2010–2015 (Hu *et al.*, 2019b).

more precipitation, wet deposition fluxes are less heterogeneous than dry deposition fluxes. The NFD dust and MED dust have reduced wet deposition over East Asia compared to the EAD dust's wet deposition. Overall, the dry deposition of the EAD dust predominates the total deposition over East Asia, while the wet deposition of the EAD dust is greater than that of the NFD dust and the MED dust, especially South of 40°N. Additionally, the total deposition over Northern East Asia is contributed by the NFD dust and the MED dust at 5–20% dry deposition and 20% wet deposition, respectively.

4.3.5 *Vertical distribution and stratosphere dust*

Figure 4.10 depicts how dust is distributed vertically during a typical dust storm according to CALIPSO measurements. On May 2, 2018, CALIPSO observed the strong and deep extinction layers over the Indian subcontinent and the Bay of Bengal throughout both the day and night. According to the location, these layers were dominated by dust, polluted dust, or dusty marine. The dust particles were raised along the CALIPSO tracks by 6 km, indicating that the Indian subcontinent and the Bay of Bengal are affected by the dust event. However, due to reduced signal-to-noise ratios for the weak extinction layer at higher altitudes, which was brought on by the contamination of sunlight, the vertical movement of dust aerosol during the day may be underestimated. Surprisingly, on May 2 during the night, CALIPSO also detected a dense layer of dust aerosols over the Southeast TP, ranging in altitude from 4 to 10 km, with a maximum extinction of 0.18 km^{-1}. The dust plume is also visible in Fig. 4.5 at a similar location. To determine the possible conveyance of the dust aerosol during the dust event at the upper level, the horizontal distribution of column-integrated ADMR from 337.5 to 525 hPa is shown in Fig. 4.6. The ThD and its surrounding areas were where the positive peak center of ADMR was most prominent, showing that strong upward movements (29°N) are responsible for the bulk of the dust aerosols hoisted above 500 hPa, as seen in Fig. 4.11(a). The downstream region and even the TP were being primarily affected by the dust aerosols continuously delivered from the ThD due to the influence of upper Northwest airflows regulated by the subtropical high over the Indian Ocean. As seen from Fig. 4.6, two typical regions over the TP had very positive ADMR. One primarily manifested at about 80°E in the Western TP, where the predominant wind direction partially changed from the Northwest to the South. Since the dominant Northwest wind was responsible

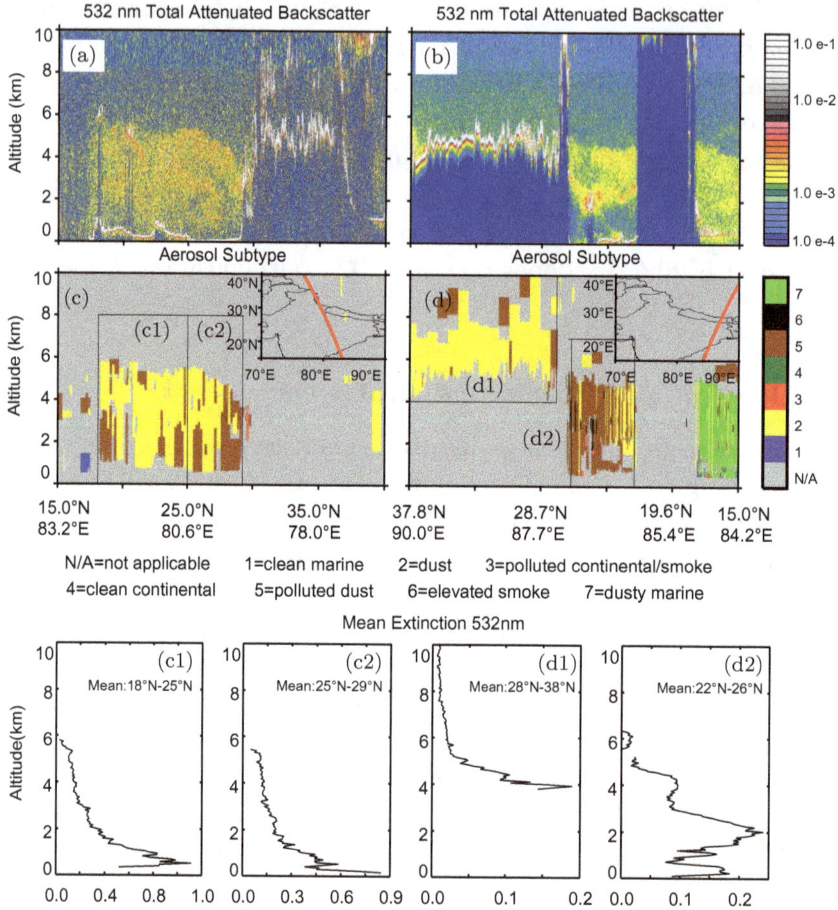

Fig. 4.10. Vertical distributions of 532 nm total attenuated backscatter (unit: $km^{-1}sr^{-1}$) (a, b), aerosol types (c, d), and 532 nm mean aerosol extinction coefficient (unit: km^{-1}) (c1, c2, d1, d2) over four specified areas along the CALIPSO tracks in the morning (left) and evening (right) on May 2, 2018 (Wang *et al.*, 2021).

for moving the majority of the dust aerosols raised by strong vertical movement (Figs. 4.11(a) and 4.11(b)), Southern India and the Bay of Bengal were affected by this dust aerosol transport. The stronger southerly wind, on the other hand, carried the remainder to the TP, causing a greater ADMR and wider areas to be influenced over the Western TP (Fig. 4.6(b) and 4.6(c)). The other was discovered at a TP in the Southeast, where a cyclonic belt curled to the Southwest and Northeast. Strong southerly winds functioned as the primary driver of the dust belt, which was observed by the CALIPSO satellite (Fig. 4.10(d)), which is consistent with other studies that examined the water vapor transport channel in the Yangtze River basin of

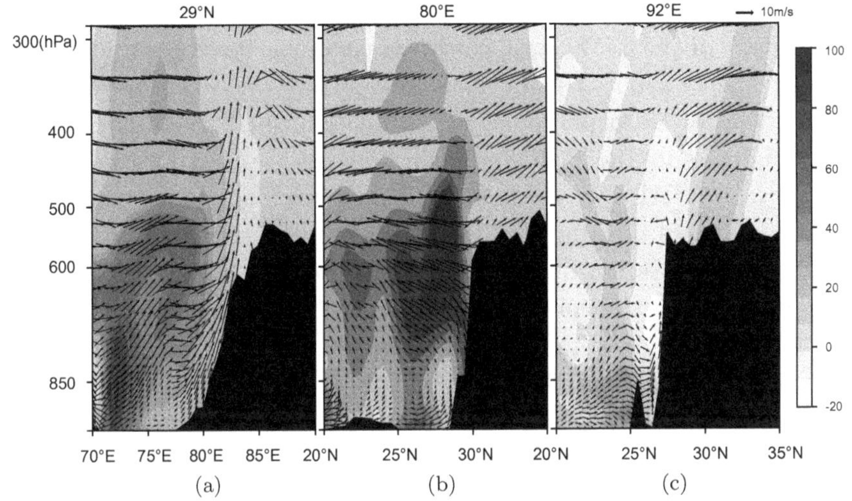

Fig. 4.11. Vertical cross-sections of the ADMR (unit: ppbm; shading) at 29° N (a), 80° E (b), and 92° E (c). The zonal and meridional circulations (unit: m/s) were superimposed on these images (Wang *et al.*, 2021).

China. The findings show that dust aerosols and other pollutants over South Asia can be vertically transported to high altitudes above 500 hPa and across the magnificent Himalayas through a variety of routes, having an impact on the weather and climate in the TP and parts (the middle and downstream regions) of the Yangtze River.

A "transfer station" for vertically spreading the accumulated dust aerosols, the TP also serves as a storage facility for the particles. *In situ* measurements have steadily revealed some strong indications of aerosol concentrations above the tropopause in recent years. Meanwhile, it was noted that the Asian Tropopause Aerosol Layer's (ATAL) dominant aerosol by mass is mineral dust, which is present in both peaks and exhibits a significant interannual fluctuation but no long-term trend (Yuan *et al.*, 2019). Dust aerosol can be transported through the TP to the upper troposphere and lower stratosphere (UTLS) (Xu *et al.*, 2018). According to Lau *et al.*'s (2010) research on the process of pollutants' vertical transfer to the stratosphere, pollution from Asia, India, and Indonesia can enter the global stratosphere through the monsoon circulation. Additionally, the anticyclone's intraseasonal east–west oscillation may be crucial in moving pollutants from the boundary layer that are convectively pumped into the UTLS. Quantitatively, it was demonstrated that the TP represents a significant transport pathway for the global tropopause dust layer (GTDL) using a three-dimensional aerosol transport-radiation model. Although the northern slope of the TP also collects dust particles, its feeble updrafts make it a poor transport route. Additionally, deep convection can also inject dust aerosols into the stratosphere. It was also established that the Himalayan-Gangetic Plain and the Sichuan Basin are ideal routes for the improved transfer to the ATAL by overshooting deep convection (Yuan *et al.*, 2019). Additionally, it was thought that the Asian summer monsoon was related to the aerosol concentration in the UTLS (Bian *et al.*, 2020).

The conclusions showing the TP is in fact a potent transport channel, particularly on its southwest slope, are summarized in Fig. 4.12 The Ethiopian Plateau and the Rocky Mountains are also used as transportation routes. The southern slope of the TP can be used to transfer dusts from IND to the GTDL, and their contribution to the GTDL at 100 hPa is the largest (Fig. 4.12(b)). The Ethiopian Plateau and the TP's southwest slope are two places where dust from ME is transported, making it the second-largest contributor to the GTDL at 100 hPa (Fig. 4.12(a)). The majority of the TP's southern slope, as well as its northern and eastern slopes, serve as transport routes for dusts coming from NWEA that are transported to the GTDL (Fig. 4.12(c)). The NAF's contribution to GTDL at 100 hPa is the smallest, but its transport routes, which include the Rocky Mountains, the Ethiopian Plateau, and the southwest slope of the TP, are the most numerous (Fig. 4.12(d)).

4.4 Climate Effects of East Asian Dust

East Asia is a significant source of dust in the world. In Asia, where there are many arid and semi-arid regions, mineral dusts in the atmosphere and their interactions with clouds and precipitation have significant effects on regional climates. This section discusses the many impacts of dust particles on clouds and precipitation, mainly over East Asian arid and semi-arid regions, as well as the usual transit routes of East Asian dust, which have an impact on regional and worldwide climates. We intend to set a standard for research in this field (Huang *et al.*, 2014).

4.4.1 *Direct radiative effect*

All anthropogenic and natural dust aerosols have a direct radiative effect (DRE), which is measured by the AOD or extinction coefficient, the SSA, and the asymmetry parameter or phase function, on the radiative flux at the TOA and at the surface, as well as on the absorption of radiation within

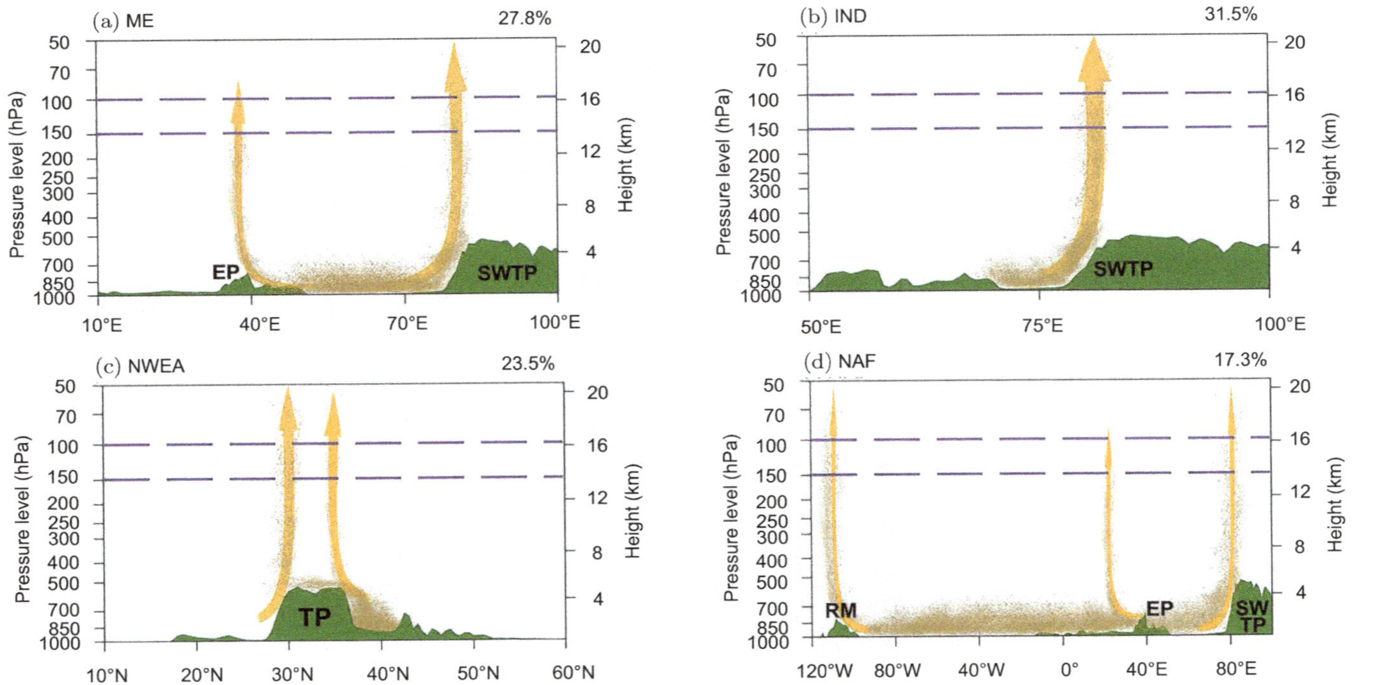

Fig. 4.12. Schematic illustration of the dust transport to the GTDL. Transport of dusts originating from Middle East (ME) (a), Indian Peninsula (IND) (b), Northwestern East Asia (NWEA) (c), and North Africa (NAF) (d) to the GTDL. The yellow arrows represent the transport pathways, and the numbers in the upper-left corner represent the contributions of the dust sources to the GTDL at 100 hPa. EP represents the Ethiopian Plateau, SWTP represents the Southwestern slope of the TP, and RM represents the Rocky Mountains. The green shadings indicate the topographic altitude. The purple lines denote the 100 and 150 hPa layers (Liu *et al.*, 2022).

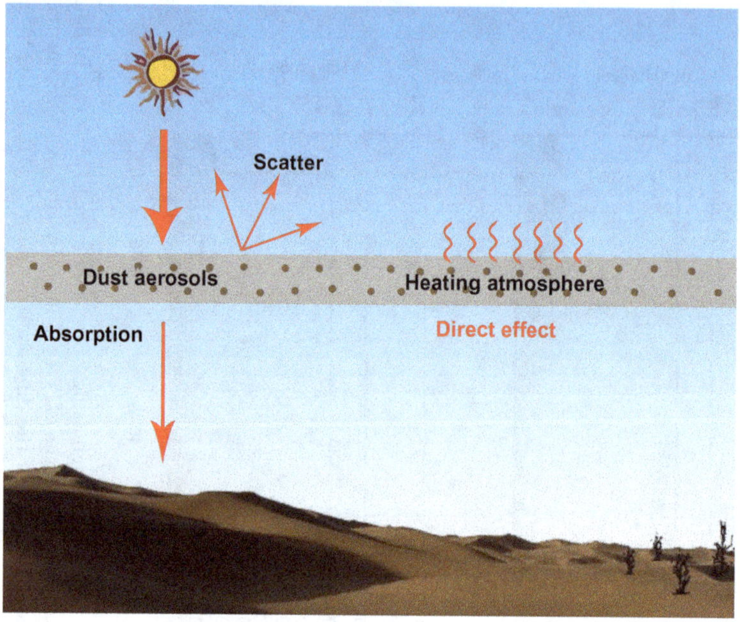

Fig. 4.13. A schematic depiction of the direct radiative effect of dust aerosols (Huang *et al.*, 2014). Radiation substantially outweighs cooling caused by the scattering of solar flux, and the magnitude of cooling at the TOA is less than that at the surface (Bi *et al.*, 2013). According to the dust DRE's diurnal cycle, the surface RF is negative during the day due to the absorption and backscattering of sunlight by dust particles, but positive at night due to the continuous heat emission from dust aerosols (Claquin *et al.*, 1998).

the atmosphere (Fig. 4.13). In order to calculate dust DRE, DRH is also employed as an index.

The SW, LW, and net DRH are each determined using the SBDART model. The seasonal geographic distributions of the decadal means of SW, LW, and net DRH averaged over the atmospheric column in the TP and its environs are shown in Fig. 4.14. As the dust aerosol absorbs SW radiation to heat the atmosphere (Figs. 4.14(a) and 4.14(d)), positive SW DRH implies a faster heating rate under the dust condition and may cause an intense vertical transport of dust aerosol from nearby sources (Lau *et al.*, 2006). The TD peaks at 9.0 K/month in the spring, whereas the ThD peaks at around 7.0 K/month in the summer. The LW DRH displayed predominantly negative values, indicating a cleaner environment in the presence of dust, even if its magnitude was much lower than that of the SW DRH (Figs. 4.14(e) and 4.14(h)). With the exception of DJF, all seasons saw the peak center of LW cooling (about −1.5 K/month) primarily near the TD. The net DRH exhibits a favorable warming effect overall. The peak centers, which were

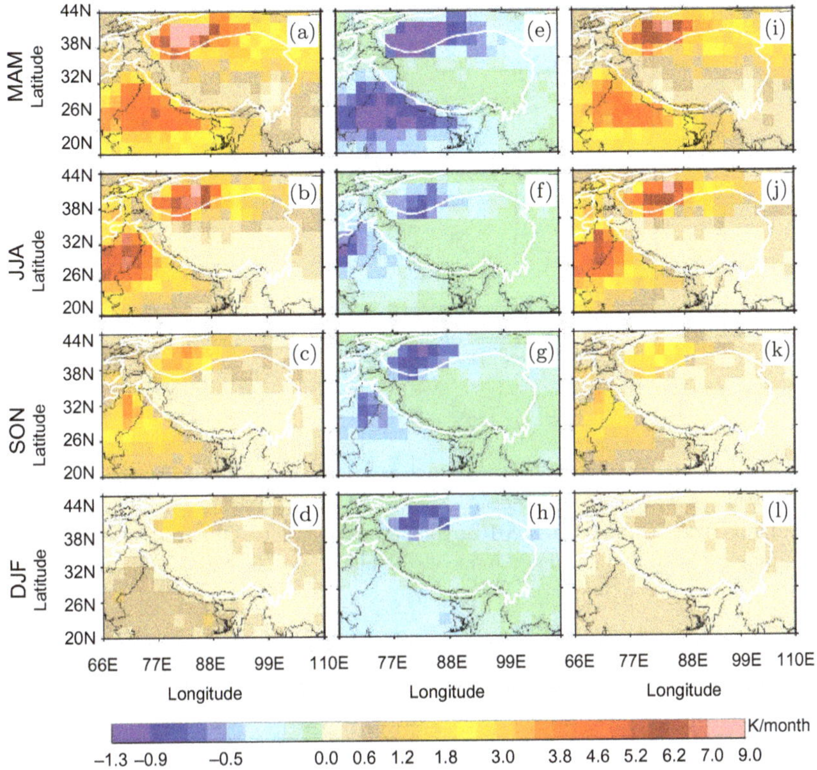

Fig. 4.14. Seasonal spatial distributions of the decadal mean of column-averaged SW (a–d), LW (e–h), and net DRF (i–l) in the atmosphere over the TPS. The core region of TP is outlined with a white solid line at an altitude of 2,500 m. The March–April–May (MAM), June–July–August (JJA), September–October–November (SON), and December–January–February (DJF) represent spring, summer, autumn, and winter, respectively (Wang *et al.*, 2020a).

7 K/month over the TD in the spring and 6.2 K/month over the ThD in the summer, were completely consistent with those of the SW DRH. In spring or summer, the effect of dust aerosol over the TD on the thermal structure of the atmosphere is more evident than it is over the ThD and nearby transport zones. The Qaidam Basin experienced a greater warming effect from dust aerosols, 2.4 K/month in spring and summer.

Dust RFs from multiple research projects in various climate areas are compiled in Fig. 4.15. According to the IPCC (2013), the annual mean DRE of mineral dust is estimated to be -0.1 W/m^2 on a global scale, with the range of the uncertainty being between -0.3 and 0.1 W/m^2. According to Zhang *et al.* (2010), the worldwide annual mean direct forcings are -0.4, -1.46, and $+1.86$ W/m^2 at the surface, within the atmosphere,

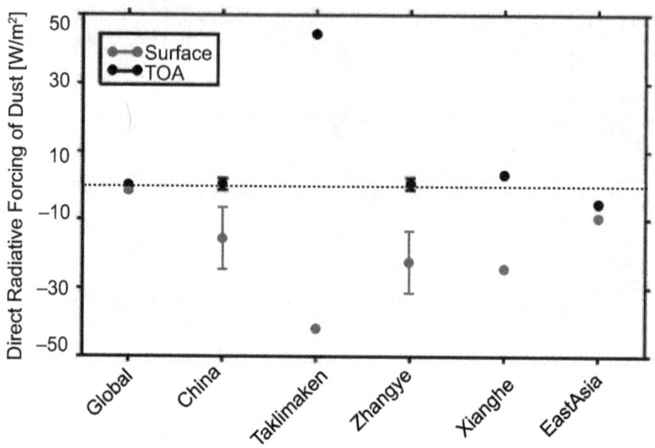

Fig. 4.15. The daily mean direct RF of dust aerosols at the TOA and surface in different regions. The TOA value for the globe is from the IPCC (2013), and the corresponding surface value is from Zhang *et al.* (2010); the values over the China region, the Taklimakan Desert, Zhangye, Xianghe, and East Asia (20°N–50°N, 100°E–150°E) (Huang *et al.*, 2014).

and at the TOA, respectively. The global annual mean all-sky SW RF by dust at the TOA is calculated to be -0.28 W/m^2. For the preindustrial, doubled-carbon-dioxide, and last glacial maximum climates, respectively, the simulated globally averaged RF values of dust at the TOA are -1.0, -0.4, and $+0.14$ W/m^2. In general, both at the TOA and the surface, the SW RF values are negative. Due to dust particles' absorption and emission of LW radiation, the LW RF is positive. However, because warming caused by the absorption of solar and thermal radiation substantially outweighs cooling caused by the scattering of solar flux, the magnitude of cooling at the TOA is less than that at the surface (Bi *et al.*, 2013). According to the dust DRE's diurnal cycle, the surface RF is negative during the day due to the absorption and backscattering of sunlight by dust particles, but positive at night due to the continuous heat emission from dust aerosols (Claquin *et al.*, 1998).

With the anthropogenic fraction of dust estimated to be as much as 30–50% of total dust production, a variety of human activities (such as land use practices and construction) can lead to significant dust loading, and its direct RF on both global and regional scales may be comparable to or greater than the forcing by other anthropogenic aerosols (Tegen and Lacis, 1996).

The sensitivity of these mineral forcings to various modifications of the aerosol complex refractive index and size distribution was investigated by Claquin *et al.* (1998). Given its rarity and uncertainty, the complex aerosol refractive index has been recognized as a crucial metric. Furthermore, if the refractive index is spectrally averaged, the imaginary portion is insufficient. Its inherent variability (related to mineralogical characteristics) causes variances in aerosol forcing calculations of up to $\pm 40\%$. Since the optical characteristics of dust are particularly sensitive to the presence of microscopic particles, a good description of the size distribution is necessary when modeling mineral aerosols. According to Wang *et al.*'s (2013) assessment of the radiative effect of a significant dust storm over Northwest China, the vertical distribution of the AOD and its importance for LW RF in the atmosphere are more essential than surface albedo, the AOD, and SSA for SW.

As it heavily depends on the source of the dust and the state of the surface, dust DRE exhibits excellent regional features. The SW RF at the TOA can be favorable over bright surfaces, such as deserts, because more radiation, which should be reflected into the atmosphere, may be absorbed by dust, particularly at shorter wavelengths. Some studies have discovered that the daily mean SW forcing for the Saharan dust outbreaks over the Mediterranean is countered by a LW forcing of about 30% at the surface and about 50% at the TOA. When Mallet *et al.* (2009) calculated the RF of dust aerosols for a significant dust event over West Africa, they discovered that the presence of dust particles caused a significant instantaneous reduction of surface incoming SW (-127 W/m^2) but that the overall effect at the TOA was only a small cooling (-12 W/m^2). Over the Taklimakan Desert, a similar positive SW RF has been detected at the TOA. For a five-day dust event that occurred over this area in July 2006, according to Huang *et al.*'s (2009) calculations, the daily mean net RF was 44.4, -41.9, and 86.3 W/m^2 at the TOA, the surface, and the atmosphere, respectively. According to these forcing estimates, solar radiation is responsible for about 90% of the atmospheric warming, while LW radiation is responsible for about two-thirds of the warming effect at the TOA. At the surface, the LW warming effect offsets about one-third of the dust solar radiative cooling effect. The dust SW direct RF varies from -4.8 to 0.4 W/m^2 at the TOA, -5.2 to -15.6 W/m^2 at the surface, and 5.2 to 10.8 W/m^2 in the atmosphere above the Badain Jaran Desert, according to research by Bi *et al.* (2013). The dust aerosol RF at the TOA and the surface in Zhangye are, respectively, 0.52 ± 1.69 and -22.4 ± 8.9 W/m^2 along the dust paths (Ge *et al.*, 2010a). In March, April, and May over SACOL, Liu *et al.* (2011) estimated the aerosol RF and discovered that the SW RFs are 0.68, -70.02, and 70.70 W/m^2, respectively, at the TOA, surface, and atmosphere. Using data from the CloudSat and CALIPSO satellites, Li *et al.* (2007b) demonstrated that the heavy dust loading combined with pollutants exerted an extraordinarily strong direct forcing of roughly −15 to −25 W/m^2 (daily mean) at the surface, whereas the force at the TOA was essentially nonexistent. This was true of Xianghe, the downward zone of the dust. The mixed dust at Taihu, in Southeast China, can decrease SW radiation by 48 W/m^2 at the surface and increase atmospheric absorption of SW radiation by 27 W/m^2. According to Li *et al.* (2010b), the daily mean net RF of dust across China is 0.3 ± 1.6, -15.7 ± 8.9, and 16.0 ± 9.2 W/m^2 at the TOA, surface, and atmosphere, respectively. A regional climate model was employed by Qian *et al.* (2003) to illustrate direct aerosol RF impacts, which were almost unfavorable over China. Seinfeld *et al.* (2004) estimated that the dust RFs at the TOA and surface were −5.5 and −9.3 W/m^2, respectively, over the area of 20–50°N, 100–150°E during April 5–15, 2001, based on data from the Asian Pacific Regional Aerosol Characterization Experiment.

The measurement of scattering efficiency to total extinction efficiency, known as SSA, has a range of 0–1. A lower value denotes a particle with a higher solar radiation absorption. When dust particles move from the source regions to remote areas, they are frequently covered with pollutants from both Asia and Europe along their pathways. As a result, these particles absorb pollutants much more strongly than dust particles from other significant deserts, such as the Sahara (Eck *et al.*, 2010). For large SSA, dust aerosols have a negative TOA forcing, whereas for small SSA, they have a positive TOA forcing.

The uncertainty of direct aerosol RF has previously been studied, and it was discovered that when the individual aerosol optical properties (AOD, SSA, asymmetry parameter, scale height, and anthropogenic fraction) are perturbed from their base values, the direct aerosol RF shows a significant sensitivity to SSA in cloudy columns. Furthermore, despite the fact that the global mean of the clear-sky direct aerosol RF is more than twice as large as the all-sky direct aerosol RF, and the direct aerosol RF uncertainty in all-sky conditions is higher than in clear-sky conditions.

Lee *et al.* (2007) determined the SSA at 25 stations throughout China, including some dust-filled regions, by integrating satellite-measured reflectance and surface measurements of transmittance. They found that the national mean of SSA at 0.5 μm is 0.89 ± 0.04 in this study. The SSA of the Asian dust varies from 0.74 to 0.84 at 0.5 μm in Northern China (Ge *et al.*, 2010a) and in India. The SSA of the Asian dust, according to Costa *et al.* (2006), can be as low as 0.76, which is substantially lower than the figure discovered for the African dust. A mean value of 0.95 for the SSA at 0.55 μm of the African dust, for instance, was reported by Fouquart *et al.* (1987). In the Saharan Dust Experiment, the SSA at 0.55 μm varied between 0.95 and 0.99, according to Haywood *et al.* (2003). Smaller SSA values for the Asian dust indicate that these particles are very aerosol-absorbing. High densities of anthropogenic species were discovered during a recent experiment in the vicinity of a significant desert in rural Western China.

The heating rate profile, which offers more data regarding dust aerosol DRE at various air altitudes than RF, is greatly influenced by the vertical distribution of aerosols. Under light, moderate, and heavy dust conditions, dust aerosols can heat the atmosphere over the Taklimakan Desert by up to 1, 2, and 3 K/d (daily mean values), respectively; the maximum daily mean of the radiative heating rate can reach 5.5 K/d at 5 km at the location of the dust layer (Huang *et al.*, 2009). According to another study (Lemaître *et al.*, 2010), dust aerosols can, depending on altitude and latitude, warm the atmosphere by an average of 0.3–4.0 K/d during the day. Within a small area of the dust plume, there was localized strong warming (i.e., a heating rate as high as 8 K/d). The range of the heating rate retrieval uncertainty was determined to be between 0.5 and 1.4 K/d in the optically thickest region of the dust plume. Less than 1 K/d of heating/cooling rate values were obtained during the night. The influence of dust particles on the surface temperature depends on the vertical extent and duration of the dust layer. Even during the day, the surface temperature might rise when a dust layer contacts the ground and remains there for several days. The surface temperature drops when there is a higher dust layer. Horizontal temperature gradients resulting from these temperature variations, brought on by dust DRE, may alter near-surface winds. Feedback on the overall emission flux might be established since the surface wind threshold controls the uptake of dust from the surface.

According to Ramanathan *et al.* (2007), local pollution by itself can increase the solar heating of the lower atmosphere by about 50%. According to another study (Mallet *et al.*, 2009), dust particles can reduce surface temperature by 0.5–10.0 K/d. The highly elevated surface air over the TP may function as an "elevated heat pump" and significantly alter the local climate through the absorption of solar radiation by dust in combination with black carbon emissions from industrial areas in Northern India, according to model studies by Lau *et al.* (2006). An early commencement and strengthening of the Indian summer monsoon may come from a tropospheric temperature anomaly that may be created in late spring and early summer over regions of Northern India and Tibet.

In conclusion, whereas dust particles chill the surface, they warm the atmosphere as a whole. As opposed to dark surfaces, Asian arid and semi-arid regions have higher surface albedo and dust aerosols that are more absorbent than those from other dust source regions, including the Sahara Desert. As a result, the total effect of dust aerosols is typically warming or nonexistent at the TOA. The highest forcing values are found over or close to places where there is a lot of dust, such as the Taklimakan Desert.

4.4.2 *Indirect radiative effect*

According to Huang *et al.* (2006c), dust aerosols can alter the macro- and microphysical characteristics of clouds, such as the liquid water content, cloud fraction, cloud top temperature, droplet number concentration, and cloud particle size, which are depicted in Fig. 4.16.

Aerosols, as we are all aware, are essential components for the creation of clouds since they act as condensation nuclei (CN) to create cloud droplets and ice crystals. One of the most frequent types of IN are dust aerosols; however, new research indicates that mineral particles may also be able to penetrate far into the upper troposphere and act as IN for the development of cirrus and mixed-phase clouds (Teller and Levin, 2006). Additionally, sea salt, man-made pollutants, and secondary organic aerosols can interact with dust to create particles that have a "core" of insoluble mineral dust and a coating of soluble material (Levin *et al.*, 2005). A soluble coating on dust particles makes them effective CCN, frequently continuing their activity as IN. The size and makeup of an aerosol particle affect its capacity to absorb water and subsequently initiate cloud condensation in microphysical processes. It has been hypothesized that CCN's size matters more than their chemical makeup.

Because they need a lower critical supersaturation, larger particles can be triggered more easily than smaller ones. Even while dust particles combine with other anthropogenic aerosols to raise CCN concentration, the activation ratio tends to quickly decline with increasing CCN during dusty events, suggesting that dust particles do not readily increase CCN concentration (Liu *et al.*, 2011b). Additionally, due to their enormous size, dust particles can function as enormous CCN, forming effective collector drops that start the development of drizzle and precipitation. The resulting cloud droplets should be more numerous and smaller, with a bigger total surface area and a higher cloud albedo (cloud albedo effect), when more aerosol particles are vying for the uptake of a fixed amount of liquid water content (Twomey,

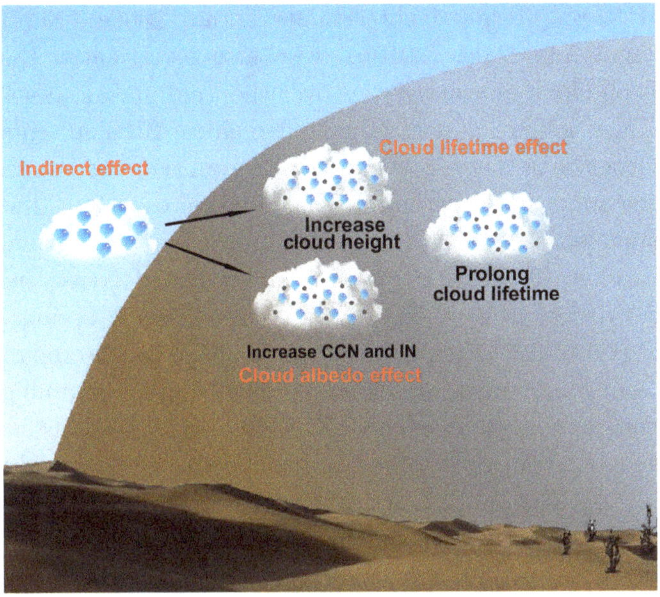

Fig. 4.16. A schematic depiction of the indirect radiative effect of dust aerosols (Huang *et al.*, 2014).

1977). The result is a negative RF at the TOA because a dust-polluted cloud reflects more solar radiation back into space (Huang *et al.*, 2006d).

Because there are more of these smaller, more numerous cloud droplets colliding with one another, the amount of precipitation produced by polluted clouds is reduced. This prolongs the life of the clouds and increases their capacity to disrupt atmospheric circulation, which causes a redistribution of clouds and precipitation (the "cloud lifetime effect"). Delaying precipitation would enable more small cloud particles to rise above the freezing level and initiate an early ice process in the warm-based mixed phase, which would release more latent heat and hence stimulate the vertical development of clouds. This is known as the aerosol invigoration effect, whose climate impact is probably a positive forcing (Koren *et al.*, 2010).

In Asia's arid and semi-arid regions, where there are few opportunities for ground-based observations, satellite products, particularly data from the A-Train constellation, are widely used to research the impacts of dust on clouds (Huang *et al.*, 2006b, 2006c, 2006d, 2010b; Wang *et al.*, 2010b). In the same meteorological environment over Northwestern China, Huang *et al.* (2006b, 2006c) compared the cloud properties under dusty and dust-free conditions. They discovered that dust aerosols injected into clouds would reduce ice cloud effective particle size and optical depth of cirrus clouds by 11% and 32.8%, respectively (Huang *et al.*, 2006c). Similar findings were made by Wang *et al.* (2010b), who also investigated the impact of dust aerosols on cloud characteristics over the Taklimakan Desert and West Pacific, the dust source, and transported regions. Similar tendencies were also seen in other data (Huang *et al.*, 2010b) from semi-arid and arid regions of Asia, which supported the notion that dust aerosols can diminish cloud effective particle size. According to a study based on simulations of general circulation models (GCMs), cirrus has an overall decrease in ice crystal number and mass of 10% and 5%, respectively, while seeing an increase in ice crystal size of 3% (Kuebbeler *et al.*, 2013). This discrepancy between observation and modeling suggests a significant degree of ambiguity in our comprehension of how dust affects cloud features. Aerosols have entirely different effects on stratiform clouds and convective clouds, as demonstrated by theoretical speculation and model simulations (Tao *et al.*, 2012).

Dust aerosols affect clouds' microphysical characteristics directly and indirectly, which has an impact on the radiative energy budget. The IPCC (2007) regarded the indirect aerosol effect as having the greatest level of uncertainty. Thus, it is crucial to calculate the indirect aerosol forcing over arid and semi-arid regions of Asia. A review of the numerous aerosol effects on clouds and climate was done by Lohmann and Feichter (2005). They discovered that the estimations provided by models for the indirect RF of all types of clouds range from -0.5 to -1.9 W/m^2. All models showed negative forcing, with a standard deviation of 0.4 W/m^2 and an average of -1 W/m^2. Averaging -0.7 W/m^2 with a standard deviation of 0.5 W/m^2, the cloud lifetime effect varies significantly among the models (ranging from -0.3 to -1.4 W/m^2). Estimates of the semidirect effect range from $+0.1$ to -0.5 W/m^2. The estimations differ because of where black carbon is located in relation to the clouds. In most areas, the indirect aerosol forcing is negative. Some research revealed that ice clouds were found in dust-laden air at substantially warmer temperatures than in dust-free settings because dust particles often function as IN (Choi *et al.*, 2010). The indirect cooling impact of aerosols would be amplified by increasing outgoing thermal radiation. According to Su *et al.*'s (2008) research, contaminated clouds had much lower absolute TOA RF levels than clean clouds. In Northwest China, Wang *et al.* (2010b) discovered that dust aerosols had an immediate net cooling effect of 43.4% on clouds. According to other studies as well (Huang *et al.*, 2006c, 2006d), dust particles had a cooling effect on the clouds at the TOA.

Two zones are chosen to represent the dust and clouds in various situations in order to detect changes in cloud properties brought on by dust aerosols. A CLD is a location where clouds form in an atmosphere devoid of dust. Partial clouds in a dusty area are part of PCOD. Even if they don't completely encircle the PCOD region, the clouds in the PCOD region are solid-deck clouds. COD is a representation of gloomy, dusty clouds. According to measurements from 701 surface meteorological stations in China and Mongolia, the CLD and COD regions were chosen (Huang *et al.*, 2006c). The four types of dust that are reported by the surface stations are storms of dust, wind-blown sand, floating dust, and no dust. If there is no dust on the surface when a cloud region is viewed by a satellite, the region is classified as a CLD; if there is a dust storm, the region is classified as a COD.

Figure 4.17 displays histograms of the pixel-level ice cloud features for each category from the 10 cases. This graph can show the potential effects of dust particles on the De, τ, and WP of ice clouds. In the COD areas, the average De decreases from 54.0 μm in the CLD regions to 48.0 m. In comparison to the CLD regions, smaller values of De occur more frequently in the PCOD and COD regions. The mean over the CLD regions (Fig. 4.17(b)) is 21.9, which is 12.8 and 7.2 more than the average τ across the PCOD and COD regions, respectively. As a result, the mean WP values for the PCOD and COD regions are similarly much lower than those from the CLD regions (Fig. 4.17(c)). For the PCOD and COD zones, the mean WP falls from 396.2 g/m^2 to 128.7 and 230.7 g/m^2, respectively. However, the Te values (Fig. 4.17(d)) for the COD regions are comparable to the values in the CLD regions, which are slightly warmer (about 5 K), suggesting that the cloud top heights in these locations differ by less than 1 km. In contrast to the other two sections, the PCOD regions (Fig. 4.17(d)) had a higher mean Te (241.3 K). It's possible that less precise retrievals of cloud characteristics near the boundaries of cloud decks in the PCOD regions are a contributing factor to the difference in cloud properties between the PCOD and COD regions. Due to the existence of dust aerosols and partially cloud-filled pixels, which cause higher reflectance and warmer brightness temperatures, the cloud parameters collected by satellite under partly cloudy conditions may be underestimated or overstated.

The aerosol–cloud interaction is often quantified as ACI $= -I_n r_e/\ln \alpha$, where r_e is the drop size and α is the aerosol burden. It can be seen from Twomey (1977) that the ACI ranges from 0 to 0.33. According to McComiskey and Feingold (2008), RF varies from -3 to -10 W/m^2 for each 0.05 rise in anthropogenic aerosol disturbance. Su *et al.* (2008) used a combination of satellite measurements and Fu–Liou model simulations for

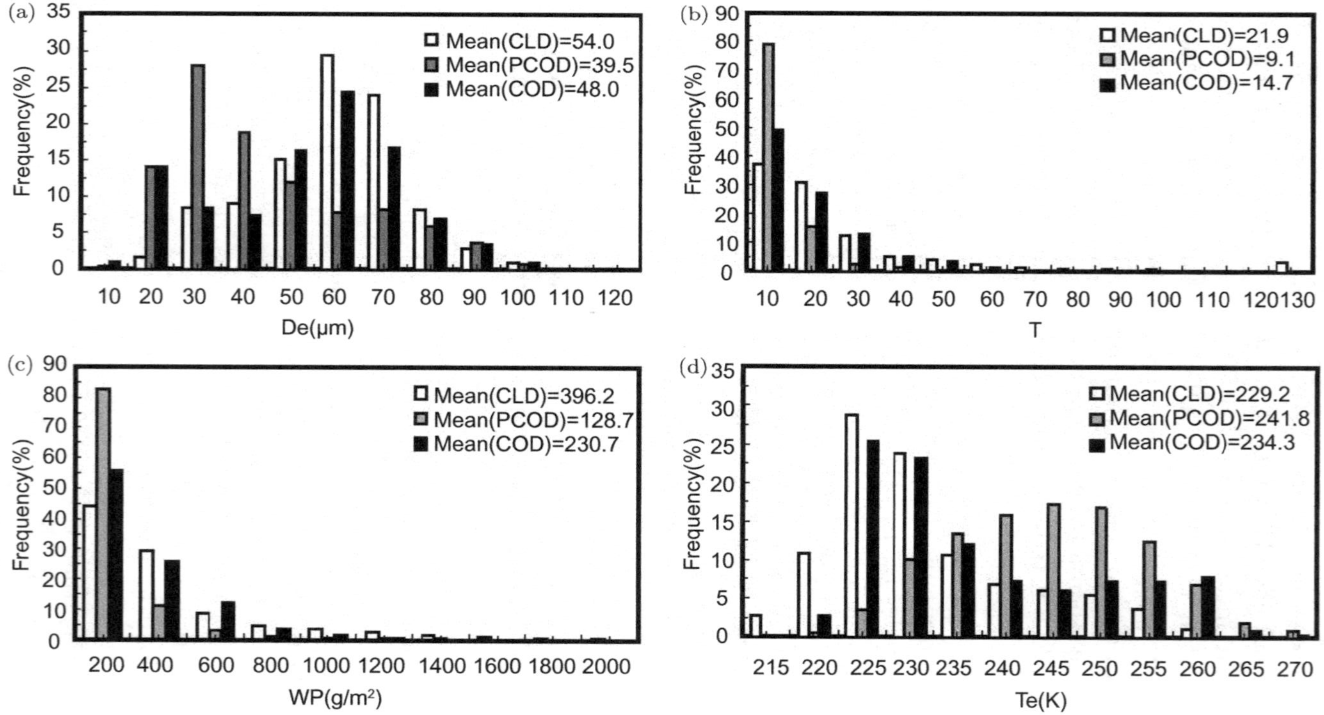

Fig. 4.17. Comparison of the cloud properties over the CLD region (open bar), PCOD region (gray bar), and COD region (black bar) for De (a), τ (b), WP (c), and Te (d). The histogram intervals are 10 μm for (a), 10 for (b), 200 g/m^2 for (c), and 5 K for (d) (Huang *et al.*, 2006c).

16 simultaneous examples during 4 years over China and Mongolia in a hazy sky to quantify the contribution of dust's direct, indirect, and semidirect impacts on cloud RF. Indirect and semidirect SW RF together have a mean value of 82.2 W/m^2, or 78.4% of the total dust effect. Only 22.7 W/m^2, or 21.6% of the entire effect, is produced directly. The dusty cloud warming effect is mostly caused by the semidirect effect of dust aerosols since both the first and second indirect effects improve the cloud cooling effect.

The growth of anvil clouds associated with DCCs can generate positive RF during the invigoration process, but the RF of the aerosol invigoration effect is still uncertain because the convective core thickening may cause cooling (Koren *et al.*, 2010). The energizing impact would also introduce positive forcing and enrich the UTLS with water vapor. The RF of the aerosol invigoration effect was marginally favorable at the TOA but negatively affected at the surface during the EAST-AIRC campaign (Li *et al.*, 2011).

It is difficult to distinguish between the cloud albedo effect and other factors. Processes that lower the size of cloud droplets for a given liquid water content also decrease the likelihood of precipitation development, extending the cloud lifetime (the cloud lifetime effect). The time-averaged cloud albedo also changes as a result of an increase in cloud lifespan. Therefore, it is still unclear which factor is more significant: the cloud lifespan or the cloud albedo effect.

4.4.3 *Semidirect effect*

Due to their ability to absorb solar energy, dust particles can affect cloud formation and growth not only by acting as CCN or IN but also by altering the environment. Figure 4.18 provides a schematic representation of the semidirect effect. Dust has the ability to absorb solar energy and cause localized heating, which alters the relative humidity and atmospheric stability and, in turn, affects the longevity and liquid water content of clouds. Depending on where it is located, the dust aerosol layer may either stabilize or destabilize the boundary layer. When dust particles develop within or above clouds over arid or semi-arid areas, it is likely that the clouds will evaporate (Huang *et al.*, 2006b, 2010b; Wang *et al.*, 2010b). This effect is an immediate reaction related to the direct effect.

Modeling studies on eddy-resolving, regional, and global scales have studied the causal relationship between aerosol atmospheric heating (and surface cooling) and cloudiness, showing a more complex picture than previously hypothesized. Due to the implicit consideration of the semidirect effect whenever absorbing aerosols are incorporated into the radiation scheme, the

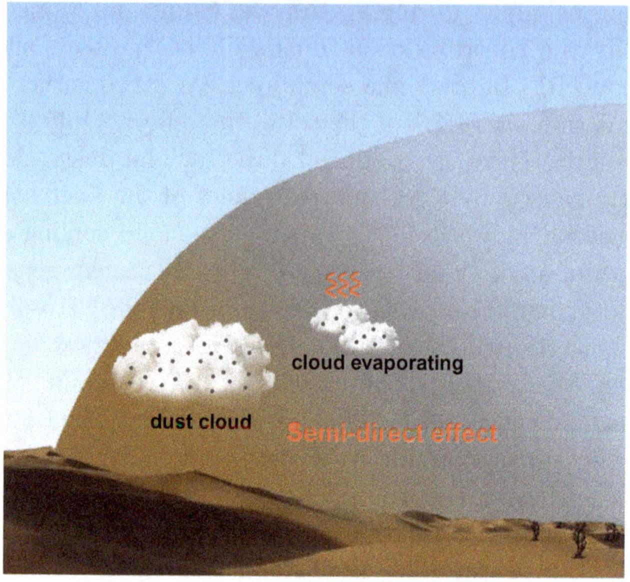

Fig. 4.18. A schematic depiction of the semidirect radiative effect of dust aerosols (Huang *et al.*, 2014).

semidirect effect has been modeled using GCMs and high-resolution cloud-resolving models (Cook and Highwood, 2004). The semidirect effect can also include cloud variations brought on by circulation and/or surface albedo effects when it is identified within a GCM framework. This is due to the GCM response's ability to accommodate large-scale circulation changes that may have important extra consequences not included in large-eddy simulation models (Johnson, 2005). Furthermore, the semidirect impact does not only apply to the absorption of aerosols. Semidirect effects were also seen in case studies of cumulus and stratocumulus, demonstrating a similar correlation between the sign of the semidirect impact and the height of the aerosol layer with respect to the cloud (Johnson, 2005).

By comparing the magnitudes of all these conflicting impacts using a global circulation model, Lohmann and Feichter (2001) discovered that the semidirect effect might be significant locally even though indirect effects predominate globally. Perlwitz and Miller (2010) investigated how dust influences cloud coverage using the Goddard Institute for Space Studies (GISS) climate model and discovered a counterintuitive decrease in low cloud cover due to heating by absorbing aerosols. In their trials, they employed aerosolized dust with different optical characteristics and a slab of water. They demonstrated that, with the exception of winter, the overall net effect of sufficiently significant dust aerosol optical depth and absorption is enhanced cloud cover. Cloud cover fell in some areas, including Eurasia, Western North America, and South America. Over the oceans, Central Africa, the Arabian Peninsula, India, and Southeast Asia, there was a significant increase in cloud cover. Over land, where the absorbing dust raised specific humidity due to increased moisture convergence driven by dust heating, the cloud cover rose. Overall, this effect is greater than the increased air evaporation resulting from the reduced humidity caused by dust absorption. Wintertime's predominant scattering of dust particles with low AOD showed poor or even reversed cloud cover enhancement effects.

Over Southern Africa, numerous calculations and observations of the semidirect aerosol effect have been made. Using the GISS climate model, Miller *et al.* (2004) discovered that dust loading increased low-level cloud cover and precipitation. Sakaeda *et al.* (2011) employed a slab ocean model and a 20-year run of the Community Atmospheric Model. Negative TOA semidirect radiative effects linked to increasing low cloud cover predominated over a weaker positive all-sky direct radiative effect over the ocean, where aerosol layers are typically found above clouds. On the other hand, reductions in cloud liquid water routes over land, where aerosols are frequently below or inside cloud layers, resulted in a favorable semidirect radiative impact that predominated over a nearly zero DRE.

The altitude of the absorbing aerosols in relation to the cloud and the type of cloud have the most effects on the amount of cloud cover. Cloud cover decreases when aerosols are embedded within the cloud layer. Convection and cloud cover may be improved by aerosols beneath a cloud layer. A cloud top is stabilized by an aerosol layer, which also tends to boost stratocumulus clouds while inhibiting cumulus cloud growth. For instance, Takemura *et al.* (2005) discovered that from a clean to a dusty environment, the areal coverage of shallow clouds rose by 20–40%. Since they facilitate deep convection and lower-level convergence as moisture moves from the ocean to the land, absorbing aerosols can also boost cloud cover in convergent areas. The majority of global model studies revealed regional differences in cloud response. Over oceans and some land areas, there is generally more cloud cover, with a net increase in lower-level cloud cover and/or a net decrease in upper-level cloud cover. As a result of absorbing aerosols, the clouds have a net adverse semidirect effect feedback.

Model simulations addressed the semidirect impact's potential significance, but observational data for the effect bolstered the conclusions by employing models. There aren't many publications that use observational data to analyze the Asian dust aerosol semidirect effect. Huang *et al.* (2006b) investigated the semidirect effect of dust particles on cloud water paths over East Asia using satellite observations. They discovered that the water path of clouds with dust contamination is much narrower than that of clouds without dust. Dusty clouds have mean ice water paths (IWP) and liquid water paths (LWP) that are, respectively, 23.7% and 49.8%, fewer than those of dust-free clouds. A substantial negative link between the dust storm index and the ISCCP cloud water path is also found in the long-term statistical relationship produced from the International Satellite Cloud Climatology Project (ISCCP) data.

The dusty cloud parameters from CERES are compared with those from dust-free scenarios in order to identify the cloud alteration brought on by dust particles. Based on surface station observations over Northwest China (30°N–50°N and 80°E–110°E), dust storms and dust-free meteorological conditions are distinguished. Based on the surface measurements, 33 dust storm cases for the 2-year CERES period from January to May were chosen. The weather was often chilly during those months, with low-level (<2 km) clouds reaching temperatures as low as about 260 K. Since dust storms typically occur near the margins of cold fronts, where surfaces are warmer than behind the fronts, the average surface skin temperature of the CLD regions for the 33 selected cases was 6.1°C lower than that for the COD region (Huang *et al.*, 2006b).

Figures 4.19(a) and 4.19(b) demonstrate the IWP and LWP histograms from the COD (gray bar) and CLD (black bar) datasets. The COD mean IWP and LWP are, on average, 23.7% and 49.8% lower than the comparable CLD values. Another distinction between these two groups is that, whereas only 22% of the CLD pixels achieve this requirement, 55% of the COD pixels have LWP $\leq$ 50 g/m^2 (Fig. 4.19(b)). The COD frequencies are lower than their CLD counterparts in the majority of LWP bins for LWP $>$ 50 g/m^2 (Fig. 4.19(b)). The moisture supply for cloud formation in CLD regions could not be greater than that in COD regions since the CLD and COD regions were in the same cold frontal systems and the COD regions were colder than the COD regions and situated in drier environments (even within deserts), except that the COD regions are found near the front's leading edge and could exhibit different characteristics from those found behind the front. In comparison to the COD region, the clouds that form in the CLD region should be thinner and contain less water. The observed results, however, diverge dramatically from the image provided by conventional meteorology. As a result, the significant decline in IWP and LWP values of COD clouds

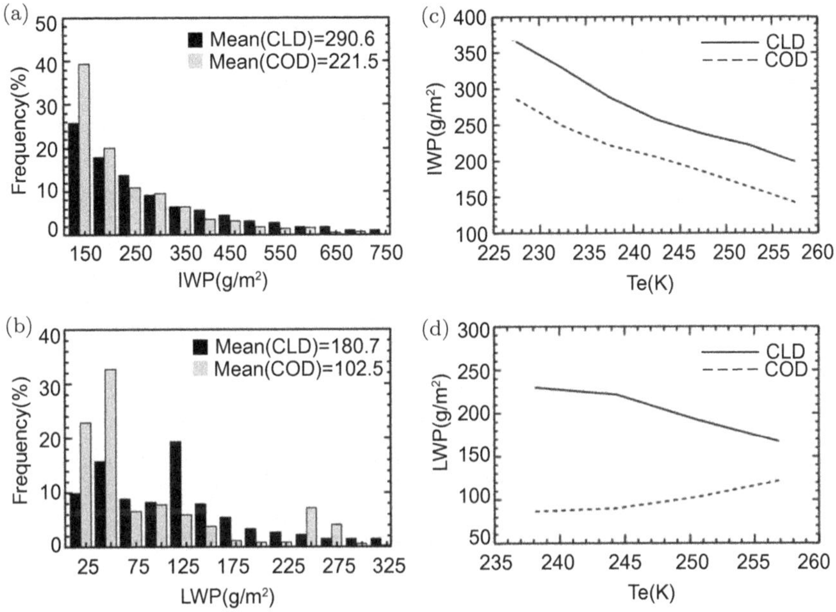

Fig. 4.19. Histogram comparison of the cloud water path over the dust-free cloud region (CLD, black bar) and overcast clouds over the dust region (COD, gray bar) for IWP (a) and LWP (b). The histogram intervals are 50 g/m^2 for (a) and 25 g/m^2 for (b). Comparison of the cloud water paths over the CLD region with the COD region as a function of effective cloud top temperature *Te* for IWP (c) and LWP (d) (Huang *et al.*, 2006b).

cannot be attributed to insufficient atmospheric water vapor or moisture transportation during cloud formation. This discrepancy shows that a key factor in the formation of clouds is the lower-layer cloud evaporation brought on by dust aerosol heating as a result of the aerosol's absorption of solar energy.

The IWP and LWP in the CLD and COD categories were displayed as functions of *Te* in order to compare cloud features under similar meteorological conditions (Figs. 4.19(c) and 4.19(d)). Over the whole range of observed cloud top temperatures ($225\,\text{K} < Te \leq 260$ K), the COD IWP (Fig. 4.19(c)) and LWP (Fig. 4.19(d)) are both lower than the comparable values for dust-free clouds. The change in IWP in Fig. 4.19(a) demonstrates that the IWP falls off as *Te* rises. All of the temperatures of the ice clouds exhibit the large IWP difference between the CLD and COD clouds. When $Te = 245$ K, for instance, the CLD IWP is about 250 g/m^2, whereas the COD value is about 200 g/m^2. Middle layer clouds (*Te* of about 250 K) are where the significant difference between dust-free and dusty CWP may be discovered for water clouds. While the fluctuations of CLD LWP are generally similar to those of CLD IWP and only decrease with increasing *Te*, the LWP values in the COD category increase with *Te*. This suggests that dust particles have a greater impact on LWP in middle-layer clouds than upper-layer clouds because high-upper-layer aerosol heating from solar radiation absorption may have a direct impact on cloud evaporation. The dusty CWPs appear to be significantly less than those of dust-free clouds in Fig. 4.19, which may be the result of cloud evaporation brought on by dust aerosol absorption or wet precipitation of dust.

The LWP and IWP for dusty clouds over Northwestern China were reported to be, respectively, 31.2% and 21.9% lower than those for unpolluted clouds by Wang *et al.* (2010b). According to these findings, dust particles might warm clouds, accelerate the evaporation of cloud droplets, and further shorten the cloud's water path. To compare the effects of aerosols, Huang *et al.* (2010b) selected two semi-arid regions, one in China and the other in the US, that were located at the same latitude and had comparable climatic circumstances. The authors discovered that the aerosols over China's semi-arid region had a greater optical depth and higher absorption, and demonstrated a more notable semidirect effect than those over semi-arid regions in the US. Over the arid and semi-arid parts of East Asia, the semidirect effect might predominate and interact with clouds in a way that reduces precipitation. (Huang *et al.*, 2006b, 2006c, 2010b).

4.5 Dust–Cloud–Precipitation Interactions

According to recent studies, these precipitation patterns were likely caused by aerosol impacts, which might prevent rain from falling in stratus clouds yet stimulate convective clouds to create heavy rain.

4.5.1 *Suppression of drizzle and light rain*

As a result, warm rain processes in stratocumulus clouds, small cumulus clouds, topographic clouds, and other clouds will be suppressed (Andreae *et al.*, 2004). Aerosol indirect and semidirect impacts are also predicted to diminish cloud particle size and potentially burn clouds. Observational data and model simulation results have been used to evaluate these theories. Yin and Chen (2007) used a two-dimensional spectral-resolving cloud model to simulate the effects of mineral dust particles on the development of cloud microphysics and precipitation over Northern China. They discovered that when dust particles acted as both CCN and IN at the same time during the development of a cloud, the heating brought on by increasing dust aerosol loadings would suppress precipitation. This is due to IN's potent suppressive impact almost completely outweighing CCN's boost. In the mid-latitude frontal zone over China, Kawamoto (2008) compared the amount of precipitation and non-precipitation water cloud properties and discovered that the relationships between precipitation and effective particle size/number density of droplets are consistent with the Twomey effect and aerosol scavenging by precipitation. It was also discovered that the reduction of precipitation caused by dust particles occurred in the other dust-prone semi-arid regions of West Africa. According to Min *et al.*'s (2009) study of precipitation in a deep convection cloud system over the Eastern Atlantic, the dust particles' microphysical impacts caused the precipitation size spectrum to shift from heavy precipitation to light precipitation.

On CCN, cloud droplets develop and lower the water vapor saturation barrier required for droplet formation. More cloud droplet embryos are fighting for available water vapor at higher aerosol concentrations. Changes in the distribution of cloud droplet sizes and numbers generate thermodynamic and microphysical feedbacks that could alter the evolution and characteristics of clouds. The formation of a positive feedback loop is a notable characteristic of dust–cloud–precipitation interactions over arid and semi-arid regions. A decrease in rainfall triggers the feedback loop, which causes a deficit in soil moisture. Dust storms appear more frequently as a result of this. As a result, dust particles in the atmosphere warm clouds, increase the evaporation of cloud droplets, and further decrease the cloud water path (the semidirect effect). Less rain falls as a result of the reduced water vapor content and low cloud cover. The frequency of dust storms would then rise, potentially resulting in even less rainfall. A schematic representation of this feedback loop is shown in Fig. 4.20.

4.5.2 *Increase of heavy rain*

The stories regarding aerosol–cloud–precipitation relationships are quite different when it comes to deep clouds or storms. Precipitation would be increased by the aerosol invigoration effect (Rosenfeld *et al.*, 2008). The hydrometeor profile of the cloud and phase shift brought on by dust–cloud interactions are additional microphysical factors that influence cloud dynamics and thermodynamics through latent heat release. According to Min *et al.* (2009), the combined effects of dynamic, thermodynamic, and microphysical cloud system processes can be seen in the vertical precipitation profile. Aerosol abundance and rain rate were compared in a study by Koren *et al.* (2012). The scientists discovered that, under a variety of circumstances,

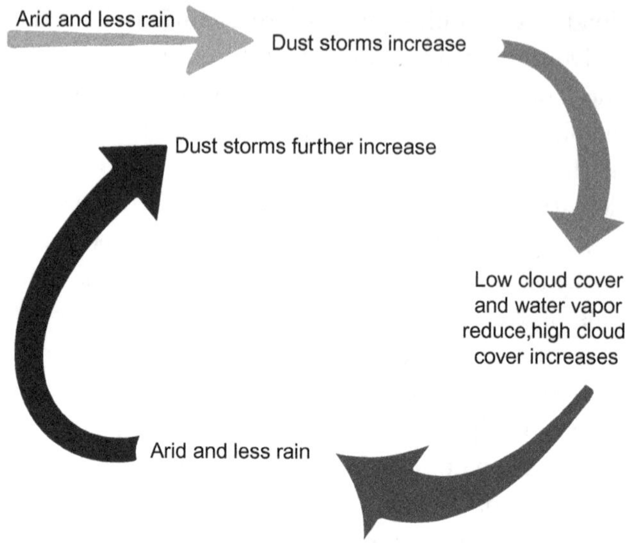

Fig. 4.20. A schematic depiction of the feedback mechanism between precipitation and a dust storm (Huang *et al.*, 2014).

a rise in aerosol abundance was linked to a local intensification of rain rate, as measured by the satellite's microwave radiometer for the TRMM. This association might be seen in the tropics, subtropics, and mid-latitudes, as well as over land and water. According to Rosenfeld *et al.*'s (2011) research, the East Asian desert dust and high levels of air pollution have the power to glaciate the tops of convective clouds as they form, resulting in ice precipitation rather than the suppression of warm rain. The researchers discovered that in deep clouds with high liquid water content, precipitation rose with aerosol concentration, while it reduced in clouds with low liquid water content. These observations were supported by simulations performed using a cloud-resolving model (Tao *et al.*, 2012).

When Qian *et al.* (2009) examined precipitation data gathered across China over nearly 50 years, they discovered a consistent pattern of declining drizzle and light rain frequency but increasing heavy rain frequency. Figure 4.21 displays the cloud microphysical characteristics from the three scenarios to demonstrate how aerosols suppress precipitation. It is qualitatively consistent with satellite measurements over EC and the distant oceanic area within the domain that cloud droplet number concentration (CDNC) increases by around 4–5 times, with a 10 times increase in aerosol concentration from the C-to-P scenarios, because collision and coalescence are substantially less effective when droplet sizes are much smaller. Figure 4.21

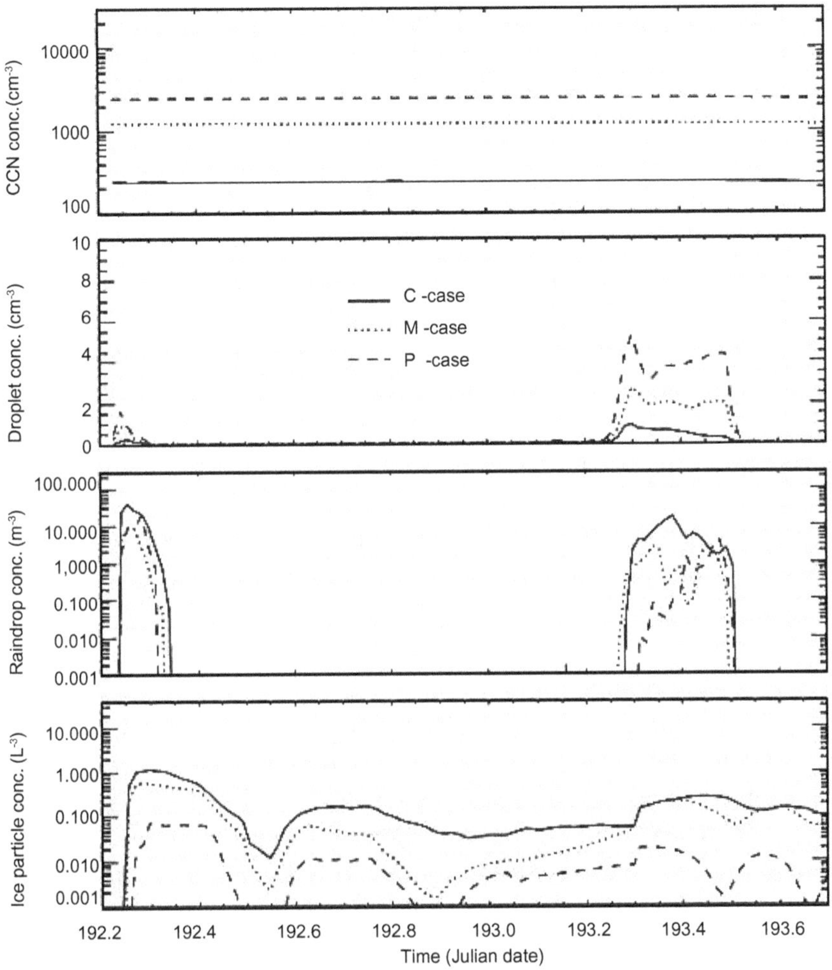

Fig. 4.21. Time series of CCN (unit: cm^3), cloud droplet (unit: cm^3), raindrop (unit: m^3) and ice particle (unit: L^3) concentrations for clean (C), moderately polluted (M), and heavily polluted (P) cases averaged over a subdomain, including the major cloud fields with a size of 450 km (Qian *et al.*, 2009).

illustrates that the raindrop concentration is significantly lowered and the generation of raindrops is delayed in the M and P cases. The P case's water drop radius (including cloud and raindrop) is reduced by up to 50% compared to the C case's, which is likewise compatible with satellite observations of CERW. According to Fig. 4.21, ice particle concentrations in the P case are roughly 10 times lower than those in the C case. Aerosols are not employed as IN directly; hence, in the simulations, ice formation is solely associated with ice supersaturation and supercooled droplets above the homogenous freezing level. Therefore, the much weaker convective strength in the M and P cases, which lowers ice production as ice supersaturation is reduced and fewer cloud droplets are carried to the upper altitudes, is primarily responsible for the lower ice particle concentrations in those cases. In fact, the greatest updraft velocity in the M case is 28.6 m/s, while it is 19.5 m/s in the P case, which is roughly two times lower than that in the C case. Since there are so many more tiny droplets, there is likely much stronger evaporation in the contaminated cases, which suppresses convection (Rosenfeld *et al.*, 2008). Additionally, the smaller size and higher hydrometeor concentration-related radiative effects of clouds may result in increased cooling and convection suppression. It should be noted that only the indirect effects of aerosols are considered in the simulations. By reducing surface heat fluxes, the inclusion of an aerosol's direct radiative effect would further restrict convection. High CCN concentrations were found to suppress convection and precipitation in both observational and modeling studies, and the significant decrease in convection and precipitation under the moderately and heavily polluted conditions is consistent with the processes discussed by Rosenfeld *et al.* (2008).

These model results raise the possibility that a 5–10-fold increase in CCN concentration can dramatically reduce low and median precipitation rates. It should be noted that some characteristics, including droplet size and precipitation rate, do not vary as much from the M case to the P case as droplet and ice number concentrations do because these variables are constrained by the total accessible water content. This is in line with earlier research, which found that in a polluted environment, cloud droplet size and precipitation rate become significantly less responsive to further increases in CCN concentrations.

4.5.3 *Dust–cloud–precipitation interactions over Tibetan Plateau*

Dust, which primarily originates from the Taklimakan Desert, is the most noticeable aerosol type over the TP and a significant component of global atmospheric aerosols (Huang *et al.*, 2007b). The TP is a significant dust source for long-distance transport when compared to other dust source regions in China because of its high altitudes, which make it easy for fine particles to be carried into the westerly jet stream. Strong ascending streams over the TP are also present during dust storms, which increases the concentration of dust particles of all sizes at all elevations. While finer

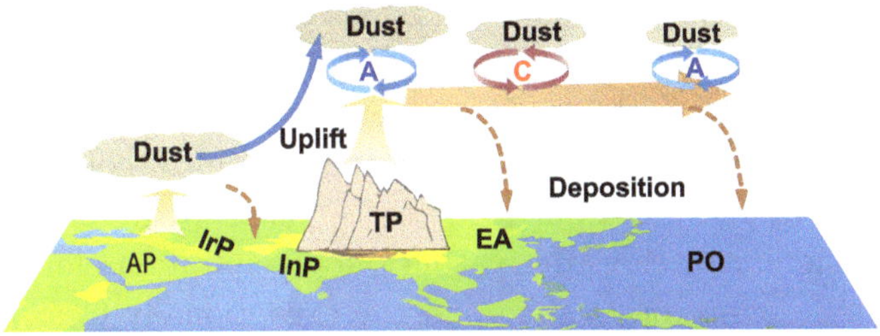

Fig. 4.22. Schematic diagram illustrating the formation of the dust belt over the Northern Hemisphere. The letters A and C denote the anticyclonic and cyclonic circulation centers, respectively (Xu *et al.*, 2018).

particles are carried by the westerly jet stream and dissipate in the Northern Pacific, lifted coarser particles mostly fall out and deposit as loess in the Eastern TP. In the spring, a dust belt that crosses the Pacific Ocean and extends to North America can be apparent at heights >6 km over the downwind direction of the TP (about 30°N–40°N) (Xu *et al.*, 2018). Such patterns are probably caused by the surface heating effect on local convection across the TP. Local convection enhances the TP as an aerosol source by controlling vertical aerosol exchange and pollution dispersion rates to the surroundings. The TP is a key region in the long-distance transfer of dust, serving as a primary conduit for dust transfer via convective ascent to the upper troposphere. Convective overshooting, which is a common spring phenomenon over the TP, can transfer dust particles from the mid-troposphere to the upper troposphere (Fig. 4.22), where they are then carried downstream by the subtropical westerly jet (SWJ). The Northern Hemisphere's long-range zonal transport of dust is thus made possible by the TP's role as a transport conduit from the lower atmosphere to the upper troposphere.

Cloud and precipitation processes can be easily influenced in relatively unpolluted atmospheric conditions, according to previous research. In comparison to other regions, the aerosol loading over the TP is significantly lower, which makes precipitation more sensitive to aerosol–cloud interactions.

During the time periods of 1961–2000 and 1800–2000, some people claimed that there were appreciable negative connections between dust aerosols and precipitation in dust source regions. It was discovered that while precipitation may not have much of an impact on dust storms, dust aerosols may have a significant impact on the hinterland of the TP. Investigations were also done into the connections between carbonaceous aerosols and precipitation. It has been proposed that the current warming may increase biogenic volatile organic compound emissions, which may increase secondary organic aerosols and lead to an increase in precipitation.

Previous research indicates that precipitation in cumulus clouds over the TP increases with aerosol concentrations, which not only promotes precipitation but also transports more ice-phase hydrometeors into the upper troposphere, as well as decreasing the precipitation efficiency. Due to the unique underlying topography and the exceptionally clean atmosphere over the TP, elevated aerosol concentrations might significantly accelerate convection. This mechanism may have an impact on the Asian summer monsoon. Additionally, earlier research has shown that absorbing aerosols over the TP may have a direct impact on monsoon rainfall via the elevated-heat-pump mechanism (Luo *et al.*, 2020). Recent regional analyses of the daily AOD during both wet and dry monsoonal years showed a significant correlation between relatively high aerosol concentrations and both cloud vertical development and precipitation. The 0°C isotherm was discovered to be higher on polluted days by 136.82 ± 18.82 m (mean $\pm$ standard deviation) than on relatively clean days. This finding may be important in influencing changes in the snow line and glacial melting, which will affect the hydroclimate of the Himalayas (Adhikari and Mejia, 2021).

Aerosol–cloud interactions above the TP affect downstream precipitation in addition to the TP's own precipitation (Fig. 4.23). Based on the analysis and simulation of an aerosol–cloud interaction event over the TP (Liu *et al.*, 2020c), we discovered that convective clouds contaminated by the Taklimakan dust over the TP's northern slope continuously move eastward and merge with convective cloud clusters along the motion path, causing significant precipitation over the TP's downstream regions. Additionally, Liu *et al.* (2020c) presented a dynamic northern drought mechanism linked to the TP in the summer. The SWJ's meridional shift is what causes the northern summertime drought. The upper-level westerly wind is diminished when the SWJ moves north, which reduces the frequency with which the water vapor, clouds, or dusty clouds over the TP are delivered to the north, reducing precipitation and increasing the frequency of droughts. On the other hand, the northern region of China sees more summer precipitation when the SWJ moves South.

4.5.4 *Case study for Indian dust-rain storm*

According to reports, on May 2, 2018, a deadly dust storm and thunderstorm ripped across Northern India (Fig. 4.24) (https://india.mongabay.com/2018/05/dust-storms-may-increase-in-india-due-to-climate-change/). The Indian

Fig. 4.23. Schematic of aerosol–cloud interactions and impacts on precipitation over the TP (Liu *et al.*, 2022).

Meteorological Department claimed that the instantaneous wind speed was high in Uttar Pradesh during this event, and the visibility in Delhi dropped steeply.

From MODIS composite images, the dust storm mainly originated from the Thar Desert; it observed a well-developed DCC (or thunderstorm) that was partially within the dust layer and covered Northwest India and Eastern Pakistan (Figs. 4.24(a) and 4.24(b)), which caused heavy rainfall. From the CALIPSO profile, the dust layer existed below 6 km over India, stretching to the northeastern margin of DCC (Fig. 4.24(c)). Dust and ice clouds coexisted over the foothills of the Himalaya (29.88°N, 79.6°E), where the convective clouds developed, and in the south, many SWCs intermingled with the dust layer. The dust storm moved to East India at night, and most of the dust mixed with anthropogenic aerosol and then lifted (Fig. 4.24(d)). Dust aerosol was transported to the TP across the Himalaya (Wang *et al.*, 2021), resulting in a dust layer appearing over the TP.

From 9:00 (15:30 local time) to 15:00 UTC, the cold air dominated this area, and the increasing incident solar radiation induced a low pressure. Being exposed to the cold front, wind speed and pressure significantly increased, bringing more dust aerosols to Delhi. When the DCC and dust

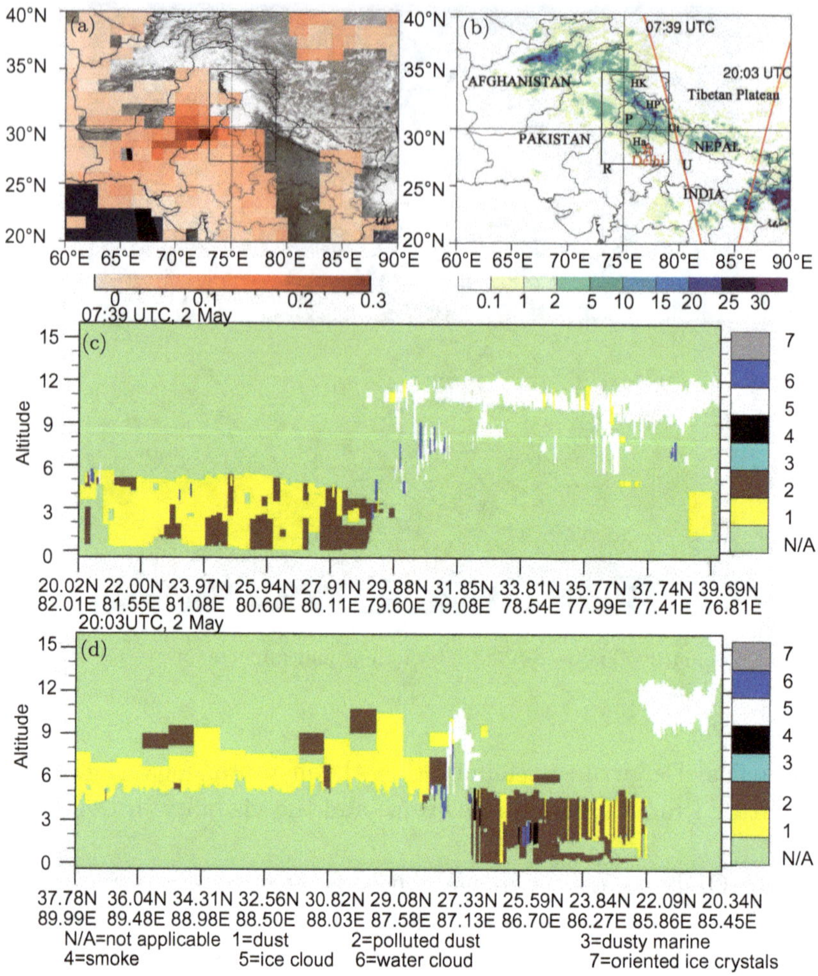

Fig. 4.24. A dust-rain storm observed by satellites. MODIS RGB composite image overlaid by OMI AAOD at 500 nm (a), GPM cumulative precipitation (unit: mm) from 00:00 to 12:00 UTC (b), CALIPSO aerosol and cloud types at 07:39 UTC (c) and 20:03 UTC (d), May 2, 2018. Orange-colored dots in (c) and (d) are SWCs selected from CALIPSO cloud products. A red star indicates the location of Delhi, red lines show the CALIPSO tracks, and a black box indicates the region this section focused on. Northern Pakistan and several states in India, including Jammu and Kashmir (JK), Himachal Pradesh (HP), Punjab (P), Uttarakhand (Ut), Rajasthan (R), Uttar Pradesh (U), and Haryana (Ha), were influenced by this event (Yuan *et al.*, 2021).

storm occur almost at the same time and area, we call it a "dust-rain storm." Here, we focus on the dust transport and development of deep convection in this area (the black box in Fig. 4.24; 73°E–79°E, 27°N–35°N, here referred to as the study region). This dust-rain storm is closely associated with the deep trough at 500 hPa over Kazakhstan. Cold air was transported from the Arctic region to Central Asia by the strong Northwesterly wind that formed behind the trough. A SW trough protruded into Afghanistan, and with the southward extension and deepening of the SW trough, sea-level high pressure increased at the Aral Sea. This weather pattern has strengthened continuously because of the strong sensible heating induced by solar radiation. Strong solar heating and a cold front provoked the dust storms, leading to high dust emissions over the Thar Desert on May 2.

Heavy precipitation is associated with proper dynamic conditions and an adequate vapor supply. The upper-level westerly jet will induce secondary circulation at the exit zone of the jet stream, providing vertical dynamics for the development of thunderstorms. There was a weak inversion layer below 850 hPa, and the cold advection started appearing below 700 hPa. After noon, the layer below 850 hPa reached a super-adiabatic state. The higher temperature dominated the Thar Desert and Northwest India; the humidity also decreased. This vertical thermal and moisture profile, combined with the uplift of secondary circulation and the cold front over the study region, triggered the convective cloud. The near-surface temperature showed a downward trend due to surface cooling at night, and then precipitation occurred. When the CAPE declines to its minimum, it represents the end of the rainfall.

The easterly wind from the Bay of Bengal increased the moisture at low levels of the atmosphere in the study region. The strong moisture input flux dominated the area; it declined over time, with the IWP of the total clouds increasing continuously, representing the development process of the cloud top. With considerable water supply, the thunderstorm occurred first in the west of the study region and then moved to the center. Then, it saw a remarkable enhancement because the increment of IWP was higher in the study region despite the decrease in moisture transport. Heavy rainfall occurred, matured, and began to move southeastward, then weakened.

The prevailing southwesterly wind in front of a low-level trough continuously delivered a large amount of dust aerosol from the Thar Desert toward the study region. The dust concentration experienced a significant increase, and some dust aerosol moved out of the study region and downstream (Fig. 4.24(d)). The dust transported to the study region can be lifted to higher altitudes by the vertical motion induced by the cold front, the upper-level westerly jet, and surface convergence (Fig. 4.25). Dust from the Thar Desert resulted in a dust layer at around 30°N. Supported by the converging ascending motion, a small portion of dust and massive moisture were lifted,

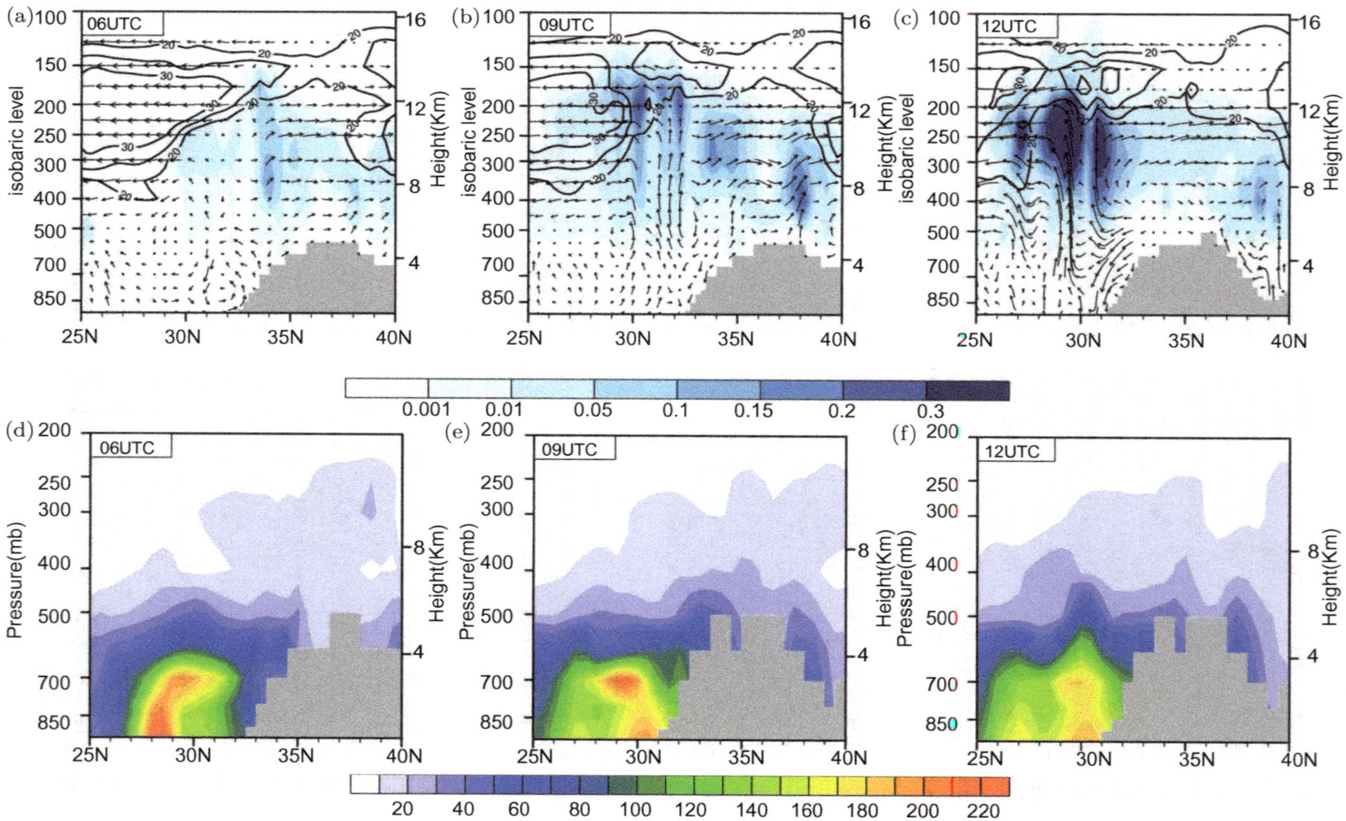

Fig. 4.25. Cross-section of cloud ice water content (color scaled, unit: kg/kg), meridional circulation (vectors, unit: m/s), and zonal wind >20 m/s (black contour lines, unit: m/s) obtained from ERA5 reanalysis data (top panel) and MERRA-2 dust concentration (bottom panel, unit: μg/m^3) along 75°E at 06:00 and 09:00 UTC and along 76.5°E at 12:00 UTC, respectively (Yuan *et al.*, 2021).

and although the dust concentration is weaker due to the wet removal, this special vertical structure of dust and IWC is still obvious. Thus, it is possible for dust to participate in the formation of convective clouds by acting as IN or CCN over the study region.

Figure 4.25 illustrates the reflectance of cirrus, the ice cloud effective radius, the supercooled cloud water path, and the AOD observed by MODIS. As the AOD and ice cloud reflectance in the southwest margin of a thunderstorm increased between two time periods, the corresponding ice particle radius decreased. The supercooled cloud water path in the center also faded with time.

As a result, the engagement of dust aerosol IN may be partially responsible for the decrease in ice radius and increase in cirrus reflectance in this DCC by changing the cloud phase. Huang *et al.* (2006c) argued that dust aerosols from the Taklimakan Desert resulted in the decrease in IWP over an arid or semi-arid region, and the semidirect effect or cloud evaporation contributed to this decline. In this study, the IWP showed an upward trend, and contrary trends may have resulted from sufficient moisture from the Bay of Bengal and strong vertical motion. Dust and anthropogenic aerosols tend to amplify convection-generated ice clouds, whereas smoke aerosols have high solar-absorbing capabilities, resulting in high evaporation and convection suppression.

4.6 Conclusions and Discussion

In this chapter, we discussed the effects of dust–cloud–precipitation interactions on a semi-arid climate. The findings indicate that the Taklimakan Desert and the Inner Mongolian Gobi Desert are the two main sources of Asian dust. Of the dust emission, 20% is transported on a regional basis, while about 30% is redeposited in deserts. A portion of the dust crosses the Loess Plateau and Hexi Corridor on its way to Southeast China. The remaining 50% travels around the globe by interacting with westerlies, moving from the continent to the ocean near Korea and Japan and even reaching Eastern North America.

Dust aerosols have a cooling impact on the surface, a nearly warming effect or no effect at the TOA and a warming influence within the atmosphere over the arid and semi-arid regions of Asia. Over or close to the dust sources are where the highest forcing values are found. Since they are more easily contaminated along their paths, dust particles from Asia are more absorbent than those from the Sahara Desert. The macro- and microphysical characteristics of clouds could be subtly altered by dust aerosols. While the invigoration effect and semidirect effect have positive forces on the climate, the first and second indirect effects are typically negative. There is still a lot of uncertainty surrounding how to quantify those opposing effects. Depending on a variety of variables, including the vertical distribution of dust particles, the amount of humidity in the air, and the kind of clouds, dust aerosols can either encourage or restrict precipitation. In deep clouds with a high liquid water content, the volume of rain rises with aerosol concentration but falls in clouds with a low liquid water content. Regarding the prevention of precipitation, interactions between clouds, dust aerosols, and precipitation have positive feedback (Fig. 4.20). In arid and semi-arid areas, dust storms can happen more frequently, which may have expedited the formation of more arid conditions over Asia's arid and semi-arid regions and contributed to the desertification that has been witnessed in recent decades. A schematic depiction of the interactions between dust, clouds, and precipitation in arid/semi-arid and downwind (wet) regions is shown in Fig. 4.26.

According to recent research, dust from arid areas may globally affect the microphysics of warm and cold cloud formation (Huang *et al.*, 2010b). It is still up for dispute how dust aerosol impacts and related feedbacks affect cloud properties, rain formation, and cloud lifetimes. While some research revealed that the system is effectively buffered, some reported that the impacts are rather significant. Owing to our limited understanding of aerosol–cloud interactions, aerosol distribution, and the physical and chemical properties of aerosols, for instance, there is significant uncertainty in existing estimates of aerosol RF. Due to a lack of data, the contribution of anthropogenic mineral dust to global dust emissions has never been explicitly assessed; there are still significant uncertainties in the estimates of the relative contributions of human and natural dust sources (Ginoux *et al.*, 2010). Some aspects of the relationships between aerosols, clouds, and climate have been qualitatively confirmed, such as the suppression of precipitation by aerosols in warm rain, but some, such as the relationship between aerosol and surface precipitation, are still incredibly uncertain in terms of their magnitude and even their qualitative relationships (the effects of aerosol on cloud fraction, cloud LWP, and cloud lifetime). Our ability to comprehend and anticipate the impacts that anthropogenic activities have on regional and global climate change is hampered by the complexity of the interactions between aerosols and clouds (IPCC, 2007).

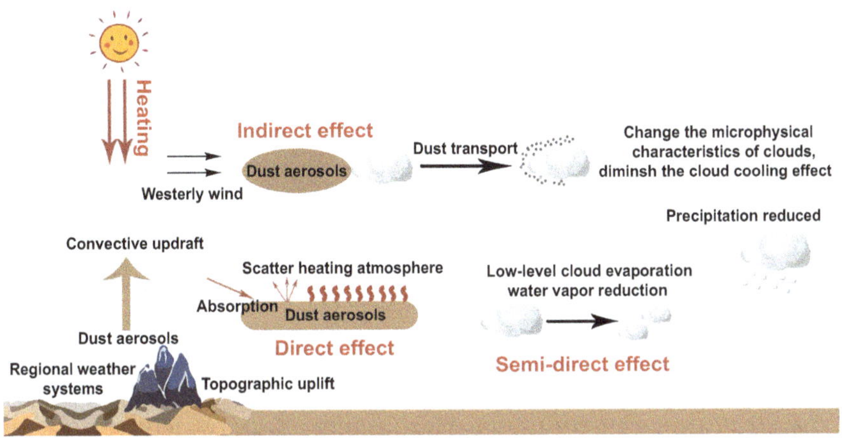

Fig. 4.26. A schematic diagram summarizing dust–cloud–precipitation interactions in arid/semi-arid and downwind (wet) regions (Huang *et al.*, 2014).

In order to better understand how South Asian dust aerosol affects DCCs, this chapter also analyzes a typical pre-monsoon dust-rain storm (a severe dust storm accompanied by a thunderstorm) that occurred in Northwest India on May 2, 2018. It does this by combining available satellite retrievals, *in situ* measurements, and reanalysis data. We concluded that South Asian dust may energize the convective clouds and boost the rainfall through IN formation based on the analysis of dust events and microphysical cloud characteristics progression. This can be explained by the fact that the SWC sharply decreased when the dust AOD, IWP, and cloud phase showed a trend in the opposite direction, which indicates that dust particles are probably acting as IN and glaciating liquid- or mixed-phase clouds. By bringing the moisture together, the latent heat generated by the phase-change process would intensify the thunderstorm and lead to significant rainfall.

Although the possible impact of dust acting as IN on DCCs was examined in this chapter, we cannot completely discount the significance of large-scale dynamics and vapor supply entrainment, which control the formation of DCC and a variety of deep cloud microphysical processes. We also acknowledge that the dust CCN may contribute to this process; however, it may not be as significant as the dust IN given the significant change in cloud phase and the rise in dust aerosol. The uncertainties resulting from the data may potentially have an impact on our findings. Although assimilating the observed AOD reduces the systematic bias between MERRA-2 and observation, the AOD assimilation is unable to constrain aerosol species or their vertical structure and has only a limited impact on systematic bias reduction. According to Li *et al.*'s (2009) hypothesis, one of the major causes of inconsistencies between MODIS's retrieval of aerosols and clouds and other active sensors, such as CALIPSO and CloudSat, is cloud filtering. CALIPSO daytime products are less dependable than their equivalents at night because of the influence of sunlight. Although CERES cloud attributes are reasonably accurate, certain significant differences in some parameters, such as liquid water path, cloud phase, and cloud-top height change, are likely to lead to ambiguity in our results (Minnis *et al.*, 2011).

This chapter summarized the various pathways through which dust affects a semi-arid climate. Due to the interaction between different temporal and spatial scales, the results in this chapter vary widely among them, and the research results are highly uncertain. We proposed a possible dust aerosol effect on a semi-arid climate, which should be better evaluated by accurate and sufficient measurements and high-resolution model simulations in a future study. With the continuous improvement of observational tools and the increasing refinement of numerical models, we believe that research in this area will continue to improve. We think that the systematic study of the dust belt from a global perspective is still insufficient. The main problems are as follows: (1) There are insufficient key observation sites in typical regions, poor knowledge of the vertical distribution of dust aerosol concentrations and their radiative properties, and limited accuracy of satellite data inversion, all of which limit the ability of numerical models to simulate dust processes, requiring in-depth research. (2) At present, there is still no international numerical model that can realize the simulation and forecasting of the sand and dust processes along the Belt and Road, mainly because of the lack of refinement of existing models, the lack of observational data, and the great uncertainty in the sand initiation scheme and dust identification. Therefore, there is an urgent need to carry out systematic research integrating observations and simulations for the sand and dust belt along the Belt and Road. To address these problems, we are building a LIDAR observation network along the Belt and Road, which combines satellite remote sensing data with the existing numerical model, and the observation data are used to optimize the model parameterization scheme. The model will provide scientific support for climate response and environmental protection and key observational data for an in-depth study of the long-range transport of dust and sand aerosols, clarifying the dust–cloud–precipitation relationship and revealing the effect of dust and sand aerosols on warming and drought in arid and semi-arid regions.

Chapter 5

Impacts of Land–Atmosphere Interactions

5.1 Introduction

One of the most fundamental facets of climate change research is the study of land–atmosphere (L–A) interactions, which regulate the exchanges of energy and mass fluxes between the surface and the atmosphere. The rates of heat, water vapor, and solar energy fluxes between the land and atmosphere have an impact on state variables that characterize climate, like air temperature, humidity, precipitation, and net radiation (Baldocchi *et al.*, 1997). They modify the near-surface climate, which includes clouds and precipitation, to affect how long extreme conditions like drought last (Section 5.2). Over a wide range of spatiotemporal scales, land surface features, such as land cover, terrain, and soil texture, are highly diverse. In their review of some of the varied traits of tropical continental climates and how land–atmosphere interactions control them in the tropics, Gentine *et al.* (2019) found that water plays a crucial role in regulating surface fluxes and land–atmosphere interactions in savanna and semi-arid climates, whereas light and energy are typically more limiting than water supply for photosynthesis and evapotranspiration (ET) in tropical rainforest climates. Increased latent heat fluxes over continuous permafrost in Eurasia cause a rise in boundary layer moisture, which in turn causes an increase in precipitation and low-level cloudiness. Weaker extra-tropical cyclones increased incident solar radiation and precipitation after the mid-1990s. The increased precipitation weakens the surface evaporation and local cooling impact, which in turn causes a warming of the land surface for the middle latitudes across East Asia (Chen *et al.*, 2020). In arid and semi-arid regions, the reduced soil moisture regulates atmospheric circulation and vertical ascent to enhance moisture transport into drylands, which benefits the increase of P–E. Therefore, soil moisture has negative feedback on P–E, which can offset about 60% of the decrease in P–E otherwise expected in the absence of soil moisture feedback over arid and semi-arid regions. In order to improve moisture transport into arid and semi-arid regions, the decrease in soil moisture will control air circulation and vertical ascent (Zhou *et al.*, 2021a).

In arid, semi-arid, and subhumid locations, the changes to the land surface brought on by droughts can result in some positive feedbacks that strengthen and prolong the droughts. Due to the considerable response of evaporation to increased soil water stress, droughts in arid and semi-arid regions are particularly prone to self-replication. In arid and semi-arid regions, upwind drought can reduce precipitation by more than 15% during a single event, and up to 30% during individual months (Schumacher *et al.*, 2022). Flash drought is a type of drought that develops quickly and is extremely destructive. It is highly influenced by the land–atmospheric interactions. On the one hand, an increase in sensible heat transfer during the commencement of a flash drought can intensify atmospheric heating and increase the lifting condensation level deficit, which effectively prevents convective precipitation. On the other hand, as the atmosphere dries, the need for evapotranspiration rises considerably, soil moisture falls, and the onset of drought is accelerated. Drought and heatwaves are strongly influenced by land-atmosphere interactions. Previous land surface conditions modify the local distribution of available energy between sensible and latent heating, which amplifies temperature anomalies during heat waves. Discovering where and when antecedent land surface conditions contribute to severe heat in Australia is made easier by understanding the spatiotemporal variations of land–atmosphere interactions.

Ecosystems of global arid and semi-arid regions, with high biomass turnover rates, have contributed the largest portions to the interannual variability (IAV) of net CO_2 flux over recent decades (Ahlström *et al.*, 2015), accounted for about 40% of the global land net primary production (NPP), and dominated the positive global land CO_2 sink trend. Variations in gross primary output have been largely responsible for this carbon sink fluctuation (GPP). Ahlström *et al.* (2015) demonstrate that the trend and IAV in CO_2 sink by terrestrial ecosystems in 1982–2011 were dominated by semi-arid ecosystems in response to changes in both precipitation and temperature. They analyzed this by using an ensemble of ecosystem and land surface models as well as an empirical observation-based product of GPP. Ecosystems of arid and semi-arid regions are highly sensitive to drought and wet events because they are water-stressed biomes, which cause abrupt, major changes in ecosystem structure and functioning and consequently high carbon exchange variability (Ponce-Campos *et al.*, 2013). Due to annual precipitation exceeding a threshold, such carbon exchange variabilities can even flip from being a net source to a net sink from year to year (Beer *et al.*, 2010). Positive arid and semi-arid regions' GPP anomalies are predominantly driven by wet events (cold and wet circumstances), whereas negative arid and semi-arid regions' GPP anomalies are strongly connected with drought events (warm and dry conditions). For instance, the increased carbon uptake by ecosystems of Australian arid and semi-arid regions during wet periods was mostly responsible for the global carbon sink anomaly in 2010–2011 (Ma *et al.*, 2014; Poulter *et al.*, 2014). Additionally, these Australian ecosystems have repeatedly experienced carbon sink episodes, indicating that under future climate change, ecosystems of arid and semi-arid regions would have a major influence on the variability of the global carbon cycle (Ma *et al.*, 2014). Here, we describe how changes in regional and subtype contributions to global arid and semi-arid regions' GPP variability, as a result of future climate change under RCP4.5, are caused by projected arid and semi-arid regions' expansion and degradation (Section 5.3). We separated the world's arid and semi-arid regions into eight regions — North America, South America, Europe, Africa, West Asia, East Asia, South Asia, and Australia — in order to calculate regional contributions to the global arid and semi-arid regions' GPP (Table 5.1).

Because of its crucial function in regulating the water and energy balance between the land surface and the atmosphere, soil moisture is widely acknowledged as a fundamental parameter in environmental processes (Seneviratne *et al.*, 2010). Although it has distinct definitions in different

Table 5.1. Latitude/longitude information for the eight regions (Yao *et al.*, 2020).

	Region	Latitude	Longitude
1	North America (NAM)	10°N–80°N	180°W–30°W
2	South America (SAM)	60°S–10°N	180°W–30°W
3	Europe (EU)	30°N–70°N	30°W–40°E
4	Africa (AF)	60°S–30°N	30°W–60°E
5	West Asia (WAS)	30°N–70°N	40°E–80°E
6	East Asia (EAS)	30°N–70°N	80°E–180°E
7	South Asia (SAS)	10°S–30°N	40°E–180°E
8	Australia (AUS)	50°S–10°S	110°E–160°E

domains, such as hydrology, meteorology, and ecology, soil moisture generally refers to the amount of water stored in the unsaturated soil zone. The planetary boundary layer regime, clouds, and precipitation are all significantly influenced by soil moisture through its effects on evapotranspiration rate, soil thermal parameters, surface albedo, and Bowen ratio (Guan *et al.*, 2009). Meanwhile, soil moisture affects plant productivity and survival as provides a direct supply of water for vegetation (Reichstein *et al.*, 2007). Therefore, to comprehend climate change and the interactions between the ground and the atmosphere, it is crucial to research soil moisture and its variability. We examine soil moisture patterns and report on the connections between soil moisture trends and climate regions in Sections 5.4.2, 5.4.3, and 5.4.4. We also investigate the processes that cause temperature and precipitation to vary soil moisture and cause the soil to dry out. The seasonal and daily soil moisture in the surface is very important to climate simulations. One of two major strategies to increase the accuracy of atmospheric numerical models' predictions is to improve the land surface parameter (Dickinson, 1995). However, uncertainties in the meteorological forcing and parameter inputs, as well as deficiencies in the model physics, have an impact on the quality of the model output. In particular, the dynamic range of the complex variable known as soil moisture is determined by the exact precipitation, evaporation, and runoff formulations used in a given mode. Inaccuracies in the structure of the model, the initial circumstances (or ensemble runs), and the applied greenhouse gas emission scenarios all contribute to model uncertainties. In Section 5.4.5, we evaluate the effectiveness of the CMIP5 and CESM in simulating soil moisture and investigate the ambiguities in the model's initial circumstances and structure.

For the advancement of land surface models, hydrological monitoring and forecasting, and water resource management, accurate observation is crucial. On the surface of the earth, there are many arid areas. The largest arid and semi-arid region in China is the Loess Plateau. Chinese loess is a common wind-deposited material in Northern China. The unique land surface physical processes in this area have an impact on China's monsoon circulation in addition to the local climate and atmospheric circulation. However, because these regions often have few ground monitoring systems and little attention has been paid to long-term observations over these areas, an adequate climatic characterization of precipitation features is absent. The Semi-Arid Climate and Environment Observatory of Lanzhou University (SACOL), which was founded in 2005 to close this gap, offers long-term and ongoing observations over the semi-arid Loess Plateau region in Northwest China (Huang *et al.*, 2008a), giving us the chance to better understand climatology in arid and semi-arid regions. Sections 5.5.2 and 5.5.3 examine the seasonal fluctuation of climatic variables and energy flux over semi-arid regions of Northwest China using a series of 2 years of continuous observations (from 1 January 2007 to 31 December 2008). Deserts are another environment that demands our attention. Despite appearing lifeless and ignored for a long time, they have silently contributed significantly to the global carbon cycle. The second-largest changing desert in the world is the Taklimakan Desert (TD). The TD has an enormous capacity for carbon storage thanks to its considerable salinization and plentiful saline–alkali groundwater. In Section 5.5.4, we consider the size of the contribution made by each internal process to the TD carbon sink as well as how these processes will adapt to climate change.

The arrangement of this chapter is as follows. Section 5.1 is the introduction to this chapter. Section 5.2 illustrates the definition and characteristics of land–atmosphere interactions. Section 5.3 discusses global and regional semi-arid regions' GPP changes, and Section 5.4 introduces variability of soil moisture on different timescales. Section 5.5 shows the data analysis of land–atmosphere interactions. Conclusions and the discussion are summarized in Section 5.6.

5.2 Land–Atmosphere Interactions

The hydrologic cycle and climate are significantly influenced by interactions between the land and the atmosphere. On spatial and temporal scales, the atmosphere and the land interact. Terrestrial components are essential to atmospheric processes, such as the rise in atmospheric humidity and the accompanying increase in the atmosphere's capacity to store water. Examples include evapotranspiration and soil moisture. These changes in the atmosphere are the root cause of extreme weather phenomena like droughts and floods. Vegetation dynamics are significantly impacted by the aerodynamic mechanisms that control surface-to-atmosphere exchange activities inside the atmospheric boundary layer. Aerosols and other large-scale atmospheric processes have a considerable impact on the planet's energy balance. Large-scale atmospheric processes as well as long-term processes of land–atmosphere interactions are crucial for understanding climate change and its future implications. Using regional and global climate models, researchers examine how solar radiation, aerosols, and greenhouse gases affect the mechanisms governing land–atmosphere interactions. We should first comprehend the definition and characteristics of land–atmosphere interactions before talking about their effects.

5.2.1 *Definition of land–atmosphere interactions*

The soil moisture and precipitation linkages have been the primary focus of the climate community's attempts to understand land–atmosphere interactions. These interactions are regarded as being between the land surface and the air column above it since they are local in space. The state variables of the land surface, the state variables of the planetary boundary layer, and the state variables of the open atmosphere that allow for the modulation of clouds and precipitation are all mutually influenced by this interaction, which can be described as local. Figure 5.1 shows a variety of physical processes that are schematically involved in these interactions and eventually result in the observed link between soil moisture and precipitation. Even though this interaction between the land and the atmosphere is schematically depicted, some understanding of this interaction exists. The diagram shows

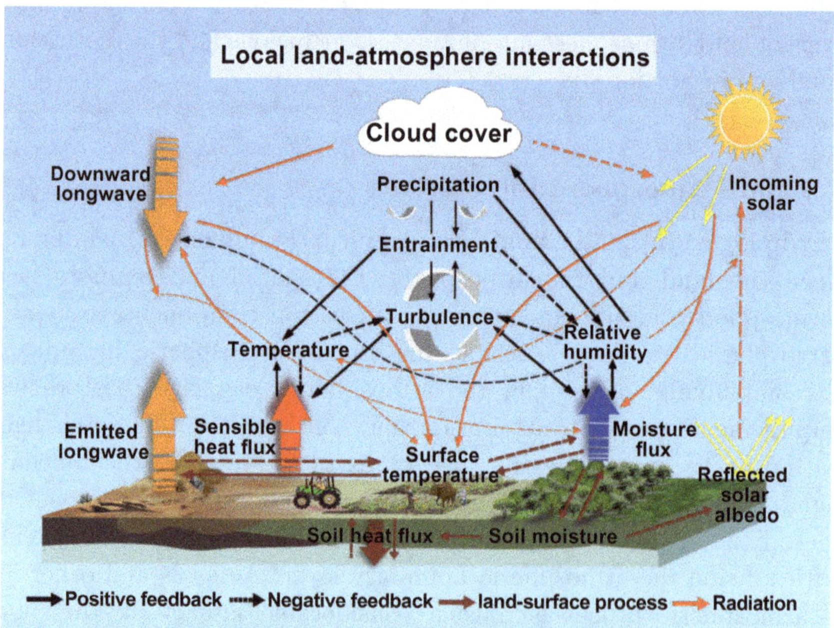

Fig. 5.1. A schematic illustration of the involved physical processes in the local land–atmosphere interactions under quiescent weather conditions. The thin red lines and dotted gray lines with arrows also represent positive feedback just as much as the solid black lines with arrows.

the process chain that connects soil moisture to precipitation and the surrounding weather. The increase in soil moisture leads to the increase in evapotranspiration, which further causes enhanced entrainment flux and accelerates the formation of clouds and precipitation. Positive and negative feedback processes can be found among the several physical processes depicted in Fig. 5.1. The difference between both feedbacks is that both variables change similarly in a positive feedback process as opposed to changing in opposite ways in a negative feedback process. For instance, a rise in sensible heat flux and/or moisture flux can cause a rise in boundary layer turbulence, or a rise in boundary layer turbulence can cause a rise in sensible heat flux and/or moisture flux. The increase in air temperature heated the atmosphere, which in turn increased the amount of longwave radiation that the atmosphere sent downward, raising the air temperature as a result. This is an analogous example of positive feedback. On the other hand, the Fig. 5.1 example of a negative feedback pathway shows how an increase in the entrainment of air from above the boundary layer can result in a decrease in the relative humidity of the boundary layer. An increase in evaporative flow that can result in a decrease in surface area is another illustration of negative feedback. Similar interpretations can be made of other feedback loops in Fig. 5.1.

In arid and semi-arid regions, land–atmosphere interactions influence the energy balance, water cycle, and carbon cycle through turbulence at the planetary boundary layer (PBL). The driving role of oceanic warming and related atmospheric processes is emphasized in many proposed mechanisms for the climate changes that occur in arid and semi-arid regions, but it should be emphasized that the aridity index response is significantly amplified by land–atmosphere feedbacks that react to climate changes (Berg *et al.*, 2016). It is obvious that understanding how the energy, water, and carbon cycles respond to global warming is essential for understanding and projecting future changes to the climate of arid and semi-arid regions.

5.2.2 *Energy balance*

In the connection between the atmosphere and the land, the energy balance must be met. The net radiative heat gain, or net radiation, must be counterbalanced by other types of heat transfer at a specific place. The expression for this thermal balance is (Burba *et al.*, 1999)

$$R_n = \mathrm{LH} + G + \mathrm{SH}, \tag{5.1}$$

in which LH is the latent heat flux, G is the conduction to or from the subsurface, and SH is the sensible heat flux (Nicholson, 2011).

Depending on the type of surface, the surface's contribution to atmospheric heating is divided into sensible and latent heat (Kool *et al.*, 2016). Due to the unique characteristics of water, the availability of surface moisture in conditions of increasing atmospheric water demand has a significant impact on this partitioning (Sherwood and Fu, 2014). Large closure errors are unacceptable since the sensible heat flow (SH) and the latent heat flux (LH), the fundamental tenet of all atmospheric models, theoretically balance the available energy ($R_n - G$). However, the land cover conditions and atmospheric evaporative demand determine how much energy is divided between SH and LH via the net radiation (R_n) (Guo *et al.*, 2016). LH normally increases more under warm temperatures, although SH generally increases with surface sun radiation in humid environments. In contrast, LH is constrained by the availability of soil water, while SH contributes a greater share of R_n in arid or semi-arid environments (Wang and Dickinson, 2013).

Low mean latent heat flow across arid and semi-arid regions is caused by low precipitation and soil moisture content, which are inputs to arid and semi-arid region water cycles and limit evaporation and transpiration associated with low vegetation cover (Yin *et al.*, 2014) (Fig. 5.2). Surface temperatures over arid and semi-arid regions must rapidly increase to provide a significant land–air temperature difference in order to release the energy from solar and infrared radiation through sensible heat fluxes. In contrast

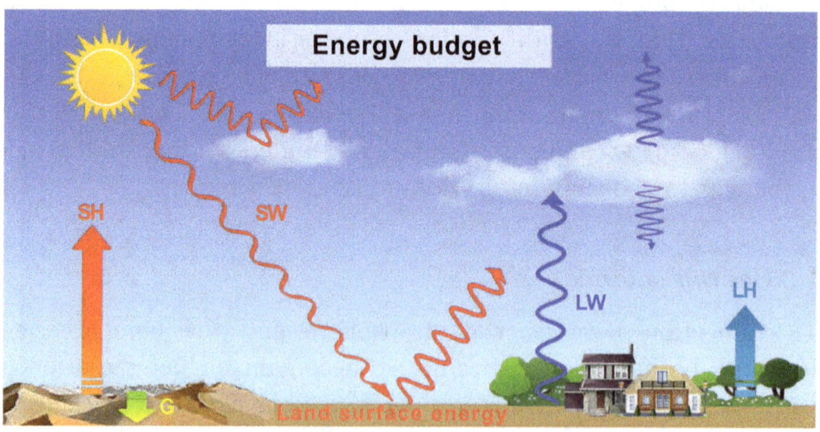

Fig. 5.2. A schematic illustration of the surface energy budget for arid and semi-arid regions. The SH, LH, SW, LH, and G represent sensible heat flux, latent heat flux, shortwave radiation, longwave radiation, and conduction, respectively.

to humid areas, this difference results in a significant sensible heat flux. The significant land–air temperature difference over arid and semi-arid regions is also indicated by the high upward net longwave radiation. Due to the absence of precipitation, arid and semi-arid regions are primarily covered with high clouds, which absorb a significant portion of the upwardly directed longwave radiation. The cloud heats up as a result of this process and begins to radiate infrared energy into the surrounding atmosphere as well as the earth. Additionally, some of the solar radiation that strikes the Earth is reflected back into the atmosphere by the ground, and some of it is absorbed by clouds. As a result, the low evapotranspiration caused by insufficient soil moisture and vegetation over arid and semi-arid regions is the main reason for the high sensitivity of surface temperatures to additional heating from rising GHGs.

5.2.3 *Water cycle*

A crucial process in the interactions between the atmosphere and the land is the water cycle, which defines the continual movement of water through the climate system. The water cycle is controlled by variations in soil moisture, ET, P, and runoff. Global ET, P, and runoff have changed due to global warming and human activity, which suggests that the global water cycle has generally intensified or accelerated (Wentz *et al.*, 2007). Asynchronous changes in these elements, however, could result in hydrological shortages and harmful ecological effects.

When water from the earth's surface rises into the atmosphere, cools, and then condenses into clouds as rain or snow, this process is known as surface water evaporation. The precipitation then falls again on the earth's surface. Rainwater that falls on land collects in lakes, rivers, soil, and porous rock strata below the surface. Most of this water then returns to the oceans, where it will once more evaporate and circulate through the atmosphere to carry water vapor to land. This is the energy exchange that impacts the climate. When water condenses, energy is released, warming the surrounding area. The energy that is taken from the environment as water vaporizes causes temperatures to drop. As a result, a crucial component of Earth's weather patterns is the cycling of water into and out of the atmosphere. Water is scarce in the arid and semi-arid regions and is mostly obtained from iceberg snowmelt and soil evaporation. The vegetation in the wetlands is abundant, and there is significant lake and river evaporation and plant transpiration. The important part of the water cycle is evapotranspiration (Fig. 5.3).

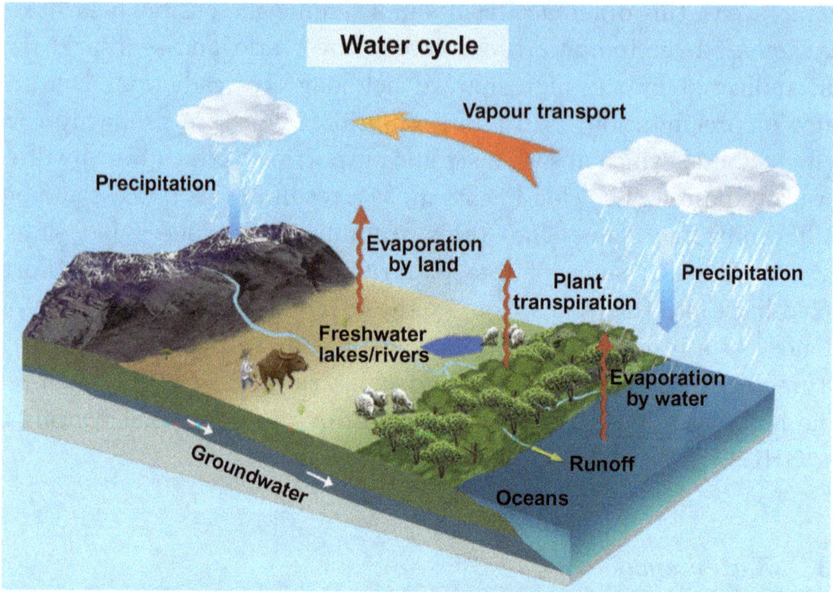

Fig. 5.3. The cycling of water. The global water cycle shows where water is stored on the Earth and the major processes that move water between the atmosphere, land, and ocean.

In many hydrologic and land–atmosphere interactions, soil moisture, a crucial component, is impacted by variations in precipitation and evapotranspiration, and from 1948 to 2010, there was a discernible decreasing trend over East Asia in comparison to the global mean (Cheng *et al.*, 2015). Transitional zones between dry and wet regions experience the most noticeable drying (Cheng and Huang, 2016). Above all, increased soil drying in these transitional regions was brought on by global warming through soil moisture–temperature feedback. In East Asia, a region with noticeable dry spells, falling precipitation is the main cause of soil drying, which is amplified nearly twofold by rising temperatures. Simultaneously, soil suction rises with falling soil moisture, making the remaining moisture less available to be absorbed by plant roots. As a result, evapotranspiration may be reduced, thus increasing the sensible heat flux and further raising temperatures. A positive feedback loop is created by the rising temperature, reduced evapotranspiration, and water loss from the soil. This loop might continue until the soil is completely dry and desertification occurs (Seneviratne *et al.*, 2010).

Numerous studies on the global land–atmosphere coupling have shown that transitional regions (such as semi-arid and semi-humid regions) exhibit the strongest coupling between soil moisture and precipitation, as measured by the omega difference, because their surface climates are most sensitive to soil moisture (Seneviratne *et al.*, 2010). Only in these transitional zones may soil moisture impact precipitation. Evaporation is appropriately high but still sensitive to soil moisture, and soil moisture can in fact cause wet convection. In areas with transitional climates, soil moisture regulates evapotranspiration and near-surface temperature simultaneously by dividing sensible and latent heat fluxes. Because of the strong linkage between soil moisture and precipitation in transitional zones, the higher local surface temperature brought on by large-scale circulations can also bring about an increase in temperature by increasing the energy partitioning to sensible heat. These processes may accelerate the drying trend and its interactions, resulting in ever drier arid and semi-arid regions. Because of this, locations with transitional climates are more severely affected by observed and anticipated warming and drying trends. Transitional areas are typically populated by a significant number of agricultural districts. Their habitats are typically delicate and vulnerable to climate change; even a small shift in the climate is likely to have disastrous effects. The ecosystem and human survival would be severely threatened by the most severe warming and drying. To lessen the severe effects of drying, strict management and logical water resource use plans are urgently required.

Despite their effects on one another, vegetation–atmosphere interactions and dust–climate feedbacks are equally significant processes in local land–atmosphere interactions. Reduced soil moisture under certain wind conditions can significantly increase dust outbreaks (Kim and Choi, 2015). The potential of atmospheric dust to diminish precipitation might help create a positive feedback loop for desertification whereby decreased vegetation cover leads to increased dust emissions and plant mortality (Huang *et al.*, 2014). The ecological environment would be destroyed, crop productivity would be decreased, and dust and drought would result. A warmer climate is expected to cause changes in the water cycle. In comparison to other parts of the water cycle, estimates of soil moisture and drought at regional and global scales remain mostly unclear.

However, expected changes in the Hadley Circulation are consistent with drying in the Mediterranean, Southwestern United States, and Southern African regions. Therefore, under RCP8.5, drying in these areas as global temperatures rise is likely to cause several degrees of warming (IPCC, 2013).

5.2.4 *Carbon cycle*

Carbon is a constituent of seawater, the atmosphere, coal, limestone, soils, and all living things. Carbon can move between these worlds as part of the carbon cycle on Earth. Carbon is taken up by plants on land from the atmosphere. Carbon dioxide (CO_2) is a gas that exists in the atmosphere and is made up of carbon and oxygen. During photosynthesis, carbon dioxide is drawn from the atmosphere to provide a substrate for plant growth. Additionally, seawater can dissolve CO_2. The two mechanisms mentioned above both consume carbon (see green arrow in Fig. 5.4).

Living beings release carbon into the atmosphere. Carbon dioxide gas is exhaled by both plants and animals during breathing. When plants and animals perish, the carbon in their bodies, wood, and leaves decomposes and is deposited in the Earth. CO_2 is also released by human activities, including burning fossil fuels to run vehicles, trucks, power plants, factories, and wildfires. These procedures produce carbon (see red arrow in Fig. 5.4).

The response of the global carbon cycle to climate change is one of the biggest sources of uncertainty for future climate predictions. Deserts and dry shrublands are examples of arid ecosystems, which are key players in the global carbon cycle. Roughly 40% of the world's net primary productivity is accounted for by arid and semi-arid regions, which make up about 41% of the terrestrial land surface (Wang *et al.*, 2012). Semi-arid regions experience

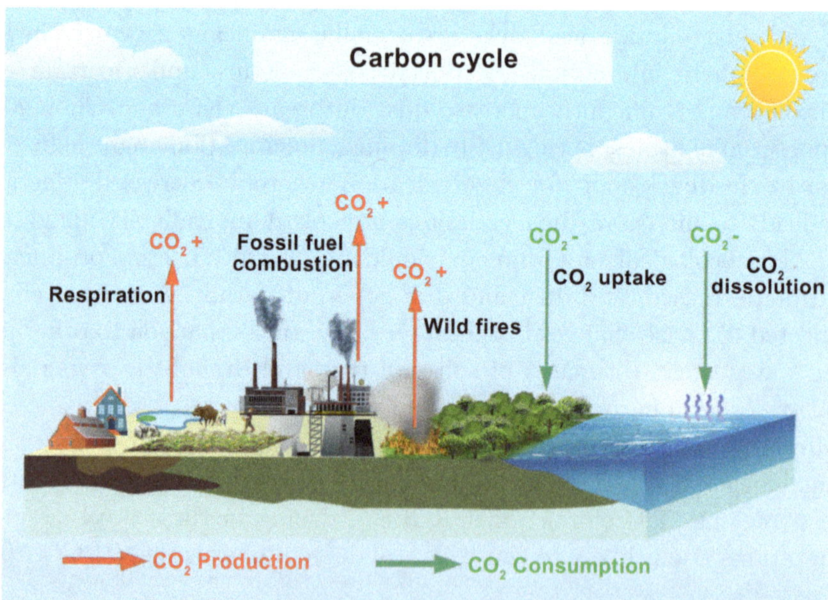

Fig. 5.4. A diagram of the generalized carbon cycle. The carbon fluxes between ocean, atmosphere, and land influence climate through major natural and anthropogenic processes.

the most pronounced climate changes resulting from rising CO_2 levels. These regions act as carbon sinks for fossil-fuel emissions. The growth of vegetation in semi-arid ecosystems contributes to 60% of carbon uptake, making it the primary driver of the trend and interannual variability in the terrestrial carbon sink (Huang *et al.*, 2017a). With annual precipitation ranging from 200 to 600 mm, semi-arid woods were able to sustain high levels of carbon sequestration by shifting the peak time for photosynthetic activity from summer to early spring. If savannas were protected from burning and grazing, estimates based on data for these ecosystems demonstrate the enormous potential for utilizing these ecosystems for carbon sequestration. An increasingly significant factor influencing the interannual variability of carbon dioxide is the greater turnover rates of carbon reservoirs in semi-arid biomes (Poulter *et al.*, 2014). The quantitative significance of terrestrial carbon cycles is now acknowledged to be equal to that of ocean–atmosphere interactions in global climate models (Osmond *et al.*, 2004). Although the combined land and ocean sinks remove about 50% of the total CO_2 emissions, coupled climate–carbon feedback models have forecast lower carbon uptake by terrestrial sinks, which is caused by declining productivity and increased respiration brought on by high temperatures. However, recent research indicates that the strength of the land carbon sink has been increasing over the past 30 years. This trend may be caused by a variety of factors, including the fertilization effect of increased CO_2 and atmospheric nitrogen deposition on plant growth, changes in growing season length, and land management (Poulter *et al.*, 2014). Increased tree cover and leaf area index (LAI) were linked to an increase in net primary productivity that was driven by CO_2 and the climate, according to Wu *et al.* (2016). However, due to the possibility of increased evapotranspiration and decreased soil moisture due to vegetation greening brought on by CO_2 fertilization in semi-arid locations, this anomaly is expected to diminish soon (Forzieri *et al.*, 2017). Due to desertification, the soil moisture will continue to decline and eventually be unable to support the current ecosystem, which would cause a sharp decline in gross primary production.

By increasing the vapor pressure deficit and evaporative demand, enhanced warming in arid and semi-arid regions may exacerbate the expansion of arid and semi-arid regions. Additionally, Seneviratne *et al.* (2010) speculate that the reduced soil moisture may cause SH to absorb more energy than LH, having a greater impact on temperature extremes (Vogel *et al.*, 2017). According to Sjögersten and Wookey (2016), climate-related changes in temperature and moisture are predicted to significantly alter the capacity of ecosystems to store carbon in high latitude regions through vegetation shifts as well as direct warming effects on photosynthesis and decomposition. Additionally, the capacity to store soil organic carbon (SOC) diminishes with rising temperature and rises with rising soil water content. Along with additional warming-induced consequences, such as an increase in soil CO_2 efflux and changes in microbial communities that would modify arid and semi-arid regions' C cycling and lower soil C stocks in the mid to long term, the ability of arid and semi-arid regions to adsorb atmospheric CO_2 will decline (Maestre *et al.*, 2013). Carbon emissions could also result from land degradation brought on by erosion (Lal, 2003). In light of this, the extension of arid and semi-arid regions will decrease SOC storage and cause CO_2 to emit into the sky. Additionally, gross primary production (GPP) is severely constrained by soil deterioration and decreased soil moisture (Peng *et al.*, 2013), which also has an impact on the rate of photosynthesis in plants that can take up CO_2 and store it as carbon. The GPP is positively connected with the AI, and it is around one-fifth as high in arid locations as it is in humid ones. Soils in arid and semi-arid regions worsen global warming by releasing more CO_2 into the atmosphere and storing less carbon through these two processes. Warming and drying processes combine to form a positive feedback cycle that reinforces each process. A recent slowdown in the increasing rate of atmospheric CO_2 during the warming pause, however, was seen (Keenan *et al.*, 2016), and it was linked to an increase in the terrestrial sink over the previous ten years (Ballantyne *et al.*, 2017). This rise was related to both the impacts of slower global warming on global respiration and the effects of rising atmospheric CO_2 on plants. Unfortunately, CMIP5 cannot replicate this pattern without significant uncertainty since certain CMIP5 models do not consider global carbon cycle processes (Huang *et al.*, 2017b). Even with the most advanced models of the global carbon cycle, there is still a significant amount of uncertainty regarding the carbon concentration. Numerous new findings about how plants react to global warming have been made, from the ecosystem scale to the leaf scale, yet the mechanisms are too complex to accurately parameterize in the model. For instance, because many deciduous trees need cold to release their dormancy and heat causes chilling to decrease, the early spring leaf unfolding of plants in response to climatic warming may be countered. Additionally, the adaptation of leaf respiration to climatic warming may prevent 80% of the anticipated increase in leaf respiration (Reich *et al.*, 2016).

5.2.5 *Oxygen cycle*

One of the most important biogeochemical cycles on earth, the oxygen cycle, is required for all heterotrophic life, including human life. It took several hundred million years for atmospheric oxygen to reach its current level, along with the development of the lithosphere and biosphere (Huang *et al.*, 2021). The modern oxygen cycle, however, has undergone significant changes compared to the cycle in the geologic timescale since industrialization because of anthropogenic disturbances, such as land-use change and fossil fuel combustion (Huang *et al.*, 2018), which are characterized by the continuous decline in atmospheric oxygen concentration over the past decades (Keeling and Manning, 2014). As a result, the modern oxygen cycle offers a novel perspective on researching how the Earth system responds to climate change and human activity (Fig. 5.5), which has raised concerns in the academic community.

Terrestrial oxygen activities, such as oxygen production from terrestrial ecosystems and oxygen consumption from burning fossil fuels, are all crucial parts of the current oxygen cycle. The biggest current source of

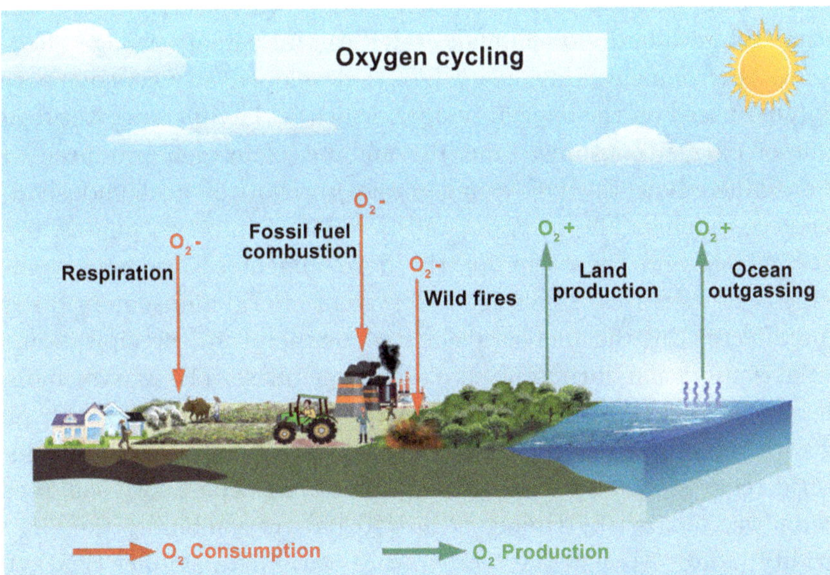

Fig. 5.5. The schematic diagram of the modern global oxygen cycle. The green arrows denote the production from land vegetation and outgassing from the ocean. The red arrows represent consumption by fossil fuel combustion, human and livestock respiration, natural fires, and residuals.

oxygen is the terrestrial ecosystem, which can create 7.10 (± 0.38) Gigaton (Gt, 1 Gt = $10^9 t$ = 10^{12} kg) oxygen for the atmosphere per year and accounts for more than 90% of all oxygen production (Ding *et al.*, 2022). According to Huang *et al.* (2018), the main factor contributing to the observed drop in atmospheric oxygen is fossil fuel burning, which uses around 38.45 (±0.61) Gt per year and accounts for more than 70% of the world's total oxygen consumption. Other terrestrial sources and sinks of oxygen include animal and human respiration, natural fires, and physical and chemical weathering. The estimated consumption from human and livestock respiration is approximately 2.48 (±0.16) and 1.73 (±0.13) Gt, respectively. 5.87 Gt could be consumed each year as a result of the natural wildfires. The primary oxygen sources and sinks are found on Earth; hence, the terrestrial oxygen budget is a predictor of changes in atmospheric oxygen content.

Furthermore, determining the ecological security and capacity for sustainable growth depends on the volatility of the terrestrial oxygen budget. As was already established, anthropogenic activities, the lithosphere, and the biosphere are all included in the terrestrial oxygen budget. Indicators generated from the terrestrial oxygen budget can therefore be used to quantitatively evaluate how human activities and climate change affect the biological environment. Han *et al.* (2021) developed an Oxygen Footprint Index (OF) based on the oxygen budget, which is the difference between the amount of oxygen consumed and the amount of oxygen produced. They then examined how the OF changed the function of arid and semi-arid regions' ecosystems. The response can be divided into two phases, one for climate change and the other for the development of arid and semi-arid regions as OF increases. The fragile ecosystem in arid and semi-arid regions will be affected by the increased surface temperature, precipitation, and potential evaporation during the climate change phase. The ecosystem in arid and semi-arid regions is then devastated during the regions' growth phase, along with the severe temperatures and low soil moisture, and the arid and semi-arid regions start to spread to adjacent land. The Ecological Security Indicator was created by Huang *et al.* (2020a) by combining the OF with the aridity index (AI) and the temperature warming magnifier (ESI). With ESI, there are four categories of land security: secure land, severe-dangerous, light-dangerous, and semi-secure zones. Land that is extremely dangerous to humans is typically found near deserts and areas that are heavily impacted by human activity, such as industrialized and densely populated areas. Since deserts and arid and semi-arid regions are mostly unstable terrain, they are expected to grow under various future scenarios with rising ESI levels.

Numerous studies have also suggested that an imbalance in the terrestrial oxygen budget may lead to the ecosystem's demise. The atmospheric oxygen level in the Biosphere 2 experiment falls from 21% to 14% over a 16-month period, resulting in the experiment's complete failure (Severinghaus *et al.*, 1994). This decline is caused by the inability of the vegetation to provide enough oxygen to offset soil respiration's use (Severinghaus, 1995). Additionally, the decrease in atmospheric oxygen directly affects solar optical depth through its effect on air density, which heightens greenhouse forcing and increases land surface temperature (Poulsen *et al.*, 2015). Ecosystems of arid and semi-arid regions may be more vulnerable to changes in atmospheric oxygen and more susceptible to dangers associated with an imbalanced oxygen budget (Han *et al.*, 2021). As a result, research into the fluxes of terrestrial oxygen processes can help us better comprehend the present oxygen cycle and can also serve as a theoretical foundation for developing development plans.

5.3 Global and Dryland Regions' GPP Changes

Terrestrial ecosystems play a crucial role in the global carbon cycle. Gross primary productivity (GPP), which is the amount of carbon absorbed by terrestrial ecosystems through plant photosynthesis, is the largest global CO_2 flux and the principal driver of many ecosystem processes. Understanding the spatiotemporal variability of GPP is essential in order to acquire precise estimates of terrestrial ecosystems' capacity to act as significant carbon storage, especially considering that atmospheric carbon dioxide concentrations are predicted to increase for the rest of this century. Northern ecosystems' GPP has grown as a result of CO_2 fertilization, extended growing seasons, and nitrogen deposition brought on by greater seasonal CO_2 exchange due to global warming. However, there are significant regional differences in GPP, thus further research is necessary.

5.3.1 *Comparison of different GPP data and methods*

(1) *MODIS GPP data*

According to Sjöström *et al.* (2013), GPP refers to the ability of vegetation to store carbon and energy during photosynthesis. From 2000 to 2014, we used monthly MODIS global terrestrial GPP products with a $0.5° \times 0.5°$spatial resolution. The first ongoing satellite-driven dataset to track the productivity of all vegetation is MODIS17. The MOD17 algorithm's principles involve using radiation conversion efficiency to predict daily GPP. The light-use efficiency (LUE) model, created by Monteith (Sjöström *et al.*, 2013), is specifically used to estimate the MODIS GPP. In this model, gross photosynthesis is proportional to the amount of photosynthetically active radiation (PAR) absorbed by plants. The LUE model is revised to become (Running *et al.*, 2004; Tang *et al.*, 2015)

$$\mathrm{GPP} = \varepsilon_{\max} \times 0.45 \times \mathrm{SW}_{\mathrm{rad}} \times f\mathrm{PAR} \times f\mathrm{VPD} \times fT_{\min}, \qquad (5.2)$$

in which $\varepsilon_{\max}$ is the maximum LUE, $\mathrm{SW}_{\mathrm{rad}}$ is the downward solar radiation, of which 45% is PAR, fPAR is the fraction of PAR being absorbed by plants, and fVPD and $fT_{\min}$ are the reduction scalars from water stress (high daily VPD) and a low daily minimum temperature, respectively. Accessing the ambiguous MODIS GPP data is difficult due to the varied plant functional categories and fluctuating climatic conditions. Lack of a precise definition, particularly for complex and diverse ecosystems where the same could lead to significant inaccuracies in estimating GPP, is one of the main sources of uncertainty. GPP can be overstated for low-productivity ecosystems and underestimated for high-productivity ecosystems (Turner *et al.*, 2006). Accounting for the effect of soil moisture in the algorithm in arid and semi-arid regions may reduce the uncertainty in GPP across arid and semi-arid regions (Kanniah *et al.*, 2009). Additionally, such an LUE model is known to have problems simulating fluxes in arid environments, especially when soil moisture and VPD are uncoupled and VPD is not a reliable indicator of water availability and stress. Nevertheless, due to its advantage of continuous spatial and temporal coverage, the MODIS GPP product continues to be the most popular global GPP product.

MODIS products (Levels 1, 2, 3, and 4) are a new type of integerized sinusoidal projection data. We converted the MODIS GPP data formats and map projections to the WGS84/geographic system using MODIS Re-projection Tools (Justice *et al.*, 2002). This study made use of the global GPP datasets from 2000 to 2014.

(2) *FLUXNET GPP data*

We contrasted the GPP obtained from MODIS with the GPP from FLUXNET. The MODIS GPP has been extensively used to validate the FLUXNET GPPs, which are based on eddy flux data (Falge *et al.*, 2002). More than 800 long-term eddy covariance flux tower sites, located between

30°S and 70°N and spanning a variety of climatic zones and terrestrial ecosystems, have been constructed by FLUXNET (https://fluxnet.org/). To assure high-quality flux datasets with greatly decreased uncertainties related to site-to-site fluctuations in fluxes, FLUXNET uses industry-standard data quality assurance and control processes, as well as post-field data processing processes. The difference between measured net ecosystem exchange and calculated ecosystem respiration is used to determine the FLUXNET GPPs (Reco; Turner *et al.*, 2006):

$$\text{GPP} = \text{NEE} - \text{Reco}, \tag{5.3}$$

in which the partitioning for Reco uses the nighttime method (Reichstein *et al.*, 2005).

In this study, we compared flux-derived GPP from 13 FLUXNET flux locations across five different global biomes with MODIS-derived GPP (e.g., SAV, GRA, WSA, CRO, and OSH). After clearing the FLUXNET quality check, the selected sites' available flux data, which spanned the years after 2000, ranged from 1 to 14 years (Table 5.2). The findings demonstrate a strong correlation between MODIS GPP and FLUXNET GPP (Table 5.3). The usefulness of MODIS datasets for examining temporal and geographical fluctuations in the arid and semi-arid regions' GPP is confirmed by our comparisons.

(3) *FLUXCOM GPP data*

Additionally, we contrasted the spatial changes in the yearly global arid and semi-arid regions' GPP between the FLUXCOM datasets and the

Table 5.2. Relative contribution of the regional arid and semi-arid regions' gross primary production (GPP) trend to the global arid and semi-arid regions' GPP trend (RCP4.5) (Yao *et al.*, 2020).

Year/Area	North America (%)	South America (%)	Europe (%)	Africa (%)	West Asia (%)	East Asia (%)	South Asia (%)	Australia (%)
2000–2014(MODIS)	45	−10	9	18	−4	21	3	18
2011–2040	9	1	10	16	17	27	18	2
2041–2070	7	−2	8	18	13	35	16	5
2071–2100	16	8	13	5	16	21	11	10
2011–2100	15	5	10	10	9	32	12	7

Table 5.3. FLUXNET site descriptions, including locations, ecosystem types, and measurement period (Yao *et al.*, 2020).

Site-ID	Site-name	Location-Lat	Location-Long	IGBP	MAP
CA-NS7	UCI-1998 burn site	56.6358	−99.9483	OSH	483.27
CA-SF3	Saskatchewan-Western Boreal, forest burned in 1989	54.09156	−106.0052	OSH	470
US-Var	Vaira Ranch-lone	38.4133	−120.9507	GRA	559
US-Wkg	Walnut Gulch Kendall Grasslands	31.7365	−109.9419	GRA	407
US-SRM	Santa Rita Mesquite	31.8214	−110.8661	WSA	380
DE-Geb	Gebesee	51.1001	10.9143	CRO	470
ZA-Kru	Skukuza	−25.0197	31.4969	SAV	547
CN-Cng	Changling	44.5934	123.5092	GRA	–
RU-Tks	Tiksi	71.5942	128.8878	GRA	323
AU-Cpr	Calperum	−34.0021	140.5891	SAV	–
AU-Emr	Emerald	−23.8587	148.4746	GRA	–
AU-Gig	Gingin	−31.3764	115.7138	WSA	–
AU-Rig	Riggs Creek	−36.6499	145.5759	GRA	–

Table 5.4. The mean and standard deviation (kg cm^{-2}) for the difference between the MODIS-derived gross primary production (GPP) and the FLUXCOM GPP in the eight regions (Yao *et al.*, 2020).

Region	North America	South America	Europe	Africa	West Asia	East Asia	South Asia	Australia Australia
Mean ± Std	0.10 ± 0.08	0.15 ± 0.08	0.08 ± 0.04	0.15 ± 0.04	0.09 ± 0.03	0.08 ± 0.03	0.14 ± 0.05	0.07 ± 0.05

MODIS datasets (Table 5.4). The 224 FLUXNET site-level observations, satellite remote sensing, and meteorological data are combined to create the FLUXCOM GPP datasets, which may be seen at http://www.fluxcom.org/. The scientific community may now access an ensemble of machine learning-based global flux products for assessing process-based land surface models thanks to the employment of three machine learning algorithms that reduce sources of uncertainty in empirical upscaling (http://www.fluxcom.org/) (Jung *et al.*, 2017). In a nutshell, site-level measurements of the explanatory climatic and land surface variables were used to train the machine learning algorithms. The variables from the high-resolution satellite remote sensing data were fed into the thorough variable selection process in order to reliably capture variations in vegetation greenness and land surface temperature with changes related to soil moisture. In Tramontana *et al.* (2016), the machine learning algorithms, their training, and a rigorous cross-validation of the data are described. With worldwide gridded satellite data for chosen explanatory variables at 10-km spatial and 8-day temporal resolutions for the years 2000–2014, trained and verified machine learning algorithms were used to produce the global GPP. Large uncertainties still exist when utilizing various upscaling techniques to extract spatial and temporal patterns of GPP on regional and global scales from FLUXNET GPP because of the diverse topography and ecosystem processes (Schimel *et al.*, 2015). When describing CO_2 dynamics over arid and semi-arid regions, care should be taken because the 224 flux sites are not evenly distributed over the globe with varied climatic types and plant function kinds, and there are less dense flux sites over arid and semi-arid regions.

We also contrasted the GPP generated by 15 CMIP5 models with the GPP derived from MODIS. The CMIP5 simulations contain decadal prediction experiments as well as long-term (century timeline) and short-term (10–30 year) integrations (Taylor *et al.*, 2012). You can get the CMIP5 GPP datasets at https://esgf-node.llnl.gov/search/esgf-llnl/. The ensemble means of the GPP from 2010 to 2100 predicted by the 15 CMIP5 models were compared to our GPP (Table 5.3). The multi-model ensemble represents a variety of best-effort attempts to mimic the climate system, as opposed to the ensemble of simulations produced by individual models. The ensemble can be used to provide a consensus representation of the climate system and, based on the spread of model results, provide some measure of how much confidence might be placed in that consensus, provided that these attempts are at least somewhat independent and that the collection of models is not generally biased.

(4) *Statistical analysis*

We calculated the correlation coefficients using the root mean squared error (RMSE) between the two GPP products and the ordinary least square (OLS) regression to demonstrate the accuracy of the MODIS product:

$$\mathrm{RMSE} = \sqrt{\frac{\sum_{t=1}^{n} (\mathrm{GPP}_{\mathrm{flux}} - \mathrm{GPP}_{\mathrm{MODIS}})^2}{n}}, \tag{5.4}$$

in which $\mathrm{GPP}_{\mathrm{flux}}$ and $\mathrm{GPP}_{\mathrm{MODIS}}$ are the FLUXNET-derived and MODIS-derived GPP, respectively. OLS regression was used to estimate the linear trend of GPP, and a two-tailed test was used to determine the significance level of the linear correlation. The mean absolute global GPP anomaly was used to calculate the IAV of GPP.

(5) *Methods to calculate relative contributions*

The sum of the trends for each of the eight regions was used to compute the respective contributions of each region's unique GPP trend from each of the eight regions to the global arid and semi-arid regions' GPP trend. We utilized an index from a previous study (Ahlström *et al.*, 2015) to determine the relative contributions of the GPP IAV from each of the eight regions to the global arid and semi-arid regions' GPP IAV:

$$f_i = \sum_t \frac{x_{it}|x_t|}{x_t} \Big/ \sum_t |x_t|, \tag{5.5}$$

in which x_{it} is the GPP anomaly for region i at time t in years, and x_t is the global arid and semi-arid regions' GPP anomaly, so that $x_t = \sum_t x_{jt}$. The f_i is the average relative anomaly x_{it}/x_t for region i, and this definition should guarantee that $\sum_i f_i = 1$. The resulting scores for a region f_i ultimately show how much it contributes to variances in arid and semi-arid regions around the world. While regions with low scores contribute less to overall differences in arid and semi-arid regions' GPP, those with high scores are significant contributors. Negatively rated areas reduce variation.

5.3.2 *Spatiotemporal change in GPP between 2001 and 2014*

We examined the spatiotemporal changes in the arid and semi-arid regions' GPP from 2000 to 2014 using the MODIS-derived GPP data as the baseline. At 13 arid and semi-arid regions' flux sites, the MODIS-derived GPP and the FLUXNET-derived GPP exhibit good agreement and a linear relationship. The vegetation types of these flux sites over arid and semi-arid regions are SAV, GRA, croplands (CRO), woody savannas (WSA), and open shrublands, and they are distributed over North America, Europe, Africa, Asia, and Australia (OSH). The MODIS product also captures IAV, as shown by the strong agreements between the annual GPP anomalies from MODIS and those from the flux towers, which is consistent with the previous thorough GPP comparisons over worldwide land. Additionally, we contrasted temporal and spatial fluctuations in the arid and semi-arid regions' GPP between the GPP produced from MODIS and the GPP derived from FLUXCOM.

Our findings support earlier findings that the worldwide arid and semi-arid regions' GPP may be reliably quantified using MODIS-derived GPP data, although some places show underestimations of the GPP acquired from MODIS. The MODIS-derived GPP remains one of the most reliable datasets to assess the spatiotemporal variability of GPP at global scales when uncertainties and other considerations for the various datasets are considered.

From 2000 to 2014, global arid and semi-arid regions showed rising trends and spatiotemporal variability in GPP. The regions that made the biggest positive contributions to the trend of rising arid and semi-arid regions' GPP were North America (45%), East Asia (21%), and Africa/Australia (18%) (Fig. 5.6; Table 5.2), whereas the regions that made the biggest negative contributions to the trend of rising arid and semi-arid regions IAV were Australia (25%), South America (20%), and Africa (15%) (Table 5.5). In keeping with the study's focus on arid and semi-arid regions, take note that relative changes in trends, IAV, and their contributions are only presented with respect to arid and semi-arid regions; the GPP of humid regions (AI >= 0.65) is not examined further. Precipitation, one of the environmental drivers, accounted for 67% of the fluctuations in arid and semi-arid regions' GPP,

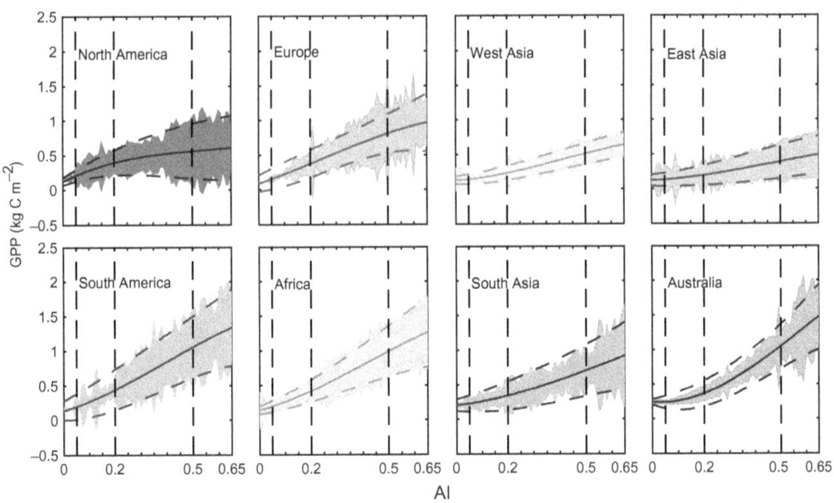

Fig. 5.6. Relationships of GPP to AI in the eight regions (2000–2014). The solid curves denote the fitted relations between gross primary production (GPP) and aridity index (AI) in the eight dryland regions, and the upper and lower dashed curves denote the 95% confidence intervals of the fitted relations. The shaded areas denote one standard deviation. The vertical dashed lines delineate the boundaries between the four dryland regions' subtypes (Yao *et al.*, 2020).

Table 5.5. Relative contribution of the regional arid and semi-arid regions' gross primary production (GPP) interannual variability (IAV) to the global arid and semi-arid regions' GPP IAV (RCP4.5) (Yao *et al.*, 2020).

Year/Area	North America (%)	South America (%)	Europe (%)	Africa (%)	West Asia (%)	East Asia (%)	South Asia (%)	Australia (%)
2000–2014(MODIS)	8	20	10	15	6	6	10	25
2011–2040	7	17	11	17	5	11	9	23
2041–2070	8	17	12	16	4	11	10	22
2071–2100	9	15	14	16	4	10	11	21
2011–2100	9	13	12	15	7	12	11	21

whereas air temperature accounted for 15% of the differences. Other factors, such as soil moisture, vapor pressure deficit (VPD), and PET, together accounted for the remaining 18% of the arid and semi-arid regions' GPP variability. According to their results, the most significant driver of arid and semi-arid regions' vegetation dynamics is precipitation. The biggest growing trend in sensitivity occurred in the most water-sensitive locations, which suffered a considerable reduction in precipitation. The LAI's sensitivity to soil moisture increases dramatically in many arid and semi-arid regions (Li *et al.*, 2022). Over 70% of vegetated land is affected by dryness stress on ecosystem output, and the main cause of this stress is soil moisture (Liu *et al.*, 2020a).

5.3.3 *Projected dryland GPP in the 21st century under RCP4.5*

Climate change is anticipated to be dominated by warming and drying trends in the 21st century, along with an increase in the frequency of extreme weather events (Ma *et al.*, 2016). Long-term trends toward warming and dryness would hasten the growth of arid and semi-arid regions by transforming humid ecosystems into arid and semi-arid regions' ecosystems and deteriorating vulnerable arid and semi-arid regions' subtypes. We first adopted the premise that the relationship between AI and GPP can be used to quantify arid and semi-arid regions' GPP variability in order to answer the questions of whether and how global arid and semi-arid regions' expansion and subtype conversions would change regional contributions to the regions' GPP and IAV as well as subtype contributions across regions.

Table 5.6. The fitting equation between the arid and semi-arid regions' gross primary production (GPP) and aridity index (AI) in each of the eight regions. One asterisk indicates that the equation exceeds the 95% confidence level, and two asterisks indicate that the equation exceeds the 99% confidence level (Yao *et al.*, 2020).

Region	Equation	R^2
North America	$GPP_{NA} = 1.70 \times AI^3 - 2.81 \times AI^2 + 1.86 \times AI + 0.12$	0.88*
South America	$GPP_{SAM} = -2.22 \times AI^3 + 2.61 \times AI^2 + 1.09 \times AI + 0.14$	0.94**
Europe	$GPP_{EU} = -4.74 \times AI^3 + 6.88 \times AI^2 - 1.81 \times AI + 0.55$	0.82*
Africa	$GPP_{AF} = -1.85 \times AI^3 + 2.31 \times AI^2 + 1.01 \times AI + 0.14$	0.98**
West Asia	$GPP_{WA} = -1.56 \times AI^3 + 1.93 \times AI^2 + 0.19 \times AI + 0.13$	0.97**
East Asia	$GPP_{EA} = -0.81 \times AI^3 + 1.14 \times AI^2 + 0.17 \times AI + 0.12$	0.96**
South Asia	$GPP_{SAS} = -0.91 \times AI^3 + 1.68 \times AI^2 + 0.39 \times AI + 0.20$	0.93**
Australia	$GPP_{AU} = -3.13 \times AI^3 + 5.49 \times AI^2 - 0.35 \times AI + 0.24$	0.92**

Note: *indicates that the equation exceeds the 95% confidence level, and **indicates that the equation exceeds the 99% confidence level.

In order to calculate the impact of arid and semi-arid regions' growth and deterioration on the regions' GPP, we looked at the link between arid and semi-arid regions' GPP and AI. We fitted one polynomial function between GPP and AI for each of the eight locations using the GPP and AI for the period of 2000–2014 (Fig. 5.6), with very good correlations (Table 5.6).

These substantial correlations support the idea that using AI to limit arid and semi-arid regions' GPP is a smart idea. To determine the uncertainty in the projected GPP, we also employed the two equations of the 95% prediction intervals of the fitted GPP–AI relations (Fig. 5.6). Substantial shifts in the regional contributions to global dryland GPP trends are observed during different periods of the 21st century. Under the RCP4.5 scenario, the three largest regional contributors to the global dryland GPP trends shift from East Asia (27%), South Asia (18%), and West Asia (17%) in 2011–2040 to East Asia (35%), Africa (18%), and South Asia (16%) in 2041–2070, and then to East Asia (21%), West Asia (16%), and North America (16%) in 2071–2100. East Asia emerges as the largest contributor to the trend, exceeding North America and South Asia. Australia, Africa, and South America remain the largest contributors to the IAV, as compared with the 2000–2014 baseline (Fig. 5.7). The largest climate-induced dryland expansion is found to occur in semi-arid regions, accounting for almost half of the total dryland expansion, covering more than one-third of the total dryland areas, and contributing to almost half of the total global dryland GPP. This explains the increased roles of semi-arid ecosystems in regulating the trends (49%) and variability (35%)

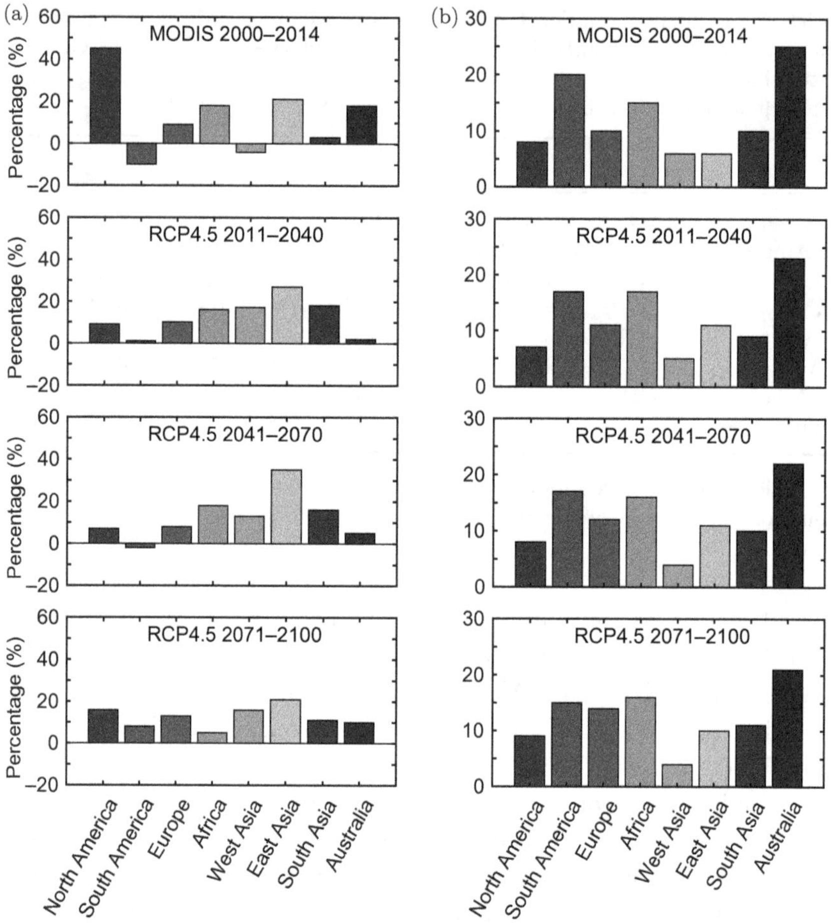

Fig. 5.7. Contributions from the eight regions to global dryland GPP. Gross primary production (GPP) trend (a) and interannual variability (IAV) (b) for 2000–2014 (MODIS) and during the three periods in 2011–2100 (RCP4.5) (Yao *et al.*, 2020).

of the global dryland CO_2 sink under the future climate change. However, the arid subtype plays an increasing role in regulating the trend (18%) and IAV (29%), whereas the subhumid subtype has a reduced impact on the trend (27%) and IAV (26%). The fitted functions for each region were then used to estimate fluctuations in the worldwide arid and semi-arid regions' GPP using the aforementioned regions' AI datasets based on the RCP4.5 forecasts for the years 2011 to 2100 (Fig. 5.8). Note that the predicted GPP over the world's arid and semi-arid regions was only calculated using these fitted functions based on data throughout the AI range of 0 to 0.65 (i.e., $0 < AI < 0.65$). For comparison, we carried out the same calculation

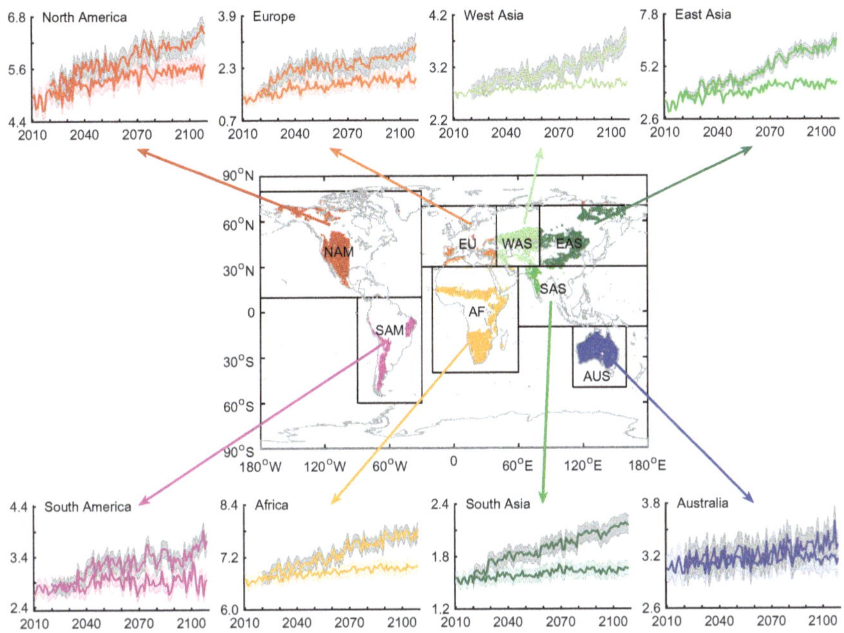

Fig. 5.8. Temporal variations of GPP in the eight regions. The curves with the small and large gross primary production (GPP) in units of Pg C year^{-1} are under RCP4.5 and RCP8.5, respectively. The shaded areas denote 95% confidence intervals (Yao *et al.*, 2020).

using RCP8.5. We carried out bias corrections on the CMIP5 model results by building mapping linkages with the 2000–2014 MODIS datasets (Taylor *et al.*, 2012).

5.3.4 *Influence of dryland regions' expansion and conversions on GPP*

Under RCP4.5, the global arid and semi-arid regions' GPP will rise by 12 ± 3% by the end of the century in comparison to the baseline of 2000–2014. The IAV values are higher, and there are increasing tendencies in the global arid and semi-arid regions' GPP. In comparison to the global arid and semi-arid regions' GPP baseline of 2000–2014, the regions' GPP trend and IAV rose by 7 ± 2 and 2 ± 1% in 2011–2040, 11 ± 3 and 4 ± 2% in 2041–2070, and 12 ± 3 and 3 ± 2% in 2071–2100, respectively. The combined effects of the accelerated arid and semi-arid regions' expansion, climate-induced shifts of the subtypes within the historical arid and semi-arid regions (e.g., primarily drought-induced degradation of arid and semi-arid regions), and conversions of arid and semi-arid regions to humid lands are responsible for these increasing trends and enhanced IAV by the end of this century. Canada, central Africa, Western Asia, and Northeastern China are the main arid regions, with a bigger area becoming drier and a smaller area becoming wetter. At the conclusion of this century, the worldwide arid and semi-arid regions' GPP will have increased by $12 \pm 3\%$ as a result of widespread expansion (0.0661 PgC year^{-1}) significantly outpacing GPP declines (−0.0211 PgC year^{-1}) brought on by arid and semi-arid regions' shrinking and degrading as a result of drought. However, the worldwide arid and semi-arid regions' GPP per unit area actually declines at a rate of −0.26 kg cm^{-2} year^{-1} due to the fact that the increased GPP in the expanding arid and semi-arid regions at a rate of 0.57 kg cm^{-2} year^{-1} is outpaced by the decreased GPP in the historical arid and semi-arid regions at a rate of −0.83 kg cm^{-2} year^{-1} (Fig. 5.9). Considering the 11% increase in global arid and semi-arid regions predicted for the end of this century under RCP4.5 compared to the global arid and semi-arid regions' baseline of 1961–1990. After 2065, under RCP4.5, the rising trends and IAV values will stabilize (Zhao *et al.*, 2015).

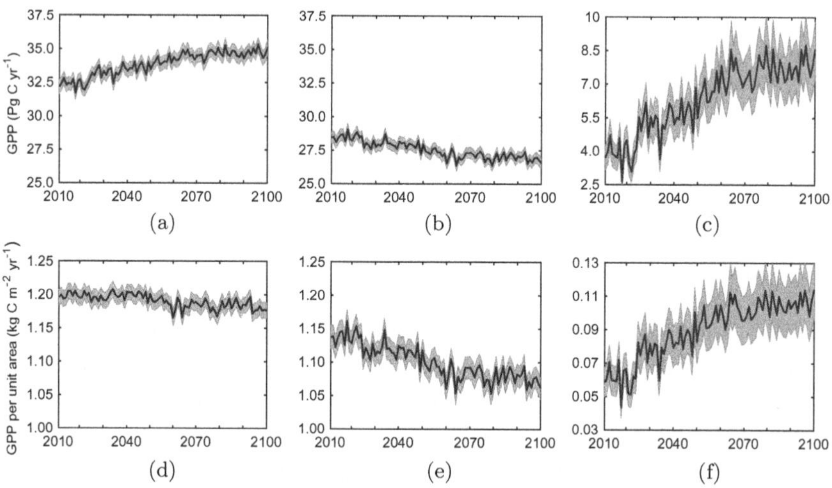

Fig. 5.9. Temporal variations in dryland GPP under RCP4.5 (2010–2100). Temporal variations in the total GPP of the global drylands, including both the historical and expanded drylands (a). Temporal variations in the total GPP of the historical drylands (b). Temporal variations in the total GPP of the expanded drylands (c). Temporal variations in the GPP per unit area in the global drylands, including both historical and expanded drylands (d). Temporal variations in the GPP per unit area in the historical drylands (e). Temporal variations in the GPP per unit area in the expanded drylands (f). The shaded areas denote 95% confidence intervals (Yao *et al.*, 2020).

At various points throughout the 21st century, there have been noticeable changes in the regional contributions to trends in the arid and semi-arid regions' GPP. Under the RCP4.5 scenario, the three largest regional contributors to the global arid and semi-arid regions' GPP trends shift from East Asia (27%), South Asia (18%), and West Asia (17%) in 2011–2040, respectively, to East Asia (35%), Africa (18%), and South Asia (16%) in 2041–2070, and then to East Asia (21%), West Asia (16%), and North America (16%) in 2071–2100. East Asia surpasses North America and South Asia to become the trend's major contributor. Compared to the baseline period of 2000–2014, Australia, Africa, and South America continue to make up a majority of the IAV's donors (Fig. 5.7(a)). The greater frequency of climate extremes and the high sensitivity of arid and semi-arid regions to droughts and precipitation in these three regions are probably to blame for the big GPP IAV (Zhang *et al.*, 2017).

It has been noted that under RCP4.5, the contributions of the various arid and semi-arid regions' subtypes to the regions' GPP trend and IAV have changed. Semi-arid regions are found to have had the most climate-induced dryland expansion, which accounts for nearly half of the total dryland expansion, more than one-third of the total dryland area, and almost half of the total world's arid and dryland GPP. This explains the increased roles of semi-arid ecosystems in regulating the trends (49%) and variability (35%) of the global arid and semi-arid regions' CO_2 sink under future climate change. However, the arid subtype plays an increasing role in regulating the trend (18%) and IAV (29%), whereas the subhumid subtype has a reduced impact on the trend (27%) and IAV (26%). The contributions of different subtypes to the trend and IAV also vary across regions. The largest contributions to the semi-arid subtype GPP trends occur in East Asia (30%) and Africa (27%), and the largest contributions to the global arid subtype GPP IAV occur in Africa (26%) and Australia (24%). The largest contributions to the global arid subtype GPP trends occur in Africa (34%) and Australia (19%), and the largest contributions to the global arid subtype GPP IAV occur in Australia (44%) and Africa (21%), whereas the largest contributions to the global subhumid subtype GPP trends occur in East Asia (31%) and North America (27%) and the largest contributions to the global subhumid subtype GPP IAV occur in East Asia (31%) and South America (16%).

During the 21st century under RCP4.5, the three prominent land conversions that led to the largest changes in the global arid and semi-arid regions' GPP were from humid to subhumid (+2.53 PgC year^{-1};

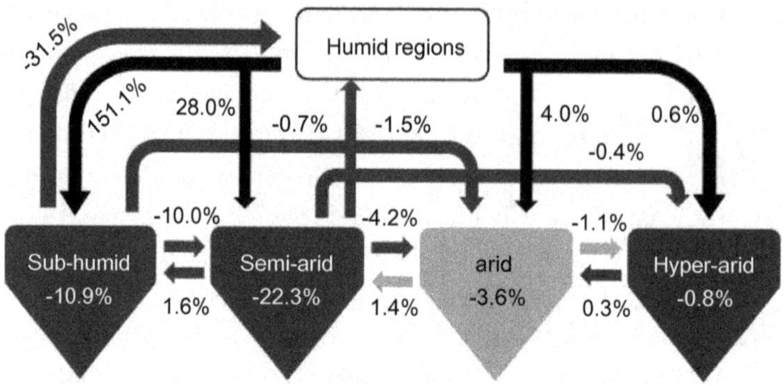

Fig. 5.10. Changes in dryland GPP due to expansion and conversions. It is estimated from our study that the global dryland gross primary production (GPP) will increase by 12% in 2085–2100 relative to the 2000–2014 baseline under RCP4.5. Taking 12% as 100 units, the magnitudes in the diagram thus denote the percentage of the contribution caused by each process to this 12% change. The sum of all the numbers in the diagram is equal to 100%. Positive values indicate increased GPP for the corresponding dryland subtypes, and negative values indicate decreased GPP for the corresponding dryland subtypes. For example, −31.5% means a reduction in the GPP contribution to the 12% change for the subhumid subtype due to its conversion to humid lands, 151.1% means an increase in the GPP contribution to the 12% change for the subhumid subtype due to its conversion of humid lands to the subhumid subtype, and −10.9% means a reduction in the GPP contribution to the 12% change for the subhumid subtype due to its degradation within its subtype. Note that changes in GPP in humid regions are not the focus of this chapter (Yao *et al.*, 2020).

151.1%), subhumid to humid (-0.53 PgC year^{-1}; -31.5%), and humid to semi-arid (0.47 PgC year^{-1}; 28.0%) (Fig. 5.10). Meanwhile, droughts also cause transitions among dryland subtypes, e.g., from subhumid to semi-arid and semi-arid to arid lands, leading to a reduction in the GPP. Degradations within semi-arid (-0.37 PgC year^{-1}; -22.3%) and subhumid (-0.18 PgC year^{-1}; -10.9%), from subhumid to semi-arid (-0.17 PgC year^{-1}; -10.0%), and from semi-arid to arid lands (-0.07 PgC y year^{-1}; -4.2%) cause substantial reductions in GPP (Fig. 5.10). Our findings show that the increased dryland GPP brought about by the subhumid and semi-arid dryland subtypes' expansion into formerly humid areas significantly outweighs the decline in dryland GPP brought about by dryland degradation due to drought and the (less frequent) contraction of the subhumid subtype because of the conversion of drylands to humid ones. As a result, regional and global dryland GPP rose both globally and locally this century.

The main factor controlling future trends and variability in arid and semi-arid regions' GPP is the regions' expansion. While Australia will continue to contribute the most to the global arid and semi-arid regions'

GPP IAV in this century, East Asia will overtake North America as the leading contributor to arid and semi-arid regions' GPP trends under RCP4.5. The arid subtype and the subhumid subtype have an increasing and a decreasing role in regulating the worldwide arid and semi-arid regions' GPP variability, respectively, despite the semi-arid subtype continuing to dominate the global dryland GPP variability. Our findings regarding global arid and semi-arid regions' GPP trends and IAV under RCP4.5 are conservative, and if future climate change follows the RCP8.5 scenario, global arid and semi-arid regions' GPP spatiotemporal variability will probably rise (Fig. 5.10). In fact, a higher rate of arid and semi-arid regions' expansion results in a 25% rise in regions' GPP under RCP8.5 by the end of this century compared to the baseline for the years 2000–2014 (Fig. 5.10). In 2041–2100, North America and East Asia, rather than Australia, will play the most significant role in regulating the global arid and semi-arid regions' GPP IAV. The main trend of the global arid and semi-arid regions' GPP hotspots shifts from East Asia to North America in 2011–2070. Our study shows that under future climate change, regional and subtype contributions to the global arid and semi-arid regions' GPP trend and IAV are prone to significant changes.

5.3.5 *Responses of terrestrial oxygen production to GPP changes*

Since the 1980s, significant advancements have been made in the observation of significant carbon fluxes and extensive vegetation cover. The phenomenon of "global greening" is characterized by an increase in vegetation greenness around the globe. This was demonstrated by both *in-situ* observation and satellite products. Additionally, it was discovered that, as a result of global greening, numerous key carbon fluxes, such as GPP and net primary production, increased during the previous few decades (Piao *et al.*, 2013; Sun *et al.*, 2018). According to global vegetation models, CO_2 fertilization is the primary cause of the planet's flora turning green (Zhu *et al.*, 2016). It is still necessary to investigate how the terrestrial ecosystem responds to global greening and the resulting rise in GPP.

The rise in the production of terrestrial oxygen cannot be matched by global greening. There are two main processes of ecosystem oxygen production: net photosynthesis and heterotrophic respiration (Rh). The oxygen produced during net photosynthesis process is offset by the cost of the Rh process. On the one hand, fertilization effects, nitrogen (N) deposition, and extending the plant growing season can all boost net photosynthesis in plants (Smith *et al.*, 2015; Whittinghill *et al.*, 2012; Piao *et al.*, 2007). Additionally, increased evaporation over the oceans and increased precipitation also help tropical plant development (IPCC, 2013). On the other hand, because temperature and soil moisture are the primary determinants of soil respiration (Yu *et al.*, 2018), increases in temperature and soil moisture under climate change may potentially promote soil respiration (Bauer *et al.*, 2008). Therefore, higher Rh also results in higher oxygen demand, whereas higher net photosynthesis increases the amount of terrestrial oxygen produced. Due to these conflicting effects, quantitative investigation is required to establish if terrestrial oxygen supplies would rise alongside GPP.

The production of oxygen over land grew historically from 5.97 to 16.01 Gt between 1900 and 2005 due to the high rise in NPP (Huang *et al.*, 2018). The main source of oxygen is found in the tropics (30°S–30°N), which produce 67.32% of all the oxygen on Earth. The principal sources in the tropics are the forests of the Amazon, Central Africa, and Indonesia (Ding *et al.*, 2022). According to the model-simulated net ecosystem production (NEP) and integrated observational oxidative ratio database, Ding *et al.*'s projection for 2022 indicated that the terrestrial ecosystem would increase its contribution of oxygen to the atmosphere. According to the SSP245 scenario and the SSP585 scenario, the amount of terrestrial oxygen emitted by 2100 will rise from the current amount of 7.10 (±0.38) to 7.51 (±0.27) Gt and 10.23 (±0.28) Gt, respectively. The Amazon forests, one of the primary sources of oxygen, will decline under the SSP245 scenario (Fig. 5.11(a)). A decline in oxygen supplies will also occur in the mid-latitudes, including Eastern North America and Western Europe. The non-oxygen-producing land on the boreal coasts will perform worse than it did historically, removing more oxygen from the atmosphere. Non-oxygen-producing land is defined as areas where the oxygen fluxes are negative, meaning that the ecosystems there consume oxygen from the atmosphere rather than producing it. Eastern South America, Central Africa, Eastern Europe, and Eastern and Southern Asia will see the largest rises. Production will rise almost everywhere under the RCP8.5 scenario, except for the high latitudes (Fig. 5.11(b)). Only in Northern Europe and Canada's coastal regions will there be decreases. In conclusion, due to extreme changes in land use and sharp increases in soil respiration, the future oxygen production trend will follow a "better gets better, worse gets worse" pattern, with more oxygen coming from the low and middle latitudes and more oxygen going into the atmosphere in the high latitudes.

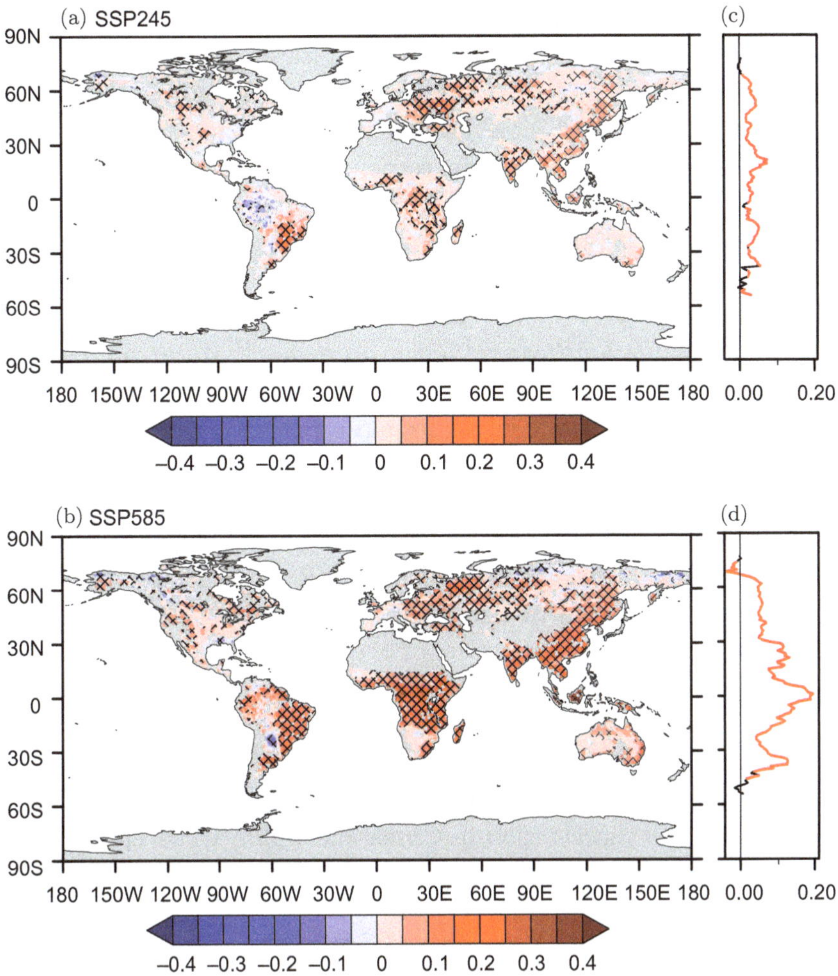

Fig. 5.11. Changes in terrestrial oxygen sources under the SSP245 (a) and SSP585 (b) scenarios and their zonal means (c) and (d). The figures show the difference between the future oxygen sources (averaged from 2087 to 2100) and the current oxygen sources (averaged from 2001 to 2014) in units of kg/m^2. The shaded areas in (a) and (b) are significant at the 95% confidence level (Ding *et al.*, 2022).

According to observations and analyses using the conditional nonlinear optimum perturbation related to parameter (CNOP-P) method (Francioni *et al.*, 2019), Rh also increases with increasing soil temperature and wetness under climate change, increasing oxygen use. There are still many unanswered questions about how anthropogenic activities will affect soil respiration. Although it is a frequent misconception that soil respiration will generally increase following fertilization, the actual reaction differs across various fertilizers and plant species. For instance, in Sweden, fertilized and unfertilized barley had identical soil respiration rates (Paustian *et al.*, 1990), whereas a wide range of mineral nutrients increased soil respiration in a pine forest. Different sites have found no change with stimulations and decreases in soil respiration in response to various land uses such as forest cutting, irrigation, and drainage. Our findings suggest that variations in soil respiration should be considered when analyzing land-use options. Soil respiration variations are frequently disregarded when discussing how anthropogenic activities and associated climate change affect ecosystems.

The model simulations for arid regions revealed that the oxygen production for the two scenarios in all drylands indicated an increase at the end of the 21st century, especially in the scenario with the highest concentration of greenhouse gases, RCP8.5 (Han *et al.*, 2021). Increased ambient CO_2 can accelerate the photosynthetic pathway by raising the rate of Rubisco's carboxylation reaction (ribulose-1,5-bisphosphate carboxylase oxygenase). Additionally, it has been proposed that variations in precipitation serve as a key driver of oxygen synthesis in ecosystems of arid and semi-arid regions with restricted access to water. One of the earliest examples of vegetation greening that enhances oxygen generation is the "greening Sahel" (Herrmann *et al.*, 2005), which is mostly driven by increases in precipitation following a severe drought. This is also true in arid and semi-arid regions where substantial afforestation is associated with a notable local increase in oxygen production, like the afforestation in China and India (Chen *et al.*, 2019a).

5.4 Soil Moisture Variability on Different Timescales

The moisture content of the soil impacts the division of net radiation into latent and sensible heat fluxes, which in turn influences the dynamics of the depth of atmospheric boundary layer and the consequent production of precipitation. It has been discovered that the interaction between soil moisture and precipitation is more pronounced in areas where soil moisture temporal variability is greater, such as in arid and semi-arid regions, as well as in transitional zones between dry and wet climates. For this reason, soil moisture variability throughout a range of time periods continues to be a focus of hydrology and climate research. This section discusses the variability of soil moisture at various scales for distinct regions.

5.4.1 *Methods*

(1) *Model and experimental design*

The Noah-MP was employed, and three experiments were conducted to examine the mechanism of soil moisture fluctuation, particularly the contributions of temperature and precipitation change to soil moisture drying. Niu and Yang assessed the Noah-MP at both local and global levels, demonstrating the benefits of using several alternative schemes in modeling simulations (Yang *et al.*, 2011). For the years 1948–2010, we drove the model using a set of $1^\circ \times 1^\circ$, 3-hourly, near-surface meteorological data processed by GLDAS. The simulation's initialization was done at random; therefore, the outcomes from the first two years were taken out to reduce the initialization's influence. We carried out three experiments: (1) the standard simulation, using the standard meteorological data processed by GLDAS as forcing data; (2) the ConTemp simulation, conducted as departures from the standard simulation but using the composite 3-hourly climatology of near-surface air temperatures over the first 5 years (1948–1952) as the forcing data for all years; and (3) the ConPrep simulation, which is the same as the ConTemp simulation but for a composite climatological precipitation.

(2) *Maximum covariance analysis (MCA)*

Singular-value decomposition analysis, also referred to as MCA, is a practical method for identifying linked patterns between two distinct geophysical domains and is frequently applied in climate research. MCA is capable of isolating pairs of spatial patterns and their corresponding time series by performing an eigenanalysis on the temporal covariance matrix between two data fields. MCA does this by removing the logical relationships between the two variables that are most tightly related to one another. The non-parametric Mann-Kendall test, which is frequently employed to identify monotonic trends in hydrometeorological time series, was utilized to calculate trends in this study. A technique was used to investigate how temperature and precipitation affected the trend in soil moisture. The soil moisture is first predicted using multiple linear regression and then regressed by the annual mean precipitation and temperature fields at each grid point. Equation (5.6) is the regression equation:

$$S' = a \times P + b \times T + C, \tag{5.6}$$

in which P and T are the time series of annual mean precipitation and temperature, respectively; a and b are the regression coefficients; and C is a constant. S' is the soil moisture predictor, which can be roughly determined as the soil moisture component caused jointly by precipitation and temperature. This concept is frequently applied to distinguish between internal and external sea surface temperature trends (Ting *et al.*, 2009). Similarly, a straightforward linear regression between soil moisture and precipitation can be used to derive the soil moisture component that is only forced by precipitation (or temperature). Equation (5.6) can be used to determine the contributions of precipitation to the trend in soil moisture based on the multiple regression (Equation 5.7):

$$\mathrm{CP} = (a \times \Delta P)/\Delta S' \times 100\%, \tag{5.7}$$

in which $\Delta S'$ and ΔP are the changes in mean S' and precipitation between the first and last 10 years. Like this, Eq. (5.7) can be used to determine how much temperature contributes to the trend in soil moisture:

$$\mathrm{CT} = (b\times)\Delta T/\Delta S'' \times 100\%. \tag{5.8}$$

It should be noted that because the contribution is relative, the total of CP and CT is 100%.

5.4.2 *Characteristics of soil moisture variability in transitional regions*

The distributions of annual mean soil moisture trends over the world from 1948 to 2010 are depicted in Fig. 5.12. High-latitude regions (above 60°S or 70°N) were disregarded because permafrost predominated there (similarly hereinafter). For 40.6% of the world's area, significant trends ($\rho = 0.05$) were found; of these, 39.8% were positive and 60.2% were negative. Eastern Australia, Northern Africa, East Asia, and Southern Europe all experienced significant drying trends. Numerous areas that already had somewhat low average soil moisture values had substantial drying trends. For central United States, South America, and Western Australia, we noticed a slight wetness trend. Given that there has been a considerable rise in precipitation in these areas over the previous few decades, this may have been precipitation induced. The general similarities between the soil moisture trend and different drought indices, such as the PDSI (Dai, 2011, 2013a) and the AI (Feng and Fu, 2013), indicate that the broad trends shown by soil moisture data are probably accurate.

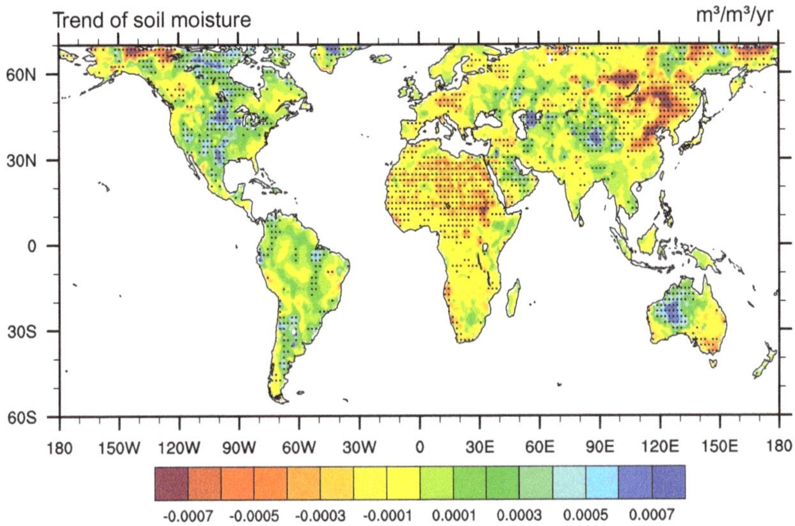

Fig. 5.12. Global annual mean soil moisture trends from 1948 to 2010. The areas marked with crosses denote significant trends at the 95% confidence level according to a two-tailed Student's t test (Cheng and Huang, 2016).

Only over the transitional zones did soil moisture drying become important, but it did so with a two-band structure. Around 1955, the apparent drying began to appear in areas with an AI between 0.65 and 0.9, which are referred to in this research as humid transitional regions. When compared to 1948, the soil moisture decreased by more than 0.004 m^3/m^3. The amplitude of the drying increased swiftly and reached extremes around the 1990s. Beginning in the 1980s, the extent of drying increased toward the arid and semi-arid regions along with its increasing intensity. The subhumid and semi-arid regions within the AI range of 0.4–0.65 (referred to as "dry transitional regions" in this section) experienced significant drying over the course of the following three decades. Furthermore, as of 2010, the amplitude of the drying was still increasing, suggesting that the soil moisture may continue to decline over the following years. A strong vertical gradient was shown for the future change in soil moisture in CMIP5, with more detrimental impacts closer to the surface. Although some northern middle-to-high latitude regions see negative annual surface changes, overall trends are positive (Berg *et al.*, 2016).

Although variations in precipitation are a major factor influencing the variability of soil moisture, the latter can also be influenced by variations in temperature (Cheng *et al.*, 2015). Greater larger vapor pressure deficit and evaporative demand are caused by rising temperatures, which could

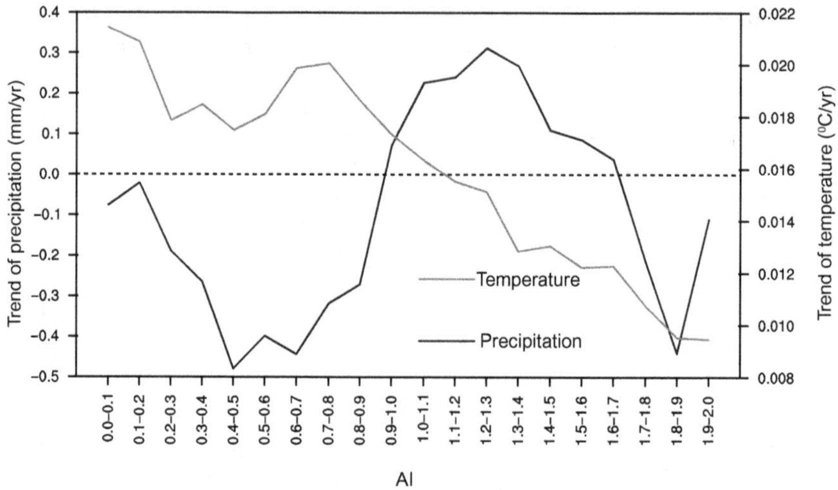

Fig. 5.13. The linear trends of global annual mean temperature and precipitation as a function of the climatological mean AI. The climatological mean AI on the x-axis is the annual mean AI for the period from 1961 to 1990 (Cheng and Huang, 2016).

increase evapotranspiration and decrease soil moisture. In Fig. 5.13, which depicts an exacerbated precipitation drop in the AI range of 0.4–0.8, we also calculated the regionally averaged precipitation trend as a function of the climatological mean AI. This shows that the increased drying over the transitional regions may be a result of the precipitation drop. The rising temperature also has a significant effect on soil drying because it can enhance evapotranspiration and intensify the soil moisture drying brought on by a decline in precipitation. Similar to the patterns in precipitation, there are regional differences in the increase in surface air temperature, which is evident in the dry regions with an AI below 0.9. This is due to the fact that soil's heat capacity diminishes as soil moisture increases, whereas water's heat capacity is noticeably higher. In these conditions, when the change in net radiation is the same, the increase in temperature on dry land is greater than that on wet land. Only in the transitional regions, when taking the variability of precipitation and temperature together, did the precipitation decrease and the temperature increase significantly. Therefore, the combined effects of reduced precipitation and elevated warmth may be to blame for the accelerated soil moisture drying in transitional locations.

5.4.3 *Mechanism analysis of enhanced soil moisture drying*

We used a straightforward linear regression to generate the soil moisture predictand regressed by annual mean precipitation or temperature fields at

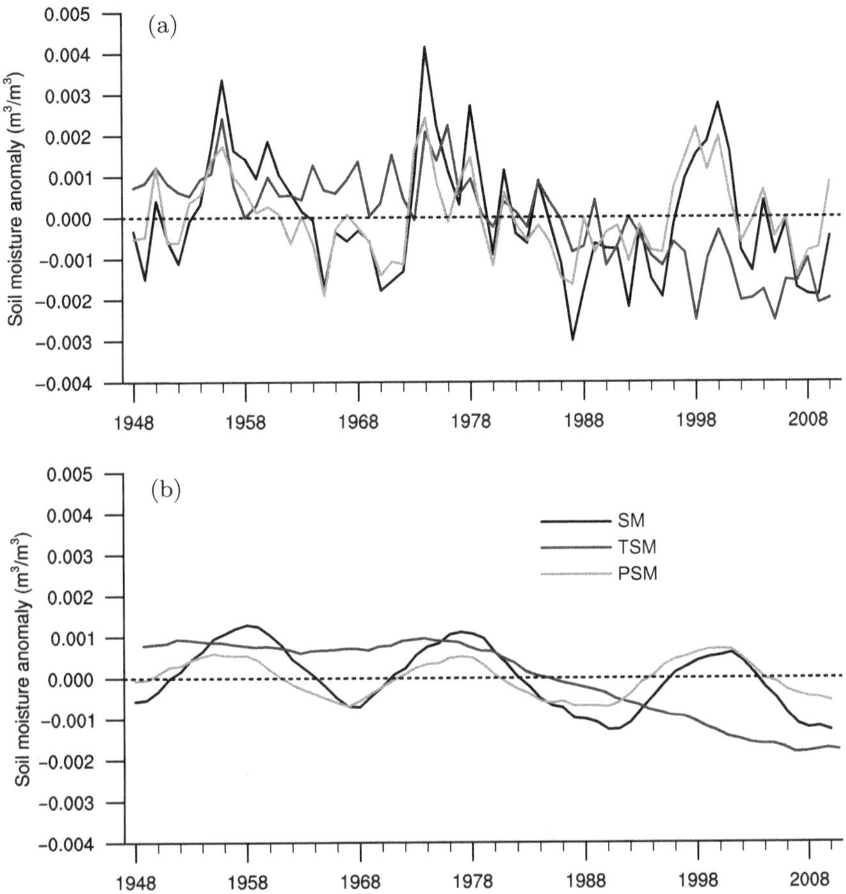

Fig. 5.14. The time series of global annual mean soil moisture (SM), TSM, and PSM anomalies from 1948 to 2010 (a). Also shown is the 7-year running mean time series (b) (Cheng and Huang, 2016).

each grid point in order to explore the effects of temperature and precipitation to the soil moisture variability. The two predictors for temperature and precipitation are designated as PSM and TSM, respectively. The PSM (TSM) can be defined as the component of soil moisture brought on by precipitation (temperature). This idea is frequently applied to distinguish between internal and external sea surface temperature trends (Ting *et al.*, 2009). The global annual mean soil moisture, TSM, and PSM anomalies from 1948 to 2010 are depicted in Fig. 5.14. The annual mean soil moisture anomaly time series exhibits a discernible decreasing trend over the past 63 years at a rate of –0.000176 $m^3/m^3/10$ yr, which is much higher than the 90% confidence level. At interannual-to-decadal timescales, there was significant heterogeneity within the long-term linear trends that were identified. From 1948 to 2010, the change in soil moisture shows a 20-year cycle period with 10 years of wetness and 10 years of drying. Despite a declining tendency over the whole time period at a rate of -0.0000284 $m^3/m^3/10$ yr, the long-term variability of the PSM is not statistically significant compared to soil moisture. However, the soil moisture time series at interannual-to-decadal timescales showed a good correlation ($r = 0.85$) with the PSM time series. Unlike the PSM, the TSM time series exhibits a perceptible long-term trend and subdued decadal fluctuation. The rate was much lower than the 99% confidence threshold at $-0.0005m^3/m^3/10$ yr.

The TSM changed slightly before 1976 but has drastically declined since then at a rate of -0.000869 $m^3/m^3/10$ yr, which is significantly higher than the 99% confidence threshold. Since the 1970s, when global warming first began, this dramatically declining tendency has been present. According to this finding, temperature mostly determines the long-term trend of soil moisture globally, while precipitation affects soil moisture variability at interannual-to-decadal time periods. The study mentioned above examines the relationship between changes in soil moisture and drought. The difference between the original soil moisture fields and the PSM (TSM) can be generally calculated as the soil moisture component minus the precipitation signal (temperature). We designated RPSM and RTSM as the two remaining components of soil moisture. The spatial distributions of the RPSM and RTSM trends are depicted in Fig. 5.15. The pattern of the RPSM and RTSM trends is strikingly like the soil moisture trend, as can be seen by comparing Fig. 5.15. Although the trends are also significant at the 95% confidence level for most places with a substantial soil moisture trend, the magnitudes are often small in comparison to the soil moisture trend. This suggests that the long-term soil moisture trend is influenced by both precipitation and temperature. However, as demonstrated in Figs. 5.15(a) and 5.15(b), the influence of temperature and precipitation on the soil moisture trend differs. In general, the RPSM's drying tendency was more notable than the RTSM's. For instance, the RPSM and RTSM drying trends over EAS were -0.00231 and -0.00210 $m^3/m^3/10$ yr, respectively. The RTSM wetness trend, in contrast, was greater than the RPSM wetting trend, particularly across Western Australia and central North America, where a modest wetting tendency occurred. The wettest RTSM trends for AUS and NAM were 0.00382 and 0.00231 $m^3/m^3/10$ yr, respectively. These trends were more than the wettest RPSM trends (0.00157 and 0.00163 $m^3/m^3/10$ yr) and even close to the wettest soil moisture trends (0.00280 and 0.00234 $m^3/m^3/10$ yr). We showed the relevance of the

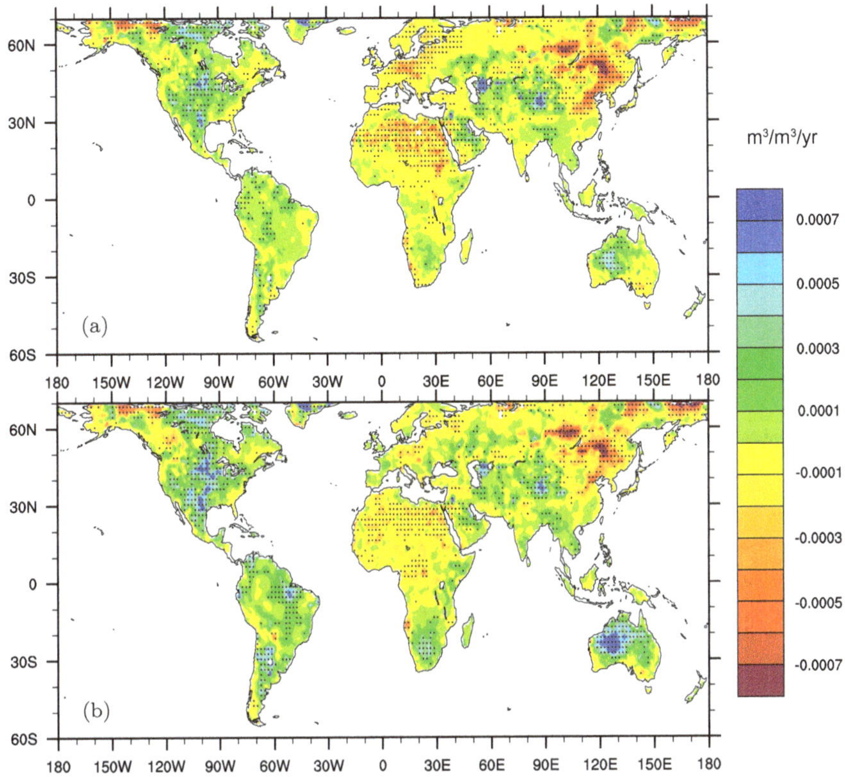

Fig. 5.15. Global trend distributions for annual mean RPSM (a) and RTSM (b) from 1948 to 2010. The areas marked with crosses denote significant trends at the 95% confidence level according to a two-tailed Student's t test (Cheng and Huang, 2016).

difference between the RPSM (RTSM) and soil moisture trends to establish the validity of the findings. For 85% (21%) of the worldwide area, a significant difference ($\rho = 0.05$) between the soil moisture trend and the RPSM (RTSM) trend was seen. This demonstrated that whereas precipitation had little effect on the long-term soil moisture pattern, temperature had a significant impact. Overall, the sign of soil moisture change is mostly influenced by changes in precipitation; wetting is caused by increased precipitation and drying by decreased precipitation. Contrarily, global warming has a negative impact on the trend of soil moisture in most locations, including the moist ones, and its impact on soil moisture drying is more pronounced than that of precipitation.

Although we used linear regression techniques to study the roles of temperature and precipitation in the soil moisture trend, the results may have several problems because the climate system is complicated and nonlinear.

We checked the outcomes of the linear regression techniques using the Noah-MP land surface model. We evaluated the temporal correlation coefficient (TCC) and relative root-mean-square difference (RRMSD) between the Noah-MP and GLDAS soil moisture data to evaluate the uncertainty of the Noah-MP soil moisture data for each spatial grid. Equation (5.8) was used to calculate the RRMSD:

$$\mathrm{RRMSD} = \frac{\mathrm{RMSD}}{\mathrm{SM}_{\mathrm{mean}}} \times 100\%, \tag{5.9}$$

in which RMSD is the root-mean-square difference between the Noah-MP and GLDAS soil moisture data. $\mathrm{SM}_{\mathrm{mean}}$ is the mean soil moisture. We used the TCC to verify the temporal consistency of long-term and interannual variability and the RRMSD to verify the deviations between the datasets.

In this section, we conducted three tests. The ConPrep (ConTemp) simulation's constant precipitation (temperature) across the course of the simulation eliminated the signal of changing precipitation (temperature) from the soil moisture. The RPSM (RTSM) field discovered by regression analysis and the ConPrep (ConTemp) simulation are equivalent in several ways. Similar to this, the PSM (TSM) field equals the difference between the standard and ConPrep (ConTemp) simulations (defined as PreSM (TemSM)). The soil moisture, TSM, and PSM correlation coefficients between the 0–200 cm and 10–40 cm GLDAS soil moisture are, respectively, 0.95, 0.99, and 0.99, which are noticeably above the 99% confidence threshold. However, with a rate of –0.000273 $m^3/m^3/10$ yr, which is significantly higher than the 95% confidence level, the downward trend of the 10–40 cm soil moisture was more significant than that of the 0–200 cm soil moisture. Additionally, the Noah-MP soil moisture anomaly's time series exhibits striking similarities to the GLDAS soil moisture at interannual-to-decadal time periods, with a correlation coefficient that is significantly higher than 99% confidence. This correlation is 0.88. The annual mean soil moisture anomaly time series exhibits a discernible downward trend over the past 63 years at a rate of –0.000642 $m^3/m^3/10$ yr, which is much higher than the 99% confidence level. The PreSM exhibits an increasing trend over the entire time period, in contrast to the PSM, at a rate of 0.00037 $m^3/m^3/10$ yr. However, at interannual-to-decadal time periods, there is a good correlation ($r = 0.89$) between the PreSM time series and the Noah-MP simulated soil moisture. The time series of TemSM has a discernible long-term trend with three periods: a mild period before 1980, fast drying from 1981 to 2001, and a drying hiatus since 2002. This is similar to the TSM. According to the data shown by the soil moisture predictor, this result indicates that precipitation has an impact on the variability of soil moisture at interannual-to-decadal timescales, but that temperature is mostly responsible for the long-term trend of soil moisture.

5.4.4 *Long-term trend soil moisture over East Asia*

Most land areas were warmed by 1°C–3°C between 1948 and 2010, with the warmest areas across Northern East Asia, particularly north of 40°N. The drop in precipitation during this time appears to be dispersed in a Northeast to Southwest belt, which coincides with the 600-mm precipitation line. Consequently, soil moisture showed a significant trend in 46% of East Asia ($P = 0.05$), and it was primarily falling (70%). The areas closest to Lake Baikal in Northeast China, North China, part of Mongolia, and Russia experienced the most dryness. The most substantial wetness, in contrast, occurred in Western China, with Southern Xinjiang having the highest rate (more than 0.01 $m^3/m^3/10$ yr) and significantly surpassing the 95% confidence level. Although there are significant regional variations, the spatial distribution of the soil moisture trend is similar to that of precipitation. For instance, while soil dryness is not considerable and is even soaking, the drop in precipitation across Yunnan Province and Southeast India is significant. On the other hand, the rains concealed the significant soil drying that had occurred close to Lake Baikal. This finding suggests that precipitation is a significant determinant of soil moisture, but it is not the only one. Given the high degree of changeability in soil moisture, it is essential to examine the variables influencing soil moisture change in order to comprehend its variability characteristics and take action to reduce the rate of soil drying. Spatial and temporal variability in soil moisture can be caused by a variety of environmental factors, including vegetation, topography, and soil texture. These processes are made more difficult by interaction with the intricate and nonlinear climate system. Although it makes intuitive sense that variations in precipitation would be the main factor influencing soil moisture, temperature variations can also have an impact. Here, we perform an MCA for rainfall and soil moisture from 1948 to 2010. The first leading MCA mode (MCA1), which can be seen in Fig. 5.16, accounts for 32% of the variability in precipitation and soil moisture. This pattern depicts precipitation and soil moisture trends that are similar to those in the trend map. We observe that the spatial patterns of precipitation and soil moisture are similar, which suggests that the long-term variability of soil moisture and precipitation trends is strongly correlated.

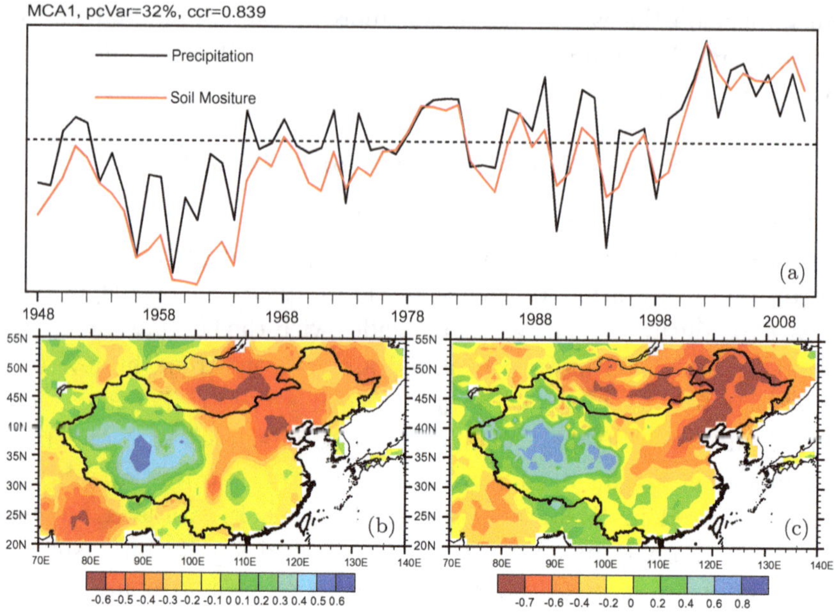

Fig. 5.16. Temporal (a) and spatial (b) patterns of the MCA1 mode for precipitation and soil moisture (c) from 1948 to 2010 (normalized by its standard deviation prior to MCA analysis). The red (blue) areas are drying (wetting) for a positive temporal coefficient on the principal component (PC) time series. In Fig. 5.16(a), pcVar is the percentage variance explained by MCA1, and ccr is the correlation coefficient between PC1 precipitation and PC1 soil moisture, which is statistically significant at the 99% confidence level (Cheng *et al.*, 2015).

To find the most notable temperature pattern impacting soil moisture, MCAs between temperature and soil moisture were also conducted. Figure 5.17 demonstrates that the major components, or MCA1 patterns, account for 75% of the total variance. The temporal coefficient is substantially linked ($r = 0.96$) with the annual mean temperature (Fig. 5.17(a)), and the spatial patterns (Fig. 5.17(b)) mirror the warming patterns over East Asia, indicating that the MCA1 patterns represent global warming. This temperature mode is accompanied by a soil moisture mode that displays a temporal evolution that is similar to that of the temperature mode (Fig. 5.17(a)), but with exceptionally intricate spatial patterns (Fig. 5.17(c)) that mirror the temperature mode's trend map. In Mongolia, Northeast China, and North China, the soil moisture MCA1 patterns are positive, but they are negative in Western East Asia. This finding shows that variations in precipitation and the terrestrial water budget were both significantly influenced by surface warming. Studies to ascertain the specific impacts of precipitation and temperature on the trend in soil moisture are

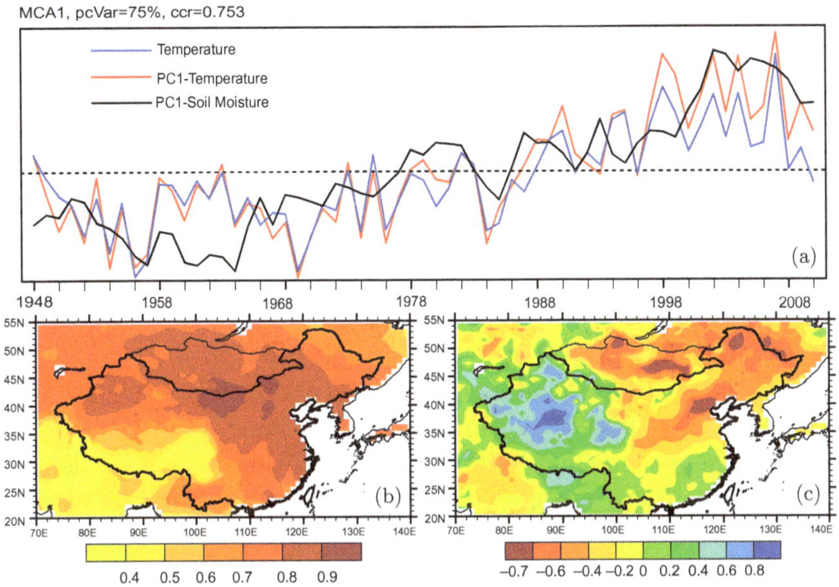

Fig. 5.17. Temporal (a) and spatial (b) patterns of the MCA1 mode for temperature and soil moisture (c) from 1948 to 2010 (normalized by its standard deviation prior to MCA analysis). The red (blue) areas are drying (wetting) for a positive temporal coefficient on the PC time series. In (a), pcVar is the percentage variance explained by MCA1, and ccr is the correlation coefficient between PC1 temperature and PC1 soil moisture, which is statistically significant at the 99% confidence level. Also shown in (a) is the area-mean temperature obtained from GLDAS (blue line). The correlation between area-mean temperature and PC1-temperature (PC1-soil moisture) time series is 0.96 (0.67), which is statistically significant at the 99% confidence level (Cheng *et al.*, 2015).

necessary because MCA cannot show the pathways between temperature, precipitation, and soil moisture.

It is possible to separate soil moisture changes into components that are influenced by temperature and precipitation. It is clear that precipitation has a key role in determining the trend in soil moisture. The direction of soil moisture change, whether drying or wetting, is determined by the precipitation trend. The wetting trend was mostly found where precipitation increased, while the drying trend was mostly found where precipitation decreased. However, it is not entirely obvious how temperature trends, particularly temperature increases, and soil moisture soaking are related. When contrasted to the temperature trend, the wetness trend's size is seen to be erratic. Only the role of temperature warming as an amplifier of soil dryness is available to us. A decrease in precipitation's impact on soil drying is amplified by an increase in temperature. It is demonstrated that the drying trend's magnitude at the top is greater than its magnitude at the bottom. This shows that while the lack of precipitation is the main cause of soil drying, the effect of the concomitantly rising temperature is amplified. The effect of rising temperatures on soil drying, however, is neither perfect nor unqualified. For instance, the most pronounced soil moisture drying trend occurs at intermediate temperature rises when the precipitation trend varies from 0 to 1 mm/yr. This issue might be caused by how warming temperatures affect changes in soil moisture. Since modifying soil evapotranspiration is how temperature impacts soil moisture, and because temperature is not the sole factor affecting evapotranspiration, it is impossible to directly link rising temperatures to increased soil moisture. In general, evapotranspiration will rise as the temperature rises, but it is not a limitless process. The soil suction will be so great if soil moisture is below a certain level that evapotranspiration will not grow. This evapotranspiration regime, which is limited by soil moisture or energy, suggests the need to analyze the changing temperature–soil moisture relationship in accordance with the climatic area.

The climatological mean precipitation, which represents the climatological state, is the yearly mean precipitation for the years 1961–1990. Semi-arid regions have an increase in temperature, with a warming peak occurring at 200–300 mm/yr of precipitation. As temperatures rise, semi-arid locations also experience a sharp decline in soil moisture, with the driest values being in the 400–500 mm/yr range. The agreement, which is an energy-limited evapotranspiration regime, after the precipitation range of 400 mm/yr is of particular importance. The amount of warming and drying in this area decreases as precipitation levels rise. While the amplitude of the warming–drying trend is fairly minimal in the wet zone (rainfall >1,000 mm/yr), it is evident in the intermediate moisture regions (precipitation ranging from 400 to 900 mm/yr). This finding implies that, at least in certain places, rising temperatures might worsen the dryness brought on by falling precipitation.

The contributions of temperature and precipitation are calculated to express the quantitative effects of each. Regions with growing soil moisture trends have been hidden since we concentrated on the mechanism of soil moisture drying in this study. The data clearly show the proportional effects of temperature and precipitation on the spatially uneven drying of soil moisture. Precipitation over Southern East Asia and temperature over North China and Northwestern East Asia make up most of the contribution in general, and their contributions over Eastern China are approximately equal. Over East Asia, the overall mean contribution of temperature to soil moisture drying is roughly 58.1%, which is marginally higher than the contribution of precipitation (41.9%). Thus, a rise in temperature nearly doubles the drying of soil moisture brought on by a fall in precipitation.

5.4.5 *Uncertainties of soil moisture in model simulations*

Studying land–atmosphere interactions also heavily relies on the accuracy of soil moisture data. A variety of historical simulations with various ESMs or starting points were made available by the CESM and CMIP5 experiments. To lessen model variation, an ensemble mean is typically used. Figure 5.18 shows the global annual mean soil moisture levels for each separate CMIP5 ESM, CESM simulation, and GLDAS data for the years 1961–1990. Only their ensemble mean is shown because there is barely any difference between the 30 CESM simulations, with values of 20.0 kg/m^2, 20.5 kg/m^2, and 22.1 kg/m^2, respectively. There are significant variations in the climatological soil moisture levels among the several CMIP5 ESMs, ranging from 8.5 kg/m^2 to 26.1 kg/m^2. In these ESMs, the maximum soil wetness was around three times higher than the minimum value. Users cannot easily transfer the soil moisture from one product to other models because of this significant variation. Since soil moisture climatology differs widely among models, and models have varying operational ranges for soil moisture, earlier research actually stressed that the fundamental information of modeled soil moisture

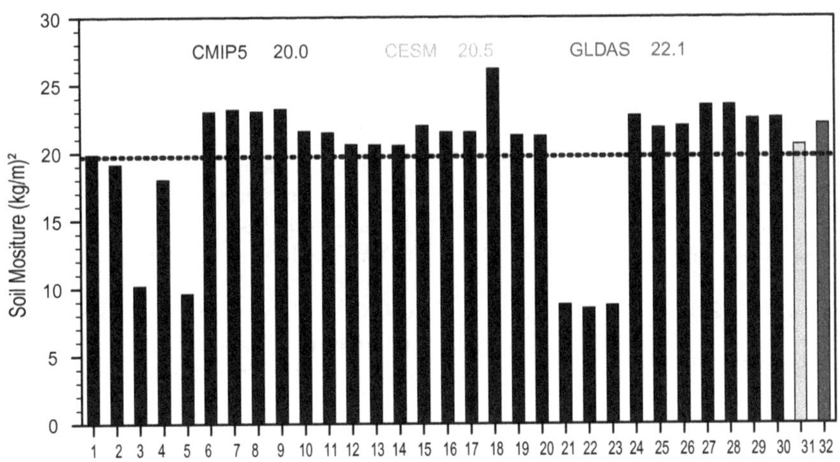

Fig. 5.18. Global annual mean soil moisture in the period of 1961–1990 for CMIP5 (nos. 1–30), CESM-EM (no. 31), and GLDAS (no. 32). The dashed horizontal line is the average value of 30 CMIP5 Earth System Models (ESMs). The numbers of CMIP5 ESMs (1–30) are the same as those in Table 5.7. The discrepancies among the 30 CESM simulations are negligible, so we present their ensemble mean here (Cheng *et al.*, 2017).

Table 5.7. List of CMIP5 ESMs used in this chapter with brief descriptions (Cheng *et al.*, 2017).

Model names	Experiments	Origin
1 bcc-csm1-1	H,R	Beijing Climate Center, China Meteorological Administration, China
2 bcc-csm1-1-m	H,R	Beijing Climate Center, China Meteorological Administration, China
3 BNU-ESM	H,R	College of Global Change and Earth System Science, Beijing Normal University, China
4 CanESM2	H,R	Canadian Centre for Climate Modelling and Analysis, Canada
5 CCSM4	H,R	National Center for Atmospheric Research, USA
6 CESM1-BGC	H,R	Community Earth System Model Contributors, USA
7 CESM1-CAM5	H	Community Earth System Model Contributors, USA
8 CESM1-FASTCHEM	H	Community Earth System Model Contributors, USA
9 CESM1-WACCM	H	Community Earth System Model Contributors, USA
10 CNRM-CM5-2	H	Centre National de Recherches Météorologiques/Centre Européen de Recherche et Formation Avancée en Calcul Scientifique, France
11 CNRM-CM5	H	Centre National de Recherches Météorologiques/Centre Européen de Recherche et Formation Avancée en Calcul Scientifique, France
12 GFDL-CM3	H,R	NOAA Geophysical Fluid Dynamics Laboratory, USA
13 GFDL-ESM2G	H,R	NOAA Geophysical Fluid Dynamics Laboratory, USA
14 GFDL-ESM2M	H,R	NOAA Geophysical Fluid Dynamics Laboratory, USA
15 GISS-E2-H-CC	H,R	NASA Goddard Institute for Space Studies, USA
16 GISS-E2-R-CC	H,R	NASA Goddard Institute for Space Studies, USA
17 GISS-E2-R	H,R	NASA Goddard Institute for Space Studies, USA
18 HadCM3	H	Met Office Hadley Centre, UK
19 HadGEM2-CC	H,R	Met Office Hadley Centre, UK

(*Continued*)

Table 5.7. (*Continued*)

Model names	Experiments	Origin
20 HadGEM2-ES	H,R	Met Office Hadley Centre, UK
21 IPSL-CM5A-LR	H,R	Institut Pierre-Simon Laplace, France
22 IPSL-CM5A-MR	H	Institut Pierre-Simon Laplace, France
23 IPSL-CM5B-LR	H,R	Institut Pierre-Simon Laplace, France
24 MIROC5	H,R	Atmosphere and Ocean Research Institute, Japan
25 MIROC-ESM-CHEM	H,R	Japan Agency for Marine-Earth Science and Technology, Japan
26 MIROC-ESM	H,R	Japan Agency for Marine-Earth Science and Technology, Japan
27 MRI-CGCM3	H,R	Meteorological Research Institute, Japan
28 MRI-ESM1	H	Meteorological Research Institute, Japan
29 NorESM1-ME	H	Norwegian Climate Centre, Norway
30 NorESM1-M	H	Norwegian Climate Centre, Norway
31 ACCESS1-0	R	Commonwealth Scientific and Industrial Research Organization (CSIRO) and Bureau of Meteorology (BOM), Australia
32 ACCESS1-3	R	CSIRO and BOM, Australia
33 CESM1-CAM5-1-FV2	R	Community Earth System Model Contributors, USA
34 CSIRO-Mk3-6-0	R	CSIRO in collaboration with Queensland Climate Change Centre of Excellence, Australia
35 FGOALS-g2	R	LASG, Institute of Atmospheric Physics, Chinese Academy of Sciences and CESS, Tsinghua University, China
36 GISS-E2-H	R	NASA Goddard Institute for Space Studies, USA
37 inmcm4	R	Institute for Numerical Mathematics, Russia

Note: The symbols "H" and "R" in the experiments indicate the historical run (1920–2005) and two future scenario (RCP4.5 and RCP8.5) runs (2006–2080) used in this study. The first ensemble run is used if a model has multiple ensemble runs.

lay not in its absolute amount but in its change over time. The soil moisture anomaly (SMA) was computed in order to reduce the impact of significant variations in absolute magnitude. The study's SMA is the annual mean soil moisture anomaly compared to the base of averaged soil moisture over a 30-year period (average of annual mean soil moisture during 1961–1990).

The climatology and average soil moisture value are shown in the previous paragraph, although variations in soil moisture are typically of interest. The general similarities between the soil moisture trend and different drought indices, such as the Palmer Drought Severity Index (Dai, 2011, 2013) and the Aridity Index (Feng and Fu, 2013; Huang *et al.*, 2016a), indicate that the general trends displayed by the GLDAS soil moisture data are probably reliable. The magnitudes of the CMIP5-EM and CESM-EM SMA trends, however, were significantly lower than the GLDAS SMA trend. When only evaluating the spatial distribution and ignoring the magnitude, the SMA trends of CMIP5-EM and CESM-EM are very comparable. While there were slight wetting tendencies in central Africa and Western Asia, the most noticeable drying patterns were recorded in the Southern USA, Northern South America, Europe, and Southeast Asia. Particularly in locations with noticeable drying tendencies, the geographical distributions were very different from those of the GLDAS. This illustrates that while CMIP5 and CESM simulations are effective at describing the geographical and temporal distributions of soil moisture on a global scale, they are not accurate at recreating variance in soil moisture between regions.

The relative standard deviations of SMA for the CMIP5 and CESM simulations, for the period of 1920–2005, are calculated and shown in Fig. 5.19 to quantitatively indicate the uncertainties in model structure and initial circumstances at each grid point. The ratio of standard deviation to mean soil moisture is used in this study to establish the relative standard deviation. It can reduce the impact of various environmental circumstances, making it a better way to convey uncertainty about soil moisture than the usually used standard deviation.

The Sahara, Arabian Peninsula, Iranian Plateau, Northern China, and Western Australia all have high values in Fig. 5.19(a), showing that these regions are the most susceptible to the structure of the climate model. The relative standard deviations of the CESM simulations display comparable spatial patterns to the CMIP5 simulations (Fig. 5.19(b)). Over most regions, the relative standard deviation values of the CMIP5 are slightly larger than those of the CESM models, but in the Sahara, the CMIP5 value is significantly higher. The regionally averaged relative standard deviations, as a function of climatological mean soil moisture, are shown in Fig. 5.19(c). As can be seen in Figs. 5.19(a), 5.19(b), the relative standard deviation is highly dependent on the climatic area. The yearly mean soil moisture for the years 1961 through 1990 is the climatological mean soil moisture, and it depicts the climatological state. As the climatological mean soil moisture increases, the relative standard deviation declines. This suggests that humid locations are where the SMA agrees the most. The comparison between CMIP5 and CESM simulations in various climate zones is also represented visually in Fig. 5.19(c), where very dry locations have the highest model structural uncertainty while other regions exhibit a somewhat higher initial uncertainty.

Although standard deviation and relative standard deviation were used to examine the quantitative uncertainty in Fig. 5.19, standard deviation can only capture the variation among individual models and simulations. The standard deviation cannot be used to determine the temporal consistency of long-term and interannual variability. The standard deviation is a state

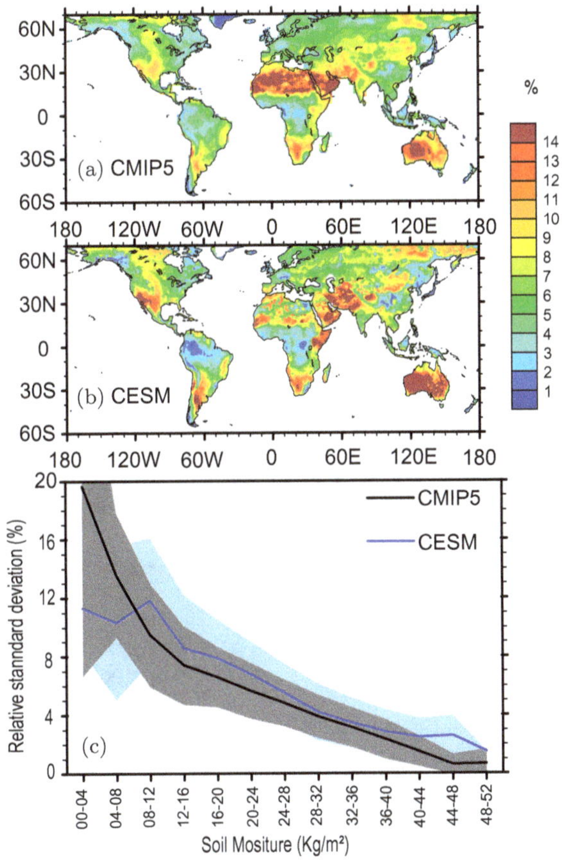

Fig. 5.19. Spatial distributions of the averaged relative standard deviation of SMA in the period of 1920–2005 for 30 CMIP5 ESMs (a) and 30 CESM simulations (b). Relative standard deviation as a function of climatological mean soil moisture (c). The climatological mean soil moisture on the x-axis is the annual mean soil moisture for the period 1961–1990. Uncertainty (shading) is one standard deviation (Cheng *et al.*, 2017).

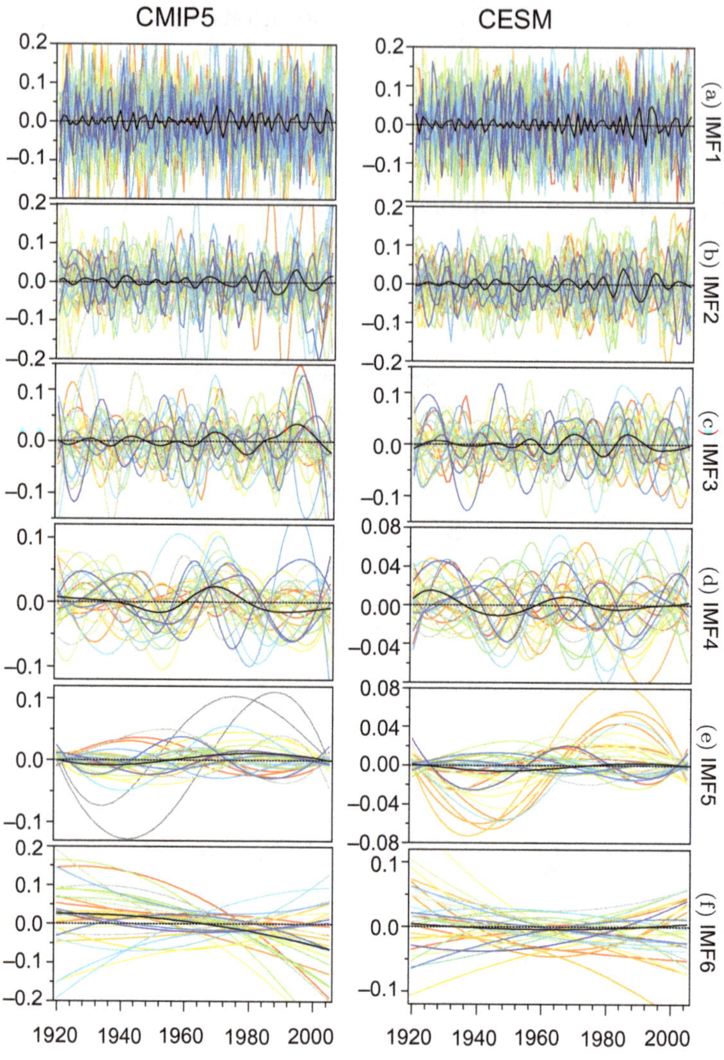

Fig. 5.20. Time series of intrinsic mode functions (IMFs) of the annual mean SMA based on ensemble empirical mode decomposition (EEMD). The colors show 30 individual members, and the black curve is their ensemble mean (Cheng *et al.*, 2017).

average that could omit some crucial information. The EEMD was used to segregate non-uniform variability at several time periods for the CMIP5 and CESM SMA in order to examine the uncertainty at various timescales. Figure 5.20 displays the time history of the 30 individual CMIP5 and CESM simulation members' EEMD IMF1 to IMF6 for the years 1920 to 2005. IMF1 to IMF5 are the signals of SMA from high to low frequency, and IMF6 can be seen as a linear trend in the time series. The largest cycle powers for IMF1 through IMF5 were found by wavelet analysis using the Morlet wavelet, and they are 3–5, 7–10, 14–20, 40–45, and 78 years. The periodic signal at the interannual timescale is primarily reflected by IMF1. IMF1's CMIP5 standard deviation has temporal variability and runs from 0.062 to 0.13 (0.066 to 0.11 for CESM). IMF6 displays the time series' linear trend. By using the formula TrendEEMD(t) = IMF6(t) – IMF6 (1920), which represents the accumulated trend since 1920, we were able to calculate the increase in the EEMD trend at a specific time from the reference time of that period (Ji *et al.*, 2014). The trend of the SMA series from 1920 to 2005 is represented by Trend EEMD (2005).

The linear trend is well supported by the fact that, out of the 30 CMIP5 simulations, 22 exhibit a declining trend and 8 exhibit an increasing trend. Nine out of the 30 CESM simulations show a rising trend, while 21 show a decreasing trend. Like IMF1 and IMF6, IMF2 through IMF5 also show a significant difference, indicating that model structure and initial conditions can both lead to uncertainty at all timescales. When we compare the ensemble mean to its individual simulations, we find that for all IMFs, the ensemble mean fluctuation is substantially lower than the individual simulations. This suggests that the ensemble mean can successfully eliminate both the internal variability fluctuations and the signal at interdecadal time periods, which can occasionally be crucial and beneficial. Therefore, it is not fully suitable to use "ensemble mean" in this situation.

In the 21st century, as predicted by CMIP5 RCP4.5 (RCP8.5), there is a decreasing trend at a rate of –0.045 (–0.074) kg/m^2 per 10 years, which is much higher than the 95% confidence level for both scenarios. The ensemble mean SMA of the CESM simulation exhibits a considerable drying trend in comparison to the CMIP5, at a rate of –0.026 (–0.044) kg/m^2 per 10 years for RCP4.5 (RCP8.5). According to the CMIP5 (CESM) model simulations, the annual mean SMA will fall by an ensemble mean of 0.25 (0.14) kg/m^2 for RCP4.5 and 0.40 (0.25) kg/m^2 for RCP8.5 for the periods of 2006–2025 and 2061–2080, respectively. Although there are substantial variances between these trajectories, almost all the CMIP5 and CESM individual simulations for the 21st century show a strong linear decreasing tendency. The findings suggest that model structure and initial condition uncertainty have a growing tendency over time. Before 2045, there is no discernible difference between the standard deviations for the four sets of simulations, but after that point, there is a substantial difference. The CESM simulations for the years 2046–2080 show little change, with rates of 0.0073 and 0.0035 kg/m^2 per 10 years for RCP4.5 and RCP8.5, respectively. However, the rates for the CMIP5 models are much higher than the 95% confidence level of 0.016 and 0.068 kg/m^2 per 10 years for RCP4.5 and RCP8.5, respectively. The averaged relative standard deviation of SMA throughout the period of 2006–2080 exhibits spatial patterns that are strikingly comparable to those of historical simulations, which exhibit significant errors in the semi-arid regions. Although the difference is not very big, the relative standard deviation for RCP8.5 is significantly higher than that for RCP4.5. This shows that model structure and beginning condition uncertainties are less significant than those of greenhouse gas emission scenarios.

5.5 Observation Analysis of Land–Atmosphere Interactions

Land–atmosphere interactions, which are made up of intricate processes and feedbacks involving radiative, hydrological, and biogeochemical processes, essentially affect the entire earth system and its climate. It is crucial to observe land–atmosphere interactions with all their components in order to accurately calculate the different fluxes and components of various interactions between land and atmosphere. Additionally, it helps with researching and forecasting the effects of land–atmosphere interactions on local and global extreme climatic events, which is a huge scientific challenge and needs international cooperation across numerous scientific communities and technological advancements. The examination of observational data on the interactions between land and air in arid and semi-arid environments is mostly covered in this part.

5.5.1 *SACOL site description and data collection*

One of the reference sites for the global Coordinated Energy and Water Cycle Observations Project is SACOL, which is situated in Northwest China's Gansu Province at 35°57′N, 104°08′E, 1965.8 m above sea level, roughly 48 km from Lanzhou's center (CEOP). The primary soil type of the parent soil is sierozem, which is an aeolian loess from the Quaternary. Short grass species, including Stipa bungeana, Artemisia frigida, and Leymus secalinu, which are typically less than 15 cm tall and cover less than 80% of the surface in summer and autumn, make up the majority of the vegetation on the surface. The site is situated on a generally north–south-oriented mesa with an average fetch length of around 120 m. The mesa is only a few hundred meters wide from east to west and roughly 600 meters long from north to south.

To the west and east of the site, respectively, are two valleys in the shape of "Vs". SACOL was created in 2005 on the Yuzhong campus of Lanzhou University and began continuous observations in May 2006. The observatory includes an array of instrumentation that measures (1) boundary layer meteorological parameters, (2) surface radiation, (3) surface flux, (4) soil heat and water moisture, (5) ambient air quality, (6) aerosol optical properties, (7) temperature and water vapor profiles, and (8) sky conditions. All the data used in this section were averaged over 30-min intervals, except for the raw eddy covariance.

The first Sino-American dust storm observation after SACOL's founding was made in March 2008. Three observatories — SACOL, the Jingtai mobile observatory, and the Zhangye mobile observatory — contributed to the measurements. The Jingtai mobile observatory was set up in a semi-arid area (37°20′N, 104°08′E) and approximately 300 kilometers to the north of SACOL. In this area, two studies were conducted, one at the Jingtai field (JF, 1592 meters above sea level) and the other at the Jingtai hill (JH, 1617 meters above sea level). It is roughly a mile between these two locations. Stipa bungeana, Artemisia frigida, and JF, a corn field irrigated with water from the Yellow River, covered the underside of JH during the trial. Data in flux have been gathered in JH from March 21 to March 31 and in JF from April 17 to May 30, 2008.

SACOL was located on the semi-arid Loess Plateau of China. It is significant since SACOL is only about 25 km from Xinglong Mountain, a remote oasis surrounded by a semi-arid region. A temporal experiment site was established at Xinglong Mountain (abbreviated as XM, 35°46′N,104°03′E, 2481m above sea level) in June 2008 to better understand the differences in the climatic components between these two surface types. The XM experiment field was situated in a transitional zone between a humid monsoon climate zone and a dry non-monsoon climate zone, comparable to SACOL, although the vegetation cover at XM was extremely vast. From July 1 to September 30, 2008, flux data were gathered, and at XM during this time, the underlying surface was benched.

5.5.2 *Variation in radiation components from SACOL site*

The distinctive underlying surface of Loess Plateau provides a source of dust aerosols for arid and semi-arid regions in Northwest China. These regions are also referred to as the "transitional zone" of climate and

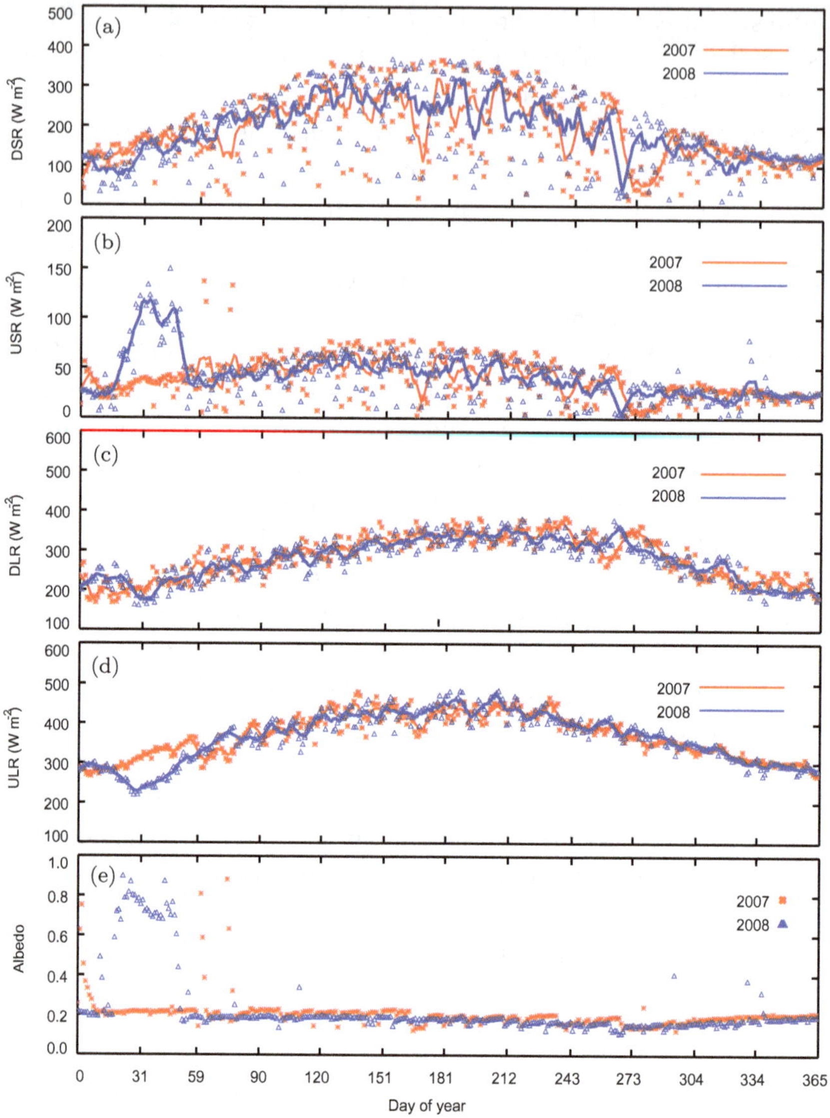

Fig. 5.21. Variations of the daily mean for downward shortwave radiation (DSR) (a), upward shortwave radiation (USR) (b), downward longwave radiation (DLR) (c), upward longwave radiation (ULR) (d), and surface albedo (e); the solid line represents the moving average smoothed by 7 days at SACOL from January 2007 to December 2008 (G. Wang *et al.*, 2010).

ecosystem change since they have experienced a severe aridity trend in recent decades. Field data gathered at the Semi-Arid Climate and Environment Observatory of Lanzhou University (SACOL, 35°57′N, 104°08′E,

Elev. 1965.8 m) between January 2007 and December 2008 were systematically analyzed, with a particular focus on the energy balance and energy partitioning of the land surface. This analysis was done to better understand the fundamental characteristics of the land surface energy budget and the seasonal and diurnal variations of moisture and heat flux over this region. The measurements (SACOL) were plotted against equivalent historical data gathered from a meteorological station from January 1956 to December 2008 to see whether the readings were indicative of the arid and semi-arid regions' climate (Fig. 5.21). About 7 kilometers from SACOL, the meteorological station was situated in Yuzhong (35°52′N, 104° 09′E), at a height of about 1875 meters. As part of normal observations, the historical data were gathered at 2:00, 8:00, 14:00, and 20:00 hours (local time). The findings demonstrate that our observations accurately reflect the weather during the past 53 years (1956–2008) and are thus regarded as indicative of the climate in this area.

The daily average surface albedo is shown in Fig. 5.22 and displays the seasonal and interannual variations of the four radiation components at SACOL: (1) downward shortwave radiation (hereafter, DSR), (2) upward shortwave radiation (USR), (3) downward longwave radiation (DLR), and (4) upward longwave radiation (ULR). A 7-day moving average was produced for these four components to account for diurnal fluctuation. The largest seasonal and interannual variability was found in ULR and DSR, particularly in the summer. Clouds and aerosols had a significant impact on both the DSR and USR, which sustained high values throughout the wet season. The highest daily mean value of DSR, which was determined in June 2008, was 369.2 W m^{-2}. However, due to the high surface albedo in winter, the highest

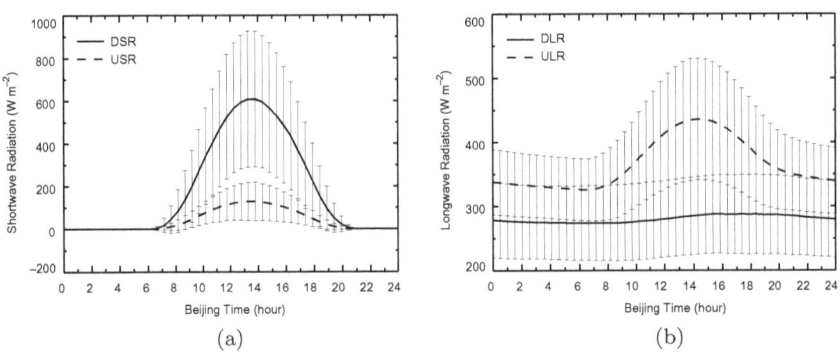

Fig. 5.22. Diurnal cycle and standard deviation of downward shortwave radiation (DSR) and upward shortwave radiation (USR) (a) and downward longwave radiation (DLR) and upward longwave radiation (ULR) (b) at SACOL from January 2007 to December 2008 (Wang *et al.*, 2010).

daily mean value of USR (150.1 W m^{-2}) did not arrive at the same time as that of DSR, as was anticipated. Cloud, aerosol, or snow cover had little to no impact on the seasonal changes in DLR and ULR, which both maintained high levels throughout the growth season. The highest daily mean values of DLR and ULR, which occurred in August 2007 and July 2008, respectively, were 386.8 W m^{-2} and 484.2 W m^{-2}. In January 2008, DLR (162.6 W m^{-2}) and ULR (220.9 W m^{-2}) reached their lowest levels.

We employed a composite analysis technique to look at the diurnal change of the radiation components. Figure 5.22 displays the yearly means of diurnal variation for the DSR, USR, DLR, and ULR from January 2007 to December 2008. The short lines represent error bars. Except for DLR, all four radiation components showed significant diurnal fluctuations. At 13:30 hours local time, the maximum DSR and USR distributions, 608.4 W m^{-2} and 129.1 W m^{-2}, respectively, are symmetrical. ULR followed the anticipated warming and cooling of the surface and displayed a pattern resembling that of DSR. It is likely that changes in clouds and aerosols were responsible for the high variability of DSR and USR.

The daily means of (1) net radiation (R_n), (2) sensible heat flux (H), (3) latent heat flux (LE), and (4) surface soil heat flux (GTDEC, 0), as determined by the TDEC method, are calculated along with their seasonal and interannual variations. A 7-day moving average was computed for the four energy components to account for diurnal fluctuations. With the exception of the surface soil heat flux GTDEC, 0, the R_n, H, and LE all exhibited significant seasonal change. R_n displayed a pattern resembling that of DSR, which exhibited the greatest seasonal and interannual fluctuation, particularly during the summer. During the dry season, H was the primary consumer of R_n, but LE was the primary consumer during the wet season. In GTDEC, 0, there was no discernible difference. The Bowen ratio is the ratio of the daily averages of H and LE. Over the course of the investigation, the Bowen ratio showed significant seasonal change. In general, the Bowen ratio in 2007 was higher than it was in the 0–151 days of 2008. The drought trend that began in 2006 and continued through the summer of 2007 is primarily to blame, as is the increased soil moisture content that appeared in 2008 during these times. The Bowen ratio is significantly below 1.0, and the latent heat flux percentage was significantly higher in 2007 due to the higher precipitation throughout the growing season. But in addition to regulating albedo fluctuation, soil moisture has a significant impact on the ratio of net radiation partitioning into latent and sensible heat fluxes. The soil moisture at the root depth is primarily responsible for evapotranspiration, particularly transpiration from vegetation. The latent heat flow in the growth season of 2007 was larger than that of 2008, particularly during days 175–225, since weak precipitation events in 2008 did not have a chance to recharge soil moisture at 80 cm depth.

5.5.3 *Surface energy balance closure over the Loess Plateau*

Large closure errors are unacceptable since the concept of surface energy balance is the foundation of all atmospheric models. A comparison of H+LE and $R_n - G_0$ is shown in Fig. 5.23. As the number of flux stations expanded to research the closure of the surface energy balance, we were concerned with the accuracy and validity of the data. Since the sonic anemometer showed significant inaccuracies before, during, and after rainfall episodes, the investigation of the surface-heating rate in this study concentrated on data obtained during dry periods. Large transient fluctuations, equipment modifications, and data gathered during calibration procedures were also disregarded. Because the measurement errors would be severe when the energy components were close to zero, analyses were also restricted to scenarios in which H, LE, and L were all greater than zero.

We ignored the soil heat storage in Fig. 5.23(a) and substituted G_0 with the soil heat flux measured at 5 cm (Gobs,5) by the soil heat flux plates. The surface heating rate was 0.78, and the correlation between H+LE and R_n Gobs,5 was 0.83. The surface heating rate is calculated as the ratio of

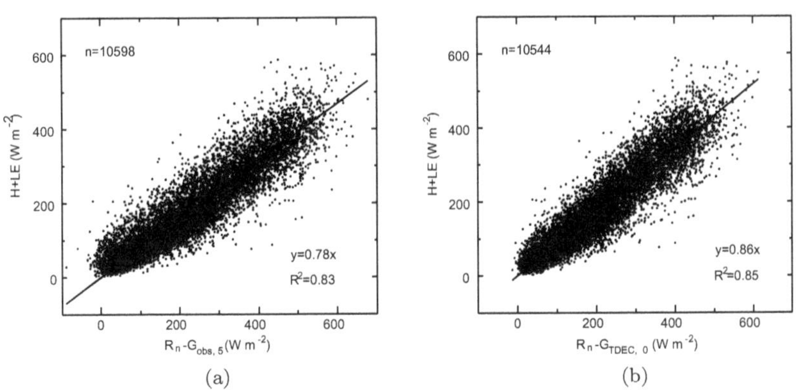

Fig. 5.23. Intercomparison of the measured sensible and latent heat fluxes (H+LE) against available energy (R_n–$G0$) at SACOL from January 2007 to December 2008. The data represent half-hourly averages. Gobs,5 is the soil heat flux observed at 5 cm depth (a). $G_{TDEC,0}$ is the surface soil heat flux calculated from the TDEC method (b) (Wang *et al.*, 2010).

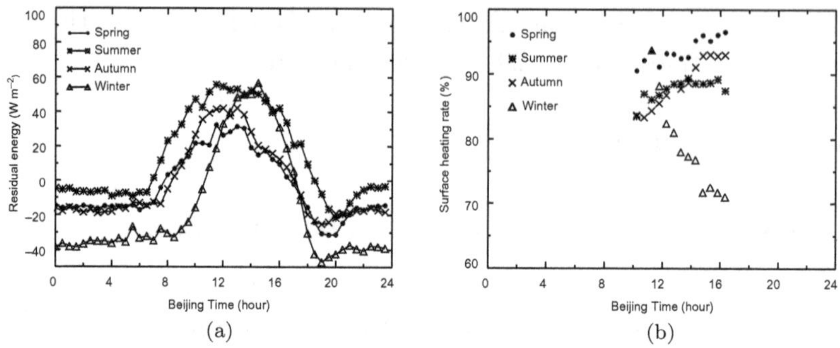

Fig. 5.24. Diurnal cycle of the residual energy ($R_e = R_n$ – H – LE – $G_{TDEC,0}$) (a) and the surface heating rate during daytime (10:00–16:00) (b) for the four seasons at SACOL from January 2007 to December 2008 (Wang *et al.*, 2010).

measured (H+LE) to accessible energy ($R_n - G_0$). We substituted $G_{TDEC,0}$ for G_0 in Fig. 5.23(b). The correlation between H+LE and $R_n - G_0$ was 0.85 in this instance, while the surface heating rate was 0.86. It is obvious that almost 8% of the energy was stored in the top 5 cm of soil, demonstrating the importance of considering soil heat storage. The seasonal mean ratio of the measured energy (H+LE) to the available energy ($R_n - G_0$) was determined to better understand the SACOL energy closure issue. The surface heating rates were calculated as 0.83, 0.78, 0.75, and 0.67 for spring, summer, autumn, and winter, respectively, without accounting for soil heat storage. The surface heating rates were calculated to be 0.92, 0.85, 0.85, and 0.73 in the spring, summer, autumn, and winter, respectively, when the soil heat storage was considered. In other words, surface heating rates increased by 0.09, 0.07, 0.10, and 0.06 in spring, summer, autumn, and winter, respectively, if surface heat storage was considered. Notably, in spring, the surface heating rate was 0.92, which was extremely close to the expected value of 1.0, while in summer and fall, it was only 0.85. The surface heating rate ought to be very near 1.0 in theory, but our measurements revealed an energy imbalance. Such energy imbalances were also experienced by other significant field operations, which complicated ensuing climate applications. According to earlier studies, the causes of the energy budget's imbalance were typically tied to measurement mistakes and uncertainties of the individual energy components as well as the impact of various footprints on those components.

For the four seasons at SACOL from January 2007 to December 2008, the seasonal mean of the diurnal variation of residual energy ($R_e = R_n$ – H – LE – G_0) and the rate of surface heating during the daytime (10:00–16:00) are presented in Fig. 5.24. Both R_e and R_n had significant diurnal variation, and their patterns of diurnal variation were similar. In Fig. 5.24(a), R_e was smallest in the spring and greatest in the summer and winter. Figure 5.24(b) shows how the surface heating rate is lower in winter and is associated with a higher R_e. Spring experienced the highest rate of surface heating, which is typically higher than that during summer and fall. The diurnal fluctuation of the surface heating rate during the day was comparable to the seasonal energy balancing conditions ((H+LE)/ (R_n−$G_{TDEC,0}$)), when the soil heat storage was considered, when the surface heating rate was computed.

5.5.4 *Oxidative ratios observations from SACOL site*

Previous research has demonstrated the strong ties between the carbon cycle and oxygen (Huang *et al.*, 2018; Liu *et al.*, 2020a). The most important link between the carbon and oxygen cycles is the oxidative ratio (OR), which is defined as the molar ratio of oxygen production (consumption) to carbon generation (production) for a specific activity. This ratio illustrates the quantitative relationship between oxygen fluxes and carbon sinks (Keeling and Shertz, 1992). OR can be used to calculate the oxygen fluxes from the observed in-situ carbon fluxes (Ding *et al.*, 2022). However, there has only ever been one commonly accepted estimate of OR (Li *et al.*, 2021), and that estimate originated from Severinghaus (1995), who determined that the value was 1.10 ± 0.05. This estimate cannot accurately reflect the OR of the entire terrestrial ecosystem because it is based on the OR of a single soil sample. Studies have already demonstrated that large differences in observational ORs occur, which are closely related to environmental conditions and plant kinds (Gallagher *et al.*, 2017). Therefore, more measurements of OR are required in order to better understand the terrestrial oxygen budget.

A field study was conducted in a semi-arid grassland environment at Lanzhou University's Semi-Arid Climate and Environment Observatory. Samples of vegetation, litter, roots, and soil were taken, and their ORs were examined. The Loess Plateau, which consists of a plain, ridge, and mound, is what gives the area around the site its distinctive topology. The primary soil type is sierozem, and the parent soil is primarily quaternary aeolian loess. The area is a typical semi-arid grassland with Tripolium vulgare, Artemisia frigida, Leymus secalinus, and Stipa bungeana as its predominant vegetation. Precipitation and the average annual air temperature are 381.8 mm and 6.7°, respectively. The surface temperature reaches its highest point in July (19°) and its lowest point in January (–8°). August is when precipitation is at its highest (Huang *et al.*, 2008a).

Each month in 2022, five distinct quadrats (1 m × 1 m) were randomly formed, and the spaces between them were greater than 5m. Soil, vegetation, litter, and roots were all tested at each quadrat when they were present. Secateurs were used to carefully cut away the vegetation above ground, and rakes were used to gather the debris. Using a soil auger, samples of soil were taken at depths of 0–20 cm and 20–40 cm. Using a large-bore root auger, samples of the roots were taken at the same depths. To lower the moisture content and the potential for oxidation, all samples were bagged in the field and air-dried (65°) in the lab. Electronic balances are used to weigh samples of vegetation and litter in their raw and dry forms. After being cleaned and allowed to dry naturally, the root samples are weighed. Samples of soil were cleared of any stones, visible roots, and plant remains. The samples were all then reduced to a fine powder (less than 0.15 mm in diameter) for elemental analysis.

The composition of each sample's elements was examined. On an elemental analyzer, samples were examined for the amounts of carbon (C), hydrogen (H), nitrogen (N), and oxygen (O). The carbon oxidation status (C_{ox}), which may be determined from the elemental compositions of organic matter as follows (Masiello *et al.*, 2008), can be used to determine the value of OR:

$$C_{ox} = \frac{2 \times [O] - [H] + 3 \times [N]}{[C]}, \tag{5.10}$$

in which the [X] represents the molar concentration of the element X. Furthermore, sulfur is not included in this equation as its concentration is low in vegetation samples. The sulfur concentrations of vegetation samples vary from 0.096% to 0.244%, which is consistent with the result that the sulfur is assumed to form 0.25% of biomass.

The following formula can be used to compute the corresponding OR for the three probable sources of N, N_2, ammonia (NH_3), and nitrate (HNO_3) (Masiello *et al.*, 2008):

$$OR = 1 - \frac{C_{ox}}{4} + \frac{3 \times [N]}{4 \times [C]}(N_2) \tag{5.11}$$

$$OR = 1 - \frac{C_{ox}}{4}(NH_3) \tag{5.12}$$

$$OR = 1 - \frac{C_{ox}}{4} + \frac{2 \times [N]}{[C]}(HNO_3) \tag{5.13}$$

Because there is little information about fertilizing at the site of our observation, N_2 is frequently used as the major source of N. The ecosystem OR can then be determined using NPP and Rh from the OR of the vegetation, roots, litter, and soil. As indicated by the formula,

$$\mathrm{OR}_{\mathrm{terra}} = \frac{\mathrm{OR}_{ab} \times \mathrm{NPP} - \mathrm{OR}_{ba} \times \mathrm{Rh}}{\mathrm{NPP} - \mathrm{Rh}}, \tag{5.14}$$

in which $\mathrm{OR}_{\mathrm{terra}}$ represents the ecosystem's OR, OR_{ab} is the OR of the live vegetation, and OR_{ba} is the weighted average of root, litter, and soil's OR. Similar methods were used in previous studies (Gallagher *et al.*, 2014).

The results showed that there were large seasonal fluctuations in ORs and the yearly mean of the ecosystem OR could reach 1.14 ± 0.04. This value exceeds what was observed in Germany and Australia (Müller *et al.*, 2004; Hockaday *et al.*, 2009). The variance in OR closely tracks the growth of the vegetation, peaking in August–September and troughing in December–January. Temperature and soil moisture fluctuations are major environmental influences on OR.

5.6 Conclusions and Discussion

We carefully go over the interactions between land and atmosphere in this chapter. The energy balance and the three main cycles — the carbon cycle, the water cycle, and the oxygen cycle — are first introduced. The three cycles are strongly tied to each other and to interactions between the land and the atmosphere (Fig. 5.25). Both carbon and oxygen, which are necessary for life and are present in the atmosphere naturally in the forms of CO_2 and O_2, are recycled and consumed by the biosphere and all of its creatures. Living matter is continually replenished because of this. Green plants are the main source of oxygen in the air. Green plants take carbon dioxide from the atmosphere to create living matter and release oxygen into the air during the day when sunshine is present. This process is called photosynthesis. The respiration of plants and animals results in the release of carbon dioxide after oxygen consumption. The carbon fixed in the organism is also released into the atmosphere as carbon dioxide when biomass is broken down or burned. Photosynthesis and the dissolution of CO_2 in seawater serve to offset the release of CO_2 into the atmosphere. Water participates in the carbon–oxygen cycle because respiration releases it while photosynthesis absorbs it. Convection causes water released into the atmosphere to ascend. As the latitude of air mass rises, its temperature falls, causing water vapor to condense into clouds and eventually generate precipitation that falls back

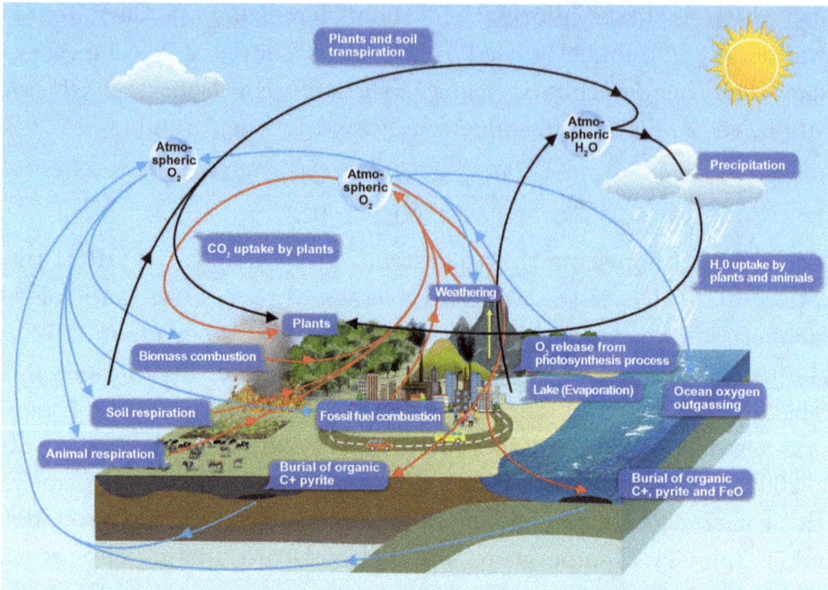

Fig. 5.25. Understanding the coupling relationships between the carbon, water, and oxygen cycles in the earth. Living organisms influence the cycling of carbon, water, and oxygen through the environment. The blue line represents the consumption and production of atmospheric O_2. The red line and the black line are the same as the blue line, but for CO_2 and water, respectively.

onto the land and oceans. The cycle continues when water continues to evaporate from lakes, soils, or vegetation, releasing water vapor into the atmosphere.

On geologic timescales, the burial and weathering of organic matter (also known as "geo" photosynthesis and "geo" respiration) also directly couple the oxygen cycle and carbon cycle. In subduction zones, a sizeable fraction of organic carbon can evade decarbonation and instead be isolated from the surface and buried deep below the mantle, resulting in the oxygenation of the atmosphere. The sulfur cycle and the oxygen cycle are linked in a manner similar to the carbon cycle. FeS_2 escapes the reaction with O_2 if it is buried in sediments and left unreacted. Pyrite burial thus constitutes an additional net source of oxygen for the atmosphere. This mechanism has the potential to cause O_2 to build up over extended periods of time. When sulfur is oxidized in the environment, reduced sulfur will weather and consume O_2, while oxidized sulfur (FeO_2) will weather and cause an increase in O_2 (Fig. 5.25). It is thought to have been formed by recycling goethite and/or H_2O downward to depths close to the core–mantle boundary in order to have stable, pyrite-structured FeO_2. The buried pyrite and organic carbon would undergo reoxidation in the rock cycle after being brought back to the earth's surface by tectonic uplift. Thus, the reoxidation processes serve as an oxygen sink to finish the cycle.

Human actions have a significant impact on the cycles of carbon, water, and oxygen in the modern environment. On the one hand, human activities significantly increase water consumption and run the risk of causing groundwater depletion. On the other hand, factory emissions and the burning of large amounts of fossil carbon tend to accumulate in the atmosphere, causing a rapid increase in CO_2 concentration that causes a greenhouse effect and global warming, as well as a decline in oxygen levels.

GPP has a direct role in the exchange of materials in the terrestrial environment and is a significant component of the carbon cycle. Under future climate change, there will be significant changes in the structure and functioning of arid and semi-arid regions' ecosystems, which will affect GPP. Global arid and semi-arid regions will also see significant expansion, deterioration, and conversions between dryland subtypes. In fact, throughout the previous few decades, alterations in ecosystems of arid and semi-arid regions have already taken place. For instance, in the past 30 years, vegetated arid and semi-arid regions have experienced extensive conversions to bare ground (Song *et al.*, 2018). There are many reasons for these transitions; for instance, in the Southwest United States, these alterations are likely a result of the increased dominance of exotic species. Reduced vegetation in Australia's arid and semi-arid regions is a result of prolonged decreases in precipitation during the growing season (Donohue *et al.*, 2009). However, the significant GRA deterioration in the Mongolian steppe was ascribed to a combination of overgrazing, declining rainfall, and rising air temperatures. Between 1982 and 2015, there was a discernible rise in the extra-tropical Northern Hemisphere's vegetation water constraint. While water surplus areas are considerably reducing, vegetation water-deficit areas are significantly expanding (Jiao *et al.*, 2021). The main factor controlling future trends and variability in arid and semi-arid regions' GPP is the regions' expansion. While Australia will continue to contribute the most to the global arid and semi-arid regions' GPP IAV in this century, East Asia will overtake North America as the leading contributor to arid and semi-arid regions' GPP trends under RCP4.5. The arid subtype and the subhumid subtype have an increasing and a decreasing role in regulating the worldwide arid and semi-arid regions' GPP variability, respectively, despite the semi-arid subtype continuing to dominate the global arid and semi-arid regions'

Table 5.8. Relative contribution of the regional arid and semi-arid regions' gross primary production (GPP) interannual variability to the global arid and semi-arid regions' GPP interannual variability (RCP8.5) (Yao *et al.*, 2020).

Year	North America (%)	South America (%)	Europe (%)	Africa (%)	West Asia (%)	East Asia (%)	South Asia (%)	Australia (%)
2011–2040	14	11	12	15	5	19	8	16
2041–2070	13	10	12	15	13	16	9	12
2071–2100	15	13	12	14	12	15	8	11
2011–2100	15	9	10	18	9	18	7	14

GPP variability. Our findings regarding the worldwide arid and semi-arid regions' GPP trends and IAV under RCP4.5 are conservative, and if future climate change follows the RCP8.5 scenario, global arid and semi-arid regions' GPP spatiotemporal variability will probably rise. In fact, a higher rate of arid and semi-arid region expansion results in a 25% rise in the regions' GPP under RCP8.5 by the end of this century compared to the baseline for the years 2000–2014. North America and East Asia, rather than Australia, will play the most significant role in regulating the worldwide arid and semi-arid regions' GPP IAV in 2041–2100, with the predominant trend of arid and semi-arid regions' GPP hotspots shifting from East Asia to North America in 2011–2070 (Table 5.8). Global arid and semi-arid regions will see increased GPP variability as a result of the semi-arid subtype found in East Asia and Africa. Our study shows that under future climate change, regional and subtype contributions to the global arid and semi-arid regions' GPP trend and IAV are prone to significant changes.

It should be noted that this study does not take into consideration the potential dynamic responses of arid and semi-arid regions' ecosystems to future climate change because it is based on historical MODIS GPP information and the derived GPP. Under multiple elevated climatic and environmental stresses like droughts, heat waves, grazing, invasions of exotic species, CO_2 fertilization, woody encroachment, adaptations, and disturbances, ecosystems of global arid and semi-arid regions would probably experience more drastic changes in their structures and functioning (e.g., changes in water use efficiencies) (Milly *et al.*, 2008; Fensham *et al.*, 2005). The relationships between arid and semi-arid regions' ecosystems and climate are all impacted in a lagging and nonlinear way by all of these mechanisms. Evidence suggests that complex interacting feedbacks, such as the lagged respiration of soil biota, carryover effects of antecedent climate responses (such as water deficits), and modifications to nutrient cycles, are to blame for lag effects (Arnone *et al.*, 2008; Anderegg *et al.*, 2015). In water-limited arid and semi-arid regions where precipitation events have significant effects on the variability in aboveground NPP, soil N availability, and drought sensitivity, variation in lag effects may be particularly important (Griffin-Nolan *et al.*, 2018). It was discovered that in ecosystems of arid and semi-arid regions, lag effects can account for between 18% and 28% of the response variables. However, existing climate–vegetation models only replicate minimal or no lag effects (Anderegg *et al.*, 2015). A new climatic regime with more frequent and powerful extreme events could prolong the lag effects, resulting in a substantial bias in anticipating climate–carbon feedbacks. This lack of quantitative understanding of lag reactions to climate extremes is of particular concern (Arnone *et al.*, 2008). However, our study does not quantify or go into the lag effects. Thus, it is assumed that projected spatiotemporal variabilities in the global arid and semi-arid regions' GPP will have uncertainties and constraints. But the variations in our MODIS-derived GPP over the world's arid and semi-arid regions are comparable to those found in earlier research, model results (such as CMIP5), and empirical observation-based datasets. Additionally, other environmental factors, such as changes in ecosystem types, have very minor roles in regulating arid and semi-arid regions' GPP fluctuation because precipitation and air temperature account for the bulk (79%) of arid and semi-arid regions' GPP variability. In light of future climate change, our projected GPP variability for the years 2010–2100 ought to include a majority of its spatiotemporal characteristics. Understanding ecosystem carbon cycle–climate feedbacks of long-term global arid and semi-arid regions requires dynamic Earth system models that couple the physical, biophysical, social, economic, and biogeochemical systems that control these interactions.

An overview of the modifications made to the carbon cycle as a result of global warming has been presented. Water supplies are significantly impacted by climate change. The water cycle is accelerated by global warming, and the spatial heterogeneity of precipitation is increased. Both of these factors directly affect soil moisture. In Section 5.4, we first look at the fluctuation of soil moisture for the years 1948–2010. In recent decades, there has been a long-term decline in soil moisture. The soil moisture trend, however, varied widely across space, with the most pronounced wetting trends occurring over the central United States, South America, and Western Australia, and the most pronounced drying trends occurring over Northern Africa,

East Asia, Eastern Australia, and Southern Europe. Among the nine subregions identified by the relative drying acreage and contributions to global drying, the drying over the NAF and EAS regions was the most severe. The most notable drying was observed across transitional regions between dry and wet areas, between the range of AI of 0.4 and 0.8, according to the regionally averaged soil moisture trend as a function of the climatological mean AI. Additionally, the percentage of places that were drying out over transitional zones was also the most pronounced, with more than 60% of those areas displaying a downward trend in soil moisture. The evolution of the soil moisture trends found by MEEMD revealed that the two-band structure of soil moisture drying was only relevant across the transitional zones. Around 1955, the humid transitional zones of the AI range of 0.65–0.9 began to noticeably dry, and they peaked in the 1990s. Beginning in the 1980s, the scope of drying grew into the dry transitional zone (within the AI range of 0.4–0.65), along with the increasing amplitude of drying.

The long-term trend of soil moisture was demonstrated by linear regression analysis to be strongly correlated with variations in precipitation and temperature. However, the function of each climate variable is unique. The sign of a change in soil moisture is primarily influenced by changes in precipitation; wetness is caused by increased precipitation and drying by decreased precipitation. Contrarily, global warming has a negative impact on the trend of soil moisture across most regions, including in wet regions, and this impact is more pronounced for soil moisture drying than for precipitation. Precipitation dominates the variability of soil moisture at interannual-to-decadal timescales for the world average, whereas temperature drives the overall long-term soil moisture trend. Thus, the increased soil moisture drying in the transitional regions is primarily due to temperature. Three tests were conducted based on the Noah-MP land surface model to confirm the outcomes of linear regression techniques that were consistent with the outcomes shown by the soil moisture predictor.

Over East Asia, there has been a persistent decline in soil moisture from 1948 to 2010. The soil moisture trend, however, varied widely over space, with the most pronounced wetting trend occurring over Western East Asia and the drying trend occurring over Northeast China, North China, part of Mongolia, and Russia close to Lake Baikal.

According to the MCA study, there is a clear connection between global warming, lessening precipitation, and the drying of soil moisture. Nevertheless, each person's specific function varies. The specific function of falling precipitation is what mostly causes the soil's moisture to dry out. However, the climatic location has a significant impact on how trends in soil moisture are impacted by rising temperatures. The degree of soil drying brought on by declining precipitation is amplified significantly by rising temperatures over the 400 mm/yr precipitation range, which is a regime of energy-limited evapotranspiration. By defining a contribution, it is possible to quantify the impact of temperature and precipitation on the soil moisture trend. The results are strikingly similar to those of the qualitative analysis and show that rising temperatures almost double the drying of soil moisture brought on by falling precipitation. As a matter of fact, a recent study has focused heavily on drying, drought, and examination of the underlying mechanisms, particularly over the East Asian and African Sahel regions, where the most severe drying occurred (Zhang and Zhou, 2015). Recent studies have frequently proposed that large-scale SST anomalies and favorable vegetation–atmosphere feedbacks worked together to cause drought in the African Sahel. Previous research has demonstrated that warming of the equatorial Atlantic and Pacific/Indian Oceans on an interannual timescale, as well as warming over the tropics on a decadal timescale (Caminade and Terray, 2010), can lead to a decrease in rainfall and a drought over the Sahel. Because Africa has the greatest influence of vegetation–atmosphere interactions in the world, the local vegetation–atmosphere interactions were also significant contributors to the Africa Sahel drought arising from the remote effects of SST anomalies (Kucharski *et al.*, 2013). Drought and dryness have a significant impact on East Asia as well. Monsoonal circulation permeates most of East Asia, and most of the region's yearly precipitation falls during the summer monsoon. Because of the interdecadal weakening of the East Asian summer monsoon circulation, which is caused by the Pacific Decadal Oscillation's phase transition, the drying trend across East Asia is a regional manifestation of a long-term or interdecadal change in precipitation.

Temperature, as previously established, can modify soil evapotranspiration, which in turn can impact soil moisture. Evapotranspiration may therefore be decreased with decreasing soil moisture, which might potentially increase the sensible heat flux and further raise the temperature. Positive feedback is created by an increase in temperature, a decrease in evapotranspiration, and an increase in soil moisture. If the feedback loop is not broken, this positive feedback can last until the soil is completely dried out and desertification occurs. This would result in the biological environment being destroyed, decreased crop yield, the presence of dust, and drought (Huang *et al.*, 2006b, 2009). Because the surface climate in transitional climatic regions, which are neither very dry nor very wet, is most susceptible to soil moisture, this effect is most severe there (Seneviratne *et al.*, 2010). The key element regulating the distribution of total surface energy into sensible and latent heat fluxes, as well as the near-surface temperature and evapotranspiration, in a location with a transitional climate is soil moisture. Additionally, transitional climate regions have the worst warming and drying. The drying trend may be further enhanced and the semi-arid regions may become even drier as a result of the strongest positive feedback between extreme warming and drying. Transitional areas are typically populated by a significant number of agricultural districts.

We also looked at how the observed data were represented in arid areas. Here are the specifics: the historical data for the years from January 1956 to December 2008 were compared to the measurements of wind speed, air temperature, water vapor pressure, air pressure, and precipitation made at SACOL. It was concluded that our measurement site at SACOL was typical of the local climate. Apart from the downward longwave radiation's daily variation, each of the four radiative components had different seasonal and nocturnal cycles. Due to the snow cover on the ground, the greatest surface albedo value (0.9) was recorded during the winter. Due to the high percentage of vegetation cover and the wetter soil, surface albedo was at its lowest during the rainy season. Surface albedo exhibited notable variances between wet and dry years due to the interannual variations in soil moisture. Multiple ecological parameters underwent sudden and systemic alterations as a result of aridification. These alterations happened in three waves that followed one another and were marked by sharp declines in plant productivity, soil fertility, plant cover, and richness (Berdugo *et al.*, 2020). Their habitats are typically delicate and vulnerable to climate change; even a small shift in the climate is likely to have disastrous effects. The ecosystem and human survival would be severely threatened by the most severe warming and drying. Future predictions for the growth of arid regions include the intensification of differences between wet and dry regions. Across all moisture levels, runoff has a general tendency to rise. However, runoff increases in wet areas are far greater than those in dry areas, particularly during the rainy season. More than radiative forcing, physiological factors contribute to the suppression of evapotranspiration in wetter regions compared to dry ones, amplifying the runoff differential (Cui *et al.*, 2021). The distinction between the wet and dry seasons within the same region has also grown more pronounced. For instance, from 1982 to 2019, the dry season in Central Asia got drier and the rainy season got wetter (Ren *et al.*, 2022). Decreases in winter ice cover, elevated lake surface temperatures, and accelerated lake evaporation are all effects of global warming. Higher evaporation rates will favor a decline in lake level and the expanse of the surface water when they are not counterbalanced by an increase in mean precipitation or input (Woolway *et al.*, 2020). Therefore, strict management and intelligent water resource use plans are urgently required to lessen the significant effects of the drying.

Studies of the oxygen cycle range in magnitude from daily to geologic (>106 years) ranges. The atmospheric O_2 is mostly controlled by the oxygen cycles of the appropriate time frames. The imbalance between photosynthesis and respiration, specifically the failure of the rate of photosynthesis to balance the rate of respiration of organic matter (specifically, the microbial oxidation of organic matter in the soil), was the main cause of the abrupt O_2 decrease from approximately 21% to 14% in the first 16 months, which was enough to cause health problems, in Biosphere 2 (an artificially sealed habitat) (Severinghaus *et al.*, 1994). Human interference with oxygen generation and consumption in a natural setting also has the potential to cause a discernible change in atmospheric O_2.

With an emphasis on heavily populated urban areas and arid or semi-arid regions at high risk of desertification, various measures should be put in place to conserve and promote local ecological services. Sparse vegetation and poor biodiversity in these places contribute to their restricted capacity to absorb and emit oxygen as well as their low capability to store and recharge groundwater flow and surface runoff. It is generally agreed that one of the most efficient ways to prevent soil erosion and restore local ecological services is to increase vegetation cover and diversity (or, in other words, to promote O_2 generation). Reducing O_2 consumption brought on by human activity is also crucial. The burning of fossil fuels in the energy sector is the largest source of atmospheric CO_2 and the largest sink of atmospheric O_2. Research and development of renewable fuels, such as carbon-neutral biofuels (with no impact on CO_2 or other GHGs) and carbon-negative biofuels (with a net reduction in GHGs but no impact on atmospheric O_2 concentration), should therefore receive priority. The decisions we make in the coming decades will determine the fate of Earth, the only planet in the universe that is currently known to be habitable. A catastrophic outcome that is essentially irreversible and nearly permanent on the timescale of humans may happen if the current trajectories of biodiversity loss and O_2 decline are allowed to continue (Ceballos *et al.*, 2015). This is because it may take millions of years to restore Earth, which is too long for humans to wait.

With the significant exception of high northern latitudes during the winter in the CMIP5 model, research on the future trend in land–atmosphere interactions indicates a general drying tendency over land (less soil moisture, less evaporation) across most of the world. Most of the world's future land–atmosphere coupling is predicted to increase, which means soil moisture changes will have more control over surface fluxes and the lower troposphere. Additionally, there is a strong consensus for a deepening atmospheric boundary layer and decreasing gradients across the entrainment zone at the top of the boundary layer, suggesting that the influence of the land surface on the atmosphere should grow both in absolute terms and relative to the free atmosphere's conditions. However, because this would establish a mechanistic connection between soil moisture and plant transpiration, simulating diversity in plant hydraulic transport and its mechanistic effects on stomatal conductance in land surface models may be essential for capturing drought intensification and climate extremes (Anderegg *et al.*, 2019). In the context of global warming and polluting emissions, the fluxes of various gases and particulate matter between the terrestrial biota and the atmosphere are directly or indirectly coupled to vegetation and soil carbon, nutrient, and water balances; therefore, proper consideration of the underlying biological processes is necessary when quantifying their geographic patterns or changes over time. Our ability to analyze how climate change or air pollution might affect human land use is, however, constrained by the lack of a mechanistically based depiction of human response processes in ecosystem models that are coupled to climate models. A more holistic viewpoint is required, and it should be a focus of ongoing research that unites the socioeconomic and biophysical communities. Santanello (2018) gives an overview of the LoCo endeavor, metric and model applications, along with scientific and programmatic developments and obstacles, and stresses the significance of comprehending, quantifying, and assessing these processes in nature in models.

Three dangers will affect developing nations: protracted drought, accelerated warming (Huang *et al.*, 2012; Ji *et al.*, 2014), and rapid population increase. Positive feedback in the climate–aridity cycle may cause these terrible scenarios and their interactions to compound nonlinearly, making the survival environment more susceptible to even little human disruptions. Degradation and desertification will pose a much greater threat than is now anticipated. Degradation and desertification will soon present a threat to the global environment and human life because developing countries make up more than 70% of the world's land area and population. To prevent future desertification and reverse current global desertification issues, a worldwide action plan is urgently required. This plan must include strict management and rational use of water resources, as well as the restoration of soils and plants to lessen ecosystem vulnerability.

Chapter 6

Impacts of Ocean–Atmosphere Interactions

6.1 Introduction

In arid and semi-arid regions, multiple factors contribute to climate change. As the anthropogenic effect is increasing, the natural effect from ocean activities and large scale of atmospheric circulation on the climate change in semi-arid regions receives less attention. Studying how oceans affect climate change in arid and semi-arid regions is crucial to understanding climate change and its causes in these regions since oceans play a significant role in the Earth's climate system. Even though most semi-arid regions are located far from the ocean, it nonetheless has a bigger impact on them. In addition to providing boundary forcing and starting and maintaining specific circulation anomalies in the atmosphere at the corresponding timescales, the decadal variations and long-term trends in sea surface temperature (SST) of the world oceans play a crucial role in the widespread drying over semi-arid regions (Kushnir *et al.*, 2002; Alexander, 2013). Evidence suggests that ocean–atmosphere interactions are also in control of the extensive drying that has taken place over the decades in semi-arid regions (Wang *et al.*, 2014). Previous research has demonstrated that oceanic activities, particularly sea surface temperature (SST) fluctuations in the Pacific, Atlantic, and Indian Ocean Basin, can have a variety of timescale effects on temperature, precipitation, and dry and wet changes in global arid and semi-arid regions.

According to studies, the past 100 years have seen a considerable intensification of warming in global arid and semi-arid regions (see Chapter 3), which clearly exhibits the interdecadal dry/wet variations. The oscillatory components of the oceanic interdecadal scale are directly tied to the nature of these changes. The land–sea thermal contrast is greatly affected by different phase combinations of oceanic oscillation factors, which in turn influences the westerly jet, planetary wave, and blocking frequency, changing the temperature and dry/wet characteristics of arid and semi-arid regions. The impact of marine activities on these regions will reveal new characteristics in the future as climate change in arid and semi-arid regions intensifies. This will enhance the uncertainty of future climate change in these regions and heighten their influence on the global climate. The Pacific Decadal Oscillation (PDO), the Atlantic Multidecadal Oscillation (AMO), and the Indian Ocean Basin Mode (IOBM) all have strong SST indications in the Pacific, Atlantic, and Indian Oceans, respectively, on the interdecadal timescale. The summertime precipitation in East Asia is influenced by each ocean's variations. Expect for the single-phase effect from typical ocean oscillation, recent studies show a "combined influence" effect from three major interdecadal ocean signals, PDO, AMO and IOBM, and through their different phase combinations, the studies have well explained the formation mechanism of the observed interdecadal variability of seasonal precipitation/aridity over the globe and region in the past 100 years (Wang *et al.*, 2014, Zhang *et al.*, 2018b). For example, the precipitation in East Asia is impacted by ocean interannual variability like ENSO and the Indian Ocean Dipole (IOD). The East Asian summer monsoon system's features can be influenced by the Indian Ocean Dipole through the atmosphere, which affects precipitation in East Asia, including China. When the positive-phase event happens, abnormal anticyclones carry more water vapor to Southern China, resulting in heavy precipitation, and abnormal signal transmission occurs in the Rossby wave air column from the Indian Ocean to the South China Sea. By affecting the strength of the East Asian summer monsoon, the cold and warm phases of ENSO also have an impact on East Asian precipitation.

The global typical semi-arid region in Northern Africa is inextricably linked to the West African monsoon (WAM) system, which blows from the tropical Atlantic Ocean across the Sahel and brings rain to this region (Nicholson, 2013). During the austral summer, rainfall is heaviest in the semi-arid region of Southern Africa, primarily because of tropical/extra-tropical influences and related convection that are heavily influenced by the nearby oceans, especially the Indian Ocean. Since 1970, the Indian Ocean has warmed noticeably. Thus, the regional circulation and SST have changed, inhibiting convection, and causing a drying trend over the semi-arid region in Southern Africa. Moreover, the influence of ENSO on SST changes in the Western Indian Ocean also affects rainfall in Southern Africa (Reason and Rouault, 2002; Misra, 2003; Kane, 2009). The Pacific Decadal Oscillation

(PDO) index increased over the last half of the 20th century, which made El Niño occurrences stronger, more frequent, and more stable in their link with South Africa. This worsened the semi-arid climate of Southern Africa.

Particularly since 1970, the overlying atmospheric circulation that produces anomalous ascending motion west of 140°E and anomalous subsidence east of 140°E possibly caused by the noticeable changes over the Indian Ocean. These changes are to blame for the wet conditions over Central/West Australia (promoting convection) and the dry conditions observed over East Australia (inhibiting convection). ENSO may be responsible for the drying trend over East Australia. About 1976, the PDO entered a warm phase, with an increasing frequency of El Niño events and a drying trend over East Australia.

The interplay between the ocean and the atmosphere, which was discussed earlier, determines the global dry and wet changes on an interdecadal scale (Guan *et al.*, 2017). The influence of oceans on climate change in arid and semi-arid regions has also been extensively studied from the perspectives of various phase changes, the combination of oceanic oscillation components, as well as the land–sea temperature contrast (Huang *et al.*, 2017a, 2017b). However, many questions remain unresolved including how the interaction of various sea areas affects climate change in arid and semi-arid regions as well as how to measure the relative contribution of marine activities to climate evolution in arid and semi-arid regions and the resulting response processes. In order to fully understand the mechanisms underlying climate change and prevent desertification in arid and semi-arid regions, this chapter summarizes the most recent findings regarding the impact of oceans on climate change in arid and semi-arid regions using data from previous studies conducted in China and other countries. It also discusses future development trends.

The arrangement of this chapter is as follows. Section 6.1 is the introduction to this chapter. Section 6.2 illustrates ocean oscillations, including El Niño and Southern Oscillation, North Atlantic Oscillation, Pacific Decadal Oscillation, Atlantic Multidecadal Oscillation, and Indian Ocean Dipole. Section 6.3 discusses impacts of ocean oscillations. Section 6.4 introduces impacts of land–sea thermal contrast. The conclusions and discussion are summarized in Section 6.5.

6.2 Ocean Oscillations

Walker published six papers in *World Weather* from 1924 to 1937, which provided the definitions of three major oscillations. The basic idea is that a

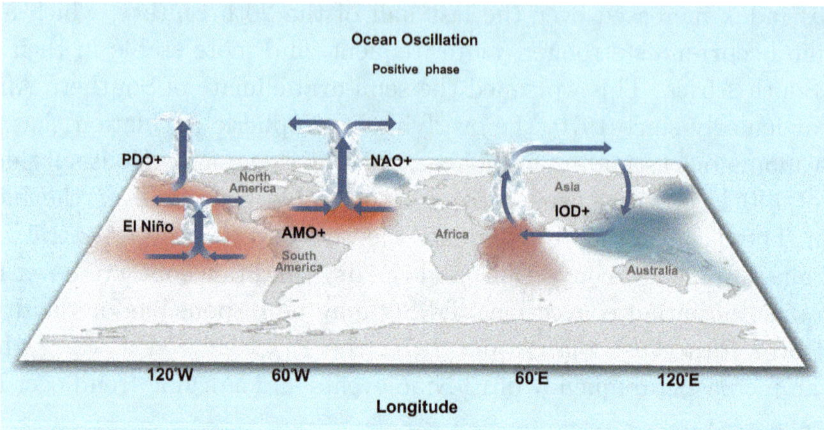

Fig. 6.1. A schematic depiction for the global distribution of ENSO, PDO, NAO, AMO, and IOD (Positive phase).

stable high-pressure or low-pressure area can be seen on the monthly mean sea level pressure map. These high-pressure or low-pressure areas are called the atmospheric activity centers (see Fig. 6.1.) However, the changes in the centers of atmospheric activity are not isolated. The pressure of two centers of atmospheric activity can occasionally occur simultaneously. For instance, the change in the central pressure in an Azores high is frequently in opposition to the Icelandic low pressure in the NAO. When the pressure over the Azores is high, the pressure over Iceland is low, and vice versa. Walker referred to this seesaw change as an oscillation. The North Atlantic Oscillation is a variation in two activity centers in the North Atlantic Ocean (NAO). Similarly, the North Pacific Oscillation refers to the oscillation between the Aleutian Low and Hawaii High in the North Pacific Ocean (NPO). Walker also mentioned that the pressure change in the Pacific Ocean is opposite that in the Indian Ocean. It is called the Southern Oscillation (SO), because it mainly affects the Southern Hemisphere. This book shows that these three oscillations can stimulate a secondary oscillation at the global scale, which modifies the global pattern and phase of surface temperature warming. DMO is the term used to describe this secondary oscillation (See Chapter 8 and Huang *et al.*, 2017b). In addition, Wallace *et al.* suggested the Pacific North American type (PNA) (Wallace and Gutzler 1981) and Arctic Oscillation (AO) (Thompson and Wallace, 1998). However, these two types of oscillations might be primarily due to NPO and NAO activity in the troposphere. The following discussion provides the fundamental ideas in this regard.

6.2.1 *El Niño–Southern Oscillation (ENSO)*

The SO refers to the reverse change of air pressure in the Indian Ocean and South Pacific, which is an anomaly of large-scale atmospheric circulation. The SO was named by Walker and Bliss (1932) and it is the most noticeable and significant phenomenon of the interannual change in tropical circulation. It alludes to the antiphase pressure vibration that occurs primarily between the tropical Indian Ocean and the Eastern Pacific Ocean. This phenomenon has been recorded since the 1920s. The term "Southern" is relative to the changes in the Northern Hemisphere, and the term "Oscillation" refers to a vibration, because this antiphase barometric vibration phenomenon will recur over an approximately 3-to-7-year period. The Southern Oscillation Index (SOI), which measures the difference in surface pressure between the East Pacific and Indian Oceans, is frequently used to define the Southern Oscillation. At present, the most used SOI is the standard sea level pressure difference between Tahiti (148°05′W, 17°53′S) (France) and Darwin (130°59′E, 12°20′S) (Australia). A negative SOI indicates that the Eastern Pacific is lower than the Indian Ocean, and a positive SOI indicates that the Eastern Pacific is higher than the Indian Ocean. A negative SOI is often associated with persistent positive SST anomalies in the equatorial Eastern Pacific (see Fig. 6.2(c)).

The El Niño phenomenon mainly refers to the abnormally persistent warming of the sea water temperature in the tropical ocean in the Eastern and Central Pacific Ocean, which changes the climate pattern worldwide, resulting in drought in some areas and excessive precipitation in other areas. Later observations found that when an El Niño event occurred, the SST would continue to rise abnormally in the coastal waters of Peru and in the entire equatorial Eastern Pacific region. Therefore, the SST of the equatorial Eastern Pacific Ocean is generally represented by the average sea surface temperature in the region covering 0–10°S, 180–90°W.

The El Niño (Fig. 6.2(a)) and SO phenomena are collectively called ENSO. Many studies have shown that there is a strong correlation between the equatorial Eastern Pacific SST anomaly event (El Niño) and the SO. El Niño refers to the continuous abnormal warming of SST in the equatorial Eastern Pacific, which can be considered an abnormal ocean phenomenon. The SO refers to the reverse change in air pressure in the Indian Ocean region and the South Pacific region, which is an anomaly of large-scale atmospheric circulation. The close relationship between El Niño and the SO is a prominent reflection of the large-scale air–sea interaction, especially the tropical large-scale air–sea interaction. Therefore, ENSO results from large-scale air–sea interactions. The eastern equatorial Pacific experiences the ENSO, which is an oscillation of the wind field and SST. The El Niño–La Niña (Fig. 6.2(b)) transition in the ocean and the SO in the atmosphere are two manifestations of the low-latitude ocean–atmosphere interaction known as ENSO. Surface waters in a significant portion of the Tropical Pacific Ocean warm or cool by anywhere from 1°C to 3°C in comparison to normal cycles lasting between two and seven years. The thermocline, which separates the warm surface water from the cold deep water in the Pacific Ocean, is one ocean layer that is affected by ENSO. Across this transition zone, the upward currents closest to the equator are strongest. Water surface temperatures are directly correlated with the thermocline's depth. Cooler water surface temperatures result from the upwelling of cold, nutrient-rich deep water from the bottom layers when the thermocline is near the surface of the ocean. Due to its ability to change the global atmospheric circulation, which in turn affects global temperature and precipitation, ENSO is one of the most significant climate occurrences on Earth. El Niño and La Niña are the extreme phases of the ENSO cycle; between these two phases is a third phase called ENSO-neutral.

El Niño events occur when SSTs are higher than average in the Central and eastern Tropical Pacific Ocean. Rainfall tends to increase over the central and eastern Tropical Pacific Ocean while decreasing throughout Indonesia. Instead of strengthening or beginning to blow in the other direction (from west to east or "westerly winds") in some cases, low-level surface winds typically blow from east to west along the equator. El Niño tends to be greater in general when ocean temperature anomalies are warmer (and vice versa). SSTs in the central and eastern Tropical Pacific Ocean drop below average during La Niña. While rainfall tends to decrease over the central and eastern Tropical Pacific Ocean, it tends to increase over Indonesia. Along the equator, the typical easterly winds become significantly stronger. In general, the stronger the La Niña, the cooler the temperature anomalies in the ocean (and vice versa). Neutral means that the phenomenon is neither La Niña nor El Niño. Tropical Pacific SSTs frequently fall just below average. There are times when the water may appear to be in an El Niño or La Niña state, but the weather is not cooperative.

The MEI makes it easier to evaluate ENSO because it incorporates both oceanic and atmospheric components. In particular, it provides real-time signals of ENSO intensity, and through historical analysis, establishes a framework for insightful comparative analysis of changing conditions.

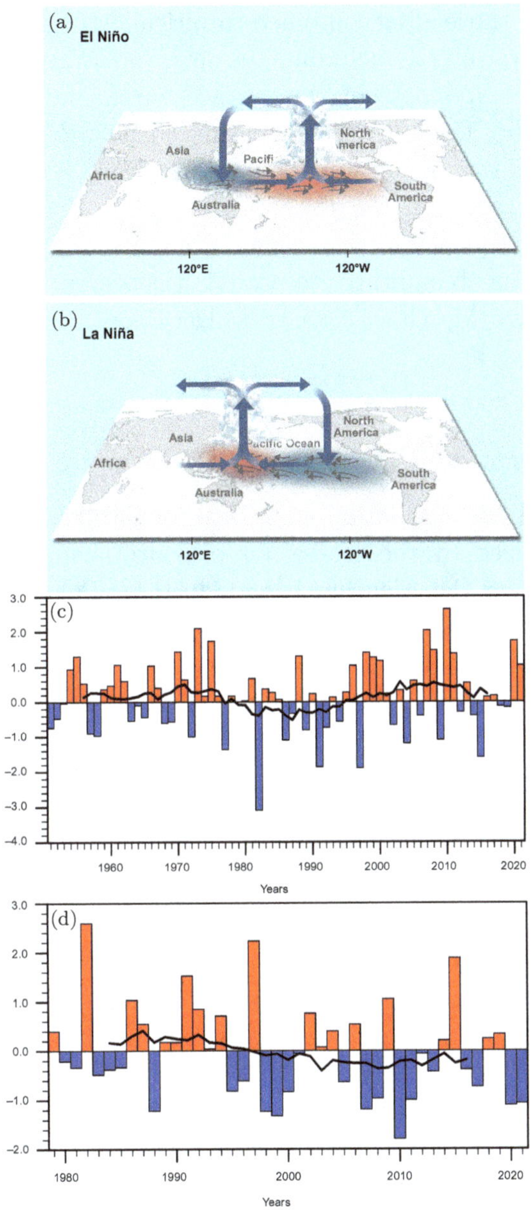

Fig. 6.2. Schematic sketch for the mechanism of positive phase of El Niño–Southern Oscillation (ENSO) (a) and negative phase of ENSO (b) (red color represents positive sea surface temperature (SST) anomalies, blue color represents negative SST anomalies). Time series of Southern Oscillation index from 1950 to 2021 (black curves indicate the 11-year running mean time series) (c). Time series of Multivariate El Niño/Southern Oscillation index (MEI.v2) from 1854 to 2021 (d) (black curves indicate the 11-year running mean time series).

The time series of the leading combined empirical orthogonal function (EOF) of five different variables, including sea level pressure (SLP), SST, zonal and meridional components of the surface wind, and outgoing longwave radiation, over the Tropical Pacific basin (30°S–30°N and 100°E–70°W), is represented by the bimonthly multivariate ENSO index (MEI.v2) (see Fig. 6.2(d)). To account for the seasonality of ENSO and lessen the effects of higher-frequency intraseasonal variability, the EOFs are estimated for 12 overlapping bimonthly "seasons" (December–January, January–February, February–March,..., November–December) (https://psl.noaa.gov/enso/mei/).

6.2.2 *Pacific Decadal Oscillation (PDO)*

Another atmospheric oscillation identified by Walker is the NPO. It is mainly characterized by the seesaw air pressure change between the low-pressure center near the Aleutian Islands in the NPO (called the Aleutian Low) and the high-pressure center near Hawaii in the South Pacific Ocean (called the Hawaii High). The pressure at the center of low pressure is opposite that at the center of high pressure, as in the NAO. Thus, two basic states of strong oscillation and weak oscillation are formed. Rossby proposed the atmospheric longwave theory in the 1930s, linking the wavelengths of planetary waves in the middle troposphere with the intensity of the westerly wind. Therefore, the intensity of the westerly wind is the basic feature of atmospheric circulation. On the ocean, the atmospheric circulation is quasi-barotropic, which means that the upper circulation has the same geographical distribution as the lower circulation. Therefore, strong oscillation and weak oscillation correspond to a high index (strong westerly wind) and a low index (weak westerly wind) in the middle troposphere. At present, the SLP of the Aleutian low (30°N–65°N, 160°E–140°W) in the Gulf of Alaska is generally used as the NPO index NPI (Trenberth and Hurrell, 1994). NPO had two developments in the late 20th century, namely, the PNA pattern and the PDO. PNA refers to the teleconnection that spreads along the Great Circle from the South of the North Atlantic through the Aleutian Islands to Northwestern North America (Wallace *et al.*, 1981, also see Chapter 8).

The PDO is the main year-round pattern of monthly SST anomalies in the North Pacific determined through EOF analysis, which was first described by Mantua *et al.* (1997). When the global mean SST anomaly is subtracted from the climatological annual cycle, the SST anomalies are defined as deviations from that cycle (as a way of removing any long-term trends that might be associated with externally forced climate change). According to Mantua *et al.* (1997), it is a long-lived variety of Pacific climate variability that resembles the ENSO type. The PDO is positive when SSTs are anomalously cool in the interior North Pacific and warm near the Pacific Coast, and when SLPs are below average over the North Pacific. When the patterns of climate anomalies are reversed, such as when there are above-average SLPs over the North Pacific or warm SST anomalies in the interior and cool SST anomalies along the North American coast, the PDO has a negative value. The phases of the PDO can be categorized as cold and warm (or cold and warm PDO events). The "cool" phase is characterized by a warm horseshoe pattern of higher-than-normal SST connecting the North, West, and South Pacific (see Fig. 6.3(b)) and a cool wedge of lower-than-normal SST in the eastern equatorial Pacific (see Fig. 6.3(a)). The Western Pacific Ocean cools during the "warm" or "positive" phase, while the wedge in the east warms. A "cool" phase occurred from 1947 to 1976 (29 years), and a "warm" phase occurred from 1977 to 1999 (22 years). However, recently, the "warm" and "cool" phases have lasted for shorter periods of time.

The spatial patterns of the PDO and ENSO are extremely similar, as noted by Zhang *et al.* (1997), except for the relative magnitudes of SST anomalies in the tropics and extra-tropics of the Pacific. ENSO prioritizes the equatorial region, whereas the PDO is concentrated in the North Pacific. The time series of ENSO and the PDO are different, although their spatial patterns are similar. ENSO is essentially an interannual phenomenon, whereas the PDO is primarily an interdecadal phenomenon. The dynamics of the PDO are still being studied, even though the physical causes behind ENSO are fairly well established. According to a recent assessment by Newman *et al.* (2016), the PDO is a confluence of atmospheric and oceanic phenomena that span the tropics and extra-tropics rather than a single event. Tropical ENSO dynamics, atmospheric teleconnections from the tropics to the North Pacific, the inherent plus tropically driven atmospheric forcing of North Pacific SST anomalies via surface energy fluxes, Ekman currents, and oceanic Rossby waves are a few of the phenomena. The low-frequency temporal nature of the PDO is influenced by the thermal and dynamic inertia of the ocean, whereas its large-scale spatial coherence is caused by atmospheric processes.

The "Interdecadal Pacific Oscillation" (IPO) is another pattern of decadal fluctuation related to ENSO (Power *et al.*, 1999; Henley *et al.*, 2015). Although there are many other ways to define the IPO, it is

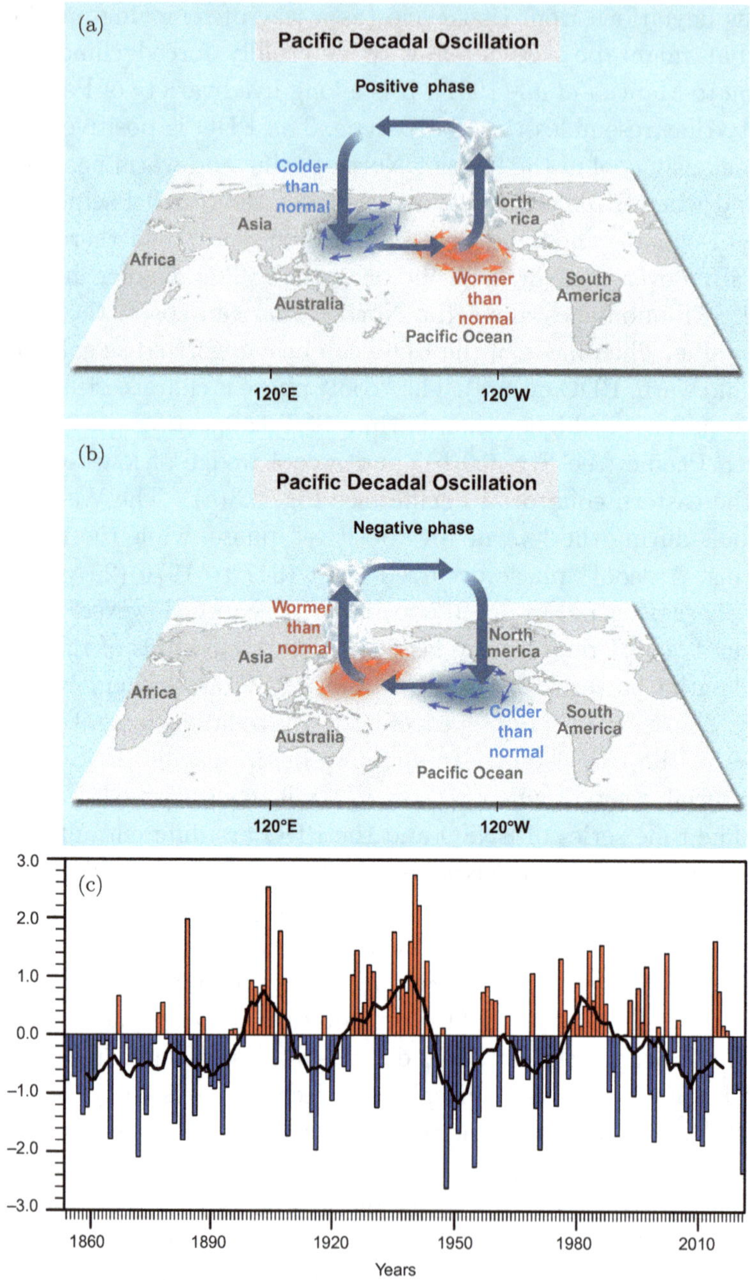

Fig. 6.3. Schematic sketch of the mechanism of positive phase of Pacific Decadal Oscillation (PDO) (a) and negative phase of PDO (b) (red color represents positive sea surface temperature (SST) anomalies, blue color represents negative SST anomalies). Time series of PDO index from 1854 to 2021 (c) (black curves indicate the 11-year running mean time series).

commonly determined to be the leading EOF of decadal low-pass filtered SST anomalies over the whole Pacific basin or the entire world. The PDO and IPO have extremely similar spatial patterns, but the former places more emphasis on the North Pacific than the latter. The monthly PDO and IPO time series are correlated at 0.74 (Newman *et al.*, 2016). Over North America, the hydroclimate, which includes precipitation, temperature, snowpack, streamflow, and drought, is connected with the PDO (McCabe *et al.*, 2012). Instead of being caused directly by the PDO SST anomalies, these consequences are mostly influenced by related variability in the large-scale atmospheric circulation.

By modifying ENSO events, the PDO can also have an indirect impact on climate. Power *et al.* (1999) noted that ENSO occurrences at various PDO phases had varied effects on precipitation in Australia, while Zhu and Yang (2003) discovered that the PDO modulates ENSO effects on summer climate anomalies at various PDO phases in China. According to Wang *et al.* (2014), the PDO phase influences how the ENSO affects worldwide dry–wet changes.

The extended reconstruction of SSTs by NOAA provides the foundation for the NCEI PDO index (ERSST Version 5). To create a PDO regression map for the North Pacific ERSST anomalies, the ERSST anomalies are regressed on the Mantua PDO index over their overlap period. The NCEI index is then calculated by projecting the ERSST anomalies onto that map. Following closely behind the Mantua PDO index is the NCEI PDO index (https://www.ncei.noaa.gov/access/monitoring/pdo/).

The time series of PDO from 1950 to 2011 is shown in Fig. 6.3(c). The PDO time series demonstrates that it did not regularly vary in the past 60 years. The PDO started out in its negative phase in the 1950s and transitioned into its positive phase in approximately 1978. Until the 2000s, the PDO was in a positive phase. It is currently in a negative phase. In addition, the PDO's negative phase is invariably linked to more frequent La Niña episodes. This phenomenon was independently modeled using CCSM46 throughout the hiatus period in warming scenarios for the 21st century.

6.2.3 *North Atlantic Oscillation (NAO)*

The NAO is another atmospheric oscillation defined by Walker (Walker and Bliss, 1932). The NAO is a long-lasting inverse oscillation of SLP in the north–south direction. The opposite pressure change in the NAO for large regions of relatively high pressure is centered over the Azores islands (West of Portugal, known as the subtropical or Azores high) and low pressure is centered over Iceland (the subpolar or Icelandic low). The NAO was first identified by Sir Gilbert Walker in the 1920s and depicts the relative changes in pressure between these two regions (Azores minus Iceland) (Walker and Bliss, 1932). This has a strong influence on winter weather and climate patterns in Europe and North America, and can extend further into Northern Asia if phases are prolonged. The North Atlantic jet stream, which constitutes ribbons of extremely rapid winds high in the atmosphere that affect the movement of regions of low pressure (depressions) and the storms that accompany them, is affected by fluctuations in the NAO, which have the effect of a large seesaw. The winter weather in Europe, Greenland, Northeastern North America, North Africa, and Northern Asia is significantly influenced by the NAO, an irregular fluctuation of air pressure over the North Atlantic Ocean. It is referred to as an "oscillation" because the changes in atmospheric pressure are essentially a back-and-forth switching between two dominant patterns, or modes: a "positive mode (see Fig. 6.4(a)) in which a strong subtropical high is located over the Azores Islands in the Central North Atlantic while a strong low-pressure system is centered over Iceland, and a "negative mode (see Fig. 6.4(b)) in which weaker high- and low-pressure systems are found over the same locations.

The NAO index, which is derived from air-pressure readings taken at sea level at stations in Iceland and the Azores (though readings collected at Lisbon have been used to represent the Azores at times), is used to determine the phase of the oscillation (i.e., whether it is in the positive or negative mode) and the relative strength of the pressure gradient (i.e., the difference in pressure between the high and low systems). Calculating the pressure difference between the centers of the Icelandic low and the Azores high yields the most accurate index values. Since the centers of these pressure cells may actually be located some distance from the recording stations, it is occasionally possible to infer the pressure differences between the cells regardless of the location of the cell centers using a method known as "principal components time series analysis." Daily index values are often graphed to track day-to-day changes, but data can also be aggregated into monthly and yearly values in order to illuminate long-term patterns.

Many scientists contend that significant changes in the climates of Europe and Eastern North America have been linked to the NAO's "locking" into one mode or another over periods of several years. Geologically, recent

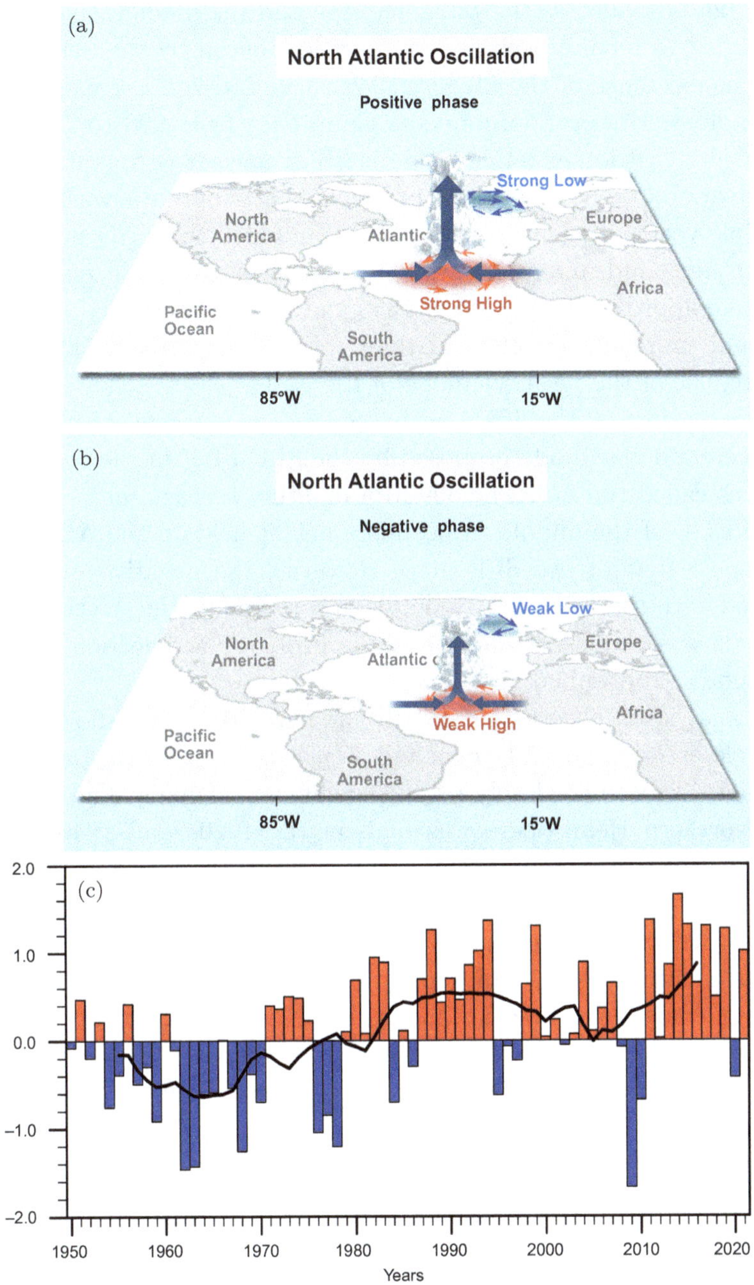

Fig. 6.4. Schematic sketch for the mechanism of positive phase of North Atlantic Oscillation (NAO) (a) and negative phase of NAO (b) (red color represents high-pressure center, blue color represents low-pressure level center). Time series of NAO index from 1854 to 2021 (c) (black curves indicate the 11-year running mean time series).

climatic periods, such as the Little Ice Age and the Medieval Warm Period, are regarded as having been significantly influenced by the behavior of the NAO. The existence of the Medieval Warm Period is a hotly debated topic in climatology. However, proponents claim that from 900 to 1300 CE, the North Atlantic region experienced a period of warmer temperatures and less climatic variability. Some of these proponents attribute the warm conditions to multiple-year periods when the positive mode of the NAO predominated. On the other hand, many scientists believe that the Little Ice Age, which lasted from approximately 1300 to 1850 and was characterized by somewhat colder and more erratic climate than today, was brought about by the predominance of the negative mode of the NAO.

An extension of the NAO is based on the concept of the Northern Hemisphere annular mode proposed by the SLP EOF in the north of 20°N, sometimes called the AO. The 500-hPa monthly average map demonstrates that the EOF of the middle troposphere height field on the Atlantic Ocean is similar to that of the SLP map. However, the negative height change associated with the Icelandic low extends throughout the Arctic. Therefore, recently, it was proposed that the AO is probably a reflection of the NAO in the middle troposphere.

The analysis of observational data and substitute data shows that NAO has a cycle of 24, 8, and 2.1 years and a cycle of 70 years. In particular, the last cycle is likely to be closely related to climate warming. The temperature in the Northern Hemisphere was high in the 1920s and 1940s. However, the temperature dropped in the 1960s and 1970s. After the 1980s, the temperature increased again. The change rule of the NAO index is generally consistent with the temperature in the Northern Hemisphere. According to the calculations, NAO can explain 34% of the variance in temperature change. Of course, this does not mean that NAO changes cause warming or vice versa. However, NAO change may be a mechanism for climate warming.

Using monthly mean 500 millibar height anomaly data from 1950 to 2000 over 0°N–90°N latitude, a rotated EOF analysis has determined the NAO loading pattern as its first mode (https://www.ncei.noaa.gov/access/monitoring/nao/).

The NAO entered its positive phase in the 1950s, and transitioned into its negative phase in the 1960s, as depicted in Fig. 6.4(c) From the 1970s to 2000, there was a particularly significant upswing. The most noticeable transition from the positive to the negative phase took place in 2000, and the NAO has stayed in its negative phase ever since then.

6.2.4 *Atlantic Multidecadal Oscillation (AMO)*

The AMO is a multi-decadal oscillation phenomenon (50–80 years is the most obvious timescale according to observational data) that affects the North Atlantic region's SST at the basin scale. The North Atlantic SST's periodic oscillations may be caused by natural internal variation in the strength of the thermohaline circulation and pertinent meridional heat movement in the ocean (McCabe *et al.*, 2004). The AMO, a mode of the climate system that operates on a nearly global scale, is one of the most prominent oscillations of decadal climate variability. It is distinguished by alternating warm and cold periods over large regions of the Northern Hemisphere. AMO warm phases occurred between 1860–1880 and 1940–1960 (see Fig. 6.5(a)), while cool phases occurred between 1905–1925 and 1970–1990 (see Fig. 6.5(b)). The signal is widespread and exhibits a co-oscillation with a positive correlation in some regions of the North Pacific, but it is strongest in the North Atlantic and spans the whole basin in that area. Numerous studies have demonstrated the connection between the warm AMO phase and local climate changes, including increased precipitation over the Sahel, less-than-normal precipitation over the Northwest Pacific, increased summer monsoon rainfall over India, wetting over Central America in summer, and drying over Northeastern Brazil in winter. This finding suggests that the multi-decadal volatility of the SST in the NAO may be largely responsible for the long-term predictability of drought frequency (McCabe *et al.*, 2004).

The major atmospheric link between the AMO and Siberian warm season (May–October) precipitation is the Atlantic–Eurasian (AEA) teleconnection, which exhibits decadal variability. Its pattern on the decadal timescale is known as the Eurasian multi-decadal teleconnection (EAMT). The AMO can stimulate an eastward downstream propagating Rossby wave train response of the EAMT, causing in-phase decadal variability in Siberian warm season precipitation. Therefore, the decadal-scale fluctuations in Siberian warm season precipitation may be indirectly caused by the AMO. In fact, the decadal changes in summer precipitation in East Asia can be included. By enhancing the EASM and the southward intrusion of cold air from the north, the eastward propagation of wave trains can be stimulated by the warm phase of the AMO throughout all of Eurasia and ultimately increase summer rainfall in Northeast Asia. Further research is necessary to determine how much the AMO contributes, in relation to other factors, to the decadal variability in summer precipitation in East Asia.

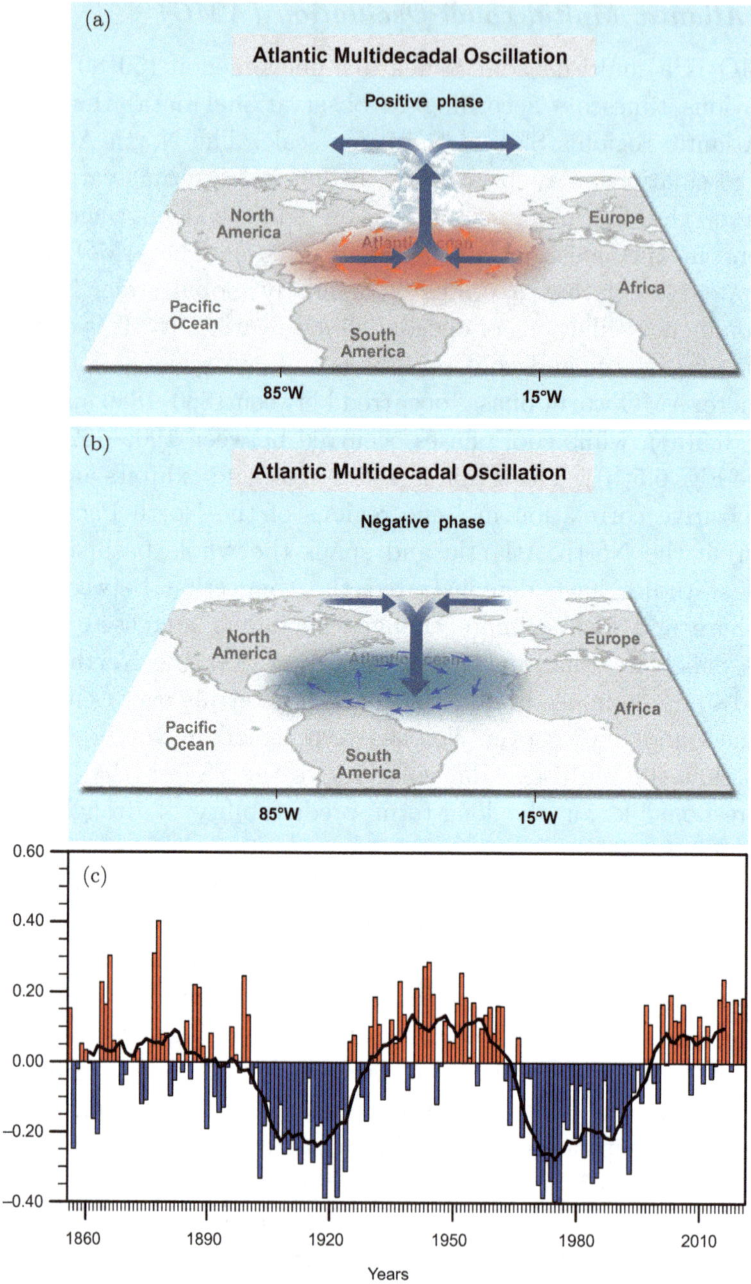

Fig. 6.5. Schematic sketch for the mechanism of positive phase of Atlantic Multidecadal Oscillation (AMO) (a) and negative phase of AMO (b) (red color represents positive sea surface temperature (SST) anomalies, blue color represents negative SST anomalies). Time series of AMO index from 1854 to 2021 (c) (black curves indicate the 11-year running mean time series).

The AMO can influence the PDO and the related PNA pattern at the multi-decadal timescale. Effects of AMO on PDO also suggested that a North Pacific regime shift (in contrast to the 1976–1977 shift) may take place several years after the observed AMO's mid-1990s shift. The equatorial Pacific decadal variability is also reported to have been significantly influenced by the AMO, and the ENSO variability has been controlled at low frequency by the AMO. According to observational and modeling studies, the AMO modulates multi-decadal variations in the atmospheric NAO, with the positive AMO phase typically occurring a few years before the negative winter NAO phase. By resolving the stratosphere and improving stratosphere/troposphere coupling, recent modeling studies have demonstrated that the predicted multi-decadal winter NAO response to the AMO can be greatly enhanced.

The area-weighted average of monthly SST anomalies in the North Atlantic is known as the AMO index (Enfield *et al.*, 2001). The index data are available at https://psl.noaa.gov/data/timeseries/AMO/ and span the years 1901 to the present (see Fig. 6.5(c)), showing that the AMO has been largely stable since 2000. Since the multi-decadal scale is what fundamentally distinguishes the standardized AMO, Huang *et al.* (2017a) separate its decadal to multi-decadal components. A period spanning more than 60 years is clearly displayed in the filtered AMO index (Fig. 6.5(c)).

6.2.5 *Indian Ocean Dipole (IOD)*

The Indian Ocean Dipole (IOD) is the difference between SSTs in two regions (i.e., poles, hence a dipole). They are the western pole in the Arabian Sea (Western Indian Ocean) and the eastern pole in the Eastern Indian Ocean, south of Indonesia. During the periodic oscillation of SSTs, known as the IOD, the Western Indian Ocean becomes warmer (positive phase) (see Fig. 6.6(a)) and subsequently becomes colder (negative phase)(see Fig. 6.6(b)) than the Eastern Indian Ocean. When a positive IOD event occurs, the SST in the Western Indian Ocean is high, and the SST in the Eastern Indian Ocean is low. This difference will cause the SST in the Indo-Pacific Warm Pool to be low and the Walker circulation to weaken, which will change the climate in Asia, Australia, and Africa (such as the heavy rain in East Africa and the drought in Indonesia and Southern Australia). The IOD significantly contributes to the variability in rainfall in Australia and has an impact on the climate of this area and other nations that border the Indian Ocean Basin. Like ENSO, the change in temperature gradients over

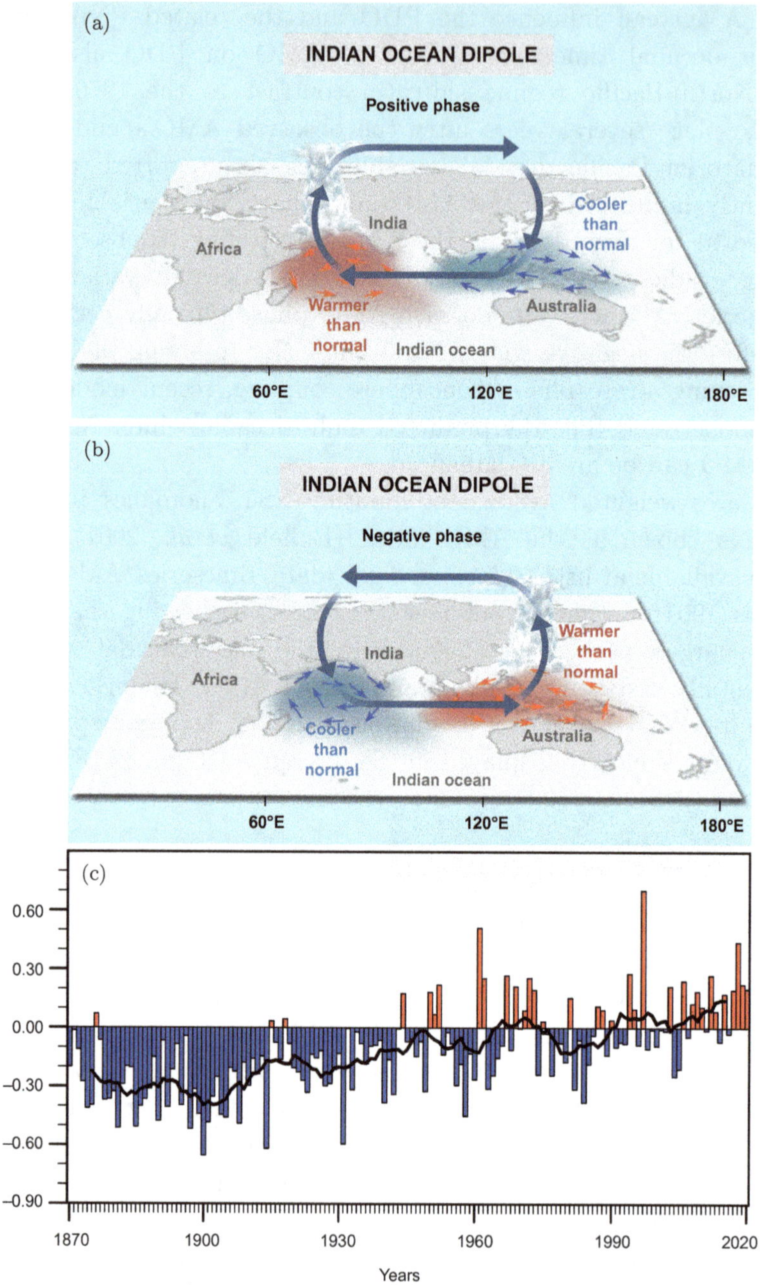

Fig. 6.6. Schematic sketch for the mechanism of positive phase of Indian Ocean Dipole (IOD) (a) and negative phase of IOD (b) (red color represents positive sea surface temperature (SST) anomalies, blue color represents negative SST anomalies). Time series of IOD mode index (DMI) from 1854 to 2020 (c) (black curves indicate the 11-year running mean time series).

the Indian Ocean causes variations in where air and moisture tend to rise and drop. The IOD is an ENSO-like coupled ocean and atmosphere phenomenon that occurs in the equatorial Indian Ocean. The IOD and ENSO occurrences are believed to be connected by an extension of the Walker Circulation to the west and related Indonesian throughflow (the flow of warm tropical ocean water from the Pacific into the Indian Ocean). On decadal timescales, ENSO and IOD activities were inversely related. Hence, positive IOD events are frequently linked to El Niño and negative IOD events are linked to La Niña. The effects of El Niño and La Niña events are frequently the greatest over Australia when the IOD and ENSO are in phase, whereas the effects can be lessened when they are out of phase. In addition, a significant proportion of IOD events and a sizable portion of ENSO events occurred independently of one another. Notably, compared to the group of IOD events that co-occurred with ENSO, the group of IOD events that occurred without ENSO accounted for a greater proportion of Indian Ocean variability.

The IOD is one of the major factors influencing Australia's climate and has a significant influence on agriculture. Thus, events often coincide with the growing season for winter crops. Neutral, positive, and negative are the three stages of the IOD. In most cases, events begin in May or June, reach their height between August and October, and then quickly subside as the monsoon reaches the Southern Hemisphere at the end of spring.

The Eastern Indian Ocean, close to Java and Sumatra, experiences cooler-than-average sea surface conditions during the positive phase of the IOD (such as what occurred in the Southern Hemisphere winter and spring of 2019). This difference contrasts with the western tropical Indian Ocean, which experiences warmer-than-average conditions. The wind, temperature, and rainfall patterns in the area are altered by this east–west contrast in ocean conditions (a "dipole"), which frequently causes floods in Eastern Africa and droughts and bushfires in Eastern Asia and Australia, among other ecological and socioeconomic effects. During the negative IOD phase, the locations of the warm/cool pattern and related effects are reversed. The growth of these IOD conditions requires collaboration between the atmosphere and ocean, like ENSO.

The anomalous SST gradient between the southeastern and western equatorial Indian Oceans (50°E–70°E and 10°S–10°N) and the equatorial Indian Oceans represents the intensity of the IOD (90°E–110°E and 10°S–0°N). This gradient is known as the dipole mode index (DMI; see Fig. 6.6(c)) (https://psl.noaa.gov/gcos_wgsp/Timeseries/Data/dmi.had.long.data). If the DMI is positive, the phenomenon is known as the positive IOD; if it is negative, it is known as the negative IOD.

6.3 Impacts of Ocean Oscillations

The section introduces the mechanisms by which various oceanic oscillation factors synergistically affect climate change in arid and semi-arid regions, with a focus on the effects of the ENSO, Pacific Decadal Oscillation (PDO), Northern Atlantic Oscillation (NAO), Atlantic Multidecadal Oscillation (AMO), and Indian Ocean Dipole (IOD). These oceanic oscillation components have a tight connection to the interdecadal scale of climate change. The surface temperature, precipitation, and dry/wet characteristics of arid and semi-arid regions are all dramatically altered by different phase combinations of oceanic oscillation components. The impact of marine activities on these regions will likely reveal new characteristics in the future as climate change in arid and semi-arid regions intensifies. This will increase the uncertainty of future climate change in these regions and intensify the effect and feedback of these arid and semi-arid regions on global climate.

6.3.1 *Impacts on temperature*

There have been two global warming slowdown periods in the past 100 years: the period from 1940 to 1975 and the period since the end of the 20th century. This is due to the transformation of the positive and negative phases of the oscillation indexes, such as the PDO, the NAO, and the AMO (Kosaka and Xie, 2013; Huang *et al.*, 2017b). Among these, the era of warming that slowed down toward the close of the 20th century sparked significant worries. According to Huang *et al.* (2017b), the contribution from land cooling to the NH was up to 66% in the cold season during the phase of warming slowdown, and the major cooling areas were found in Eurasia and North America, with large regions of arid and semi-arid regions. Furthermore, Guan *et al.* (2015a) used a dynamic adjustment approach to break down the original temperature change into the radiative forced temperature and dynamically induced temperature changes (DIT). According to Fig. 6.7, the RFT denotes temperature changes brought on by human activities like CO_2 emissions, whereas the DIT denotes temperature changes mostly brought on by the internal variability of the climate system. The DIT displayed a growing tendency during the period of increased warming (while RFT exhibited a warming trend). During this time, warming was increased by the convergence of the rising DIT and RFT, but warming was slowed down by the cooling DIT that counterbalanced the rising RFT. In particular, the interaction of the NAO, PDO, and AMO in the interdecadal scales resulted in the cooling

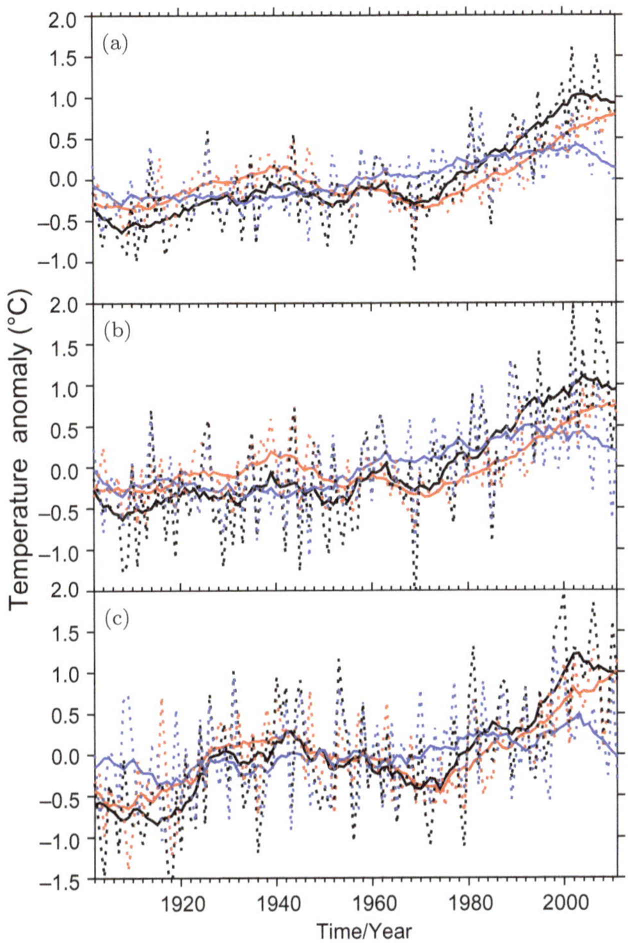

Fig. 6.7. Regional mean time series for the raw temperature (black), DIT (blue), and RFT (red) over the NH (a); Eurasia (b); and North America (c). Monthly NASA Goddard Institute for Space Studies (GISS) SAT dataset is used in Fig. 6.7 (Guan *et al.*, 2015a).

of DIT (See Chapter 8 and Huang *et al.*, 2017b). Furthermore, research by Guan *et al.* (2015a, 2015b) and Huang *et al.* (2017b) has demonstrated that the interdecadal signals were responsible for the change between the accelerated warming and warming slowdown. Due to its apparent conflict with the notion of human-induced global warming, the phenomenon of the WTS has been referred to as a global warming pause or hiatus and has garnered attention on a global scale. The observed SAT deviation throughout the WTS period casts doubt on the contribution of GHGs in global warming in the past century.

The annual mean SAT increased consistently between the two hiatus periods from 1970 to 2000, as illustrated in Fig. 6.7(a). During the cold season, the mean SAT declined over Eurasia (Fig. 6.7(b)) and North America, but this was countered by a consistently increasing SAT throughout the warm season. This concludes that the yearly mean SAT slowing that happened over the Northern Hemisphere for about 10 years after 2000 (when the WTS began) was caused by the declining SAT in the cold season. The SAT hiatus has only occurred in the cold season over the Northern Hemisphere.

The raw SAT shows a continuous growth that began in the 1970s and continued until the beginning of the hiatus in around 2000, after which there is a steady trend in the average SAT over the Northern Hemisphere (Fig. 6.7(a)). The time series of SAT anomalies (Fig. 6.7) is adjusted by 11-year running period. Interestingly, the adjusted SAT time series shows a linear increase from 1970 to 2000, follows a relative slower rate of warming during the hiatus, and is accompanied by the clear cooling trend of dynamically induced SAT, whereas it has been demonstrated that dynamically induced variability plays a role in the WTS. In the period from 1902 to 2011, the interannual to interdecadal variability was represented by the dynamically induced SAT, which is a component of raw SAT. It contributed positively to accelerating warming between 1980 and 2000 when this component was in the upswing phase (warming). However, when the dynamic temperature was in the downtrend phase (cooling), it lessened or even balanced the radiatively caused warming in the recent WTS (hiatus) in the period of 2000–2010. Over Eurasia and North America, the regional average SAT time series in Figs. 6.7(b), 6.7(c) showed a notable slowing in the raw SAT and an increase in the adjusted SAT. It suggested that the effect of adjusted SAT in the SAT shift is mostly a warming one. The first hiatus, which occurred between 1940 and 1970, likewise showed a contribution from both dynamically and radiatively produced SAT, but the most recent hiatus is different because of the balance between dynamically induced SAT and radiatively imposed SAT. The dynamically generated SAT assumed the dominant role in the raw SAT variation, and the radiatively forced SAT (adjusted SAT) was much weaker in the first hiatus period than it is in the present hiatus period. As a result, during the two hiatus periods, the dynamically and radiatively produced SATs contributed differently to the raw SAT change. Further research is required to investigate the mechanism of the hiatus in the years 1940–1970.

The impact of the PDO, NAO, and AMO on WTS has been examined in previous results. Therefore, in this book, we individually estimated the correlative coefficients of the PDO (Fig. 6.8(a)), NAO (Fig. 6.8(b)), and AMO (Fig. 6.8(c)) using dynamically generated SAT. The correlation coefficient between PDO and dynamically induced SAT is depicted in Fig. 6.8(a). Over Northwest North America and the nearby Pacific Ocean coast, the correlation coefficient is positive. The PDO's negative phase causes cooling through changes in the dynamically produced SAT, as seen by the positive correlation coefficient between the PDO and the SAT.

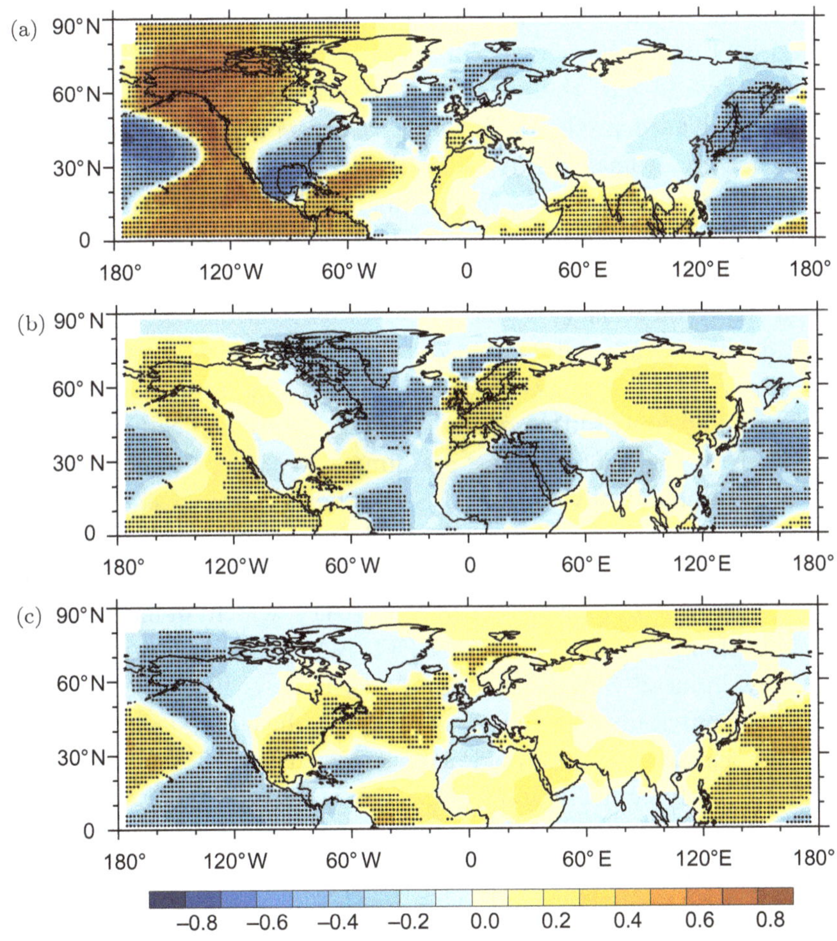

Fig. 6.8. Spatial distribution of the correlation coefficient between the PDO (a), NAO (b), and AMO (c) (detrended and 11-year running means) and the detrended dynamically induced SAT during the cold seasons of 1950–2011. The stippling indicates a 95% confidence level according to a two-tailed Student's t-test (Guan *et al.*, 2015a).

The distribution of the correlation coefficient between the dynamically produced SAT and the interdecadal NAO is shown in Fig. 6.8(b). Over the mid-to-high-latitude regions of Eurasia, Northwestern North America, and close to the Pacific Ocean coast, there is a positive pattern. This research shows that the dynamically generated SAT over the Eurasian and North American continents, which is normally reported to exhibit decadal fluctuation, is strongly influenced by the NAO. The positive correlation coefficient represents the cooling effect on the dynamically induced SAT changes over the mid-to-high-latitude regions of Eurasia, Northwestern North America, and the nearby coast of the Pacific Ocean because the interdecadal NAO time series was negative during the recent hiatus period. The dynamically generated SAT across northern North America, Africa, and the Atlantic is increasing, according to the negative correlation coefficient. The NAO has been suggested as a predictor for SAT change projections in addition to its significant involvement in dynamically driven SAT.

AMO is an efficiency metric for monitoring dynamical activities in addition to NAO and PDO. Over the mid-to-high-latitude regions of Eurasia, Western North America, and the neighboring Pacific Ocean coast, there is a negative connection (Fig. 6.8(c)). The correlation coefficient between AMO and the dynamically induced SAT in Fig. 6.8(c) shows that the relationship between the two is positive over the rest of the Northern Hemisphere. The contribution of the AMO to the dynamically induced SAT is confirmed by Fig. 6.8(c). The "Stadium wave" theory, the effect of the AMO on the dynamically produced SAT, shows an extensive impact over the entire Northern Hemisphere. We displayed the regional distribution of contributions in Fig. 6.9 to summarize the contributions of PDO, NAO, and AMO to the dynamically induced SAT, which demonstrates how the contributions distribution from PDO, NAO and AMO on the WTS's dynamically induced SAT reduction. According to Fig. 6.9(a), the PDO significantly contributes to the dynamically induced SAT across Western North America and the nearby Pacific Ocean coast. The contribution of the NAO to the dynamically induced SAT over mid-to-high-latitude parts of Eurasia and Northern Africa is shown in Fig. 6.9(b), where it is considerably more visible than in the other regions. The figure also shows the existence of a comparatively weaker center along the coast of Greenland. The impact of the NAO on dynamically induced SAT virtually exclusively affects these two centers, as evidenced by the contribution exceeding 60% over the mid-to-high-latitude regions of Eurasia and Northern Africa. Figure 6.9(c) shows how the AMO contributed to the dynamically induced SAT. There are no

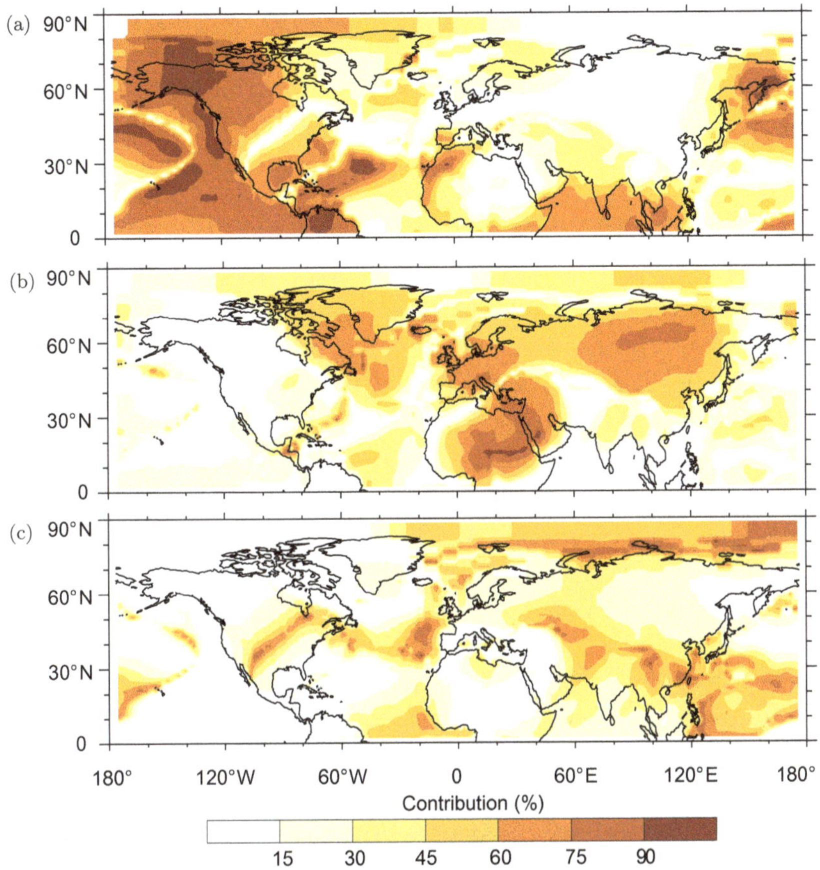

Fig. 6.9. The contribution of the PDO (a), NAO (b), and AMO (c) to the dynamically induced SAT (Guan *et al.*, 2015a).

obvious local maxima in the AMO, which is a comprehensive dynamical factor.

Teleconnection allows the AMO to have an impact on local temperature as well. The Africa–Asia multi-decadal teleconnection pattern (AAMT) (Sun *et al.*, 2017), which originates in North Africa and spreads to East Asia roughly along 30°N, where the North African–Asian jet is located in the winter, is another southern atmospheric link between the AMO and the climate in East Asia during the cold season (November–April). Over Northwest Africa, the Arabian Peninsula, and Central China, decadal shifts in surface and tropospheric air temperatures can result from the AMO's ability to excite the AAMT Rossby wave train during the cold season as it travels along the North African–Asian jet stream (Sun *et al.*, 2017).

According to earlier research, anthropogenic forcing is primarily to blame for the long-term trends in extreme high temperatures (EHT). However, the amount of natural forcing that contributes to the decadal to multi-decadal variance in temperatures and EHT varies significantly across different regions. For example, the AMO and natural external forcing (solar forcing, volcanic eruptions, and sulfate aerosols) could affect the summer temperature/EHT over Europe (Ghosh *et al.*, 2017). According to Ruprich-Robert *et al.* (2018), heat waves across North America may be influenced by the moisture content of the soil. Mega-El Niño–Southern Oscillation and AMO have been reported to have an impact on the decadal variation in the frequency of EHT over Eurasia (Zhou and Wu, 2016). Furthermore, it is demonstrated that the footprints of the AMO may be demonstrated in the huge decadal to multi-decadal changes in EHT over the four hotspots (the Mongolian Plateau, Eastern Siberia, Europe, and Western North America–Mexico) (Gao *et al.*, 2018). The AMO explains 57.6% of the overall variance of the decadal to multi-decadal variabilities in temperature averaged over these domains and contributes to the coherent fluctuation in temperature over the four hotspots through a barotropic circumglobal teleconnection (Gao *et al.*, 2018). The AMO was also crucial in the natural decadal variation of T_{min} temperature across North America and was the primary factor for the decline of in T_{min} temperature during the global warming hiatus (Gan *et al.*, 2019). The spatiotemporal changes in temperature T_{min} and the AMO are inextricably linked. At night, when the AMO is in the positive (negative) phase, there are two low-pressure (high-pressure) systems in the Northeastern Pacific and the North Atlantic, and warm (cold) advection in North America is primarily from the south (north). Due to the synchronized drop of the AMO phase, the temperature T_{min} decreased throughout the hiatus (Gan *et al.*, 2019).

6.3.2 *Impacts on precipitation*

We determined the correlation coefficient between summer precipitation and sea surface temperature in order to better understand the variables influencing summer precipitation over semi-arid regions of East Asia. The SST exhibits negative correlation patterns in the tropical East-Central Pacific, Western Indian Ocean, and Northwestern Pacific, whereas positive correlation patterns in the Barents Sea and East China Sea were 90% significant. These oceanic activities may influence precipitation in semi-arid regions in East Asia through atmospheric circulation, and the SST factors are primarily related to the monsoon route and the sea basin where ENSO occurs, so the monsoon and ENSO have a great impact on the precipitation over semi-arid regions in East Asia.

We determined the correlation coefficients among precipitation, the Niño 3.4 index, and the monsoon index to expose their linkages in order to investigate the impact of ENSO and the monsoon on precipitation. The monsoon index is defined as the U850 in (5°N–15°N, 90°E–130°E) minus U850 in (22.5°N–32.5°N, 110°E–140°E) (Wang and Fan, 1999); it can capture the total variation of precipitation over East Asia in three dimensions. The monsoon index has an obvious negative effect ($r = -0.86$) on the precipitation in the wet years and passed the significance test when the monsoon edge crossed the semi-arid region of East Asia by 99%. When it does not make a distinction between dry and rainy years, the correlation coefficient falls to −0.21 and is not significant. Accordingly, the monsoon cannot affect the precipitation in semi-arid regions until its edge crosses through those areas. The same formula can be used to determine how ENSO affects precipitation. Although there is no statistically obvious correlation between Nino 3.4 and precipitation, Nino 3.4 clearly has a favorable impact ($r = 0.30$) on the monsoon index, which has the power to alter precipitation across semi-arid regions. As a result, when the edge of the monsoon passes the semi-arid regions of East Asia, ENSO has a detrimental impact on the precipitation there.

Pacific decadal oscillation (PDO) occurrences have a significant impact on weather and climate, not just in the Pacific and North America but also throughout the entire Northern Hemisphere (Gutzler and Kann, 2002). Regional variations in temperature and precipitation in China are all correlated with PDO, with regional differences in the strength of this correlation. On the interdecadal scale, SST variability plays a crucial role in the formation of drought in North China, with low precipitation and significantly higher temperature in North China during the warm phase of PDO (Ma and Shao, 2006). Ma and Shao (2006) found a negative association between the dry and wet variability in the eastern half of Northern China, which is bordered by 100°E, and the PDO, while a positive correlation was found in the western section.

By modifying ENSO events, the PDO can also have an indirect impact on climate. Power *et al.* (1999) pointed out that ENSO events at various PDO phases have varied effects on precipitation in Australia, while Zhu and Yang (2003) discovered that ENSO effects on summer climate anomalies at different phases in China are obviously modulated by the PDO. According to Wang *et al.* (2014), the PDO phase influences how the ENSO affects worldwide dry–wet changes.

PDO, also known as the "decadal El Niño phenomena," is an oscillated SST that occurs in the North Central Pacific and lasts between 20 and 30 years (Mantua *et al.*, 1997; Mantua and Hare, 2002). The PDO is generally considered to be a part of IPO in the North Pacific. As interdecadal signals in the Pacific Oceans, they both affect the dry and wet changes in the semi-arid regions. Previous research indicated a tight connection between the PDO transition from the cold to the warm phase in 1976 and the drought in the North China region from the middle to the end of the 1970s (Ma and Shao, 2006). In the central and western regions of the United States, the cold and warm phases of the IPO correspond to decreases and increases in precipitation, respectively (Dai, 2013b).

During winters when the NAO is in its positive mode, strong high-pressure and strong low-pressure systems produces warmer, wetter conditions over Northern Europe and most of Northeastern North America. This occurs because the polar front jet stream tends to be free of large undulations (Rossby waves), and the jet stream's westerly winds funnel storms over the Mid-Atlantic states, between the strong North Atlantic pressure cells, and over Northern Europe. Additional sea ice forms in Hudson Bay, Baffin Bay, and off Western Greenland under the positive NAO mode, which also causes colder weather to prevail over areas of Quebec, Newfoundland and Labrador, and Greenland. The Mediterranean region, in contrast, experiences a cool, dry winter.

During winters governed by the negative mode of the NAO, colder conditions are brought to Eastern North America and Northern Europe mostly by more frequent intrusions of Arctic air. Europe experiences below-average precipitation while North America experiences more snow than usual. The weak state of the pressure cells over Iceland and the North Atlantic during the NAO's negative mode leads to the drier conditions over Northern Europe. The reduced pressure gradient over the region slows the pace of westerly winds, allowing cold, dry air to be drawn into Northern Europe from Northern Russia and the Arctic. Warmer weather can prevail from Hudson Bay to Western Greenland during these years due to a prominent northward-reaching arc in the polar front jet stream, which is facilitated in part by blocking anticyclones that direct the jet stream northward. The jet stream, after skirting the reduced low-pressure cell over Iceland, arcs South over the North Atlantic to bring moisture and warm air to Southern Europe.

The effect of AMO on precipitation focuses on the multi-decadal scale, as AMO can control climate variability over a long timescale. Fig. 6.10(a) shows uneven spatial distributions of precipitation in the semi-arid regions

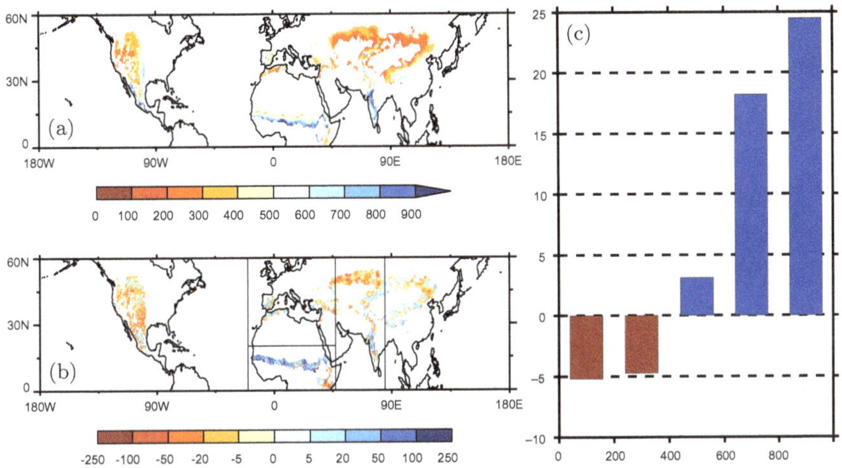

Fig. 6.10. Spatial patterns of climatological mean precipitation (a) (units: mm a^{-1}) from 1987 to 2016. Regression coefficient (b) (units: mm a^{-1}) between annual mean precipitation and AMO from 1901 to 2016 in the semi-arid regions of the NH. The semi-arid regions are divided into five parts (black lines). Area-weighted average regression coefficient (c) (units: mm) between annual mean precipitation and AMO under different semi-arid precipitation (units: mm a^{-1}) areas of the NH (Zhu *et al.*, 2021).

of the NH. The average precipitation exceeds 500 mm per year in the semi-arid regions of North Africa, India, and the southern edge of North America, while the annual precipitation in the semi-arid regions of mid-latitude Asia is generally less than 200 mm. In addition, there is a definite relationship between rainfall in the Eastern Hemisphere and latitude, with lower latitudes experiencing greater precipitation. We analyzed the patterns of regression coefficients between the two with climatological mean precipitation in order to investigate the relationship between precipitation over NH semi-arid regions and AMO. Figure 6.10(b) displays the patterns in the regression coefficients between precipitation and AMO. The negative anomalies are mainly concentrated in the semi-arid regions of North America and higher latitudes of the Eastern Hemisphere. In comparison, positive anomalies are distributed in lower latitudes of the Eastern Hemisphere. In addition, Fig. 6.10(c) shows that the area-weighted average regression coefficients between mean annual precipitation and AMO are increased along with the local precipitation amount and are, respectively, −5.19, −4.72, 3.13, 18.20, and 24.48 in the 0–200 mm, 200–400 mm, 400–600 mm, 600–800 mm, and >800 mm precipitation area. For instance, during the positive phase of the AMO in 1987–2016, the regions with annual precipitation below 400 mm experienced negative anomalies, while those with annual precipitation more than 400 mm experienced positive anomalies. This indicates that annual precipitation below 400 mm decreased and above 400 mm increased (Fig. 6.10(c)). In particular, the aforementioned patterns demonstrate more regularity toward the Atlantic Ocean. Such patterns suggest that the AMO has exerted a significant influence on the precipitation in NH semi-arid regions.

The AMO index recorded by the devices clearly comprises two cycles between 1901 and 2016. Based on the cycles, the modern AMO can be divided into two periods: one from 1901 to 1963 and the other from 1964 to 2016. According to a comparison of the relationship between AMO and precipitation in these two eras, there is a long-term and stable connection between AMO and precipitation, which can contribute to future precipitation predictions in the semi-arid regions (Sutton and Dong, 2012). The spatial distributions of average precipitation anomalies (relative to climatological mean) for the positive phases and negative phases of the AMO are also demonstrated in the CRU TS v. 4.05 dataset. The composite precipitation patterns for AMO+ (positive AMO phase) and AMO− (negative AMO phase) exhibit a high degree of consistency between the two periods, especially around the North Atlantic. In the positive phase, most of the semi-arid regions in North America show decreased rainfall (the reduced value is more than 20 mm) except in some marginal areas (precipitation increased in the Northernmost areas), while the semi-arid regions in North Africa present strongly increased precipitation, with variations ranging from 20 mm to 50 mm. During the same phase, more precipitation is distributed in the Iberian Peninsula and Asia Minor Peninsula, but the variations are only 10 mm. The increased pattern (more than 50 mm) on the Indian subcontinent represents the most substantial changes in precipitation in the positive phase in the semi-arid parts of Asia. In the negative phases, a different pattern of change in precipitation is observed. The correlation coefficients between AMO and precipitation during the two periods are greater than 0.6 or less than −0.6, and pass the 95% significance test, further supporting the idea that there is a strong correlation between AMO and precipitation in the semi-arid regions. Given the similarity in the two periods' relationships between AMO and semi-arid precipitation, it is possible that similar processes underlie the semi-arid regions' AMO-modulated precipitation and that these mechanisms have similar effects on the climate (Sutton and Dong, 2012).

As established earlier, the SST over the Pacific Ocean combined with the influence of interdecadal fluctuations in the IOBM (Indian Ocean Basin mode) has an effect on the climate of WNWC (Western Northwest China).

The 21-year running correlation coefficients between January–March IOBM and April–June precipitation over the WNWC are examined to investigate the varying relationship between IOBM and the precipitation variation over the WNWC. Even though there were fewer station-observed rainfall data before 1961, the analysis in this case is extended to an earlier period beginning in 1951 to highlight the interdecadal influence of IOBM on WNWC precipitation. As seen from Fig. 6.11, the findings show that this relationship was not robust during 1970s–1990s, despite the positive

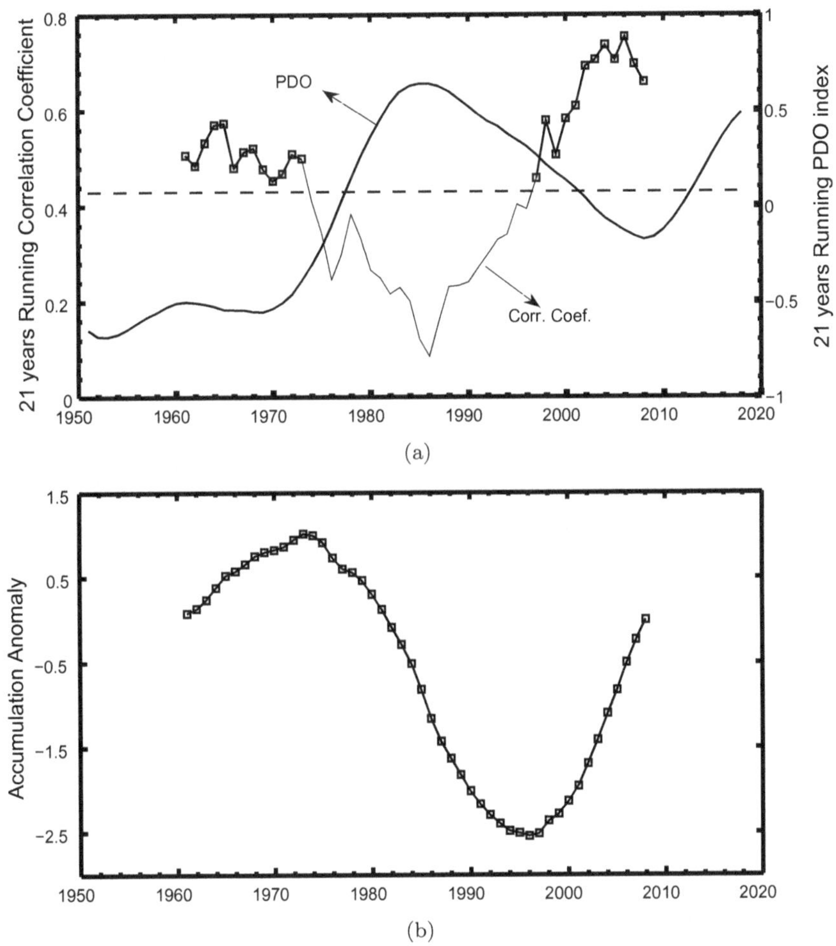

Fig. 6.11. The evolution of 21-year running correlation coefficient between January–March IOBM and April–June precipitation over the WNWC (a) and its accumulation anomaly. The straight line is the critical value (0.42) of the significant test at 0.05 level (b). The black curve in (a) is the decadal PDO index based on 21-year low-pass filter. The shift occurred in 1974 and 1997 (Wang *et al.*, 2022b).

correlation over the entire studying period. Before 1974 and from 1997 onward, these coefficients were more than 0.43, which is the threshold of the 0.05 significance level. Recent decades display a high correlation (0.75) between the January–March IOBM and the subsequent April–June WNWC precipitation (Fig. 6.11(a)). Additionally, the accumulative anomaly of the correlation coefficient demonstrates that an interdecadal shift took place around 1974 and 1997 (Fig. 6.11(b)). This result is in line with other studies on the appearance of large-scale SST anomalies in the Tropical Pacific in the middle of 1970s and late 1990s, which show an interdecadal transition (Wang *et al.*, 2014). This indicates that the PDO may affect the interdecadal lagged relationship between WNWC precipitation and IOBM, which is linked to the El Niño-like pattern of Pacific climate variability around 1976 and La Niña-like pattern in the late twentieth and early 21st centuries (Newman *et al.*, 2016).

6.3.3 *Impacts on dry and wet changes in semi-arid regions*

The global oceans are important forcing sources for the formation of global dry and wet distributions. At the same time, semi-arid regions are sparsely precipitated, ecologically fragile, and are highly vulnerable to factors regulating the global ocean. Numerous studies have shown that oceanic activities, especially SST anomalies in the Pacific and Atlantic basins, can substantially impact the dry and wet fluctuations in semi-arid regions (Dai, 2013a; Trenberth and Hurrell, 1994; Trenberth *et al.*, 2014a). The characteristics of the dry and wet shifts in semi-arid regions will be affected through teleconnections as the SST anomaly signals at the interannual and interdecadal scales of various basins regulate atmospheric circulation anomalies in the corresponding timescales.

The ENSO signal, the largest global interannual variation signal, can influence the upper atmospheric circulation, activate and maintain atmospheric circulation anomalies, and further initiate global dry and wet changes. The main cause of tropical drought is El Niño, which occurs in the equatorial Pacific (Trenberth *et al.*, 2014a). The global dry land range is twice as large when EI Niño oscillations are strong as opposed to when they are weak (Lyon, 2004). In different regions and seasons, the impact of ENSO on wet and dry changes in the global semi-arid regions varies (Ropelewski and Halpert, 1987). A study found that, when El Niño struck, drought would usually occur in South Africa, since East Africa's semi-arid regions saw rainy weather from the end of one year to the beginning of the next. Due to the increase in the SST in the eastern equatorial Pacific and enhancement of the Hadley Cell, the Pacific subtropical high in East Asia was strengthened and moved southward. As a result, the summer monsoon was weak, the northward strength of the warm and humid airflow was weak, and the cold air moved South. Due to the stagnation of the monsoon rain zone caused by the convergence of cold and warm air in the middle and lower Yangtze River, East Asia experienced constant rain and flooding in the summer.

The impact of the Atlantic SST anomalies on global dry and wet climate changes cannot be ignored (Sutton and Hodson, 2005; Knight *et al.*, 2006). The precipitation in Southeast South America is impacted by SST anomalies in both the equatorial Pacific and Atlantic, but the effects take longer to manifest. While the equatorial Atlantic primarily operates on the interdecadal or a longer timescale, the equatorial Pacific primarily operates on the interannual timescale. Corresponding to the cold phase of the AMO in the 20th century, the Southeastern part of South America became continuously wet (Seager *et al.*, 2010b). Ting *et al.* (2011) reported that the Atlantic equatorial convergence zone moved northward when the AMO was in the warm phase, increasing precipitation in Western Sahel of Africa, the equatorial North Atlantic, and the Central United States while the decreasing it in Western South Africa, the equatorial South Atlantic, and Eastern South America. Many places in the Eastern Hemisphere experience greater precipitation when the AMO is in the positive phase, except for Central America and Northern South America, as the precipitation in the Western Hemisphere decreases. In comparison to the PDO, the degree of precipitation anomalies induced by AMO in the global area is relatively small and the high-impact areas are mainly found in Africa and Northern South America.

During the warm/cold AMO period, a phase difference in the precipitation at various timescales appears between the Central Asian dryland (CAD) and East Asian dryland (EAD), revealing distinct precipitation changes in these two regions. In particular, the rebuilt IAV represents the variation in high-frequency signals from initial precipitation, and the rebuilt MDV of precipitation varies on a multi-decadal timescale. The MDV cycles of the CAD and EAD are approximately 77a and 58a, respectively, so it makes sense that the MDV of precipitation over these two regions can be represented by the sum of C4 and C5 (Fig. 6.12). Moreover, the MDV time series of the annual precipitation of the CAD exhibits obvious negative (positive) anomalies in the warm (cold) phase of the AMO, identifying a robust out-of-phase relationship conforming to the subsequent composite analysis.

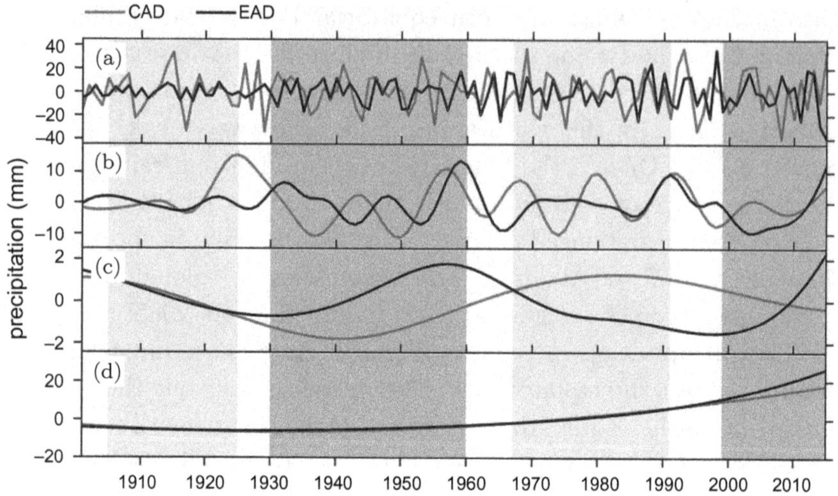

Fig. 6.12. Interannual variability (a), interdecadal variability (b), multi-decadal variability (c), and secular trend (d) for precipitation in the CAD and EAD. The dark (light) shaded areas denote the AMO-positive (negative) periods (Guo *et al.*, 2022).

Furthermore, it is believed that the tropical Indian Ocean (TIO) sea surface temperature (SST) anomaly has a crucial influence on the Asian monsoon (Xie *et al.*, 2002; Yang *et al.*, 2007). The Indian Ocean basin mode (IOBM), the dominant mode of SST variability in the TIO basin, demonstrates basin-wide warming that peaks in late winter and lasts into the spring and summer after an El Niño event (Yang *et al.*, 2007). The Indo-West Pacific region may see significant climatic anomalies as a result of the IOBM, which could prolong the effects of El Niño after the tropical East Pacific SST returns to normal (Xie *et al.*, 2002). Nevertheless, the IOBM is not just a passive response to El Niño, but an important driver of summer climate variability in the East Asian region, via forcing a Matsuno-Gill pattern in the upper troposphere with a strengthened South Asian High (SAH) (Schott *et al.*, 2009; Yang *et al.*, 2007). According to earlier research, the extra-tropical reaction to the tropical heating from the Indian Ocean is what causes the wave-like circulation that is connected to the summer precipitation in the WNWC (Chen and Huang, 2012; Li and Ma, 2018). The circulation pattern connected to the Indian summer monsoon (ISM) may cause low-level water vapor from the Indian Ocean to reach WNWC along the Tibetan Plateau's eastern edge. Recently, Yang *et al.* (2015) found that the persistent anomaly of the IOBM in the previous winter and spring is significantly correlated with the subsequent May precipitation change in ENWC by analyzing the relationship between the IOBM and precipitation in the eastern part of Northwest China (ENWC). However, it is known that the geographical distribution of climate responses over the arid and semi-arid regions of Northwest China has been distinctly different (drying in the east and soaking in the west) in recent decades (Wang *et al.*, 2020c).

The seasonally evolving patterns of IOBM-related SST anomalies for the sub-periods of 1974–1996 and 1997–2018 are shown in Fig. 6.13 to explore the potential mechanism by which the interdecadal IOBM's affect the WNWC precipitation. Compared with 1997–2016, the positive SST anomalies are relatively weaker in intensity and scope during 1974–1996. In addition, the strongest SST anomalies lie in the southern TIO near the equator from 1974 to 1996, while two additional warm centers are observed from 1997 to 2018, which are the Arabian Sea and the southwestern TIO.

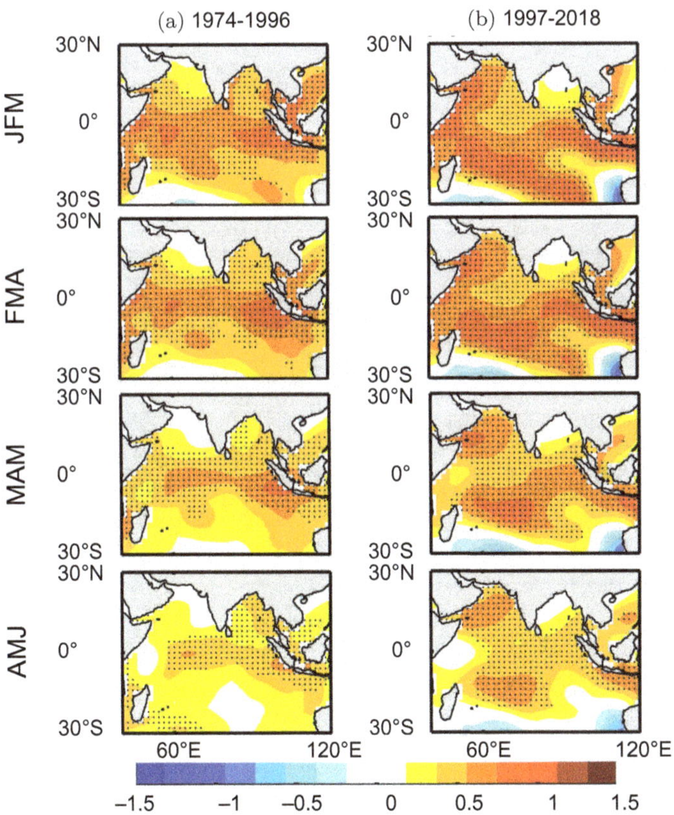

Fig. 6.13. Seasonally evolving patterns of composited SST difference (unit: °C) between the warm and cold IOBM events for the sub-periods of 1974–1996 (a) and 1997–2018 (b). The dot sign denotes the IOBM-related SST anomalies are significant at a 0.05 level (Wang *et al.*, 2022b).

Particularly over the Arabian Sea, the averaged SST over 10°N–30°N, 50°E–74°E increases by 0.8 times and 1.5 times for January–February and April–June, respectively. Similar results are obtained once ENSO is removed. Based on the diagnostic analysis and numerical experiments, the decadal variability of TIO SST in intensity and location induces significant climate anomalies over the northern tropical zone and exerts a very significant impact on the Asian summer climate (Qu and Huang, 2012; Sun *et al.*, 2019). Further research is conducted on the relationship between the geopotential height fields and the IOBM in the January–March preceding the two sub-periods of 1974–1996 and 1997–2018. (Fig. 6.13). In order to remove the impact of the remote oceanic and atmospheric forcing of ENSO, a partial correlation analysis is performed by removing the signal of the El Niño 3.4 index during the two epochs. Over the sub-period 1997–2018, the warm IOBM SST anomalies, which act as a heating source near the equator, can induce

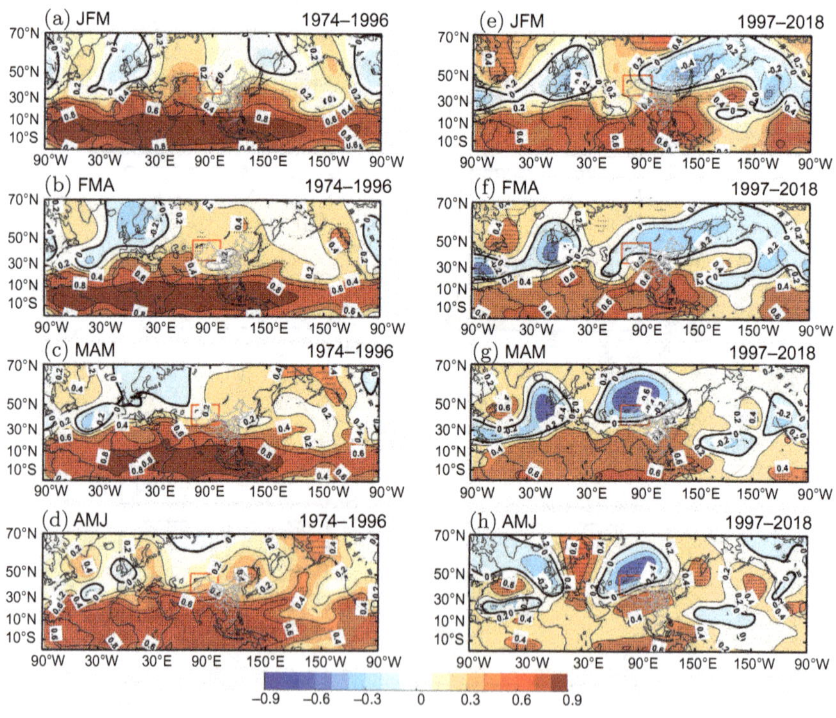

Fig. 6.14. Seasonally evolving patterns of the partial correlation coefficient between 200-hPa geopotential height field and IOBM in January–March without the ENSO signal for the sub-periods of 1974–1996 and 1997–2018. The red square box indicates the WNWC region. The dot sign indicates the correlation is significant at the 0.05 level (Wang *et al.*, 2022b).

atmospheric responses of the like "Matsuno-Gill Pattern" (Hoerling *et al.*, 2004). This anomalous atmospheric circulation appears to be an anomaly high center over the region around the Indian Peninsula and Tibetan Plateau (Fig. 6.14). The atmospheric circulation anomaly, therefore, seems to be an abnormally high pressure center over the region along 30°N around the Indian Peninsula and Tibetan Plateau in the subsequent April–June. Eastern Europe is dominated by significant high-pressure anomalies as well. Then, such anomalies further spread downstream to form a remote wave train along the jet axis in the mid-high latitude of the North Hemisphere (Figs. 14(e–h)). This wave train spans from Europe to the Western North Pacific, with an anomalously negative pressure center around Baikal Lake (termed as Baikal Low) and a significantly positive pressure center over the Northwestern Pacific, as the IOBM is warmer in the previous January–March (Fig. 6.14).

6.3.4 *Combined Impact*

In tropical regions, the most prominent feature of the ENSO signal is the striking disparity in the Walker Circulation-related vertical wind velocities between the two phases of the PDO. Note that the positive omega anomaly suggests either strengthened descending motion or weakened ascending motion. Evidently, the ENSO-induced Walker Circulation anomaly is more prominently accompanied by stronger divergence/convergence of water vapor when in phase with the PDO, but relatively weaker when out of phase with the PDO. The dry–wet changes over these areas can be explained by the descending motions over the Central and Northeastern Pacific, weakening during the warm PDO phase relative to the cold PDO phase, and the ascending motions associated with El Niño enhancing from Indonesia to Australia, especially along the eastern coast and Northeastern South America. For example, when El Niño is associated with the warm phase of the PDO, there is significant drying from Indonesia to Australia, especially to the east, and as well as from Northern Brazil to Colombia and Venezuela.

The ENSO-induced dry–wet distributions when combined with the PDO confirm the results obtained previously and reveal many important new features. When ENSO and PDO are in phase, the El Niño/La Niña-induced dry/wet anomalies are not only intensified over the canonical regions influenced by a typical ENSO event but also expand poleward (Fig. 6.15(a)). If the ENSO and PDO are out of phase, then the associated dry/wet anomaly is dampened or disappears (Fig. 6.15(b)). Generally, during the warm phase

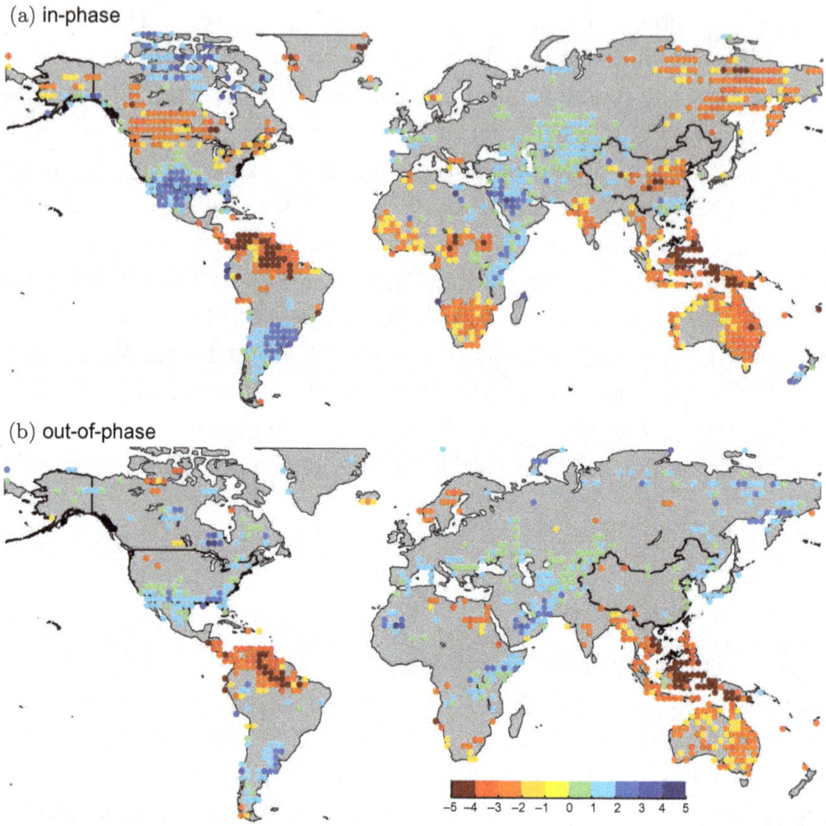

Fig. 6.15. Differences in sc_PDSI_pm between El Niño and La Niña. In-phase combination (a). Out-of-phase combination (b). Colored circles indicate aridity changes greater than the 90% confidence level. Maps and plots were produced using licensed Matlab (Wang *et al.*, 2014).

of the PDO, El Niño induces much broader and more severe droughts over land compared with the cold PDO phase. For example, the arid and semi-arid climate over the Sahel and Southern Africa worsens. Correspondingly, during the cold phase of the PDO, more rain occurs over land during La Niña winters than during the warm PDO phase. However, there are a few exceptions. The amplitude of the dry–wet variation is larger over Northern Europe and the Mediterranean during the out-of-phase condition, while the variation over the Horn of Africa tends to be stronger in the cold phase of the PDO.

Stippling in Fig. 6.16 denotes the percentages of the world's land area that experienced considerable ENSO-induced dryness and wetness during each of the two PDO periods. About 18% of the world's land area is affected

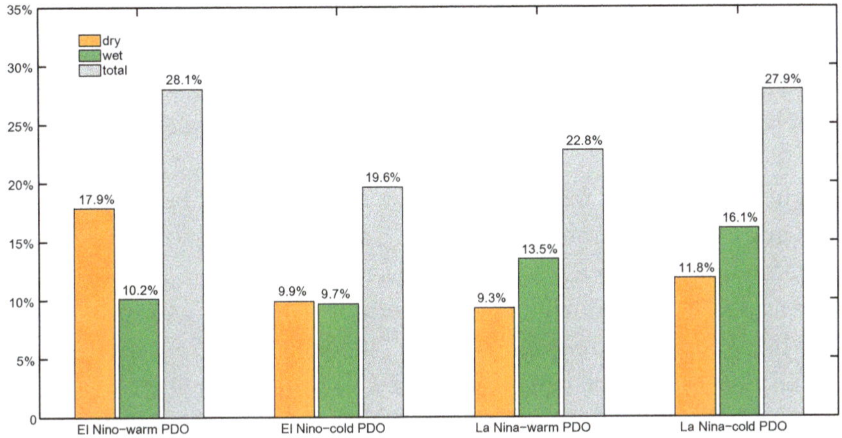

Fig. 6.16. Fractions of the global land area with significant drying and wetting induced by ENSO during warm and cold PDO regimes. Maps and plots were produced using licensed Matlab (Wang *et al.*, 2014).

by El Niño-induced dryness when the PDO and El Niño are in phase, about twice as much as when they are not. The intensity of the El Niño-induced wetness is much more severe, compared to the induced changes in the cold phase; yet, the most seriously affected areas are minor. When La Niña occurs during the cold PDO phase, approximately 16% (12%) of the global land area suffers abnormal wetness or flooding (significant dryness or drought), which is approximately 20% (27%) more than that in the warm PDO phase. In general, in phase with the PDO, the ENSO-induced dry–wet changes are magnified, and approximately 28% of the global land area is affected by the considerably abnormal dryness and wetness.

Additionally, composites of the detrended Climatic Research Unit (CRU) precipitation anomalies and the Global Land Data Assimilation System (GLDAS) surface soil moisture (0–10 cm) are examined. The composites largely match the sc_PDSI_pm changes, but there are a few small regional variations. The sc_PDSI_pm represents a cumulative departure in surface water balance (Sheffield *et al.*, 2012), influenced not only by the current precipitation changes but also by evaporation and the previous precipitation anomaly, which accounts for the small regional differences between the sc_PDSI_pm and the precipitation. The El Niño (La Niña) signal tends to be stronger during the warm (cold) phase of the PDO, which is similar to the findings for the sc_PDSI_pm, when ENSO-induced precipitation anomalies are compared with the two PDO phases. In addition, the coherence of the sc_PDSI_pm and precipitation is relatively high during the in-phase condition. It is slightly different in a few regions when the ENSO and the PDO are out of phase. The differences primarily occur over the northern high latitudes in the soil moisture because the GLDAS dataset does not adequately describe snow and other processes that affect soil moisture (Rodell *et al.*, 2004).

A crucial subject is why the dry–wet anomaly distribution linked with ENSO varies when combined with the different phases of the PDO. The ENSO-related SST anomalies intensify over the equatorial Pacific during the in-phase state and spread from the equatorial Pacific to the upper latitudes of the North Pacific via the eastern ocean. The typical SST anomalies over the equatorial Eastern Pacific become weaker and more concentrated around the equator as El Niño (La Niña) occurs against the backdrop of the interdecadal varying cold (warm) PDO, whereas the secondary signal across the North Pacific is considerably weaker and even the opposite. The corresponding atmospheric circulation at 500 hPa and water vapor transport at 850 hPa, including the geopotential height (HGT), the vertical velocity (omega), and the water vapor flux, are further examined. The positive Pacific/North America pattern, which extends from Central Canada to Greenland, and abnormally low pressure in the Aleutian Islands and the Southeast US make up the El Niño signal throughout the North Pacific–North America region (PNA). Further compositing with the warm PDO phase reveals a relatively larger HGT anomaly; however, the Aleutian Low is substantially stronger and a stronger high-pressure center is concentrated over Canada. According to earlier research, this heightened positive PNA causes two branches of abnormally high water vapor concentrations over North America. One branch occurs along the southern flank of the Aleutian Low, transporting the water vapor from the Pacific to the southwest, and the other branch brings the water vapor from Atlantic to the southeast. Therefore, North America is favored for increased precipitation and dry–wet changes (Fig. 6.16). In contrast, during the cold phase of the PDO, El Niño displays a comparable, largely inconsequential HGT anomaly pattern over the North Pacific (weaker positive PNA) with less water vapor transport to the US, which results in relatively weaker aridity changes over North America. As for La Niño winters, especially those occurring during the cold PDO phase, the PNA is in its negative phase, while the water vapor transport is anomalously divergent around Southern US and Mexico.

ENSO is also related to dry–wet variations over the Middle East, such as wetter (drier) conditions during El Niño (La Niña) winters over Irania (Nazemosadat and Ghasemi, 2004). When sub-composited by the warm PDO phase, El Niño-induced wetting is substantially intensified over the Middle East and Central Asia, abating the dry conditions over these arid regions. The deeper trough of low pressure that spans from Eastern Europe to the middle east, which speeds up the middle east jet stream and carries water vapor there, is the key contributor to this phenomenon. When El Niño occurs during the cold PDO phase, these wetting conditions are relatively weak or reverse to the east of the middle east. Correspondingly, accordingly, the cold PDO phase is characterized by significant dryness that is associated with La Niña throughout the Middle East and Central Asia.

In addition, ENSO is the most dominant perturbation responsible for interannual climate variability over Eastern and Southern Africa (Nicholson and Kim, 1997). Due to pressure anomalies over Southern Africa and the Indian Ocean, Southern Africa is known to be affected by the ENSO, with drier than normal weather during El Niño and wetter circumstances during La Niña (Reason and Rouault, 2002). However, by comparing the composites of the sc_PDSI_pm under different conditions, this relationship is only stable and significant when ENSO and the PDO are in phase. The high-pressure (low-pressure) anomalies over Southern Africa and the Indian Ocean are stronger when El Niño (La Niña) occurs with the warm (cold) PDO phase, thereby encouraging tropical convection and rainfall. In the out-of-phase circumstance, this anomalous atmospheric circulation pattern is rather weaker and more unstable. Unlike Southern Africa, the typical rainfall anomaly over the Horn of Africa (Eastern Africa) has a favorable association with ENSO (Nicholson and Kim, 1997). According to the current study, the Horn of Africa specifically dries up during El Niño winters and gets wetter during La Niña winters. When the PDO is in the cold phase, particularly since 1950, extra compositing by the PDO phase has exposed additional information, including a robust relationship between ENSO and the dry–wet variations across the Horn of Africa; however, the relationship is insignificant in the warm PDO phase.

Most of the rainfall in Western Africa, including the Sahel, is caused by the West Africa monsoon circulation regime; however, its association with ENSO is more ambiguous (Nicholson and Kim, 1997; Nicholson, 2013). From the observed sc_PDSI_pm composites, a typical El Niño (La Niña) is associated with a slightly drier (wetter) belt over the Sahel. However, the Sahel to the Gulf of Guinea, especially to the east, undergoes grave drying (wetting) when ENSO and the PDO are in phase. This is mainly due to the enhanced hot and dry Harmattan winds in the lower troposphere that do not occur during the out-of-phase condition.

During El Niño winters, especially during the cold PDO phase, the negative North Atlantic Oscillation (NAO), which comprises an anomalously high pressure that extends over Eastern Canada, Greenland, and Iceland and an anomalously low pressure that extends over the Eastern US, the North Atlantic, and the Mediterranean region, is present; this finding was previously mentioned by Gershunov and Barnett (1998). However, when the warm PDO phase is present, this NAO pattern is barely perceptible. As an important factor responsible for anomalous winters in Europe and the Mediterranean regions (Huang *et al.*, 1998, 2006a), the negative NAO reduces the pressure gradient between the weak subtropical high and the weak Icelandic low and results in fewer and weaker winter storms crossing on a more West–East pathway, transporting moist air into the Mediterranean region and cold air into Northern Europe. These changes induce moistening over Southern Europe and the tip of Northwestern Africa and drying over the Scandinavian Peninsula (Fig. 6.14(c)). The opposite is true for La Niña Niña winters; however, the effects are less pronounced, i.e., the positive NAO is not very strong. Therefore, the corresponding dry–wet variations are insignificant over the Mediterranean region and Europe.

6.4 Impacts of Land–Sea Thermal Contrast

In the second half of the 20th century, as a result of global warming, the warming of surface air temperature (SAT) increased significantly during the cold season than during the summer season (Huang *et al.*, 2012). With a rise of 1.53°C, the boreal winter warming trend is most pronounced in semi-arid regions and contributed to 44.46% of the global annual mean land surface temperature increase (Huang *et al.*, 2012). Cohen *et al.* (2012) observed an apparent flattening of the warming trend in the Northern Hemisphere since 1998; however, there is insufficient data to assess this most recent trend modification.

During the second half of the 20th century, the global temperature change was closely related to changes in atmospheric circulation throughout the winter. Gong and Wang (1999) investigated changes in the intensity of the Siberian high based on a monthly mean sea level pressure (SLP) dataset for the Northern Hemisphere obtained from the Climate Research Unit (CRU) at the University of East Anglia in the UK. They found that the Siberian high had weakened in recent decades, which led to decreased cold surges from the Arctic in the East Asia in winter. The East Asian Trough (EAT) axis was examined at 500 hPa and it was found that both the trough intensity index and the trough axis index had linear downward tendencies. The East Asian winter monsoon tended to take a southern pathway when the EAT tilt was smaller than usual, resulting in less cold air moving to the Central North Pacific. During the positive phase of the North Atlantic Oscillation (NAO), the blocking in the North Atlantic consistently tended to decrease and weaken (Huang *et al.*, 2006a). Therefore, over the last few decades of the 20th century, a significant winter warming trend was typically accompanied by a weakening of the cold surge. Attention was, however, drawn to the influence of sea surface temperature (SST); for example, the North Atlantic SST can modulate the phase of the interdecadal component of the NAO; the wavelet cross-spectrum displays the correlation between El Niño/Southern Oscillation (ENSO) and Asian monsoon (Hudgins and Huang, 1996; Huang *et al.*, 1998). Even though the land surface temperature is more sensitive to climate change, few studies have examined the atmospheric circulation response to land temperature changes. Studies are also required to ascertain whether feedback exists between the temperature change in the underlying surface and the circulation anomaly above.

Numerous studies have demonstrated a strong relationship between blocking activity in the Northern Hemisphere and zonal mean flow (Egger, 1978; Kaas and Branstator, 1993). The studies proposed that the linear resonance of planetary waves with surface thermal force and topographical forcing can account for the emergence and growth of blocking. Shabbar *et al.* (2001) found that blocking frequency and persistence are sensitive to the phase of the NAO. The "cold ocean/warm land" (COWL) pattern describes how the SAT anomaly is distributed during the NAO's positive phase, which lessens or eliminates the resonant forcing of topography. Although it has been demonstrated that the thermal forcing pattern over the Northern Hemisphere is relatively consistent with this theory, much less effort has been devoted to a dynamical explanation of the increased warming in the boreal winter.

6.4.1 *Land–sea index (LSI)*

A land–sea index (LSI) was defined to measure asymmetric zonal thermal forcing. This chapter concentrates on winter since it is the season with the greatest land–sea thermal contrast in order to examine the effect of this contrast on enhanced warming during the cold season in the Northern Hemisphere. The second empirical orthogonal function (EOF) corresponds to the second principal component of 500-hpa Geopotential height (GPH) anomalies over the extra-tropics (20°N–80°N), which accounts for 16% of the total variance. This pattern has been discussed by many authors (e.g., Wallace *et al.* 1996; Wu and Straus 2004, hereafter as WS04), with broadly congruent conclusions. The "cold ocean–warm land" (COWL) pattern, which accounts for 65% of the SAT variance during the cold season, was mentioned by Wallace *et al.* (1996). They pointed out that the substantial positive bias of the COWL pattern index was consistent with the unusual warming during the 1980s winters. WS04 compared the pattern using traditional and modified COWL definitions; the latter was defined using the second EOF of 500-hPa GPH. They discovered that the COWL pattern was primarily responsible for the trend in the mid-troposphere temperature, with the Arctic Oscillation's (AO) role limited to local cooling over Greenland.

It is crucial to quantify the intensity of the asymmetric thermal forcing because the surface thermal pattern can affect how two potential dynamic equilibria (wave-like or zonal component) alternate. Molteni *et al.* (2011) defined it as the zonal wave-2 component of the net surface heat flux averaged in four sectors of 90° longitude each. Although the land–sea thermal contrast can be represented by this definition, it does not correspond to the region of maximum land–sea contrast. From the "COWL" pattern induced by the land–sea distribution, we clearly identifies a wave-2 distribution. Four pattern centers are selected as key areas:

$$
\begin{aligned}
\mathrm{LSI} = {} & T\_\mathrm{ano}[80^\circ\mathrm{E}\text{–}120^\circ\mathrm{E}, 40^\circ\mathrm{N}\text{–}60^\circ\mathrm{N}] \\
& - T\_\mathrm{ano}[170^\circ\mathrm{E}\text{–}150^\circ\mathrm{W}, 40^\circ\mathrm{N}\text{–}60^\circ\mathrm{N}] \\
& + T\_\mathrm{ano}[130^\circ\mathrm{W}\text{–}100^\circ\mathrm{W}, 40^\circ\mathrm{N}\text{–}60^\circ\mathrm{N}] \\
& - T\_\mathrm{ano}[70^\circ\mathrm{W}\text{–}40^\circ\mathrm{W}, 57.5^\circ\mathrm{N}\text{–}77.5^\circ\mathrm{N}] \qquad (6.1)
\end{aligned}
$$

where T_ano is an area-averaged surface temperature anomaly for the period of 1960–1990. The LSI time series are shown in Fig. 6.17. In the 1980s, the index saw a positive anomaly, changing from a negative anomaly. The 500-hPa geopotential height grew in the mid-latitudes and declined in the polar areas during the positive phase of the LSI. The tropospheric planetary wave activity also became weaker and exerted less easterly forcing on the westerly wind. These circulation changes were favorable for westerly wind acceleration and reduced blocking.

In contrast to the Molteni's COWL index, which is expressly based on a regular wave structure, He *et al.* (2014) defines the LSI index as a linear

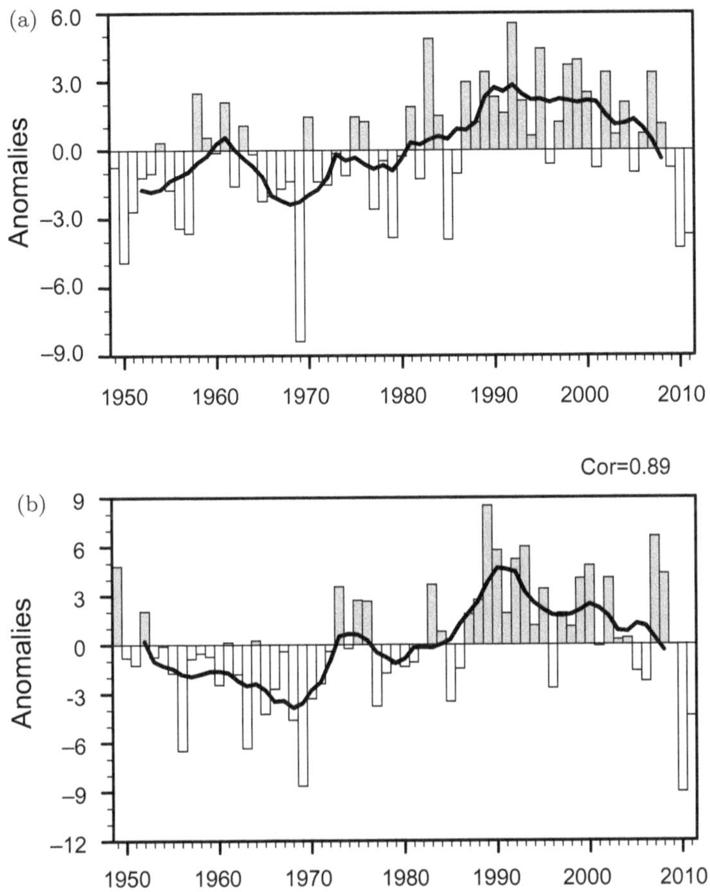

Fig. 6.17. Time series of landsea index in winter (a). Winter (DJF) mean time series (1948/1949–2008/2009) of the westerly wind index (b). The black curves indicate 7-year Gaussian-type filtered values (He *et al.*, 2014).

combination of the average surface temperature anomaly in the important regions of the COWL pattern, in order to represent the maximum land-sea thermal contrast. We chose the LSI defined by the temperature anomaly to depict the land–sea thermal contrast in order to keep things simple because the LSI defined by net surface heat flux in the four major areas is identical to that defined by that LSI. The warming on a continent in winter under global warming is substantially stronger than that over the ocean because the heat capacity of the land is much smaller than that of the ocean. The smaller land–sea temperature difference and warmer climate are thus suggested by the positive LSI. Wintertime anomalous warming over the land causes the LSI to transition from a negative to a positive phase, which signifies a reduction in the land–sea thermal disparity.

It is obvious that the LSI increased in the second half of the 20th century and turned from a negative to a positive anomaly in the 1980s. However, the LSI index displayed a decreasing trend after the 2000s. All the seasons, apart from winter, showed rapid and significant warming; the linear temperature trend during this time was essentially zero. Large regions with quite strong cooling tendencies, however, were observed across Europe, Northern and Central Asia, and a portion of Central and Eastern North America, in contrast to the nearly neutral wintertime trend (Cohen *et al.*, 2012).

The thermal differential between land and ocean has shifted since the 1980s as a result of the increasing winter warming in Eurasia and North America in the second half of the 20th century. According to Charney and DeVore's theory, the atmospheric circulation would switch from one equilibrium to another if the zonal asymmetric thermal contrast exceeded a specific threshold. This gives rise to the question of whether there is a relationship between the land–sea thermal contrast and the COWL pattern. Although the COWL pattern is important for understanding the internal variability of atmospheric circulation, there were few studies on the mechanism of the COWL pattern. As advocated in Kucharski *et al.* (2005), heat fluxes generated in the North Pacific may induce SST anomalies in the Subtropical Pacific on a decadal timescale and embed a COWL-like pattern to support the NAO through the cooling of the subtropical West Pacific and a consequent enhancement of the west Tropical Pacific SST gradient.

6.4.2 *Atmosphere circulation associated with LSI*

The westerly wind index in Fig. 6.17(a) is an obvious indicator of the intensity of the mid-to-high-latitude westerly wind. The index is computed using a modified definition by Li and Wang (2003), namely, the difference in zonal mean SLP between 35°N and 65°N from the monthly NCEP/NCAR reanalysis data, which contains a larger signal-to-noise ratio than the traditional zonal index suggested by Rossby (1939). After using a 7-year Gaussian filter, the westerly wind index anomaly's correlation coefficient with the LSI increased from 0.49 to 0.89, changing its sign in 1980. The 99% level of confidence is significantly exceeded by both coefficients. The outcome suggests that the westerly wind became stronger, as the LSI changed from a negative to a positive phase.

The 500-hPa GPH anomaly and the SLP anomaly patterns must be taken into consideration in order to examine the responses of atmospheric circulation to the change in the thermal forcing associated with the land–sea distribution. Here, a composite analysis is conducted to measure the impact of land–sea thermal forcing on the circulation. When the absolute normalized value of the LSI was more than 1.0 in DJF, we chose the positive and negative LSI examples for our analysis. The negative LSI cases are 1950, 1951, 1956, 1957, 1969, 1979, 1985, 2010, and 2011 and the positive LSI cases are 1983, 1987, 1989, 1992, 1995, 1998, 1999, 2002, and 2007. Similar to earlier research, the anomaly pattern shows that 500-hPa GPH decreased in the Arctic during the positive LSI phase, while the negative anomaly increased over North America and the Eurasian Continent in the mid-latitudes and extended into the North Atlantic and Northwest Pacific. This increased the meridional gradient of GPH between the mid and high latitudes (Fig. 6.18(a)). As shown by the shaded areas, these differences between the positive and negative phases significantly exceed the 99% confidence level for the two-sided Student's t test (Fig. 6.18(c)). He *et al.* (2014) looked at the structures of the monthly mean SLP anomaly field for positive and negative LSI phases and their differences in order to better understand the variation in 500-hPa GPH. In several phases of the LSI, semi-permanent pressure centers were discovered to have noticeable changes in strength and position. The positive and negative SLP anomaly centers, which represent the Azores high and the Icelandic low, respectively, were more pronounced in the positive LSI. The Aleutian low differed from that of Favre and Gershunov (2006) in that it was deeper and moved eastward. It transported warmer and more humid air masses along North America's West Coast. With a negative anomaly in the 500-hPa GPH and SLP in the Siberian area, the Siberian high weakened (Panagiotopoulos *et al.*, 2005). The semi-permanent centers were induced by the land–sea distribution and modulated by planetary waves, so the increase in the LSI may be the reason for these variations in strength and position.

Chen *et al.* (2005) pointed out that the quasi-stationary planetary waves may play a role in bridging the AO and climate anomalies. The amplitudes of the planetary waves in the lower troposphere over mid-to-high latitudes are affected by the impact of the AO or westerly wind intensity on the vertical propagation of planetary waves. It is of interest to determine whether the magnitude and propagation of planetary waves are correlated with the geographic pattern of thermal forcing as an external force. The tropospheric interannual oscillation index (TIO) was developed by Chen *et al.* (2002) to gauge the activity of planetary waves in order to study this link. The index is defined as the difference in the Eliassen–Palm (EP) flux divergence of planetary waves between 50°N at 500 hPa and 40°N at 300 hPa. According to Chen *et al.* (2002), a positive TIO index anomaly denotes increased equatorward planetary wave propagation and increased planetary wave energy trapped in the troposphere. Due to increased equatorward propagation and weaker upward wave refraction into the polar region, Fig. 6.17(b) demonstrates that planetary wave activity in the mid-latitudes has become less active in recent decades. Additionally, the correlation coefficient between TIO and LSI is 0.49, which is above the 99% confidence level. Due to land–sea thermal forcing, planetary wave activity also propagated more strongly equatorward while the LSI was in a positive phase. In latitude–height cross sections for positive phase, negative LSI phase, and the difference between the two composites, Fig. 6.19(d–f) depicts the divergence of the zonally averaged EP flux caused by planetary wave activity. In both positive- and negative-phase winters, the EP flux was convergent in most of the extra-tropical troposphere, resulting in an easterly wind forcing on the westerly wind exerted by planetary waves. In the difference map, in the mid-latitudes, positive values appeared along the polar wave guide and negative values arose along the equator wave guide. The positive values indicate that during winters of the positive LSI phase the forcing exerted by planetary wave activity on the westerly wind was weaker than that during winters of the negative LSI phase, which helped to accelerate the westerly wind. The shaded areas in Fig. 6.19(f) show where the discrepancies between the positive and negative LSI phases significantly exceeded the 95% confidence level.

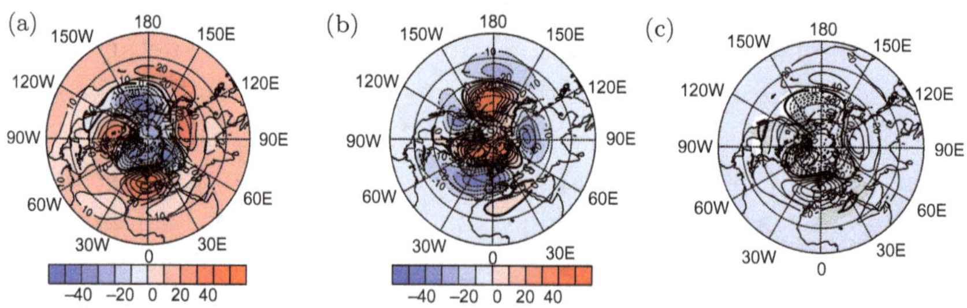

Fig. 6.18. Composites of 500-hPa geopotential height anomalies in winter (DJF) for positive land–sea index (a) and negative land–sea index (b). Difference (c) between positive and negative LSI phases (positive minus negative). The contour intervals are 10 m in (a) and (b), and 20 m in (c). The shaded areas in (c) indicate significant values at the 99% confidence level (He *et al.*, 2014).

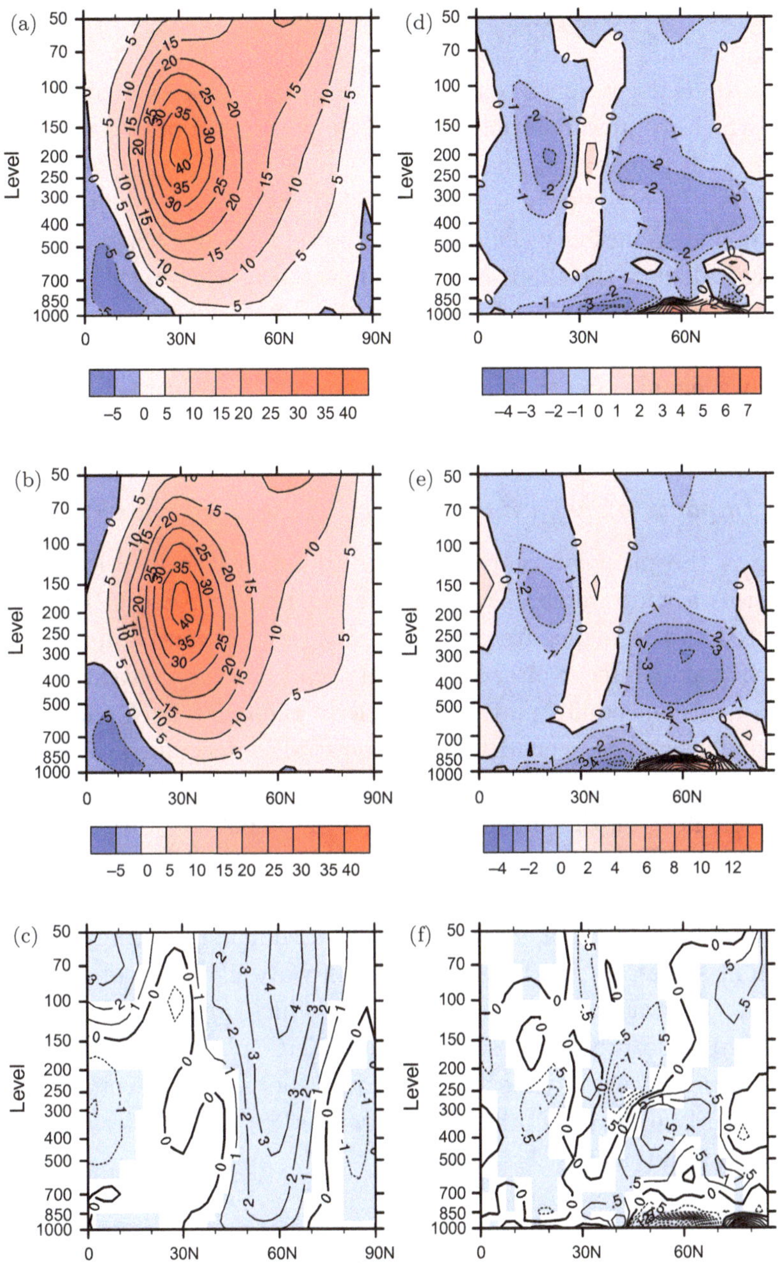

Fig. 6.19. Left panel composites of globally averaged zonal wind in winter (DJF) for positive LSI phase (a) and negative LSI phase (b). The difference (c) between positive and negative LSI phases, i.e., (a) minus (b). The contour intervals are 5 m/s in (a) and (b) and 1 m/s in (c). Right panels are same as the left panels, except for the DJF EP flux divergence induced by planetary waves. The contour interval is 0.50/ms/day and the shaded areas in bottom panels indicate significant values at the 99% confidence level (He *et al.*, 2014).

Fig. 6.19(a–c) shows the composites of zonally averaged wind for winters with positive and negative LSI phases and the difference between the two phases. The clearly reduced (enhanced) westerly wind appeared around 30°N (60°N), which resulted in a poleward shift of the subtropical jet stream. This is in line with previous studies that showed a positive correlation between the activity of the westerly wind or the AO and the interannual variability of planetary wave activity (Chen *et al.*, 2005). Through a higher TIO index, the rising LSI since the 1980s may be connected to a stronger westerly wind. During the winters with a larger TIO index, the upward wave propagation from the troposphere into the stratosphere became weaker, and the low-latitude waveguide prevailed, along with an apparently weaker East Asian trough and Siberian high, which is consistent with observations.

6.4.3 *Impacts of asymmetric thermal contrast on blocking*

Blocking is the primary weather system following the invasion of cold air and induces a wide range of cooling effects. The decrease in Ural blocking contributed to the higher frequency of warm winters in China. It has been suggested that nonlinear interactions between planetary waves or large-scale flow and transient eddies may play a role in the emergence and persistence of blocking events (Egger, 1978). Many studies recognized that blocking may be the result of the adjustment of planetary waves to deviations in the zonal mean flow or the linear resonance of planetary waves with surface forcing (Kaas and Branstator, 1993). Shabbar *et al.* (2001), who developed a straightforward conceptual model based on Charney and DeVore's theory that blocking is a metastable equilibrium alternating between two equilibrium states of high-index flow (weak-wave-like component) and low-index flow (strong-wave-like component), which are associated with the NAO phases, recently proposed that the NAO plays a significant role in the occurrence of wintertime blocking over the Atlantic sector. Therefore, the NAO can significantly influence the winter SAT by moderating blocking frequency and intensity.

Table 6.1. Blocking sectors defined in this chapter (He *et al.*, 2014).

Attributes	Euro-Atlantic		Pacific	
Sector	ATL	EUR	WPA	EPA
Domain	(100°W, 0°)	(0, 90°E)	(90°E, 180°)	(180, 100°W)

As with many other studies, blocking events were shown to be relatively more frequent over the Central Pacific Ocean and the Eastern North Atlantic Ocean in cold season, whereas blocking activity was lower over the continents. These results support the existence of four different blocking sectors, as shown in Table 6.1 (Barriopedro *et al.*, 2006). An objective blocking indicator developed by Tibaldi and Molteni (1990) is applied. In order to make our discussion more efficient, a simple definition is provided here. For each longitude, the GHGS and GHGN 500-hPa GPH gradients are calculated:

$$\text{GHGS} = \frac{Z(\emptyset_0) - Z(\emptyset_S)}{\emptyset_0 - \emptyset_S} \tag{6.2}$$

$$\text{GHGN} = \frac{Z(\emptyset_n) - Z(\emptyset_0)}{\emptyset_n - \emptyset_0}, \tag{6.3}$$

where $\emptyset_n = 80°\text{N}+\Delta$, $\emptyset_0 = 60°\text{N}+\Delta$, $\emptyset_S = 40°\text{N}+\Delta$, and $\Delta = -2.5°, 0°, 2.5°$. Since the reanalysis data are on a 2.5° grid, in this book, we choose the spacing between latitudes to be 2.5° as well. A given longitude is defined as blocked on a specific day if both of the following conditions are met, GHGS >0, and GHGN ≤10 m per degree latitude. The local blocking index is further constrained by imposing additional space and time constraints to obtain a regional blocking signature. A regional blocking event is defined as a blocking-like pattern that covered three or more nearby longitudes and persisted for five days or more. The blocked days are calculated by simply counting the number of days considered as blocked (5 days or longer) by the blocking index for December-February.

Only the blocking days in Europe and the Western Pacific significantly reduced, with linear trends of −13% and −4.9%, respectively, at the 99% confidence level. The blocking days appear to exhibit interannual

Table 6.2. Composite analysis of blocking parameters (columns) for the Northern Hemisphere and blocking sectors (rows) defined in Table 6.1 (He *et al.*, 2014).

Sectors	Trend (day/ year)	Mean days (day)		Duration (day)		Event number			
						5–9 days		≥10 days	
		Pos	Neg	Pos	Neg	Pos	Neg	Pos	Neg
ATL	0.05	18.1	29.4	10.1	11.8	7	9	10	14
EUR	−0.13	21.6	35.7	9	11.8	16	11	7	15
WPA	−0.049	14.3	20	11.3	11.5	5	14	8	7
EPA	0.02	18.1	21	10.1	9.7	16	13	4	7

and interdecadal variability in all four regions. The North Atlantic region exhibited significant interdecadal variation and experienced a climate shift in the mid-1980s, but its averaged blocked days did not decrease. A composite analysis is carried out to investigate the connection between regional blocking activity and the LSI. Table 6.2 lists the blocking parameters, including duration, event counts, and the typical number of blocked days per winter. In each of the four sectors, the number of blocked days in winter was higher during the negative LSI phases by 11.3, 14.1, 5.7, and 2.9 days, respectively. From the linear trend of blocked days, it is also reasonable to infer that the mechanism controlling the blocking activity was different in the Atlantic compared with that in the Pacific. During the positive and negative phases, blocking episodes in the Atlantic region lasted an average of 10.1 and 11.8 days, respectively. It was discovered that fewer lengthier blocking events were to blame for the fewer blocked days in the positive phase compared to the negative phase in the Atlantic region by comparing the number of occurrences lasting 5–9 days with those lasting more than 10 days. However, in the European region, the duration of blocking in the two phases did not change much but the total number of blocking events in the negative LSI phase was much higher than that in the positive LSI phase. For example, there were 16 blocking events lasting 5–9 days in the positive LSI phase compared to 11 in the negative LSI phase, and there were 7 occurrences lasting more than 10 days in the positive phase compared to 15 in the negative. These results show that, in comparison with the negative phase, it was difficult for the blocking to remain in the positive phase, so that blockings could not last long. Therefore, the environment during the negative LSI phase is considered more conducive to the blocking persistence. Fewer short-duration blocking occurrences occurred in the Western Pacific region when the LSI transitioned from the negative to the positive phase; the opposite was true in the Eastern Pacific region. Other processes might have an impact on blocking in these areas. According to Molteni *et al.* (2011), the zonal thermal differential over the North Atlantic sector is mostly to blame for the circulation pattern there and the NAO, whereas the meridional temperature advection may be the dominant force that modifies the circulation pattern in the Pacific sector. After the 2000s, the surface temperature has appeared to flatten out in the Northern Hemisphere winter (Cohen *et al.*, 2012); however, the blocked days in winter have an obvious upward trend only in the Atlantic sector.

6.4.4 *Impacts of zonal thermal difference on cold events*

The cold events of 2020–2021 caused high-frequency and stronger cooling (Fig. 6.20(a)). Cold events were reported on 12–14, 27, and 29–30 December in 2020, and on 1–5 and 14–16 January in 2021, and reached the historically lowest record for the last 67 years in some areas. As seen from the associated circulation pattern, the geopotential height at 1,000 hPa exhibits a larger change around the Ural Mountains (Fig. 6.20(b)), and East Asia was dominated by a prolonged anticyclone centered over the Ural Mountains at 500 hPa during this period (Fig. 6.20(c)). Meanwhile, both upstream (North Africa and Western Europe) and downstream (other areas) regions saw unique low-pressure centers (East Asia).

This circulation pattern caused Northeast China to be in front of the strengthened Siberian cold high and behind the deepening East Asian trough, which led to a descending movement and upper-level Northwesterly that were conducive for cold occurrences (Zheng *et al.*, 2021). East Asia experienced extremely cold winters thanks to the ensuing northerly cold air. When the polar vortex weakened or even split, it allowed frigid air to escape and push southward toward East Asia and cooled East Asia, inducing the cold events that occurred in the winter of 2020–2021. The areas affected by the cold events showed clear zonal differences (Fig. 6.20(d)). The most widespread area was located near 50°N with the largest area of 5.71×10^7 km^2, where the decrease was even over 10°N near 40°N; however, the largest cooling area was concentrated at -3°C (Fig. 6.20(d), Table 6.3.)

According to reports on the cold events in 2020–2021, the cold event frequency in the winter of 2020–2021 was five times greater than usual. Traditionally, temperatures below the 10th percentile of the average temperature that last for five days straight are known as "cold episodes" in China. This definition indicates that the frequency of the most recent cold occurrences was two times higher, which conflicts with CMA report data (Fig. 6.21(a)). However, cold events with the definition of 5% and 3 days occurred five times, which is consistent with the CMA's data.

The raw and thermodynamic temperatures increased sharply after 1980, especially North of 40°N. The temperatures were maintained low before 1980. Since the 1990s, there has been a noticeable increase in raw temperature, which was mostly brought on by increased thermodynamic temperature rising. In 2020–2021, the dynamic temperature was substantially lower than the thermodynamic temperature. When the thermodynamic temperature was in an upward (warming) phase in 2020–2021, it balanced or reduced the

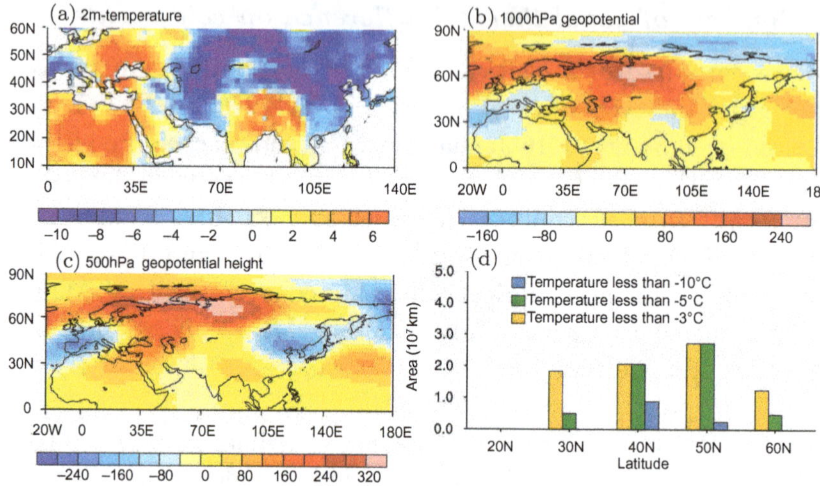

Fig. 6.20. Spatial distribution of average temperature anomalies during Jan. 5–7 (a) (units: °C). Spatial distributions of average geopotential height anomalies during Jan. 5–7 at 1,000 (b) and 500 hPa (c) (units: m). The areas of cold events over East Asia (100°E–140°E, 20°N–60°N) under different definitions during Jan. 5–7 (d) (units: 10^7 km) (Guan *et al.*, 2022b).

dynamic cooling and resulted in a relatively high value of raw temperature. Therefore, the 2020–2021 years were ideal for exploring the effect of regional warming on cold events.

The recurrence of cold events in the winter of 2020–2021 resulted in a weak meridional circulation, a fall in temperatures throughout the mid–high latitudes of the Northern Hemisphere, and a narrowing of the temperature gap between the high and mid-latitudes (Fig. 6.20(a)). The weak meridional circulation encouraged the reduction of westerly wind and resulted in Arctic cold air. The meridional circulation turned positive (histogram in Fig. 6.20(a)), ending the cold events and cutting the length of each cold event while the temperature adjusted by anthropogenic factors was at a high level in 2020–2021. This quickly offset the cooling from the Arctic (Guan *et al.*, 2015b). Such a mechanism in the equilibrium of a low index resulted in stronger and shorter cold events in the past winter.

We undertake sensitivity analysis using the outputs from 23 global climate models (See Chapter 3) in order to confirm the feedback between surface thermal differences and large-scale circulation. The outputs were retrieved from the CMIP5 multi-model data archive (https://pcmdi.llnl.gov/mips/cmip5/). These are known as pre-industrial control simulation (picontrol) and transient 1% annual rise in CO_2 over the simulation period in

Table 6.3. The area influenced by cooling of −3°C, −5°C, and −10°C over East Asia. Units: 10^7 km^2 (Guan *et al.*, 2022b).

Latitude	SAT less than −3	SAT less than −5	SAT less than −10
20°N	0	0	0
30°N	1.84	0.50	0
40°N	2.07	2.077	0.89
50°N	2.73	2.73	0.25
60°N	1.25	0.48	0

accordance with CMIP5 nomenclature guidelines (1pctco2). The picontrol experiment ran for a 500-year simulation by imposing non-evolving, pre-industrial conditions and served as the baseline for the analysis of future scenario runs with prescribed concentrations. Under rising greenhouse gas concentrations, the 1pctco2 experiment simulates a warm temperature state. The 1pctco2 experiment was initialized from the pre-industrial control and CO_2 concentration was prescribed to increase at 1% per year for 140 years (reaching 4 × pre-industrial levels at the end). In Taylor *et al.* (2012), the boundary conditions for each experiment were detailed.

Figure 6.22 demonstrates that while the LSI fluctuated around zero in the picontrol run, it gradually climbed in the 1pctco2 run along with rising CO_2 concentrations and a general warming of the planet. It is assumed that the CO_2 concentration was low and the LSI oscillated close to the mean condition of the climate in the picontrol run. When the LSI increased and was accompanied by global warming, the system departed from the climate mean state and was in the positive anomaly, which triggered the positive feedback mechanism. At that point, the large-scale circulation shifted from one equilibrium (wave component) to the other (zonal component). The LSI and the altered atmospheric circulation would create a positive feedback loop that sped up the warming process.

The comparison between two experiments in the 100–140 years based on the ensemble average of the 23 models confirms the response of large-scale circulation to changes in the LSI. Here, the picontrol experiment represents a low-LSI scenario and the 1pctco2 experiment represents a high-LSI scenario. In the Northern Hemisphere, the change in GPH in the COWL pattern is quite obvious, with greater GPHs across two continents and lower GPHs over two seas. Due to the CO_2 concentration being quadrupled in the 1pctco2 experiment, the level of global warming is more intense than would be otherwise. The pattern illustrating the differences in the oceans — land

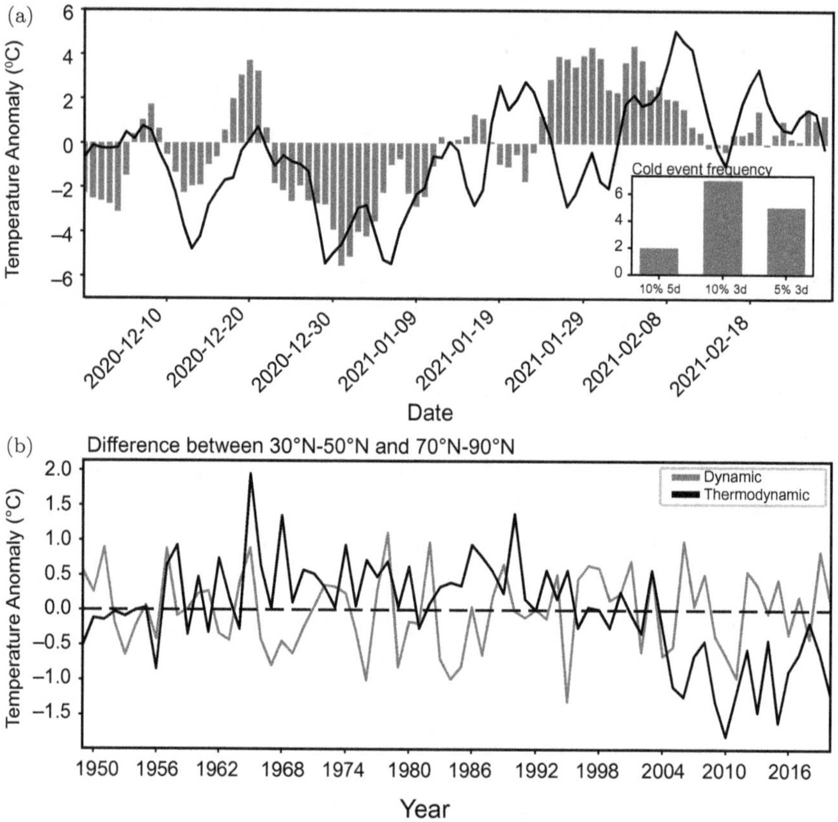

Fig. 6.21. Time series of temperature anomaly over East Asia (100°E–140°E, 20°N–60°N) in the winter of 2020–2021 and the anomaly of the meridional thermal difference between 30°N–50°N and 70°N–90°N (histogram) in the winter of 2020–2021. Subfigure is the frequency of cold events defined as temperatures lower than the 10th or 5th percentile of the winter temperature during 1948–2021, persisting for five or three consecutive days in the winter of 2020–2021 (a) (Units: times). Time series of the meridional thermal difference between 30°N–50°N and 70°N–90°N during 1949–2020 cold seasons (b). (Guan *et al.*, 2022b)

contrast between the 1pctco2 and picontrol — is similar to the reanalysis, even though the amount of change is smaller in the former than that in the latter. Because other forcing factors, such as the ENSO, the NAO, or polar vortex, have changed and interacted under the increasing CO_2 concentration scenario (Huang *et al.*, 1998, 2006a), perhaps another sensitivity experiment with the other forcing components is required, in which the other forcing factors are prescribed.

When the LSI was in the positive phase, the westerly wind was stronger and less blocking activity occurred. A heat advection pattern that strengthens the COWL pattern is produced by the large-scale circulation alteration. To verify this, we calculated the temperature advection between different phases of the COWL index. And their difference, based on NCEP reanalysis data, shows that the temperature advection was mainly distributed at the junction of continents and oceans and in inland Eurasian areas. As seen from the different fields, the warm advection over eastern part of continents and cold advection over western part of oceans in the positive phase of the COWL index were stronger than those in the negative phase.

6.5 Conclusions and Discussion

The impact of the seas on climate change in semi-arid regions is mostly evident in the characteristics of temperature, precipitation, and dry and wet variations on the interdecadal scale. The oceans are a significant part of the Earth's climate system. Climate change in semi-arid regions can be greatly impacted by SST anomalies in various sea basins (Fig. 6.23). Because the long-term upward trends in temperature are influenced by the interdecadal modulation of oscillations, which exhibit upward or downward

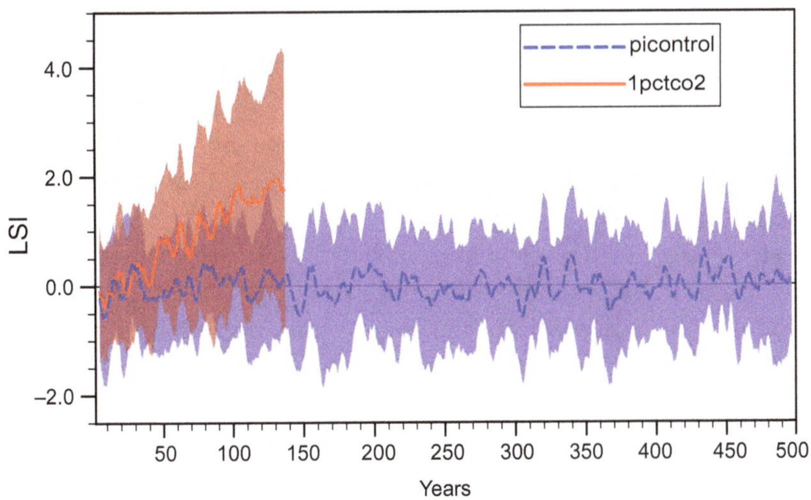

Fig. 6.22. Time series of the LSI in the 1pctco2 and picontrol experiments. The red solid curve represents the ensembles mean of the 1pctco2 experiment, and the blue dashed curve represents the ensembles mean of the picontrol experiment. The shading denotes one standard deviation of the 23 models from the picontrol/1pctco2 simulations. A 7-year running smoothing was applied to emphasize the climate change (He *et al.*, 2014).

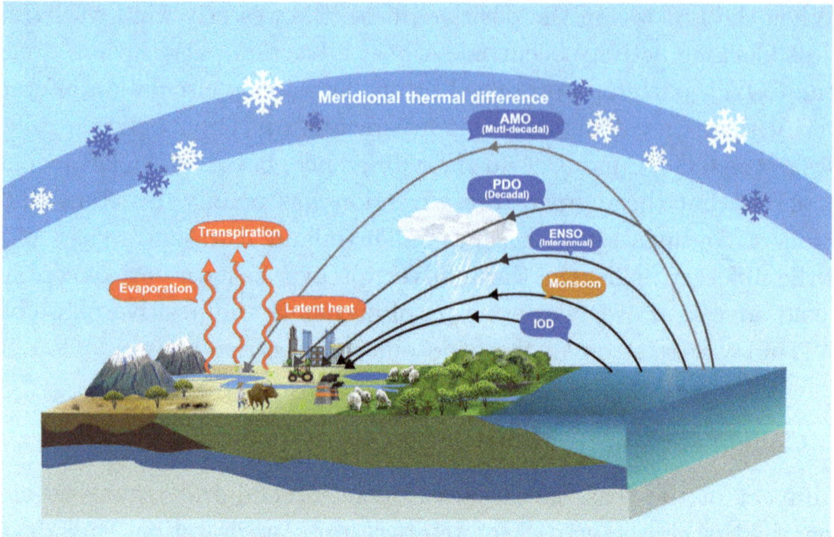

Fig. 6.23. The effect from ocean activities (AMO, PDO, ENSO, Monsoon, and IDO) on semi-arid region for different timescales (the snowflake represents the cooling from Arctic by decreasing meridional thermal difference).

trends through various phase transitions, the oceanic oscillation factors cause the temperature in semi-arid regions to exhibit interdecadal variation characteristics of accelerated warming or warming slowdown. At the same time, the oceanic oscillation factors such as ENSO, PDO, AMO, and NAO change the land–sea thermal contrasts through different phase transitions. It leads to atmospheric circulation anomalies and global temperature changes, which in turn affects the precipitation and dry/wet changes in semi-arid regions.

These oceanic oscillation elements affect the climatic change across semi-arid regions both individually and collectively. It was discovered that the AMO has an impact on precipitation across semi-arid regions of the Northern Hemisphere for the single effect. Over the past three decades, the AMO underwent a phase transition from a negative phase to a positive phase. The semi-arid regions of North America, except for some marginal areas (where precipitation increased), experienced less precipitation (the reduced value is more than 20 mm) during both positive phases between 1901 and 1963 and between 1964 and 2016; most semi-arid regions of North Africa experienced more precipitation (the range from 20 mm to 50 mm), and the Iberian Peninsula and Asia Minor Peninsula experienced slightly more precipitation with only 10 mm of recorded variation. The most significant changes of precipitation in the positive phase are the increased pattern (more than 50 mm) in the subcontinent of India. In the negative AMO phase, patterns that were roughly opposite each other appeared. The positive anomalous SST during the summer's positive AMO phase decreased the SLP throughout the Atlantic and Eastern Pacific. The anomalous SLP altered the regional atmospheric activities over the semi-arid regions around the Atlantic. Then, three cyclonic circulation anomalies developed over North Atlantic, North America, and Eastern Pacific. The North Atlantic cyclone strengthened the African monsoon and even had an impact on precipitation in the semi-arid areas surrounding the Atlantic as the cyclonic circulation anomalies intersected over the semi-arid parts of North America and induced low-level divergence. A roughly opposite flow field occurs during the negative phase of the AMO.

The dynamics of the PDO remain very complex and climate models can't predict the future evolution of the PDO, especially the shift from one PDO phase to another. The PDO entered a cold phase around 2000, which increased the frequency of La Niña episodes and increased rainfall over land (Trenberth *et al.*, 2014a). Moreover, areas of La Niña-induced precipitation became wetter and La Niña-induced drought areas became drier; however, these areas are smaller as a result of the modulation of the cold PDO phase. For instance, since the PDO entered a cold phase around 2000, contiguous US, particularly the Southwest US, has experienced severe drought (US Drought Monitor, Svoboda *et al.*, 2002; https://www.drought.unl.edu/Monitoring.aspx). Numerous studies have linked this phenomenon to the Tropical Pacific Ocean (Trenberth *et al.*, 2002, 2014a; Dai, 2013a). Furthermore, unless the interdecadal PDO transitions back to a warm phase, the American drought is not going to get better. China has also undergone a dry–wet pattern turnaround over the previous ten years; the pattern of "Southern flood and Northern drought" has changed to "Southern drought and Northern flood," and it appears that this pattern will last for another ten years. In addition, floods and abnormally high rainfall are more likely in India, from the Indonesian islands to Eastern Australia, and in equatorial South America. Meanwhile, Southern Brazil and Uruguay have been prone to drought in recent decades. All of these effects of ENSO on global dry–wet changes as the interdecadal PDO transitions into a cold phase should be taken seriously in the coming decades, especially over the fragile arid and semi-arid regions. A new type of ENSO, ENSO Modoki, has become more common since the late 1970s (Ashok *et al.*, 2007).

When combined with the PDO, the ENSO-induced dry–wet distributions were shown to validate earlier findings and reveal several significant new aspects for the combined effect. The El Niño/La Niña-induced dry/wet anomalies are not only amplified over the traditional regions influenced by a typical ENSO event but also expand poleward, when ENSO and PDO are in phase. If ENSO and the PDO are out of phase, then the associated dry/wet anomaly is attenuated or vanishes. Generally, during the warm phase of the PDO, El Niño induces much broader and more severe droughts over land compared with the cold PDO phase. For example, the arid and semi-arid climate over the Sahel and Southern Africa worsens. Correspondingly, during the cold phase of the PDO, more rain falls over land in La Niña winters than during the warm PDO phase. However, there are a few exceptions. The amplitude of the dry–wet variation is larger over Northern Europe and the Mediterranean during the out-of-phase condition, while the variation over the Horn of Africa tends to be stronger in the cold phase of the PDO.

Since the middle of the 1980s, the continental temperatures have been rising more rapidly, and the LSI has changed from a negative to a positive phase. According to composite analysis, when the LSI was higher, the planetary waves in the mid-latitudes were weaker due to larger equatorward propagation. Weaker easterly wind force was applied to the westerly wind by planetary waves, which increased the westerly wind. Thus, a stronger westerly wind would prevail in the positive LSI phase and apparently reduced the meridional heating exchange because of a weaker meridional flow. Another possibility is that a strong westerly wind may advect warm air masses from the oceans to the west of the continents, causing warming in these areas. On the western side of the continents, annular-mode oscillations are predicted to cause a similar link between zonal wind and temperature advection (Thompson and Wallace, 1998). Much attention was given to the annular-mode variation, but coupled simulations including anthropogenic effects have failed to account for the observed trend in the Northern Annular Mode (NAM). Meanwhile, a recent increase in meridional thermal circulation has contributed to shorter and more intense extreme cold events. The composed effect by LST and meridional thermal circulation contributed to the change of equilibrium state, which in the low index by thermodynamic temperature decreased the duration of each cold event and led to stronger and shorter cold events. According to Huang *et al.* (2017b), between 2000 and 2010, the natural oscillation was greater than the anthropogenic meridional thermal temperature and had a high index of equilibrium, leading to the characteristic high-frequency cold episodes.

In the winter of 2020–2021, although the natural oscillation was abnormally strong, the thermodynamic meridional thermal index was also powerful and dominated equilibrium in the low index. Therefore, the cold events in 2020–2021 were typical of stronger and shorter characteristics due to anthropogenic warming. Therefore, it is impossible to deny how the ocean is influencing climate change in dry and semi-arid areas. Against the backdrop of global warming, the total area of arid and semi-arid regions has expanded dramatically, and extreme drought events have occurred frequently. During the process of climate change in the past 100 years, the semi-arid regions have shown significant periods of heightened warming and aridification, and there have been prominent spatial differences, which deteriorate the local environment. The SST anomalies in different sea basins can significantly affect climate change in arid and semi-arid regions. The ocean effect on semi-arid region always exhibits a longer timescale. Different mechanisms of ocean oscillation over semi-arid regions need more investigation. In the future, different ocean oscillations will undergo different changes, and these changes will result in new characteristics of the mechanism. For example, over the semi-arid region of East Asia, the monsoon edge tends to extend westward with the increased emission of GHGs. Westward spread of the monsoon edge will take place in the high-emissions scenario. Such westward extension of the edge of the monsoon will result in more precipitation. Moreover, as the frequency of El Niño increases and the intensity of El Niño-induced atmospheric anomalies under greenhouse warming increases, the climate in the semi-arid regions of East Asia may be greatly changed by ENSO. In addition, monsoon expansion will turn the semi-arid regions of East Asia wet and green. As a typical underlying surface of the semi-arid region of East Asia, the Loess Plateau will undergo great ecological changes in the future, and more investigation is needed.

For the multi-decadal period, AMO affects climate change over semi-arid regions by teleconnection. In the previous result, the AMO effect on regional $T_{\max}$ and $T_{\min}$ was enhanced and offset by anthropogenic effects. However, the anthropogenic force contribution to the change of $T_{\max}$ and $T_{\min}$ should be studied in the future. By comparison, significant decreasing trends in $T_{\min}$ occurred to the Southeastern area of inner East Asia, which may be associated with the downstream propagation of the cyclonic–anticyclonic teleconnection wave train. However, the AMO effect on precipitation over the semi-arid regions of North America and Africa is clear. Such an effect from the AMO on the precipitation over the semi-arid areas of Asia has not been greatly explored, and the AMO effect on drought and aridity is also obvious.

In addition to these direct effects from ocean oscillation on sensitive semi-arid regions, the indirect and semidirect influence that oceans exert on climate change in semi-arid regions receives less attention. The significance of disparities between the various regions is still not fully understood. In addition, there are relatively few studies on how the interaction of different sea areas affects climate change in arid and semi-arid regions. At present, it is still not possible to exactly quantify the relative contributions of different ocean oscillation factors to the climate evolution of arid and semi-arid regions and their response processes. Moreover, the understanding of the climate evolution mechanisms of arid and semi-arid regions affected by air–sea interactions is not sufficient. Future research on these mechanisms should combine diverse observational data sources with numerical simulations to be thorough and methodical. There is major uncertainty regarding the future of climate change in arid and semi-arid regions. Ocean oscillation changed the local temperature and precipitation at the same time, increasing the complexity of local energy and water cycle changes, which may receive more attention in the future.

Chapter 7

Impacts of Human Activity

7.1 Introduction

By altering emissions and the resulting atmospheric concentrations of radiatively significant gases and aerosols, as well as by altering land surface features, human activities continue to have an impact on the Earth's energy balance. Multiple lines of evidence from earlier analyses have already demonstrated that human activity is largely to blame for the climate change that is occurring throughout our world. There is no doubt that human activity has caused the atmosphere, ocean, and land to warm. The atmosphere, ocean, cryosphere, and biosphere have all undergone widespread and quick changes (IPCC, AR6, 2021). In addition to raising the average global surface temperature and land precipitation, human activity is already having an impact on numerous weather and climate extremes, including the global retreat of glaciers and the rise in the world's mean sea level (IPCC, AR6, 2021).

Anthropogenic activities are responsible for climate change in arid and semi-arid regions. Anthropogenic activities are partially to blame for salinization, desertification, the disappearance of vegetation cover, the decline in biodiversity, and other types of environmental degradation (Huang *et al.*, 2016a, 2017a). More fossil fuels have been consumed as economic development has accelerated, resulting in significant GHG emissions and energy production (Barnett and O'Neill, 2010). The regional distribution of temperature has been significantly impacted by the released GHGs and energy in recent years, particularly in developing countries where economic policies result in the inefficient use of resources and, consequently, increasing energy waste. The consequences of human activity, particularly in high-density GHG scenarios (Huang *et al.*, 2017b), are closely related to the enhanced warming that has occurred in arid and semi-arid regions (Huang *et al.*, 2012). Increased GHG emissions, anthropogenic aerosols, land use, and urbanization have all had an impact on the climate of arid and semi-arid regions.

The most significant challenges, according to earlier studies, have been rising temperatures, area expansion, extreme events, water scarcity, and desertification. These issues are directly tied to human activity, including increased GHG emissions, anthropogenic aerosols, and land use. Recent research has demonstrated that arid and semi-arid regions have seen the most warming over the past century (Huang *et al.*, 2012; Ji *et al.*, 2014; Guan *et al.*, 2015). Using a recently established method that can identify and separate the temperature change into dynamically generated and radiatively caused changes (Wallace *et al.*, 2012), Guan *et al.* (2015) studied the driving elements of the enhanced semi-arid warming (ESAW). The findings demonstrated that the radiative forced temperature (RFT) changes brought on by GHGs, land cover change, clouds, aerosols, and human activities are primarily responsible for the warming trend on the global and regional scales. Additionally, Zhou *et al.* (2016) discovered that the processes are probably mostly due to increased longwave radiative forcing, which is connected to stronger water vapor feedback over drier ecoregions in response to the positive global-scale GHG forcing. The phenomenon of enhanced warming has significant implications for understanding the patterns of global warming and evaluating the effects of climate change. As a result, it should be clear what the spatial and temporal features of enhanced warming and the effect of human activity are.

Climate models have reliably predicted an increase in global mean precipitation in response to rising GHG emissions in relation to the role that GHGs play in climate change (Stephens and Ellis, 2008; Pendergrass and Hartmann, 2014). According to numerous studies (Feng and Fu, 2013; Huang *et al.*, 2017b), global arid and semi-arid regions, as defined by the AI, have grown over the past 60 years. These studies also predicted that there will be significant increases in GHG emissions and drying trends over land during the next century. Increased aridity, growing arid and semi-arid regions, rising temperatures, intensive land usage, and other human activities could all contribute to desertification — which would be extremely dangerous for ecosystems and populations of arid and semi-arid regions — and intensify regional tensions. Therefore, it is important to improve our understanding of how human activity affects the spread of arid and semi-arid regions.

With the publication of IPCC AR6 (2021), the evidence of observed increases in severe events, especially those attributed to human influence, such as heat waves, heavy precipitation, droughts, and tropical cyclones, has strengthened since AR5. Numerous weather and climatic extremes are already being impacted by human-induced climate change on every continent. With a high degree of certainty that human-induced climate change is the primary driver of these changes, it is virtually certain that hot extremes (including heat waves) have increased in frequency and intensity across most land regions since the 1950s, whereas cold extremes (including cold waves) have decreased in frequency and severity. Arid and semi-arid regions would experience more intense temperature episodes due to their accelerated warming. Increased extreme heat occurrences may cause more calamities for the citizens of arid regions in the context of the existing challenging ecological environment and living environment in arid and semi-arid regions. Therefore, it is critical to have a better understanding of how human activities affect extreme events in arid and semi-arid regions.

Additionally, two key factors influencing socio-environmental transformations in the 21st century are urbanization and climate change (Chapman *et al.*, 2017). Urban areas are home to more than half of the world's population, 80% of the global gross domestic product (GDP), and more than 70% of the world's greenhouse gas emissions, which are hot spots of environmental problems. They are also key zones for supporting sustainable development (IPCC, 2013). Rapid urbanization and the resulting alteration in land use over the past few decades have created a number of significant environmental and socioeconomic problems for global sustainable development, particularly the decline in vegetation's ability to produce terrestrial oxygen. Historical anthropogenic carbon emissions have been rising since the Industrial Revolution, and in recent years, this growth has been more rapid. This has been accompanied by a global drop in atmospheric oxygen. The Scripps O_2 Program's observations show that atmospheric O_2 content has been falling since the late 1980s, and the oxygen concentration in the atmosphere has been dropping (by about 4 ppm/yr) at a rate that is almost twice as fast as the rate of rising carbon dioxide (IPCC, 2013). However, oxygen has received much less attention from the scientific community than carbon dioxide up to this point. As a result, it is crucial for sustainable development to quantify the oxygen cycle's current state in the Earth's system and consider its influence.

Based on a series of typical disaster characteristics such as enhanced warming, area expansion, decreased water storage, and increased extreme events in arid and semi-arid regions, this chapter mainly introduces the impact of human activities on these phenomena in arid and semi-arid regions. The arrangement of this chapter is as follows. Section 7.1 is the introduction to this chapter. Section 7.2 illustrates the phenomenon of enhanced warming and impacts of GHG emissions. Section 7.3 discusses the impacts of human activities on the expansion of arid and semi-arid regions. Section 7.4 introduces the responses of extreme temperature events in different GHG emission pathways. Section 7.5 presents the impacts of human activities on the oxygen cycle of the Earth system and projects future oxygen cycle change. Conclusions and discussion are summarized in Section 7.6.

7.2 Energy Budget in Arid and Semi-Arid Regions

Policymakers decided to set a target to keep global mean surface warming (GMSW) to no more than 2°C over pre-industrial levels after the Paris Climate Agreement was signed in April 2016 and ratified by the USA and China at the G20 meeting in September 2016. The pursuit of a warming limit as low as 1.5°C was also suggested, and the Intergovernmental Panel on Climate Change (IPCC) was asked to provide a special report by 2018 on the effects of global warming of 1.5°C and on the associated global greenhouse gas emission pathways. Before making a new choice, it is important to quickly assess the benefits and necessity of a half-degree less warming as part of an ambitious and urgent plan.

Concerns have been raised since warming over land is not uniform and there are significant geographical variations (Seneviratne *et al.*, 2016). Increased warming over arid and semi-arid regions has been seen in observations (Huang *et al.*, 2012, 2016a; Guan *et al.*, 2015). Nearly 50% of the planet's geographical surface is made up of arid and semi-arid regions, which also contain about 38% of the world's people. Due to their delicate ecosystems, arid and semi-arid regions are particularly vulnerable to desertification and land degradation. Examining variations across arid and semi-arid regions under the 2°C and 1.5°C objectives is crucial.

7.2.1 *Historical radiation balance*

In Fig. 7.1(a), historical warming trends as a function of mean precipitation (P) are examined across three observational datasets, GISS, CRUTEM4, and MLOST, as well as historical simulations driven by the Fifth Coupled Model Intercomparison Project (CMIP5). Since there are few observations from many areas before 1920, we concentrate on the 1920–2015 time frame in this chapter. The three observational datasets all demonstrate declining warming with rising mean precipitation, suggesting that warming has been more pronounced in arid regions. This finding holds true across multiple precipitation datasets and time periods. The distribution of arid and semi-arid regions, which are defined as land areas with a mean annual precipitation (P) below 600 mm and a mean precipitation to potential evapotranspiration (P/PET; Feng and Fu, 2013) ratio below 0.65 (Fig. 7.1(b)), is consistent with regions that have seen rapid warming. The mean temperature trend for arid and semi-arid regions from 1920 to 2015 is $1.28 \pm 0.18°C/96\,yr$ in MLOST ($1.30 \pm 0.19°C/96\,yr$ and $1.25 \pm 0.19°C/96\,yr$ in CRUTEM4 and GISS, respectively). The mean temperature trend for humid lands,

which are regions with annual $P > 600\,mm$ and $P/PET > 0.65$, is $0.92 \pm 0.13°C/96\,yr$ in MLOST ($0.81 \pm 0.14°C/96\,yr$ and $1.04 \pm 0.14°C/96\,yr$ in CRUTEM4 and GISS, respectively), which is between 60% and 80% of the trend for arid and semi-arid regions.

However, CO_2 emissions from anthropogenic sources over various land areas show a geographical pattern that is opposed to the warming patterns (Fig. 7.1), even though anthropogenic CO_2 and other GHG emissions are thought to be the primary cause of the recent global warming.

The Carbon Dioxide Information Analysis Center's time series of annual CO_2 emissions from anthropogenic sources, such as the burning of fossil fuels, the production of cement, the flaring of gas in oil fields, and the trade of energy, were used in this study. They covered the years 1751–2013 and were plotted on a grid of $1° \times 1°$ (Carbon dioxide information analysis center, CDIAC). Since there are several sources of observational temperature data for the years 1920–2013, we focused on that time frame for this analysis.

Between 1920 and 2013, arid and semi-arid regions produced only a little amount of CO_2 emission totaling about $4.58\,kg\,m^{-2}$, whereas humid lands produced a large amount of CO_2 emission totaling about $15.43\,kg\,m^{-2}$, or around 3.5 times that of arid and semi-arid regions, primarily from Europe, Eastern North America, and Eastern Asia (Fig. 7.1(c)). Additionally, this pattern holds up well across time periods. Surface temperatures over arid and semi-arid regions are more susceptible to CO_2 levels than other locations because atmospheric CO_2, a highly mixed GHG after emissions, is distributed uniformly over the world. Most of the worldwide CO_2 emissions in this example were from humid lands (about 750 GtC from 1920–2013; 1 Gt=1,015 g, compared to the emissions of about 230 GtC from arid and semi-arid regions), although arid and semi-arid regions saw 20%–40% higher

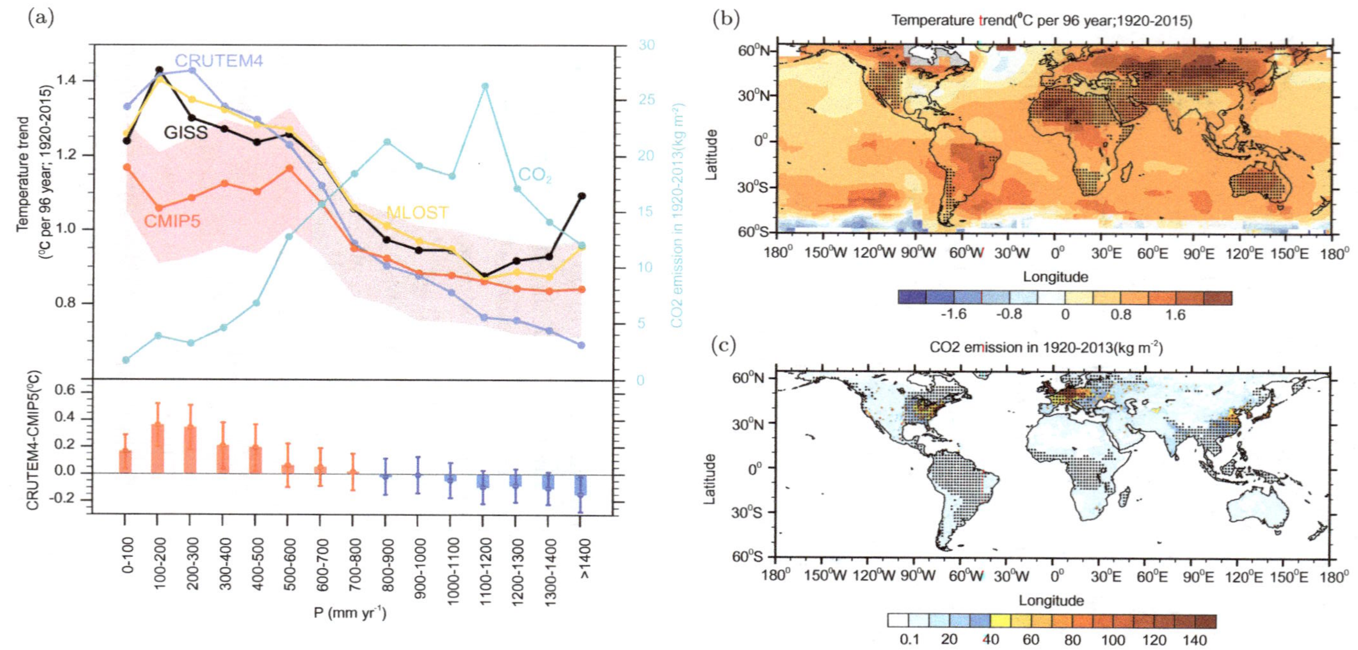

Fig. 7.1. Temperature trends and historical CO_2 emissions for drylands and humid lands. Variations in regionally averaged temperature trends from 1920–2015 (unit: °C per 96 yr) calculated from the MLOST (yellow line), CRUTEM4 (blue line), and GISS (black line), and the ensemble mean of the CMIP5 simulations (red line), together with regionally averaged total CO_2 emissions flux during 1920–2013 (unit: kg m^{-2}) as a function of the climatological (1948–2005) mean precipitation from GPCC V6 over land surfaces. Bars denote the difference between regionally averaged CRUTEM4 temperature trends and the CMIP5 ensemble mean (CRUTEM4 minus CMIP5), and error bars and red shading denote the 95% confidence intervals for the 18 models. CMIP5 all-forcing historical simulations (for 1920–2005) and the RCP4.5 projections (for 2006–2015) were used (a). The global distribution of surface temperature trends for 1920–2015 from the MLOST, with the gray shading denoting the identified drylands (b). The global distribution of total CO_2 emissions flux during 1920–2013, with the gray shading denoting the identified humid lands (c) (Huang *et al.*, 2017c).

warming than humid lands (Fig. 7.1(a)). Not much attention has been paid to the significant disparity between the geographic distributions of CO_2 emissions and warming rates.

Surface and atmospheric mechanisms can provide an explanation for the increased surface heat in arid and semi-arid regions (Fig. 7.2). Cloud radiative impacts were evaluated using the monthly averages of all-sky fluxes and clear-sky fluxes at the top of the atmosphere (TOA) from the CERES Energy Balanced and Filled (EBAF-TOA) Ed2.8 dataset (CRF). We calculated CRF data for January 2001 to December 2015 using a 15-year average. First, the radiation differences between clear-sky and all-sky circumstances were calculated in order to get the shortwave (SW) and longwave (LW) CRFs. The total of the SW and LW CRFs, or the net CRF, was then calculated:

$$\begin{aligned} \mathrm{CRF_{SW}(TOA)} &= (\mathrm{SW}^{\downarrow} - \mathrm{SW}^{\uparrow}_{\mathrm{all}}) - (\mathrm{SW}^{\downarrow} - \mathrm{SW}^{\uparrow}_{\mathrm{clear}}) \\ &= S\mathrm{W}^{\uparrow}_{\mathrm{clear}} - \mathrm{SW}^{\uparrow}_{\mathrm{all}} \end{aligned} \tag{7.1}$$

$$\begin{aligned} \mathrm{CRF_{LW}(TOA)} &= (\mathrm{LW}^{\downarrow}_{\mathrm{all}} - \mathrm{LW}^{\uparrow}_{\mathrm{all}}) - (\mathrm{LW}^{\downarrow}_{\mathrm{clear}} - \mathrm{LW}^{\uparrow}_{\mathrm{clear}}) \\ &= \mathrm{LW}^{\uparrow}_{\mathrm{clear}} - \mathrm{LW}^{\uparrow}_{\mathrm{all}} \end{aligned} \tag{7.2}$$

$$\mathrm{CRF_{Net}(TOA)} = \mathrm{CRF_{SW}(TOA)} + \mathrm{CRF_{LW}(TOA)} \tag{7.3}$$

In order to examine the cloud properties over arid and semi-arid regions and humid lands, cloud area fractions and liquid water pathways from the SYN1deg Ed3A datasets were also employed. The MODIS and CERES cloud retrievals were merged. Since SYN1deg Ed3A's monthly cloud parameter data end in November 2015, the averaged cloud fraction and liquid water path were calculated from December 2000 to November 2015.

Limited soil moisture content limits evaporation over drylands, and poor vegetation cover leads to low transpiration rates, causing a low mean latent heat flux of $16.12\,\mathrm{W\,m^{-2}}$ over drylands compared to $66.49\,\mathrm{W\,m^{-2}}$ over humid lands (Fig. 7.2(b)). As a result of transpiration and photosynthesis, vegetation can reduce air temperature. This reduces the additional heating caused by elevated GHGs and leads to slower warming rates. Figure 7.2(c), which depicts a decrease in surface warming rates with increased vegetation cover, appears to support this. Surface temperatures over arid and semi-arid regions must rapidly increase to produce a significant land–air temperature gradient in order to release the heat from solar and infrared radiation through sensible heat fluxes. Due to this, there is a significant sensible heat flux of $66.44\,\mathrm{W\,m^{-2}}$ as opposed to $36.02\,\mathrm{W\,m^{-2}}$ over humid areas

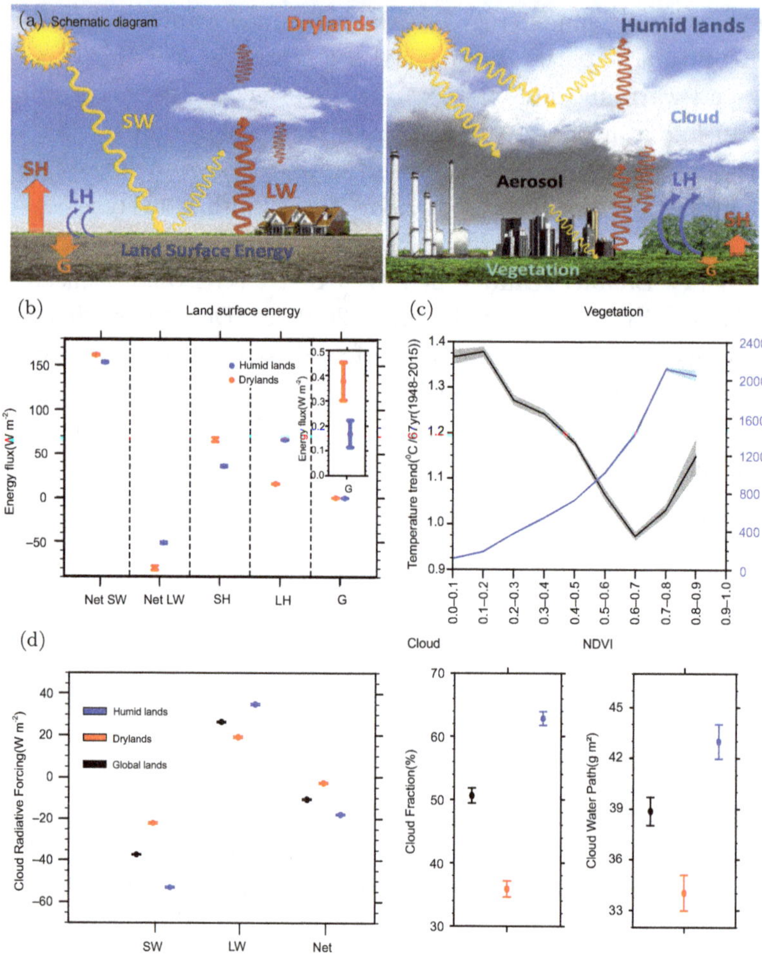

Fig. 7.2. The thermodynamic mechanisms of dryland-enhanced warming. Schematic diagram of local thermodynamic processes in drylands and humid lands (a). Comparison between the mean surface energy fluxes for drylands and humid lands based on GLDAS data, including net shortwave radiation (Net SW, downward positive), net longwave radiation (Net LW, downward positive), sensible heat flux (SH), latent heat flux (LH), and ground heat flux (G). The error bar denotes the ± 1 standard deviation range of temporal variations from 1948 to 2010 (b). Variations in regionally averaged temperature trends over 1948–2015 (black) and climatological precipitation for 1948–2005 (blue) with respect to the climatological (2001–2015) mean Normalized Difference Vegetation Index (NDVI). The shading denotes the ± 1 standard deviation range of the temporal variations in the NDVI for 2001–2015 (c). The mean cloud radiative forcing, cloud fractions, and liquid water paths during 2001–2015 averaged over drylands (red), humid lands (blue), and all land surfaces (black). The error bar denotes the ± 1 standard deviation range of temporal variations from 2001 to 2015 (d) (Huang *et al.*, 2017c).

(Fig. 7.2(b)). A significant difference in land–air temperature over arid and semi-arid regions is also shown by the high upward net longwave radiation over arid and semi-arid regions of 79.33 W m^{-2} (compared to 50.82 W m^{-2} over humid lands). Because there is little soil moisture and vegetation, surface temperatures over arid and semi-arid regions are highly sensitive to additional heating from increased GHGs, and any additional heating will cause the surface temperature to rise. In contrast, a portion of this additional energy could be used to increase evapotranspiration over humid land, reducing the subsequent surface warming. The same amount of heating will result in greater warmth over arid and semi-arid regions than over humid land because a dry surface has a lower specific heat than a wet surface. The aforementioned physical processes might also be active in the future under warming brought on by GHGs.

Additionally, there are several kinds of clouds over wet and dry countries. In humid areas, optically thick low clouds are common. Although they effectively reflect sunlight, their longwave warming effect is constrained by their low cloud tops. Low clouds like these keep humid land surfaces cool. In contrast, low humidity results in fewer clouds over arid and semi-arid regions, which reduces the amount of sunlight that is reflected and scattered. The most common type of cloud over arid and semi-arid regions is thin cirrus clouds, which let most sunlight through but collect infrared radiation from the atmosphere's surface and lower troposphere and radiate at considerably lower temperatures (Hartmann *et al.*, 1992). Cirrus clouds thus have a tendency to warm the surface of arid and semi-arid regions. Over arid regions, shortwave cloud radiative forcing is -21.92 W m^{-2}, while over humid lands it is -52.87 W m^{-2}. The difference results in a net (longwave plus shortwave) cloud radiative forcing over arid and semi-arid regions of -2.63 W m^{-2} and over humid lands of -17.68 W m^{-2} (Fig. 7.2(d)). Thus, clouds have a considerably greater tendency to chill humid land surfaces than dry land surfaces. Nevertheless, it will depend on how the clouds develop in a warmer climate whether or not they will contribute to the varied sensitivity to GHG forcing over the two distinct regions.

Last but not least, aerosol optical depth (AOD) data from satellites reveal that anthropogenic aerosols are more prevalent over wet lands than over dry ones, although dust aerosols are more prevalent in extremely arid places (Fig. 7.3). The retrievals from the Level-3 Deep Blue AOD of the MODIS (moderate-resolution Imaging Spectroradiometer) were used. As part of the Terra and Aqua platforms, MODIS instruments offer a large spectral range (0.412–14.240 m), a high spectral resolution, a high horizontal

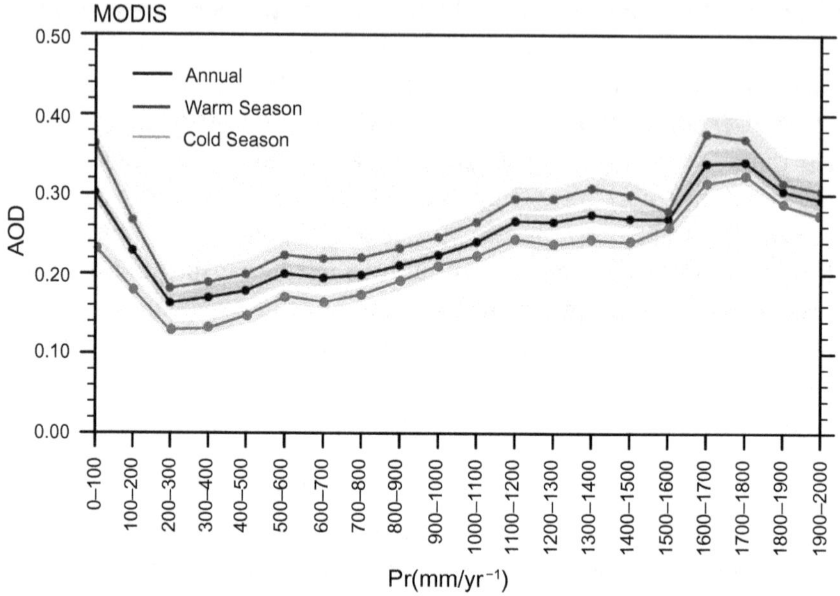

Fig. 7.3. Variations in the regionally averaged AOD (2001–2015) with respect to the climatological mean precipitation level (1948–2005) for warm seasons (May–September), cold seasons (November–March), and the annual mean. The shading denotes the standard deviation of temporal change for 2001–2015 (Huang *et al.*, 2017c).

resolution (10 km), and virtually worldwide coverage (36 channels). The dataset is accessible at daily, eight-day, and monthly temporal resolutions on a 1° × 1° grid at https://ladsweb.modaps.eosdis.nasa.gov/. These man-made aerosols scatter and absorb sunlight, cooling the surface in the process. Additionally, the aerosols can raise cloud water content and, consequently, cloud albedo, which will lower surface temperatures over humid terrain even more. This might be a factor in the regional variations in past warming rates (Fig. 7.1).

7.2.2 *Impacts of GHG emissions on warming*

How are the three abovementioned warming amplifications produced? We investigate the solution by looking at the regional energy balance at the surface. We start by looking at how surface energy sources alter as GHG levels rise (Fig. 7.4). The increase in downward thermal infrared radiation caused by more GHGs results in a significant increase in energy on average (Fig. 7.4), but solar radiation contributes little to changes in the global mean. Changes in energy sources appear to favor polar and terrestrial amplifications in terms of geographic patterns (Fig. 7.4, left). The findings also point to

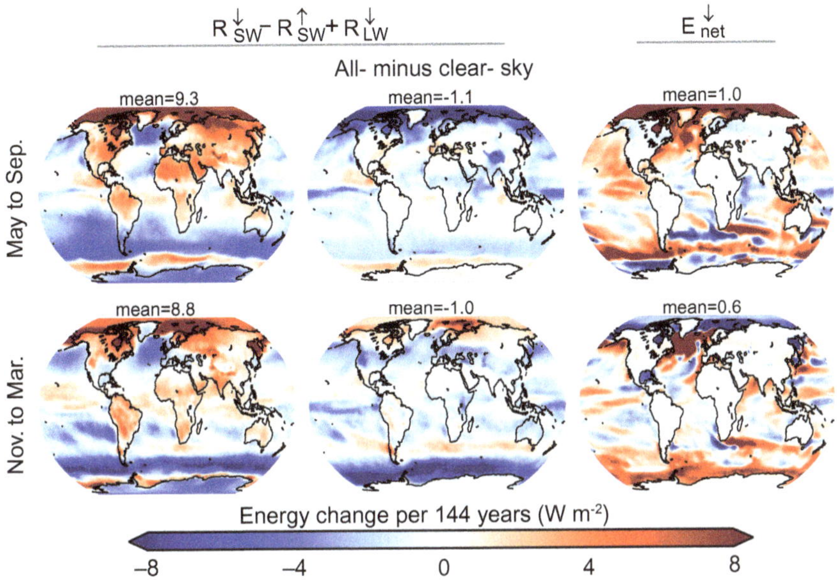

Fig. 7.4. The energy source ($R^{\downarrow}_{SW} - R^{\uparrow}_{SW} + R^{\downarrow}_{LW}$) and net energy ($E^{\downarrow}_{net}$) change at the surface for the period from 1861 to 2004. $R^{\downarrow}_{SW}$, $R^{\uparrow}_{SW}$, and $R^{\downarrow}_{LW}$ are incident and reflected solar radiation and incident thermal infrared radiation, respectively. The results are all ensemble means from historical greenhouse gas simulations. The linear trend of $R^{\downarrow}_{SW} - R^{\uparrow}_{SW} + R^{\downarrow}_{LW}$ during May to September, and November to March, respectively, in units of W/m^2 per 144 years (left). The effect of clouds is represented by the differences using all-sky results (left) minus clear-sky results (middle). (Right) Same as (left), but for $E^{\downarrow}_{net}$ (Xie *et al.*, 2019).

an accelerated increase in incoming energy over arid and semi-arid regions (such as over Africa and Asia), suggesting that changes in the energy source also contribute to the amplification of arid and semi-arid regions.

Considering the clear-sky results, we demonstrate that during local winter, clouds encourage the increased energy gain in polar locations, whereas during local summer, the reverse effect is shown (Fig. 7.4, middle). As a result, changes in surface albedo brought on by the polar ice/snow loss are likely what are responsible for the dramatic energy increase over the polar regions throughout the summer. The very significant decrease in clear-sky reflected shortwave radiation in the polar regions during local summer, when the rate of sea ice loss is at its greatest, strongly suggests this origin of the energy increase (Notz and Stroeve, 2016). However, summertime polar amplification is essentially nonexistent due to the physical requirement of a constant temperature of the ice–water mixture (IPCC, 2013). As a result, polar amplification is most prevalent during the winter. The reported negative effect of clouds on energy sources in the summer has no direct effect on simultaneous temperatures in the Arctic, which suggests that the direct effect of clouds on polar amplification is positive when the results are considered as a whole. Later, quantifiable research into the relative effects of clouds and other factors on polar amplification was shown.

The final warming pattern depends on the energy source as well as the way the fluctuating incident energy is discharged. Due to the net energy term's inclusion of both surface heat uptake and horizontal energy transmission, the net energy in our stated budget will not be zero (Fig. 7.4, right). In the polar oceans, strikingly positive and negative net energy shifts are seen in the local summer and winter, respectively (Fig. 7.4, right). This finding implies that extra energy collected by melting sea ice and open ocean during the summer is later released during the following winter (Screen and Simmonds, 2010). As a result, the pronounced polar amplification that occurs during local winter rather than summer appears to be caused by this phenomenon.

The surface energy budget changes for the three warming amplifications are further investigated in order to reach quantification (Fig. 7.5). First, it is evident that the increasing downward thermal infrared radiation dominates the Arctic amplification throughout the winter, with contributions from sea ice change being roughly twice as large (Fig. 7.5, top-left, $R^{\downarrow}_{\mathrm{LW}}$ and $-E^{\downarrow}_{\mathrm{net}}$). Additionally, the contribution to Arctic amplification from clouds is around one point five times more than that from the sea ice changes during the summer. Again, take note that the consistent temperature of the ice–water mixture means that the cloud's negative impact on midsummer net solar radiation (Fig. 7.5, top-left) has minimal direct impact on concurrent Arctic amplification. Therefore, rather than the generally cited sea ice change, clouds play the primary role in the observed Arctic amplification (Screen and Simmonds, 2010). The previous study (Screen and Simmonds, 2010) only looked at changes over a 20-year period (1989–2008); therefore, decadal variability must have had a significant impact (Trenberth, 2015). Their conclusion is therefore inappropriate for long-term global warming. Because of this, our conclusion diverges significantly from that study. The Arctic and Antarctic oceans are both in the same condition. Therefore, the general case for polar amplification is covered by our result.

Amplifications of terrestrial and arid and semi-arid regions do not exhibit a significant seasonal change, in contrast to polar amplification. Additionally, their energy budgets are largely constant throughout the year (Fig. 7.5, middle and bottom). The distribution of newly acquired energy between

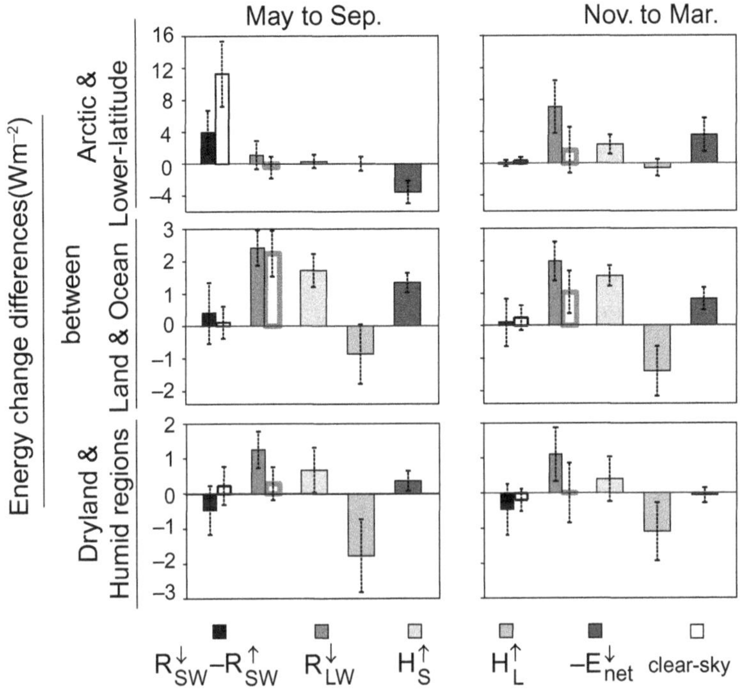

Fig. 7.5. The energy budget changed at the surface for the period from 1861 to 2004. The differences in the regional mean linear trends of the energy budget components between the Arctic and lower-latitude regions, in units of W/m^2 per 144 years (top). The means for the Arctic and lower latitudes are averaged over 60–90 and 0°N–60°N, respectively. (middle) Same as (top) but for differences between land and ocean. (bottom) Same as (top) but for differences between dryland and humid regions. For May to September, dryland and humid regions are represented by precipitation less than or more than 300 mm, respectively, while the cutting line of precipitation is 200 mm for November to March. $R^{\downarrow}_{SW} - R^{\uparrow}_{SW}$, $R^{\downarrow}_{LW}$, $H^{\uparrow}_{S}$, $H^{\uparrow}_{L}$, and $E^{\downarrow}_{net}$ are the net solar radiation, incident thermal infrared radiation, sensible, latent heat flux, and net energy, respectively. The open column represents the clear-sky result (Xie *et al.*, 2019).

latent and sensible heat is undoubtedly the primary cause of terrestrial and arid and semi-arid regions' amplifications (Fig. 7.5, middle and bottom). Although the enhanced downward thermal infrared radiation over land is about half as much as it is over the ocean (Fig. 7.5, middle), the net energy shift also indicates that surface heat intake and horizontal energy transmission may potentially have some impact. The difference in net energy changes between land and ocean regions is mostly driven by ocean heat uptake carried to the deeper ocean, and the influence of horizontal heat transport is minimal because the net energy change over land is essentially balanced (zero). In order to encourage terrestrial amplification, the deeper ocean's improved capacity to absorb heat also plays a significant role. Furthermore, enhanced downward thermal infrared radiation for the amplification of arid and semi-arid regions reveals a major contribution from clouds (Fig. 7.5, bottom), but clouds have a negligible effect on terrestrial amplification (Fig. 7.5, middle). Despite the inevitability of model differences, our primary conclusions related to energy budget studies are solid (Fig. 7.5).

7.2.3 *Warming of arid and semi-arid regions*

In addition to historical data, it is anticipated that the heightened warmth over arid and semi-arid regions will continue in the future. Under the high RCP8.5 (low–moderate RCP4.5) emissions scenario, surface warming over arid and semi-arid regions is anticipated to reach about 6.5° (about 3.5°) by the end of this century (Fig. 7.6). Almost all of the CMIP5 models show higher warming over arid and semi-arid regions than humid land, despite the fact that there are significant differences between them. The mean warming over arid and semi-arid regions should be around 2.7° (around 2.0°) when GMSW hits 2° (1.5°). The CMIP5 ensemble mean arid and semi-arid regions' warming for the period of 1920–2015 is only about $1.05 \pm 0.17°/96\,yr$, which is significantly less than the $1.28 \pm 0.19°/96\,yr$ average obtained from the

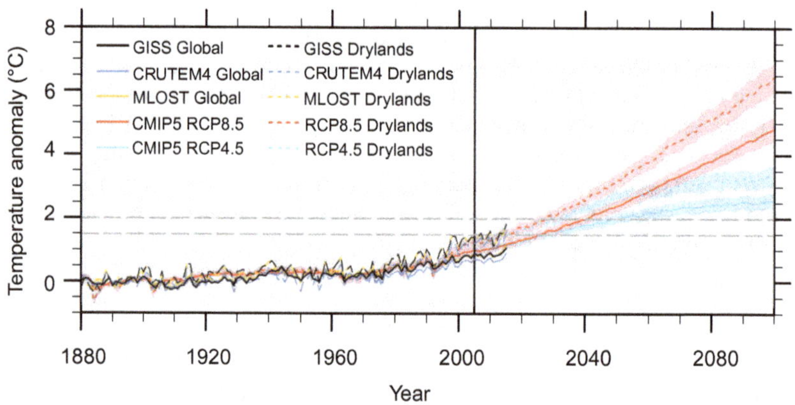

Fig. 7.6. Time series of global mean surface air temperature anomalies for the CMIP5 (relative to the 1861–1900 mean) and three observational datasets (relative to the 1880–1899 mean). The GISS, CRUTEM4, and MLOST are denoted by black, blue, and yellow lines, respectively. Red (green) lines denote the ensemble mean of the CMIP5 from historical simulations and RCP8.5 (RCP4.5) projections. The solid (dashed) lines are the mean of global (drylands) grids. The shading denotes the 95% confidence intervals of the 18 models, and gray dashed lines denote the 2.0°C and 1.5°C warming limitation conditions (Huang *et al.*, 2017c).

three observational datasets, as shown in Fig. 7.1(a). As a result, according to Fig. 7.1(a), the CMIP5 models overestimate the historical warming over arid and semi-arid regions, and the anticipated future warming over arid and semi-arid regions may exceed that of CMIP5.

According to the aforementioned analyses, warming of arid and semi-arid regions would be significantly higher under a GMSW limit of 2° than it would be under a goal of 1.5°. The effects of the temperature difference between these two objectives on arid and semi-arid regions are considerable in terms of the implications for agriculture, hydrology, drought, and health (Schleussner *et al.*, 2016). Four sectors were used to evaluate the effects of climate change on the 1.5° and 2° warming targets: agriculture (maize yield), hydrology (total runoff), drought (P/PET), and health (climatic suitability for malaria transmission). The ISI-MIP Fast Track database was used to generate the impact datasets (https://www.isimip.org/protocol/fast-track-simulation-protocol/). The four bias-corrected CMIP5 models (GFDL-ESM2M, HadGEM2-ES, MIROC-ESM-CHEM, and NorESM1-M) used for the climate simulations had a spatial resolution of 0.5° × 0.5°, a temporal range of historical simulations from 1971 to 2005, and future projections under RCP8.5 with a temporal range from 2006 to 2100. The global mean temperature and precipitation ranges of the other CMIP5 models can be well represented by the climate predictions of these four models, and the related effect findings are anticipated to be similar to those of the other CMIP5 models. The 11-year smoothed mean annual temperature was projected to attain the 1.5° (2.0°) warming target relative to that of 1861–1900 for each climate model. This was used to determine the 1.5° (2.0°) warming condition. Each sector has its own impact model, and the most popular ones were selected for this article. The 11-year time slices' estimated 1.5° (2.0°) warming condition was compared to the simulation's start (1861–1900).

In contrast to the increases in yields over humid lands, a majority of which are found in the humid lands of Europe, maize yields would decrease slightly over arid and semi-arid regions if GMSW were to reach 1.5°, but the yield reduction would be more significant with 2.0° global warming (about $-0.04\,\text{t}\,\text{ha}^{-1}\,\text{yr}^{-1}$ or 1.1% compared with the 1971–1981 level) (Fig. 7.7(a)). However, this rise is only transitory, and by the end of the 21st century, the world's maize yields would decline as the detrimental impacts of an increase in evaporative demand and a reduced growing season took control. With a GMSW between 1.5° and 2.0°, total runoff (including surface and subsurface runoff) will drop across all land surfaces, but most notably in arid and semi-arid regions (Fig. 7.7(b)), suggesting that the water deficit

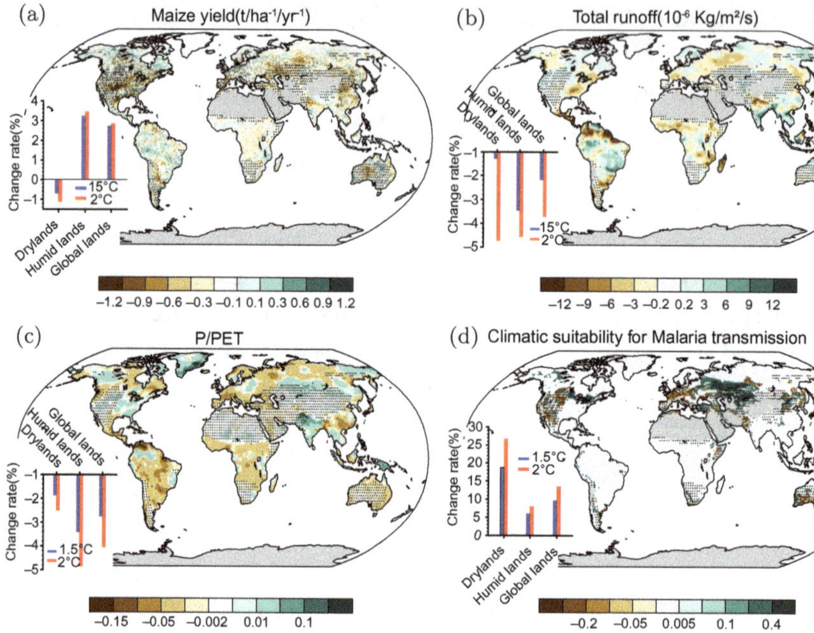

Fig. 7.7. Differences in climate change impacts between the GMSW of 1.5°C and 2.0°C. Global distributions of future additional changes under the RCP8.5 for maize yields (a), total runoff (b), P/PET (c), and climatic suitability for malaria transmission (d). Gray shading denotes barren land or ice cover and the black points denote the defined drylands. The inserted graphs show percentage change rates for drylands, humid lands, and all land surfaces (60°S–65°N) compared with historical mean values for 1971–1981 under a GMSW of 2.0°C (red bars) and 1.5°C (blue bars) (Huang *et al.*, 2017c).

in arid and semi-arid regions could worsen. In all land areas, a GMSW of 1.5°–2.0° would result in a drop in the P/PET ratio (Fig. 7.7(c)). The increased likelihood of both long-lasting drought over arid and semi-arid regions and flash drought over humid lands is indicated by this, which shows that increases in precipitation would not be adequate to offset the higher PET under rising temperatures. With a GMSW of 1.5°–2.0°, the climatic appropriateness for malaria transmission will increase over all land surfaces, with the largest increases happening in arid and semi-arid regions, from 19% to 27% greater than that of 1971–1981 (Fig. 7.7(d)). These impact assessments indicate that ecosystems of arid and semi-arid regions could be more vulnerable if GMSW rose from 1.5° to 2.0°.

7.2.4 *Differences in impacts of GHG emissions on Warming*

Here, the monthly results of historical simulation (1850–2005) and future predictions (2006–2100) under four scenarios (RCP45, RCP85, LW1.5deg and LW2deg) in the CESM experiment were used. The LW1.5deg and LW2deg scenarios are designed such that the global mean temperatures never exceed 1.5° and 2° above the pre-industrial levels. In this section, the ensemble mean were calculated to eliminate the internal climate variability, and the model results were modified to have the same climatology for the period 1961–1990 as the observations (Feng and Fu, 2013).

The scheduled warming targets are typically examined using one of three main approaches: the transient climate response (passing through 1.5° or 2° global warming), the short-term stabilization response (late 21st century output of simulations driven with emissions that stabilize the mean global warming to 1.5° or 2° by 2100), and the long-term equilibrium stabilization response (output of simulations at 1.5° and 2° once climate equilibrium is reached after several millennia). Studies on climate change at a century scale often employ the first two techniques. In this section, we employed both of them to examine and contrast the variations in the arid and semi-arid regions' climate response to global warming. The following are more detailed introductions.

(1) *Stabilized 1.5°C and 2°C warmer worlds*

Short-term stabilization measures for climate change are based on low-warming runs of the CESM model in which the global mean temperature will reach 1.5° (LW1.5deg) or 2° (LW2deg) at the end of the 21st century. In those circumstances, a target for the rate of global temperature rise has been set, and emission reductions are modified to change the equivalent carbon dioxide in order to achieve the target. RCP85 is followed by emissions in LW1.5deg (LW2deg) until 2017, at which point they are meant to rapidly decrease and reach 50% of 2017 levels by 2027 (2042). In 2038, total carbon emissions from land use and fossil fuels will be zero (carbon neutral) (2078). In 2065, CO_2 emissions will peak at a net negative flux of about −1.8 (−0.8) GtC/year (2120). Negative emission fluxes are subsequently decreased (Sanderson *et al.*, 2017). LW1.5deg and LW2deg are represented as stabilized warmer worlds in the following. The two warmer worlds are represented by an 11-year period spanning 2090 and 2100.

(2) *Transient 1.5°C and 2°C warmer worlds*

The 1.5°C and 2°C warming time periods for the RCPs, on the other hand, occur in the model runs prior to the temperatures rising to exceed the warming targets and are transitory, not stabilized climates. The temporary warmer worlds are represented by RCP45 and RCP85 in the following debates. The 11-year running average of the global mean surface air temperature (TAS) for the ensemble mean is used to calculate when the 1.5°C and 2°C warming above pre-industrial levels (1850–1920) will occur.

Numerous earlier studies from the preceding century provided support for the higher warming in arid and semi-arid regions (Huang *et al.*, 2012). According to Huang *et al.* (2012), the global annual mean temperature increase over land was 1.1°C, but the warming trend during the semi-arid regions' boreal cold season revealed a temperature increase of 1.5°C. Furthermore, worldwide arid and semi-arid regions have experienced surface warming of 1.2°C–1.3°C, which is 20%–40% higher than the surface warming of humid lands (0.8–1.0°). Figure 7.8 displays the time series of annual mean temperature over arid and semi-arid regions and the world to compare the warming in these regions when the global warming reaches 1.5°C and 2°C relative to pre-industrial levels. It is clear that, particularly in the long term, the temperature anomaly in arid and semi-arid regions is larger than the globe's averaged temperature anomaly. Global arid and semi-arid regions' temperature increases will be 1.6°C and 2.3°C when global warming slowly approaches 1.5°C and 2°C at the end of this century, respectively (Fig. 7.8). By the end of the century, the RCP85 scenario predicts an increase in average temperatures of 5.0°C for the world and 6.4°C for the arid and semi-arid regions. The global averaged TAS warming will also reach 1.5°C and 2°C in 2023–2033 (2031–2041) and 2035–2045 (2050–2060), respectively, according to the RCP85 (RCP45) scenario. However, the average TAS warming in the arid and semi-arid regions will be 1.5°C and 2°C in the years 2018–2028 (2023–2033) and 2028–2038 (2038–2048), respectively, about 10 years before the average TAS warming worldwide (Fig. 7.8).

Figure 7.9(a–c) shows the differences in warming of arid and semi-arid regions between a 1.5°C warmer world and the 1961–1990 baseline era.

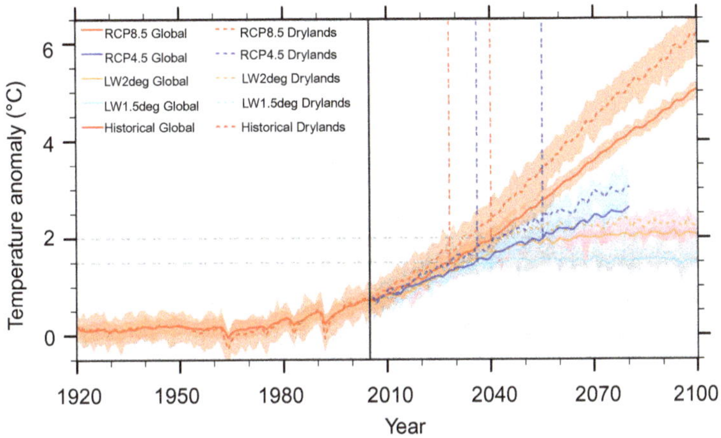

Fig. 7.8. Time series of annual mean surface air temperature (TAS) anomalies for the CESM (relative to the 1850–1920 mean). The solid colored lines represent global (90°S–90°N) averaged TAS anomalies, and the dashed colored lines are for drylands (60°S–65°N). Thick dotted lines show the timing of 1.5°C (2°C) warming over the globe, and the timing of 1.5°C and 2°C warming for RCP45 (RCP85) scenarios is 2036 and 2055 (2028 and 2040), respectively. The shading denotes the 5th to 95th percentiles for the ensemble members (Wei *et al.*, 2019).

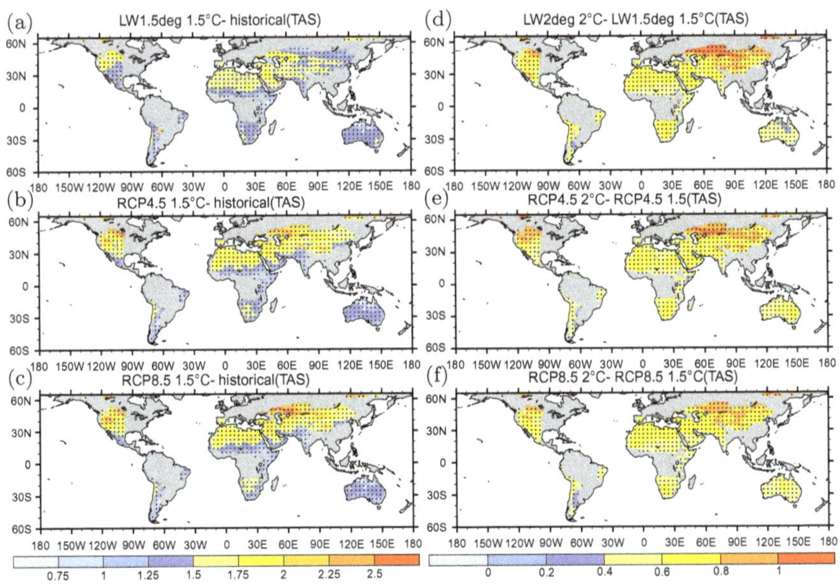

Fig. 7.9. Multi-model ensemble mean changes in annual mean TAS (units: °C) over drylands between a 1.5°C warmer world and the 1961–1990 baseline period (a–c); (d–f) same as (a–c), but for the changes between a 2°C and a 1.5°C warmer world. Dots denote where the changes are passing a 95% significance test (Wei *et al.*, 2019).

Global arid and semi-arid regions will experience a markedly higher temperature than the baseline in a 1.5° warmer world, with Eurasia, North Africa, and North America experiencing the most pronounced warming and up to 1.5°C warming. The contrast in temperature change between stabilized (Fig. 7.9(a)) and transient (Figs. 7.9(b) and 7.9(c)) 1.5°C warmer worlds is particularly notable in all three regions relative to the baseline. Drylands generally warm faster in a world that is 1.5°C warmer temporarily than it does in a world that is 1.5°C warmer stabilized. The widespread warming in the extra-tropical Northern Hemisphere, which is in line with several earlier studies (Wallace *et al.*, 2012), is more intriguing. Global drylands would see favorable temperature changes in a 2°C warmer world compared to a 1.5°C warmer world (Figs. 7.9(d–f)). The distinction is that this extra 0.5°C of warming causes temperature rises that are most noticeable in the northern drylands of North America and, particularly, in the Eurasian continent, where increases may reach 1.0°C (Fig. 7.9(d)). In contrast to Northern North America and Australia, the difference between temporary 2°C and 1.5°C warmer worlds is greater on the Eurasian continent (Figs. 7.9(e) and 7.9(f)).

According to the AI, we classified drylands into four subcategories. Figures 7.10(a) and 7.10(b) illustrates the warming in drylands and other dry subtypes in stabilized and transient 1.5°C and 2°C warming environments relative to pre-industrial. The warming in drylands is 1.6°C (2.3°C) when the worldwide averaged surface air temperature is progressively growing to 1.5°C (2°C) relative to pre-industrial levels at the end of hyper-arid regions, reaching up to 1.8°C (2.4°C) in a stabilized 1.5°C (2°C) warmer world, whereas the warming in semi-arid regions, at 1.5°C (2.2°C), is minor. A lower mean latent heat flow compared to wetter drylands is the fundamental

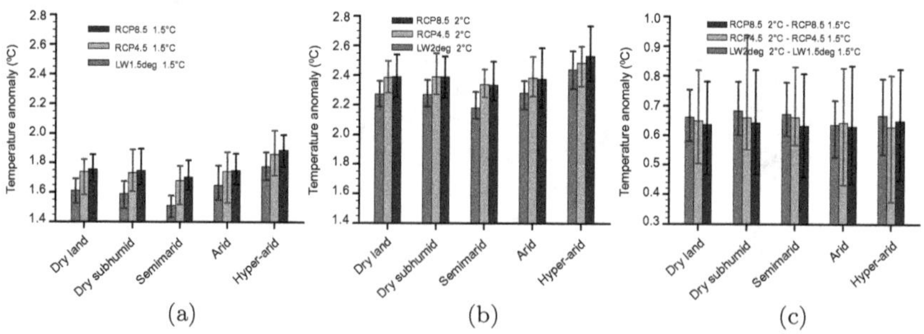

Fig. 7.10. Temperature anomalies in drylands and various dry subtypes are relative to pre-industrial in a 1.5°C (a) and a 2°C (b) warming world. The difference in warming in drylands and various dry subtypes between a 1.5°C and a 2°C warming world (c). The error bars denote the 5th to 95th percentiles for the ensemble members (Wei *et al.*, 2019).

reason why warming is generally more pronounced in drier locations. Other contributing factors include decreased precipitation over drier drylands and the reduced evaporation and transpiration associated with lower vegetation cover (Fig. 7.10). To release the warmth from solar and infrared radiation, the surface temperature over drier drylands must rise dramatically in order to provide a greater sensible heat flow (Fig. 7.10). This can be explained by the fact that the radiative forcing values in the RCP45 and RCP85 scenarios are higher than in the LW scenarios for the same temperature increase range (Wei *et al.*, 2019), regardless of whether the warming is greater for the total dryland area or the four subtype areas. The dryland warming is anticipated to increase by about 0.7°C when comparing a stabilized 2°C warming world to a stabilized 1.5°C warming world, with the greatest additional warming occurring in the dry subhumid regions (about 0.7°C) and the least additional warming occurring in the arid regions (about 0.6°C), whereas the difference in warming in the drylands for transient 2°C warming compared with a transient 1.5°C warming is smaller than for the stabilized scenarios (Fig. 7.9(c)). This outcome might be explained by the fact that in the transient scenario, the difference in radiative forcing in the world between 2°C warming and 1.5°C warming is smaller than that in the stable scenario (Wei *et al.*, 2019).

(3) *Underestimated warming of arid and semi-arid regions*

The technique utilized to distinguish between the internal variability and the externally imposed component in the observed temperature time series is based on that proposed by Dai and colleagues (Dai *et al.*, 2015). In order to concentrate on the anomalies, the mean for the whole analysis period (1920–2015) was first removed. The annual mean temperature was used in each analysis. The measured temperature time series at each grid point was linearly regressed using the global mean temperature anomaly time series of the CMIP5 ensemble mean:

$$T(n,i) = b_F(i)\overline{T}_m(n) + T'(n,i), \tag{7.4}$$

in which $T(n,i)$ is the surface temperature anomaly at grid point i for year n from observations and $n = 1, 2, \ldots, 96$ for 1920, 1921, …, 2015, respectively. $\overline{T}_m(n)$ is the global mean surface air temperature anomaly for year n from the CMIP5 ensemble mean of historical simulations (for 1920–2005) and RCP4.5 projections (for 2006–2015). $\mathrm{b_F}(i)$ is the regression slope at grid point i, $\mathrm{b_F}(i)\overline{\overline{T}}_m(n)$ is defined as the externally forced component in the observations at grid point i for year n, and the residual $T'(n, i)$ is defined as the unforced component due to internal variability. Here, the externally forced component averaged over the global (90°S–90°N) surfaces (including oceans) and the arid and semi-arid regions is denoted by $\overline{T}_F^{\text{global}}(n)$ and $\overline{T}_F^{\text{dry}}(\mathrm{n})$, respectively, and the amplification factor of the externally forced component over arid and semi-arid regions, $\mathrm{A}_{\text{drylands}}^{\mathrm{F}}$, is defined as follows:

$$A_{\text{drylands}}^{F} = \frac{\overline{T}_F^{\text{dry}}(n)}{\overline{T}_F^{\text{global}}(n)} = \frac{\sum_{i=1}^{N_{\text{dry}}} \frac{b_F(i)\overline{T}_m(n)\cos\varphi_i}{\sum_{i=1}^{N_{\text{dry}}}\cos\varphi_i}}{\sum_{i=1}^{N_{\text{global}}} \frac{b_F(i)\overline{T}_m(n)\cos\varphi_i}{\sum_{i=1}^{N_{\text{global}}}\cos\varphi_i}}$$

$$= \frac{\sum_{i=1}^{N_{\text{dry}}} b_F(i)\cos\varphi_i \cdot \sum_{i=1}^{N_{\text{global}}}\cos\varphi_i}{\sum_{i=1}^{N_{\text{global}}} b_F(i)\cos\varphi_i \cdot \sum_{i=1}^{N_{\text{dry}}}\cos\varphi_i}, \tag{7.5}$$

in which N_{global} and N_{dry} denote the number of grid points over the globe and the arid and semi-arid regions, respectively, and φ_i is the latitude of grid i. Note that $\mathrm{A}_{\text{drylands}}^{\mathrm{F}}$ does not depend on time. Furthermore, regional means of unforced temperature anomalies over arid and semi-arid regions and over the globe are, respectively, denoted as follows:

$$\overline{T}'^{\text{dry}}(n) = \frac{\sum_{i=1}^{N_{\text{dry}}} T'(n,i)\cos\varphi_i}{\sum_{i=1}^{N_{\text{dry}}}\cos\varphi_i}, \tag{7.6}$$

$$\overline{T}'^{\text{global}}(n) = \frac{\sum_{i=1}^{N_{\text{global}}} T'(n,i)\cos\varphi_i}{\sum_{i=1}^{N_{\text{global}}}\cos\varphi_i}, \tag{7.7}$$

in which the amplification of the unforced component over the arid and semi-arid regions, $\mathrm{A}_{\text{drylands}}^{\mathrm{I}}$, is defined as the regression slope between time series of $\overline{T}'^{\text{dry}}(n)$ and $\overline{T}'^{\text{global}}(n)$:

$$\overline{T}'^{\text{dry}}(n) = A_{\text{drylands}}^{\mathrm{I}}\overline{T}'^{\text{global}}(n). \tag{7.8}$$

Due to the removal of the period's mean, the intercept in this case is zero. Other regions' amplification factors (for humid lands, all land surfaces, and seas) are derived in a similar way. The findings indicate that the amplification factors of the unforced internal components are about unity and vary less across different locations. For instance, using GISS temperature data, the unforced amplification factors for arid and semi-arid regions, humid lands, lands, and seas are 1.01, 1.04, 1.08, and 1.01, respectively. This indicates that the spatial distribution of the unforced temperature signal is largely uniform.

The amplification of the raw observational anomalies over arid and semi-arid regions, A^{I}_{drylands}, is defined by the following regression:

$$\overline{T}^{\text{dry}}(n) = A^{R}_{\text{drylands}}\overline{T}^{\text{global}}(n), \tag{7.9}$$

in which $\overline{T}^{\text{drylands}}$ and $\overline{T}^{\text{global}}$ denote regional mean temperature anomalies over the arid and semi-arid regions and the globe, respectively:

$$\overline{T}^{\text{dry}}(n) = \frac{\sum_{i=1}^{N_{\text{dry}}} T(n,i)\cos\varphi_i}{\sum_{i=1}^{N_{\text{dry}}}\cos\varphi_i}, \tag{7.10}$$

$$\overline{T}^{\text{global}}(n) = \frac{\sum_{i=1}^{N_{\text{global}}} T(n,i)\cos\varphi_i}{\sum_{i=1}^{N_{\text{global}}}\cos\varphi_i}. \tag{7.11}$$

The amplification factors calculated from the raw data and those based on the predicted forced components are contrasted. The lack of a discernible difference between the two scenarios shows that the unforced component's influence is minimal.

We compare (Fig. 7.11) the mean arid and semi-arid regions' temperature increases with the GMSW level for both data and CMIP5 models to further quantify the amplified warming over arid and semi-arid regions. The CMIP5 ensemble mean clearly shows that the arid and semi-arid regions' warming rate is roughly 1.35 times that of the GMSW for both historical and future time frames (Fig. 7.11(a)). However, as shown in Fig. 7.1(a), the CMIP5 models overestimate the historical warming over arid and semi-arid regions and are likely to overestimate the future warming over arid and semi-arid regions as well. Therefore, we are unable to estimate the future climate's amplification factor using the future heat predicted by the model over arid and semi-arid regions.

However, because the observational data contain both internally generated variations and externally forced changes, whereas the CMIP5 ensemble mean primarily represents forced changes, the amplification in the three observational datasets cannot be directly compared with that of the CMIP5 ensemble mean. To estimate the forced component for each of the three observational datasets, we utilized the global mean temperature series from the CMIP5 ensemble mean to regress on the observed temperature series at each grid point, with the residual taken as the unforced component (Dai *et al.*, 2015).

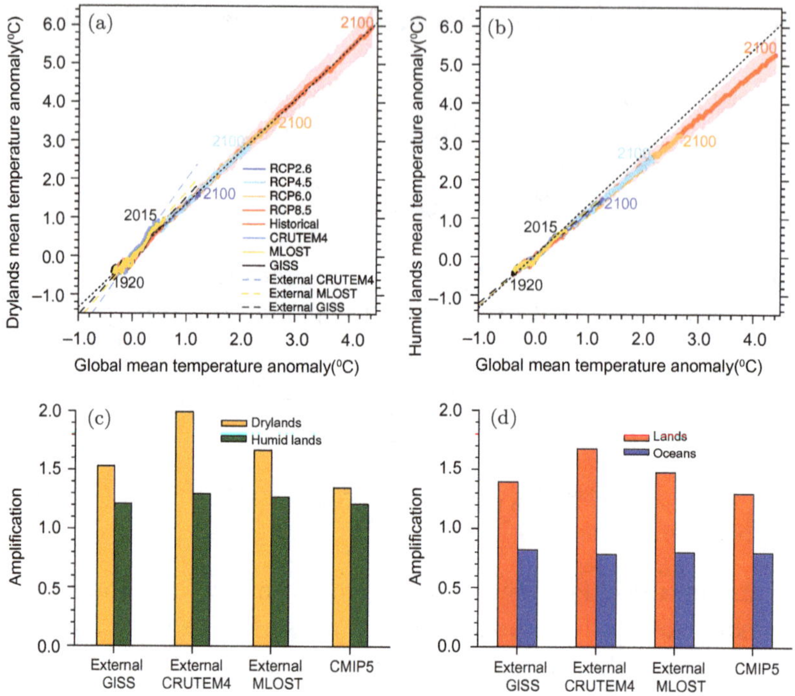

Fig. 7.11. The comparison of warming amplification over different regions is based on CMIP5 and three observational datasets. Anomaly relationship between the mean global surface air temperature and that of drylands (a) and humid lands (b) based on the GISS, MLOST, and CRUTEM4 datasets (after 5-year running averaging). Also shown are the estimated externally forced components in the datasets (dashed lines) and the ensemble mean of the CMIP5 historical simulations, RCP8.5, RCP6.0, RCP4.5, and RCP2.6 21st century projections (2006–2100). The shading denotes the 95% confidence interval for the 18 models. The black dotted line is the linear regression of the CMIP5 over drylands. (c, d) The amplification factor of drylands and humid lands (c) and all land and ocean surfaces (d) from the ensemble mean of CMIP5 and the estimated externally forced components of the GISS, CRUTEM4, and MLOST datasets (Huang *et al.*, 2017c).

The externally forced amplification, which is about 1.7 for arid and semi-arid regions based on MLOST (about 2.0 and about 1.6 using CRUTEM4 and GISS, respectively), is the ratio of the externally driven component averaged over arid and semi-arid regions (humid lands) to that averaged across the globe (Fig. 7.11(c)). A sensitivity test is performed to determine whether the predicted forced amplification would be trustworthy for future projections. If data from 1920 to 2000 or later years were taken into consideration, Fig. 7.12 demonstrates that the predicted amplification factor over drylands is essentially constant. This means that estimates for the future can be made using the amplification factor calculated using data from 1920

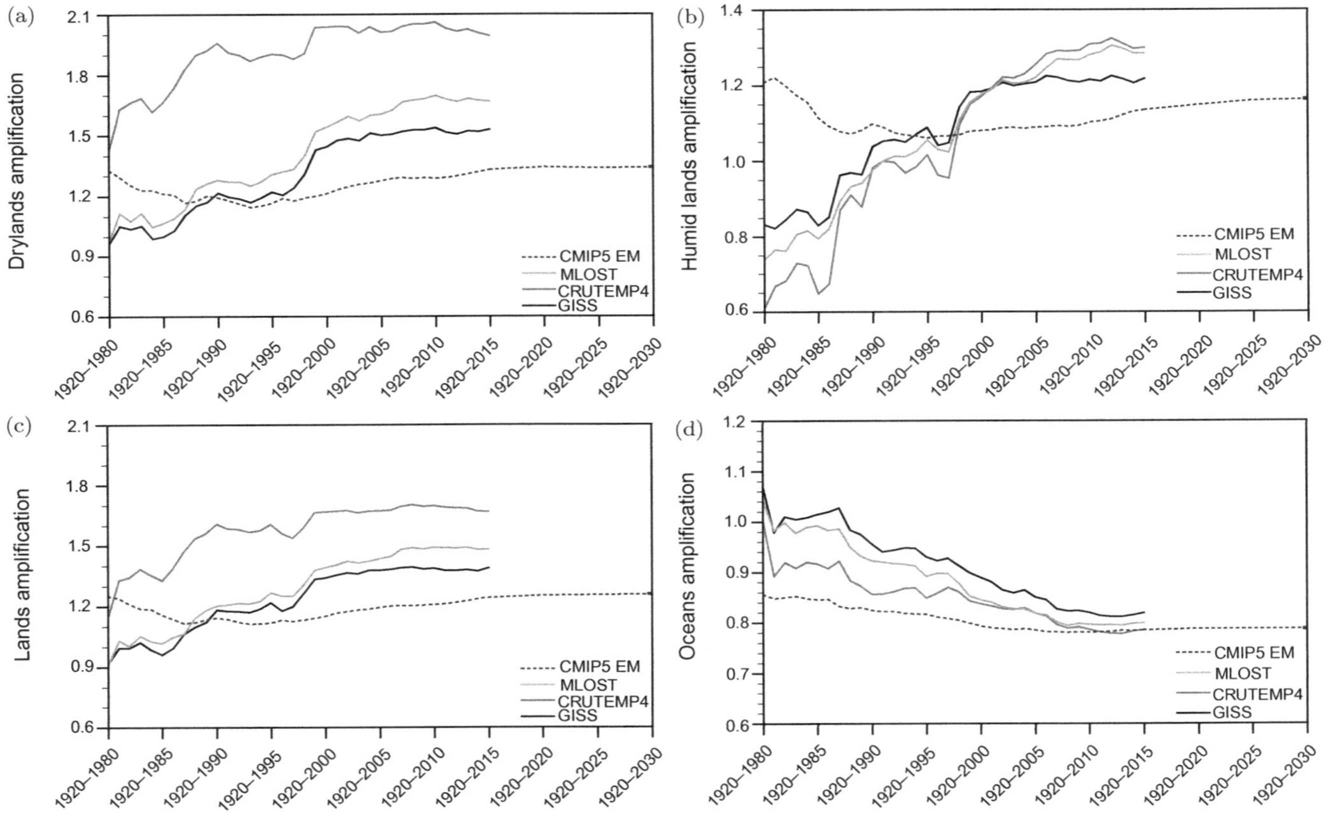

Fig. 7.12. The amplification factors of the forced changes over different regions are estimated using data over different time periods from 1920–1980 to 1920–2030 based on the MLOST, CRUTEM4 and GISS datasets and the CMIP5 18-model ensemble mean (Huang *et al.*, 2017c).

to 2015. This aids in correcting the CMIP5 models' underestimation of the warming over drylands for the next time frame.

By extrapolating this 1.6–2.0 amplification factor into the future, warming over arid and semi-arid regions might reach 3.2°C–4.0°C (2.4°C–3.0°C) when the global mean surface temperature exceeds 2°C (1.5°C). The externally imposed amplification for the MLOST in humid lands is also about 1.3 (around 1.3 and around 1.2 for the CRUTEM4 and GISS, respectively), whereas the amplification for the CMIP5 in humid areas is about 1.2 (Figs. 7.11(b) and 7.11(c)). As a result, warming in humid regions may be between 2.4°C and 2.6°C (1.8°C–2.0°C) when the GMSW hits 2°C (1.5°C). Amplification by the unforced component can be determined in a manner similar to that described earlier, and the difference is negligible when compared to other areas or datasets. If more model runs are used to derive the forced signal, the outcomes are remarkably similar.

7.3 Expansion of Arid and Semi-Arid Regions

Ecosystems of arid and semi-arid regions are far more vulnerable due to most of the soil being very poor and the scant vegetation cover (Reynolds *et al.*, 2007). Due to global warming, as well as the results of rapid economic development, population increase, and urbanization, desertification and degradation are ubiquitous in arid and semi-arid regions. Studies have also shown that the lengthening of dry spells and shorter, less frequent, and less widespread precipitation events are two additional effects of warming that will grow more pronounced in the 21st century (Giorgi *et al.*, 2014). These patterns could lead to the growth of arid and semi-arid regions and raise the proportion of people whose lives are affected by water scarcity and land degradation (Reynolds *et al.*, 2007). It is crucial to understand how future climate change may impact arid and semi-arid regions in order to protect them and develop adaptation methods (Nicholson, 2011). The spatial and temporal evolution of arid and semi-arid regions in future GHG emission pathways is clearly explained in this section using climate model simulations. The findings might spur policymakers to act quickly and wisely to stop the impending worldwide desertification.

7.3.1 *Dry/wet change of arid and semi-arid regions*

Drought is typically understood to indicate that the soil is experiencing a water deficit, and the specific manifestation is that there is not enough precipitation to make up for PET and runoff (Feng and Fu, 2013). Precipitation and PET are key factors in drought. The CESM model ensemble mean nearly captures these characteristics, except for dry subhumid regions, where the increase in PET during the most recent several decades has been more substantial than any minor trend in precipitation (Fig. 7.13). To forecast future precipitation and PET changes, we use forecasts from the CESM model. Additionally, two sets of observed temperature data, three sets of observed precipitation data, and two sets of observed PET data were used to validate the climate model simulation. The Goddard Institute for Space Studies surface Temperature (GISS) dataset, created by NASA, is one temperature dataset. It may be accessed at https://data.giss.nasa.gov/gistemp/. The Merged Land–Ocean Surface Temperature Analysis (MLOST) V3.5.4 created by NOAA is another temperature data collection that can be found at https://climatedataguide.ucar.edu/climate-data/global-surface-temperature-data-mlost-noaa-merged-land-ocean-surface-temperature. MLOST is a global dataset that spans the years 1880 to the present and includes monthly

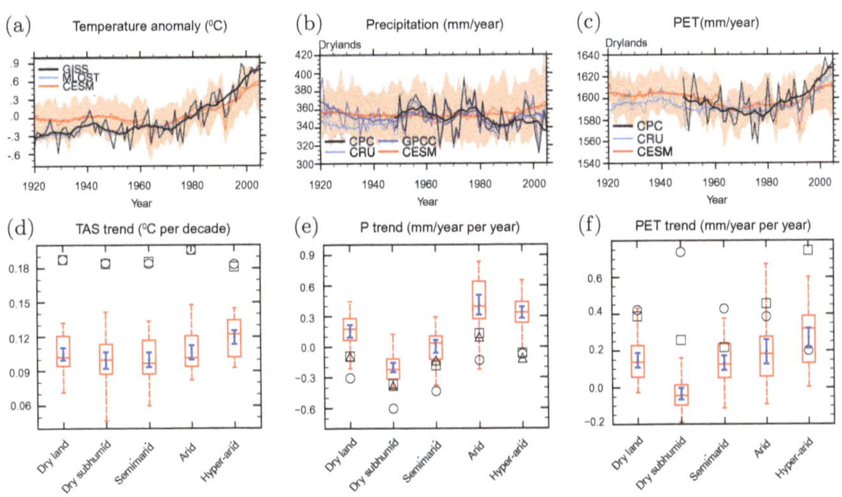

Fig. 7.13. Comparison of annual mean surface air temperature (TAS) anomalies, annual precipitation, and PET in drylands and their trends in drylands and four dry subtypes from CESM historical simulations and observations. Thick colored lines indicate the 11-year running mean. The shading denotes the 5th–95th percentiles for the ensemble members (a–c). Linear trend of TAS, P, and PET in model ensembles and observations for the 1948–2005 period. The red box plot represents the minimum of the trend, the 25% quartile value, the median, the 75% quartile value, and the maximum value for 30 ensemble members. The blue dots denote the CESM ensemble mean trend, and the blue error bar represents ± one standard deviation for the ensemble mean (d–f). The circles and squares in (d) represent the TAS trends from GISS and MLOST observations, respectively. The circles and squares in (e, f) represent CPC and CRU, respectively. The triangles in (e) are from GPCC (Wei *et al.*, 2019).

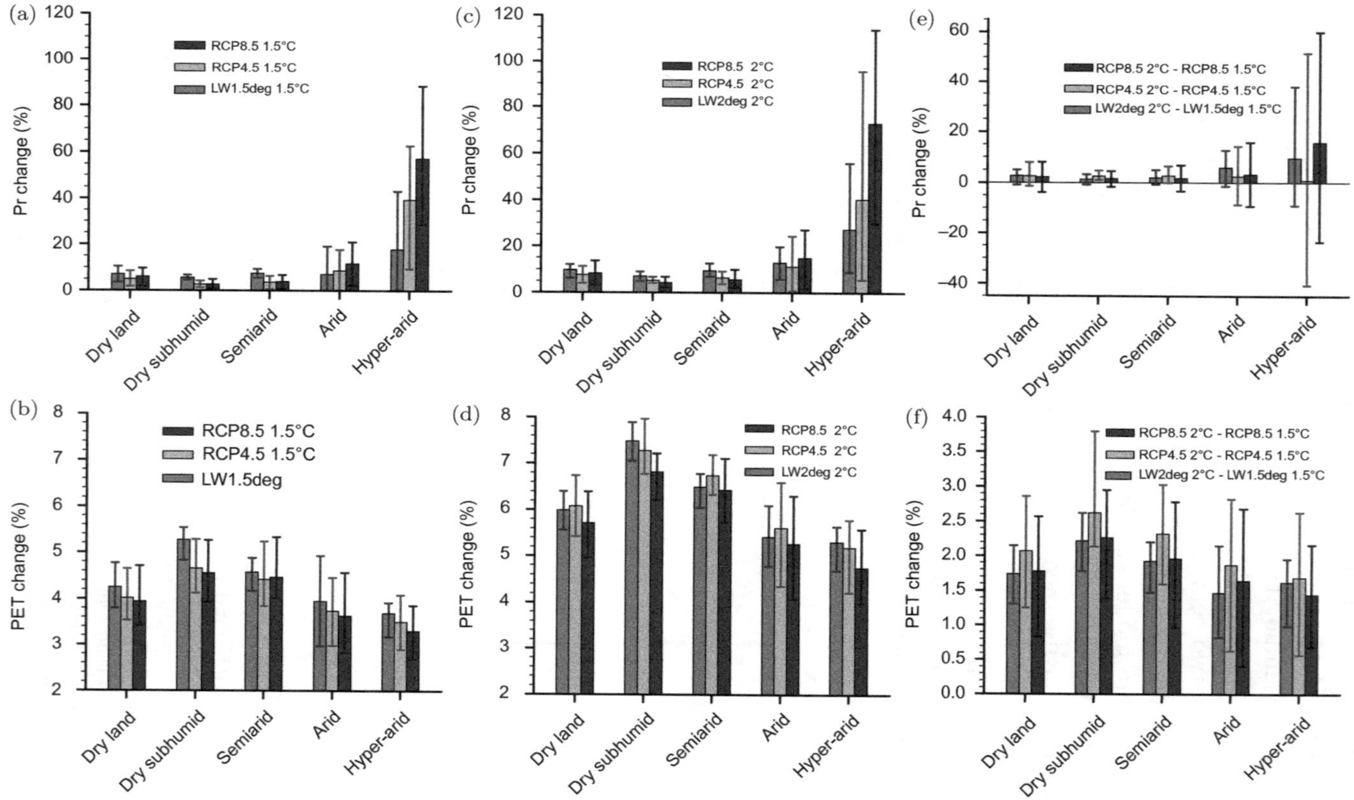

Fig. 7.14. Annual precipitation (a) and PET (b) changes as a fraction of the baseline period 1961–1990 in drylands and four dry subtype regions between a 1.5°C warmer world and the 1961–1990 baseline period; (c, d) same as (a, b), but for a change in a 2°C warmer world relative to the 1961–1990 baseline. (e, f) The difference of change in annual precipitation (e) and PET (f) between a 1.5°C and 2°C warmer world. The error bars denote the 5th–95th percentiles for the ensemble members (Wei *et al.*, 2019).

anomalies of air (over land) and sea surface temperatures (over ocean). Three precipitation datasets are CRU TS 3.25 (1901–2016, website at https://crudata.uea.ac.uk/cru/data/hrg/), the Global Precipitation Climatology Centre (GPCC) combined V7 and V4 data product (1901–2015, website at https://psl.noaa.gov/data/gridded/data.gpcc.html), and the PRECipitation REConstruction over Land (PREC/L) developed by the Climate Prediction Center (CPC) (1948–2005) and retrieved from https://www.esrl.noaa.gov/psd/data/gridded/data.precl.html. Two PET datasets are sourced from CRU TS 3.25 (1901–2016) and CPC (1948–2005) (Feng and Fu, 2013).

Both precipitation and PET indicate a sharp increase in the coming years, regardless of whether one is looking at the LW1.5deg and LW2deg forecasts or the RCP45 and RCP85 emission scenarios. But the rise in precipitation (12.9–37.0 (the 5th to 95th percentile for ensemble members) (21.8–42.6) mm/year in a stabilized 1.5°C (2°C) warmer world) is not enough to make up for the rise in PET (60.3–75.7 (88.3–101.7) mm/year in a stabilized 1.5°C (2°C) warmer world), which can worsen aridity. According to Fu and Feng (2014), the rise in temperature is mostly to blame for the rise in PET, which also causes the vapor pressure deficit (VPD) to rise (Ficklin and Novick, 2017). While the increase in PET does not significantly differ between the two scenarios, the increase in precipitation in stabilized scenarios is greater than in transitory scenarios (Fig. 7.14). The arid and semi-arid regions in a world with intermittent warming are anticipated to be drier than those in a world with stabilized warming, in accordance with the concept of aridity (P/PET) in this study (Fig. 7.15).

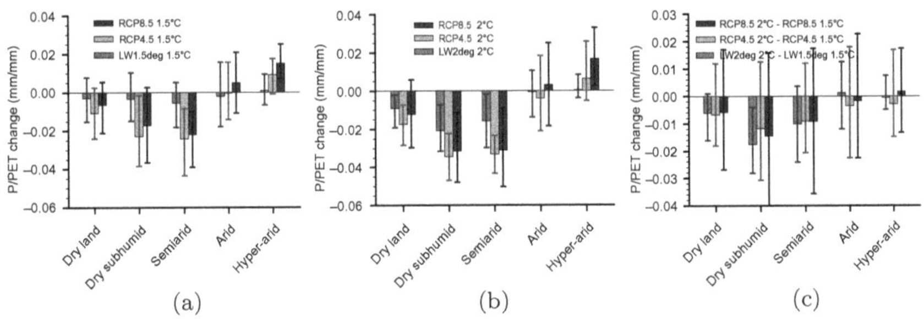

Fig. 7.15. Changes of annual P/PET in drylands and four dry subtype regions in a 1.5°C warmer world and in the 1961–1990 baseline period (a). (b) Same as (a) but for a change in a 2°C warmer world relative to the 1961–1990 baseline. Difference of change in annual P/PET between a 1.5°C and 2°C warmer world (c). The error bars denote the 5th to 95th percentiles for the ensemble members (Wei *et al.*, 2019).

7.3.2 *Impacts of GHG emissions on area expansion*

It is crucial to compare data from the same time period with historical simulations of arid and semi-arid regions' variability from the CMIP5-EM model (1948–2005) to assure the accuracy of future estimates (2006–2100). Table 7.1 compares the global observed AI and CMIP5-EM values over the course of 58 years. The areas with a drying tendency account for up to 66% of the world's land area, and the reported AI fell dramatically, with a mean net trend of −0.050 per 58 yr (Table 7.1). While arid regions only make up 59% of the world's surface area, the mean trend of CMIP5-EM is just −0.012 per 58 yr, or around one-fourth of the observed trend. To demonstrate the temporal changes, a portion of the AI data from the past 15 years (1991–2005) is contrasted with the prior 15 years (1948–1962) (Table 7.1). The observed areal increases in hyper-arid, arid, semi-arid, and subhumid land types from nearby wetter subtypes are 0.62%, 1.16%, 2.32%, and 3.32% of the global land area, respectively. By contrast, the increases predicted by CMIP5-EM are 0.05%, 0.14%, 0.37%, and 0.50%, respectively. This is similar to how the declines in subtype regions between neighboring wetter and drier subtypes in CMIP5-EM are roughly one-third that of the observations.

The aforementioned contradiction may be caused by natural variability, the uncertainty of model design, or the uncertainty of external forcing in models because the observed changes are prone to both driven and natural

Table 7.1. Comparison of the AI for the observed data and the CMIP5-EM simulation (Huang *et al.*, 2016b).

Items	**Category change**	**Observed data**	**CMIP5-EM simulation**
Percentage of global land area with drying trend during 1948–2005		66%	59%
Linear trend of global mean AI during 1948–2005		−0.050 per 58 yr	−0.012 per 58 yr
Neighboring subtype changes during 1991–2005 relative to 1948–1962	Arid to hyper-arid	0.62	0.05
	Semi-arid to arid	1.16	0.14
	Subhumid to semi-arid	2.32	0.37
	Humid to subhumid	3.32	0.50
	Hyper-arid to arid	0.27	0.14
	Arid to semi-arid	0.78	0.27
	Semi-arid to subhumid	0.68	0.23
	Subhumid to humid	0.95	0.34

Note: Unit: percentage of global land area.

variability. Therefore, modifications are required because CMIP5-EM by itself is insufficient for estimating future aridity changes. We performed bias corrections on CMIP5-EM by developing a mapping relationship with historical observations (Yu *et al.*, 2022). The simulated aridity changes are more compatible with observations after rectification, according to a posteriori independent validation. This finding shows that the historical link can be transferred to the future, and correcting the CMIP5-EM forecasts will be helpful for assessing changes in aridity and aridity patterns around the world (Yu *et al.*, 2022).

It is impossible to build a mapping link between the CMIP5 and the observed data across the historical period because they both contain too many degrees of variation. The CMIP5 and observed datasets are both reduced to a lower-dimensional space using principal component analysis (PCA), which then decomposes them into eigenvectors and related principal components (PCs). A matrix of regression coefficients is created by regressing each PC in the CPC dataset by the leading PCs of CMIP5. The future observed PCs can be anticipated by the simulated PCs of CMIP5 in accordance with this connection. Leading PCs are considered as correction factors not only because they account for the majority of mathematical variance but also because they exhibit strong correlations with patterns brought about by both internal and external climate variability. The connection may show the potential physical implications underlying each PC, even though it does not establish causality. In this study, we used this method to correct the long-term changes in the CMIP5 projections. It has previously been used to adjust seasonal predictions (Qin *et al.*, 2011; Wei *et al.*, 2021b).

The following introduces the practical procedures:

The climate drift, which is calculated as the difference between projections and observations averaged over climatic timescales, is subtracted from the simulated fields before they are adjusted. When m is the number of spatial grid points and n is the length of the time series, the simulated field $Z_{\mathrm{sim}}(t)$ and observed field $Z_{\mathrm{obs}}(t)$ can both be written as matrices of size $m \times n$. Usually, m is a lot bigger than n. By using PCA, we increase $Z_{\mathrm{sim}}(t)$ and $Z_{\mathrm{obs}}(t)$ to condense the spatial dimensions:

$$Z_{\mathrm{sim}}(t) = \sum_{i=1}^{n} b_i(t) X_i, \tag{7.12}$$

$$Z_{\mathrm{obs}}(t) = \sum_{j=1}^{n} a_j(t) Y_j, \tag{7.13}$$

in which X_i and Y_j are spatial patterns with a dimension of m; b_i and a_j are PCs corresponding to each of X_i and Y_j with a dimension of n. Using multiple linear regression, each $a_j(t)$ can be regressed by the K leading $b_i(t)(i = 1, 2, \ldots, K)$:

$$a_j(t) = \sum_{i=1}^{K} b_i(t)c_{i,j} + e_j, \tag{7.14}$$

in which $c_{i,j}$ is the regression coefficient and e_j is the regression residual. Projecting $Z_{\text{sim}}(n + 1)$ (the predicted field at the time of $n + 1$) onto each spatial pattern X_i will provide the corresponding time coefficient $b_i(n + 1)$:

$$b_i(n + 1) = X_i^T Z_{\text{sim}}(n + 1). \tag{7.15}$$

By inserting $b_i(n+1)$ into Eq. (7.14), we can obtain a prediction of the observed time coefficient:

$$\tilde{a}_j(n + 1) = \sum_{i=1}^{K} b_i(n + 1)c_{i,j} + e_j. \tag{7.16}$$

Then, the adjusted prediction of Z_{obs} at a time of $n + 1$ can be obtained by combining the spatial pattern:

$$\tilde{Z}_{\text{obs}}(n + 1) = \sum_{j=1}^{K} \tilde{a}_j(n + 1)X_j. \tag{7.17}$$

Generally speaking, high-frequency variability is difficult for climate models to reproduce, and keeping higher-order PCs could result in more noise. In this situation, it is important to consider the ideal number of leading PCs. To choose the total number of top PCs, the commonly utilized leave-one-out cross-validation over the historical period is applied.

The revised AI is thought to be more accurate than the initial forecasts since it uses data from both observations and CMIP5. The results show that the correction method is reliable and robust, and *a posteriori* independent validation is used to demonstrate the method's validity. Based on this, the same PCA adjustment procedure is used to adapt for the future. The linear regression is derived when PCA decomposes the 58-year record of observations and simulations during the historical period. To get a time coefficient, the simulation field for each year of the future period is projected onto the decomposed spatial patterns. A "predicted" time coefficient for the observations will be produced by applying this time coefficient to the regression relationship, which will also enable a "prediction" of the observed field. By obtaining the adjusted projections under RCP8.5 and RCP4.5, the adjustment is carried out for each year between 2006 and 2100.

Clearly, CMIP5-EM grossly underestimates the growth of arid and semi-arid regions over the historical era. This underestimate and its spatial variability were lessened by the modified CMIP5-EM, making the simulation equivalent to the observations. Regarding future projections, the original CMIP5-EM drops more slowly and achieves values of 0.72 and 0.75, but the corrected mean AI decreases monotonically and reaches values of 0.67 and 0.72 in 2100 under Representative

Concentration Pathway 8.5 (RCP8.5) and RCP4.5, respectively. In accordance with the definition of arid and semi-arid regions, the corrected area coverage in 2100 under RCP8.5 is 56% of the world's land area, which is 23% bigger than the observed climatology (1961–1990). Dry subhumid, semi-arid, arid, and hyper-arid dryland subtypes, respectively, make up 8.3%, 20.3%, 14.9%, and 12.6% of the world's land area. The comparable subtypes under RCP4.5 account for 8.9%, 19.0%, 14.4%, and 8.4% of the world's land area, for a total coverage of 50%. The growth of drylands is most pronounced in semi-arid regions, which account for more than one-third of total drylands and nearly half of total dryland expansions in both RCP8.5 and RCP4.5.

To better understand the temporal evolution of the arid and semi-arid regions, the future era is split into three 30-year intervals. The expansion of drylands is not noticeable in the early 21st century (Fig. 7.16(a)), and the distribution is dispersed; new semi-arid and dry subhumid areas typically develop around the edges of the historical drylands, while more arid and hyper-arid regions develop in Alaska, NW Canada, and Eastern Siberia. The subtype alterations become increasingly important in the middle of the 21st century (Fig. 7.16(b)). Eastern Siberia, Northeastern China, Western Asia, Central Africa, and Canada are the primary regions that experience drying, while the western half of the continents experiences a majority of the continents' wetter regions. There is little doubt that the drier subtypes cover a bigger area than the wetter subtypes. According to Fig. 7.16(c), which shows the spread of drier regions in the late 21st century, drylands now encompass nearly all of the continents of Africa, Eurasia, and Western North America between 30°N–60°N and 15°S–50°S. As a result, drylands now make up a majority of the world's land surface. The aridity variations over land have not simply followed the "dry gets drier, wet gets wetter" paradigm, according to earlier studies (Greve *et al.*, 2014). These results are in agreement with those studies. According to Fig. 7.16, emerging nations account for 78% of the increasing drylands, but the area transitions to

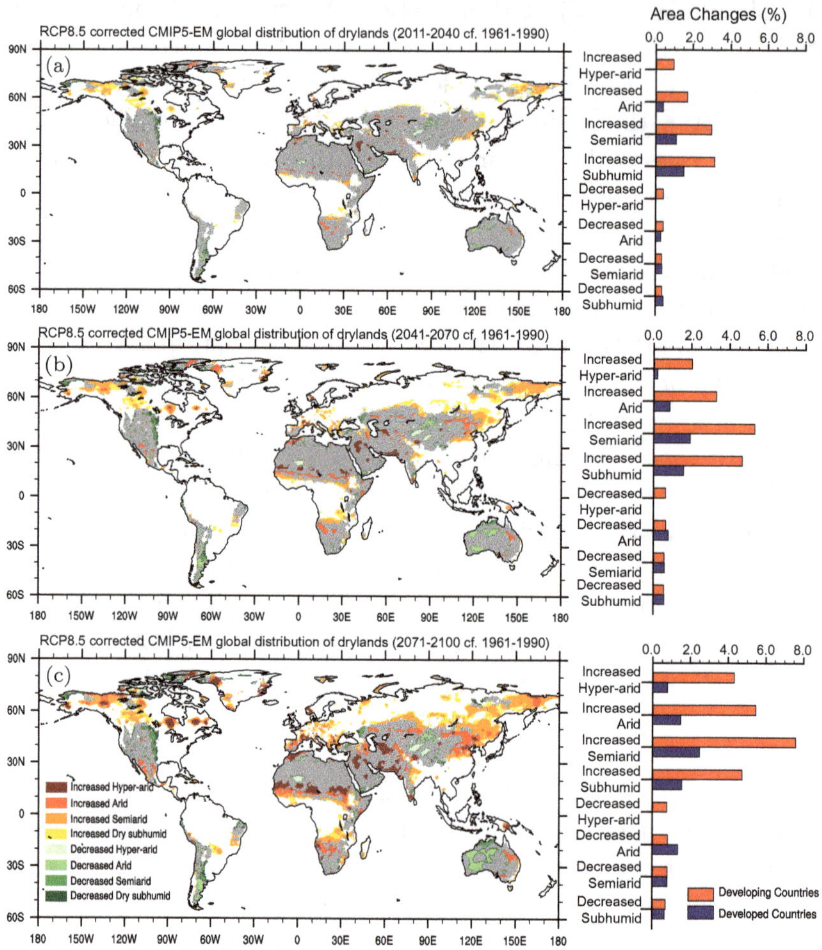

Fig. 7.16. Global distribution of future changes in the dryland subtypes. Projections of subtype changes from the corrected CMIP5-EM and RCP8.5 are shown relative to the baseline period (1961–1990) for 2011–2040 (a), 2041–2070 (b), and 2071–2100 (c) with the corresponding area changes (unit: percentage of global land area) in developing and developed countries. The gray shading denotes the baseline arid and semi-arid regions in 1961–1990. Changes include any transition between adjacent and non-adjacent subtypes. For example, the "increased" category means the indicated regions transitioned from any of the subtypes, for example, semi-arid, dry subhumid, and humid, to a drier subtype; the "decreased" category refers to transitions from drier to wetter subtypes (Huang *et al.*, 2016b).

wetter subtypes are significantly less frequent, and the percentages between developing and developed countries are equal. These findings suggest that the survival environment in underdeveloped countries will be more fragile.

7.3.3 *Differences in impacts of GHG emissions on area expansion*

To better understand the changes in various places, we classified drylands into four categories. Figures 7.14(a–d) displays the deviations from the baseline in the rate of precipitation and PET in drylands and the four subtype regions in 1.5°C and 2°C warmer worlds. In static and transitory 1.5°C (2°C) warmer worlds, the changes in PET for drylands are comparable (Figs. 7.14(b) and 7.14(d)). In global drylands, precipitation and PET rise, although there are regional variations. The changes in precipitation in transiently warmer worlds are noticeably higher than those in stabilized warmer worlds, and there are significant disparities between the changes in precipitation over hyper-arid regions in stabilized and transiently 1.5°C (2°C) warmer worlds (Figs. 7.14(a) and 7.14(c)). Global drylands or precipitation changes are comparable in the transitory warming and stabilized warming scenarios (Figs. 7.14(a) and 7.14(c)). For stabilized 1.5°C (2°C) warmer worlds, the precipitation change is greater in drier regions than in wetter drylands (Figs. 7.14(a), 7.14(c) and 7.14(e)), but the PET change is greater in wetter drylands than in drier regions (Figs. 7.14(b), 7.14(d) and 7.14(f)), which may mean that drier regions will become wetter and wetter regions will become drier, and the differences in drought across global drylands will diminish.

The expansion of drylands could cause desertification, endangering the ecosystems and people who live there. According to the RCP45 and RCP85 scenarios, the global drylands will increase by 11% and 23%, respectively, at the end of the century (Feng and Fu, 2013; Huang *et al.*, 2016b), and the increase in drylands in a stabilized 2°C warmer world is double that of a stabilized 1.5°C warmer world. In this section, we examined the changes in drylands due to 1.5°C or 2°C global warming under four scenarios, including stabilized (LW1.5deg and LW2deg) and transient (RCP45 and RCP85) scenarios (Fig. 7.17(a)). The area coverage of global drylands continuously increases from the baseline to the transient period of 1.5°C (2°C) warming. Drylands' area coverage variations under the RCP45 and RCP85 scenarios are similar when global warming hits 1.5°C (2°C). In stabilized conditions, there is less growth in dryland area than in transitory ones. This finding indicates that, in comparison to the transient warming scenario, the stabilized low-warming scenarios can successfully stop the future growth of drylands by reducing greenhouse gas emissions (RCP85).

The temporal change of area coverage for the four dry subtype regions (Figs. 7.17(b–e)) reveals distinct variances. The areas of dry subhumid and

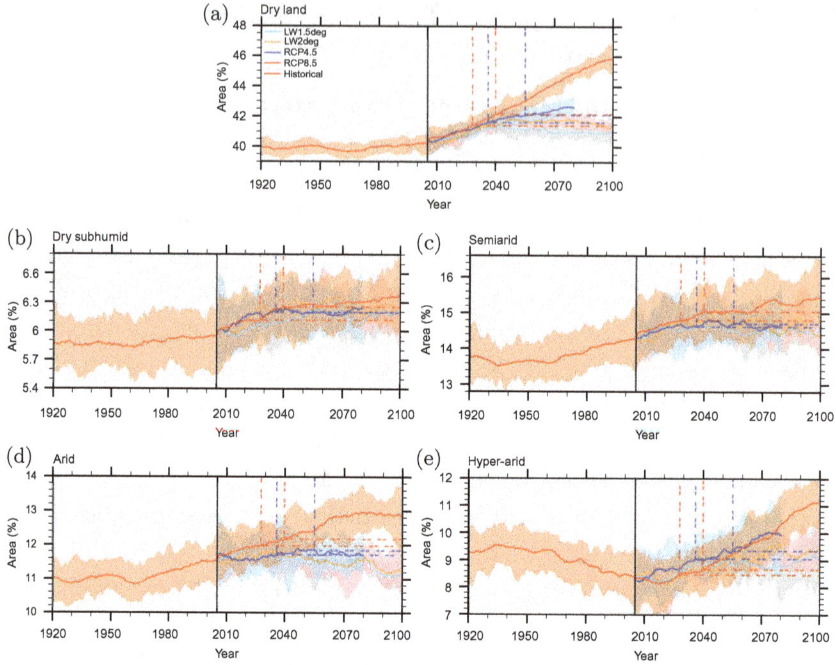

Fig. 7.17. Temporal variation in areal coverage for different land cover types. Areal coverage (percentage of global land area) is shown for total drylands (a), dry subhumid regions (b), semi-arid regions (c), arid regions (d), and hyper-arid regions (e). Eleven-year smoothing was applied. Solid colored lines show the multi-model ensemble mean, and shading denotes the 5th to 95th percentile for the ensemble members. Dotted lines show the timing of 1.5°C (2°C) warming over the globe and the corresponding changes in the multi-model mean (Wei *et al.*, 2019).

hyper-arid regions will increase under the stabilized 1.5°C and 2°C scenarios, the areas of semi-arid regions will initially increase and then decline, and the areas of arid regions will decrease. All four subtypes of drylands, however, will greatly expand under the RCP85 scenario (Figs. 7.17(b–e)). Numerous studies have connected the poleward extension of the Hadley circulation to the subtropical expansion of drylands (Lau and Kim, 2015). The dry, subhumid, and semi-arid regions are mainly found in the subtropical regions, where the Hadley Cell expansion could lead to the expansion of drylands, according to the spatial distribution map of the various dryland classes. The largest P/PET decrease (increase) (Fig. 7.15) occurs in semi-arid (hyper-arid) regions, which may be reduced (increased) to between 0.05 and 0.2, which will be classified as arid regions. This may be the cause of the expansion of arid regions. Since they are mostly found in the tropics, the hyper-arid regions would not be impacted by the Hadley Cell extension.

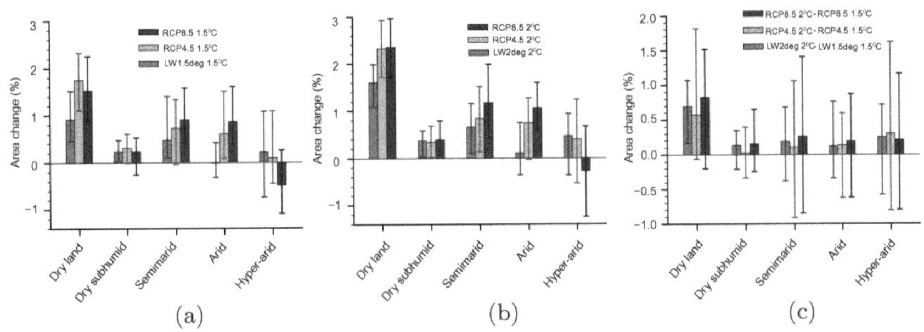

Fig. 7.18. Comparison of areal coverage (percent of global land area) changes for the total drylands and four dry subtype regions between a 1.5° C (a) and a 2° C (b) warmer world and the 1961–1990 baseline period. The difference in area coverage (percent of global land area) of drylands and four subtype regions between a 1.5° C and a 2° C warmer world (c). The error bars denote the 5th to 95th percentiles for the ensemble members (Wei *et al.*, 2019).

The area coverage of dryland subtypes other than arid will grow relative to the baseline in a stabilized 1.5°C warmer world, and the area coverage of the total drylands will increase by about 1% of the global land area (Fig. 7.18). The four dryland subtypes' area coverage in a stabilized, 2°C warmer world is comparable to that in a 1.5°C warmer world. The only alteration is the expansion of drylands and their four subtypes into a greater area (Fig. 7.18). In a stabilized 2°C warmer world compared to a 1.5°C warmer future, the area coverage of drylands will expand by 0.7% more, with hyper-arid areas increasing the most and arid regions the least (Fig. 7.18). Comparing the area coverage of drylands to the stabilized and transient 1.5°C (2°C) warmer worlds reveals a significant difference. Under a temporary warmer world, the dryland expansion from the baseline to a 1.5°C warmer world is more important, particularly for arid regions. Under a transient warmer world, the entire dryland area might fluctuate by a factor of two greater than under a stabilized warmer world. Additionally, the RCP85 scenario has a greater impact on the area covered by semi-arid and arid drylands than the RCP45 scenario (Figs. 7.18(a) and 7.18(b)). The expansion of drylands can be effectively resisted by stabilized warming scenarios, particularly in arid regions.

7.3.4 *Impacts of expansion of arid and semi-arid regions*

The inhabitants and ecosystems of arid and semi-arid regions may be threatened by desertification, which could result from the expansion of arid and semi-arid regions. Desertification is a growing concern for arid and semi-arid regions, although assessments to inform management and policy solutions are constrained by uncertainty surrounding the term's definition and the processes involved. Kassas (1995) distinguished between desertification and drought, a natural danger (degradation of land resources). Desertification, which includes both biophysical and social elements, is defined as "land deterioration in arid, semi-arid, and dry sub-humid areas resulting from many variables, including climate fluctuations and human activities" (Reynolds *et al.*, 2007). The loss or decline of arid and semi-arid regions' biological or economic productivity is referred to as land degradation. It is due to a persistent inability to maintain a balance between the supply and demand for ecosystem services in arid and semi-arid regions.

Improving global food production and managing ecosystems sustainably in the face of climate change require an assessment of the state of desertification today. It was estimated that 78.5 million people lived in areas that were directly impacted by recent severe desertification at the time of UNCOD (United Nations Conference on Desertification, 1977). The Gridded Population of the World, version 3 (GPWv3), which can be found at http://sedac.ciesin.columbia.edu/gpw/index.jsp, provided the population estimates. Population values for 2000 are assigned to grid cells with a geographic resolution of $0.5° \times 0.5°$ using a proportional allocation gridding technique that makes use of more than 300,000 national and subnational administrative units. The Country-level Population and Downscaled Projections Based on the Special Report on Emissions Scenarios (SRES), B2 Scenario (1990–2100), data collection are used to construct the population counts for 2025; the basis map is CIESIN's Gridded Population of the World, version 2 (GPWv2).

The United Nations Plan of Action to Combat Desertification's (PACD) 1984 review verified that land continued to be lost irretrievably to desertification or deteriorated to desert-like conditions at a pace of 6 million ha annually. According to the International Convention to Combat Desertification (CCD) secretariat, desertification poses a threat to about 30% of the Earth's surface and already affects about 70% of its agricultural arid and semi-arid regions, which are home to more than 250 million people in over 110 countries, and is at risk of displacing another 1 billion people, as well as causing potential economic losses in the neighborhood of $42 billion annually (Winslow *et al.*, 2004). Bestelmeyer *et al.* (2015) advocated that assessments of land transformation and desertification be placed within a state change-land use change (SC-LUC) framework and discussed how the SC-LUC approach might lead to more effective assessments of land transformation and management of arid and semi-arid regions.

According to Warren *et al.* (1996), significant changes are predicted to occur over arid and semi-arid regions in the coming decades. Increased

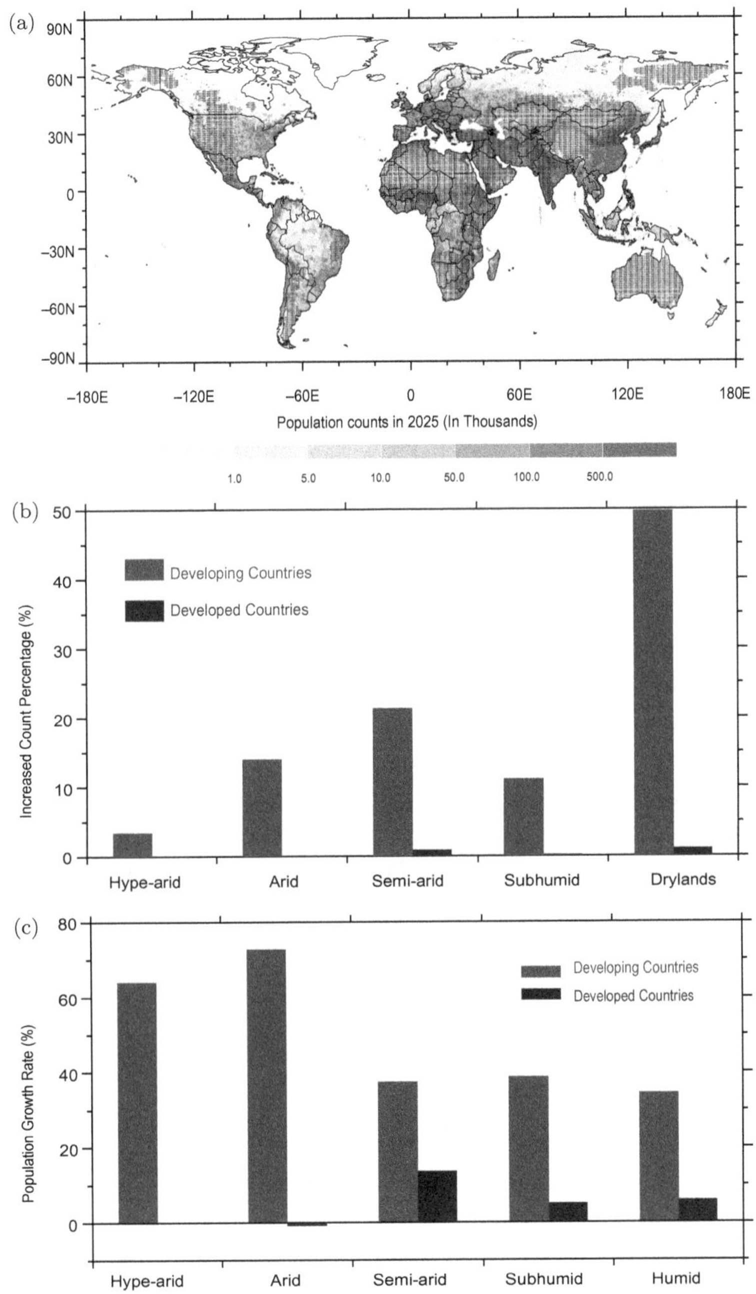

Fig. 7.19. The distribution of the projected population in 2025 overlaid with drylands (AI < 0.65) with corrected CMIP5 and RCP8.5 (gray shaded area) (a); the percentage of the regional population increase to the global increase from 2000 to 2025 for each dryland subtype (b); the regional population growth rate from 2000 to 2025 for each dryland subtype (c) (Huang *et al.*, 2016b).

human demands of all kinds, including population growth in urban and rural regions, ongoing oil development, and an increase in the number of large-scale water infrastructure projects, will cause the most significant changes. There may not be enough land left to cultivate in many rural areas, which could have an effect on the soil and vegetation. The most vulnerable regions of the continent are the Kalahari–Namib region in Southern Africa, the Sudano–Sahelian region, and Mediterranean Africa. Notably, one-third of Africa suffers from desertification, and 73% of all arid and semi-arid regions exploited for agriculture have deteriorating conditions (Darkoh, 1998).

Additionally, due to rapid population growth, future expansion of drylands will affect the ability of additional people to survive. The population estimates for 2025 are depicted in Fig. 7.19(a). Drylands can be found where there are dense populations, such as in Northern China, Western Asia, India, and Eastern Africa. The environment necessary for human survival will be threatened by the escalation of droughts since ecosystems of drylands are delicate. 50% of the increase in world population since 2000 took place in drylands of developing countries, where the proportions of hyper-arid, arid, semi-arid, and subhumid regions to the total land area are, respectively, 4%, 14%, 21%, and 11% (Fig. 7.19(b)). In developed countries' drylands, the rise is only 1%. Since the population bases for the various dryland subtypes vary greatly, another indicator that is utilized is the population growth rates, which are determined by dividing the increase by the corresponding population base in 2000. The forecast population growth for the dryland subtypes in developing countries is 65%, 73%, 38%, and 38%, respectively. These increases are significantly larger than the 35% predicted for humid regions (Fig. 7.19(c)). In hyper-arid and arid regions, the growth rate is significantly higher given the lower population base and population increase.

7.4 Extreme Temperature Events

Analysis of extreme climate change within particular defined global warming targets has greatly benefited from the work of some academics. Keep in mind that most prior research focused on the spatiotemporal alterations of severe climate extremes (e.g., heat waves and floods). Overall, the studies showed that as the world warmed, climate extremes would become more intense, frequent, and long-lasting. Furthermore, when the temperature rises, so will the likelihood of connected dangers. Due to the enormous sea–land thermal variations and the difficult topography, East–Central Asia is a sensitive and vulnerable region (Zhou and Zou, 2010). Additionally, the risk of extreme temperature occurrences and the intensity of any given event rising increase when the surface air temperature rises quickly (as discussed previously). This part primarily introduces the impact of GHG emissions on extreme temperature events with an emphasis on East–Central Asia.

7.4.1 *Extreme cold events*

The Expert Team for Climate Change Detection Monitoring and Indices (ETCCDI) developed the indices used in this section (Sillmann *et al.*, 2013). The daily maximum temperature, daily lowest temperature, daily mean temperature, and daily precipitation are used to calculate these indices. You may find a detailed description of each of these indices at http://etccdi.pacificclimate.org/list_27_indices.shtml. There are five categories in which these 27 indices might be categorized (Alexander *et al.*, 2006). The threshold indices, which represent annual days over a certain threshold [frost days (FD), icing days (ID), summer days (SU), and tropical nights (TR)], as well as the absolute indices, which represent maximum or minimum values within a season or year (TX_X, TN_X, TX_N, and TN_N), could intuitively show future extreme climate features. The diurnal temperature range (DTR) is a significant indicator since it has the potential to significantly affect both human health and crop growth. These indices could make it easier to forecast extreme temperature changes.

Our study area was determined to be within the latitude and longitude ranges of 10°N–55°N and 60°E–150°E, encompassing East and Central Asia. It has a big population and a variety of climate zones, making it particularly vulnerable to severe weather disasters. Research on extreme climate change under 1.5°C and 2°C warming targets can be done in the arid and semi-arid areas (ASAs), which is located east of the Afro-Asian arid zone and close to the East Asian monsoon region.

Along with geographical plots and line charts, box plots and probability distribution functions (PDFs) were employed to visualize the research findings. Using 10%, 25%, 50%, 75%, and 90% level presentations, box plots might provide more details regarding extreme climatic indices, while PDFs could more precisely characterize the data distribution.

In addition, the probability ratios PR and FAR, with definitions of PR=P1/P0 and FAR $= 1 -$ P0/P1, were employed for attributions. P0 denotes the historical period's extreme temperature indices, and P1 denotes the same under 1.5°C and 2°C conditions. Both of these metrics could be used to identify the contributions caused by changes in external forcing.

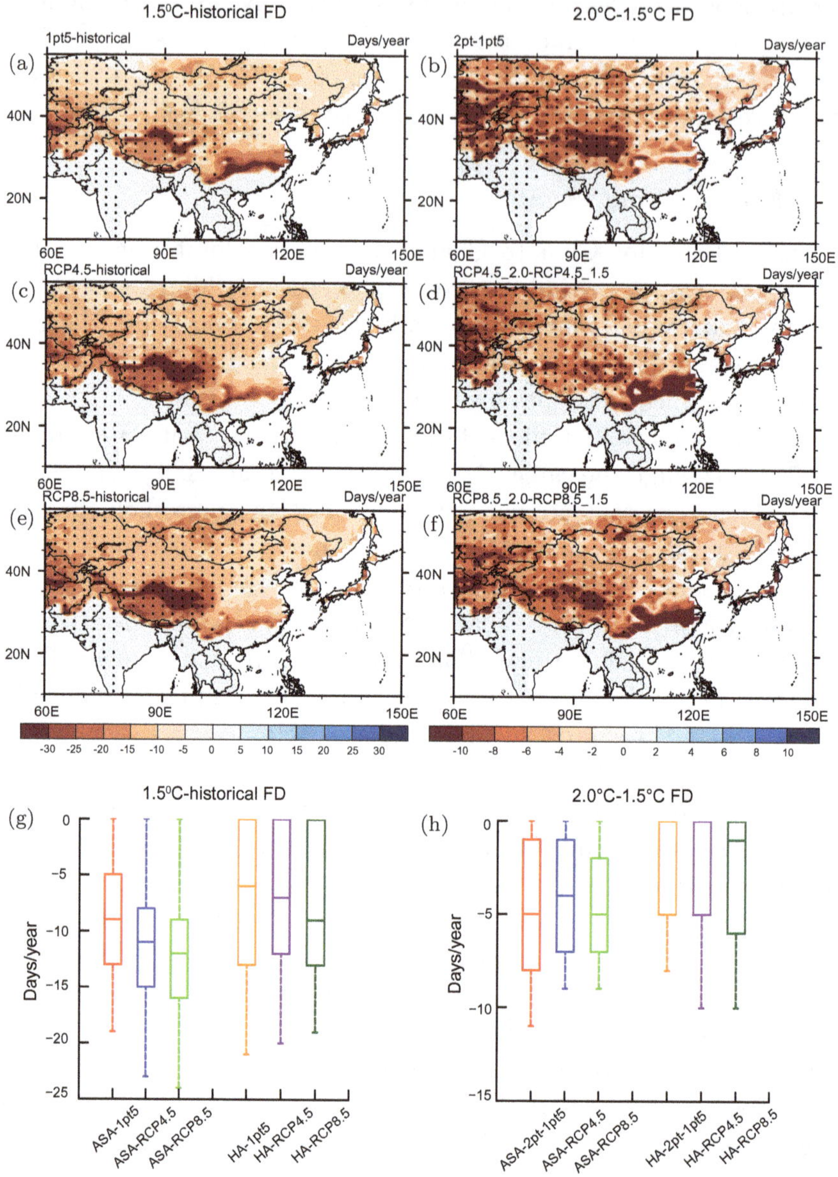

Fig. 7.20. Spatial changes in FD in the 1.5°C scenario relative to the historical period (a, c, e) and the 2°C scenario relative to the 1.5°C scenario (b, d, f). For the box plots, subplot (g) refers to the 1.5°C scenario relative to the historical plots. Subplot (h) refers to the 2°C scenario relative to the 1.5°C scenario. The box whisker plots describe the 10th, 25th, 50th, 75th, and 90th intervals. Red represents LWR in ASA, blue represents the RCP4.5 in ASA, green represents the RCP8.5 in ASA, dark orange represents LWR in HA, purple represents the RCP4.5 in HA, and dark green represents the RCP8.5 in HA (Zhang *et al.*, 2019).

As the name implies, there are two cold indices: FD denotes the number of days per year when there is conditional frost on the ground and ID denotes the number of days per year when there is conditional ice on the ground. No matter the situation, Fig. 7.20 demonstrates that declining areas are essentially present north of 25°N, where the ASA (Figs. 7.20(a–f)) predominates. It is important to remember that the zone between 30°N and 40°N would have the highest value. For instance, all simulation results show a quick decline in the middle and lower parts of the Yangtze River and the Tibetan Plateau, whereas the southern regions remain steady because their temperature rarely falls below 0°C. The ASA median change in FD for the box plots under the 1.5°C scenario relative to the historical era would be close to 10 days/year, which is 2–3 days greater than the humid areas' (HAs) change (Fig. 7.20(g)). Similarly, the difference between the 1.5°C and 2°C scenarios would be around 5 days every year, which is 2–3 days more than the HA change (Fig. 7.20). However, due to the Yangtze River basin's contributions, the amplitude of the HA variation in the 90th interval would be quite similar to that of the ASA.

Similar to the FD, the boreal regions, which are primarily part of the ASA, would similarly exhibit the spatial distribution of high-value zones in ID. Due to its tougher temperature requirements, changes in the ID take place further north than those in the FD. Given its high altitude and extreme cold, the Tibetan Plateau and its surroundings would be an intriguing area to study. All of the subplots indicate that the wet–dry border would overlap with the southern boundary of the ID, indicating that the demarcation line might be crucial to ID alteration. When comparing 2°C–1.5°C, the magnitude of ID changes under LWR is greater than those under RCPs. The box plots show that the ASA has a higher value than the HA regardless of the median or 90th interval.

7.4.2 *Extreme warm events*

Extreme temperature variations during the cold season may be represented by changes in the FD and ID. Similarly, the warm-season circumstances might be represented by the SU and TR. In contrast to the latitudinal trend depicted in Figs. 7.20 and 7.21 shows a more complicated distribution. The HA would make up a sizable portion of this increase, which would mostly affect South Asia and the Eastern and Central Areas of China. These regions are regarded as conventional hot spots. In contrast, the absence of SU days would lead the expansive ASA, such as the Tibetan Plateau and Northern Mongolian Plateau, to change significantly and even have a lot of open space

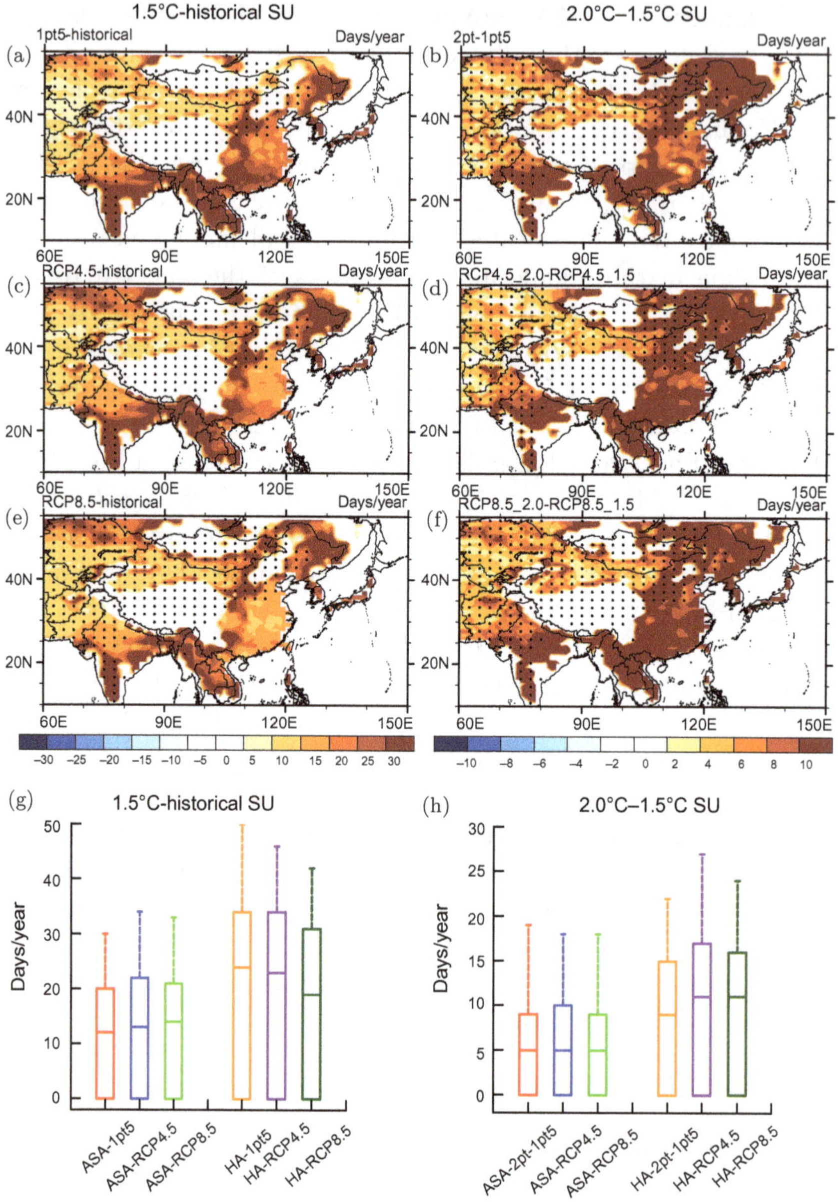

Fig. 7.21. Same as Fig. 7.20, but for SU (Zhang *et al.*, 2019).

(Fig. 7.21(a–f)). These inconsistencies correspond to those in the box plots as well. In the 1.5°C scenario relative to the historical period (Fig. 7.21(g)), the median changes in the SU in the ASA would be about 10 days per year, and they would be cut in half in a world that warmed by an additional 0.5°C.

All simulation findings indicate that the increment in the HA is nearly two times greater than that in the ASA (Fig. 7.21(h)).

The magnitude of changes in TR is less pronounced than those in SU, which is similar to the connection between the FD and ID (Figure not shown). In particular, the growth would stop and remain stable North of 50°N (particularly in Northeast China). The box plots showed considerable drops as a result of improved inclusion index criteria. According to the median in the box plots, the change in TR would drastically decrease by 10 days in HA and 5 days in ASA when compared to SU (Figure not shown). Similarly, under the additional 0.5°C warming, there would be a slight increase (less than 5 days), but it is not immediately apparent (figure not shown).

7.4.3 *Extreme temperature events*

As a result, under 1.5°C and 2°C warming targets, we could describe the changes in the chosen extreme temperature indices. Figure 7.22 provides an explanation of the temporal fluctuations in the LWR from 1971 to 2100. In both the ASA and HA, the FD exhibits a gradual decline of about 15% in 2100 compared to the original (Fig. 7.22(a)). The fundamental FD days between the ASA and HA also have a significant gap. For most features, the ID change is comparable to the FD change, and the intensity difference is minimal (Fig. 7.22(b)). The inflection periods in the two warm indices are essentially identical to those in the cold, reaching a peak about 2040 and then remaining steady (Figs. 7.22(c) and 7.22(d)).

Figure 7.22 illustrates the strong relationships between the threshold extreme temperature indices and the mean temperature rise. To offer confirmation, attributions and detections are also required. To measure the warming effect on the occurrence of extreme cold or warm temperature indices, we employed the FAR and PR. As was previously determined, the cold indices are adversely correlated with the rise in mean temperature. PR in both the FD and ID drops from about 1 in 2006 to roughly 0.9 in 2100, a 10% reduction (Figs. 7.23(a) and 7.23(b)). Due to threshold variations, the HA and ID in the ASA see the highest change in FD. The same downward trend in the FAR suggests that the frequency of cold index readings will decline in the future (Figs. 7.23(e) and 7.23(f)). The two warm indices reach a plateau and undergo a mild increase (0.15–0.2) in different directions (Figs. 7.23(c), 7.23(d), 7.23(g) and 7.23(h)). Extreme temperature indices will be significantly impacted by global warming in the future.

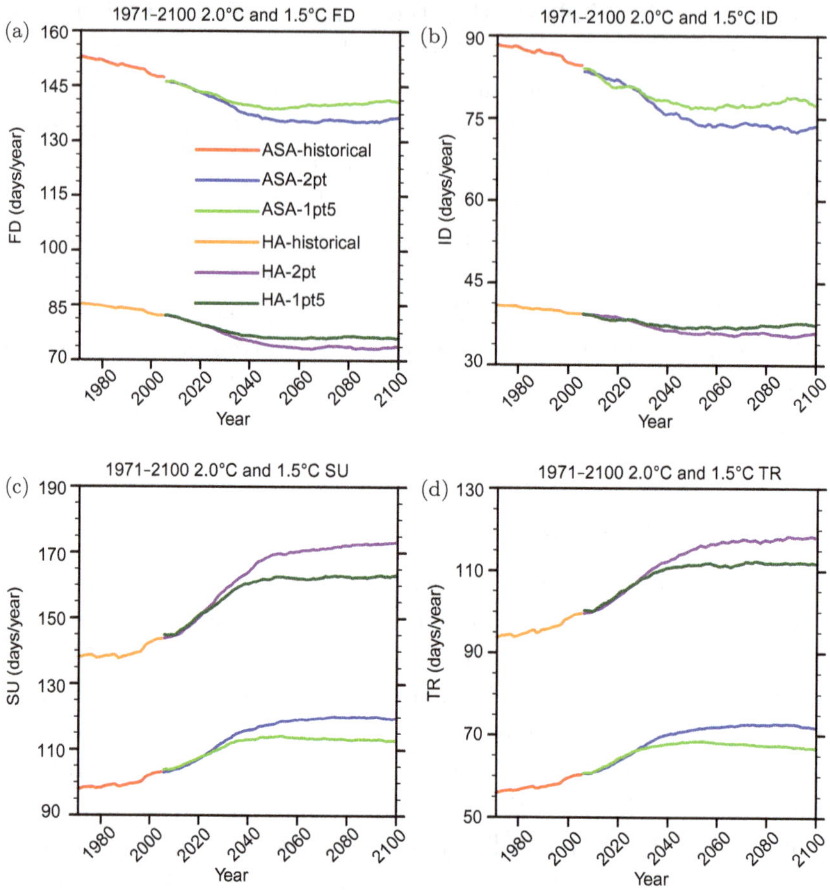

Fig. 7.22. Temporal changes during 1971–2100 for FD (a), ID (b), SU (c), and TR (d). Red lines represent the historical period in ASA, blue lines represent 2 pt in ASA, green lines represent 1 pt5 in ASA, the dark orange lines represent the historical period in HA, purple lines represent 2 pt in HA, and the dark green lines represent 1 pt5 in HA (Zhang *et al.*, 2019).

7.4.4 *Impacts of GHG emissions on extreme events*

The FD, ID, SU, and TR threshold extreme climatic indices correspond to the frequency of extreme temperature events, while TX_X, TX_N, TN_X, and TN_N could represent the intensities (Alexander *et al.*, 2006). TX_X in the winter would have the biggest amplitude and peak at 1.5°–1.8° in December for a scenario of 1.5°C warming relative to the historical period. It is anticipated that the ASA will see a summertime temperature increase that is around 0.2°C–0.3°C greater than the HA. The differences between the ASA and HA are less noticeable in the other three seasons than they are in the summer.

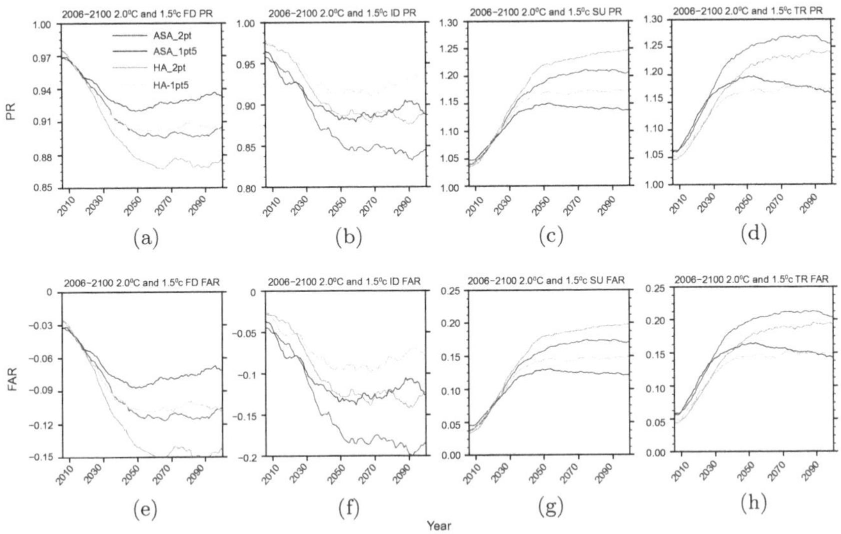

Fig. 7.23. FD (a, e), ID (b, f), SU (c, g), TR PR (d, h), and FAR values for the 2°C scenario and the 1.5°C scenario relative to the historical period in both ASA and HA (Zhang *et al.*, 2019).

However, the condition changes when there is a 0.5°C difference. In nearly all experiments, the annual temperature exceeds 0.5°C (figure not shown). The PDF graph might provide a comprehensive view of the temperature ranges, which include additional information about climate change (Fischer and Knutti, 2014). The median value in the two 1.5°C scenarios, compared to the historical scenario, is similar to 1.5°C in the ASA experiments, which generally reach a higher degree than the HA experiments (Figs. 7.24(a) and 7.24(i)). On the other hand, both extra 0.5°C scenarios depict graphs in the average TX_X that are only slightly different from the 0.5°C global mean climb (Figs. 7.24(e) and 7.24(m)). Additionally, compared to summer, the TX_X increase in winter is greater. With a less-than-average value under 1.5° warming limits and a more-than-average value under 2.0°C warming limits, TX_N is comparable to TX_X in terms of concrete data variances. Therefore, variations in the maximum temperature may follow similar trends. Similar to monthly adjustments, the consistency of the PDFs in the TX_N is excellent. The distributions shown in all four figures (Figs. 7.24(b), 7.24(f), 7.24(j) and 7.24(n)) are comparable. Less space is devoted to the HA experiments than to the ASA experiments. The distribution forms for PDFs are nearly identical to those for TX_X. The difference in TN_X between the ASA and HA is depicted in Fig. 7.24(c), with median values of roughly 1.2 (HA) and 1.7°C (ASA). The greatest value of the four indices under the 1.5°C warming target is

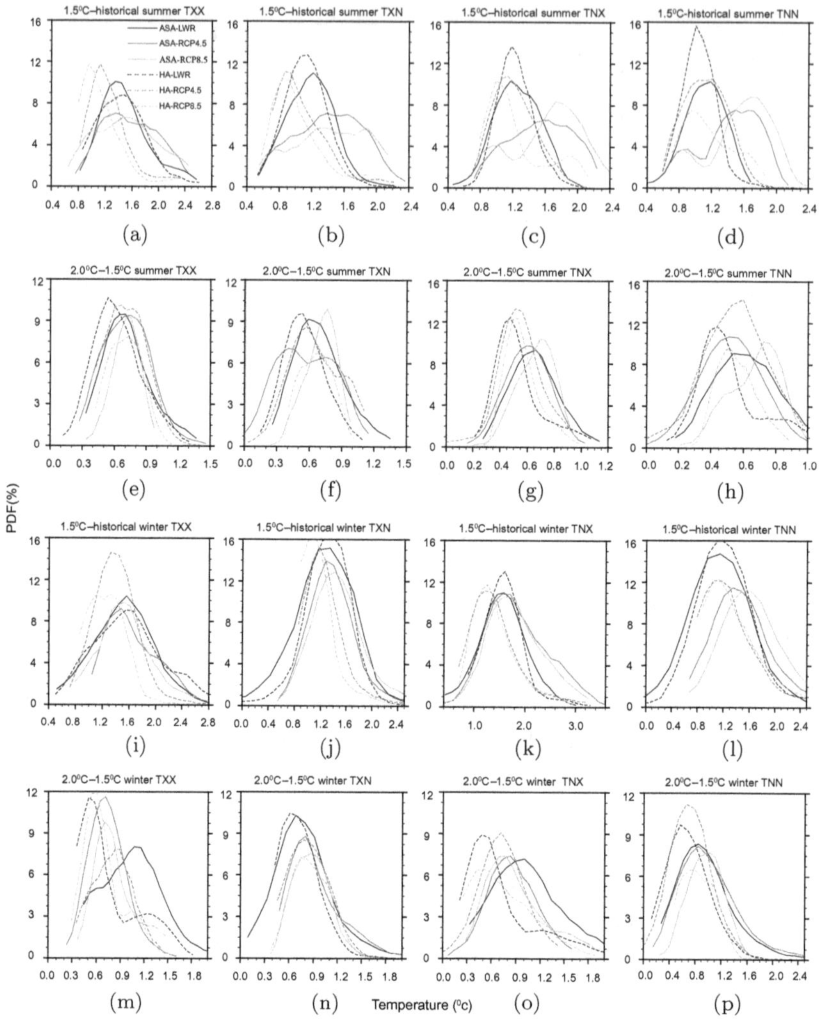

Fig. 7.24. Temperature PDF distribution about TX_X (a, e, i, m), TX_N (b, f, j, n), TN_X (c, g, k, o), and TN_N (d, h, l, p) in both summer and winter under the 1.5°C scenario relative to the historical period and the 2°C scenario relative to the 1.5°C scenario (Zhang *et al.*, 2019).

found in the right-tail distribution of the PDFs in TN_X in Fig. 7.24(k), which exceeds 3.0°C. The average value has not changed significantly, though. When compared to 1.5°C in the summer, 2°C is warmer, and the growth rate of TN_X returns to normal (Fig. 7.24(g)). Compared to summertime, TN_X has a little edge in wintertime conditions (Fig. 7.24(o)). Recent research has shown that, particularly in the mid- and high-latitude regions of the Northern Hemisphere, the minimum temperature increases at a faster rate than the maximum temperature (Huang *et al.*, 2012; Guan *et al.*, 2015). This assertion may have some support from Fig. 7.24(p). In some sensitive places, the extreme value rises by an extra 0.5°C throughout the winter, to about 2°C–2.5°C. In conclusion, it is extremely concerning that extreme temperatures, particularly the minimum temperature, are increasing so quickly.

Under future climate change, the DTR might display the relative changes in the maximum temperature and minimum temperature. The health of the flora and animals also depends on it. For instance, high daytime temperatures encourage the production of organic matter through vegetation photosynthesis, whereas low nighttime temperatures may decrease vegetation respiratory loss and accelerate the storage of organic matter. As a result, plants may grow and develop more easily the larger the DTR is in general. The DTR in Fig. 7.25 indicates seasonal mean differences as opposed to monthly

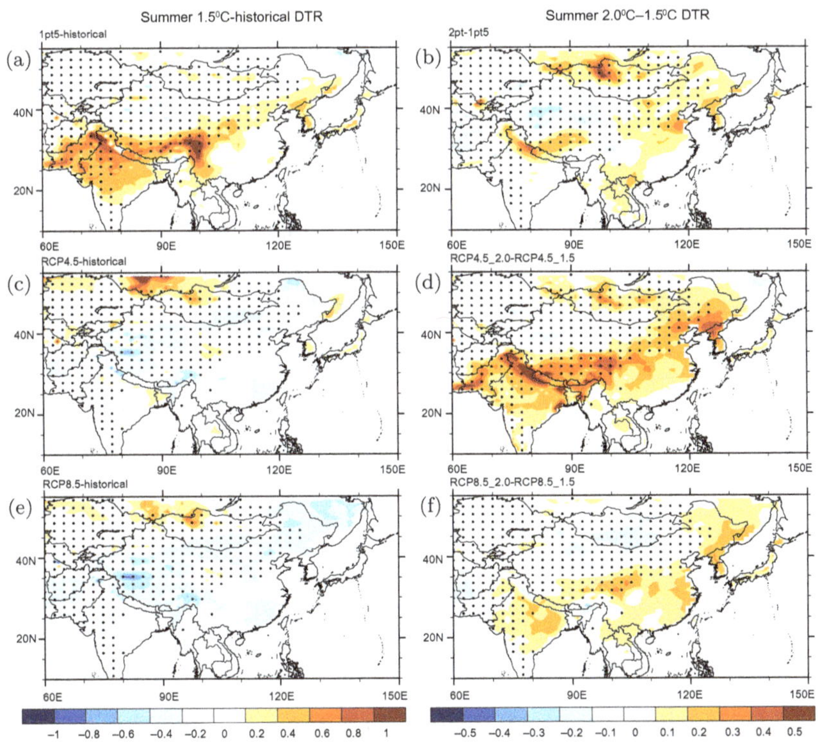

Fig. 7.25. Spatial changes of the DTR in the 1.5°C scenario relative to historical (1971–2000) for 1 pt5 (a), RCP4.5 (c), and RCP8.5 (e). 2°C scenario relative to the 1.5°C scenario for 2 pt (b), RCP4.5 (d), and RCP8.5 (f). The dotted parts represent ASA (Zhang *et al.*, 2019).

mean differences between maximum and minimum temperatures. Regarding the historical era, there are sizable positive regions in Northern India, the Tibetan Plateau, and a Northeastern Plain extension of the Loess Plateau (Fig. 7.25(a)). The three extra 0.5°C conditions make these zones even more evident. On the Northern Mongolian Plateau, the DTR exhibits either stability and steadiness or a small decline (Figs. 7.25(b), 7.25(d) and 7.25(f)). However, under the 1.5°C warming limits in RCP4.5 and RCP8.5, the entire research area experiences a modest reduction (Figs. 7.25(c) and 7.25(e)). In Fig. 7.25, the distributions are entirely different. The transition zone between the ASA and HA is marked by a significant gap in every subplot. In most of the ASA and most of the HA, particularly in Southeast China, the DTR decreases. In the 1.5°C scenarios, the increase is roughly 0.6°C–1.0°C in comparison to the historical period (Figs. 7.25(a–c)) in the HA. There is an increase of 0.3°C–0.6°C for the additional 0.5°C of warming (Figs. 7.25(d), 7.25(e) and 7.25(f)). Additionally, an average drop of 0.3°C–0.6°C was seen in the ASA. According to further studies, the maximum temperature has a greater impact in the summer, and the minimum temperature has the greatest influence in the winter (figures not shown). The most likely explanation is that, throughout the summer, as the number of clouds decreases, solar energy is reflected less and has a greater positive impact on maximum temperature than minimum temperature in the dry–wet transition zone, increasing DTR. In comparison to summer, the association between cloud coverage and DTR is less in the winter (Liu *et al.*, 2018). Most East–Central Asian regions have a drop in DTR as the lowest temperature rises. In Southeast China, the maximum temperature, however, predominates. And DTR indicates growth in certain areas.

7.5 Oxygen Cycle

7.5.1 *Ocean deoxygenation*

The current volume fraction of O_2 in the modern atmosphere has essentially stabilized at 20.95% after a billion years of atmospheric O_2 evolution. The precise measurement of atmospheric O_2 at the part-per-million scale is still difficult, despite the fast global improvement of atmospheric CO_2 measurements during the past few decades. It is difficult to detect relative changes in atmospheric O_2 since the background concentration of O_2 is much higher than that of CO_2 (210,000 ppm for O_2 and 400 ppm for CO_2). As a result, only a small number of organizations worldwide, such as the Scripps Atmospheric Oxygen Program (http://scrippso2.ucsd.edu/), are engaged in

O_2 observations. O_2 levels have steadily decreased (by about 4 ppm yr^{-1}) according to high-precision observations since the late 1980s because of an imbalance between production and consumption fluxes (Fig. 7.26(a)).

Ocean deoxygenation has posed a substantial danger to marine ecosystem productivity and biodiversity since the middle of the 20th century, despite the fact that significant atmospheric O_2 depletion is unlikely to occur in the near future (Fig. 7.26(b)). There has been a fourfold increase in oxygen-depleted waterways during the previous 50 years. 40% of the oxygen in some maritime regions has been depleted (Schmidtko *et al.*, 2017). Physical and biogeochemical processes work together to control the amount of oxygen in the ocean. Oceans absorb oxygen through photosynthesis by autotrophic organisms and through O_2 dissolution from the atmosphere in undersaturated waters. Oceans lose oxygen through oxygen outgassing in oversaturated zones, aerobic organism respiration, and the oxidation of reduced compounds. The warming of ocean waters and eutrophication, two human-caused factors, are primarily blamed for the diminishing oxygen levels in the ocean. This trend is thought to be mostly driven by ocean warming, both directly (solubility effect) and indirectly (changes in circulation, mixing, and respiration). A significant amount of oxygen is used as a result of eutrophication in coastal areas brought on by rising human inputs of nutrients (nitrogen and phosphorus) from land runoff (Diaz and Rosenberg, 2008). The ocean is currently losing oxygen and has even turned into a net source of atmospheric O_2 despite the little marine oxygen store (about 0.6% of that in the atmosphere).

Li *et al.* (2020) identified the ocean oxygen budget using CMIP5 models, highlighting the significant contribution of sea surface oxygen outgassing to marine deoxygenation. Deoxygenation can have an impact on the physiology, behavior, and ecology of marine creatures as well as various elements of marine ecosystem services. If marine invertebrates are exposed to decreased oxygen supply, their vision may suffer noticeably (by 60%–100%) (McCormick *et al.*, 2019). Numerous important zooplankton species that dwell close to their physiological limitations can be directly affected by minute changes in ocean oxygen (Wishner *et al.*, 2018).

Deoxygenation can have a severe impact on a variety of fish species' traits, such as growth, survival, and reproduction. This can lead to changes in population distribution and the food chain. The future is expected to see a worsening of ocean deoxygenation due to the effects of global warming and increased fertilizer inputs to coastal regions. According to CMIP5 models, deoxygenation in the surface and deep oceans is predicted to get worse in the

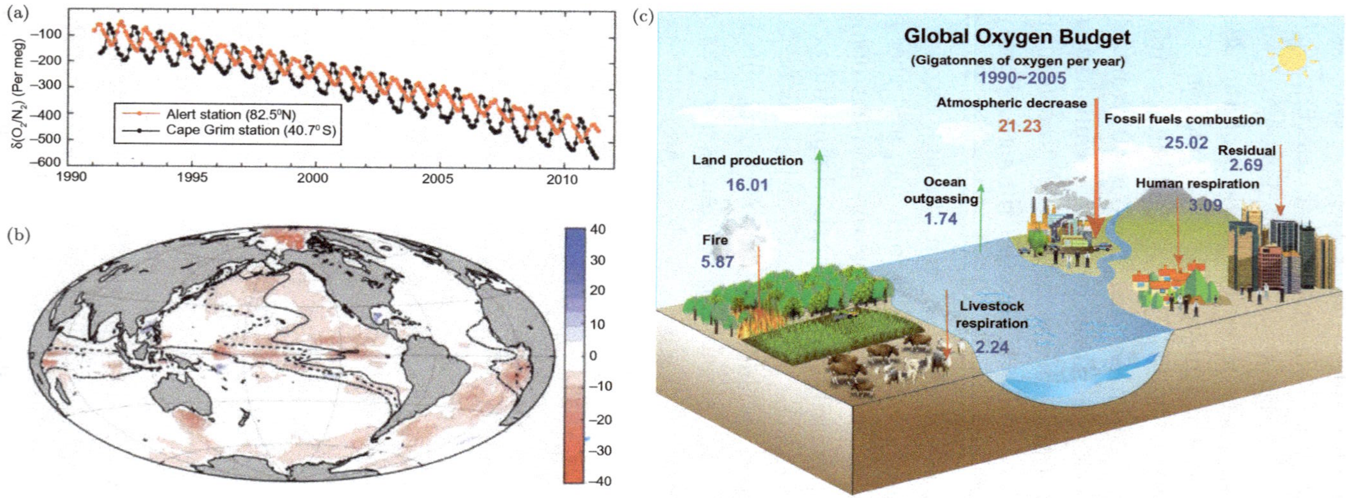

Fig. 7.26. Global oxygen cycle in the modern Earth system. Time series of the O_2 concentration (ppm) at Alert Station and Cape Grim Observatory (a). Changes in the oxygen content of the global ocean in mol O_2 m^{-2} $decade^{-1}$ (Schmidtko *et al.*, 2017) (b). Average global oxygen budget from 1990 to 2005 in 10^{12} mol yr^{-1} (Huang *et al.*, 2018) (c), where the green arrows denote the production from land vegetation and outgassing from oceans and the red arrows represent consumption by fossil fuel combustion, human and livestock respiration, fires, and residual processes (Huang *et al.*, 2021).

21st century under the RCP2.6 (with a drop of 1.81 ± 0.31%) and RCP8.5 (3.45 ± 0.44%) scenarios (Bopp *et al.*, 2013). Despite the danger that O_2 depletion might pose, significant research efforts into the global oxygen cycle have focused on improving the estimation of the carbon budget for the entire planet using the various inorganic interactions of CO_2 and O_2 (carbonate reactions) with seawater. We have only recently begun to understand the contemporary oxygen cycle. To definitively define a contemporary oxygen cycle, the specifics of the fluxes involved in O_2 production and consumption processes must be more constrained.

7.5.2 *Hypoxia related to urbanization*

A crucial gas for Earth systems, oxygen serves as a link between the hydrosphere, biosphere, and atmosphere. Recent research has demonstrated that the oxygen cycle cannot continue in its current form (Huang *et al.*, 2018). The concentration of atmospheric oxygen has been declining at an accelerated rate in recent decades due to increased human activity and rapid industrialization worldwide (Huang *et al.*, 2018; Liu *et al.*, 2020b), and in some densely populated places, the decline may be more than 50%. By assessing the difference between oxygen consumption and production in the city, Yin *et al.* (2014b) and Peng *et al.* (2015) calculated the oxygen balance in Beijing and the Pearl River Delta, China. According to their findings, the oxygen flows in the two major cities were wildly out of equilibrium. Although the possibility of an emerging oxygen crisis in cities around the world has received little attention so far, there are no publicly available observations of the oxygen content in urban regions around the world and no publicly available converted high-resolution global oxygen consumption data. Here, the oxygen production and consumption at global grid points are calculated to suggest an oxygen index (O_I), or the ratio of oxygen production to consumption, which is closely related to vegetation, population growth, economic development, and global warming — all factors crucial for sustainable development. The O_I can quantify human oxygen consumption from nature as a result of intricate interactions between environmental and socioeconomic systems at various spatial and temporal scales.

The following equation can be used to calculate O_I:

$$O_I = \frac{O_C}{O_P} = \frac{O_{\mathrm{CFF}} + O_{\mathrm{CFI}} + O_{\mathrm{CL}} + O_{\mathrm{CH}}}{O_P}, \tag{7.18}$$

in which O_{CFF}, O_{CFI}, O_{CL}, and O_{CH} represent oxygen consumed by fossil fuel, fire, livestock, and human respiration, respectively.

In non-urban locations, flora produces atmospheric oxygen primarily through photosynthesis. In urban areas, human activities like burning fossil fuels and breathing use atmospheric oxygen. Urban areas only make up around 3.8% of the earth's land surface, yet our findings demonstrate that they were responsible for 39.3% of the planet's annual terrestrial oxygen consumption (36.4 ± 1.5 Gt/yr) between 2001 and 2015 (Fig. 7.27(a)). However, oxygen consumption (14.30 ± 0.6 Gt/yr) was far higher than the amount of oxygen produced in urban areas by natural terrestrial ecosystems, which was just 0.02 ± 0.03 Gt/yr (Fig. 7.27(a)). The production of oxygen in non-urban regions must be carried to urban areas to make up for the loss through air circulation in order to maintain the O_2 levels there when an oxygen deficit ($O_I \sim 59$, Fig. 7.27(b)) exists. However, before taking into account supplying oxygen to highly urbanized areas, the value of oxygen generation of 8.0 ± 0.6 Gt/yr (Fig. 7.27(a)) in non-urban areas was not enough to complement their own oxygen consumption (22.1 ± 0.9 Gt/yr).

Urban regions in Asia, Europe, and North America consumed more oxygen per area than those on the other three continents (Fig. 7.27(c)),

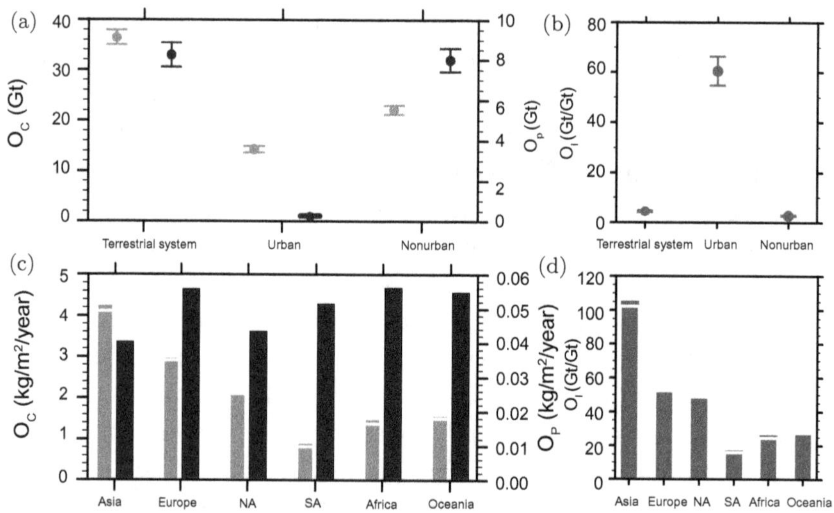

Fig. 7.27. Oxygen balance over urban and non-urban regions. (a, b), Comparison of oxygen consumption and production (a) and index [O_I, (b)) in natural terrestrial systems and urban and non-urban regions. The error bars show the 95% confidence intervals for the average oxygen consumption (O_C), production (O_P), and O_I during 2001–2015. The mean (2001–2015) O_C and O_P in urban areas of 6 continents (c). The average O_I during 2001–2015 in urban areas of 6 continents (d). In (c, d), each bar is divided into two sections by a white horizontal line; the part below the white horizontal line represents fossil fuels, and the part above it represents non-fossil fuels (Wei *et al.*, 2021a).

and these continents also had a worse oxygen deficit than the other three continents (Fig. 7.27(d)). The highest oxygen deficit ($O_I \sim 105$) was found in metropolitan areas of Asia, where oxygen production and consumption were 4.24 kg/m^2/yr and 0.04 kg/m^2/yr, respectively (Fig. 7.27(c)). Despite having the highest oxygen production per unit area among the six continents, Europe's metropolitan regions had the second-highest oxygen consumption per unit area (2.96 kg/m^2/yr), resulting in an O_I of up to 53 (Figs. 7.27(c) and 7.27(d)). Higher O_I values are mostly caused by the burning of fossil fuels in urban areas, which accounted for 97.5% of the total oxygen consumption (contributions varied between 92.7% and 98.4% for different continents, Fig. 7.27(d)).

The O_I in urban areas around the world fell by 4.6 between 2001 and 2015, but it began to climb quickly after 2010. The rise in the burning of urban fossil fuels, which increased O_I by 51.6 and increased oxygen generation, which decreased O_I by 26.5, caused the urban areas of Asia's cities to see an increase in O_I of 25.5. The increased oxygen production in metropolitan areas on all six continents could lower the O_I on each continent between 2001 and 2015. Only metropolitan areas in Europe and North America, where O_I declined by 5.5 and 6.7, respectively, demonstrated a significant (prob. < 0.05) fall in O_I. This was mostly because of the lower oxygen consumption of fossil fuel burning. However, in South America and Africa, this could not be offset by the decrease in O_I caused by the increase in oxygen production, which ultimately led to the weak uptrend of the O_I in urban areas on these two continents. In urban areas on the other three continents of Asia, South America and Africa, the increased oxygen consumption from fossil fuel combustion would increase the O_I in these urban areas. In metropolitan areas over Oceania, the decreased O_I brought on by less oxygen being burned in fires may balance out the increased O_I brought on by more oxygen being burned in fossil fuels.

The problem of oxygen imbalance in some cities may be more significant even though the oxygen balance in urban areas around the world has been badly affected (the oxygen consumption in urban areas was roughly 59 times the area's oxygen output). Wei *et al.* (2021a) also displayed the oxygen balance status for all major cities with more than one million inhabitants. In 2000, 1.3 billion people lived in these cities, or 48% of all people living in urban areas. Large cities in the Middle East and West Asia were typically the only places where a perennial oxygen imbalance occurred. For instance, Shanghai had an urban population of 13.7 million people in 2000, and from 2001 to 2015, its O_I was 2,051.0. There were 26.3 million people living in

Shenzhen, and its O_I was 457.9. Furthermore, the O_I in Beijing was 339.1, which was higher than the O_I of 166.2 estimated by Yin *et al.* (2014). This difference may be explained by the fact that the statistics by Yin *et al.* (2014) covered the months of July to September, which are considered to be the summer in New Hampshire and the relative best time of year for vegetation growth. The O_I in Al-Qahirah, Egypt, reached 2430.6, which is significantly higher than the O_I in New York (161.0) and Los Angeles (240.4).

O_I values are higher in cities with more people. The cities with O_I greater than 50/100 account for 78%/53% of all large cities and 98%/75% of cities with a population of above 5 million people, where 574 million residents living there in 2000 (Wei *et al.*, 2021a). In cities with more than 5 million inhabitants, the O_I had an average value of 228.6 from 2001 to 2015 (Wei *et al.*, 2021a). O_I is influenced by a number of factors, not just the population. The amount of urban vegetation and the energy structure both affect the O_I value. For instance, despite having a higher population than Los Angeles, New York's O_I value was lower, which was caused by the city's higher vegetation coverage, which provided more oxygen. Although Paris and Moscow had similar populations, Moscow's O_I was far larger than Paris's, at 647.7 compared to Paris's 111.2 (Wei *et al.*, 2021a). The primary factor was that Paris had far lower oxygen consumption per unit area than Moscow. Therefore, the issue of a significant oxygen imbalance in cities cannot be resolved by merely controlling urban population growth. The structure and energy use of cities must be optimized in order to enhance oxygen production by urban landscapes and decrease oxygen consumption.

In addition to increasing the use of natural resources, high population density and extensive urbanization can weaken ecosystem stability. In the context of global warming, risks related to exceptionally high temperatures that afflict urban populations are also rapidly rising. Large cities with higher O_I are more likely to experience longer and more frequent heat waves, as well as higher temperatures for their residents. The frequency, intensity, and duration of high-temperature heat waves all rose as the O_I steadily climbed in major cities, indicating that residents of larger cities were more likely to be subjected to severe heat waves (Wei *et al.*, 2021a). The minimum temperature at night and the entire period from 2001 to 2015 support this pattern. Cities with higher O_I have less vegetation cover and produce less oxygen, which cause less latent heat flux in the region and more sensible heat net flow to heat the atmosphere, which in turn cause more frequent high-temperature heat waves in the region.

Similar to how excessive groundwater use will make it challenging to restore groundwater quickly, this could endanger the lives and well-being of future generations. With a rise in O_I, the amount of water storage steadily changed from growing to decreasing. When the O_I reached 500, the water storage fell at a rate of 8.3–9.2 cm/10 years, which aggravated the issue of water shortage in big cities (Wei *et al.*, 2021a). The water storage capacity reduced as the O_I increased beyond 10. The pressure on the environment brought on by human activity in the area has increased, which is the main cause of the increasing rate of decline in water storage in cities with large O_I (containing people, roads, and structures). Consequently, as a result of urbanization in recent decades, human activity intensity has greatly increased in cities. Due to increased carbon emissions that are directly linked to urban economic development, urban oxygen consumption has also increased even though there is a lack of natural water. In the end, these circumstances have caused towns with higher O_I values to extract more water.

Assuming a city with an O_I greater than 1000 and there is no external oxygen transmission, the atmospheric oxygen content below 300 m of the city would drop from 20.9% to 19.5% in just 10 days (volume percentage in the air), which would cause serious health issues for people even though the current level of oxygen on earth is at a relatively healthy level (20.9%), assuming a large city with an O_I over 1,000 and no external oxygen transmission (Stefana *et al.*, 2015) (Fig. 7.28).

The following formulas determine the amount of time needed to reach a hypoxic environment.

Assuming that there is no horizontal advection of air, that the inversion layer is above H (m), and that the air below H can mix vertically, the loss of oxygen (m_o) when the oxygen content drops from the current level (O_1 = 20.946%) to a harmful level (O_2 = 19.5%) was calculated from the following equation:

$$m_O = (P(0) - P(H))/g \times r \times (O_1 - O_2), \tag{7.19}$$

in which $P(H)$ is the atmospheric pressure (Pa) at H; g is the gravitational acceleration (9.8 m/s^2); and r is the ratio of oxygen density to air density ($r = 32/28.959$). Given the oxygen consumption rate O_C (kg/m^2/day) and oxygen production rate O_P (kg/m^2/day), the time required for the oxygen content to decrease from O_1 to O_2 (hereinafter referred to as time) was as follows:

$$\text{Time} = m_O/(O_C - O_P), \tag{7.20}$$

in which the unit of time is a day.

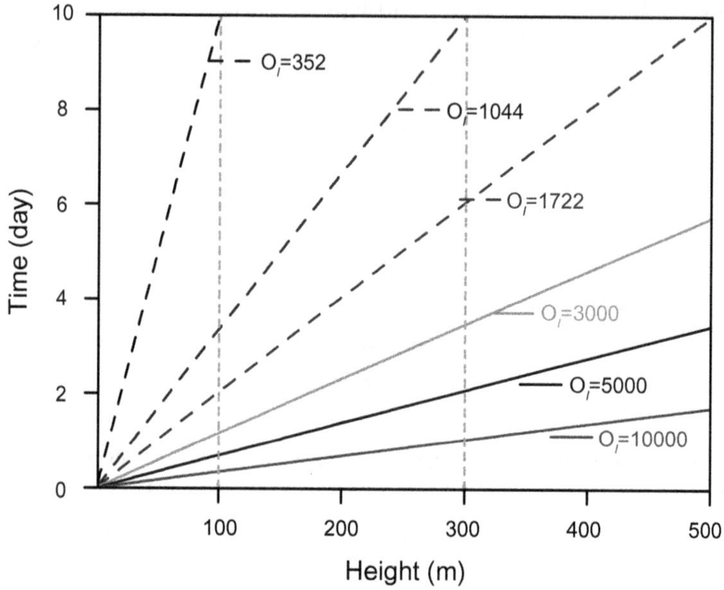

Fig. 7.28. Time required for cities with different O_I values to reach a hypoxic environment. Under the assumption of no horizontal and vertical exchange, the time required for cities with different O_I values to reach a hypoxic environment (the oxygen content is less than 19.5%, which would lead to significant health problems for humans) from the surface to a certain distance (abscissa: height) (Wei *et al.*, 2021a).

The relationship between time and O_I was as follows:

$$\text{Time} = \frac{m_0}{O_p}/(O_I - 1) \quad \text{or} \quad \text{Time} = \frac{m_0}{O_C}/(1 - 1/O_I), \tag{7.21}$$

in which O_I is the oxygen index, which is defined as the ratio of oxygen consumption to oxygen production. Here, the maximum O_P ($O_P = 0.20\,\text{kg/m}^2\text{/yr}$ at the city of Yaounde, Cameroon, Africa) from the 391 large cities was used to estimate the time required for a given change in O_I.

According to Fig. 7.28, when a city's O_I reaches 1722, the range where oxygen content drops to 19.5% of the air volume in 10 days would extend to a height of 500 m, which is roughly the height of the atmospheric mixing layer during periods of severe haze, which are occurring more frequently as the climate warms and the global atmospheric circulation changes. The long-term hypoxic environment would cause immediate and severe harm to the urban population if humans continued to unrestrainedly expand the size of cities and raise the density of urban people and buildings. Additionally, because of a lack of oxygen, urban dwellers may experience extreme heat waves and water shortages more frequently.

7.5.3 *Global oxygen budget*

The primary source of atmospheric O_2 in the modern atmosphere comes from terrestrial and marine ecosystems, while sinks include autotrophic and heterotrophic respiration, wildfires, the weathering of organic matter and minerals, the oxidation of volcanic gases, and most significantly, human activities (such as the burning of fossil fuels and other industrial processes) (Fig. 7.26(c)). The depletion of atmospheric oxygen is largely attributed to the combustion of fossil fuels, according to scientific agreement. According to a quantitative global O_2 budget estimate by Huang *et al.* (2018) based on CMIP5 models, the average O_2 consumption flux from 1990 to 2005 was 36.22 Gt O_2 yr^{-1} $(1,131.9 \times 10^{12}\,molyr^{-1})$ and the average compensation from land and ocean systems over the same period was 17.75 Gt O_2 yr^{-1} $(554.7 \times 10^{12}\,mol\,yr^{-1})$. Over the past century, O_2 levels have steadily decreased as a result of an imbalance between O_2 production and consumption. X. Liu *et al.* (2020b) tried to quantify the land O_2 flux and took the first step toward quantitatively describing the anthropogenic effects on the global O_2 budget on a grid scale from 1975 to 2018. They discovered that the burning of fossil fuels was the primary cause of O_2 fluxes in Central Africa, while wildfires were the main cause in East Asia, India, North America, and Europe. Wei *et al.* (2021a) conducted an extensive analysis of urban districts around the world and discovered that, between 2001 and 2015, 75% of cities with a population $>$ 5 million consumed at least 100 times as much oxygen as the local flora could produce. Researchers looked at the possibility that under highly sluggish weather for several days, the mixing layer's O_2 content could fall below 19.5% (20.9% in the natural background), which could increase the risk of hypoxia-related health impacts in susceptible populations.

High-resolution observations have successfully recorded both the long-term variability in atmospheric O_2 and the shorter-term variations (from the nocturnal to seasonal scales). With opposite variations in CO_2, O_2 is shown to rise in the spring and summer and fall and winter in both hemispheres, reflecting the productivity of the biosphere and air–sea exchange on the same timescale. On Hateruma Island, Japan, Minejima *et al.* (2012) looked at the O_2 concentrations during pollution events and discovered that mixed pollutants from China, South Korea, and Japan caused the O_2 content to drop by about 10 ppm. Wildfires raging 70 km to the east of the station caused the O_2 content to drop by 20 ppm, according to Lueker *et al.*'s detection (2001). There are variations in the spatial distribution of the atmospheric O_2 content. Ishidoya *et al.*'s observations of fluxes in urban

Tokyo allowed them to quantitatively separate O_2 and CO_2 variation into combustions of liquid and gaseous fuel and calculate O_2 and CO_2 flux owing to human respiration based on population data and an expected OR of 1.2.

7.5.4 *Future projection of oxygen cycle*

The Earth system has been gradually directed away from its natural tendencies as a result of human impacts on global habitats that have grown more severe (Lenton *et al.*, 2008). Anthropogenic changes to the globe have already left global stratigraphic signatures and are severe enough to alter long-term geologic processes. As a result, the "Anthropocene" was designated as a new geological epoch that is distinct from the Holocene both stratigraphically and functionally. A clear and widespread consensus has emerged that human modifications of the global environment are occurring at unprecedented rates, which is gradually moving us toward the sixth mass extinction. Although the beginning of this new epoch has been hotly debated and several thresholds have been invoked to demarcate the onset of the Anthropocene, Ceballos *et al.* (2015) employed incredibly cautious assumptions to determine if human activity is causing a mass extinction and came to the conclusion that biodiversity has been disappearing extraordinarily quickly over the past few centuries (for vertebrate species, the current extinction rate is as much as 100 times that of the background rate). To avert a dangerous situation, we must act collectively and quickly. By doing so, we may be able to stop the loss of ecosystem functioning and biodiversity resulting from this decline. But as time goes on, our chances will reduce greatly. Even if anthropogenic emissions are controlled, studies have shown that positive feedbacks within the Earth system can be a possible trigger. When a particular threshold is reached, the Earth system may shift uncontrollably toward an unstable condition.

Understanding and even forecasting anthropogenic impacts on the Earth system can be done with the help of geological analogs from past climatic events. However, because the Anthropocene's perturbations are much more profound than those in the geological past, scientists find it difficult to compare the present climate with a historical analog. For instance, it is generally agreed that the Paleocene–Eocene Thermal Maximum (PETM, about 56 Ma), which was linked to a significant carbon release and a temperature of 5°C, is the greatest analog for forecasting future trajectories. Would the global oxygen cycle undergo significant alterations in the future as a crucial component of the Earth's system? In the 21st century, a decrease in O_2 content is predicted. The O_2 concentration would decrease from its present level of 20.946%–20.825% under the Representative Concentration

Pathways (RCPs) RCP8.5 scenario (Huang *et al.*, 2021). Approximately 100 Gt of O_2 would be lost from the atmosphere each year until 2100 (Huang *et al.*, 2018).

Various possibilities for future O_2 projections are depicted in Huang *et al.* (2021). In RCP8.5, RCP6.0, and RCP4.5, respectively, it is predicted that the O_2 concentration will finally stabilize at 20.61%, 20.85%, and 20.90%. Persistent O_2 drops are possible in no-stop RCPs, where fossil fuel burning is unaffected. The O_2 concentration will fall below 20.0% in the 26th century and reach 19.5% at the start of the 29th century in no-stop RCP8.5, which may cause potential health issues. Livina *et al.* (2015) projected the future O_2 concentration based on known data for atmospheric O_2 and discovered that the reduction in O_2 will result in health-threatening issues for people in 5,000–10,000 years if no action is taken. According to Martin *et al.* (2017), humanity might exist in the atmosphere without protection for about 3,600 years. These projections have shown that the atmospheric O_2 concentration is closely related to the total amount of fossil fuels consumed over a certain period of time, and they imply that ongoing human interference with the global oxygen cycle will eventually pose a serious threat to our planet's ability to support life. But it is important to remember that fossil fuels have a limited supply. These resources change over time and are not fixed. Due to limits in knowledge and boundary conditions, estimates of the global supply of fossil fuels are very uncertain (IPCC, 2014). Of course, before exhausting all the resources on our planet, we must find alternatives to fossil fuels (such as bioenergy, wind energy, and solar energy), and we might still have a chance to maintain atmospheric O_2 at a somewhat healthy level. Our mitigating measures will have a significant impact on how the atmospheric O_2 concentration evolves in the future.

7.6 Conclusions and Discussion

The impacts of rising GHG emissions, anthropogenic aerosols, land use, and urbanization on the climate change of arid and semi-arid regions are discussed in this chapter, with an emphasis on the impacts of increased GHG emissions. Several studies (Huang *et al.*, 2016b; Feng and Fu, 2013; Sherwood and Fu, 2014) have demonstrated that arid and semi-arid regions could increase in size and become increasingly drier during the 21st century as a result of GHG-induced global warming. This shows that future warming of arid and semi-arid regions may have a greater impact on a larger population (Huang *et al.*, 2016b). Additionally, global warming will affect most regions of the world under the stated global mean warming target, with some being more vulnerable than others. One method to comprehend regional differences is to contrast arid and humid places. Conditions would be worse for localized, extreme temperature swings. Thus, while establishing such a global mean warming target, it is crucial to take into account these regional variations.

To sum up, we find that regional differences should be fully taken into account when discussing CO_2 emission reduction plans because, despite humid (arid) areas having contributed roughly 58% (19%) of historical global CO_2 emissions, drylands have experienced warming that is 20–40% greater than humid lands; agricultural, hydrological, and drought- and health-related problems are predicted to grow dramatically across drylands if the GMSW climbs from 1.5°C to 2.0°C. A 1.5°C GMSW target is necessary to limit mean dryland warming to 3.0°C; since drylands are the most vulnerable to GHG-induced climate change, they should be given more consideration when establishing GMSW targets.

By the end of the 21st century, drylands will cover 56% and 50% of the world's land surface under RCP8.5 and RCP4.5, respectively, expanding by 23% and 11% above the observed baseline (1961–1990). According to RCP8.5, drylands will make up around 61% of the total area of developing countries by the end of the 21st century, with about 78% of the newly enlarged drylands areas being found in these nations. Drylands in developing countries are more susceptible to climatic and environmental changes, as well as disturbances brought on by humans, than drylands in developed countries. Additionally, the warming trends over drylands, especially in arid areas, are twice as strong as those over humid areas. The same temperature rise will undoubtedly have a higher effect on the poor and vulnerable communities inhabiting drylands, leading to increased poverty, degradation of the land and ecosystems, soil erosion, and additional desertification. Last but not least, by 2025, drylands may make up 48% of the world's land surface and support 51% of the population growth that will occur between 2000 and 2025, 50% of which will take place in developing countries versus only 1% in developed ones. Any dryland subtype has a higher population growth rate in developing countries than it does in humid areas. The increased agricultural requirement due to population growth would need tillage, pasturage, and other human activities that can swiftly degrade sensitive arid and semi-arid regions.

The projection skill was enhanced by this correction technique, but it should be noted that the enhancement is still only modest because the correction method is reliant on the statistical mapping relationship between the observation and model output. When results are only based on the projections of the original CMIP5, the risk of arid and semi-arid regions' expansion that was proposed in our study is still present. This correction method also suggests that the situation may get worse, especially in developing countries, a fact that is typically overlooked by the general public. As a result, three threats will affect developing countries: prolonged drought, increased warming (Huang *et al.*, 2012; Ji *et al.*, 2014), and rapid population growth. Positive feedback in the climate aridity cycle may cause these terrible scenarios and their interactions to compound nonlinearly, making the survival environment more susceptible to even little human disruptions. Degradation and desertification will pose a much greater threat than is now anticipated. Degradation and desertification will soon present a threat to the global environment and human life because developing countries make up more than 70% of the world's land area and population. To prevent future desertification and reverse current global desertification issues, a worldwide action plan is urgently required. This plan must include strict management and rational use of water resources, as well as the restoration of soils and plants to lessen ecosystem vulnerability.

Drylands generally warm more quickly in a warming world than they would in one that has stabilized. However, in the stabilized scenario, regulated emissions or even negative emissions will allow 1.5°C or 2°C warming to occur in the latter half of this century. During these two times, there will be variations in land use, aerosol forcing, and greenhouse gas emissions. Regional precipitation and PET can be significantly impacted by aerosols (Fu and Feng, 2014), particularly for the arid regions of Central Asia, Africa, and Australia. The difference in precipitation between the transient and stabilized scenarios was not primarily due to particles, according to comparisons of aerosol impacts on precipitation over various time periods. The aerosol forcing in the transient and stabilized scenarios can be consistent, and the decrease in aerosol will also cause an increase in temperature, if the aerosol effect from the 1.5°C (2°C) warming period to the end of the 21st century is added to the 1.5°C (2°C) warming world in transient scenarios. Currently, the transient scenario and the stabilized scenario have different levels of global warming. Therefore, in this work, rather than taking into account aerosol forcing and land use, we focus primarily on the variation in CO_2 emissions under the same warming situation in different scenarios.

Under the 1.5°C and 2°C scenarios, as well as the RCP45 and RCP85 emission scenarios, both precipitation and PET show considerable future increases. The rise in PET (60.3–75.7 (88.3–101.7) mm/year in a stabilized 1.5°C (2°C) warmer world can result in an increase in aridity because the rise in precipitation (12.9–37.0 (the 5th to 95th percentile for ensemble members) (21.8–42.6) mm/year in a stabilized 1.5°C (2°C) warmer world) is insufficient to make up for it. Reduced precipitation in North America, South Africa, and North Africa will be caused by the additional 0.5°C warming that shifts the world from 1.5°C to 2°C stabilization, while the PET will increase more visibly, thus increasing the probability of aridity in these drylands.

It is important to note that the P/PET utilized in this study is just a measurement of aridity variations, which introduces certain uncertainties. For example, sample anomalies of precipitation, soil moisture, runoff, and net primary productivity (NPP), as well as the Palmer Drought Severity Index (PDSI) and the Aridity Index (P/PET), have all been created to monitor and quantify variations in aridity over worldwide land in earlier research. The coverage of drylands is likely to grow in the future, according to our AI data. The outcome is in line with the results drawn from the PDSI and soil moisture data (Cook *et al.*, 2015). However, the chosen variables have a significant impact on variations in local land aridity. For instance, the "drying" pattern shown by predicted declines in AI and PDSI in CMIP5 models is in contrast to the pattern of "greening" over the bulk of the world's land, which manifests itself as increases in precipitation, runoff, and NPP (Scheff, 2018). Additionally, it is predicted that a significant portion of the world's land regions will see both increased vegetation growth and a decrease in soil moisture. As a result, it is challenging to utilize one index to accurately capture all aspects of aridity around the world. Furthermore, various components of the relationship between the land surface, plants, and atmosphere must be taken into account for a more complete understanding of aridity variations. These components include soil moisture, runoff, NPP, and vapor pressure deficit (VPD). Aridity trends in the past were also greatly underrepresented by the models when compared to data (Huang *et al.*, 2016b). According to the updated model simulations, the future aridity trend might be worse than we originally thought (Huang *et al.*, 2016b).

Numerous earlier studies (Huang *et al.*, 2016b; Feng and Fu, 2013) have highlighted the accelerated dryland expansion. The worldwide dryland area coverage will continue to rise from the baseline to the transient timing of 1.5°C (2°C) warming. The extent of dry subhumid and hyper-arid regions will rise under the stabilized 1.5°C and 2°C scenarios, semi-arid regions will initially expand and subsequently decline, and arid regions will decrease. In a stabilized 2°C warmer world compared to a 1.5°C warmer future, the area coverage of drylands will grow by 0.7% of the global mean land area, with semi-arid and hyper-arid places expanding the most. Under a momentarily warmer climate, the dryland expansion is more substantial, especially for arid regions.

Under various definitions, the climate of arid and semi-arid regions generally reacts to warming differently. The radiative force at any given change in the global mean temperature is larger than in a stabilized experiment due to the time delay of surface warming in transient circumstances. Therefore, in a world that is temporarily warmer than one that is stabilized, the temperature response and area coverage shift for global drylands are higher. The additional 0.5°C of global warming in transient warming scenarios results in smaller differences in temperature and precipitation over drylands than in a stabilized warmer world. As they are closer to the Paris Agreement's proposal, the stabilized warming scenarios are better suited for examining the effects of 1.5°C and 2°C warming targets. Controlling global warming to 1.5°C by the end of this century, as opposed to the 2°C objective, will cut dryland warming by 0.7°C and dryland expansion relative to the baseline by about 44%. Drylands like Southern North America and North Africa could experience less deterioration.

Both warming and drying trends that have been present historically are expected to continue into the future. Thus, dynamics and climate may link these two variables. By increasing the vapor pressure deficit and evaporative demand, the warming may exacerbate dryland expansion. Additionally, the decreased soil moisture may increase the sensible heat flux while decreasing the latent heat flux, further affecting temperature extremes. With a heated peak of 1.17°C per 58 yr, which is around 2.1 and 1.5 times more than the increase in humid regions and in the world mean, respectively, the average temperature increase is most noticeable in drylands. Additionally, the storage of soil organic carbon (SOC) increases with soil water content and decreases with temperature. Carbon emissions could result from the degradation of the land caused by erosion. In light of this, the expansion of drylands will decrease SOC storage and cause CO_2 emissions into the sky. Additionally, the gross primary production (GPP) is severely constrained by soil deterioration and decreased soil moisture, which also has an impact on the rate of photosynthesis in plants that can take up carbon dioxide and store it. According to Fig. 7.29(b), the GPP and AI have a positive correlation, and the GPP in arid areas is roughly one-fifth that in humid areas. Dryland soils worsen global warming by releasing more CO_2 into the atmosphere and storing less carbon through these two processes. Warming and drying

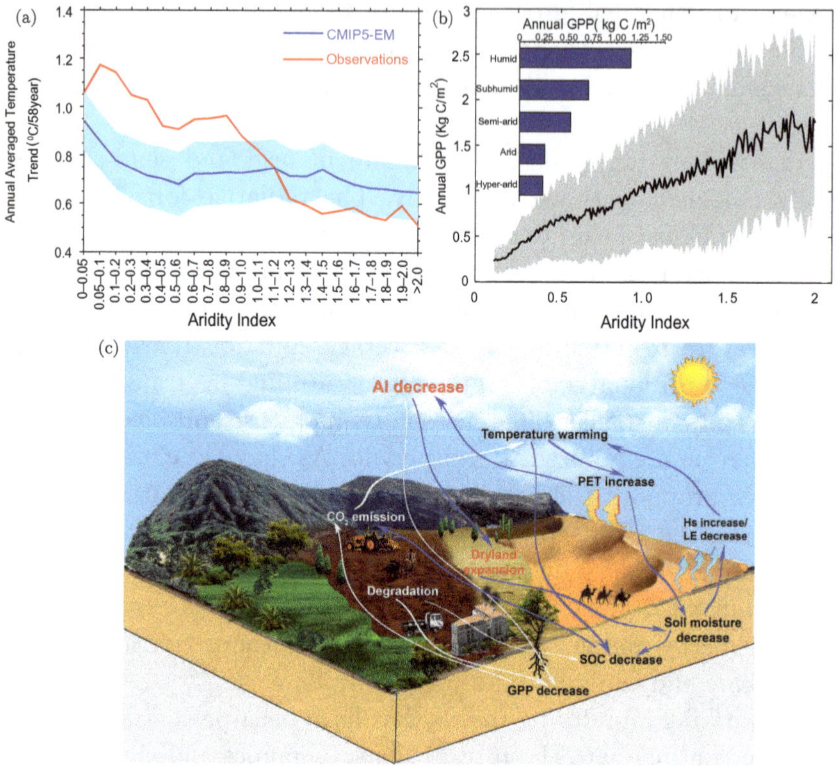

Fig. 7.29. Warming and drying feedback and associated impacts related to AI. Variation in the regionally averaged annual temperature trends (1948–2005), as calculated from CRU observations (red) and CMIP5-EM simulations (blue); the shading denotes the 95% confidence intervals (a). Annual GPP as a function of the AI; the black line denotes the regional mean from 2000 to 2010, with the gray shading denoting one standard deviation. The regional mean annual GPP of each subtype is shown in the inset (b). A schematic diagram of positive feedback cycles, including processes of global warming, dryland expansion, and carbon emissions from soils (c) (Huang *et al.*, 2016b).

processes combine to form a positive feedback cycle that reinforces one another (Fig. 7.29(c)). Unfortunately, because certain CMIP5 models do not take into account global carbon cycle processes, they are unable to replicate this trend without significant uncertainty (Fig. 7.29(a)).

Similar to the previous study, we find that for every extra 0.5°C of warming, the warm extreme indices greatly increase, whereas the cold extreme indices noticeably decrease. The trends of potential risks (FAR and PR) show that the frequency of extreme climate occurrences is predicted to increase with global warming. Despite using a small number of data points in the CESM model, our research clearly shows how temperature extremes vary under various warming circumstances. The ASA and dry–wet dividing zone, particularly on the Tibetan Plateau, undergo a little decline in the two indices (FD and ID) that indicate the cold season. Due to an additional 0.5°C warming, both the ASA and HA experience a decline of about 5 days every year. For warm indices, the SU changes significantly in the HA, including East China, Southeast Asia, and South India. In comparison to the historical period, the SU is 20 days/year greater under the 1.5°C scenario and 10 days/year larger under an extra 0.5°C warming from 1.5°C to 2°C. The TR follows a similar tendency to the SU, but under more challenging circumstances, the deviations are smaller in both time and space. According to Sui *et al.* (2018), within 2°C warming limits, the incidence of SU and TR lasts nearly the entire summer in China. If we keep global warming at 1.5°C, more than 6% and 11% of the increases in SU and TR might be decreased (Li *et al.*, 2018a). Additionally, if the target temperature was 2°C, East Asia would prevent around 20% of severe heat stress (Lee and Min, 2018).

Four key aspects, with variations under the 1.5°C and 2°C warming targets, are identified in the intensity indices TX_X, TX_N, TN_X, and TN_N. (1) Under a warming of an additional 0.5°C, the decreasing range for the minimum temperature is somewhat larger than that for the maximum temperature. (2) The indices in winter are 0.2°C–0.4°C larger than those in summer. (3) As a whole, there are no noticeable differences between the ASA and HA. (4) These chosen indicators have a nonlinear growth tendency; an additional 0.5°C warming will have a significant impact on the rise in severe temperatures. The 0.5°C less warming in the 1.5°C warming limits over East Asia helped avoid roughly 35% of consequences in TX_X and TN_X, according to Li *et al.* (2018b).

Under the assumption that the world gets 0.5°C warmer, the DTR is predicted to warm up by 0.3°C–0.5°C in the summertime in and around the dry–wet transition zone. However, there is a significant variance in winter due to positive values in East and Central China and negative values in other areas of East–Central Asia. Previous research revealed that the DTR has a markedly declining tendency over the course of history over a majority of the world, including East–Central Asia (Liu *et al.*, 2018). The fact that the minimum temperature rose more quickly than the highest temperature can be attributed to this phenomenon. We discovered that the DTR in some subregions increased while it decreased in other subregions under 1.5°C or 2°C warming future, in contrast to the uniform decreasing trend during the historical era. And its association with the cloud cover is still very strong.

Therefore, keeping global warming below 1.5°C will benefit mankind by lessening the severity, frequency, and duration of climate extremes over East–Central Asia. We believe that these scientific advances will provide policymakers with the assistance they need. To further understand the changes in climatic extremes, additional research should include more global and regional climate models as well as 1.5°C and 2°C warming experiment samples.

Anthropocene has been a new epoch for mankind since the Industrial Revolution. Due to the intensifying influence of humans on nature, some planetary boundaries have now been crossed (Rockström *et al.*, 2009a). Urban areas make up only 3.8% of the earth's land surface, but from 2001 to 2015, they consumed 39.3% of the world's terrestrial oxygen (36.4 Gt/yr). Our study shows that there was already a significant oxygen imbalance in many large cities around the world. As a result of the oxygen deficit, urban residents in cities with higher O_I would experience more frequent high-temperature heat waves and water shortages, which would have a negative impact on the health status of residents (Ginzburg *et al.*, 2014) (Fig. 7.30). O_I could be used as a measure of urban sustainability because its value is strongly related to numerous socioeconomic and environmental variables in urban areas around the world.

The average O_I value in metropolitan areas between 2001 and 2015 was around 60, and it was higher in Asia than on the other 5 continents. The O_I in metropolitan areas around the world has fallen by 4.6 over the past 15 years, but it began to climb quickly after 2010. Only in cities in Europe and North America did the O_I show a significant lower trend during the past 15 years, mostly as a result of the twin effects of increased oxygen production by vegetation and decreased oxygen consumption from the burning of fossil fuels. While the O_I in urban areas in Asia increased by 25.5, this was primarily due to the interaction of rising urban fossil fuel burning, which increased the O_I by 51.6, and rising oxygen generation, which decreased the O_I by 26.5. Our findings indicate that if a city's O_I is greater than 352, the atmospheric oxygen content below 100 m could decrease from its current level of 20.95% to a deficiency level (19.5%) for humans living there in 10 days. However, this result should be understood under the presumption of very calm weather and persistent, typical meteorological conditions (including surface pressure, temperature, and humidity). According to Ginzburg *et al.* (2014), if the temperature and absolute humidity of the surface air are both high, there is little atmospheric oxygen present and humans may have hypoxic symptoms. This suggests that the oxygen concentration is

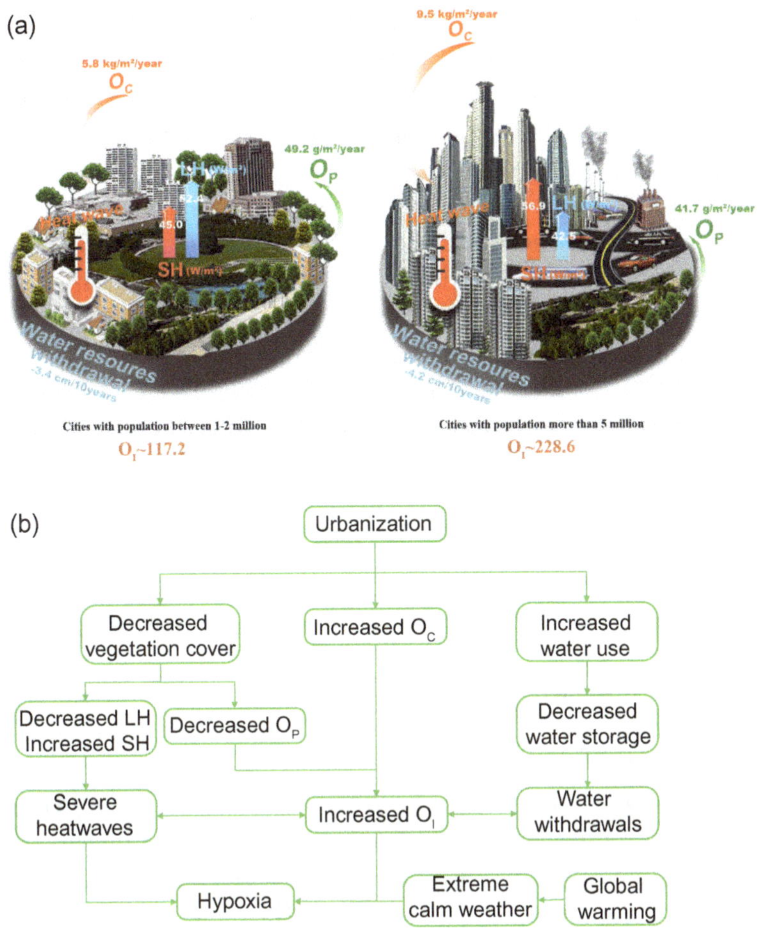

Fig. 7.30. Schematic of the oxygen balance and its related risks in large cities. Oxygen balance and its related heat wave, water shortage, and hypoxia risks affect cities with populations between 1 and 2 million and more than 5 million (a). The differences in numbers in (a) all passed a 99% t-test for significance. Schematic framework (b); the boxes denote the related causes or results, whereas the arrows denote a possible correlation or causality between variables in this chapter. SH and LH represent the net sensible and latent heat flux, respectively (Wei *et al.*, 2021a).

directly tied to external weather conditions. Additionally, local pollution events are directly associated to a reduction in the oxygen concentration. Accordingly, residents of major cities with higher O_I values would experience serious health issues during periods of extreme calm, and these issues would exacerbate when extreme heat wave and pollution events coexist.

A typical phenomenon that is spreading and intensifying globally is a decrease in oxygen concentration. The Northern Hemisphere, where human activities are more concentrated, is experiencing a faster decline in oxygen concentration, raising the alarming possibility that in the near future, people living in large cities could experience severely hypoxic environments during exceptionally calm weather conditions. The expected rate of oxygen loss is predicted to increase to 18 ppm/yr by the end of the 21st century from its current pace of about 4 ppm/yr. A hypoxic environment in very calm weather, more frequent heat waves, and the expansion of water shortages to more cities would all result from O_I's unchecked proliferation. By that time, the oxygen reduction will probably be a widespread occurrence all across the planet, and residents of big cities may need to relocate to oxygen-rich places or may not be able to find oxygen refuges. In order to prevent the continuous depletion of oxygen (Huang *et al.*, 2018), as well as frequent heat waves and water shortages in big cities in the future, we must focus on limiting urbanization, which is brought on by an increase in urban population, buildings, and access to necessities like water. We also need to increase the proportion of urban vegetation cover to offset this depletion of oxygen, as well as create new energy sources to offset industrial oxygen consumption. Smart, sustainable, and healthy urbanization has a long way to go before it is fully achieved.

Arid climate factors like GHGs, aerosols, and urbanization have all been the subject of numerous studies, in addition to the implications on humans as already mentioned. According to Sherwood and Fu's (2014) research, a warming temperature will result in an overall drying of the terrestrial climate. Fu and Feng (2014) evaluated the terrestrial mean aridity index, which is defined as the ratio of P to PET (i.e., P/PET), in response to GHG forcing using CMIP5 transient CO_2 increase to $2 \times CO_2$ models. They discovered that although PET increased at a rate of 5.3%/°C, P increased at an average rate of 1.7%/°C over land. As a result, P/PET drops by roughly 3.4%/°C, resulting in a drier environment. Lin *et al.* (2016) computed the global mean (ocean mean) SAT rise using the Community Earth System Model (CESM) ensemble simulations to isolate the impact of GHGs as a 2.7%/°C (about 3.2%/°C) drop in the terrestrial mean aridity index. Cook *et al.* (2014) examined projected 21st century drying and wetting trends under the RCP8.5 scenario according to PDSI SPEI, using the GCM output from CMIP5. Their findings revealed that robust cross-model drying occurs in Western North America, Central America, the Mediterranean, Southern Africa, and the Amazon, whereas robust wetting occurs in the NH high latitudes and East Africa (PDSI only). Particularly in arid areas like the Sahara and the Middle East, the SPEI is more susceptible to PET changes than the PDSI. In their analysis of 15 coupled climate models, Seager *et al.* (2010a) found that as GHG emissions rise, the climate in Southwestern North America will grow drier over the course of the 21st century. The regional differences seen in this earlier research are in line with the findings reported by Scheff and Frierson (2015). They discovered that, with the exception of high latitudes, land P does not routinely increase or decrease with greenhouse warming. As a result, in most models, P/PET decreases (drying) are significantly more common than increases (wetting) in the tropics, subtropics, and mid-latitudes; this conclusion supports and extends previous studies' findings.

Greve and Seneviratne (2015) employed an ensemble of 30 CMIP5 models to examine the future changes in hydroclimatological conditions using P-Ep (i.e., precipitation–potential evaporation) as a measure of aridity rather than P/PET, which was used in the previous studies. The expansion of dryland was further supported by the discovery of a significant increase in aridity in a number of humid subtropical regions and their surrounding subtropical regions. Increases in GHGs, such as CO_2, may further damage the vegetation of arid and semi-arid regions by influencing the precipitation and the fertilization effect. By influencing plant water-use efficiency and soil water availability, Lu *et al.* (2016) postulated that increased CO_2 levels cause global dryland greening. To test this hypothesis, they used meta-analytic approaches to examine 1705 field observations from 21 different sites. In arid and semi-arid regions, there was a noticeable reaction between the increasing CO_2 and the soil water availability. According to Ukkola *et al.* (2015), who used NDVI and evapotranspiration data from 1982 to 2010, an increased atmospheric CO_2 level had an impact on the vegetation and hydrology across Australia. Future research should thus assess the biological effects of vegetation responses to GHG increases.

Aerosols are another important forcing agent linked to human activities in addition to GHGs (IPCC, 2013). Aerosol decreases relative to current levels that are anticipated in the 21st century would result in extra system warming, which could exacerbate the effects of GHG-induced warming on the climate. Therefore, it is crucial to look into and comprehend how aerosols affect terrestrial aridity.

Fu *et al.* (2016) revealed that although anthropogenic aerosols have minimal effects on terrestrial mean aridity index, they might largely switch the attributions from the temperature change to precipitation change. This might make the identification and attribution of aridity change problematic owing to GHG rises when aerosols also exert a forcing because the precipitation has considerably more natural variability than temperature. Using CMIP5 model results, C. Li *et al.* (2015) examined how anthropogenic aerosols affected temperature variations in China's arid/semi-arid and humid/semihumid regions. In comparison to the arid/semi-arid regions, humid/semihumid regions saw lower temperatures and showed stronger contributions from human aerosols.

Meehl *et al.* (2012) analyzed time series of annual mean globally averaged SAT data from 1850 to 2005 for sulfate (SO_4) aerosol forcing using the Community Climate System Model, versions 4 (CCSM4). According to their findings, the temperature was lowered by 0.25°C–0.40°C by the end of the 20th century due to the negative radiative forcing (RF) of SO_4 aerosols. Based on the fully coupled chemistry–climate model CM3 from the Geophysical Fluid Dynamics Laboratory (GFDL), Levy *et al.* (2013) concluded that SO_4 aerosols' direct effects dominate the strong precipitation response and that their indirect effects significantly strengthen the impacts of aerosols on surface temperature.

Positive BC RF increases global mean SAT and direct radiative heating of the atmosphere (Bond *et al.*, 2013). According to Held and Soden (2006), the warmer surface tends to promote longwave cooling, which causes more precipitation. However, P is suppressed by the atmospheric solar absorption that BC causes (O'Gorman *et al.*, 2012). The suppression of P causes a net drop in the global mean precipitation in the CESM and GFDL models, outweighing the effects of BC's surface warming (Lin *et al.*, 2016). Additionally, BC lowers the amount of solar radiation that reaches the surface, which results in a reduction in PET. On the other hand, surface warming brought on by BC tends to raise PET. Intriguingly, the effect of surface warming on PET overcomes the effect of radiation, leading to a net rise in the global mean PET, which, when combined with the aforementioned P suppression, lowers the AI (increasing aridity) [Lin *et al.*, 2016).

Lin *et al.* (2016) used CESM simulations to further investigate how the terrestrial aridity index responded to anthropogenic BC and SO_4 aerosols. Positive BC RF causes a 1.9%/°C warming (i.e., drying) drop in P/PET, with P decreasing by 0.9%/°C and PET rising by 1.0%/°C. Due to the cancelling effects of decreased P (6.7%/°C) and decreased PET (6.3%/°C), negative SO_4 forcing has a negligible impact on terrestrial mean P/PET (drying of 0.4%/°C cooling). Despite having opposing impacts on the world's average temperature, both BC and SO_4 played a role in the 20th century's drying.

This section focuses on the effects of anthropogenic dust, while Section 4 explored the effects of natural dust aerosols on climate change. According to Guan *et al.* (2015), regional warming and anthropogenic dust particles in arid and semi-arid regions have a positive association. Huang *et al.* (2015) found that local anthropogenic dust aerosols originating from human activities, such as agriculture, industrial activity, transportation, and overgrazing, account for about 25% of the global continental dust load using data from CALIPSO dust, PBL height retrievals, and a land-use dataset. Furthermore, PET is reduced over a majority of these areas, mostly as a result of changes in net radiation flux. Overall, aridity and the geographic breadth of arid and semi-arid regions are not made worse by dust aerosols. Additionally, a series of GCM experiments revealed that dust aerosol forcing and changes to the land surface were probably responsible for the US's intensity and persistence. The Central Plains had megadroughts because dust increases the shortwave planetary albedo, thereby reducing energy inputs to the surface and boundary layer. Energy deficits also increase atmospheric stability, inhibiting convection and reducing cloud cover and precipitation over the Central Plains.

By changing the energy, momentum, and moisture exchange between the atmosphere and the ground surface, land use/cover change (LUCC) has a significant effect on the local climate and environment (IPCC, 2013). Negative LUCC, especially over semi-arid regions, can dry out the climate and cause soil degradation. Positive feedbacks between degraded land and the local climate, for instance, have been observed in the Sahel. Drought would be made worse by dry soil, decreased leaf area, and an increase in surface albedo. Olson *et al.* (2008) discovered similar positive feedback across arid and semi-arid regions in East Africa and demonstrated that the region would experience drier conditions as a result of the predicted massive expansion of farming into natural savannas.

Avila *et al.* (2012) used a climate model in conjunction with a land surface scheme to show that land-use-driven land cover change (LULCC) may regionally disguise or magnify the impact of increasing CO_2 on anthropogenically generated variations in temperature extremes. However, compared to the effect of doubling CO_2, the effects of LULCC on extreme temperature indices is lesser in scope and more localized. According to Junkermann *et al.* (2009), changes in surface qualities are important contributors to the reported decrease in precipitation.

SAT, P, PET, P/PET, and other meteorological variables averaged over land were explored by Fu *et al.* (2016) using CESM simulations induced with the transient evolution of land-use circumstances. A minor drop in precipitation (non-significant) and PET (significant) as a result of the land usage led to non-significant changes in P/PET. The combined impacts of declining SAT and accessible energy and rising RH and surface wind led to a fall in PET related to land usage.

More precise vegetation and management data are also required to continuously improve the simulations of the climatic impacts of land cover change, according to Mahmood *et al.*'s (2014) recommendation that several observational platforms be used to better detect the climate impacts of land-use change.

Ecosystems may be disproportionately impacted by extreme climate events. A region's biogeophysical climate impacts from deforestation may be amplified by drought-induced forest die-off, creating significant uncertainty in regional climate–ecosystem interactions and carbon cycle feedbacks on the global climate. Land cover change is a significant factor in the Earth system, as shown by Sterling *et al.*'s (2013) research, which also showed that it reduces yearly terrestrial evapotranspiration and changes annual runoff to a similar or higher level than other main factors.

Uncertainty surrounds human influences on drought and potential societal reaction from drought. Because the feedback between drought and human activity is not fully understood, drought management is ineffective. We need to update the definition of drought in this era of human influence to consider how humans are amplifying and mitigating drought (Van Loon *et al.*, 2016). In fact, Cook *et al.* (2016) hypothesized that changes to the land's surface could cause megadroughts in North America.

Chapter 8

Dynamics of Semi-Arid Climate System

8.1 Introduction

A long-term drought trend and regional wet and dry oscillation are the outcomes of the coupling of interdecadal oscillations controlled by oceanic processes with long-term trends brought on by external forcing in the semi-arid environment. Globally speaking, climate change has sped up the water cycle, which has accelerated water evaporation in arid and semi-arid areas. This has resulted in a significant drop in terrestrial evapotranspiration and water storage, as well as a general trend of aridity intensification and expansion in arid and semi-arid areas. Oceanic processes provide a modulating function at the regional level. Anomalies in the atmospheric circulation caused by various sea surface temperature (SST) phase combinations result in interdecadal oscillations and substantial regional variances in the dry and wet changes of arid regions. The dynamic mechanisms of semi-arid climate change will be theoretically examined in this chapter.

Over the past 100 years, the amount of carbon dioxide in the Earth's atmosphere has steadily increased (IPCC, 2013), and it is thought to be the primary factor contributing to global warming. The global mean surface air temperature (SAT) has stabilized since 1998, a phenomenon known as a warming hiatus, despite the fact that this rise in atmospheric carbon dioxide has been linked to rising temperatures (IPCC, 2013; Kosaka and Xie, 2013; Trenberth *et al.*, 2014b; Guan *et al.*, 2015a). Because it appears to defy the notion of human-induced global warming, the warming hiatus has garnered a lot of interest globally (IPCC, 2013). Global warming will continue in the first two decades of the 21st century, according to a previous scientific prediction (IPCC, 2007). But the sudden cooling off was interpreted as a failure of the forecast and cast doubt on the veracity of climate science. The scientific community has previously offered numerous hypotheses for the warming hiatus, greatly advancing our understanding of how the hiatus process works. Previous research has revealed that internal climate variability (ICV), particularly the oceanic ICV, was primarily responsible for the warming pause. According to Steinman *et al.* (2015), with just a minor contribution from Atlantic variability, the hiatus has been predominantly caused by a substantial negative trend in the Pacific multi-decadal variability. However, the definition of the Pacific multi-decadal variability that was put forward by Steinman *et al.* (2015) did not consistently match the traditional ICV modes that are typically employed to depict Pacific variability. A negative phase of the Pacific Decadal Oscillation (PDO), or (more generally) the Interdecadal Pacific Oscillation (IPO) (Trenberth *et al.*, 2014b), and the La-Niña-like cooling caused by accelerated trade winds (Kosaka and Xie, 2013) were the two main examples of the suggested classic Pacific variability related to the recent warming hiatus. The Northern Hemisphere (NH) or global mean temperature is significantly influenced by the Atlantic variability, which is symbolized by the Atlantic Multidecadal Oscillation (AMO) (Tung and Zhou, 2013). A series of atmospheric and lagged oceanic teleconnections result in the opposite signal in the NH temperature roughly 30 years after the AMO signal. These analyses show a strong correlation between the AMO signal and the previous warming hiatus (about 1940–1975). The North Atlantic Oscillation (NAO) typically serves as a proxy for the atmospheric ICV over the Atlantic (Huang *et al.*, 1998). According to Guan *et al.* (2015a), this is thought to have contributed to the warming hiatus.

A majority of these studies have mainly concentrated on specific oceanic ICV modes and their influence on global cooling without thoroughly discussing or describing how these ICV modes have caused the warming hiatus (Chen and Tung, 2018). We shall examine the dynamics of the warming hiatus from a global standpoint in Section 8.2 of this chapter, as well as the combined effects of ICV modes on SAT over the NH land. We found that these ICV modes can modify the asymmetric meridional and zonal thermal forcing (MTF and ZTF) and ignite a global decadal modulated oscillation (DMO) (Huang *et al.*, 2017b). The DMO's modulation effect on SAT was strongest when both weaker (stronger) MTF and stronger (weaker) ZTF occurred simultaneously. The current hiatus is the result of a downward DMO combined with a weaker MTF and stronger ZTF, which stimulate a weaker polar vortex and westerly winds along with amplified planetary waves, and facilitate an Arctic cold air invasion and blocking formation. Global warming will be accelerated when the DMO swings upward.

High-index flow, which has strong zonal westerlies and relatively weak wave perturbations, and low-index flow, which has relatively weak westerlies with large-wave amplitudes and typically develops into blocking, are two distinct patterns of large-scale atmospheric circulation over middle–high latitudes (Rossby, 1939; Namias, 1950). The multiple-flow equilibria theory was put forth by Charney and DeVore (1979) to explain the two distinct flow patterns. When topographic and thermal forcings are present, they found that multiple equilibrium states may exist using a low-order (also known as "highly truncated") spectral barotropic channel model. Two different equilibrium states, known as the high- and low-index flows, were stable among the multiple equilibrium states. Using a two-layer baroclinic model, Charney and Straus (1980) expanded the research of Charney and DeVore (1979) to examine the instabilities that generate and feed on multiple equilibrium states. They suggested that topographic instability just serves as a catalyst for the generation of multiple equilibria, and the energy required to maintain the wave-like equilibria is produced by the conversion of mean flow potential energy.

Numerous studies in the multiple-flow equilibria theory and low-order spectral model have been sparked by Charney's groundbreaking study. Using a two-layer low-order spectral model, Zhu and Zhu (1982) and Zhu (1985) found that some stable equilibrium states had features that were typical of actual blocking. They stressed that the main factors in blocking dynamics were the zonally asymmetric thermal and topographic forcings as well as the nonlinearity of flow. When Reinhold and Pierrehumbert (1982, 1985) added synoptic-scale waves to the Charney and Straus (1980) model, they found two different weather regime states. They proposed that the model flow might be transferred from one regime equilibrium to another via wave–wave interactions. A higher-order barotropic spectral spherical model was employed by Legras and Ghil (1985), who reported that the model may display properties of an index cycle. Due to the deterministic nature of Charney's model, stochastic forcing was incorporated, and the model flow then manifested transitions between high- and low-index states (Egger, 1981). Additionally, some research investigated the physical causes of abrupt changes in flow patterns across mid-latitudes and subtropical regions by utilizing low-order spectral models (Li and Luo, 1983; Liu and Tao, 1983; Miao and Ding, 1985; Luo, 1987). In several recent studies, the roles of the high- and low-index flow patterns in the interdecadal variability of the continental temperature were demonstrated using Charney's multiple-flow equilibria theory (He *et al.*, 2014, 2018; Huang *et al.*, 2017a, 2017b). We shall go over multiple equilibria in a land–atmosphere coupled system in Section 8.3.

Amplification of Arctic warming has been the most noticeable trend over the past few decades (IPCC, 2013; Xie *et al.*, 2019), especially during the 1990s (Xie *et al.*, 2022a), in addition to the previously stated PDO, IPO, AMO, and NAO. Studies have shown that Arctic warming has a significant impact on Eurasian cooling (Xie *et al.*, 2020, 2022a), whereas other studies have shown that the Arctic has little impact on mid-latitudes (Dai and Song, 2020). Therefore, it is crucial to understand how the land is affected by the Arctic's ongoing warming in order to make the issue more understandable.

We shall go into more detail on the dynamic connections between the Arctic and Eurasian climatic changes in Section 8.4. In particular, a thorough methodology based on potential vorticity (PV) dynamics is proposed and discussed. Because PV is essential for atmospheric circulation, such as PV gradients for the occurrence of Rossby waves, PV dynamics are used (Hoskins *et al.*, 1985). In particular, PV has illuminated the dynamic relationships between the Arctic and mid-latitudes, including mechanisms involving PV generation of Arctic warming and further PV advection to mid-latitudes (Xie *et al.*, 2020) and the mediation of Arctic warming on atmospheric blocking via meridional PV gradients (Luo *et al.*, 2018).

Teleconnection is an essential component of the climate system, which refers to the climate variability links between geographically separated areas. The remote region need not exhibit fluctuations of the same sign in order to be "teleconnected." In fact, the teleconnections often involve contemporaneous variations of opposite signs. Climate analysis is facilitated by the construction of a teleconnection map, which describes the linkage between a region of interest (a base point) and all other points in the domain that are farther than the de-correlation length scale of the variable. Teleconnection maps thus provide information about the structure of recurrent low-frequency variability that is characterized by the correlation-at-a-distance feature. The maps are useful because climate variability often manifests itself with such structure (Wallace and Gutzler, 1981). Climate teleconnections are built using observations that have been time averaged over a duration long enough to minimize daily weather variations but brief enough to preserve the seasonal-to-interannual component of climate variability. Because a month is greater than the duration of most large-scale synoptic waves in the troposphere, monthly averages have traditionally been employed. Climate teleconnections emphasize the "standing" component of low-frequency variability, or the nodes and anti-nodes that are fixed spatially. The teleconnection maps' depiction of the interconnection of far locations suggests the possible importance of remote forcing in the production of regional climate anomalies. The teleconnection map based on contemporaneous correlations, however, is unable to identify the driving and response regions separately. The wave-like shape of the jet stream traps a different kind of teleconnection. These teleconnections include the circum-global teleconnection (CGT) in boreal winter (Branstator, 2002) and summer (Ding and Wang, 2005), and the Silk Road pattern (SRP) in boreal summer (e.g., Lu *et al.*, 2002; Enomoto *et al.*, 2003). Tibetan Plateau-modulated coupling teleconnection (TPTTCT) (Huang *et al.*, 2023) is another example. Since the North Atlantic Oscillation and the Southern Oscillation (Walker and Bliss, 1932) were covered in Chapter 6, in Section 8.5, we will introduce the PNA, SRP, CGT, and TCT teleconnection patterns.

The arrangement of this chapter is as follows. Section 8.1 is the introduction to this chapter. Section 8.2 illustrates the dynamics of surface temperature oscillation. Section 8.3 discusses multiple equilibria in a land–atmosphere coupled system, and the impacts of the Arctic on the Eurasian climate are demonstrated in Section 8.4. Section 8.5 introduces the teleconnection in the atmosphere. Conclusions and discussion are summarized in Section 8.6.

8.2 Dynamics of Surface Temperature Oscillation

8.2.1 *Surface temperature change and causes*

The temperature changes have varied based on the season, land versus sea, and hemisphere, as shown in Fig. 8.1 (Wu *et al.*, 2011; Huang *et al.*, 2012; Ji *et al.*, 2014). Across land, the mean SAT changes significantly more quickly than it does over ocean. Although both the land and the ocean exhibited a hiatus between 1998 and 2012, it is clear that the land SAT increased following the heightened warming in the 1980s and 1990s of the previous centuries. Before the most recent hiatus phase, the SAT and SST of the ocean continued to warm at a relatively moderate rate compared to land (Fig. 8.1(a)). The SAT over NH had the strongest warming trend before the 1998–2012 hiatus and the most pronounced hiatus in recent decades (Fig. 8.1(b)). In recent decades, the SAT of land over NH has clearly cooled during the boreal cold season, with a slight increase in the mean SAT of the ocean (Fig. 8.1(b)).

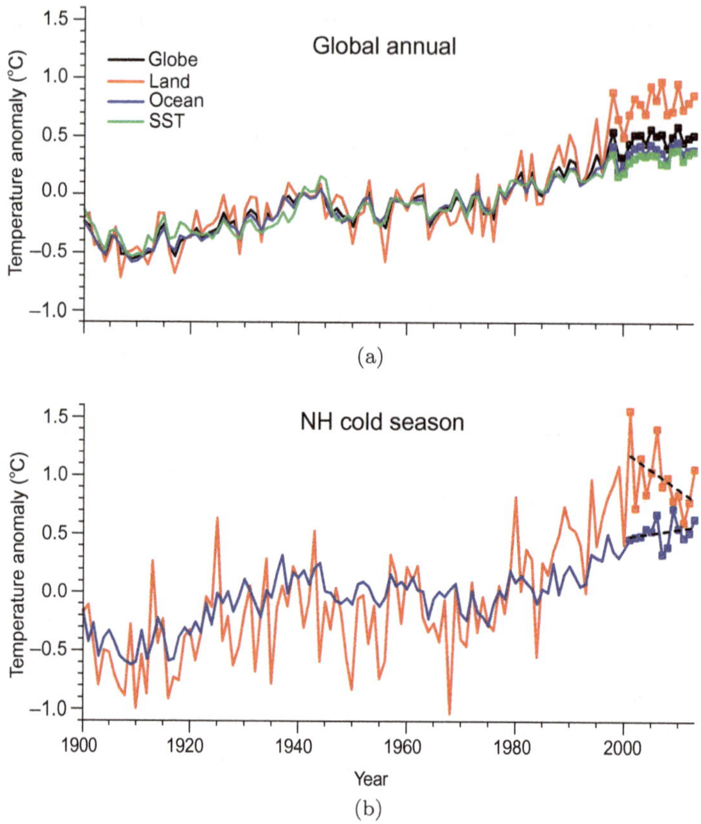

Fig. 8.1. The SAT anomaly time series relative to the 1961–1990 global annual mean for land and ocean (black), land (red), ocean (blue), and SST (green) (a). The marks at the line indicate the recent warming hiatus period. (b) Same as (a) but for the NH cold season; only the SAT anomalies over land (red) and ocean (blue) are shown, and the black dashed lines are the linear trend for 2001–2013. Note that the boreal warm season runs from May to September, while the cold season runs from November to the following March (Huang *et al.*, 2017b).

Figure 8.2 displays the linear trends of SAT during the warm and cold boreal seasons from 2001 to 2013. Figure 8.2 depicts the apparent variations in the spatial patterns of SAT changes over the most recent break between the warm and cold seasons. During the most recent hiatus, there were more cooling locations and bigger cooling trends in the boreal cold season than in the warm season. During the warm season, cooling centers are mostly found in small regions over Eurasia continents and some regions over Pacific (Fig. 8.2(a)). The majority of NH land still exhibits a warming trend throughout the hotter months. However, a sizable portion of NH territory has a clear cooling trend during the winter (Fig. 8.2(b)). Additionally, cooling

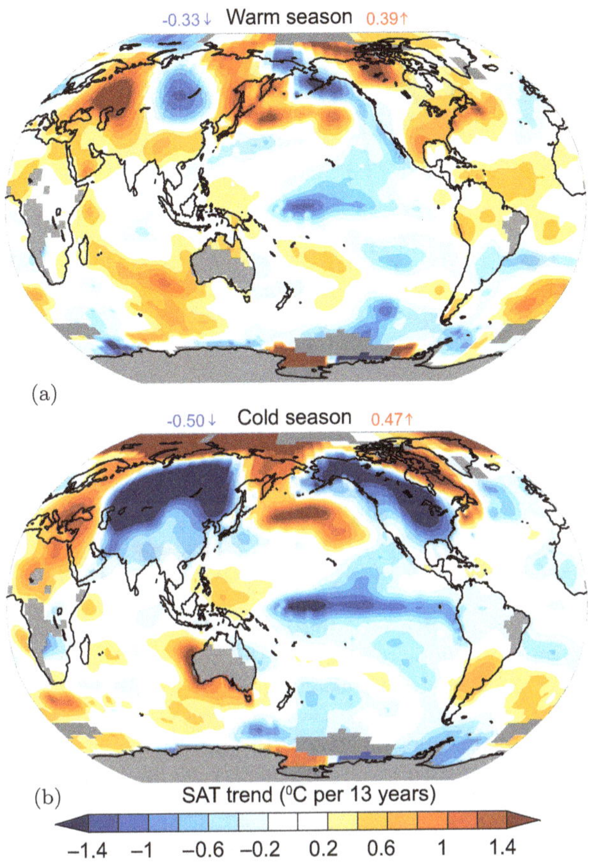

Fig. 8.2. The linear trend in the boreal warm season SAT for 2001–2013 (a). The blue (red) number in the left (right) corner is the global mean cooling (warming) trend in °C per 13 years. (b) Same as (a) but for the boreal cold season (Huang *et al.*, 2017b).

over the equatorial central and Eastern Pacific (ECEP) (Kosaka and Xie, 2013) is also more pronounced in the cold season (Fig. 8.2(b)). We limit our research to SAT variations in the NH during the boreal cold season as a result.

We estimated the areal weighted relative contribution of the SAT for each grid cell to the global mean cooling or warming trend during the boreal cold season (Fig. 8.3) in order to evaluate the spatial characteristics of the recent hiatus. Using Eq. (8.1), the relative contribution seen in Fig. 8.3 was computed. N refers to the entire number of grids with negative SAT trends and $\sum_i^N K_i\omega_i$ was used to compute the global mean cooling SAT trend. The global mean warming trend is determined using the same process. The total relative contribution of all grids in an area in Table 8.1 represents

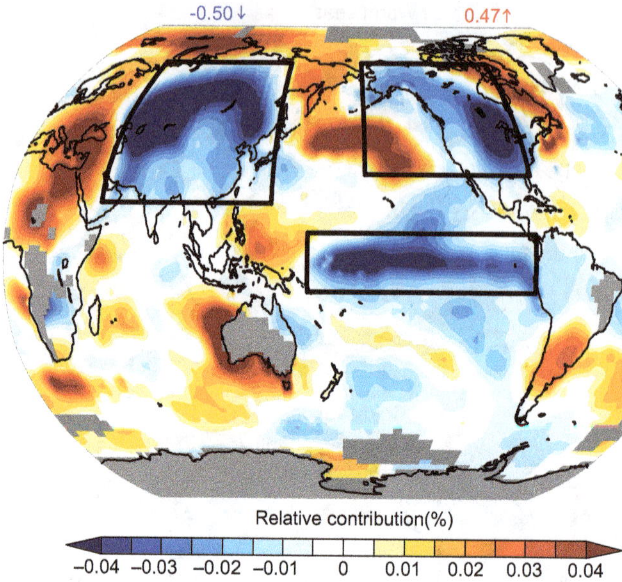

Fig. 8.3. Areal weighted relative contribution of the SAT trend in each grid unit to the global mean warming/cooling trend during the boreal cold season in 2001–2013. The relative contribution was calculated by using Eq. (8.1). The blue (red) number in the left (right) corner is the global mean cooling (warming) trend in °C per 13 years. The black squares represent the North American (NA), the Eurasian (EUA), and the Equatorial Central and Eastern Pacific region (ECEP) (Huang *et al.*, 2017b).

the regional mean cooling trend's relative contribution to the global mean cooling trend (Fig. 8.3). The ratio of the regional mean relative contributions to the average relative contributions over NH, land, or ocean was used to determine the relative contributions of the regional mean cooling trends to each of these regions:

$$\gamma_i = \frac{K_i \omega_i}{\sum_i^N K_i \omega_i}, \tag{8.1}$$

where K_i is the SAT trend of grid i, $\omega_i = \cos(\theta_i \cdot \pi/180.0)$, θ_i is the latitude of the grid, and N is the number of grids whose SAT trend is negative (cooling) or positive (warming). For example, N is the number of all grids with a negative SAT trend when the SAT trend of grid i is negative.

Figure 8.3 demonstrates that the recent warming hiatus's cooling centers are primarily found above the Eurasian (EUA) and North American (NA) continents as well as in the ECEP regions. Over the Arctic, some areas of the Mediterranean, and the North Pacific Ocean, new warming centers have been found. The worldwide cooling trend, which was −0.50°C per 13 years, was

Table 8.1. For the boreal cold season, the relative contribution (%) of the regional mean cooling trend to the averaged Northern Hemisphere (NH), global, land, and ocean cooling trend (Huang *et al.*, 2017b).

	NA	EUA	ECEP	Land	Ocean
NH	18	43	N/A	66	34
Globe	12	29	15	48	52
Land	25	60	N/A	N/A	N/A
Ocean	N/A	N/A	29	N/A	N/A

Note: N/A, not applicable.

slightly higher than the areal weighted global mean warming trend, which was 0.47°C per 13 years. The EUA and NA continents, respectively, made for 43% (60%) and 18% (25%) of the NH (terrestrial) mean cooling (Table 8.1). Although the oceanic ICV was a significant factor in the explanation of the hiatus, the contribution of the ocean to local cooling was only 52% (Table 8.1), suggesting that 48% of the explanation is provided by terrestrial cooling, a number that should not be disregarded.

8.2.2 *Decadal modulated oscillation*

The decadal modulated oscillation is the name we give to the ICV-modulated components of SAT variation on the decadal to multi-decadal scale (DMO). We employed the EEMD approach, which can separate the evolution of SAT into long-term trends and oscillation components, to evaluate the DMO components of SAT (Wu *et al.*, 2011; Ji *et al.*, 2014). The oscillation component is believed to be mostly caused by ICV (Wu *et al.*, 2011; Wallace *et al.*, 2012; Guan *et al.*, 2015a), while the long-term warming trend is primarily attributable to radiative forcing (Wu *et al.*, 2011; Guan *et al.*, 2015a). The DMO components of the global annual mean SAT and NH cold season mean SAT are presented in Fig. 8.4(a) and 8.4(b) respectively. The DMO component is the sum of IMF 3, 4, and 5 from the EEMD, and the long-term trend is the IMF 6. The DMO enhances or suppresses the long-term trend on the decadal to multi-decadal timescale. The DMO's magnitude is comparable to that of the long-term trend at the decadal scale (Fig. 8.4). The DMO contributes to an accelerated warming trend when it is in an upward (warming) phase, as it was from 1985 to 1998. The recent global warming hiatus appears to be the result of a downward swing in the DMO that has been going on since around 2000, which has balanced or reduced the radiatively caused warming.

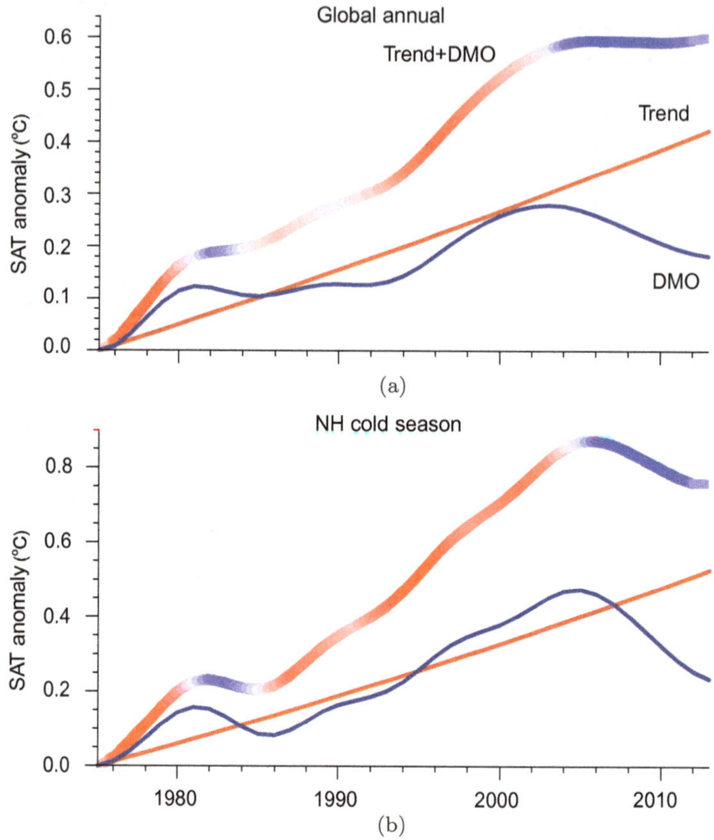

Fig. 8.4. The EEMD decomposed global annual mean SAT anomalies for the long-term trend component, DMO component, and the trend plus DMO (a). (b) Same as (a) but for NH cold season (Huang *et al.*, 2017b).

The mean SAT over the NH shows a clear cooling during the boreal cold season, indicating that the DMO's cooling tendency is stronger than the long-term warming trend (Fig. 8.4(b)).

The DMO index was produced using multiple linear regression (MLR) and EEMD techniques in order to investigate the modulated impact of the oceanic ICV modes on the NH SAT during the cold season. First, the EEMD method was applied to the indices of three classic oceanic modes and an atmospheric mode: the PDO, Niño3.4, AMO, and Arctic Oscillation (AO). Then, the DMO component of the NH SAT during the cold season (Fig. 8.4(b)) was regressed with standardized IMF 3, 4, and 5, for each index using a stepwise MLR. The regressed DMO component in Fig. 8.5 is expressed in Eq. (8.2). The regression-based approximation of DMO

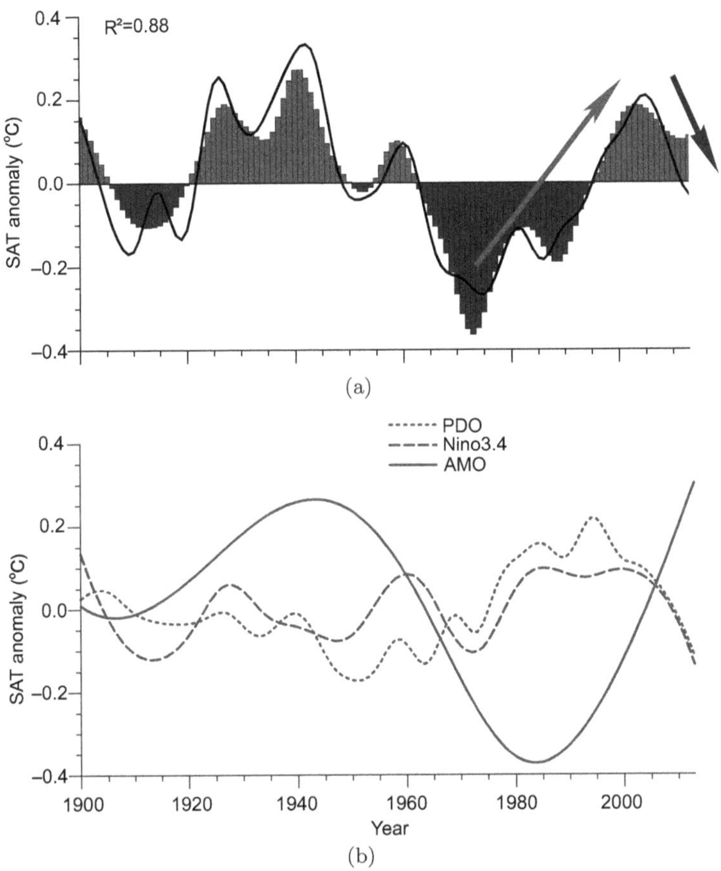

Fig. 8.5. The DMO component of the NH SAT anomalies during the cold season (black line) and the regression using the decadal variability of the PDO, Niño3.4, AMO, and AO indices (bar) (a). The contribution of PDO, Niño3.4, and AMO to the regressed DMO component in (a), (b). Note that the regression in (a) illuminated 88% of the variance in corresponding SAT anomalies, and the PDO, Niño3.4, AMO, and AO dedicated 22%, 18%, 56%, and 4%, respectively. As the contribution of AO was small, the line representing AO's contribution was not shown in (b) (Huang *et al.*, 2017b).

component for NH SAT using PDO, Niño3.4, AMO, and Arctic Oscillation (AO) indices can explain 88% of its variance during the boreal cold season (Fig. 8.5). The three classic oceanic ICV modes — PDO, Niño3.4, and AMO — contributed 22%, 18%, and 56%, respectively:

$$\begin{aligned}\mathrm{DMO} = {} & 0.008 + 0.032\mathrm{PDO}_3 + 0.094\mathrm{PDO}_5 + 0.063\text{Niño}_4 + 0.015\text{Niño}_5 \\ & + 0.206\mathrm{AMO}_5 + 0.016\mathrm{AO}_3 - 0.042\mathrm{AO}_4,\end{aligned} \tag{8.2}$$

$$\begin{aligned}\text{DMO} &= 0.008 + 0.018\text{EOF1}_3 + 0.012\text{EOF1}_4 + 0.104\text{EOF1}_5 \\ &\quad + 0.063\text{EOF2}_4 + 0.095\text{EOF2}_5 + 0.031\text{EOF3}_3 + 0.05\text{EOF3}_5 \\ &\quad + 0.024\text{AO}_3 - 0.037\text{AO}_4, \qquad (8.3)\end{aligned}$$

$$\text{DMO} = 0.002 + 0.044\text{EOF}_1 + 0.092\text{EOF}_2 + 0.114\text{EOF}_3 + 0.029\text{AO}. \qquad (8.4)$$

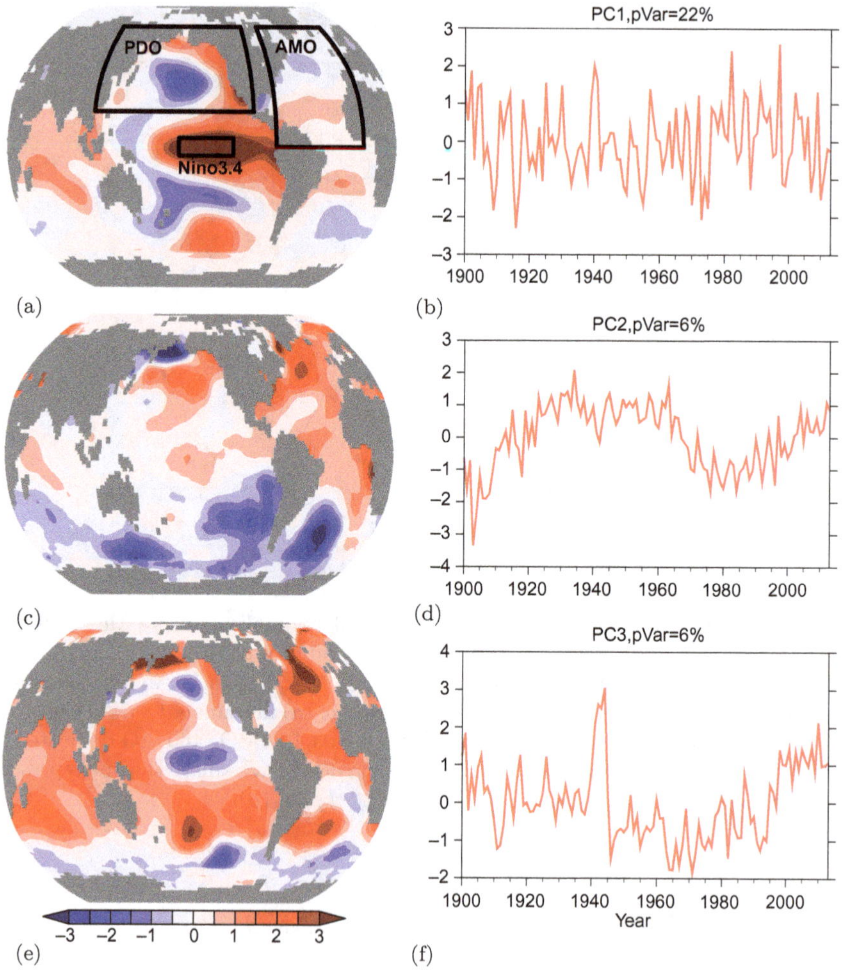

Fig. 8.6. The spatial pattern and temporal coefficient of the first EOF mode for the linear detrended global SST anomaly over the 1900–2013 period with respect to the 1961–1990 reference period for the boreal cold season, (a, b) respectively. (c–f) The same as (a, b) but for the second and third EOF modes, respectively. Note that the three black rectangles in (a) indicate the ocean areas defined by the PDO, AMO, and Niño3.4 indices (Huang *et al.*, 2017b).

Furthermore, the temporal coefficients in the first three empirical orthogonal function (EOF) modes of the detrended global SST (Fig. 8.6) and the AO index were also treated using a similar methodology. The outcomes are represented in Eqs. (8.3) and (8.4) and displayed in Fig. 8.7. The EOF analysis was initially performed to the linear detrended global SST for the boreal cold season (Fig. 8.6) to produce the results displayed in Fig. 8.7(a). The temporal coefficients of the first three EOF modes and the AO index were then calculated using techniques identical to those used to create Fig. 8.5. Similar techniques were used to produce the results in Fig. 8.7(c), but the EEMD approach was swapped out for the 11-year running mean method, which is represented by Eq. (8.4). The PDO and Niño3.4 indices have strong correlations with the first EOF mode of the detrended global SST, with correlation coefficients of 0.69 and 0.93, respectively (Table 8.2). The correlation coefficients for the second and third SSTs' EOF modes with the AMO are 0.61 and 0.64, respectively (Table 8.2). The relative contribution of each classic

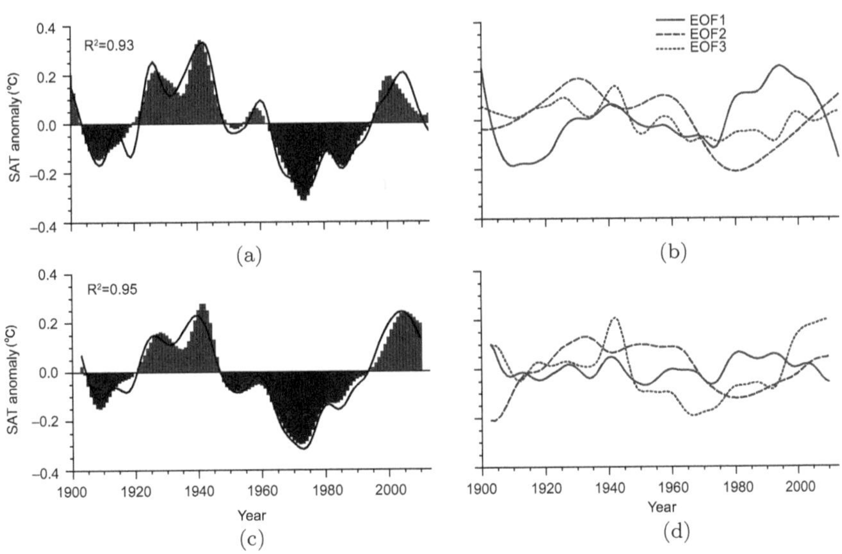

Fig. 8.7. The DMO component for the NH cold season mean SAT anomalies (black line), and its regression using decadal variability of EOF modes for global SST, and AO indices (bar) (a). The contribution of first, second, and third EOF modes to the regressed DMO component in (a), (b). (c, d) same as (a, b) but using the 11-year running mean instead of the EEMD method. Note that the regression in (a, c) explained 93% and 95% of the variance in corresponding SAT anomalies, respectively. The contribution rates of EOF1, EOF2, EOF3, and AO in (b) were 33%, 45%, 15%, and 7%, and those in (d) were 11%, 42%, 43%, and 4%, respectively. The lines that represent AO's contributions are not shown in (b, d) (Huang *et al.*, 2017b).

Table 8.2. The correlation coefficients between the temporal coefficients in Figs. 8.6(b), 8.6(d) and 8.6(f) and the PDO, Niño3.4, and AMO indices (Huang *et al.*, 2017b).

	EOF1	EOF2	EOF3
PDO	0.69	−0.25	0.19
Niño3.4	0.93	0.08	−0.12
AMO	0.16	0.61	0.64

oceanic ICV mode or SSTs EOF mode and AO to the regressed DMO component was calculated following the method described by Huang and Yi (1991).

According to Figs. 8.5 and 8.7, the decadal variability of the Atlantic Ocean contributes the most to the DMO, whereas the decadal variability of the North Pacific and Tropical Pacific has a relatively smaller impact. However, the combined output from the North Pacific and Tropical Pacific is comparable to that of the Atlantic. For decadal timescales, the North Pacific contributes more to the SAT variability of NH than the Tropical Pacific. Nevertheless, the Pacific Ocean's decadal variability played a major role in the recent warming hiatus (Kosaka and Xie, 2013; Trenberth *et al.*, 2014b). According to the data, which revealed that the relative contributions of the AO were 4%, 7%, and 4% in Figs. 8.5 and 8.7(b), 8.7(d), respectively, it appears that the AO makes little contribution to the DMO on the decadal scale. However, numerous studies have shown that the Arctic sea ice and increased Arctic warming have had a major impact on the SAT variability on mid-latitude NH continents in recent decades (Cohen *et al.*, 2014; Mori *et al.*, 2014; Screen and Simmonds, 2014). Therefore, the impact of the Arctic on the recent warming hiatus will be further explored in the next section. Since, the majority of NH is land, the next section will also look at how the suggested oceanic modulation affects the SAT variability of NH land.

8.2.3 *Thermally forced atmospheric circulation regime*

During the boreal cold season, the main factor influencing the decadal variability of terrestrial SAT over the NH is air circulation change (Wallace *et al.*, 2012; Guan *et al.*, 2015a), which is influenced by oceanic ICV via oceanic thermal forcing (Latif and Barnett, 1994; Rodwell *et al.*, 1999). To study the thermally forced atmospheric circulation change, the surface thermal forcing fields were examined first. In contrast to the recent slowdown in warming over the mid-to-high NH latitudes for the boreal cold season, the warming in the Arctic increased during the same time frame, which is consistent with Cohen *et al.* (2014). The SAT over NA and EUA demonstrated notable downward tendencies, but the Western EUA (areas near the Mediterranean) and Pacific regions continue to exhibit the rising trend.

The meridional temperature gradients between middle and high latitudes are represented by the MTF. The ZTF depicts temperature asymmetries between extra-tropical large-scale warm and cold zones in a zonal direction, particularly the thermal differential between sea and land. During the recent warming hiatus, the various SAT shifts across the mid-latitudes and the Arctic caused a rapidly diminishing meridional temperature gradient over the NH extra-tropics, which in turn caused an accelerated weakening of the extra-tropical MTF. However, the asymmetric ZTF, which is connected to the land–sea thermal contrast, intensified, changing from a major falling trend to a significant increasing trend. The difference between the two cooling centers over the EUA and NA continents and the two warming centers over the Western EUA and the North Pacific, which have been rapidly expanding, is what caused this heightened extra-tropical ZTF.

"Warm Arctic and Cold Land" (WACL) was a pattern that the weaker MTF followed (Overland *et al.*, 2011; Mori *et al.*, 2014), and this pattern was linked to a clear positive anomaly at 500-hPa geopotential height (GPH) over the Arctic. Additionally, a negative 500-hPa GPH anomaly was present in the mid-latitudes over the Eastern EUA, NA, Atlantic, and Mediterranean. Besides, the weakening of the meridional GPH gradient and the slower westerly winds will result in slower planetary wave speeds and larger amplitudes (Screen and Simmonds, 2014), which will facilitate the invasion of extremely cold air at the mid-latitudes and cause cooling over the EUA and NA during the recent hiatus. Additionally, the Arctic showed cooling trends during the previous warming hiatus (1940–1975), which intensified the polar vortex in the region. As a result of the stronger MTF in high latitudes, the Eurasian extra-tropical zones (poleward of 45°N) displayed warming tendencies. However, because of the weaker MTF over these areas, cooling trends were observed in the Eurasian mid and low latitudes (southward of 45°N). Because the stronger polar vortex stretched to high latitudes over North America during the preceding warming pause period, the high latitudes of North America displayed significant cooling trends.

The enhanced ZTF normally followed a "warm ocean and cold land" (WOCL) pattern (Wallace *et al.*, 1995; He *et al.*, 2014), but one warm center was over the Western EUA regions rather than over the Atlantic Ocean. One of the most noticeable warming areas was in the North Pacific, which coincided with a positive 500-hPa GPH anomaly and a negative PDO phase. Sea level pressure (SLP) fields, which are used to characterize the Aleutian Low, show quite a weaker Aleutian Low during the most recent warming hiatus. The air over NA was forced southward by the positive GPH or SLP anomalies over the North Pacific, which resulted in cooling throughout NA (Fig. 8.3). Additionally, during the previous warming pause period, the positive GPH anomalies over the North Pacific were also seen, which significantly influenced the cooling of the NA during that time. The obvious positive GPH anomalies and the weaker Aleutian Low across the North Pacific during the previous warming hiatus period were also suggested by earlier studies (Klyashtorin and Lyubushin, 2007). These studies investigated changes in atmospheric circulation across a time span (about 1915–1975) that contains the entire evolutions from the accelerated warming (about 1915–1940) to the warming hiatus (1940–1975). The findings suggested that the GPH and Aleutian Low over the North Pacific changed significantly during both the accelerated warming (about 1915–1940) and the previous warming hiatus period by analyzing the Pacific Circulation Index, Atmospheric Circulation Index, which represent the large-scale winds and atmospheric mass transfer, and Aleutian Low Pressure Index (Klyashtorin and Lyubushin, 2007) Large portions of the Eastern EUA received cold Arctic air from the southward airflow and the cyclone anomaly, which caused obvious cooling in those areas. The cyclone-like negative 500-hPa GPH anomaly over the Eastern EUA further accelerated the flow of air southward across EUA (Fig. 8.3) Furthermore, the same cyclone anomaly was present during the previous warming hiatus, but it was located over the mid and low latitudes of Eurasian. Additionally, a negative-phase NAO-like pattern was indicated by the opposite SLP or GPH anomaly at 850 hPa between Greenland and the mid-latitude Atlantic, which also contributed to the recent warming hiatus (Guan *et al.*, 2015a). However, the earlier warming hiatus was accompanied with a positive NAO-like pattern that most likely influenced the warming trends in extra-tropical regions across the Eurasian continent at that time.

8.2.4 *Abrupt shift of atmospheric circulation*

It was confirmed by a theoretical dynamic model (Källén, 1981; Charney and DeVore, 1979) that the circulation patterns responded as predicted to the MTF and ZTF variations over the extra-tropical Northern Hemisphere. The equation for a sphere's non-dimensional quasi-geostrophic barotropic vorticity governs the model (Källén, 1981; Charney and DeVore, 1979):

$$\frac{\partial}{\partial t}\nabla^2\psi = J(\nabla^2\psi + h, \psi) - 2\frac{\partial \psi}{\partial \lambda} + k\nabla^2(\psi^* - \psi), \tag{8.5}$$

where ψ is the non-dimensional stream function, ψ^* is the non-dimensional stream function forcing, t is non-dimensional time, ∇^2 is the non-dimensional horizontal Laplacian operator, J is the non-dimensional Jacobian operator, λ is the longitude, μ is the sine of the latitude ($\mu = \sin(\varphi)$), h is the non-dimensional topographic height parameter, and k is the non-dimensional dissipation rate.

For the sake of simplicity, a purely zonal component and two wave components are considered in the low-order model. Therefore, we choose three spherical harmonics,

$$F_A = P_2^0(\mu), \quad F_K = \cos(2\lambda)P_3^2(\mu), \quad F_L = \sin(2\lambda)P_3^2(\mu) \tag{8.6}$$

and truncate the expansions for ψ, ψ^*, and h as follows:

$$\psi = \psi_A F_A + \psi_K F_K + \psi_L F_L, \tag{8.7}$$

$$\psi^* = \psi_A^* F_A + \psi_K^* F_K, \tag{8.8}$$

$$h = h_o F_K. \tag{8.9}$$

The standardized associated Legendre function is defined as

$$P_n^m(\mu) = \sqrt{\frac{(2n+1)(n-m)!}{2(n+m)!}}\frac{(1-\mu^2)^{\frac{m}{2}}}{2^n n!}\frac{d^{n+m}}{d\mu^{n+m}}(\mu^2-1)^n$$

$$(m = 0, 1, 2, \ldots, n; \quad n = 0, 1, 2, \ldots) \tag{8.10}$$

so that

$$\int_{-1}^{1} P_n^m(\mu)P_{n*}^{m*}(\mu)d\mu = \begin{cases} 1, m = m^* \text{ and } n = n^* \\ 0, m \neq m^* \text{ or } n \neq n^* \end{cases}. \tag{8.11}$$

The flow of the atmosphere is confined in the Northern Hemisphere in the model (i.e., $\lambda \in [0, 2\pi]$, $\mu \in [0, 1]$). By substituting the expansions (8.6)–(8.9) into Eq. (8.5), we can get

$$\begin{cases} \frac{d\psi_A}{dt} = -\frac{1}{6}\alpha h_0 \psi_L + k(\psi_A^* - \psi_A) \\ \frac{d\psi_k}{dt} = \alpha\psi_A\psi_L + k(\psi_k^* - \psi_k) + \frac{1}{3}\psi_L \\ \frac{d\psi_L}{dt} = -\alpha\psi_A\psi_K + \frac{1}{6}\alpha h_0\psi_A - \frac{1}{3}\psi_k - k\psi_L \end{cases}, \tag{8.12}$$

where $\alpha = 105/128 \cdot \sqrt{10}$.

Solving for the equilibrium solution of the system given by Eq. (8.12), and we obtained the following equation using ψ_A:

$$6k(\psi_A^* - \psi_A)\left[\left(\alpha\psi_A + \frac{1}{3}\right)^2 + k^2\right] + \alpha h_0 k\left[\psi_A^*\left(\alpha\psi_A + \frac{1}{3}\right) - \frac{1}{6}\alpha h_0 \psi_A\right] = 0 \tag{8.13}$$

and

$$\psi_L = \frac{6k(\psi_A^* - \psi_A)}{\alpha h_0} \tag{8.14}$$

$$\psi_k = \frac{\alpha\psi_A + \frac{1}{3}}{k}\psi_L + \psi_K^* \tag{8.15}$$

According to the cubic Eq. (8.13), there can only be three equilibria for certain values of the forcing parameters. The stability of an equilibrium solution is hinged on the characteristic values of the three-order coefficient matrix of the linear perturbation equations obtained from Eq. (8.12), i.e., from

$$\begin{vmatrix} -(k+\sigma) & 0 & -\frac{1}{6}\alpha h_0 \\ \alpha\bar{\psi}_L & -(k+\sigma) & \alpha\bar{\psi}_A + \frac{1}{3} \\ -\alpha\bar{\psi}_K + \frac{1}{6}\alpha h_0 & -\alpha\bar{\psi}_A - \frac{1}{3} & -(k+\sigma) \end{vmatrix} = 0 \tag{8.16}$$

Two stable equilibrium states stand out among the other multiple equilibrium states: the first is a low-index flow with a strong wave component and a weaker zonal component (Fig. 8.8(e)), and the second is a high-index flow with a lower wave component and a stronger zonal component (Fig. 8.8(c)). Additionally, the change in MTF and ZTF causes an abrupt shift between the low-index and high-index equilibrium (Fig. 8.8(a)).

We first individually look at MTF and ZTF to analyze each pattern's unique impact profile before assessing the thermally induced atmospheric circulation alterations that result from the simultaneous modifications of both MTF and ZTF. Figure 8.9 depicts the equilibrium bifurcation associated with the changes to the parameter of MTF (ψ_A^*; ψ_A^* is opposite to MTF). $\psi_A^* < 0$ demonstrated that thermal forcing reduces as latitude increases. $\psi_A < 0$ indicates that the large-scale flow is a westerly wind, whereas $\psi_A > 0$ implies that the large-scale flow is an easterly wind. When the absolute value of ψ_A is higher than the absolute value of ψ_K and ψ_L, the zonal component is stronger than the wave component. We can notice that

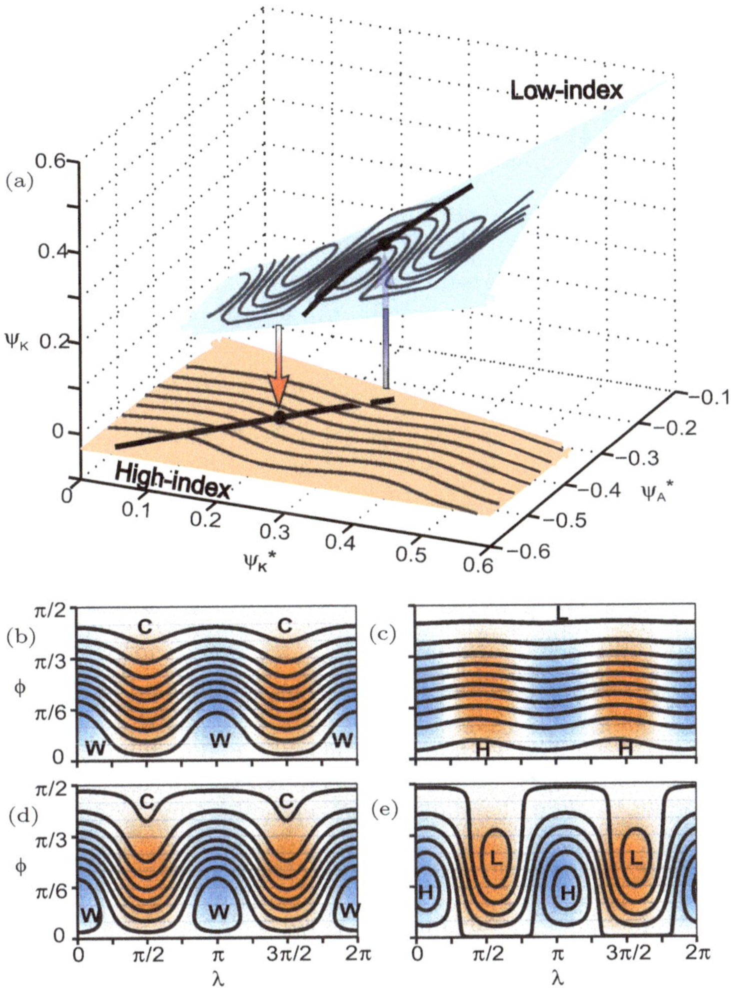

Fig. 8.8. The equilibrium bifurcation associated with the change both in the parameter of MTF (ψ_A^*) and the parameter of ZTF (ψ_K^*) (a). Note that ψ_A^* is the opposite of MTF, and ψ_K^* is same phase with ZTF. All the points in the orange and sky blue surface illustrate high-index and low-index equilibrium states, respectively. All the points with unstable equilibria are not shown. The black line is a possible track of equilibrium states that correspond to the change of both ψ_A^* and ψ_K^*. The thin black lines over the sky blue and orange surfaces in (a) represent the stream function field corresponding to the two black points. The heating field and stream function field correspond to the black point on the orange surface, respectively (b, c). (d, e) same as (b, c) but for the *black point* on the sky blue surface. The colored background shows the topography distribution in the model, and the warm tone indicates the "land" the cool tone demonstrates the "ocean." "C" and "W" in (b, d) represent the cold and warm center, respectively. "H" and "L" in (c, e) represent the high- and low-pressure center, respectively. The contour intervals are all 0.1 units. Parameter values are $k = 0.05$, $h_0 = -0.15$, $\psi_A^* = -0.42$, $\psi_K^* = 0.2$ in (b) and $\psi_A^* = -0.32$, $\psi_K^* = 0.3$ in (d) (Huang *et al.*, 2017b).

when the MTF is large ($\psi_A^* < -0.42$), there is only one feasible equilibrium state for the large-scale flow, which is a high-index flow with a large zonal component (Fig. 8.9(a)) and minor wave components (Figs. 8.9(b) and 8.9(c)). When the MTF is small ($-0.3 < \psi_A^* < 0$), the large-scale flow only has one conceivable equilibrium state, which is a low-index flow with a minor zonal component (Fig. 8.9(a)) and a significant wave component (Fig. 8.9(b) and (c)). When the MTF falls between the two cases outlined ($-0.42 < \psi_A^* < -0.3$), a high-index flow and a low-index flow are the two stable equilibrium solutions. Depending on the flow's initial state and how the MTF evolves over time, the atmosphere will be drawn to one or more stable flow regimes (Shabbar *et al.*, 2001).

The equilibrium bifurcation associated with the changes to the parameter of ZTF (ψ_K^*) is shown in Fig. 8.10. $\psi_K^* > 0$ indicates that the warming of the ocean is stronger than that of the land, whereas $\psi_K^* < 0$ means

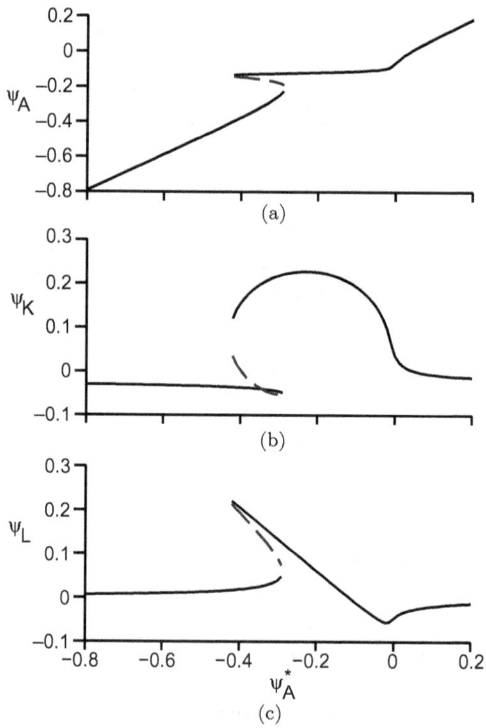

Fig. 8.9. The equilibrium bifurcation associated with the changes in the parameter of MTF (ψ_A^*). The ordinate gives the equilibrium values ψ_A in (a), ψ_K in (b), and ψ_L in (c). A numerical evaluation of the eigenvalues illustrates stability properties as demonstrated by the solid (stable) and dashed (unstable) lines. Parameter values are $k = 0.05$, $h_0 = -0.15$, and $\psi_K^* = 0.2$ (Huang *et al.*, 2017b).

that the cooling of the ocean is stronger than that of the land. When the ZTF with a "warm ocean and cold land" pattern (WOCL, Fig. 8.8(d)) is large ($\psi_K^* > 0.8$), there is only one possible equilibrium condition for the large-scale flow, which is a low-index flow with a minor zonal component (Fig. 8.10(a)) and a big wave component (Figs. 8.10(b) and 8.10(c)).

When the ZTF is small ($\psi_K^* < 0.2$) or if the ZTF exhibits a "cold ocean and warm land" pattern (COWL, $\psi_K^* < 0$), there is only one possible stable equilibrium state for the large-scale flow, which is a high-index flow with a large zonal component (Fig. 8.10(a)) and small-wave components (Figs. 8.10(b) and 8.10(c)). When the ZTF falls between the two conditions ($0.2 < \psi_K^* < 0.8$), there are also two stable equilibrium solutions. According to Figs. 8.9 and 8.10, a low-index flow was more likely to occur when MTF was lower and ZTF was higher. Figure 8.8(a) depicts the equilibrium bifurcation brought on by the simultaneous modification of the MTF (ψ_A^*) and ZTF (ψ_K^*) parameters.

Additionally, during the current hiatus, blocking corresponding to the low-index flow occurred more frequently over the EUA and NA (Fig. 8.11).

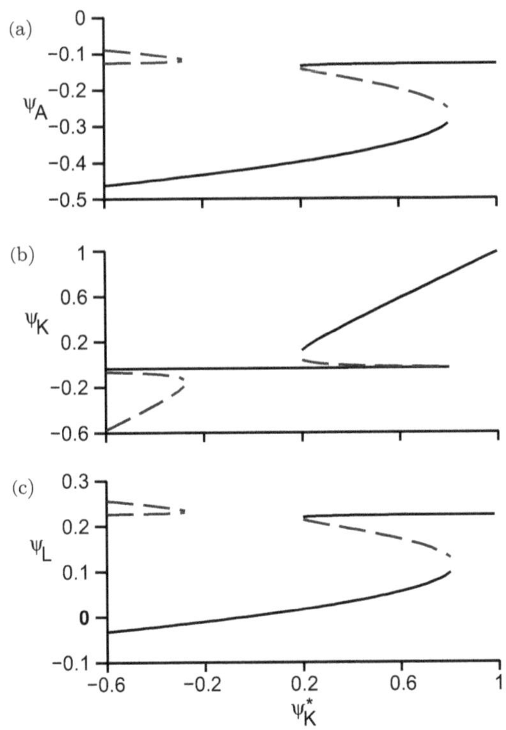

Fig. 8.10. Same as Fig. 8.9, except that the abscissa all give the parameter of ZTF (ψ_K^*) and have a parameter value $\psi_A^* = -0.42$ (Huang *et al.*, 2017b).

The EUA (50°E–140°E) and NA (170°E–80°W) regions' decadal variations of sector blocking frequency are depicted in Fig. 8.11(a). Figure 8.11(b) contrasts the beginning of the hiatus (1999–2003) with the current in terms

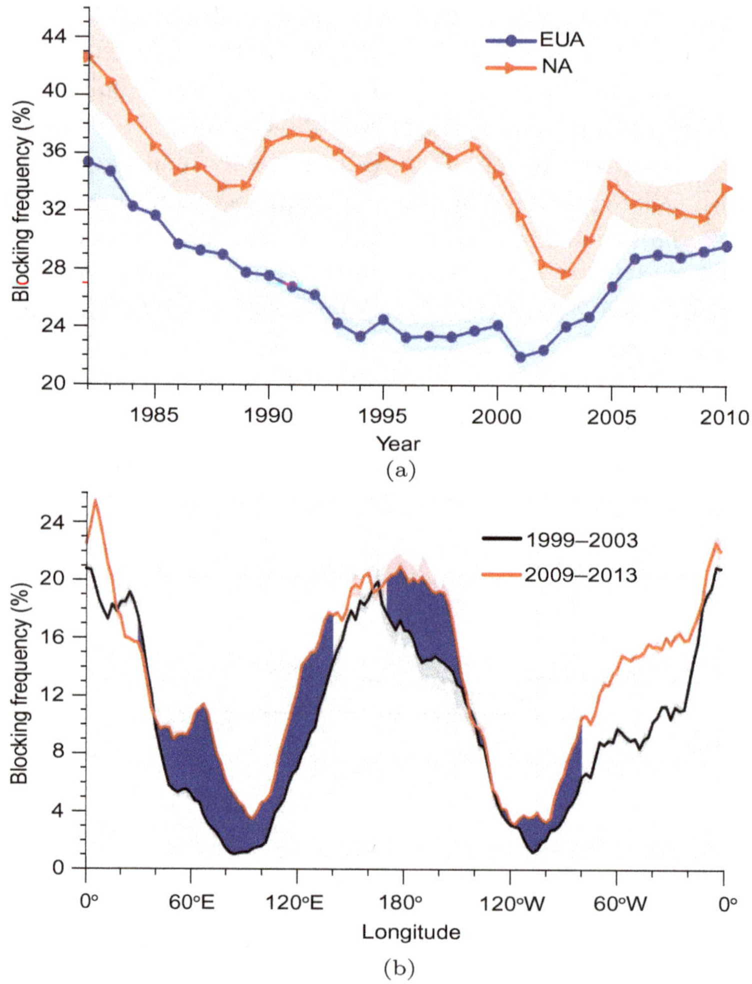

Fig. 8.11. The 11-year running mean sector blocking frequency for the boreal cold season over EUA (blue line) and NA (red line) (a). The shading represents the one standard deviation range of results from the ERA-interim, NCEP I, and II reanalysis products. The local blocking frequency for the boreal cold season average from 1999 to 2003 (black line) and from 2009 to 2013 (red line) (b). The pink shading and grey shading refer to the one standard deviation range of results from the ERA-interim, NCEP I, and II reanalysis products. The blue area stands for the positive difference between the average of the results for 2009–2013 and 1999–2003 over the EUA and NA regions. Note that the sector and local blocking was calculated following the procedure outlined in D'Andrea *et al.* (1998), Huang *et al.* (2017b).

of the 5-year mean difference of local blocking frequency (2009–2013). The method provided in D'Andrea *et al.* (1998) was used to compute the sector and local blockage. According to Fig. 8.11(a), since 2001, the blocking frequency over EUA increased by 8% in boreal cold season since 2001, which is 27% of its climatology. A cooling of the EUA during the recent hiatus will undoubtedly result from such a sharp rise in blocking events. Additionally, the blocking frequency over NA increased by 6%, which also caused the NA to cool (Fig. 8.11(a)). Besides, before the recent warming hiatus, the blocking frequency over EUA and NA dramatically dropped (Huang *et al.*, 2012; He *et al.*, 2014). The enhanced warming was also influenced by the frequency of blocking events. The increased frequency of the local blocking over the EUA and NA longitude zones also pointed to a recent rise in blocking events (Fig. 8.11(b)). In conclusion, the Arctic amplification (Cohen *et al.*, 2014; Mori *et al.*, 2014), land–sea thermal contrast (Wallace *et al.*, 1995; He *et al.*, 2014), the negative phase of PDO, and negative phase of NAO (Guan *et al.*, 2015a) jointly induced the weaker MTF and stronger ZTF, and then made the frequent blocking events in the recent warming hiatus.

8.3 Multiple Equilibria in Land–Atmosphere Coupled System

8.3.1 *Theoretical model*

The "thermal forcing," which refers to the radiative equilibrium temperature field as determined by Charney and Straus (1980) and the direct forcing of the flow wave field in Charney and DeVore (1979), is always artificially specified, which is a drawback of the traditional Charney model. As a result, there is no feedback from the atmospheric flow to the "thermal forcing," which means that in Charney's model, the atmospheric flow can only be adapted to the "thermal forcing" and cannot change the thermal distribution. It is possible that Charney's model overstates the impact of "thermal forcing" on large-scale atmospheric motions. A new model that couples the flow and temperature fields should be created to address this flaw. Some fundamental physical processes, such as the generation of atmospheric motions from horizontally inhomogeneous temperature fields, which then alter the distribution of temperature, should be included in the coupled model. The uneven solar heating, which is zonally symmetric and decreases from low to high latitudes, should be the external energy input to make up for the energy dissipation caused by friction. We establish such a simple coupled model in this section. We find that in the presence of topography, there are still multiple equilibria with distinct wave amplitudes (i.e., the high- and low-index flow). It is interesting to note that some stable equilibria have lower-layer stream functions that are either in phase with the topography or out of phase with it, i.e., their lower-layer ridges or troughs are over the mountains; we refer to these as ridge- or trough-type equilibria. In our coupled model, the multiple-wave phase equilibria associated with ridge and trough types are more noticeable than the multiple-wave amplitude equilibria associated with high- and low-index types. Besides, the coupled model exhibits more remarkable multiple-wave phase equilibria than the uncoupled model. There have not been as many investigations of multiple-wave phase equilibria as there have been of multiple-wave amplitude equilibria. This section investigates multiple-wave amplitude equilibria associated with high- and low-index types as well as multiple-wave phase equilibria associated with ridge and trough types using a low-order coupled land–atmosphere model.

The atmospheric model is a two-layer quasi-geostrophic flow limited by a periodic β plane channel with zonal walls at $y = 0$ and πL, similar to Charney and Straus (1980). The equations under pressure coordinates are as follows:

$$\frac{\partial}{\partial t}(\nabla^2\psi^1) + J(\psi^1, \nabla^2\psi^1) + \beta\frac{\partial\psi^1}{\partial x} = -k_d'\nabla^2(\psi^1 - \psi^3) + \frac{f_0}{\Delta p}\omega, \tag{8.17}$$

$$\frac{\partial}{\partial t}(\nabla^2\psi^3) + J\left(\psi^3, \nabla^2\psi^3 + f_0\frac{h}{H}\right) + \beta\frac{\partial\psi^3}{\partial x} = k_d'\nabla^2(\psi^1 - \psi^3) - \frac{f_0}{\Delta p}\omega - k_d\nabla^2\psi^3, \tag{8.18}$$

where x and y are eastward and northward coordinates, respectively; t is time, ∇^2 is the horizontal Laplace operator, and J is the Jacobian operator; ψ^1 and ψ^3 are the geostrophic stream function fields at $p_1 = 250$ and $p_3 = 750\,\text{hPa}$; $\omega = dp/dt$ is the vertical velocity; f_0 is the Coriolis parameter at central latitude $\phi_0 = 45°N$, with $\beta = df/dy$ as its meridional gradient; $\Delta p = 500\,\text{hPa}$ is the pressure difference between the two layers; H is mean depth of each layer; $h(x, y)$ is the lower-boundary topographic height; and we assume that $h \ll H$. The constants k_d and k_d'multiply the surface friction term and the internal friction between layers, respectively.

We define

$$\psi = (\psi^1 + \psi^3)/2, \quad \theta = (\psi^1 - \psi^3)/2, \tag{8.19}$$

then the equations of atmospheric motion become as follows:

$$\frac{\partial}{\partial t}(\nabla^2\psi) + J(\psi, \nabla^2\psi) + J(\theta, \nabla^2\theta) + \beta\frac{\partial\psi}{\partial x} = -0.5J\left(\psi, f_0\frac{h}{H}\right) + 0.5J\left(\theta, f_0\frac{h}{H}\right) - 0.5k_d\nabla^2(\psi - \theta), \tag{8.20}$$

$$\frac{\partial}{\partial t}(\nabla^2\theta) + J(\psi, \nabla^2\theta) + J(\theta, \nabla^2\psi) + \beta\frac{\partial\theta}{\partial x} = 0.5J\left(\psi, f_0\frac{h}{H}\right) - 0.5J\left(\theta, f_0\frac{h}{H}\right) + 0.5k_d\nabla^2(\psi - \theta) - 2k_d'\nabla^2\theta + \frac{f_0}{\Delta p}\omega. \tag{8.21}$$

In the equation for temperature of the baroclinic atmosphere, a radiative scheme and a heat flux scheme are introduced to reflect the energy exchanges among the land, atmosphere, and space (Barsugli and Battisti, 1998; Vannitsem *et al.*, 2015; De Cruz *et al.*, 2016):

$$\gamma_a\left(\frac{\partial T_a}{\partial t} + J(\psi, T_a) - \sigma\omega\frac{p}{R}\right) = -\lambda(T_a - T_g) + \varepsilon_a\sigma_B T_g^4 - 2\varepsilon_a\sigma_B T_a^4 + R_a, \tag{8.22}$$

where T_a and T_g are atmospheric and land temperature, respectively; σ is the static stability with p being the pressure; R is the gas constant for dry air; ω is the vertical velocity in pressure coordinates; γ_a is the heat capacity of the atmosphere for a 1,000-hPa deep column; λ is the heat transfer coefficient between the land and atmosphere; σ_B is the Stefan-Boltzmann constant; and ε_a is the longwave emissivity of the atmosphere. $\varepsilon_a\sigma_B T_g^4$ is the longwave radiation emitted from the land that is absorbed by the atmosphere; $-2\varepsilon_a\sigma_B T_a^4$ is the longwave radiation emitted from the atmosphere to the land and space; and R_a is the shortwave solar radiation directly absorbed by the atmosphere.

The land temperature equation is similar to the atmospheric temperature equation:

$$\gamma_g\frac{\partial T_g}{\partial t} = -\lambda(T_g - T_a) - \sigma_B T_g^4 + \varepsilon_a\sigma_B T_a^4 + R_g, \tag{8.23}$$

where γ_g is the heat capacity of the active layer of the land for a mean thickness of 10 m (Monin, 1986); $-\sigma_B T_g^4$ is the longwave radiation emitted from the land; $\varepsilon_a\sigma_B T_a^4$ is the longwave radiation emitted from the atmosphere absorbed by the land; and R_g is the shortwave solar radiation absorbed by the land.

Akin to Vannitsem *et al.* (2015), the quartic terms in the radiative fluxes are linearized. The system of equations is closed by the thermal wind relation:

$$\theta = \frac{R}{2f_0}\ln\left(\frac{p3}{p1}\right)T_a \approx \frac{R}{2f_0}T_a. \tag{8.24}$$

Let x and y be scaled by L, t by f_0^{-1}, ψ and θ by $L^2 f_0$, T_a and T_g by $L^2 f_0^2/R$, and let $h' = h/H$, $\omega' = \omega/(f_0\Delta p)$, $\beta' = \beta L/f_o$, $2k = k_d/f_0$, $k' = k_d'/f_o$, $\sigma' = \sigma(\Delta p)^2/(2L^2 f_0^2)$.

Other non-dimensional coefficients are

$$\left.\begin{aligned} &R_g' = R_g R/(\gamma_g f_0^3 L^2), R_a' = R_a R/(2\gamma_a f_0^3 L^2) \\ &\lambda_g' = \lambda/(\gamma_g f_0), \lambda_a' = \lambda/(\gamma_a f_0) \\ &\sigma_{B,a} = 8\varepsilon_a \sigma_B T_{a,0o}^3/(\gamma_g f_0), \sigma_{B,g} = 4\sigma_B T_{g,0}^3/(\gamma_g f_0) \\ &S_{B,a} = 8\varepsilon_a \sigma_B T_{a,0o}^3/(\gamma_a f_0), S_{B,g} = 4\varepsilon_a \sigma_B T_{g,0}^3/(2\gamma_a f_0) \end{aligned}\right\}. \tag{8.25}$$

We get the non-dimensional equations of the model as

$$\begin{aligned} &\frac{\partial}{\partial t'}(\nabla^2\psi') + J(\psi', \nabla^2\psi') + J(\theta', \nabla^2\theta') + \beta'\frac{\partial\psi'}{\partial x'} \\ &\quad = -0.5J(\psi', h') + 0.5J(\theta', h') - k\nabla^2(\psi' - \theta') \end{aligned} \tag{8.26}$$

$$\begin{aligned} &\frac{\partial}{\partial t'}(\nabla^2\theta') + J(\psi', \nabla^2\theta') + J(\theta', \nabla^2\psi') + \beta'\frac{\partial\theta'}{\partial x'}, \\ &\quad = 0.5J(\psi', h') - 0.5J(\theta', h') + k\nabla^2(\psi' - \theta') - 2k'\nabla^2\theta' + \omega', \end{aligned} \tag{8.27}$$

$$\frac{\partial\theta'}{\partial t'} + J(\psi', \theta') - \sigma'\omega' = -\lambda_a'(\theta' - 0.5\delta T_g') + S_{B,g}\delta T_g' - S_{B,a}\theta' + \delta R_a', \tag{8.28}$$

$$\frac{\partial\delta T_g'}{\partial t'} = -\lambda_g'(\delta T_g' - 2\theta') - \sigma_{B,g}\delta T_g' + \sigma_{B,a}\theta' + \delta R_g'. \tag{8.29}$$

Note that Eqs. (8.28) and (8.29) have been linearized (see Appendix A in Li *et al.*, 2018c). The non-dimensional atmospheric temperature, T_a', is already replaced by non-dimensional θ' in Eq. (8.28) according to Eq. (8.24); $\delta T_g'$is the non-dimensional land temperature anomaly; and $\delta R_a'$ and $\delta T_g'$ represent the non-dimensional meridional differential shortwave solar radiation absorbed by the atmosphere and the land, respectively. All the variables are now dimensionless unless otherwise specified. Next, for simplicity, we omit the primes of the non-dimensional variables ψ', θ', $\delta T_g'$and x', y', t', but others are retained to avoid confusion.

In accordance with the work of Charney and Straus (1980), we truncate the expansions for ψ, θ, δT_g, ω', and h' as

$$\left.\begin{array}{l}\psi = \sum_{i=1}^{3}\psi_i F_i, \quad \theta = \sum_{i=1}^{3}\theta_i F_i, \quad \delta T_g = \sum_{i=1}^{3} T_{g,i} F_i \\ \omega' = \sum_{i=1}^{3}\omega_i F_i, h' = h_2 F_2\end{array}\right\}. \tag{8.30}$$

We choose

$$\left.\begin{array}{l}F_1 = \sqrt{2}\cos y \\ F_2 = 2\cos(nx)\sin y \\ F_3 = 2\sin(nx)\sin y\end{array}\right\}. \tag{8.31}$$

Here, the zonal wavenumber n may be chosen freely, and it is related to the planetary zonal wavenumber $m = na\cos(\phi_0)/L = 2.83n$, where a is the radius of the earth. Notice that the channel is periodic in direction over the scale $2\pi/n$.

The dimensional boundary topography is given by

$$h = Hh_2F_2 = 2Hh_2\cos(nx/L)\sin(y/L). \tag{8.32}$$

In our model, we set $h_2 = 0.1$. Thus, the dimensional amplitude of the topography is fixed at $0.2H = 1.46\,\text{km}$.

The non-dimensional meridional differential shortwave solar radiation absorbed by the land and the atmosphere is given by

$$\delta R_g' = C_g' F_1, \quad \delta R_a' = C_a' F_1, \tag{8.33}$$

where $C_g' = C_g R/(\gamma_g f_0^3 L^2)$ and $C_a' = C_a R/(2\gamma_a f_0^3 L^2)$. The dimensional forms are

$$\delta R_g = \sqrt{2}C_g\cos(y/L), \quad \delta R_a = \sqrt{2}C_a\cos(y/L), \tag{8.34}$$

and we set $C_a = 0.4C_g$. Thus, the dimensional meridional differences in solar heating absorbed by the land and atmosphere between the southern wall $y = 0$ and the northern wall $y = \pi L$ of the channel are $2\sqrt{2}C_g$ and $2\sqrt{2}C_a$, respectively. The variable C_g is a dimensional parameter, which is an indicator of the meridional difference in solar heating absorbed by the land between the walls, and it is the most important variable parameter in our coupled model. Based on the observations, a typical value of C_g is 15 W m^{-2} for boreal summer, 55 W m^{-2} for boreal winter, and 35 W m^{-2} in spring or autumn.

We can obtain 12 spectral equations and by eliminating ω', the number of equations can be down to 9. The final spectral equations are given as follows:

$$\dot{\psi}_1 = -k(\psi_1 - \theta_1) - c\tilde{h}(\theta_3 - \psi_3), \tag{8.35}$$

$$(n^2+1)\dot{\psi}_2 = -cn^2(\psi_1\psi_3 + \theta_1\theta_3) + \beta n\psi_3 - B_1(\psi_2 - \theta_2), \tag{8.36}$$

$$(n^2+1)\dot{\psi}_3 = c[n^2(\psi_1\psi_2 + \theta_1\theta_2) + \tilde{h}(\theta_1 - \psi_1)] - \beta n\psi_2 - B_1(\psi_3 - \theta_3), \tag{8.37}$$

$$C_1\dot{\theta}_1 = c[\psi_2\theta_3 - \psi_3\theta_2 - \sigma'\tilde{h}(\psi_3 - \theta_3)] - B_3\theta_1 + k\sigma'\psi_1 - d_1\theta_1 + d_2T_{g,1} + C'_a, \tag{8.38}$$

$$(n^2+1)C_2\dot{\theta}_2 = c(A_1\psi_3\theta_1 - A_2\psi_1\theta_3) + \beta n\sigma'\theta_3 - B_2\theta_2 + B_1\sigma'\psi_2 - d_1\theta_2 + d_2T_{g,2}, \tag{8.39}$$

$$(n^2+1)C_2\dot{\theta}_3 = c[A_2\psi_1\theta_2 - A_1\psi_2\theta_1 + \sigma'\tilde{h}(\psi_1 - \theta_1)] - \beta n\sigma'\theta_2 - B_2\theta_3 + B_1\sigma'\psi_3 - d_1\theta_3 + d_2T_{g,3}, \tag{8.40}$$

$$\dot{T}_{g,1} = -d_3T_{g,1} + d_4\theta_1 + C'_g, \tag{8.41}$$

$$\dot{T}_{g,2} = -d_3T_{g,2} + d_4\theta_2, \tag{8.42}$$

$$\dot{T}_{g,3} = -d_3T_{g,3} + d_4\theta_3, \tag{8.43}$$

where the coefficients used here are

$$\left.\begin{aligned} c &= \tfrac{8\sqrt{2}n}{3\pi}, \quad \tilde{h} = \tfrac{h_2}{2} \\ d_1 &= \lambda'_a + S_{B,a}, \quad d_2 = \tfrac{\lambda'_a}{2} + S_{B,g} \\ d_3 &= \lambda'_g + \sigma_{B,g}, \quad d_4 = 2\lambda'_g + \sigma_{B,a} \\ A_1 &= 1 - \sigma' n^2, \quad A_2 = 1 + \sigma' n^2 \\ B_1 &= (\mathrm{n}^2+1)\mathrm{k}, \quad B_3 = (2k' + k)\sigma' \\ B_2 &= (\mathrm{n}^2+1)(2k'+k)\sigma' \\ C_1 &= \sigma' + 1, \quad C_2 = \sigma' + \tfrac{1}{n^2+1} \end{aligned}\right\}. \tag{8.44}$$

8.3.2 *Multiple equilibrium solutions*

We shall demonstrate the equilibrium solutions (i.e., stationary solutions) of Eqs. (8.35)–(8.43) and stabilities of the equilibrium solutions.

In Eqs. (8.35)–(8.43), we set all the time derivatives and wave components to zero, and then we obtain a specific equilibrium state:

$$\left.\begin{array}{l}\bar{\theta}_1 = \frac{D_2}{2k'\sigma' - D_1} \\ \bar{\psi}_1 = \bar{\theta}_1 \\ \bar{T}_{g.1} = \frac{d_4\bar{\theta}_1 + C'_g}{d_3}\end{array}\right\}, \tag{8.45}$$

where

$$D_1 = \frac{d_2 d_4}{d_3} - d_1, \quad D_2 = \frac{d_2 C'_g}{d_3} + C'_a. \tag{8.46}$$

This equilibrium state is referred to as "Hadley circulation" in Charney and Straus (1980). The $\bar{\psi}_1 = \bar{\theta}_1$ indicates that there is no lower-layer zonal flow, i.e., horizontally motionless in the lower layer, while there are strong westerlies without any meridional perturbations in the upper layer. It should also be pointed out that the topography has no effect on the Hadley solution.

Refer to Appendix B in Li *et al.* (2018c) for the procedure to derive the general equilibrium solution of Eqs. (8.35)–(8.43) and to ascertain the stabilities of the equilibrium solutions.

Next, the outcomes of calculations of the equilibrium solutions and their stabilities are then displayed. In this chapter, we favor the planetary zonal wavenumber $m = 3.7$ ($n = 1.3$), which is comparable to the "demonstration case" in Reinhold and Pierrehumbert (1982). Additionally used for comparison is wavenumber $m = 6$. We set $2k = 0.02$ and $k' = 0.005$ (same as Yoden, 1983). Therefore, the dissipation time of dimensional surface and internal friction is 5.6 days and 22.4 days, respectively.

The results of the equilibrium solutions and their stabilities for $m = 3.7$ are present in Table 8.3. In Table 8.3, for a given realistic value of C_g, there may be one ($C_g \leq 45$ W m^{-2}) or three ($C_g > 45$ W m^{-2}) equilibrium states, and some of them are stable. There are high-index equilibria and low-index equilibria among the stable states. The presence of at least one closed streamline for a low- or high-pressure center in the upper layer, as well as a stream function magnitude of 10^7 m^2 s^{-1} in the upper layer and 10^6 m^2 s^{-1} in the lower layer, is a requirement for low-index equilibria. High-index equilibria are those that do not match the aforementioned requirements. Specifically, the Hadley solution is high-index equilibria. For $C_g = 20$ W m^{-2}, the only one equilibrium state is stable. This equilibrium state has zero values for each of the wave components, making it a Hadley solution. For $C_g = 30$, 40, and 45 W m^{-2}, the results are the same as for $C_g = 20$ W m^{-2}. For $C_g = 50$ W m^{-2}, only the latter two of the three equilibrium states

Table 8.3. Non-dimensional equilibrium solutions for $m = 3.7$ (Li *et al.*, 2018c).

C_g (W m^{-2})	ψ_1	ψ_2	ψ_3	θ_1	θ_2	θ_3	$T_{g,1}$	$T_{g,2}$	$T_{g,3}$	S	Character
20	0.0258	0	0	0.0258	0	0	0.0603	0	0	—	Hadley
30	0.0387	0	0	0.0387	0	0	0.0905	0	0	—	Hadley
40	0.0516	0	0	0.0516	0	0	0.1207	0	0	—	Hadley
45	0.0580	0	0	0.0580	0	0	0.1358	0	0	—	Hadley
50	0.0644	0	0	0.0644	0	0	0.1509	0	0	N	High 2
	0.0620	—	—	0.0633	—	—	0.1487	—	—	—	High 1
	0.0643	0.0025	0.0071	0.0618	0.0030	0.0069	0.1461	0.0053	0.0123	—	—
	—	—	0.0111	—	—	0.0109	—	—	0.0194	—	—
	—	0.0035	—	—	0.0026	—	—	0.0047	—	—	—
55	0.0709	0	0	0.0709	0	0	0.1659	0	0	N	High 2
	0.0620	—	—	0.0661	—	—	0.1575	—	—	—	Low 1
	0.0643	0.0115	0.0092	0.0612	0.0126	0.0087	0.1487	0.0225	0.0155	—	—
	—	—	0.0187	—	—	0.0183	—	—	0.0328	—	—
	—	0.0124	—	—	0.0104	—	—	0.0186	—	—	—
60	0.0773	0	0	0.0773	0	0	0.1810	0	0	N	Low 1
	0.0804	0.0227	0.0012	0.0711	0.0177	0	0.1699	0.0316	0	N	—
	0.0650	—	0.0219	0.0610	—	0—	0.1517	—	0.0383	—	—
	—	0.0201	—	—	0.0171	0214	—	0.0305	—	—	—
70	0.0902	0	0	0.0902	0	0	0.2112	0	0	N	Low 1
	0.0816	0.0437	0	0.0685	0.0324	—	0.1724	0.0579	—	N	—
	0.0664	—	0.246	0.0607	—	0.0017	0.1583	—	0.0031	—	—
	—	0.0329	—	—	0.0279	0.0238	—	0.0498	0.0426	—	—
80	0.1031	0	0	0.1031	0	0	0.2414	0	0	N	Low 1
	0.0816	0.0563	—	0.0674	0.0412	—	0.1776	0.0736	—	N	—
	0.0676	—	0.0013	0.0605	—	0.0031	0.1652	—	0.0055	—	—
	—	0.0433	0.0257	—	0.0362	0.0248	—	0.0648	0.0443	—	—

The column S describes the stability of the equilibria. Sign "–" denotes a stable equilibrium and "N" denotes an unstable equilibrium.

are stable. The Hadley solution, which was the first equilibrium state, is now unstable. Figure 8.12 shows the temperature and stream function fields of the second and third equilibrium stages.

Due to the presence of both strong zonal westerlies and weak meridional perturbations in the upper layer, the second and third equilibrium states are both high index (Figs. 8.12(a) and 8.12(e)). However, in the second equilibrium state, there are wavy easterlies in the lower layer (Fig. 8.12(b)), while in the third equilibrium state, there are wavy westerlies in the lower layer (Fig. 8.12(f)). The isotherms in the two equilibrium states' atmospheric and land temperature fields are all quite flat (Figs. 8.12(c), (d), (g) and 8.12(h)) and almost in phase with the corresponding upper-layer stream function fields. Both of the two equilibrium states have a characteristic baroclinic structure, i.e., the waves of stream function fields demonstrated westward phase shifts with height. However, they have different wave phases

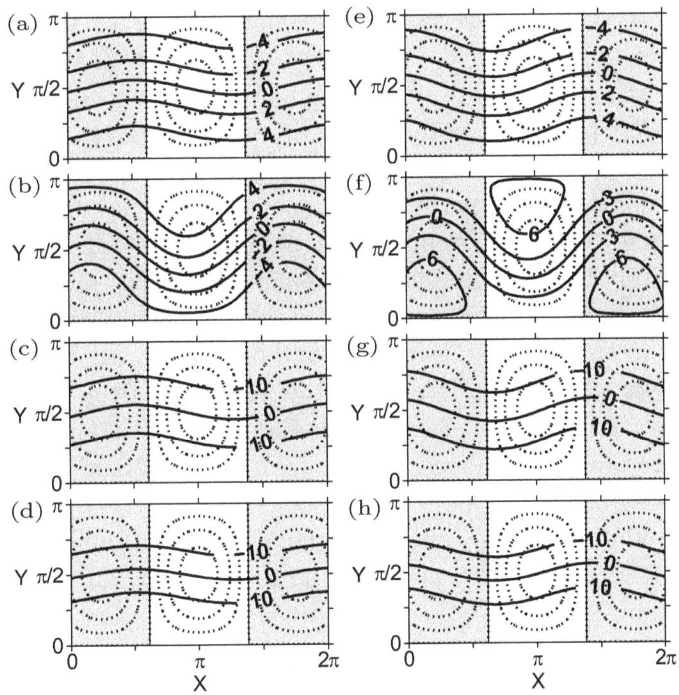

Fig. 8.12. The second one (left panels) and third one (right panels) of the three equilibrium states for $m = 3.7$ at $C_g = 50\,\mathrm{W\,m^{-2}}$. They belong to "High 2" and "High 1" equilibria, respectively. The stream function fields of the upper (a, e) and lower (b, f) layers. The temperature fields of the atmosphere (c, g) and the land (d, h). The contour intervals are (a, e) 2.0×10^7 m^2 s^{-1}, (b) 2.0×10^5 m^2 s^{-1}, (f) 3.0×10^5 m^2 s^{-1}, and (c, d, g, h) 10 K. The background dotted lines show the topographic heights in the model, with negative regions shaded (Li *et al.*, 2018c).

relative to the topography. With respect to the second equilibrium state, its lower-layer stream function is close to the topography, the lower-layer inverted ridges (anticyclonic flow) cross the mountains (positive topographic heights), lying slightly west of the mountain crests (Fig. 8.12(b)), and the upper-layer ridges are located to the west side of the mountains (Fig. 8.12(a)). This equilibrium is referred to as "ridge type." In contrast, for the third equilibrium state, the lower-layer stream function is almost out of phase with the topography, the lower-layer low-pressure centers and troughs are over the mountains, also lying just west of the mountain crests (Fig. 8.12(f)), and the upper-layer troughs are located to the west of the mountains (Fig. 8.12(e)). This is called "trough-type" equilibrium. We simply refer to the two equilibrium states' characteristics as "High 2" and "High 1," respectively. In this case, "High" stands for "high-index," "2" refers to "ridge type," and "1" means "trough type."

For $C_g = 55\,\mathrm{W\ m^{-2}}$, the last two of the three equilibrium states are stable as well. While the second one becomes low-index equilibrium, the first one remains in "High 2" high-index equilibrium (Table 8.3). Figure 8.13 depicts the temperature and stream function fields of this low-index equilibrium (left panel). In both the upper- and lower-layer stream function fields (Figs. 8.13(a) and 8.13(b)), there are relatively weak westerlies with strong meridional flow, especially closed streamlines in the former. Both the atmosphere and land temperature fields (Figs. 8.13(c) and 8.13(d)) exhibit relatively large meridional disturbances, with closed isotherms in the latter. This low-index equilibrium is a trough-type equilibrium because the lower-layer stream function is almost out of phase with the topography, the upper-layer troughs are located to the west side of the mountains, and the lower-layer troughs are placed above the mountains. The nature of this equilibrium condition is referred to as "Low 1," where "Low" stands for "low index."

At $C_g = 60\,\mathrm{W\ m^{-2}}$, only the third one of the three equilibrium states is stable, that is, the "Low 1" equilibrium. The second one becomes unstable. For $C_g = 70$ and $80\,\mathrm{W\ m^{-2}}$, the results are the same as for $C_g = 60\,\mathrm{W\ m^{-2}}$.

For the sake of comparison, for wavenumber 6, we have calculated the equilibrium solutions. We found that for a given value of C_g (not shown), there may be one, three, or five equilibrium states. Some of them are stable. Besides the stable "High 1", "High 2", and "Low 1" equilibrium, a new stable low-index equilibrium may exist. As illustrated in Fig. 8.13 (right panel), this low-index equilibrium has strong meridional perturbations in both the upper layer stream function field (Fig. 8.13e) and temperature

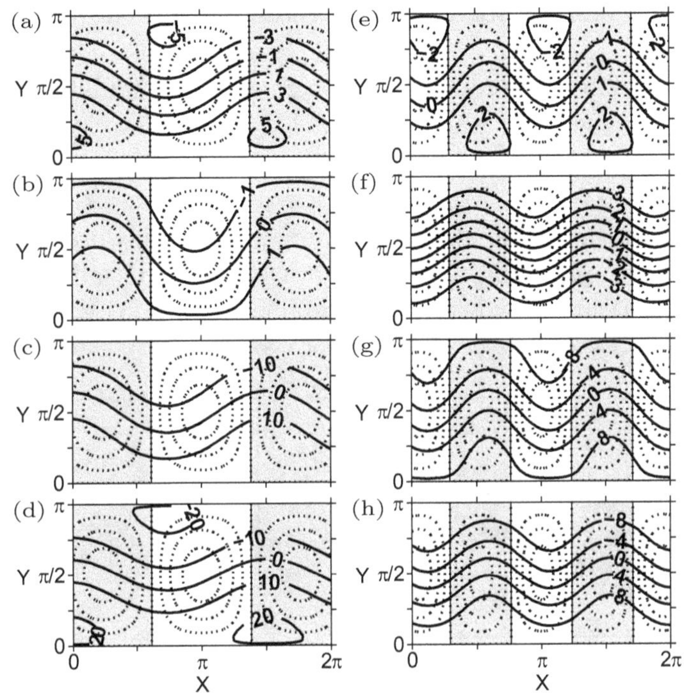

Fig. 8.13. Same as Fig. 8.12, but for the third one of the three equilibrium states for $m = 3.7$ at $Cg = 55$ W m^{-2} (left panels) and for the third one of the five equilibrium states for $m = 6$ at $Cg = 30$ W m^{-2} (right panels). They belong to "Low 1" and "Low 2" equilibria, respectively. The "Low 2" equilibria have non-dimensional solutions with $(\psi_1, \psi_2, \psi_3, \theta_1, \theta_2, \theta_3, T_{g,1}, T_{g,2}, T_{g,3}) = (0.0216, -0.0079, -0.0051, 0.0311, -0.0098, -0.0043, 0.0771, -0.0175, -0.0077)$. The contour intervals are (a) 2.0×10^7 m^2 s^{-1}, (e) 1.0×10^7 m^2 s^{-1}, (b, f) 1.0×10^6 m^2 s^{-1}, (c, d) 10 K, and (g, h) 4 K (Li *et al.*, 2018c).

fields (Figs. 8.13g and 8.13h). However, the lower-layer stream function field has wavy easterlies (Fig. 8.13(f)). It should be noted that this is a ridge-type equilibrium since the lower-layer stream function is almost in phase with the topography, the lower-layer inverted ridges are positioned above the mountains, and the upper-layer ridges are situated on the west side of the mountains. The nature of this equilibrium condition is referred to as "Low 2."

As a further demonstration of the multiple equilibrium states and their stabilities for wavenumbers 3.7 and 6, we show several simple bifurcation diagrams in Fig. 8.14. The zonal component ψ_1^1, the wave component ψ_2^1, and ψ_3^1 of the upper-layer stream function ψ^1 are given by $\psi_1^1 = \psi_1 + \theta_1$, $\psi_2^1 = \psi_2 + \theta_2$, and $\psi_3^1 = \psi_3 + \theta_3$, respectively. To show the existence of equilibrium solutions, we denote the 2 W m^{-2} interval of the parameter C_g.

With regard to wavenumber 3.7, there are four equilibrium branches (Fig. 8.14, left panel). For small values of C_g, the Hadley circulation (black) is the only equilibrium and it is stable. As C_g is gradually increased, around $C_g = 50\,\mathrm{W\ m^{-2}}$, the Hadley circulation becomes unstable, and two new equilibria (blue and red) emerge. The blue branch represents a trough-type equilibrium, and it is always stable. It includes "High 1" ($50 \le C_g \le 52\,\mathrm{W\,m^{-2}}$) and "Low 1" equilibria ($C_g \ge 54\,\mathrm{W\,m^{-2}}$). The red

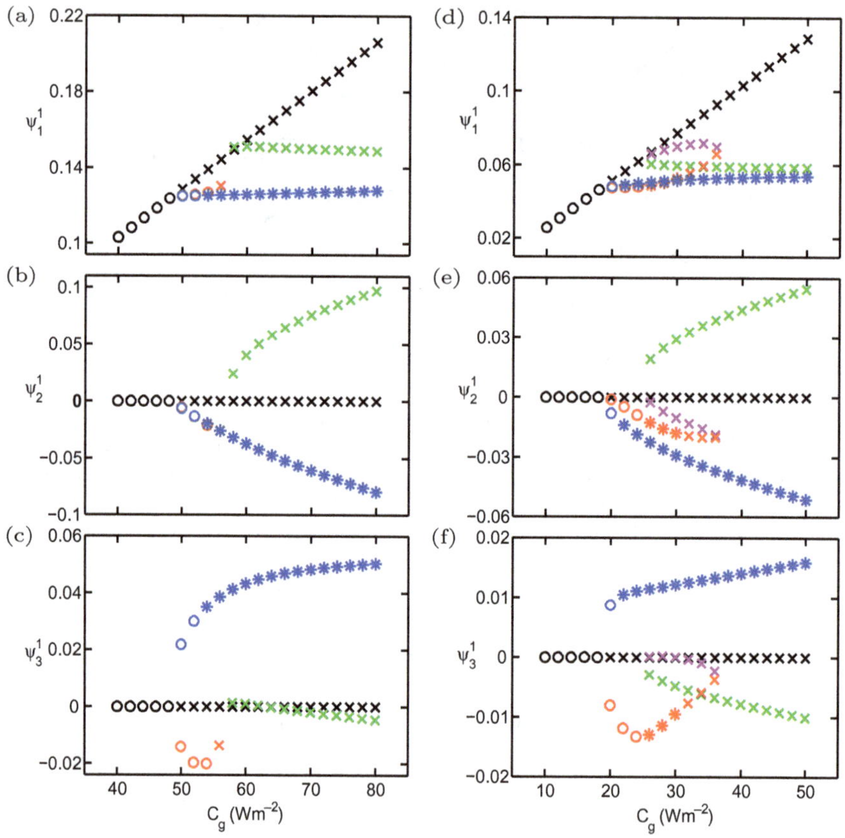

Fig. 8.14. The equilibrium bifurcation associated with the change in meridional differential solar heating parameter C_g for $m = 3.7$ (left panels) and $m = 6$ (right panels), respectively. The ordinate shows the nondimensional equilibrium values of (a, d) the zonal component and the wave components (b, e) and (c, f), respectively. Different branches of equilibrium solutions have different colors. The crosses denote unstable equilibria, the circles denote stable high-index equilibria, and the asterisks denote stable low-index equilibria.

branch refers to a ridge-type equilibrium, and it includes stable "High 2" equilibrium ($50 \leq C_g \leq 54$ W m^{-2}). It becomes unstable around $C_g = 56$ W m^{-2}, then it disappears and appears as a new equilibrium (green) when $C_g > 56$ W m^{-2}. The green branch equilibrium is always unstable.

Five equilibrium branches exist for wavenumber 6 (Fig. 8.14, right panel). For small values of C_g, the stable Hadley circulation (black) is still the only equilibrium. As C_g is increased to about $C_g = 20$ W m^{-2}, the Hadley circulation becomes unstable, and generates two new equilibria (blue and red). The blue branch represents a trough-type equilibrium and is always stable. It includes "High 1" ($C_g = 20$ W m^{-2}) and "Low 1" equilibria ($C_g \geq 22$ W m^{-2}). The red branch stands for a ridge-type equilibrium, and it includes stable "High 2" equilibrium ($20 \leq C_g \leq 24$ W m^{-2}) and stable "Low 2" equilibrium ($26 \leq Cg \leq 30$ W m^{-2}). The ridge-type equilibrium is unstable within $32 \leq C_g \leq 36$ W m^{-2} and it disappears when $C_g > 36$ W m^{-2}. At around $C_g = 26$ W m^{-2}, there are two more unstable equilibria (green and magenta). The magenta branch disappears when $C_g > 36$ W m^{-2}.

According to the aforementioned findings, the coupled model contains multiple equilibrium states with different wave phases and amplitudes. For an extensive range of C_g values ($50 \leq C_g \leq 54$ W m^{-2} for $m = 3.7$ and $20 \leq C_g \leq 30$ W m^{-2} for $m = 6$), different wave phase relatives to the topography, i.e., ridge- and trough-type equilibria, may exist simultaneously (Fig. 8.14). However, it is only for a small range of C_g values ($C_g = 54$ W m^{-2} for $m = 3.7$ and $22 \leq C_g \leq 24$ W m^{-2} for $m = 6$) that a significant wave amplitude, i.e., high- and low-index equilibria, may coexist (Fig. 8.14). Therefore, ridge- and trough-type multiple-wave phase equilibria are more noticeable than high- and low-index types' multiple-wave amplitude equilibria.

Next, we explore the origin of the multiple equilibria. Figure 8.15(a) shows the stability curves of the Hadley circulation in the coupled model. The orographically unstable region is surrounded by blue lines. The variable α changes from having a negative real value to having a positive real value (pitch-fork bifurcation) when crossing the blue lines between the sides of stability and instability. In the presence (absence) of topography, the baroclinically stable and unstable zones are distinguished by the red (black dashed) lines. In crossing these lines from the stable to unstable sides, the real part of the complex α changes from negative to positive while the imaginary part is not zero (Hopf bifurcation). The orographic instability of the Hadley circulation is only present when the topography is present. Moreover, when topography is present, there is no overlap between orographic instability and baroclinic instability. Besides, it can be observed

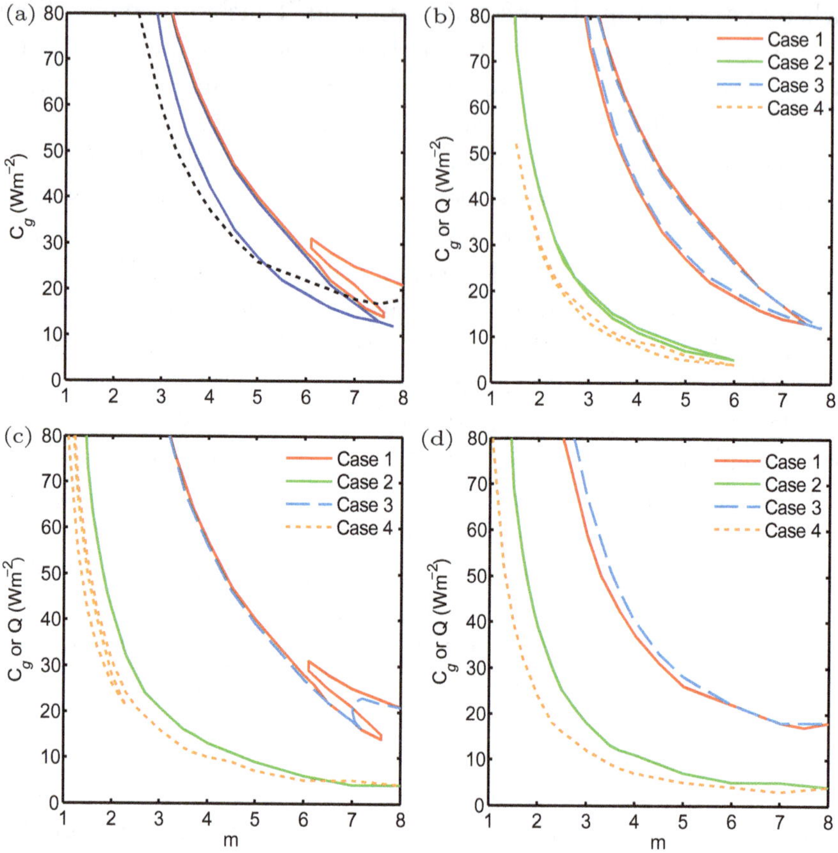

Fig. 8.15. Stability curves of the Hadley circulation in the coupled land–atmosphere model (Case 1) (a). The blue solid lines enclose the region of orographic instability. The red solid (black dashed) lines and the top x-axis and the right y-axis enclose the region of baroclinic instability in the presence (absence) of topography. Comparison of the regions of orographic instability (b), baroclinic instability in the presence of topography (c), and baroclinic instability in the absence of topography for the four experiments (d) (Cases 1–4) (Li *et al.*, 2018c).

from comparing the baroclinic stability curves with and without topography that topography stabilizes the Hadley circulation for most wavenumbers. In the case of topography, due to the baroclinic instability of the Hadley circulation (which is a Hopf bifurcation), there is only the Hadley circulation or the traveling wave when topography is absent. For example, for $m = 3.7$ at $C_g = 50$ W m^{-2} without the topography, using any initial conditions as a starting point, numerical integration converges to a periodic solution with a period of 18 days (Fig. 8.16). Here, the zonal component ψ_1^3 and the wave component ψ_3^3 of the lower-layer stream function ψ^3 are given by

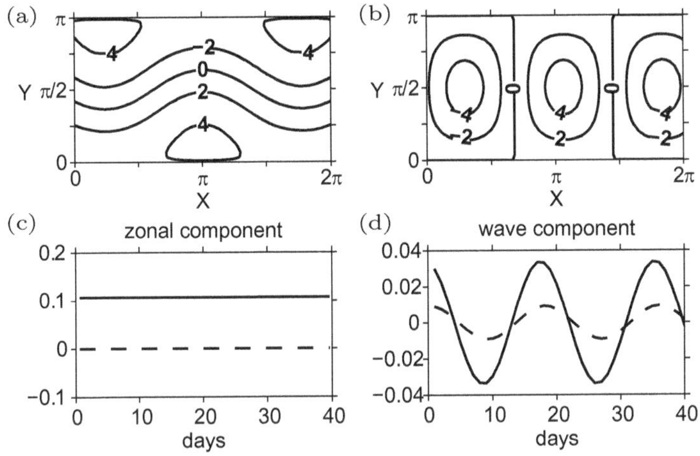

Fig. 8.16. A traveling wave solution in the absence of topography for $m = 3.7$ at $C_g = 50\,\mathrm{W\,m^{-2}}$ with $(\psi_1, \psi_2, \psi_3, \theta_1, \theta_2, \theta_3, T_{g,1}, T_{g,2}, T_{g,3}) = (0.0534, 0.0214, 0.0016, 0.0534, 0.0125, -0.0008, 0.1311, 0.0014, 0.0048)$ at this moment. The stream function fields of the upper and lower layers (a, b), respectively. The contour intervals are (a) $2.0 \times 10^7\,\mathrm{m^2\,s^{-1}}$ and (b) $2.0 \times 10^6\,\mathrm{m^2\,s^{-1}}$. Temporal evolution for the non-dimensional equilibrium values of the zonal component ψ_1^1 (solid line) and ψ_3^1 (dash line) (c). Temporal evolution for the non-dimensional equilibrium values of the wave component ψ_3^1 (solid line) and ψ_3^3 (dashed line) (d) (Li *et al.*, 2018c).

$\psi_1^3 = \psi_1 - \theta_1$ and $\psi_3^3 = \psi_3 - \theta_3$, respectively. In this periodic solution, the upper-layer stream function exhibits blocking-like flow (Fig. 8.16(a)), whereas the lower-layer stream function exhibits wave train rather than zonal flow (the zonal component ψ_1^3 remains zero in Fig. 8.16(c)) (Fig. 8.16(b)). Compared with the wave components of upper- and lower-layer stream function (ψ_3^1, ψ_3^3), which change frequently over time (Fig. 8.16(d)), the zonal components of upper- and lower-layer stream function (ψ_1^1, ψ_1^3) remain constant (Fig. 8.16(c)). In particular, this traveling wave moves westward.

Multiple equilibrium states may exist when topography is present. It appears from a comparison of Figs. 8.12 and 8.16 that topography has the effect of splitting the traveling wave into two different types of stationary waves. In fact, the first bifurcation (Fig. 8.14, around $C_g = 50$ W m^{-2} for $m = 3.7$ and around $C_g = 20\,\mathrm{W\,m^{-2}}$ for $m = 6$) results from the orographic instability of the Hadley circulation (Fig. 8.15(a), around $C_g = 50$ W m^{-2} for $m = 3.7$ and around $C_g = 20\,\mathrm{W\,m^{-2}}$ for $m = 6$), and it is a (supercritical) pitch-fork bifurcation. Because it controls the occurrence and coexistence of trough- and ridge-type equilibria, this bifurcation is significant. It should be noticed that the occurrence of baroclinic instability of the Hadley circulation (Fig. 8.15(a), around $C_g = 64$ W m^{-2} for $m = 3.7$ and around

$C_g = 28$ W m^{-2} for $m = 6$) is unrelated to the disappearance of the ridge-type equilibria (Fig. 8.14, around $C_g = 56$ W m^{-2} for $m = 3.7$ and around $C_g = 36$ W m^{-2} for $m = 6$).

In light of this, the Hadley circulation's orographic instability, which is a pitch-fork bifurcation, is the cause of the multiple-wave phase equilibria associated with the ridge and trough types.

8.3.3 *Role of land–atmosphere coupling*

The existence and properties of the equilibrium states are examined in this section in combination with the role of the land–atmosphere coupling. Four experiments are designed with different diabatic heating terms. For simplicity, here we refer to the coupled land–atmosphere model as Case 1. The diabatic heating terms in Case 1 consist of three terms, heat flux, longwave radiation, and shortwave radiation, according to Eqs. (8.22) and (8.23). For Case 2, we eliminate all the terms on the right side of Eq. (8.22) and replace with specified heating Q^* and delete Eq. (8.23) ($Q^* = \sqrt{2}Q\cos(y/L)$, where Q is a heating parameter that is similar to C_g). As a result, Case 2 is the traditional uncoupled model. We eliminate the heat flux terms of Eqs. (8.22) and (8.23) (or set $\lambda = 0$ W m^{-2} technically) for Case 3. Equations (8.22) and (8.23) (or put $\sigma_B =$ W m^{-2}K^{-4} technically), which deal with longwave radiation, are removed for Case 4.

The orographic instability of the Hadley circulation in the four studies is compared in Fig. 8.15(b). The thresholds of orographic instability in Cases 2 and 4 are obviously much lower for wavenumbers 1–6 than they are in Case 1. In Cases 2 and 4, the orographically unstable regions are also both very narrow. Unexpectedly, Cases 3 and 1 have nearly overlapped orographically unstable regions. The Hadley circulation's baroclinic instability with and without topography is shown in Fig. 8.15(c) and 8.15(d), respectively. Similar to Case 1, Cases 2 and 4 both have significantly lower thresholds for baroclinic instability for wavenumbers 1–8. However, Cases 3 and 1's baroclinic stability curves roughly overlap. The findings show that, in comparison to the uncoupled model (Case 2), the coupling between the land and the atmosphere (Case 1) considerably stabilizes the Hadley circulation. This stabilizing effect is primarily attributed to the presence of longwave radiation fluxes, but not heat fluxes.

Furthermore, Case 4 has lower thresholds than Case 2 for all orographic instability and baroclinic instability, both with and without topography (Figs. 8.15(b–d)). It implies that the Hadley circulation is severely destabilized by the presence of heat fluxes, whether topography is present or not. However, the destabilizing effect of the heat fluxes on Hadley circulation is almost entirely suppressed in Case 1, which includes both the heat fluxes and the longwave radiation fluxes.

Next, the equilibrium bifurcation in the four experiments is compared (see Li *et al.* (2018c) for more detailed results). We have analyzed the bifurcation in Case 3 for wavenumber $m = 3.7$ and $m = 6$. Cases 3 and 1 have almost overlapped orographic and baroclinic instability curves (Fig. 8.15), but their equilibrium bifurcation is not identical. When the heat flux is missing in Case 3, as opposed to Fig. 8.14(c), we found that practically all ridge-type equilibrium states become unstable $m = 3.7$. However, when $m = 6$, Case 3 does not significantly alter the equilibrium bifurcation. These indicate that the ridge-type equilibrium is more or less stabilized (destabilized) by the presence (absence) of heat fluxes.

Furthermore, we get the bifurcation diagrams in Cases 2 and 4 for wavenumber $m = 3.7$. The Hadley circulation first splits at remarkably low heating parameter values in Cases 2 and 4 (Fig. 8.15(b)), because of the low thresholds of orographic instability of the Hadley circulation. However, because the orographically unstable regions in Cases 2 and 4 are so narrow (Fig. 8.15(b)), the ridge-type equilibrium eventually vanishes at low heating parameter values. The ridge- and trough-type equilibria in this situation may coexist, but only for a very narrow range of heating parameter values. The ridge- and trough-type equilibria may coexist in Case 1 (Fig. 8.14) for a large range of heating parameter values, primarily because of the relatively extensive orographically unstable regions (Figs. 8.15(a) and 8.15(b)). As a result, the multiple wave phase equilibria associated with the ridge and trough types in the coupled model (Case 1) are more remarkable when compared to the uncoupled model (Case 2).

Then, we compare the stream function and temperature fields of equilibrium states in the four experiments for $m = 3.7$ at the same heating parameter values: $Q = 50$ or $C_g = 50$ W m^{-2}. In Case 2, the stream function and temperature fields both exhibit blocking-like large amplitude perturbations (Fig. 8.17, left panel), and there is only one stable equilibrium state. It is clear that the meridional perturbations in stream function and temperature fields of the equilibrium state in Case 2 are substantially stronger than those of the two equilibrium states in Case 1 (Fig. 8.12). This result is partially explained by the Case 2 Hadley circulation's extremely low threshold of orographic instability (Fig. 8.15(b)). The result in Case 4 is similar to Case 2, but the meridional perturbations in the stream function and temperature fields are stronger than they were in Case 2

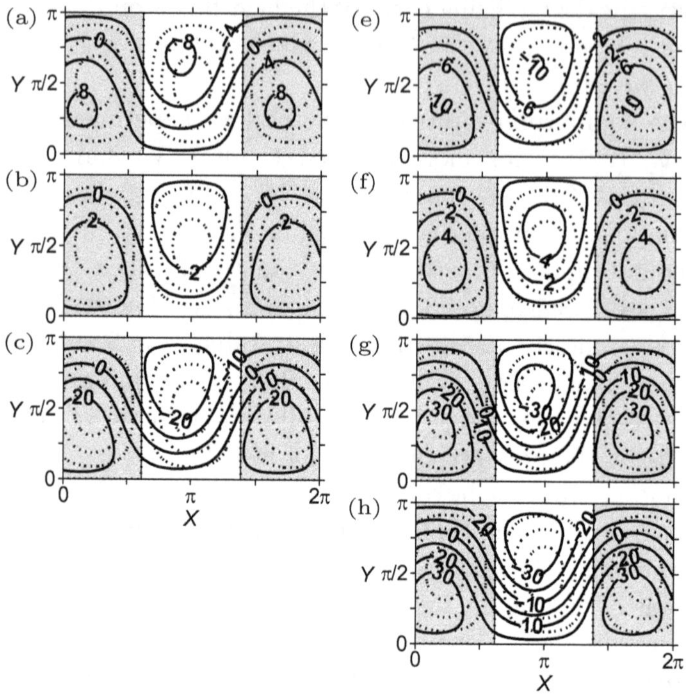

Fig. 8.17. Same as in Fig. 8.12, but for the only stable equilibrium state for $m = 3.7$ at $Q = 50$ W m^{-2} in Case 2 (left panels) and for the third one of the three equilibrium states for $m = 3.7$ at $C_g = 50$ W m^{-2} in Case 4 (right panels). The former has non-dimensional solutions with (ψ_1, ψ_2, ψ_3, θ_1, θ_2, θ_3) =(0.0656, −0.0679, 0.0259, 0.0613, −0.0616, 0.0253). The latter has non-dimensional solutions with (ψ_1, ψ_2, ψ_3, θ_1, θ_2, θ_3, $T_{g,1}$,$T_{g,2}$,$T_{g,3}$) = (0.0662, −0.0863, 0.0278, 0.0615, −0.0776, 0.0272, 0.1762, −0.1551, 0.0544). They both belong to "Low 1" equilibria. The contour intervals are (a, e) 4.0×10^7 m^2 s^{-1}, (b, f) 2.0×10^6 m^2 s^{-1}, and (c, g, h) 10 K. Note that there is no land temperature field in Case 2 (left panel) (Li *et al.*, 2018c).

(Fig. 8.17, right panel), most likely because Case 4's Hadley circulation has a lower threshold for orographic instability than Case 2 (Fig. 8.15(b)).

These findings suggest that, in comparison to the uncoupled model (Case 2), the land–atmosphere coupling may reduce the atmospheric response to thermal and topographic forcing, with the presence of longwave radiation fluxes playing a major role in this reduction. However, in the coupled model that combined the heat fluxes and longwave radiation fluxes, the heat fluxes only strengthened the response of the lower-layer flow and only slightly decreased the meridional gradient of the land temperature. The presence of heat fluxes greatly strengthens the atmospheric response to the thermal and topographic forcing.

To further understand the reason for the different results in the four experiments, we should compare the heating fields in the four experiments. Li *et al.* (2018c) tell us that the presence of longwave radiation fluxes reduces the meridional gradient of the net diabatic heating field, resulting in a more stable atmosphere flow. On the contrary, the heat fluxes decrease from low to high latitudes; thus, the presence of heat fluxes increase the meridional gradient of the net diabatic heating field, resulting in a less stable atmosphere flow. In addition, it implies that the land–atmosphere coupling, as compared to the uncoupled model (Case 2), reduces the meridional gradient of the net diabatic heating, with the main cause of this effect being the presence of longwave radiation fluxes. The meridional gradient of the net diabatic heating is significantly increased by the existence of heat fluxes. However, the heat fluxes only moderately increase the meridional gradient of the net diabatic heating in the coupled model that combines the heat fluxes and longwave radiation fluxes.

In conclusion, the multiple-wave phase equilibria associated with the ridge and trough types in the linked model are more notable than those in the uncoupled model, primarily because the land–atmosphere coupling widens the Hadley circulation's orographic instability region. Besides, the Hadley circulation is significantly stabilized by the land–atmosphere coupling, which also reduces the atmospheric response to topographic and thermal forcing. In particular, the longwave radiation fluxes, which rise from low to high latitudes and lessen the meridional gradient of the net diabatic heating, are principally responsible for these impacts of the land–atmosphere coupling. The effects of longwave radiation fluxes are somewhat modified by the existence of heat fluxes.

8.3.4 *Ridge- and trough-type equilibria*

In this section, we address the wave phases of ridge- and trough-type equilibria in relation to the topography. It is evident that both the ridge- and trough-type equilibria's wave components ψ_2^1 are negative (Figs. 8.14(b) and 814(e)). The negative sign of ψ_2^1 indicates that this wave component of upper layer stream function for the two types of equilibrium is out of phase with the topography. The ridge- and trough-type equilibria have negative and positive wave components ψ_3^1, respectively (Figs. 8.14(c) and 8.14(f)). The negative (positive) sign of ψ_3^1 indicates a 90° phase lag (lead) in relation to the mountain crests for this wave component of the upper-layer stream function of the ridge-type (trough-type) equilibria. As a result, the west side of the mountains is where the upper-layer ridges (troughs) of ridge-type (trough-type) equilibria are found.

Table 8.4 displays the results of our calculations of the wave phase of the stream function with respect to the mountains for wavenumbers 3.7 and 6. The equilibrium states of the ridge type (High 2 and Low 2) have lower-layer ridges over the mountains, upper-layer ridges on the west side of the mountain crests, and lower-layer easterlies (Mean_U3 is negative, see Table 8.5 for definition of Mean_U3). Lower-layer troughs over the mountains, upper-layer troughs on the west side of the mountain crests, and lower-layer westerlies (Mean_U3 is positive) are characteristics of the trough-type (High 1 and Low 1) equilibrium states. We found that the ridge-type (trough-type) equilibria have lower-layer easterlies (westerlies). Therefore, the distinct characters of ridge- and trough-type equilibria are robust.

The forced topographic Rossby wave theory can practically explain the aforementioned phenomena. Based on the barotropic potential vorticity equation (Smith, 1979; Holton and Hakim, 2012), the forced topographic Rossby wave solution is given by Eq. (8.47), where Re [ ] denotes the real part; $\tilde{k}$ and $\tilde{l}$ are zonal and meridional wavenumbers, respectively; h is the boundary topography; H_0 is the height of the homogeneous atmosphere;

Table 8.4. Wave phase of the equilibrium states relative to the mountains (Li *et al.*, 2018c).

m	C_g (W m^{-2})	Character	Phase	Δ Phase(°) Lower	Δ Phase(°) Upper	Mean_U^3 (m s^{-1})	g_1 (*10^{-11} m^{-2})	g_2 (*10^{-11} m^{-2})
3.7	50	High 2	Ridge	−12	−84	−0.18	9.10	−1.49
	50	High 1	Trough	−9	−57	0.23	−6.93	1.17
	55	High 2	Ridge	−18	−108	−0.61	2.76	−0.44
	55	Low 1	Trough	−9	−45	0.44	−3.57	0.61
	60	Low 1	Trough	−6	−36	0.60	−2.59	0.45
	80	Low 1	Trough	−6	−27	1.03	−1.46	0.26
6	20	High 2	Ridge	0	−42	−0.20	8.31	−1.68
	20	High 1	Trough	0	−24	0.26	−6.01	1.29
	26	Low 2	Ridge	−6	−60	−0.77	2.32	−0.44
	26	Low 1	Trough	0	−12	0.49	−3.09	0.69
	28	Low 2	Ridge	−6	−66	−1.06	1.74	−0.32
	28	Low 1	Trough	0	−12	0.52	−2.90	0.65
	32	Low 1	Trough	0	−6	0.55	−2.73	0.61
	40	Low 1	Trough	0	−6	0.56	−2.67	0.60

Notes: "Ridge (trough)" in the "Phase" column indicates that the ridges (troughs) of lower-layer stream function are over the mountains. The subsequent column "ΔPhase" gives the phase of the ridges (troughs) of the lower-layer and upper-layer stream function relative to the mountain crests, and negative values indicate that the ridges or troughs are located to the west side of the mountain crests.

$\bar{u}$ is the mean zonal wind speed; and ε is the dissipation factor. In this model, $\tilde{k} = n/L$, $\tilde{l} = 1/L$, $h = 2Hh_2 \cos(nx/L) \sin(y/L)$, $\varepsilon = k_d$, and we might assume $H_0 = H$. The boundary topography has little influence on the upper-layer flow. Thus, we select the mean lower-layer zonal wind speed, i.e., Mean_U3, as $\bar{u}$:

$$\Psi_{(x,y)} = Re\left[\frac{f_0 h/H_0}{\tilde{k}^2 + \tilde{l}^2 - \beta/\bar{u} - i\varepsilon(\tilde{k}^2 + \tilde{l}^2)/(\bar{u}\tilde{k})}\right]. \tag{8.47}$$

We might write the boundary topography as

$$h = \mathrm{Re}\{2Hh_2[\cos(nx/L) + i\sin(nx/L)]\sin(y/L)\} \tag{8.48}$$

and set

$$g_1 = \tilde{k}^2 + \tilde{l}^2 - \beta/\bar{u}, \tag{8.49}$$

$$g_2 = \varepsilon(\tilde{k}^2 + \tilde{l}^2)/(\bar{u}\tilde{k}), \tag{8.50}$$

then Eq. (8.47) becomes the following:

$$\begin{aligned}\Psi_{(x,y)} &= \mathrm{Re}\left\{\frac{2f_0h_2}{g_1 - ig_2}[\cos(nx/L) + i\sin(nx/L)]\sin(y/L)\right\}\\ &= \mathrm{Re}\left\{\frac{2f_0h_2(g_1 + ig_2)}{(g_1 - ig_2)(g_1 + ig_2)}[\cos(nx/L) + i\sin(nx/L)]\sin(y/L)\right\}\\ &= \frac{2f_0h_2}{g_1^2 + g_2^2}[g_1\cos(nx/L) - g_2\sin(nx/L)]\sin(y/L),\end{aligned} \tag{8.51}$$

Table 8.4 demonstrates examples of calculated values of g_1 and g_2 . The absolute values of g_1 are almost always substantially higher than that of g_2. Thus, the sign of g_1 is the major factor that determines the wave phase of the stream function Ψ in relation to the topography. In other words, ridges (troughs) should be over the mountains if g_1 is a positive (negative) value and the stream function Ψ is approximately in (out of) phase with the topography. Obviously, this rough theory's prediction of the wave phase for the lower-layer stream function of these equilibrium states is absolutely accurate.

In fact, the zonal wind direction and horizontal scale of the topography influence the wave phase of equilibrium states. The absolute vorticity decreases over the mountains because the potential vorticity is conserved. For westerly flow ($\bar{u} > 0$), in the case $g_1 > 0$, i.e., $\tilde{k}^2 + \tilde{l}^2 > \beta/\bar{u}$ [also called "long-wave" case (Smith, 1979)], the creation of negative relative vorticity, which leads to the formation of ridges over the mountains, is what principally causes the decrease in absolute vorticity. By contrast, in the case $g_1 < 0$, i.e., $\tilde{k}^2 + \tilde{l}^2 < \beta/\bar{u}$ [also called "ultralong-wave" case (Smith, 1979)], troughs are produced over the mountains as a result of the decrease in planetary vorticity, which is linked to the southward movement of air parcels and the decrease of absolute vorticity. However, for easterly flow ($\bar{u} < 0$), there is always $\tilde{k}^2 + \tilde{l}^2 > \beta/\bar{u}$, inverted ridges are created above the mountains as a result of the decrease in absolute vorticity, which is caused by the development of negative relative vorticity as well as the drop in planetary vorticity brought on by the southward motion (Holton and Hakim, 2012). In our model, the lower-layer easterlies are solely responsible for the occurrence of the "long-wave" case linked to ridge-type equilibria.

In other words, the equilibrium states with the ridge type (trough type) over mountains have lower-layer ridges (trough) and easterlies (westerlies). The wave phases of equilibrium states relative to the topography depends on the direction of lower layer zonal wind and horizontal scale of the topography.

8.3.5 *High- and low-index equilibria*

The high- and low-index equilibria are investigated in this section. We define a few dimensional physical variables, as shown in Table 8.5, to further explore the distinctions between these two types of equilibrium. All of the variables are defined over the domain ($0 \leq x \leq 2\pi L$, $0 \leq y \leq \pi L$) of the channel. The upper-layer geopotential height field's amplitude of wave component is

Table 8.5. List of variables and notations used in this section (Li *et al.*, 2018c).

Notation	Variable
U^1 (m s^{-1})	Zonal mean upper-layer u-component (east-west) wind
U^3 (m s^{-1})	Zonal mean upper-layer u-component wind
Mean_U^1 (m s^{-1})	Channel-average upper-layer u-component wind, i.e., mean of U^1
Mean_U^3 (m s^{-1})	Channel-average lower-layer u-component wind, i.e., mean of U^3
Mean_U^2 (m s^{-1})	Middle-level u-component wind, mean of Mean_U^1 and Mean_U^3
AH (gpm)	The amplitude of wave components of upper-layer geopotential height field
ΔT_a(K)	Meridional gradient of atmospheric temperature, mean atmospheric temperature at the southern wall minus that at the northern wall
ΔT_g(K)	Meridional gradient of land temperature, mean land temperature at the southern wall minus that at the northern wall
AT$_a$(K)	The amplitude of wave components of atmospheric temperature field
AT$_g$(K)	The amplitude of wave components of land temperature field

defined as

$$\mathrm{AH} = \frac{L^2 f_0^2}{g_0}\sqrt{(\psi_2^1)^2 + (\psi_3^1)^2}. \tag{8.52}$$

The definition of the wave components' amplitude of the atmospheric and land temperature fields is

$$\mathrm{AT}_a = \frac{2L^2 f_0^2}{R}\sqrt{(\theta_2)^2 + (\theta_3)^2}, \tag{8.53}$$

$$\mathrm{AT}_g = \frac{L^2 f_0^2}{R}\sqrt{(T_{g,2})^2 + (T_{g,3})^2}, \tag{8.54}$$

respectively. The zonal asymmetry of atmospheric and land temperature fields is represented by these two variables.

As expected, results show that the wave amplitude AH of the low-index equilibrium state is always greater than that of high-index equilibrium state at the same value of C_g. This phenomenon also occurs in wave amplitude of the atmospheric temperature field AT_a. Meanwhile, at the same value of C_g, the meridional atmospheric temperature gradient $\Delta\,\mathrm{T}_a$ of low-index equilibrium state is always smaller than that of high-index equilibrium state.

The mean upper-layer zonal wind speed is not remarkably different between these two types of equilibrium: The Mean_U^1 of the low-index equilibrium state is smaller than that of the high-index equilibrium state at $C_g = 54$ W m^{-2} for $m = 3.7$. By contrast, the former is greater than the latter at $C_g = 22$, 24 W m^{-2} for $m = 6$. Besides, the distinctions in value of Mean_U^1 between high- and low-index equilibrium states are no more than 0.5 m s^{-1} (details see Figs. 11 and 12 in Li *et al.* (2018c)). In addition, these two equilibrium types have no obvious changes in non-dimensional zonal component ψ_1^1 (Figs. 8.14(a) and 8.14(d)), which implies that the high- and low-index equilibria are similar in upper-layer zonal wind speed.

Our findings on the differences between high- and low-index equilibria in zonal wind are notable because they diverge from those of earlier studies (e.g., Charney and DeVore, 1979) based on barotropic models, in which the zonal component ψ_A (i.e., zonal wind) of high-index equilibria was significantly greater than that of low-index equilibria. One could contend that the low-index equilibrium states of the trough type (Fig. 8.14(d)) in the baroclinic model, which can be compared to the outcomes based on the barotropic model, have a higher upper-layer zonal component ψ_1^1 than the magenta branch equilibrium states, which are characterized by small-wave amplitude (Figs. 8.14(e) and (f)). However, both in this work and earlier investigations (e.g., Charney and Straus, 1980; Reinhold and Pierrehumbert,

1982), the magenta branch equilibrium states in the baroclinic model are consistently unstable. It should be emphasized that the barotropic model results do not correspond with the observations, e.g., some research has indicated that the zonal wind's probability density distribution is unimodal (Benzi *et al.*, 1986; Sutera, 1986).

In fact, as emphasized in Charney and Straus (1980), the wavelike equilibrium is maintained not by the conversion of mean flow kinetic energy, but by the mean flow potential energy in baroclinic atmosphere. Therefore, in our baroclinic model, the meridional atmospheric temperature gradient of low-index equilibria does decrease since the mean flow potential energy is consumed and the low-index equilibria have larger wave amplitudes. In our baroclinic model, there are no dramatic variations between the low- and high-index equilibria in zonal wind speed. However, in the barotropic model, the only source of energy for the wavelike equilibria was the mean flow kinetic energy. Because of this, the low-index equilibria with large-wave amplitude in the barotropic model unquestionably have a lower zonal wind speed than the high-index equilibria with small-wave amplitude.

According to Figs. 8.12 and 8.13, each upper-layer stream function field is almost always in phase with the equilibrium temperature fields of the atmosphere and the land. The meridional perturbations of the stream function field would continue to develop or degrade as a result of temperature advection if the atmospheric temperature field lagged behind or led the stream function field in phase. Thus, there would be no stationary waves, i.e., equilibrium states. The atmospheric temperature field is undoubtedly in phase with the stream function field because we have only got equilibrium solutions from Eqs. (8.35)–(8.43). The interactions between the land and atmospheric temperature fields through radiative and heat exchange are the cause of the zonal asymmetric structure of the land temperature field. As a result, there is excellent agreement between the stream function field of the upper layer and the variations in wave amplitude of the atmospheric and land temperature fields. This finding also suggests that the mean flow potential energy is converted in order to preserve the wavelike equilibrium.

The results in this section demonstrate that the meridional atmospheric temperature gradient is smaller (larger) and the wave amplitude is larger (smaller) in the low-index (high-index) equilibrium states, but the zonal wind speeds are not significantly different between the two types of equilibrium states. These results are attributed to the wavelike equilibrium that is maintained by the conversion of the mean flow potential energy in the baroclinic atmosphere.

8.4 Impacts of the Arctic on Eurasian Climate

8.4.1 *Eurasian cooling and Arctic warming*

One of the most predominant phenomena relating global temperature change over the past few decades is enhanced Arctic warming (Xie *et al.*, 2019). Meanwhile, in the context of overall global warming, notable regional cooling on decadal or multi-decadal scales occurs, such as the Eurasian cooling (Cohen *et al.*, 2014; Xie *et al.*, 2022a). Internal climate variability is typically attributed to this kind of decadal variability. On the one hand, previous studies have suggested that internal climate variability associated with Atlantic and Pacific Oceans accounts for a majority of multi-decadal temperature variability (Kosaka and Xie, 2013; Huang *et al.*, 2017b). On the other hand, Arctic warming is one of the strongest signatures during the past several decades, which overwhelms the signals of internal climate variability directly reflected by the temperature change over the Atlantic and Pacific Oceans.

In addition to the well-known spatial pattern, Fig. 8.18 depicts the vertical structure of Eurasian cooling and Arctic warming, i.e., a vertical–latitudinal cross section of the changes in temperature and geopotential height averaged over 60°E–120°E. Apart from the surface, Eurasian cooling affects nearly the entire troposphere (Fig. 8.18(a)). At about 500 hPa, the troposphere experiences the strongest cooling. Meanwhile, due to the Arctic warming, the whole troposphere is also warming. According to Screen and Simmonds (2010) and Cohen *et al.* (2014), the surface of Arctic experiences the strongest warming. A dominant cooling or warming also happens in the stratosphere. In Section 8.4.4, we will examine the stratospheric influences on the troposphere.

Under geostrophic approximation, circulation could be expressed by geopotential height in pressure coordinates. In pressure coordinates, the hydrostatic relation is $\partial\Phi/\partial p = -(R/p)T$, and the relation holds for both the mean state and change in the pressure coordinates. Therefore, the temperature change should be proportional to the vertical gradient of geopotential change at the arbitrary pressure level. The results indicate that regions where the absolute value of temperature changes is bigger for the same pressure level have larger absolute changes in the vertical gradient of geopotential. Thus, it is clear that changes in atmospheric circulation have a direct impact on regional temperature.

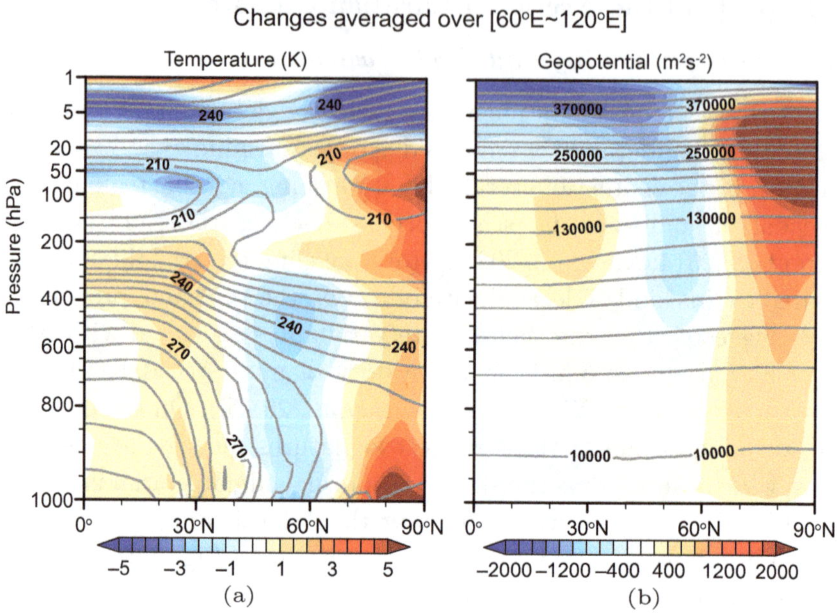

Fig. 8.18. Vertical–latitudinal cross section of temperature changes (linear trend multiplied by the time span) for the period of 1979–2017 averaged over the region of 60°E–120°E during boreal winter (a). The gray contour line indicates the corresponding climatology for the period of 1979–2017. (b) As in (a), but for geopotential (Xie *et al.*, 2020).

8.4.2 *Theory of potential vorticity dynamics*

Because Potential vorticity (PV) naturally illustrates seamless coupling between dynamic and thermodynamic elements, it is utilized to analyze the dynamic linkages between Eurasian Cooling and Arctic Warming (Ertel, 1942; Yeh and Chu, 1958; Charney and Stern, 1962). In other words, any observed circulation and temperature change can, on a fundamental level, be attributed to PV. Piecewise PV inversion and the PV equation are employed (Xie *et al.*, 2020). While the PV equation explains how the PV change occurred, piecewise PV inversion (PPVI) quantifies the impact of any arbitrary PV change on the entire atmosphere. Moreover, although it is frequently utilized in synoptic cases, PV inversion can also be used in the climate change realm.

The basic concept of PPVI is similar to the electric charge in the electric field (Hoskins *et al.*, 1985; Xie *et al.*, 2020). Here, PV corresponds to the electric charge, while temperature or circulation corresponds to the properties of the electric field. Here, the electric charge is represented by PV, while the characteristics of the electric field are represented by temperature or circulation. The simplest explanation for PV inversion is $\Phi = L^{-1}(q)$, where Φ and q stand for the geopotential and pseudo- (or quasigeostrophic) PV (Charney and Stern, 1962), respectively. The operator L is a three-dimensional Laplacian-like operator, while L^{-1} is the inverse operator. Notably, L is an operator rather than a coefficient, which ensures that the three-dimensional coupling among varied vertical levels and horizontal locations can be described by PV inversion more than just correspondence between Φ and q. The superposition principle of a linear operator L allows piecewise PV inversion, namely, one-to-one correspondence of $\Phi_n\Phi_n$ and q_n in $\sum_1^n \Phi_n = \sum_1^n L^{-1}(q_n)$ (Xie *et al.*, 2020).

In pressure (p) coordinates, pseudo-PV is in the form

$$q = f + \frac{1}{f_0}\nabla^2\Phi + \frac{\partial}{\partial p}\left(\frac{f_o}{\bar{S}}\frac{\partial \Phi}{\partial p}\right), \tag{8.55}$$

where f is the Coriolis parameter and ∇^2 is a two-dimensional Laplacian operator (Holton and Hakim, 2012). $\bar{S} = -\bar{\alpha}(\partial \ln \bar{\theta}/\partial p)$ is just a function of pressure and indicates the static stability of the basic state. That is to say, the specific volume $\bar{\alpha}$ and potential temperature $\bar{\theta}$ are averaged over a large domain (northward of 10° N in this study). The f-plane assumption is adopted for relative vorticity, namely, f in the denominator of $(1/f)\nabla^2\Phi$ is replaced by $f_0 = 2\Omega\sin(\varphi_0)$, in which $\varphi = 45°\text{N}$ and $\Omega = 7.292\times 10^{-5}s^{-1}$ are Earth's angular speed of rotation. PPVI is utilized in this study to examine change rather than the mean state. Here, the PV change is in the form

$$q' = \frac{1}{f_0}\nabla^2\Phi' + \frac{\partial}{\partial p}\left(\frac{f_0}{\bar{S}}\frac{\partial \Phi'}{\partial p}\right) = \frac{1}{f_0}\nabla^2\Phi' + \frac{\partial}{\partial p}\left(\frac{f_0}{\bar{S}}\right)\frac{\partial \Phi'}{\partial p} + \frac{f_0}{\bar{S}}\frac{\partial^2 \Phi'}{\partial p^2}. \tag{8.56}$$

The PV equation is in the form

$$\frac{dq}{dt} = -f_0\frac{\partial Q}{\partial p} + \mathbf{k}\cdot\nabla\times\mathbf{F}, \tag{8.57}$$

where $Q = \bar{\alpha}\dot{\theta}/(\bar{S}\bar{\theta})$ indicates the diabatic heating ($\dot{\theta}$ demonstrates the material change rate of potential temperature) and $\mathbf{F}$ stands for the horizontal frictional force. The effect of friction is difficult to specifically evaluate from reanalysis data (Egger *et al.*, 2015). Physically, friction tends to reduce the relative motion between two objects at an interface. According to Smagorinsky (1953) and Yeh and Chu (1958), the vorticity perturbation is roughly proportional to the effect of boundary layer friction. In particular, the frictional term in the PV equation is expressed as $\mathbf{k}\cdot\nabla\times\mathbf{F} = -\mu(1/f_0)\nabla^2\Phi'$, where μ is the positive frictional coefficient. Thus, friction always has a tendency to cancel out changes in PV. However, friction cannot modify the sign of PV change. In short, friction causes the final change to be more modest than it would have been without it.

The Eulerian form PV equation without friction is

$$\frac{\partial q}{\partial t} = -V_g \cdot \nabla_q - f_0 \frac{\partial Q}{\partial p}, \tag{8.58}$$

where V_g is geostrophic wind. In a climate change sense, the equilibrium form should be adopted, namely, $-V_g \cdot \nabla_q - f_0(\partial Q/\partial p) = 0$, by neglecting the local tendency term. Here, the linear perturbation form of the equilibrium PV equation is $-\ (V_g \cdot \nabla_q)' - f_0(\partial Q'/\partial p) = 0$. Finally, the PV equation in the perturbation form is obtained by ignoring the high-order term

$$-V_g' \cdot \nabla_{\bar{q}} - \bar{V}_g \cdot \nabla q' - f_0 \frac{\partial Q'}{\partial p} = 0. \tag{8.59}$$

The perturbation expansion was applied to $(V_g \cdot \nabla_q)$' in Eq. (8.59). That is to say, there is $(AB)' = AB - \bar{A}\bar{B} = A'\bar{B} + \bar{A}B' + A'B' \approx A'\bar{B} + \bar{A}B'$ for $A = \bar{A} + A'$ and $B = \bar{B} + B'$ because $A'B'$ is negligible for the large-scale features of PV advection.

8.4.3 *Potential vorticity adaptation*

In the PV view, Eq. (8.56) clearly demonstrate the seamless dynamic and thermodynamic coupling, in which any change in geopotential Φ' could lead to the changes in both circulation [represented by vorticity term $(1/f_0)\nabla^2\Phi'$] and temperature [represented by static stability term $(\partial/\partial p)[(f_0/\bar{S})(\partial\Phi'/\partial p)]$ and supported by results shown in Fig. 8.18]. The intrinsic dynamic and thermodynamic coupling are depicted succinctly and vividly by the classic Ertel PV conservation picture in Fig. 8.19(a). The equivalent pseudo-PV conservation picture, which is similar to Fig. 8.19(a) but substitutes the isobaric surface for the isentropic surface, is shown in Fig. 8.19(b). In contrast to the Eulerian viewpoint typically used in discussions of climate change, the PV conservation picture is presented in a Lagrangian perspective. As shown by Eulerian-form Eq. (8.58), local PV conservation is extremely difficult to achieve because PV advection occurs widely and affects local PV even in the absence of diabatic heating and friction.

However, the picture of PV conservation still aids in providing understanding. Here, two ideal cases that are expanded from the PV conservation picture are employed for some instructive discussions. When applying PV

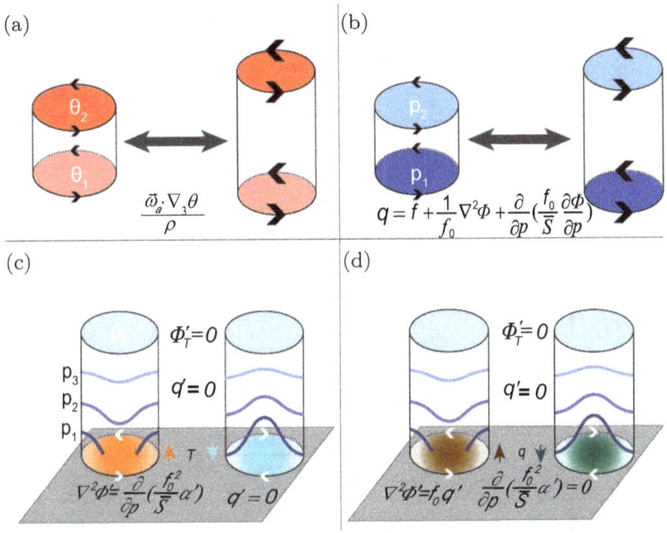

Fig. 8.19. Schematic views of (top) PV conservation for Ertel PV (EPV) (a) and pseudo-PV (PPV) (b) and (bottom) PV adaptation (c) under constant PPV but anomalous surface thermal conditions and (d) under anomalous PPV but constant surface thermal conditions. PV adaptation means the adaptation of PV (PV change) to forcing and coherent dynamic–thermodynamic adaptation to PV change. Subscript T indicates the top of the atmosphere. The gray parallelogram indicates Earth's surface. The visual illustration in (c) and (d) is similar to the traditional PV conservation illustration in (b) but with a fixed top and bottom (Xie *et al.*, 2020).

dynamics to climate change, the following two questions came to light: (1) What is the effect of PV change? (2) Why did the PV change? PV inversion and the PV equation can provide answers to the two problems, as mentioned in Section 8.1.

According to the superposition principle, the ideal case in Fig. 8.19(c) can be quantified by using PPVI, which needs to maintain only the change in surface thermal condition in the PPVI, as illustrated in Fig. 8.20(b). The geopotential change of atmosphere will be much stronger in the ideal case (Fig. 8.20(b)) with surface heating change but maintaining a constant PV compared to the real case (Fig. 8.20(a)). In other words, the motion of atmosphere is moderated by a change in PV in reaction to surface diabatic heating. The friction as illustrated in Section 8.4.2 supports this conclusion as well. Therefore, in a basic sense, PV change functions as an atmospheric stabilizer. This is a response to the following question: "Why should the PV change?"

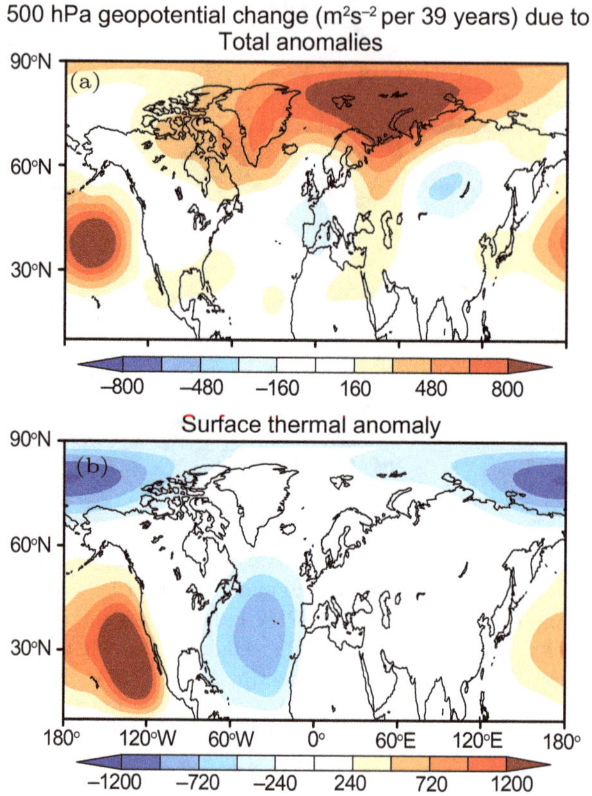

Fig. 8.20. Inverted geopotential changes ($m^2\ s^{-2}$) (linear trend multiplied by the time span) from PPVI with three pieces, at the 500-hPa level, during boreal winter from 1979 to 2017. The three pieces are (b) surface thermal condition change, surface and tropospheric PV change (not shown), and stratospheric total change. The sum of all the three pieces (a). Note that the ranges of the two color bars are different (Xie *et al.*, 2020).

Here, let us consider the impact of diabatic heating on the production of PV, which has the following straightforward mathematical expression: $\partial q/\partial t \propto -f_0(\partial Q/\partial p)$. Thus, diabatic heating generates positive, negative, and zero PV change when Q is increased, decreased, or unchanged with height, respectively. Here, in Fig. 8.19(c), an ideal scenario that is expanded from the PV conservation is presented. In this ideal case, the surface initially experiences a temperature change.

Then, the atmosphere makes a significant adjustment to the change in surface temperature, so that the diabatic heating of the atmosphere (caused by the surface temperature change and corresponding vertical motion of atmosphere) is even in the vertical direction (i.e., $\partial Q/\partial p = 0$). In this case, the change in surface temperature does not result in a PV change, but the geopotential significantly changes, and the balance relation is $\nabla^2\Phi' = (\partial/\partial p)[(f_0^2/\bar{S})\alpha']$ according to Eq. (8.56), where the hydrostatic relation, $\partial\Phi'/\partial p = -\alpha'$, is introduced.

In contrast to the ideal situation in Fig. 8.19(c), the other ideal case is presented in Fig. 8.19(d), which prescribes a PV change at the surface but without surface thermal condition change $\{(\partial/\partial p)[(f_0^2/\bar{S})\alpha'] = 0\}$ rather than a zero-temperature change. In Fig. 8.19(d), the balance relation is $\nabla^2\Phi' = f_0 q'$. Thus, geopotential must have a negative change in response to a positive PV change (according to $\Phi' \propto -\nabla^2\Phi'$), and vice versa. Regarding the ideal case in Fig. 8.19(c), for surface warming stronger at the surface than above, $(\partial/\partial p)[(f_0^2/\bar{S})\alpha']$ is positive under the current climatological $\bar{S}$. Therefore, the geopotential change must be negative according to $\nabla^2\Phi' = (\partial/\partial p)[(f_0^2/\bar{S})\alpha']$, and vice versa.

The real case is the sum of Figs. 8.19(c) and 8.19(d). That is to say, for surface warming stronger at the surface than above, the balance in Fig. 8.19(c) tends to generate negative geopotential, while the stronger diabatic heating at the surface than above also induces negative PV, which tends to generate positive geopotential. Thus, the eventually observed geopotential change depends on the net effects after summing Fig. 8.19(c) and 8.19(d). In other words, the reality is the sum of one rule "warming is cyclonic and cooling is anticyclonic" and another rule "a positive PV change is cyclonic and a negative PV change is anticyclonic." In summary, two important adaptation processes are involved from a "forcing" [i.e., diabatic heating and friction in Eq. (8.57)] to the final observed alterations in atmosphere. One is a direct adaptation to forcing shown by a change in PV [Eq. (8.57)]. The other is the commonly observed change to PV change in the atmosphere (e.g., temperature and circulation). Therefore, briefly speaking, "PV adaptation" refers to the adaptation of PV to forcing and coherent dynamic–thermodynamic adaptation to PV change. PV adaptation is characterized by the seamless dynamic–thermodynamic coupling.

PPVI can be used to quantify the idea of PV adaptation and further comprehend the vertical coupling between various levels (Xie *et al.*, 2020). A new three-piece PPVI scheme shows that the geopotential change is dominated by surface total change at 850 hPa. The tropospheric PV change shows a tiny contribution to the geopotential change at 850 hPa, and even the 850-hPa level itself is located in the troposphere. The dominant surface contribution means that the change at 850 hPa is determined by the surface PV change via PV adaptation. In contrast, the geopotential change at 500 hPa is significantly influenced by both the surface total change and the tropospheric

PV change. Both the overall change in the surface area and the PV change in the troposphere contribute to the positive geopotential change at 500 hPa that corresponds to Arctic warming. The tropospheric PV change, rather than the surface total change, has the greatest influence on the negative geopotential change associated with Eurasian cooling (Fig. 8.18). Due to the dominating role of tropospheric PV change, increasing Eurasian cooling at 500 hPa is primarily driven by PV adaptation rather than other factors.

8.4.4 *Vertical and horizontal coupling of atmosphere*

Figure 8.21 also displays the outcomes of the two PPVI methods. The pieces at 800–250 hPa and 400–250 hPa reflect the influences from upper levels, while Fig. 8.21(a) and 8.21(c) represents the effects of a PV change within

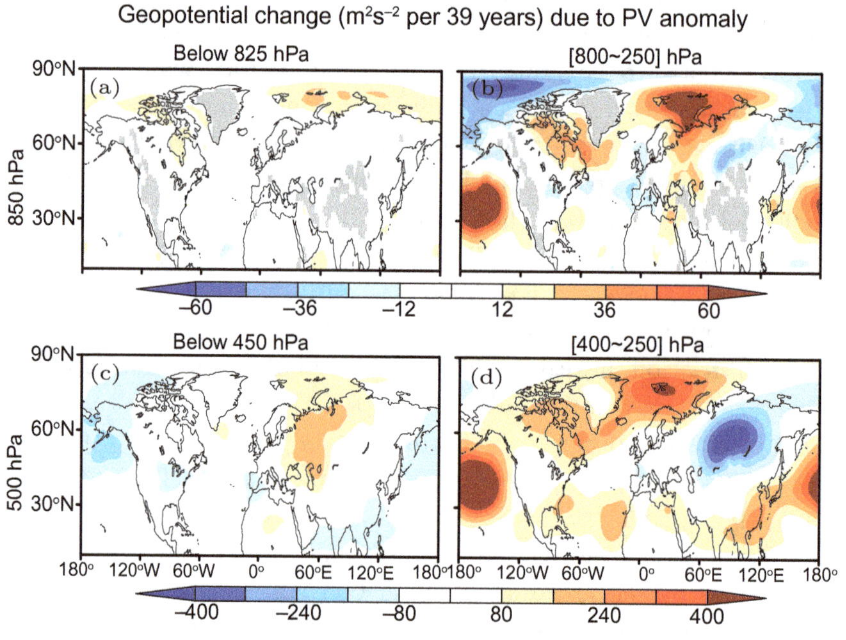

Fig. 8.21. Inverted geopotential changes (m^2 s^{-2}) (linear trend multiplied by the time span) from PPVI with five pieces, at the (top) 850- and (bottom) 500-hPa levels, during the boreal winter from 1979 to 2017. Two of the five pieces are surface and stratospheric total change. For 850 hPa, the remaining three pieces are tropospheric PV change from the surface to 900 hPa, 875–825 hPa, and 800–250 hPa (b). For 500 hPa, the remaining three pieces are tropospheric PV change from the surface to 600 hPa, 550–450 hPa, and 400–250 hPa (d). The sum of two pieces (surface–900 hPa and 875–825 hPa) is shown (a), because the surface–900-hPa piece has a very small value. (c) As in (a), but the sum of two pieces is shown (surface–600 hPa and 550–450 hPa). Again, please notice the range of the color bar (Xie *et al.*, 2020).

the vertical levels of 850 and 500 hPa, respectively. At both the 850- and 500-hPa levels, the influences of the upper-level PV change are greater than those of local (vertical levels) PV change on geopotential change. In short, the influences of tropospheric PV change on both the 850- and 500-hPa levels are basically top-down in the vertical direction.

Notably, there is little the top-down influence of PV change on 850 hPa. However, on the 500-hPa geopotential change, the top-down influence of PV change is equal to or even greater than the influence of the surface total change (Xie *et al.*, 2020). In particular, the top-down influence of PV change is nearly entirely responsible for the negative geopotential change at 500 hPa over the Eurasian cooling region. Thus, the top-down effect of PV change via PV adaptation is the fundamental cause of the accelerated Eurasian cooling at 500 hPa. Stratospheric total change shows a tiny direct influence on both the 850- and 500-hPa levels, while the stratosphere may indirectly affect 850 and 500 hPa via a top-down mechanism, such as a gradual influence through the upper troposphere to 500 and 850 hPa.

The linkage between the Eurasian climate and the Arctic must also have a horizontal channel in addition to the vertical coupling. In order to access the horizontal connection, one should ascertain whether the local PV change of the target domain is a result of horizontal PV advection. Regarding climate change, the horizontal PV advection term is linearized as $-\mathrm{V}_g' \cdot \nabla\bar{q} - \bar{V}_g \cdot \nabla_q{}'$ according to Eq. (8.59). The PV change itself in the equilibrium PV equation is directly reflected in the horizontal advection term. This is very different from the synoptic case, in which the PV change is reflected in the local tendency term [$\partial q/\partial t$ in Eq. (8.58)]. In terms of physical understanding, the nature of the difference between climate and the synoptic case is a complete or ongoing PV adaptation.

In Fig. 8.22, the wind change and climatology for both 850 and 500 hPa are superimposed with the PV climatology and change. The combinations of wind change/PV climatology and wind climatology/PV change represent $-\bar{V}_g' \cdot \nabla\bar{q}$ and-$\bar{V}_g' \cdot \nabla q'$, respectively. Figure 8.22(a) demonstrates, over the Eurasian cooling–Arctic warming zone, how the wind change crosses the isoline of climatological PV (with a large crossing angle). The importance of PV advection via wind change/PV climatology is clear. Similar conditions exist at 500 hPa (Fig. 8.22(c)).

According to Eq. (8.59), the balance relation is $-\mathrm{V}_g' \cdot \nabla\bar{q} + \bar{V}_g \cdot \nabla_q' = 0$ when diabatic heating and friction are not considered. This ideal case is more closely resembled by the situation at 500 hPa than that at 850 hPa. Therefore, we make use of the included figures for PV and wind at 500 hPa.

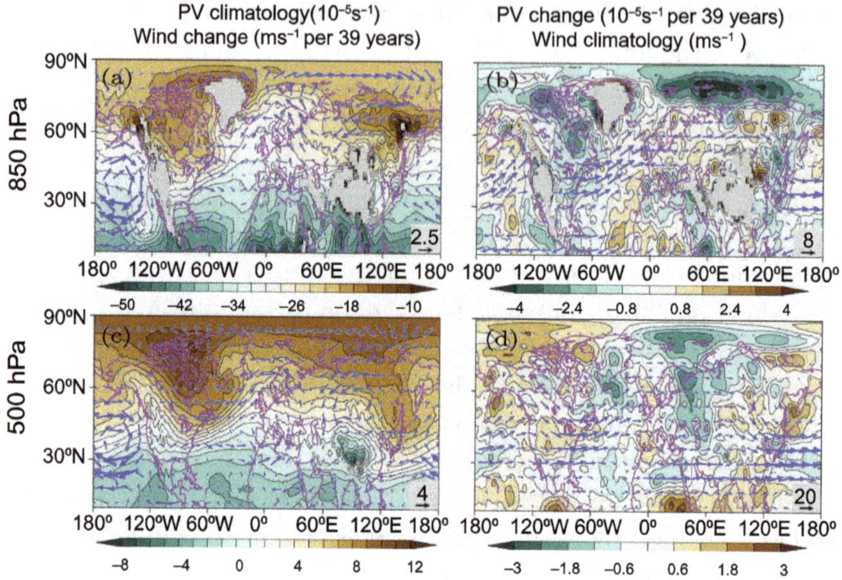

Fig. 8.22. PV climatology (filled; unit: 10^{-5} s^{-1}) and horizontal wind change (vectors; linear trend multiplied by the time span; unit: m s^{-1}) during the boreal winter from 1979 to 2017 (a), (c). (b), (d) As in (a) and (b), but for PV change and horizontal wind climatology. Results are shown at the (top) 850- and (bottom) 500-hPa levels (Xie *et al.*, 2020).

Symbolically, the domain is set to 0°E–60°E and 40°N–80°N, and the featured values in Fig. 8.22 (bottom) are used to roughly establish the magnitudes of the variables in climatology and change. For simplicity, the climatological wind is purely zonal (uniformly 12 m s^{-1}), the wind change is purely meridional (with anticyclonic zonal shear of 5 to -5 m s^{-1} from west to east, and northward is positive), and PV is zonally uniform but with a meridional gradient (1–8 × 10^{-5} s^{-1} from south to north). This ideal situation is depicted in Fig. 8.23 (top), where the overall patterns do not change with, although magnitudes will change with, the different values of selected parameters.

According to the PV climatology shown in Fig. 8.23 (top left), the southward wind change has positive PV advection, but the northward wind change has negative PV advection. The PV advection change induced by $-V'_g \cdot \nabla \bar{q}$ for a given zonal shear value because of the wind change is shown in Fig. 8.23 (top center). Then, the term $-\bar{V}_g \cdot \nabla q'$ is equal to $-V'_g \cdot \nabla \bar{q}$. The PV change is illustrated in Fig. 8.23 (top right) after the balance relation has been resolved, with the PV change at the western boundary set as zero. The balance between $-V'_g \cdot \nabla \bar{q}$ and $-\bar{V}_g \cdot \nabla q'$ will eventually make the distribution

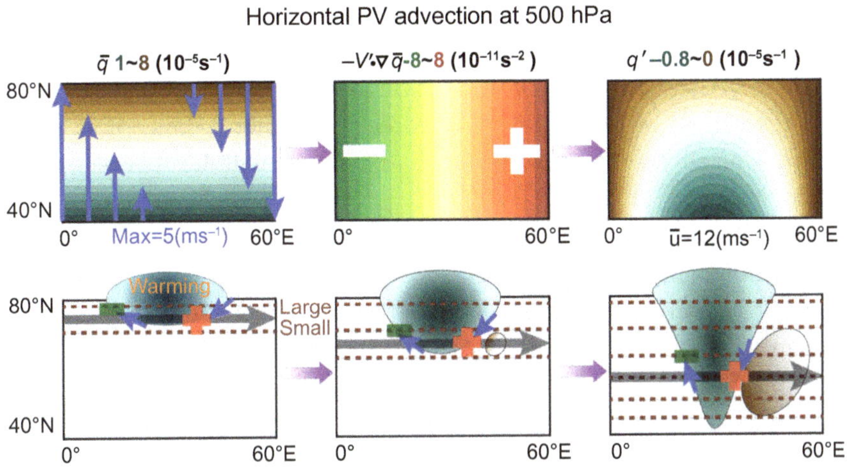

Fig. 8.23. (Top) An ideal case of zero-net horizontal PV advection (i.e., $\bar{V}' \cdot \nabla \bar{q} + \bar{V} \cdot \nabla q' = 0$), with prescribed climatology of PV ($\bar{q}$) and wind ($\bar{V}$) and change of wind (V′). (Top left) The prescribed $\bar{q}$ and V′ are shown as filled and blue arrows. Note that $\bar{q}$ is zonally uniform, and V′ is purely meridional (with a maximum value of 5 m s^{-1}); $\bar{V}$ is a uniform zonal wind ($\bar{u}$) of 12 m s^{-1}. These prescribed values approximately correspond to the situation at 500 hPa (Fig. 8.22, bottom, within the domain of approximately 40°N–80°N, 0°E–60°E). (Top center) Distribution of $-\text{V}' \cdot \nabla \bar{q}$ (10^{-11} s^{-2}) under prescribed values. (Top right) Distribution of $q\prime$ calculated from $\text{V}' \cdot \nabla \bar{q} + \bar{V} \cdot \nabla q' =$, in which $q\prime$ at 0° longitude is prescribed as zero. (Bottom) Schematic view of the evolution of horizontal PV advection under a situation with enhanced regional warming over the Arctic, a meridional PV gradient (larger to the north, brown dashed lines indicate isoline of PV), and purely westerly wind (gray arrow). The bottom row is based on results shown in the top row. As in the top row, the blue arrow indicates wind anomaly, "small" and "large" indicate the relative value of $\bar{q}$, and the minus and plus symbols indicate the sign of $-\text{V}'_g \cdot \nabla \bar{q}$. The filled symbol indicates a change in PV ($q\prime$) (Xie *et al.*, 2020).

of PV change induced by $-\text{V}'_g \cdot \nabla \bar{q}$ move downstream. Thus, implied by the circulation change ($-\text{V}'_g \cdot \nabla \bar{q}$), the eventual PV change does not directly overlap with the PV advection change.

Understanding an actual shift at the 500-hPa level (Figs. 8.22(c) and 8.22(d)) is not difficult with the aid of the ideal situation (Fig. 8.23, top). The phenomena that took place in the vicinity of the Eurasian cooling region (Figs. 8.22(c) and 8.22(d)) are comparable to the ideal case (Fig. 8.23, top). The climatological wind downstream (Fig. 8.22(d)) advected PV generated by the wind change upstream of the Eurasian cooling area (Fig. 8.22(c)). Eventually, the horizontal PV advection change generated a positive PV change over the Eurasian cooling region and a negative PV change over the upstream regions (Fig. 8.22(d)). In contrast, the situation at 850 hPa (Figs. 8.22(a) and 8.22(b)) seems different from that at 500 hPa and in the ideal case (Figs. 8.22(c), 8.22(d) and 8.23, top). This discrepancy is anticipated given the importance of the diabatic heating effect in the lower troposphere. Although the compensation between $-V' \cdot \nabla \bar{q}$ and -$\bar{V}_g' \cdot \nabla q'$ persists regardless of the presence of force, the general concept is that PV advection caused by the upstream wind change will be advected by the climatological wind downstream.

In extra-tropical regions, large PV gradients are prevalent over the extra-tropical regions (Figs. 8.22(a) and 8.22(c)), making it inevitable for an anti-cyclonic (or cyclonic) circulation change to cross the isoline of climatological PV. Therefore, through horizontal PV advection and PV adaptation, the anticyclonic circulation change brought on by Arctic warming will spread southward. The extension of the anticyclonic circulation change brought on by Arctic warming is depicted schematically in Fig. 8.23 (bottom). According to the bottom right panel of Fig. 8.23, the local anticyclonic circulation modification brought on by Arctic warming initially advects positive PV to the nearby Southeast region and negative PV to the nearby Southwest region. Due to the very small PV gradient in the very northern regions (Figs. 8.22(a) and 8.22(c)), no PV advection is depicted on the northern side of the anticyclonic circulation anomaly (Fig. 8.23, bottom).Then, the PV change brought about by meridional wind advection is carried downstream by climatological wind. As shown in Fig. 8.23(top), the negative PV change on the southern side of the anticyclonic anomaly is amplified once a temporal balance between meridional and zonal PV advection is attained. In turn, the enhanced PV change causes the anticyclonic circulation anomaly to extend to the north via PV adaption. Meanwhile, PV advection caused by climatological wind results in a rise in PV in the downstream regions (east of the anticyclonic circulation anomaly). Fig. 8.23 (bottom right) depicts roughly the ultimate balanced PV change.

The schematic perspective in Fig. 8.23 (bottom) and the illustration of the horizontal PV advection at 500 hPa (Fig. 8.22, bottom) are identical. The negative PV change brought on by Arctic warming in the Arctic Ocean, its northward expansion, and the positive PV change over the downstream Eurasian cooling zone are all visible in Fig. 8.22 (bottom). The last section showed that the top-down effect dominates Eurasian cooling at 500 hPa. At pressures higher than 500 hPa, the circumstances surrounding horizontal PV advection are comparable to those at 500 hPa. In other words, both horizontal coupling (with Arctic warming) at higher levels and top-down vertical coupling control the enhanced Eurasian cooling at 500 hPa.

8.4.5 *Multi-sphere interactions*

Along with the atmospheric dynamics, it addresses a thorough understanding of the mechanisms governing the multi-sphere interactions (Xie *et al.*, 2022a) The prominent wintertime atmospheric circulation patterns, the wintertime energy budget at the Earth's surface and its summertime sources related to the sea ice, the coupling between the energy budget at the Earth's surface and the diabatic heating of the atmosphere aloft in winter, the detailed seasonality in the circulation patterns and diabatic heating of the atmosphere, the crucial role of the baroclinicity, and the situation with regard to the interannual variability are some of the key factors. The storylines from summer to winter reveal that the Arctic is a major factor in wintertime Eurasian cooling, as shown in Fig. 8.24. The following is a summary of the main findings and important points.

The positive sensible heat change in winter that began in the summertime points to the Arctic as the primary driver. The simultaneous prominent anticyclonic circulation change that links wintertime Eurasian cooling and Arctic warming can be brought on by enhanced sensible heat in the Eastern Arctic sector, which results from the summertime enhanced subsurface ocean heat uptake. The changes in wintertime sensible heat and diabatic heating, in particular, exhibit a dipole pattern in the Arctic area, with negative and positive indications in the west and east, respectively. However, the negative change is meaningless. Furthermore, the growth of circulation brought on by the positive sensible heat in the east can be used to explain the negligible negative change in the Western Arctic sector and widespread Arctic warming. Due to the significant climatological background baroclinicity in mid-high latitudes, the two-way interaction between the Arctic and mid-latitudes is essential for circulation development in winter. Consequently, both Arctic warming and downstream Eurasian cooling are favored by the feedback between temperature advection and circulation development similar to the baroclinic wave mechanism. The seasonality of the change shows that Arctic and mid-latitude links similar to the wintertime scenario can also exist in other months. In general, the region downstream of the largest Arctic warming is where mid-latitude cooling occurs. Therefore, the strongest Arctic warming over the western or middle flank to the north of the Eurasian continent is when Eurasian cooling occurs. Since the diabatic heating around the Barents–Kara seas is strongest in winter, Eurasian cooling only occurs in winter.

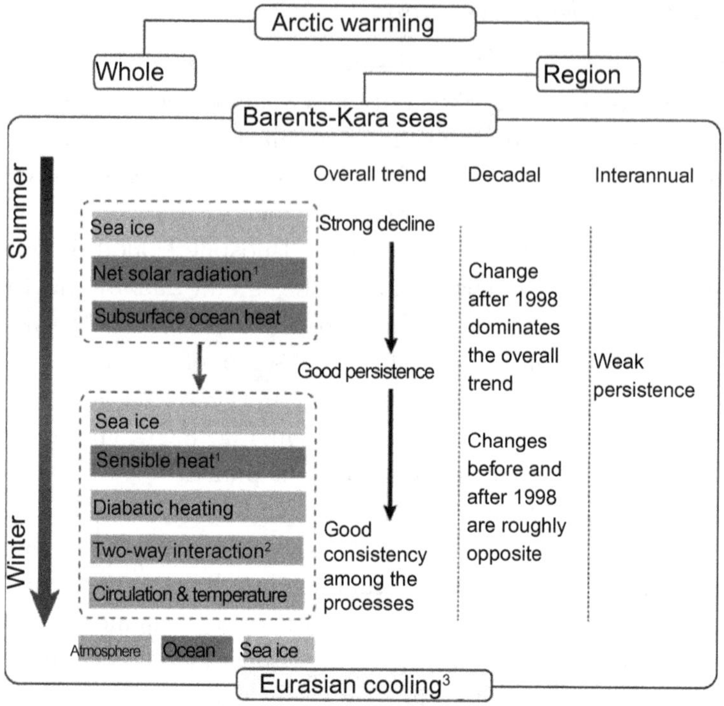

Fig. 8.24. A schematic view of the key factors of the multi-sphere interactions. The scope is (1) the regional enhanced warming in Barents-Kara seas rather than warming in the whole Arctic (top); (2) storylines from summer to winter (arrow on left); (3) focus on the overall trend and with comparisons to decadal and interannual variability (the right part). The key processes concerning the mechanisms are listed on the left part (Xie *et al.*, 2022a).
Notes: [1]At the sea surface. [2]Dynamics of atmosphere, also affects sea surface sensible heat [3]Downstream of Arctic warming

Summertime enhanced oceanic heat uptake in the Eastern Arctic sector is regulated by the strong sea ice decline and corresponding sea ice–albedo feedback. Although longwave radiation feedback can increase the magnitude of change and is essential in creating energy budget equilibrium, sea ice is critical in the energy cycle between seasons. While sea ice loss during the winter favors energy release, sea ice loss during the summer favors energy intake (Xie *et al.*, 2022b). Around the Barents–Kara seas, upper-level subsurface ocean warming is also enhanced. The seasonality of subsurface ocean warming also implies that the subsurface ocean absorbs more energy in the summer and releases it to the atmosphere in the winter, favoring the largest atmospheric warming observed in the Barents–Kara seas region. Furthermore, it is possible for the Southern Barents–Kara seas to warm, according to the examination of the interannual variability.

8.5 Teleconnection in Atmosphere

A consequence of the transient behavior of the atmospheric planetary-scale waves is that anomalies in climate on seasonal timescales typically occur over large geographic regions. Some regions may be cooler or perhaps drier than average, while at the same time thousands of kilometers away, warmer and wetter conditions are prevalent. In the literature on climate change, these simultaneous variations in climate, frequently of the opposite sign, over remote regions of the world, are referred to as "teleconnections."

8.5.1 *PNA teleconnection pattern*

A teleconnection map describes the linkage between a region of interest (a base point) and all other points in the domain that are farther from the de-correlation length scale of the variable. Teleconnection maps thus provide information about the structure of recurrent low-frequency variability that is characterized by the correlation-at-a-distance feature.

The 500-hPa height time series at a "reference grid point" and the matching time series at all grid points are correlated to create the Pacific–North American (PNA) teleconnections patterns (Wallace and Gutzler, 1981). The PNA pattern, which has the largest amplitude during the boreal winter, is arguably the most noticeable over the NH. By following Wallace and Gutzler's instructions (Wallace and Gutzler, 1981), the monthly mean PNA index can be created as PNA = $1/4[Z^*(20°N, 160°W) - Z^*(45°N, 165°W) + Z^*(55°N, 115°W) - Z^*(30°N, 85°W)]$, where Z^* denotes monthly mean 500-mb height anomaly that is obtained by subtracting the calendar mean value for each month.

Figure 8.25 show the one-point correlation maps of 500-hPa geopotential heights for boreal winter (December–February) over 1958–2020. The PNA teleconnection pattern has four centers of action. Over the North Pacific Ocean, geopotential height fluctuations near the Aleutian Islands vary out of phase with those to the South, forming a seesaw pivoted along the mean position of the Pacific subtropical jet stream. Over North America, variations in geopotential height over the Southeastern United States are positively correlated with the Subtropical Pacific center, but negatively connected with those over Western Canada and the Northwestern United States. The link between the four PNA centers' locations and individual phases and the mean atmospheric circulation is significant. Changes in the PNA pattern correspond to changes in the waviness of the atmospheric flow in the western half of the hemisphere, which in turn corresponds to

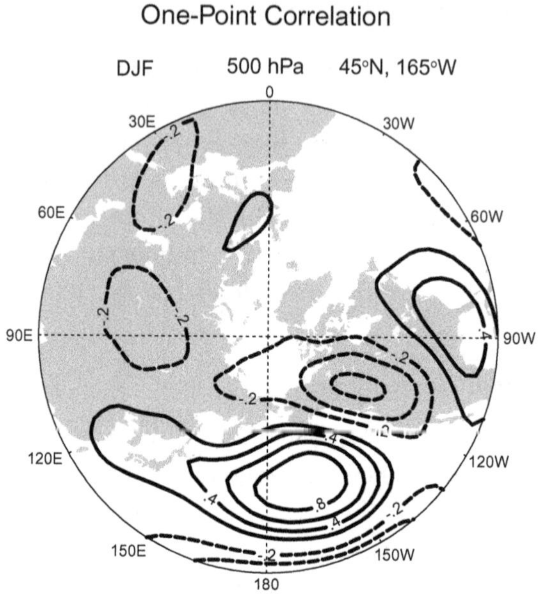

Fig. 8.25. One-point correlation maps of 500-hPa geopotential heights for boreal winter (December–February) over 1958–2020. The reference point is 45° N, 165° W, corresponding to the primary center of action of the PNA pattern (Wallace and Gutzler, 1981).

changes in the large-scale Pacific and North American air masses' north–south migration and the weather that goes along with it.

Over Western Canada and the extreme Western United States, the PNA pattern's positive phase is characterized by above-average temperatures, while the South-Central and Southeastern regions of U.S. have below-average temperatures. In summer, the PNA typically has little impact on surface temperature variability over North America. The associated precipitation anomalies include below-average totals over the upper Midwestern United States and above-average totals in the Gulf of Alaska and the Pacific Northwestern United States.

8.5.2 *Great circle path propagation of Rossby waves*

The great circle path propagation of Rossby waves (Hoskins and Koraly, 1981) refers to the fact that, under certain circumstances, Rossby waves can spread not only along the zonal direction to distant regions of different longitudes but also along the meridional direction, causing the atmospheric response to SST anomalies in the tropics to spread to middle and high latitudes.

The barotropic Rossby wave linearized perturbation equation is

$$\left(\frac{\partial}{\partial t}+\bar{u}\frac{\partial}{\partial x}\right)\nabla^2\psi'+\left(\beta-\frac{d^2\bar{u}}{dy^2}\right)\frac{\partial\psi'}{\partial x}=0, \tag{8.60}$$

where ψ' is the perturbed flow function and $\bar{u}$ is the zonal basic flow. Provided $\psi'=\psi(y)\exp[i(kx-\sigma t)]$, where k and σ are the zonal wavenumber and circular frequency, respectively, and ψ is the complex amplitude of the perturbation, after putting it into the Eq. (8.60), one can get the perturbed meridional structure equation:

$$\frac{d^2\Psi}{dy^2}+\left(\frac{\beta-\frac{d^2\bar{u}}{dy^2}}{\bar{u}-c_x}-k^2\right)\Psi=0, \tag{8.61}$$

where $c_x=\sigma/k$ is the zonal phase velocity of perturbation. For stationary Rossby waves, $c_x=0$. Provided

$$Q_k=\frac{\beta-\frac{d^2\bar{u}}{dy^2}}{\bar{u}-c_x}-k^2, \tag{8.62}$$

Q_k is the refractive index squared, which is influenced by the structure of the basic flow and the scale of the perturbation. And the concise form of Eq. (8.61) is

$$\frac{d^2\Psi}{dy^2}+Q_k\Psi=0. \tag{8.63}$$

Assuming Q_k is close to a constant, the form of the Ψ solution can be discussed qualitatively based on the theory of simple ordinary differential equation as follows.

If $Q_k<0$, then provided $Q_k=-l^2$, the Ψ solution will be exponentially decaying, i.e., $\psi(y)\sim A\exp(-ly)$, where the amplitude of the perturbation will be greatest at the latitude of the source and will rapidly decrease away from its source. This particular type of trapped wave only provokes a local reaction; it is unable to cause a meridional propagation (or a meridional teleconnection).

If $Q_k>0$, then provided $Q_k=-l^2$, the Ψ solution will have a complex exponential type, i.e., $\psi(y)\sim A\exp(\pm ily)$, resulting in the appearance of a fluctuating type in the meridional direction, known as the meridional propagation wave. l is the meridional wavenumber, and such Rossby waves can result in teleconnections in locations far from the source latitude.

Therefore, whether stationary Rossby waves can propagate in the meridional direction is determined by the positive and negative values of Q_k, and the propagation path is also determined by the distribution of the values. Hence, Eq. (8.62) indicates that in the easterly wind belt, only local responses can be generated due to $\bar{u} < 0$ and $Q_k < 0$, which cannot affect far latitudes. While in the westerly wind belt, whether the planetary waves can propagate in the meridional direction is also determined by the scale k of the waves. Only when the number of waves k in the zonal direction is smaller than a critical value k_s, i.e.,

$$k < k_s = \left(\frac{\beta - \frac{d^2\bar{u}}{dy^2}}{\bar{u}} \right)^{1/2}, \tag{8.64}$$

can waves be capable of moving in meridional directions. The smaller the k, i.e., the longer wavelength, the easier Eq. (8.64) is to satisfy, the stronger the meridional propagation of planetary waves. On the other hand, the weaker the westerly wind (the smaller $u > 0$ is), the easier Eq. (8.64) is to satisfy. Therefore, Rossby waves are more likely to propagate in the meridional direction in a region with light westerly winds.

The path of Rossby wave propagation, which forms the waveguide where the most significant teleconnections (also known as the seesaw effect) occur, is of interest for climate forecasting.

The path of wave propagation can usually be discussed in terms of the refractive index squared Q_k and the distribution of the group velocity $\vec{C}_g$ of Rossby waves (which relies on the structure of waves).

According to research, Rossby waves tend to propagate in the direction of the gradient Q_k, and the extreme value region of Q_k is conducive to the meridional (and vertical) propagation of waves, while the extreme small-value region of Q_k (e.g., westerly rapids) hinders the meridional (and vertical) propagation of waves.

Hoskins and Koraly (1981) confirmed that Rossby waves move horizontally along a kind of a great circle path through numerical studies using multi-layer linear models. An important continuous response would be produced at mid-latitudes by a non-adiabatic heat source in the tropics if the zonal flow supports the propagation of Rossby waves.

To discuss the propagation path of Rossby waves, we will calculate the group velocity C_g of barotropic Rossby waves. It is possible to assume a perturbed single-wave solution of the following kind for waves that can travel in both zonal and meridional directions, i.e., $\psi' = \psi \exp[i(kx + ly - \sigma t)]$. Adding it to Eq. (8.60) results in the frequency equation being obtained as

$$\sigma = k \left(\bar{u} - \frac{\beta - \frac{d^2\bar{u}}{dy^2}}{k^2 + l^2} \right)^{1/2}. \tag{8.65}$$

Using Eq. (8.65), one can get the zonal component C_{gx} and meridional component C_{gx} of the group velocity:

$$C_{gx} = \frac{\partial \sigma}{\partial k} = \bar{u} - \frac{(\beta - \frac{d^2\bar{u}}{dy^2})(l^2 - k^2)}{(k^2 + l^2)^2} \tag{8.66}$$

and

$$C_{gy} = \frac{\partial \sigma}{\partial l} = \frac{2(\beta - \frac{d^2\bar{u}}{dy^2})kl}{(k^2 + l^2)^2}, \tag{8.67}$$

assuming $k > 0$, and the sign of l depends on the structure waves. When the perturbed trough ridge line has a northwest–southeast orientation, $l > 0$. Conforming to Eq. (8.67), $C_{gy} > 0$, and Rossby waves have northward group velocity components. On the contrary, when the ridge line of perturbed trough faces a northeast–southwest orientation, $l < 0$, then $C_{gy} < 0$, and Rossby waves have southward group velocity components. On the other hand, there is usually $l^2 > k^2$ for Rossby waves, according to Eq. (8.66), at low latitudes in the easterly or weak westerly zone; when $C_{gx} < 0$, Rossby waves have westward group velocity components. However, at mid-latitude, $\bar{u}$ in Eq. (8.66) is larger than the latter term, then $C_{gx} > 0$, and Rossby waves have eastward group velocity components. Hence, there are several folds in the Rossby wave energy propagation channel (great circle), depending on the fundamental airflow conditions at the latitude, where the planetary wave forcing source is located, and the horizontal phase structure of the perturbation. For example, if the SST anomaly located in the equatorial Eastern Pacific Ocean generates local Rossby wave responses and the perturbation has a

Northwest–southeast phase structure, then Rossby waves energy will leave the source to propagate northwestward ($C_{gx} < 0$, $C_{gy} > 0$) first. After producing teleconnections in the central Pacific, due to increasing westerly winds at mid-latitudes, the perturbation energy will turn to propagate Northeastward ($C_{gx} > 0$, $C_{gy} > 0$) to produce teleconnections in North America. The resulting Rossby wave teleconnection in this case is similar to the PNA teleconnection pattern.

The fluctuation of wave rays and amplitudes of different waves with latitude is illustrated in Fig. 8.26 using parameters that correspond to the 300-hPa airflow sites in the Northern Hemisphere during winter and assuming that the disturbance source is at 15°N. According to the wave ray diagram, ultralong waves (1–3 waves) have rays that form a large circle to the north and east. The perturbation amplitude rises with latitude and is highest at high latitudes. For waves with a shorter wavelength, the rays

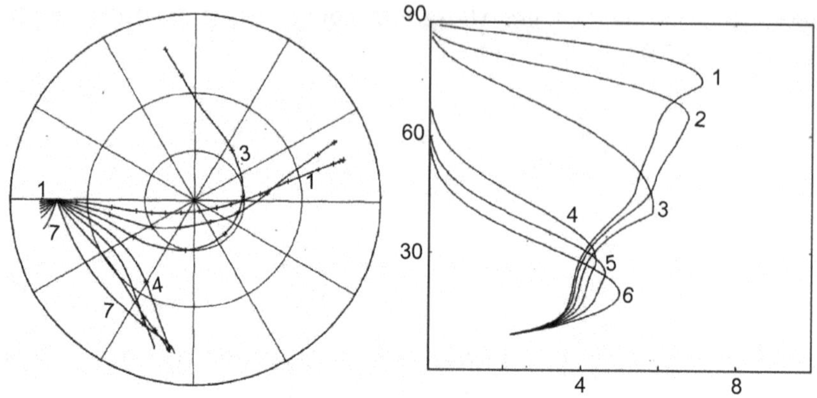

Fig. 8.26. A 15° source in the NH 300 hPa zonal flow: rays and propagation time marked by crosses (a); amplitudes of the extrema on the rays for different wavenumbers (b) (Hoskins and Karoly, 1981).

travel east and north for a certain distance before turning to the equator. The amplitude of the subtropical zone reaches its maximum.

In conclusion, the two-dimensional energy dispersion hypothesis based on Rossby waves on the sphere explains both the phenomenon of increased atmospheric response to tropical and subtropical forcing at middle and high latitudes as well as the great circle routes of large-scale perturbation wave trains. Therefore, it is regarded as one of the key theoretical foundations for atmospheric teleconnection.

8.5.3 *Tropical SST anomalies on cooling in North America*

On interannual time frames, variations in Tropical Pacific sea surface temperatures (SSTs) due to the ENSO phenomena are linked to air circulation anomalies across the North Pacific, including the PNA. The deep convection in the tropics can generally force the PNA-like teleconnection from the Tropical Pacific to North America (Trenberth *et al.*, 1998). Therefore, the PNA pattern is occasionally considered to be the extra-tropical branch of ENSO. However, the PNA shows significant variability even when ENSO is absent, proving that the PNA is an "internal" component of atmospheric variability. The PNA pattern often has two phases: a positive phase that is connected with Pacific warm episodes (El Niño) and a negative phase that is associated with Pacific cold episodes (La Niña).

According to Kosaka and Xie (2013) and Trenberth *et al.* (2014b), the ENSO transition to a cold phase, i.e., with La Niña, is responsible for the cooling of the equatorial east-central Pacific Ocean. Additionally, the La Niña phenomenon, which propagates along the Rossby wave's great circle course, coincides with the cooling of the equatorial east-central Pacific, i.e., the PNA (Pacific–North America) teleconnection pattern leading to the cooling of the North American continent (Trenberth *et al.*, 2014b).

8.5.4 *Silk Road teleconnection*

In addition to the wave-like teleconnections propagating along the great circles, such as the Pacific–North America pattern (Wallace and Gutzler, 1981), and the dipolar teleconnections, such as the Southern Oscillation and the North Atlantic Oscillation (see Chapter 6), there is a third type of teleconnection that is trapped in the jet stream with a wave-like pattern. These teleconnections include the British–Baikal Corridor (BBC) pattern in boreal summer (Xu *et al.*, 2019), the circumglobal teleconnection (CGT) in boreal winter (Branstator, 2002), and the Silk Road pattern (SRP) in boreal summer (e.g., Lu *et al.*, 2002; Enomoto *et al.*, 2003). They are a result of the jet stream waveguide effect (Hoskins and Ambrizzi, 1993). This means that specific band-like locations in the atmosphere with a local maximum of the meridional gradient of absolute vorticity or potential vorticity might act as an atmospheric waveguide and trap the propagation of Rossby waves (Hoskins and Ambrizzi, 1993). As a result, the stationary Rossby waves trapped in the waveguide develop teleconnection patterns that are sometimes referred to as waveguide teleconnection patterns and have large zonal scales and tiny meridional scales.

Along the Asian subtropical jet in the boreal summer, there are two waveguide teleconnections called the SRP and CGT (e.g., Lu *et al.*, 2002; Enomoto *et al.*, 2003; Ding and Wang, 2005). Their centers of action over Eurasia have remarkably resemblance, but the CGT may reach as far as the Pacific and North America. Due to the high similarity, SRP and CGT are frequently used interchangeably in the literature. On the one hand, the Indian Summer Monsoon (ISM) and the East Asian Summer Monsoon (EASM) precipitation can be altered by variations of the SRP/CGT from intraseasonal to interdecadal timescales (Enomoto *et al.*, 2003). On the other hand, the latent heat released from the ISM and the EASM precipitation is essential for the formation and maintenance of the SRP/CGT (Ding and Wang, 2005).

According to Chen *et al.* (2019), the Asian continent can be roughly split into "Westerlies Asia," which is mostly dominated by mid-latitude westerly circulation, and "Monsoonal Asia," which is under the control of monsoon circulation. Westerlies Asia experiences a "dislocation phase" or "antiphase" with the monsoon climate change in Asia, with a generally dry climate in the early Holocene and a relatively humid climate in the mid-late Holocene. In contrast to the monsoon zone, there is a considerable climate change in this area with "dominant modes of westerly." In terms of the spatial pattern over the last millennium, Westerlies Asia's hydrological climate variability differs greatly from Monsoonal Asia. The primary component in the production of the "dominant modes of westerly" on the suborbital scale is the external drive of the climatic system (Northern Hemisphere solar radiation). Additionally, on a centennial–decadal scale, the "dominant modes of westerly" are mostly governed by the internal variability of the climatic system (global teleconnection/silk road teleconnection).

8.5.5 *Tibetan Plateau-modulated teleconnection*

Thermal forcing in the spring and summer and mechanical forcing in the winter are the main ways in which the large-scale orography of the Tibetan Plateau (TP) influences the Asian climate (Wu *et al.*, 2007). The effects of the TP on the local atmospheric circulation, the westerly jet stream, and the ASM were thoroughly examined in previous research (Yeh, 1950). Furthermore, it has also been shown that TP forcing and its variability have the power to influence the hemispheric-scale atmospheric circulation throughout the year, with many of these effects being connected to teleconnections that are already well known. Through forced atmospheric responses, the TP can interact with distant oceans to control ocean currents and sea surface temperatures (Yang *et al.*, 2020c). These interactions may be extremely important to the global climate. The role of the TP in modifying the global climate, which involves both atmospheric circulation and air–sea interactions, is schematically summarized in Fig. 8.27.

Paleoclimate simulations have revealed that TP topography may have influenced the establishment of the AMOC, and shutting down of the PMOC, helping to determine the Earth's current climate (Yang *et al.*, 2020c). Consequently, a warmer and wetter (i.e., more habitable) Northern Hemisphere is produced by the augmentation of northward ocean heat transport and atmospheric moisture transport across the equator. The Hadley cell was weakened by 15% and the Walker circulation was increased by 40% as a result of the TP's existence, which has also had an impact on the tropical climate (Yang and Wen, 2020). While the enhanced Walker circulation correlates with the zonally asymmetric changes in the equatorial convections, which are further related to the enhanced easterlies, and cold

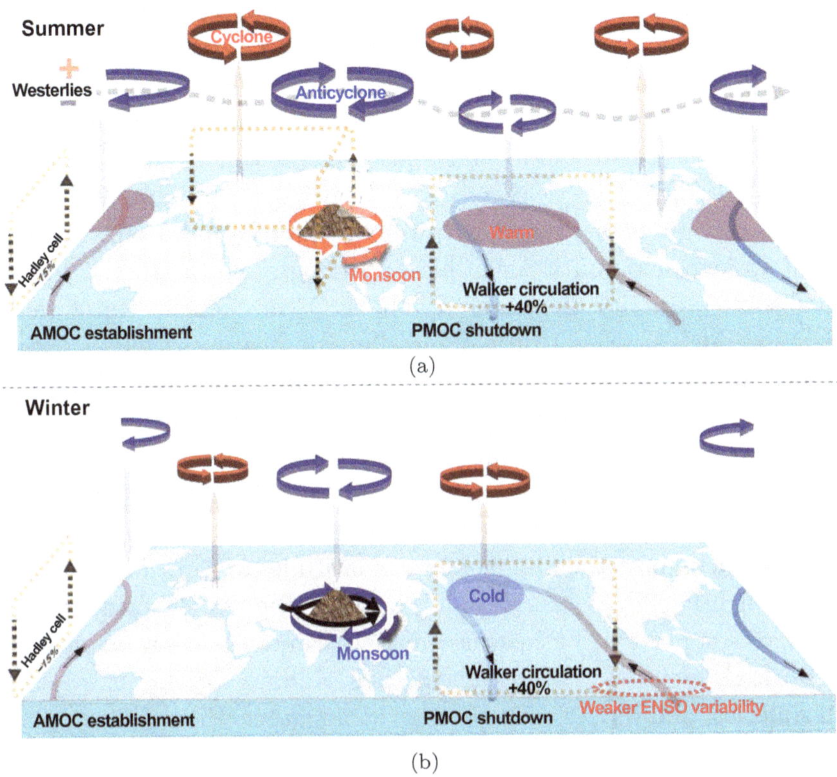

Fig. 8.27. Schematic view of the TP-modulated global circulations in summer (a) and winter (b), illustrations of the AMOC, PMOC, Hadley cell, and Walker circulation represent the annual mean climate (Huang *et al.*, 2023).

tongue in the eastern equatorial Pacific, the weaker Hadley cell correlates with the northward meridional displacement of the convective zone over the Indian–Asian–Pacific region. In addition, TP uplift lowers winter ENSO variability by causing the convection zone to migrate westward and by reducing ocean sensitivity due to the deeper mixed layer and stronger thermocline.

In addition to these paleoclimatic phenomena, seasonal dynamics of TP's influence on global climate have been summarized. The TP modulates westerlies, hemispheric Rossby wave trains, zonal–vertical cells (TP-to-upstream and TP-to-downstream), and meridional–vertical cells (TP-to-Indian Ocean) during the summer (Fig. 8.27(a)). Combining mechanisms, such as the threshold behavior theorem of thermally driven circulation and the stationary waves theorem, can theoretically explain these behaviors (Huang *et al.*, 2023) as follows: The Asian Monsoon's coupling with the

TP makes it easier for the TP to exert its wide-ranging impact. The TP induces the Northeastward migration of the East Asian summer monsoon and the northward extension of the South Asian summer monsoon, leading to substantial diabatic heating both inside and outside of its boundaries. In addition to thermal equilibrium stationary waves, angular-momentum-conserving meridional–vertical cells also form when the diabatic heating is intense enough to pass past the threshold (Plumb and Hou, 1992). The powerful anticyclonic South Asian high contributes to the threshold, which is the absolute vorticity disappeared. The T-Q_z mechanism states that the South Asian high's vertical profile and diabatic heating together define its zonal location (Wu *et al.*, 2015). Further inducing zonal–vertical cells is the movement of meridional–vertical cells. Due to the TP's subtropical location, anomalies in the upper-level circulation caused by the TP directly disrupt the westerlies and create the Rossby waves that travel with them (Fig. 8.27(a)).

A low-level cyclone, warming, anomalous ascent, and an upper-level anticyclone encompass the middle and low latitudes of Eurasia as well as the Western North Pacific as a result of TP heating (Fig. 8.27(a)). In addition to Rossby waves, a positive TP heating anomaly can cause less rain and warmer temperatures over the Mediterranean Sea, Indian Ocean, extra-tropical North America, and the Atlantic through zonal and meridional–vertical cells. Furthermore, such a positive anomaly can not only cause higher precipitation across the pan-Arctic regions of Northern Europe and North America but also decreased precipitation and warming over the Pacific through air–sea interactions. In contrast to summer, the influence of TP on winter climate is primarily mediated through Rossby waves (Fig. 8.27(b)). Theoretically, this results from winter TP-related diabatic heating only having the ability to trigger thermal equilibrium response (Plumb and Hou, 1992) in the absence of strong diabatic heating of the Asian summer monsoon. Westerly winds were redirected climatologically by the topography of the TP, while Rossby wave trains were further modulated by a quasi-barotropic anticyclonic circulation anomaly caused by negative diabatic heating over the TP (Fig. 8.27(b)). Furthermore, a distinctive PNA teleconnection pattern can also be triggered by an unusual winter TP snow cover, which will change the winter climate in North America (Liu *et al.*, 2017).

8.6 Conclusions and Discussion

The Arctic, diverse atmospheric circulation regimes, and oceanic ICV modes are a few of the many elements that affect climate variability in semi-arid regions, as elucidated in this chapter (Huang *et al.*, 2017b; Li *et al.*, 2018c; Xie *et al.*, 2020). Weaker MTF and greater ZTF from the oceanic ICV modes and the Arctic affected the terrestrial SAT, which in turn affected the polar vortex, westerly winds, planetary waves, and meridional wind, causing changes in the semi-arid climate (Huang *et al.*, 2017b). Additionally, as the DMO trended upward, an associated stronger asymmetric MTF and weaker ZTF exacerbated the warming, and vice versa. Positive feedbacks from the asymmetrical thermal forcing and the shift in terrestrial SAT could result in increased regional cooling or warming (He *et al.*, 2014).

A vital direction for future research is the interdecadal climate forecast of the semi-arid climate based on the oceanic ICV modes. Numerous studies have linked internal climate variability (ICV) to the current warming pause. But few studies have been done on the dynamics of how these ICV modes affect cooling over land in the NH. The modulating influence of these oceanic ICV modes and the Arctic on the SAT change at decadal to multi-decadal timescales is indicated by the DMO. We employed the EEMD approach, which can divide the evolution of SAT into long-term trends and oscillation components, to examine the DMO components of SAT. The oscillation component is believed to be mostly caused by ICV, while the long-term warming trend is primarily attributable to radiative forcing (Wu *et al.*, 2011; Guan *et al.*, 2015a). The DMO's magnitude is comparable to the decadal scale's long-term trend. The warming was increased when the DMO trended higher; conversely, when it is in a downward phase, the radiative forced warming is balanced or reduced, leading to the global warming hiatus. We use the warming acceleration and hiatus events to highlight the modulation effect of the DMO on SAT change. We use the multiple linear regression (MLR) and EEMD methods to create the DMO index; it not only explains the impact of the oceanic ICV on the DMO but also quantifies the contribution of oceanic ICV to DMO. The downward DMO combined with a weaker MTF and stronger ZTF, which stimulate both a weaker polar vortex and westerly winds, along with the amplified planetary waves, thereby facilitating southward invasion of cold Arctic air and promoting the blocking formation, led to the warming hiatus over the NH.

In the second portion of this chapter, we provided the idea of the DMO's modulation influence on SAT and demonstrated that the DMO causes the whole warming and warming hiatus process on the multi-decadal timescale. The dynamic mechanism of the warming hiatus over the NH is shown by discussing the interaction of oceanic ICV and its contribution to DMO changes in light of this definitive result. It is revealed that the DMO can be used to interpret both the current warming hiatus and the likelihood that global warming will resume when it swings back upward. The DMO model gives us a fresh perspective on how climate change and oceanic ICV interact. Moreover, we anticipate applying this model to research semi-arid climate change.

We employ a coupled land–atmosphere model to address the shortcoming of the classic Charney's model that the thermal forcing is always artificially specified. For a specific realistic case of uneven solar heating, we find that there are still multiple equilibrium states when topography is present. Therefore, Section 8.3 of this chapter confirmed the multiple-flow equilibria theory once more. The multiple-wave phase equilibria associated with ridge and trough types are more apparent in our coupled baroclinic model, while they are also present along with the multiple-wave amplitude equilibria associated with high- and low-index types (Fig. 8.14). The multiple-wave phase equilibria associated with ridge and trough types originate from the orographic instability of the Hadley circulation in the presence of topography, which is a pitch-fork bifurcation. Thus, the ridge- and trough-type equilibria can also coexist in the uncoupled model as long as the topography is present (Fig. 8.17(c)). But the multiple-wave phase equilibria in the uncoupled model are unremarkable, mainly due to the very narrow orographically unstable region (Fig. 8.15(b), Case 2). The land–atmosphere coupling considerably expands the orographically unstable region (Fig. 8.15(b), Case 1), and thus the multiple-wave phase equilibria in the coupled model are prominent (Fig. 8.14). We have shown that the longwave radiation fluxes are the main contributors to the effect of the land–atmosphere coupling, and that the heat fluxes more or less modify this effect. Besides, the coupled model's atmospheric response to topographic and thermal forcing is substantially weaker. In other words, the atmospheric flow is significantly stabilized by the land–atmosphere interaction.

We have thoroughly examined the wave phase and ridge- and trough-type equilibria. Topography causes the first bifurcation of the Hadley circulation. The bifurcation has two equilibrium states with opposing lower-layer zonal wind directions: the trough-type equilibrium has lower layer-westerlies, while the ridge-type equilibrium has lower-layer easterlies. Therefore, the existence of topography is still responsible for the creation of two opposing directions of the bottom-layer zonal wind. Additionally, we investigated the wave amplitude and high- and low-index equilibria. The findings demonstrate that the meridional atmospheric temperature gradient is smaller (greater) and the wave amplitude is larger (smaller) in the low-index (high-index) equilibrium states. Meanwhile, our coupled baroclinic model finds no marked differences in the zonal wind speeds of the high- and low-index equilibrium states. These findings can be explained by the fact that the mean flow potential energy in the baroclinic atmosphere is converted, maintaining the wavelike equilibrium. Therefore, it is important to carefully reevaluate the earlier barotropic model conclusion (Charney and DeVore, 1979).

Nevertheless, the low-order model we employed is oversimplified and has several drawbacks. For instance, our two-layer model's vertical resolution is still poor, the land–sea thermal contrast is not taken into consideration, and the equilibrium states' flow patterns are sensitive to the model's horizontal resolution. The study of various large-scale atmospheric phenomena, such as the recurrence of quasi-stationary planetary wave trough and planetary wave ridge over particular regions, e.g., the Ural (Dole and Gordon, 1983; Li and Ji, 2001; Holton *et al.*, 2003; Ren *et al.*, 2006; Pan *et al.*, 2009), is made easier by our conclusions on multiple-wave phase equilibria. The extent to which our findings concur with the observations requires further investigation. Future research on multiple-wave phase equilibria should make use of more realistic models.

The influence of the Arctic also has a significant role in the variability of the semi-arid climate. Based on PV dynamics, Section 8.4 investigated the broad dynamics that link the Arctic to the climate of Eurasia. In conclusion, the Arctic has an impact on Eurasia through interactions between different spheres. Specifically, in summer, sea ice, surface solar radiation, and subsurface ocean heat uptake are important factors in multi-sphere interactions, whereas in winter, sea ice, surface sensible heat, diabatic heating of the atmosphere, the interaction between the Arctic and mid-latitude, and large-scale circulation are important factors (Xie *et al.*, 2022). The atmospheric circulation regime underlying the teleconnection between the Arctic and Eurasia is determined by PV adaptation and advection mechanisms (Xie *et al.*, 2020). The relationship between the Arctic and the mid-latitudes is greatly influenced by the spatial distribution, vertical structure, and climatological PV distribution. Due to the intricacy of multi-sphere interactions, the use of the Arctic as a climate predictor for Eurasia requires more investigation.

The structure of recurrent low-frequency variability, which is defined by the correlation-at-a-distance feature, is revealed via teleconnection maps. There is a growing understanding that variations in the frequency and amplitude of teleconnection modes affect changes in climate in different places. Teleconnection, which is common in the atmosphere, controls and influences climatic parameters thousands of kilometers away. We briefly describe many teleconnection models in Section 8.5, where we concentrate on the Rossby wave's great circle path propagation (Hoskins and Koraly, 1981). The great circle pathways of large-scale perturbation wave trains and the phenomenon of an increase in atmospheric response to tropical and subtropical forcing at middle and high latitudes are both explained by the two-dimensional energy dispersion theory based on Rossby waves on the sphere. According to the study, the PNA (Pacific–North America) teleconnection pattern, which results in the cooling of the North American continent, coincides with the La Niña phenomenon, which causes the equatorial east-central Pacific to cool (Trenberth *et al.*, 2014b). Furthermore, two new types of teleconnections were discovered. One is based on the waveguide effect of jet streams (Hoskins and Ambrizzi, 1993), which trapped the wave-like pattern Silk Road pattern (SRP) and the circumglobal teleconnection (CGT). The combination of the two teleconnection models primarily controls the "dominant modes of westerly" on the centennial–decadal scale. The other teleconnection is the Tibetan Plateau-modulated teleconnection, which is also crucial for the current climate variability, and plays an important role in the global climate. In addition to showing a phase relationship with the same or opposite sign between two places, teleconnection, a type of long-distance interacting climatic mode, is crucial for the evolution of the global climate. Many potential dynamic mechanisms of climate changes are expected to be resolved by learning how to capture new teleconnection models and expose the interaction between these existing teleconnection models.

We have summarized the potential dynamic mechanisms that could contribute to semi-arid climate change in this chapter, but these theoretical studies are incomplete and do not yet form an integrated theoretical framework. The task at hand will be to develop and enhance this theoretical framework. Although the process of this theoretical system should require nonlinear interaction over several timescales and multiple cycles, this does not exclude the discovery of a simple nonlinear coupling model. A great conceptual model is the DMO. If we continue along this path, we might achieve significant theoretical advancement. We may accurately predict semi-arid climate change and learn more about the fundamental dynamic mechanism of semi-arid climate change if this theoretical system is constructed.

Chapter 9

Desertification of Semi-Arid Regions

9.1 Introduction

According to the previous chapters, the world's arid and semi-arid regions have grown over the past 60 years and will do so again in the 21st century. Increased aridity, growing arid and semi-arid regions, rising temperatures, intensive land usage, and other human activities could all contribute to desertification, which would be extremely dangerous for arid and semi-arid region ecosystems and their populations and would intensify regional tensions. Land degradation will occur in some sensitive locations, such as arid and semi-arid regions, as a result of the extension of such regions and increased human activity. According to reports, one-fourth of the world's land area is affected by desertification, and 10%–20% of arid and semi-arid regions have already experienced degradation (medium certainty; Millennium Ecosystem Assessment 2005; UNCCD 1994), which has a direct impact on 250 million people in developing countries. Given the effects of climate change and population growth (PG), this number is probably going to increase significantly (Reynolds *et al.*, 2007; Huang *et al.*, 2016a, 2017a, 2019). Global climate change is predicted to worsen desertification in the future and alter the world's biogeochemical cycles significantly. With careful evaluation of region-specific aspects of a nature–society system, proactive planning and adaptation strategies should be created, and management and adaptation activities must be performed for global arid and semi-arid regions. Understanding the causes of desertification, evaluating the hazards of desertification, and advancing research on the ecosystem of sensitive areas are therefore of utmost importance in the context of the extension of arid and semi-arid regions.

Desertification is currently a major social, economic, and environmental issue that affects all nations in the world. The French plant biologist and geographer André Aubréville coined the term desertification in his 1949 book "Climate, Forests, and Desertification in Tropical Africa." Desertification was described at the United Nations Conference to Combat Desertification (UNCCD) in Nairobi in 1977 as the loss or destruction of the biological potential productivity of land, leading to the eventual emergence of a desert state. It was then described as land degradation brought on by unsuitable human activity in arid, semi-arid, and arid subhumid environments during the Global Assessment of Desertification Conference held in Nairobi in 1990. It was further described as the process of land degradation in arid, semi-arid, and arid subhumid regions under the influence of several causes, including climate and human activities, during the United Nations Conference on Environment and Development held in Rio de Janeiro in 1992. This definition was included in the 1994 International Convention to Combat Desertification. Desertification is a process of land degradation that results from the interaction of irrational economic activity with a precarious ecological environment. It results in decreased land productivity, the depletion of land resources, and the surface appearance of a desert landscape.

There are four primary types of desertification: wind erosion desertification, water erosion desertification, freeze–thaw desertification, and salinization. (1) Among all types of desertification, wind erosion desertification has the largest area and the widest dispersion. It primarily occurs in arid and semi-arid regions. Climate, especially the amount of dryness and moisture, is a threat to it. This is because soil erosion resistance and water content have a positive effect in wind erosion desertification. The type and extent of vegetation are also influenced by changes in soil moisture and dryness. As a result, depending on the kind of climate, the degree of wind erosion desertification varies from subhumid arid to semi-arid to arid regions. The general trend of change is from mild to moderate to severe to extremely severe. To put it another way, when climate types dry up, wind erosion desertification becomes a more serious issue, and its distribution area grows in size. (2) The distribution of water erosion desertification is primarily concentrated in the middle and upper reaches of rivers and the foothills of some mountains, and it clearly demonstrates topographic characteristics. High altitude, significant shape variability, and favorable precipitation conditions are typical of these regions. Additionally, many of these locations are hubs of human activity. Vegetation is destroyed and the surface is exposed as a result of practices like overgrazing and land reclamation on steep slopes. This speeds up the process of desertification due to water erosion. (3) In high-altitude regions with significant day–night temperature differences, freeze–thaw desertification occurs. Due to strong thermal expansion and cold contraction, the soil structure is harmed. (4) One type of land degradation brought on by chemical activity is soil salinization. It is extensively dispersed in places experiencing desertification. Climate, poor drainage, a high groundwater table, and inefficient irrigation practices are its key causes. It is most severe in arid regions and only mildly so in subhumid arid regions.

Desertification is tightly linked to environmental, economic, and social development rather than being a purely natural phenomenon. According to the International Convention to Combat Desertification, numerous ecological, biological, political, social, cultural, and economic variables interact to cause desertification. The region's harsh natural environment and excessive human activity in the sensitive biological environment are the two main causes of desertification, although there are many other variables as well. By changing the spatiotemporal patterns of temperature, rainfall, and wind, climate change causes desertification.

The primary causes of desertification and land degradation are human activities such as pollution, overgrazing, and the use of natural land for agriculture (Yu *et al.*, 2018). However significant changes are predicted to occur over arid and semi-arid regions during the next few decades. Increased human demands of all kinds, including population growth in urban and rural regions, ongoing oil development, and an increase in the number of large-scale water infrastructure projects, will cause the most significant changes. In many locations, rural populations may be approaching the boundaries of arable land, which may have an effect on the soil and vegetation. In Africa, desertification is affecting 1.9 million ha of irrigated croplands (or 18% of the total area), 48.86 million ha of rain-fed croplands (or 61% of the total area), and 995.08 million ha of rangelands (or 74% of the total area) at a moderate or higher level, according to UNEP's most recent assessment. The most vulnerable regions of the continent are the Kalahari–Namib region in Southern Africa, the Sudano–Sahelian region, and Mediterranean Africa. Specifically, 73% of all agriculturally exploited arid and semi-arid regions are deteriorated, and one-third of Africa is afflicted by desertification.

Governments and organizations from throughout the world have been interested in desertification as a worldwide environmental issue. The International Year of the Desert and Desertification was declared in 2006, and the United Nations has occasionally focused on desertification and arid and semi-arid regions. Notable examples include the adoption of the Convention to Combat Desertification in 1992. On the recommendation of the Second Committee of the General Assembly (Economic and Financial), the 49th General Assembly of the United Nations adopted resolution 49/115 on December 19, 1994, designating June 17 of each year as World Day to Combat Desertification and Drought, starting in 1995. The resolution's goal was to further increase global awareness of the significance of combating desertification and to arouse people's sense of responsibility and urgency. The Conference of the Parties to the Convention was convened annually from 1997 to 2001. The Conference of the Parties has taken place every two years since 2002. Annual meetings are held by the Convention Performance Review Committee. In 2019, the Intergovernmental Panel on Climate Change (IPCC) released a Special Report on climate change, desertification, land degradation, sustainable land management, food security, and greenhouse gas fluxes in terrestrial ecosystems. Therefore, it is crucial to investigate arid and semi-arid region desertification using a consistent criterion to combine global climate change and human activity.

Vulnerability assessment is a tool for examining climate change policy research and will be used as a guide for managing of adaptation. Researching the susceptibility of desertification is also a critical concern. Policymakers could create the finest plans for slowing down desertification and implementing adequate land rehabilitation programs in sensitive areas with the identification and projection of areas that are susceptible to it. Desertification, however, is a complicated phenomenon that is typically fueled by a variety of causal agents, including biophysical and socioeconomic causes (Jafari and Bakhshandehmehr, 2016). On the basis of various indices and datasets, many mathematical models have been built to identify regions that are vulnerable to desertification. The Mediterranean Desertification and Land-Use Model, which incorporates the four primary factors that contribute to land degradation and desertification — climate, soil, vegetation, and land management — into an environmental sensitivity index to map the sensitivity of the environment to desertification, is the most widely used model (Lavado Contador *et al.*, 2009). However, in order to take into account unique features in various regions, the Mediterranean Desertification and Land-Use Model needs to be updated (Pravalie *et al.*, 2017). As a result, a majority of studies have concentrated on the degree of regional desertification; however, since diverse factors were taken into account, it is not possible to compare these assessments directly. On the basis of a reclassification of the Global Soil climate map and Global Soil map, the US Department of Agriculture

(USDA) first published a global susceptibility to desertification (Eswaran and Reich, 2003). The risk of worldwide desertification was evaluated by Spinoni *et al.* (2015) solely taking climate change into account.

With the increasing use of satellites to monitor the Earth, abundant biophysical data are available on a global scale, such as the leaf area index (LAI), which is an important factor of net primary production (NPP), water and nutrient use, and carbon balance. A growing variety of datasets, such as those on population density (PD), carbon dioxide (CO_2) emissions, and gross domestic product (GDP), characterize social factors at a worldwide level. PD denotes the level of demand for natural resources generated by human activity, which puts the ecosystem under direct stress. The economic and development status is represented by both (CO_2) emissions and GDP. The World Atlas of Desertification, third edition (Cherlet *et al.*, 2018), recommended merging a variety of other global datasets due to the growing availability of global datasets. However, determining the vulnerability of the world to desertification requires a consistent index that takes into account both human activity and climatic change. The susceptibility of the entire planet to climate change and human activity has received little attention, despite some studies (Rampone and Valente, 2019) exploring future desertification vulnerability using various methodologies. The Fifth Coupled Model Intercomparison Project (CMIP5) has produced projections under various emission scenarios and may offer the information required to project the vulnerability to desertification.

Along with identifying the area most susceptible to desertification, it is critical to gain a better understanding of the mechanisms and feedback that occur during the process. In arid and semi-arid regions, the plant decline phase precedes soil disruption, ecosystem collapse, and ecosystem degradation (Berdugo *et al.*, 2020). Actually, the destruction of ecological security is what has caused desertification. Ecological security is a condition in which stable, naturally occurring and semi-natural ecosystems can be found. It is related to the ecological environment and offers ecological guarantees for the long-term sustainability of the entire eco-economic system (Newbold *et al.*, 2015). However, ongoing global warming and increased human activity pose a serious danger to ecological security (Huang *et al.*, 2012, 2015, 2016b; Fu *et al.*, 2017). According to several studies (Glover *et al.*, 2010; Humphrey *et al.*, 2017), ecological damage has led to desertification, food shortages, and water insecurity, all of which pose a threat to human existence and development in the 21st century. Under conditions of climate change and human activity, the biological, oxygen, carbon, hydrological, and thermal cycles are closely related to ecological security. This is due to how the oxygen cycle, which is affected by changes in vegetation structure and function, regulates the flow of energy, water, and carbon between the land and the atmosphere (Piao *et al.*, 2020). However, nothing is known about how these changes may affect ecological security. To assess the security of the terrestrial ecological system through the processes of respiration, photosynthesis, and burning, it is crucial to comprehend how these indicators vary. Therefore, understanding how future changes in ecological security will be impacted by climate change and human activities is crucial for both protection and adaptation efforts.

Despite the fact that the idea of ecological security was first put forth in the 1970s (Sohn, 1973), there is still no single definition that is widely accepted because ecological security is a complicated topic that encompasses numerous factors (Feng *et al.*, 2018). The current analysis of ecological security has two drawbacks. First, although some indicators have been used to assess regional ecological security in previous studies (Feng *et al.*, 2018), those assessments have only considered specific regional issues, such as water security (Wang *et al.*, 2015) and urban ecological security. Second, because ecological security has many definitions (Hu *et al.*, 2019a), there is no standardized indicator system that is widely accepted. In order to assess ecological security, it is crucial to comprehend how human activities and variations in the climate affect the cycles of biological, oxygen, carbon, hydrological, and thermal energy. All of Earth's areas are related by the oxygen, carbon, hydrological, and thermal cycles, which are likewise linked to all biological cycles (Schlesinger *et al.*, 2017; Huang *et al.*, 2017b, 2017c, 2018). This chapter presents a framework that we have created. It considers the interactions between the biological, oxygen, carbon, thermal, and hydrological cycles and offers a quantitative assessment to gauge ecological security throughout the entire planet and forecast its changes in the future. The biological, oxygen, carbon, thermal, and hydrological cycles must be maintained in balance for humans and other animals to survive. This is what is meant by ecological security. The process of photosynthesis absorbs solar radiation and maintains carbon balance in addition to producing essential O_2 and being crucial for the food chain. Because of limited biological activity and sluggish soil organic carbon decomposition caused by high water levels, carbon accumulates in soils in significant amounts. Carbon accumulation plays an important driving role in the carbon cycle through atmosphere, green vegetation, and surface soils.

When land without photosynthesis is unable to sustain life, it exhibits similar characteristics to cancer. This type of land will invade and destroy the surrounding land and lead to biological extinction. By viewing the Earth as a living body, the concept of Earth cancer was proposed and we developed a detection method for diagnosing cancerous land. Earth cancer starts in semi-arid regions with the lowest amounts of oxygen production, urban centers, and in concentrated industrial areas with the highest amounts of oxygen consumption. For policymakers to decide where and how to carry out the combatting action against desertification, it is crucial to identify the places that are vulnerable to desertification and to uncover the mechanisms and feedback in the process. Mitigating climate change may be a more dramatic approach when compared to biological and mechanical methods of preventing and controlling desertification. Nature-based approaches have recently shown great promise in addressing a range of environmental and socioeconomic issues. Due to their great sensitivity to inter-annual climate variability, arid ecosystems play a significant role in the global carbon cycle, dominating the trend and variability of the terrestrial carbon sink. The scientific community has been particularly interested in the desert ecosystem recently because it has been shown that deserts can store enormous amounts of CO_2 to fill in the gap left by the missing carbon sink. The desert ecosystem is a typical widely dispersed severe environment in arid and semi-arid regions. The Taklimakan Desert (TD), the second-largest changing desert in the world, is crucial for the ability of desert ecosystems to act as carbon sinks. The TD has a significant capacity for storing carbon thanks to its considerable salinization and plentiful saline–alkali groundwater. Regarding the magnitude and how that will react to climate change, there is ambiguity, though. Despite the low levels of soil organic carbon (SOC), the arid and semi-arid region soils' total SOC pool is 241 Pg, or 15.5% of the global average of 1550 Pg to a 1-meter depth. The historic C loss due to desertification ranges from 20 to 30 Pg. Afforestation with the right species, soil management on crops, pasture management on grazing land, and restoration of degraded soils and ecosystems through afforestation and conversion to other restorative land uses are all examples of land-use and management strategies that sequester SOC. The entire potential of SOC sequestration over a 50-year period is 12 to 20 Pg C, assuming that two-third of the past loss can be re-sequestered. In order to counteract desertification, we shall introduce the knowledge in this chapter regarding improving carbon storage capacity through sustainable land management.

The arrangement of this chapter is as follows. Section 9.1 presents the definition and distribution of desertification, which also reviews recent progress in research on global desertification. Section 9.2 discusses global desertification vulnerability to climate change and human activities. Declines in global ecological security under climate change, impacts of differences in soil temperature on desert carbon sink, and soil conservation and desertification control are introduced in Sections 9.3, 9.4, and 9.5, respectively. Conclusions and discussion are summarized in Section 9.6.

9.2 Global Desertification Vulnerability

Desertification is the impoverishment of arid, semi-arid, and some subhumid ecosystems. To assist decision-makers in developing the best strategies for land rehabilitation and combating global desertification in vulnerable areas, it is crucial to analyze the vulnerability of desertification on a global scale to climate change and human activities. There is no map of worldwide vulnerability to desertification that takes climate change and human activity into account. The global desertification vulnerability index (GDVI), which was created by fusing human activity and climate change, offers another viewpoint on the susceptibility of desertification on a worldwide scale and projects its future development. Desertification vulnerability can be divided into four groups based on the GDVI's probability density function: very high, high, medium, and low. According to the analysis's findings, areas around deserts and barren land have a higher risk of desertification. The percentages of the global area that were subject to moderate, high, and extremely high desertification risk were 13%, 7%, and 9%, respectively. Among the representative concentration pathways (RCPs), RCP8.5 projected that the area of moderate to very high desertification risk will increase by 23% by the end of this century. In Africa, North America, and the northern regions of China and India, desertification hazards are expected to rise over time.

9.2.1 *Establishment of global desertification vulnerability index (GDVI)*

(1) *Estimation of HAI*

We created an HAI to calculate the impact of human activities based on four human indicators: PD, PG, CO_2 emissions, and GDP. The greater PD shows that more natural resources are needed and that the environment is under more stress. PG is the product of how social policy affects population, and as population grows, so does the pressure on the environment. GDP and CO_2 emissions are measures of the economy's and world's states of development, respectively. The following is a definition of the HAI:

$$\mathrm{HAI} = w_1\mathrm{PD(t)} + w_2\mathrm{PG(t)} + w_3\mathrm{CO_2(t)} + w_4\mathrm{GDP(t)}, \qquad (9.1)$$

in which w_1, w_2, w_3, and w_4 are the weighting coefficients of PD, PG, CO_2 emissions, and GDP, respectively, as determined by the criteria importance obtained through the intercriteria correlation (CRITIC) method. PG in year i was defined as the difference in PD between year i and year $i-1$:

$$\mathrm{PG}_i = \mathrm{PD}_i - \mathrm{PD}_{i-1}. \qquad (9.2)$$

Prior to creating the HAI, all four human activity indicators were non-dimensionalized.

The following procedure was used to normalize the HAI in the study:

$$x' = \begin{cases} 0 & (x < a) \\ \frac{x-a}{b-a} & (a \leq x \leq b), \\ 1 & (x > b) \end{cases} \qquad (9.3)$$

in which x represents HAI, $a = \bar{x} - 2 \times \sigma$, $b = \bar{x} + 2 \times \sigma$, $\bar{x}$ is the mean of HAI, and σ is the standard deviation of HAI. After normalization, the HAI has a range of 0 to 1. Higher HAI signifies more human activity, which suggests a greater need for natural resources and a higher likelihood of the exploitation of land and water resources (e.g., land overdevelopment, water shortages, and grassland overgrazing). Therefore, a higher HAI shows that human activity is contributing more to the likelihood of desertification.

(2) *Estimation of GDVI*

The GDVI was built on the premise that regions with a high rate of warming, high aridity, little vegetation cover, and significant human activity had a higher risk of desertification (Fig. 9.1). To create GDVI, we merged LAI, ΔT, AI, and HAI. The following formula was used to determine the GDVI values for each grid across the world:

$$\mathrm{GDVI} = \mathrm{CEI} \times \mathrm{HAI}, \qquad (9.4)$$

in which $\mathrm{CEI} = 1/\mathrm{LAI} \times \Delta\mathrm{T}/\mathrm{AI}$, and CEI is the climate environment index.

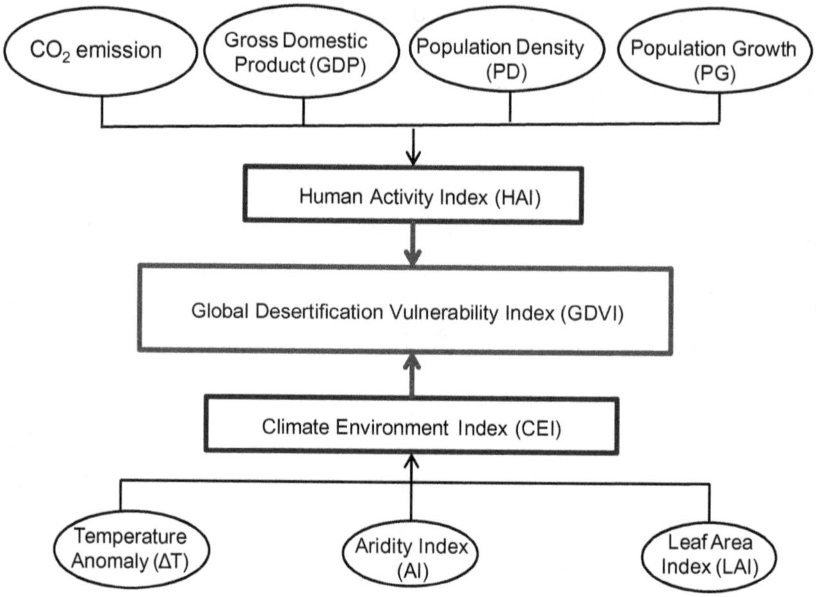

Fig. 9.1. Diagram of the global desertification vulnerability index (GDVI) (Huang *et al.*, 2020b).

Because the HAI was non-dimensional and ranges from 0 to 1, ΔT, LAI, and AI were rescaled using Eq. (9.5):

$$x' = \frac{x - x_{\min}}{x_{\max} - x_{\min}}, \tag{9.5}$$

in which x represents ΔT, LAI, and AI. $x_{\max}$ and $x_{\min}$ are their maximum and minimum values, respectively, which were defined by the 99.9% and 0.1% thresholds. An overview of the development of ecosystems and surroundings was provided by the type of indicator used to evaluate the vulnerability to desertification.

(3) *Statistical analysis*

We employed a probability density distribution function to categorize desertification susceptibility, as calculated by our GDVI, into four classes: low, moderate, high, and extremely high. This gave an intuitive depiction of the global desertification vulnerability to climate change and human activities. We compared our findings to the USDA's desertification vulnerability map in order to validate the grading results (Eswaran and Reich, 2003). In order to match the GDVI data, the USDA desertification vulnerability map data were first aggregated into 1° spatial resolution data using the majority algorithm, which determines the most prevalent desertification risk class for a 1° grid cell. This was done because the USDA data are rasterized on a 2-min grid cell. Then, we compared the spatial distribution of global vulnerability of our results and those of the USDA. The area proportion of various subtypes determined by the USDA and GDVI was also calculated. We also determined the Pearson correlation coefficients between the GDVI, NPP, and soil moisture to assess the accuracy of the GDVI data at each pixel. Using the Student's t test at the $p < 0.05$ level, the correlation's significance was examined. We compared the outcomes of a CMIP5-EM simulation with observations made during the same time period to ensure the accuracy of the future projection.

To obtain a relatively accurate result, the spatial resolution results of the CMIP5 models were interpolated into a $1° \times 1°$ spatial resolution using bilinear interpolation. The National Center for Atmospheric Research Command Language Version 6.5.0 was used to create all of the maps and plots in Figs. 9.2–9.5, in particular the function "gsn_ csm_ contour_ map." On the specified workstation, that function might produce and display a contour plot over a map (https://www.ncl.ucar.edu/Document/Graphics/Interfaces/gsn_csm_contour_map.shtml). Using MapPlot and Map Transformation (https://www.ncl.ucar.edu/Document/Graphics/Resources/mp.shtml#mpProjection), the attribute of the map was added and adjusted. The coastlines on the map are a condensed version of those in Wessel and Smith's Global Self-Consistent Hierarchical High-resolution Shoreline database, which Rainer Feistel created from the Regionally Accessible Nested Global Shorelines database.

9.2.2 *Evaluating desertification vulnerability by GDVI*

The GDVI was computed using the CEI and HAI, and values are displayed in Fig. 9.2. According to the probability distribution function, we divided the vulnerability to desertification into four groups: extremely high (GDVI >81.5), high ($31.5 <$ GDVI ≤ 81.5), medium ($10.0 <$ GDVI ≤ 31.5), and low (GDVI ≤ 10.0). In Fig. 9.2(b), the probability density function for the logarithm of GDVI has two different peaks at 1.70 (GDVI = 5.5) and 2.50 (GDVI = 12.2). The point that established the distinction between the low and medium classes was the local minimum between two peaks, where the logarithm of the GDVI is 2.3 (GDVI = 10.0). The probability density swiftly declines to the point where the GDVI logarithm is 3.45 (GDVI = 31.5) after reaching the second peak, and then it gradually declines until the GDVI

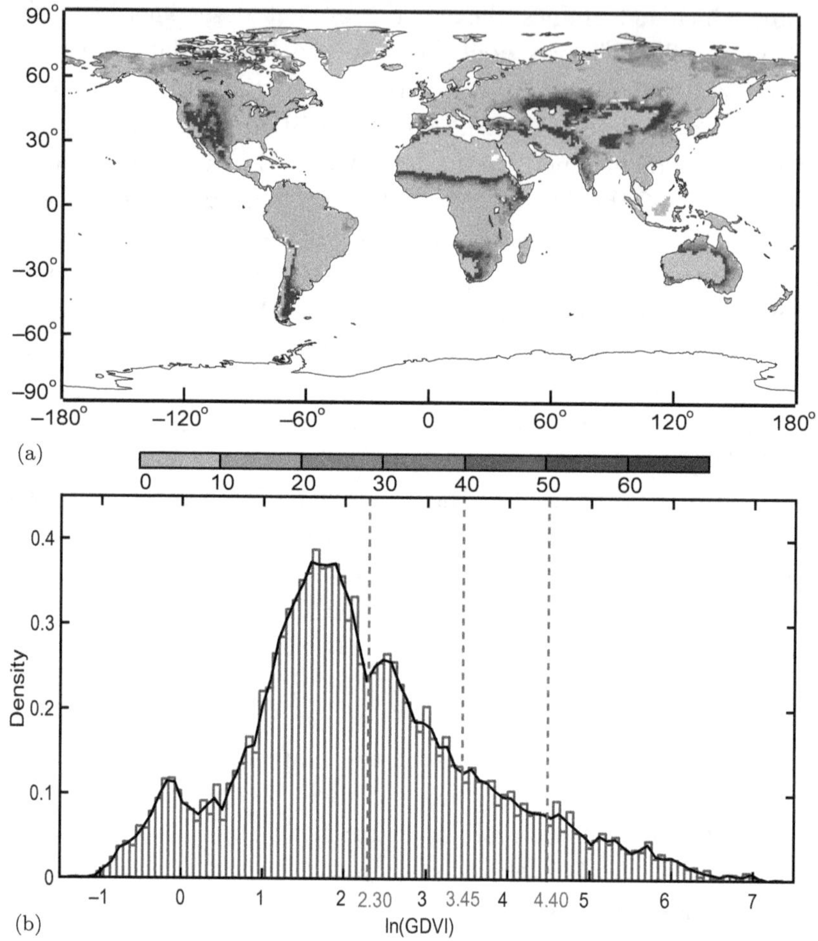

Fig. 9.2. Global distribution of the mean global desertification vulnerability index (GDVI) for 2000–2014. Areas dominated by barren land are shown in gray (a). The probability density function for the logarithm of the GDVI. Dotted line represents the threshold value used to classify the GDVI (b) (Huang *et al.*, 2020b).

logarithm is 4.40 (GDVI = 81.5). Desertification vulnerability was shown to be high in the transition region, where the GDVI is greater than 31.5 and less than 81.5, compared to very high and medium vulnerability to the right and left, respectively.

9.2.3 *Global distribution of desertification vulnerability*

The distribution of the mean GDVI from 2000 to 2014 is depicted in Fig. 9.3(a). Western North America, the Sahel, Central Asia, and Northern China are the areas with the highest risk of desertification due to their

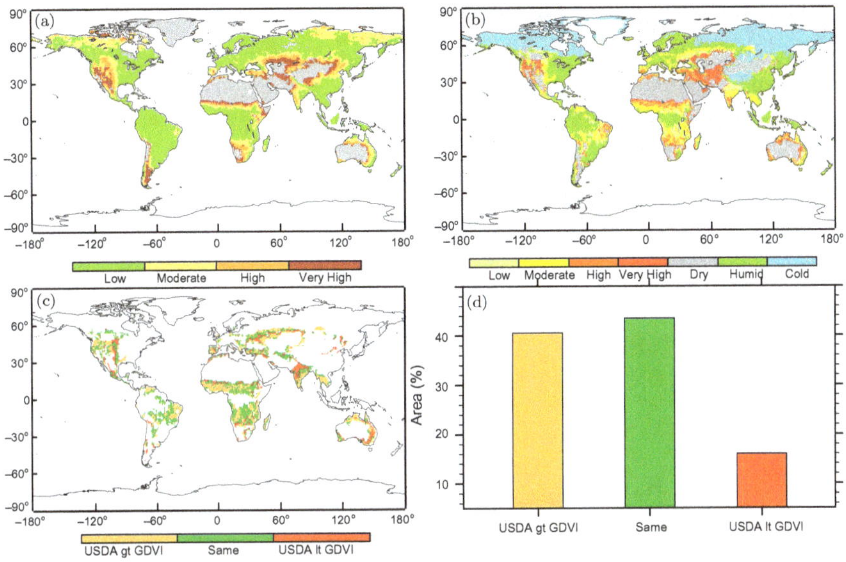

Fig. 9.3. The global distribution of desertification risk in the global desertification vulnerability index (GDVI) for 2000–2014 (a) and U.S. Department of Agriculture (USDA) data (b). The global distribution of difference between GDVI with USDA data (c); the area fraction of different risk occupy region was estimated by both two data (d). "USDA gt GDVI" indicates that the level of the USDA is higher than GDVI's, "USDA lt GDVI" indicates that the level of the USDA is lower than GDVI's, and "Same" indicates that the level of two datasets is coincident (Huang *et al.*, 2020b).

scant vegetation and arid environment. In the Indian subcontinent and North China Plain, the risk of desertification is moderate to severe due to high levels of human activity. According to earlier research (Martínez–Valderrama *et al.*, 2016), the danger of desertification is similarly moderate in Southwestern Europe.

We compared the fresh findings to the USDA's map of vulnerability to desertification (Eswaran and Reich, 2003). The UNCCD definition of desertification excludes regions with humid, arid, or cold climates. The USDA data's spatial distribution resembles that of the GDVI data (Figs. 9.3(a) and 9.3(b)). The distribution pattern is dominated by the region where two data points are coincident (Fig. 9.3(c)), primarily in Africa and South America. India and the Indo-China Peninsula are the two most erratic regions. In India, the risk is overestimated because of intensive human activity.

On the basis of USDA statistics, it is decided that the Indian subcontinent has a high danger of desertification after taking the PD into account (Eswaran *et al.*, 2009). However, in the Indo-China Peninsula, where the GDVI is low, the vulnerability to desertification is overestimated due to the high LAI. The two maps could not be compared point for point since the USDA map was based on a reclassification of the global soil climate map and the global soil map, which are different components from those utilized in the GDVI. In general, GDVI is reasonable and trustworthy.

We compared the GDVI with NPP in order to see how well the GDVI measures the vulnerability to desertification. NPP can be used to estimate a region's ecosystem and land productivity (Higginbottom and Symeonakis, 2014). Arid and semi-arid regions experience a decline or loss in productivity as a result of desertification. NPP is closely related to desertification, with a higher probability of desertification resulting in a lower NPP. The distribution of NPP differs from the GDVI's. Greater GDVI areas have lower NPP (Fig. 9.4(b)). For regions with various GDVI levels, the frequency of

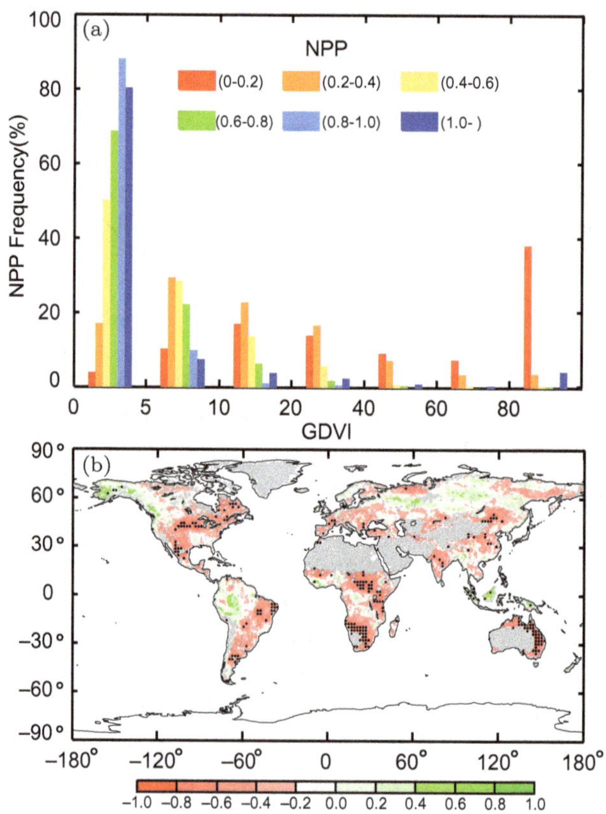

Fig. 9.4. Frequency of certain net primary productivity (NPP) values for regions with different global desertification vulnerability index (GDVI) values for 2000–2014 (a). Distribution of the correlation coefficients for the relationship between NPP and GDVI (b). The stippling pattern indicates a 95% confidence level according to a two-tailed Student's *t* test (Huang *et al.*, 2020b).

specific NPP values is shown in Fig. 9.4(a). It demonstrates that the regions with high NPP values tend to be found in areas with lower GDVI values, and that the frequency of high NPP values declines as GDVI values rise. A majority of regions with a GDVI higher than 80 are those where the high frequency of NPP values is 0–0.2. This demonstrated the GDVI's strong correlation with biological productivity. Low vegetation productivity areas are more susceptible to climate change and human activities, making them more subject to desertification (Lavado Contador *et al.*, 2009). Most places, with the exception of the extremely cold high latitudes where there is a lot of snow and ice cover, show a negative association between NPP and GDVI (Fig. 9.4(b)). The negative correlation coefficient demonstrated that areas with a low NPP have a significant probability of desertification. NPP and GDVI have a favorable association over the high latitudes. This might be due to the fact that these locations are mostly covered in ice and snow, with little flora. The ice and snow in these areas may melt as the temperature rises, making the climate more favorable for the development of flora. This contradicts our previous assumption that the desertification risk is higher in areas with high temperatures.

9.2.4 *Risk expansion of global desertification*

We compared CMIP5 simulations of the GDVI with observations for the same time span to ensure the future projection (2000–2014). The outcomes demonstrate that the area coverage and spatial distribution of various risk classes as assessed by simulation are congruent with the data (Fig. 9.5). In general, desertification is highly likely to occur in arid and hyper-arid regions. Desertification is less likely to occur in areas with dense vegetation. In Northern Canada and Eastern Russia, there is a region where the anticipated values and the observations disagree. Places with an overestimation in the ensemble mean of the CMIP5 data, or areas with an observed lower risk of desertification, are primarily found in Western Canada and Northern China. The observed area coverage of low-risk species, however, exceeds the expected area. Future desertification risk will increase globally under the RCP8.5 scenario, with a significant decline in low (−23%) risk areas and an increase in moderate(+10%), high(+5%), and very high(+8%) risk areas assessed by GDVI. Under the RCP4.5 scenario, except for low-level risk, all risk categories will slightly increase by 2100 (Fig. 9.6). According to the GDVI, the risk of desertification will rise over Europe, Western Asia, Northern China, the edge of the Sahel, and Mexico under the RCP4.5 when

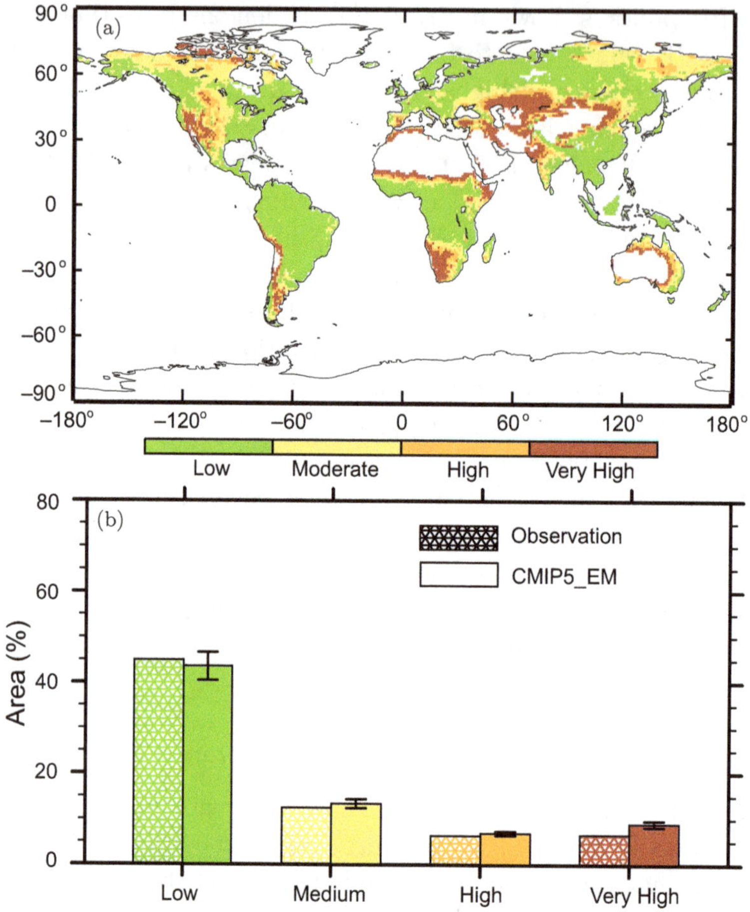

Fig. 9.5. The global distribution of desertification risk level during 2000–2014 estimated by the global desertification vulnerability index in the ensemble mean of the Fifth Coupled Model Intercomparison Project (CMIP5) (a); the area coverage [percentage of global land area (60°S–65°N)] of different levels of risk (bar charts filled with an asterisk represent the observed value, and the solid color represents the projected value) (b) (Huang *et al.*, 2020b).

comparing the means of 2001–2014 and 2086–2100 (Fig. 9.7(a)). However, in India and the Qinghai–Tibet Plateau, the probability of desertification declines. Indicating that both climate change and human activities are responsible for the rising danger of desertification in Europe and Western Asia, the AI declines while the surface air temperature and HAI both rise. Over India, the HAI declines, which lowers the likelihood of desertification. However, the AI declines and surface air temperature rises, indicating that climate change predominates the increase in the risk of desertification there. HAI likewise declines in Northern China and at the edge of the Sahel.

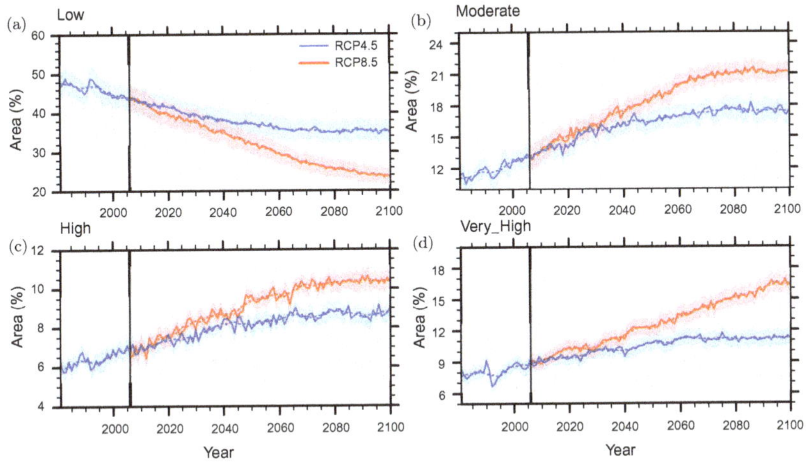

Fig. 9.6. The time series of area coverage [percentage of global land area (60°S–65°N)] of different desertification risks from 1981 to 2100. The blue lines represent the area under the representative concentration pathway 4.5, the red lines represent the representative concentration pathway 8.5, and the shading denotes the 95% confidence intervals. Low level (a), moderate level (b), high level (c), and very high level (d) (Huang *et al.*, 2020b).

Due to a decline in human activity and a growth in AI, the Qinghai–Tibet Plateau's risk of desertification has been reduced. Desertification risk over Northern China, India, Mexico, and Europe increases dramatically under RCP8.5 because of a considerable increase in HAI, a drier environment, and an increase in surface air temperature (Fig. 9.7(b)). However, in Australia, the rising risk of desertification is primarily due to climate change.

Despite the fact that vegetation coverage rises with warmth at high latitudes, the risk of desertification is increased by drier weather and greater human activity.

9.3 Assessment of Global Ecological Security

When an ecosystem maintains its stability in the face of external stress, this condition is known as ecological security. Global ecological security has been rapidly deteriorating during the 20th century due to climate change and the rise in human activity, endangering sustainable human growth. For the growth of adaptation methods to preserve sustainable development in vulnerable areas, the assessment and projection of global ecological security are crucial. However, because the linkages between many factors have not been taken into account in the existing evaluations of ecological security, future projections are subject to a great deal of uncertainty. Here, a new index of ecological security was created by taking into account the

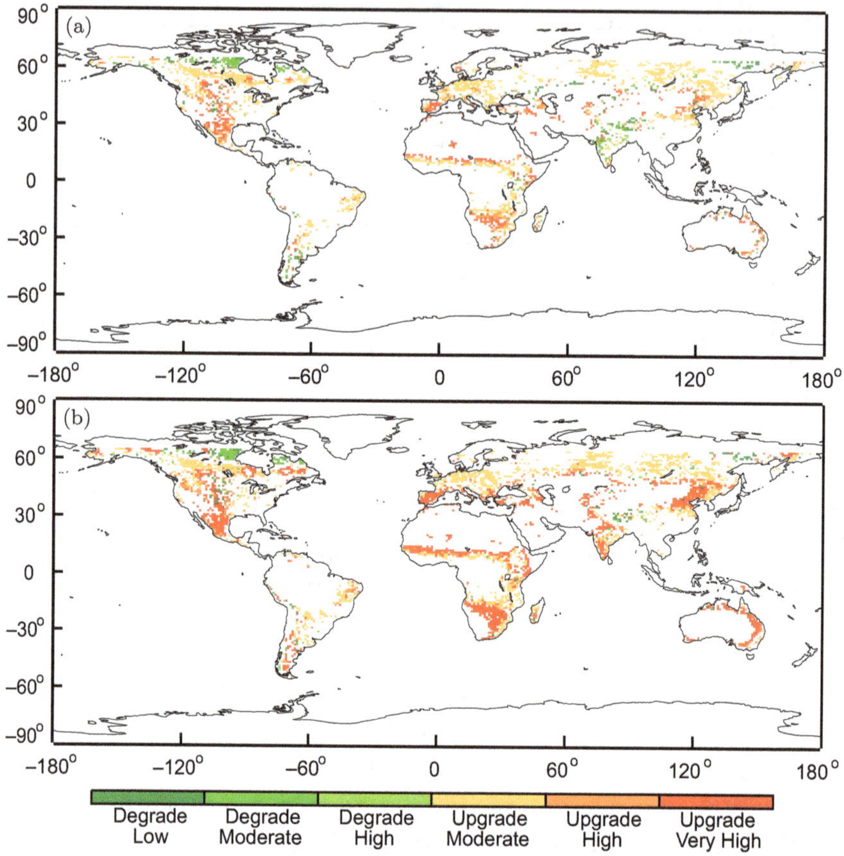

Fig. 9.7. The change in desertification risk level in 2086–2100 compared with 2000–2014 under the representative concentration pathway (RCP) 4.5 and RCP8.5. 2081–2100 under the RCP4.5 (a) and 2086–2100 under the RCP8.5 (b). "Upgrade" means the indicated risk level transitioned from lower to higher subtypes, and "Degrade" refers to the transition from original subtypes to a lower risk level (Huang *et al.*, 2020b).

cycles of life, oxygen, carbon, heat, and water as well as the effects of global climate change and human activity on the stability of ecosystems. Ecological security has been distributed globally over the last 60 years, with forecasts for the next 100 years included. Over the past 60 years, there has been a dramatic loss in ecological security in arid and semi-arid regions that has spread to neighboring regions. It is anticipated that ecological security will respond more strongly to human activity and global warming. Under a scenario with significant greenhouse gas emissions, globally insecure land would occupy more than 57% of the planet's land area by about 2100.

9.3.1 *Establishment of ecological security index (ESI)*

Combining four key variables — oxygen consumption (O_c), oxygen production (O_p), temperature warming magnification (T_m), and the aridity index (AI) — identifies ecological security. The following is a description of the data compilation for each phrase.

(1) *Oxygen consumption*

O_2 is consumed by numerous processes, some of which are insignificant or challenging to measure, such as the weathering of organic matter and sulfide minerals, and volcanic gas oxidation. There are five primary O_2 consumption processes: (1) fossil fuel combustion, (2) human respiration, (3) livestock respiration, (4) fires, and (5) heterotrophic and soil respiration (Turekian and Holland, 2013). In Huang *et al.* (2018) and X. Liu *et al.* (2020), the processes (1)–(4) with specific methodology and datasets are described in detail. The process by which soil organisms breathe out oxygen in this instance is known as heterotrophic and soil respiration ($R_h + R_d$), where R_h is the respiration of heterotrophs and R_d is the respiration of decomposers (microbes), and it is quantified by the disparity between net primary productivity (NPP) and net ecosystem productivity (NEP). The Fifth Coupled Model Intercomparison Project (CMIP5) simulation produced the simulated $R_h + R_d$ dataset (Taylor *et al.*, 2012). The Global Fire Emissions Database (GFED, http://www.globalfiredata.org) is where the observed R_h+R_d data are found (Van *et al.*, 2017).

(2) *Oxygen production*

When plants and other organisms collect carbon dioxide (CO_2) from the atmosphere and release oxygen, it is known as photosynthesis (CO_2). The chemical equation below can be used to describe photosynthesis:

$$6H_2O + 6CO_2 \rightarrow C_6H_12O_6 + 6O_2. \tag{9.6}$$

The entire quantity of CO_2 fixed by a plant during photosynthesis is known as gross primary production (GPP). NPP is the total gross primary productivity that is left over after accounting for plant respiration costs. If the known amount of carbon is fixed through photosynthesis (NPP), then we can apply the following equation to determine the net amount of O_2 created during the process, according to Eq. (9.6):

$$O_2 = \text{NPP} \times 2.667. \tag{9.7}$$

The ratio is 2.667 because the molar mass of O_2 is 32 g per mole and the molar mass of C is 12 g per mole. The observed NPP data based on MODIS from 2000 to 2015 were received from the Global Fire Emissions Database (Van *et al.*, 2017), whereas the modeled NPP dataset from 1948 to 2100 was derived from the CMIP5 simulation (Taylor *et al.*, 2012). For comparison, the NPP data from both the observed and the simulated NPP are re-gridded to a resolution of 1.0° × 1.0°. Huang *et al.* (2018) and X. Liu *et al.* (2020) both contain comprehensive methodologies and datasets related to oxygen production.

(3) *Temperature warming magnification*

The temperature warming magnification is represented by $T_m(i, n)$, which is defined as the ratio of the grid warming rate to the global warming rate as follows:

$$T_m(i,n) = \frac{T_{\text{trend}}(i,n)}{T_{\text{Gtrend}}(n)}, \tag{9.8}$$

in which $T_{\text{trend}}(i, n)$ is defined as the linear trend of the surface temperature from 1901 to year n at grid point i, and $T_{\text{Gtrend}}(n)$ is defined as the linear trend of the global average surface temperature from 1901 to year n for year n.

The surface air (over land) and sea surface (over ocean) temperature data are combined in the CRUTEMP4 dataset, which was created by the Met Office Hadley Centre and the Climatic Research Unit at the University of East Anglia (Morice *et al.*, 2012). Dai and Zhao (2017) have added the CRU TS2.3 temperature data to the CRUTEMP4 to complete the missing values. Therefore, we compute the observed T_{Gtrend} and T_{trend} using this modified version of the CRUTEMP4. 20 CMIP5 climate models are used to generate the surface air temperature data for simulations that are utilized to determine T_{Gtrend} and T_{trend} (Taylor *et al.*, 2012). To determine the ecological security index (ESI) for each model, we averaged the Tm data from 20 different models.

(4) *Aridity index*

The AI(i, n) represents the aridity index at grid i for year n, which is defined as the ratio of precipitation to potential evapotranspiration (PET) and denoted as follows:

$$\mathrm{AI}(i,n) = \frac{\mathrm{P}(i,n)}{\mathrm{PET}(i,n)}. \tag{9.9}$$

The precipitation information used for observation comes from the NOAA PRECipitation REConstruction over Land (PREC/L) dataset (Chen *et al.*, 2002a), created by the Climatic Prediction Center (CPC), and covering the years 1948 to the present on a 0.5° grid. Additionally, the PET data come from the CRU TS 3.25 dataset (Harris *et al.*, 2014), which offers 0.5° precision and spans the years 1901–2016 for all geographical areas. We maintain the climatology of observed precipitation and PET data in a consistent manner with the observed precipitation and PET data provided by Feng and Fu (2013) in order to be consistent with the AI in the models.

The simulated datasets for precipitation and PET utilized here were given by Feng and Fu (2013). These numbers are taken from CMIP5 climate model outputs of monthly mean temperature, precipitation, solar radiation, specific humidity, and wind speed (Taylor *et al.*, 2012). 20 models were used by Feng and Fu (2013) to generate AI data, but these models do not produce NPP and $R_h + R_d$. To determine the ESI for each model, we therefore summed the AI data from 20 different models.

(5) *Land datasets*

The ERA5 reanalysis from the European Centre for Medium-Range Weather Forecasts (ECMWF) on a 0.25° × 0.25° grid was used to compile the surface energy flux data for sensible and latent heat, 0–7-cm volumetric soil water, and 0–7-cm soil temperature for 2000–2015. Additionally, for 1948–2010, we explored the surface energy flux data for sensible and latent heat, 0–10-cm soil moisture content from the NASA Global Land Data Assimilation System (GLDAS) land model simulation using the Noah Land Surface Model forced by observational data on a 1° × 1° grid (GLDAS_NOAH10_M.2.0) using the CPC Soil Moisture dataset (Fan and Dool, 2014) on a 0.5° × 0.5° grid from a model for verification, which covers the period of 1948–present and was provided by the NOAA/OAR/ESRL PSD, Boulder, Colorado, USA.

(6) *Ecological security index (ESI)*

Here, biological, oxygen, carbon, thermal, and hydrological cycles were taken into consideration to measure the ecological security of the world's land because the land was disturbed by human activities and climate change. An ESI is defined here as a tool to quantify the ecological system's overall level of security, identify vulnerable areas, and forecast future changes. The following is how the ESI is built:

$$\mathrm{ESI} = \left(\frac{O_c}{O_p} \cdot T_m \cdot \frac{1}{\mathrm{AI}} \right)^{\frac{1}{3}} \times 10. \tag{9.10}$$

Here, ESI is a multi-disciplinary indicator that connects various fields and describes the stability status of the terrestrial ecosystem, taking into account factors like human-induced oxygen consumption, vegetation photosynthesis, precipitation, potential evapotranspiration, and temperature. The ecosystem activity in this instance is shown by oxygen-related indicators. The strains placed on the ecosystem would also have a significant impact on its vulnerability. Local ecological security would be impacted by temperature rise and aridity change. As a result, we evaluate the ecological security of the land under the influence of climate change and human activity by combining oxygen-related metrics, AI, and temperature magnification.

We normalized O_c/O_p, T_m, and $1/\mathrm{AI}$ to values between 0 and 1, then set the 1% and 99% thresholds as the lowest and highest values to remove the effects of extreme values. The three terms were then added together using Eq. (9.6) to define the ESI. Consequently, the range of the ESI's value is 0 to 10. The model simulations are adjusted to have the same mean-normalized O_c/O_p, T_m, and $1/\mathrm{AI}$ of 2000–2015 as the observations in order to concentrate on temporal variation and long-term climate change.

9.3.2 *Evaluating ecological security by ESI*

We assumed that each classification's ESI indices differed significantly from one another and that each classification's ESI values should be aggregated at a single location. Following that, there should be multiple peaks in the probability density distribution of the ESI. As the classification threshold, we utilize the ESI value that corresponds to the valley of the probability density distribution. The values corresponding to the three troughs of the probability density distribution of ESI, which are 1.1, 1.4, and 1.9, were utilized by us as the categorization thresholds based on these norms. The biggest distinction from other hierarchical aggregation centers is also present. The mean ESI for each categorization for the years 2000–2015 is shown in Fig. 9.8. The mean ESI values for secure, semi-secure, light-dangerous, and severe-dangerous zones, respectively, are 0.72, 1.24, 1.63, and 2.74. The significance test for the ESI difference between different classifications was passed at the 99% level. With thresholds of 1.1, 1.4, and 1.9, ESI is therefore separated into

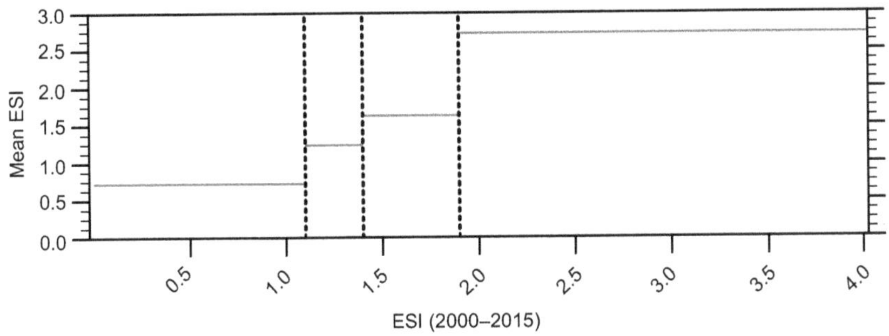

Fig. 9.8. The average value of the mean ESI during the period 2000–2015 under each classification. The abscissas corresponding to the dotted lines are the hierarchical intervals (Huang *et al.*, 2020a).

four levels. Insecure land is described as a territory with ESI $\geq$ 1.1, which can be further subdivided into zones that are semi-secure (1.1 $\leq$ ESI <1.4), light-dangerous (1.4 $\leq$ ESI <1.9), and severe-dangerous (ESI $\geq$ 1.9). Here, secure land is referred to as a region with ESI <1.1.

The growth of ecologically unstable land coincides with a decline in ecological security. A positive feedback cycle where warming and drying reinforce one another occurs when land on Earth runs out of water and loses the ability for plants to produce oxygen through photosynthesis. This land will invade and destroy the surrounding environment and gradually expand its territory to form more deserts or arid and semi-arid regions on Earth (Fig. 9.9). This kind of unstable ground is what we refer to as unstable natural land. While this is happening, population shifts lead to an increase in megacities and industrial hubs, which lead to a rise in the demand for fresh water and the extinction of flora (Cai *et al.*, 2015). If no steps are taken to adjust for and prevent these negative environmental consequences, they appear to have the potential to influence and overrun surrounding areas. These geographical areas are what we refer to as anthropogenic unstable terrain. The aforementioned natural and manmade insecure lands make up the overall insecure land. Less oxygen is created, combined with a larger loss of fresh water and plants, as insecure land increases, reducing the amount of land that is appropriate for supporting human existence. Controlling the balance between the regional oxygen, water, carbon, and thermal cycles is therefore essential for sustaining safe land; otherwise, negative consequences, such as the release of heat and soil dust, would directly influence the ambient land through a vicious cycle (Fig. 9.9). The terms in Eq. (9.5) are closely related to the oxygen, carbon, soil, water, and heat cycles, which play

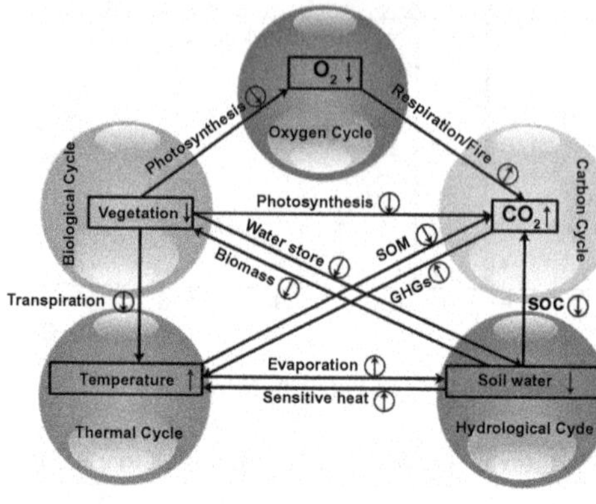

Fig. 9.9. Conceptual framework for an ecological security index. The connections among indicators from the oxygen cycle (oxygen production and consumption), biological cycle (vegetation photosynthesis and biomass), thermal cycle (soil temperature, evaporation, and sensible heat flux), hydrological cycle (soil water and TWS), and carbon cycle (SOC and CO_2) are described. The changes in indicators associated with the symbols ↓ or ↑indicate the decrease and increase in response to ecologically insecure land compared to ecologically secure land. The single arrows represent the paths of indicators. The O_2, CO_2, TWS, and SOC indicate oxygen, carbon dioxide, terrestrial water storage, and soil organic carbon, respectively (Huang *et al.*, 2020a).

indiscernible roles in global land ecosystems, and the ESI is built using the framework outlined above (Fig. 9.9).

9.3.3 *Global distribution of ESI by observations*

The regional distribution of ESI values from measurements made between 60°S and 60°N, between 2000 and 2015 is depicted in Fig. 9.10. The proportions of the various land surface types (60°S–60°N) in the semi-secure, light-dangerous, and severe-dangerous regions are 13.3%, 16.4%, and 23.3%, respectively. Insecure land covers 53.0% of the world's land area (60°S–60°N). Here, desert represents deserts like the Sahara and Taklimakan, making up 15.9% of the world's area (Fig. 9.10(a)). The severe-dangerous regions are mainly distributed at the edges of deserts and in regions highly affected by climate change, as well as in some other places with high human activities, including industrial and densely populated regions (Fig. 9.10(a)). In Australia, North America, Europe, Southern Africa, and Central Asia, there are a few arid and semi-arid regions and high plateau regions without vegetation and snow cover that fall under the light-dangerous regions (Fig. 9.10(a)). The semi-secure regions, which are primarily found in semi-arid and cattle-producing regions (Fig. 9.10(a)), are transitional areas

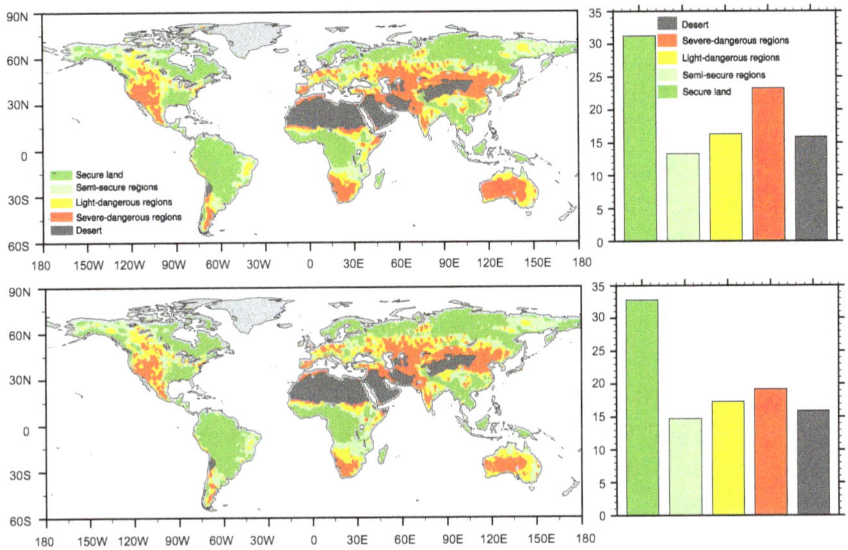

Fig. 9.10. Global distribution of ESI and area coverage of different land surface types based on observations (a–b) and CMIP5-EM (historical and RCP8.5) (c–d) from 2000 to 2015. Spatial distribution of observed ESI values for 2000–2015 (a). Area coverage (percentage) of different land surface types (60°S–60°N) based on observed ESI values from 2000 to 2015 (b). (c) and (d) are similar to (a) and (b), but for CMIP5-EM (Huang *et al.*, 2020a).

between secure and risky land where the amount of oxygen produced through photosynthesis is insufficient to make up for the oxygen consumption. Only 31.2% of the world's land area is considered secure, with the majority of this territory located in South America, Siberia, Central Africa, and the Tibetan Plateau. These regions are known for their tremendous ecological resilience and lack of human disturbance. The cryosphere (polar regions and Greenland), which includes Greenland and the polar areas but is not covered in this study, can also be categorized as safe terrain with abundant freshwater reserves and minimal human interference.

9.3.4 *Earth cancer under climate change*

The evolution of planet Earth can spontaneously establish an ecological balance fit for all forms of life and demonstrate its own capacity for self-recovery. However, due to unending economic development and the impacts of growing populations — such as deforestation, intensifying farmland production, and animal husbandry — the release of more heat, and increasing industrial emissions, including greenhouse gases like carbon dioxide and nitrogen oxides (Knohl and Veldkamp, 2011; Knohl and Cuntz, 2017), we are destroying the ecological balance of Earth at an astounding and almost irreversible rate (Groisman, *et al.*, 2017). Natural disasters and climate crises have been threatening human life more regularly in recent times; examples of these catastrophes include global warming, glacier melting, drought, flooding, the loss of biodiversity, and more powerful hurricanes and earthquakes (Huang *et al.*, 2016b, 2018). As a result of human activity, the Earth's fossil fuel, freshwater, and oxygen reserves are being dramatically depleted, and the vegetation that has grown here for billions of years is being cleared from the land. As a result, the oxygen content of the atmosphere is declining (Huang *et al.*, 2017a; Keeling *et al.*, 2010), and oxygen consumption is outpacing oxygen output through the land and oceans. If the Earth is perceived as a physical body, cancerous land is a very serious disease, which spreads rapidly like cancer cells to the surrounding normal land. The primary causes for the origin and spread of Earth's "diseases" are still unknown, though. Here, we put forth the idea of "land cancer," provide the conceptual framework for identifying illnesses on Earth, and go over the manner in which these illnesses affect the prognosis for human survival.

Because of its aggressive growth, cancer is known to have harmful impacts on human health. The cancer cells in a person will indiscriminately ingest nutrients from healthy cells and eventually put the person's life in jeopardy, maybe even resulting in death, if we cannot figure out efficient techniques to identify the cancer sooner and stop its growth (Schlesinger *et al*, 1990). In contrast to the human body, when land on Earth lacks water and plants are unable to produce oxygen through photosynthesis, this land will encroach on neighboring areas, destroy them, and slowly enlarge its territory to create more deserts or arid and semi-arid regions, sharing anaerobic respiration traits with human cancer cells. This kind of land is what we refer to as naturally diseased land. Meanwhile, as additional megacities and industrial hubs have sprung up due to population movement, fresh water and plant life are becoming increasingly scarce. If no steps are taken to adjust for and prevent these negative environmental consequences, they appear to have the potential to influence and overrun surrounding areas. These geographical locations are what we refer to as anthropogenic malignant land. The aforementioned natural and man-made cancerous lands make up the entire cancerous area. The amount of land that is suitable for supporting human existence decreases as more cancerous land grows because less oxygen is created, as well as larger losses of fresh water and vegetation.

All of Earth's areas are related by the oxygen, carbon, hydrological, and thermal cycles, which are likewise linked to all biological cycles (Piao *et al.*, 2009). In addition to producing essential O_2 and being crucial for the food chain, photosynthesis in plants also takes in solar energy, takes in CO_2, and stores water. Within the aforementioned five cycles, plant flora should coexist in a manner that ensures the existence of humans and other animals. The hydrological thermal cycle and net oxygen production are related, as shown in Fig. 9.11. With increased oxygen generation, soil moisture and terrestrial water storage (TWS) both rise (Figs. 9.11(a) and 9.11(b)). Land will progressively lose its ability to store water when it cannot produce enough oxygen to support the biological life cycle (Figs. 9.11(a) and 9.11(b)). By heating the air with a high sensible heat flow and a low latent heat flux, low soil moisture causes a temperature increase over land with low oxygen generation (Huang *et al.*, 2017a) (Figs. 9.11(c) and 9.11(d)). As a result, in order to produce oxygen, the land must hold more water (Figs. 9.11(a) and 9.11(b)) and absorb more heat (Figs. 9.11(c) and 9.11(d)). In contrast, the process of oxygen consumption causes land to lose water (Figs. 9.11(a) and 9.11(b)) and emit heat, which raises the temperature (Figs. 9.11(c) and 9.11(d)). More lands will become diseased or even cancerous, if these processes are harmed by aggressive human activity that pushes the limits of their capacity for self-recovery and adjustment. According to the findings, diseased land holds less water and releases more heat into the sky, creating a cycle where warming and drying reinforce one another. Controlling the balance between the regional oxygen, water, carbon, and thermal cycles is therefore essential for keeping healthy land; otherwise, negative consequences, such as the release of heat and soil dust, will immediately influence the surrounding land through a vicious cycle.

Given the current predictions for the rate of global warming, there are several issues with land cancer that pose scientific, management, and policy difficulties. Land will eventually lose its ability to store water and grow drier when it cannot create enough oxygen. This causes a positive feedback loop whereby warming and drying support one another and more lands will get diseases, including cancer. In order to improve our knowledge of land cancer and develop strategies to prevent or at least delay it, urgent action and an unparalleled scientific effort are therefore required now more than ever.

When taken as a whole, our suggested ecological security index (ESI) can shed more light on land cancer and suggest areas for future study. The predicted biodiversity species will have irreparable losses if we are unable to successfully examine the cancer's origins and stop it from spreading

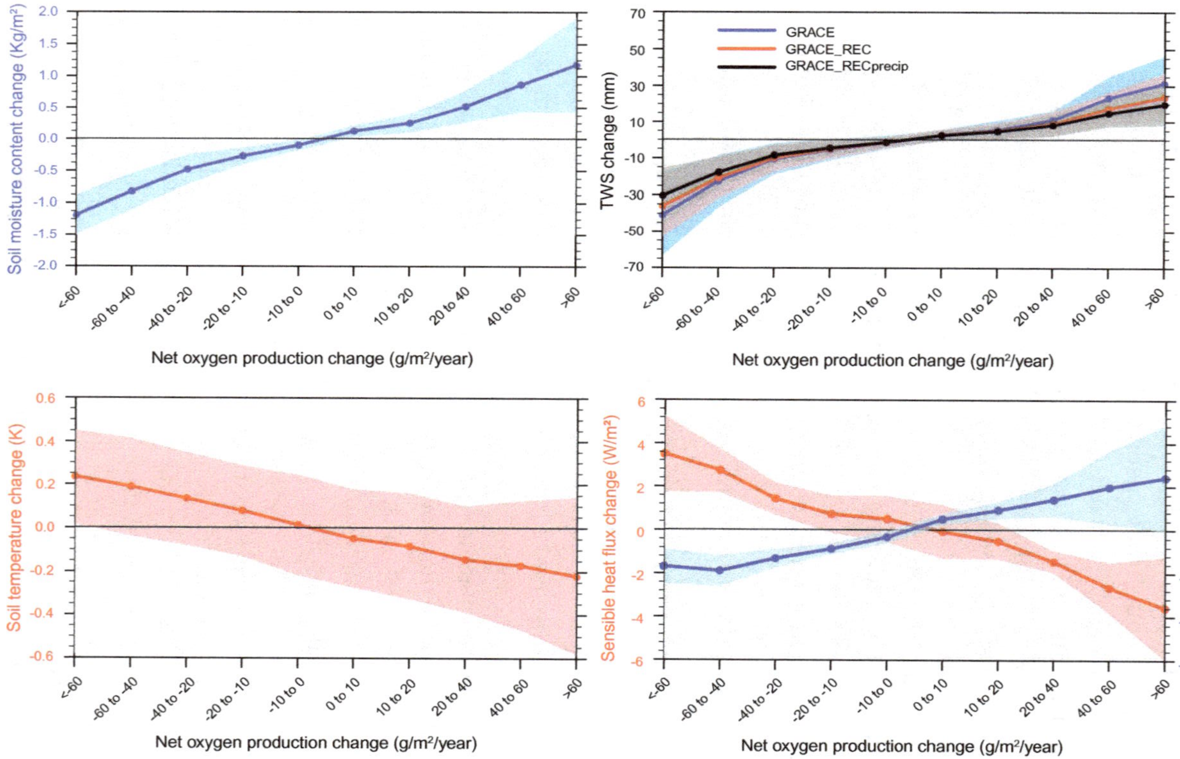

Fig. 9.11. The hydrological and thermal changes associated with net oxygen production change. The inter-annual variability of regional average soil moisture content for 2000–2010 from the GLDAS dataset (blue line) with respect to the corresponding inter-annual variability of net oxygen production derived from the MODIS dataset (a). (b–d) are same as (a), but for the terrestrial water storage (TWS) for 2002–2015 from GRACE satellite observations (blue line), and two statistical models of GRACE-REC (red line) and GRACE-RECprecip (black line) which are trained with GRACE observations (b), soil temperature (red line) within 60°S–50°N for 2000–2010 from the GLDAS dataset (c), and sensible heat flux (red, upward positive) and latent heat flux (blue, upward positive) for 2000–2010 from the GLDAS dataset (d). Shading denotes the ±1 standard deviation range of the corresponding temporal variation (Huang *et al.*, 2020a).

(Pimm *et al.*, 2014). This forecast also forewarns us to act quickly and pay more attention to the land that is cancerizing. Failure to do so, however, might result in the transformation of more areas into "land cancer," the extinction of rare species as a result of this transition, and other serious social and economic repercussions, particularly in developing nations. Therefore, it is important to keep in mind that land cancer is a threat to human well-being worldwide and is intimately related to existing ecological security and biodiversity. If all nations worked together, concerns about preventing land cancer and preserving biodiversity might be addressed with newfound optimism.

9.3.5 *Future changes of global ecological security*

It is crucial to look into how ecological security will alter in the future based on the aforementioned climatology. The future regional and temporal evolution of ecological security may be traced back to forecasts made by the CMIP5 using a variety of emission scenarios.

The CMIP5-EM simulations of ESI over 2000–2015 (Figs. 9.12(c) and 9.12(d)) are compared to observations to confirm the accuracy of future predictions. The findings demonstrate that CMIP5 is suitable for estimating future ESI changes because the spatial distribution and area coverage of insecure land determined by CMIP5 are compatible with those suggested by observations. For the CMIP5-EM historical and RCP8.5-simulated ESI values from 2000 to 2015, insecure land covered 51.2% of the world's land area, whereas the percentages of secure, semi-secure, light-dangerous, and severe-dangerous land surface types (60°S–60°N) were 32.8%, 14.7%, 17.3%, and 19.2%, respectively. 15.9% of the earth's surface is covered by deserts, primarily in the Sahara and Taklimakan; this figure is consistent with what has been observed (Figs. 9.12(c) and 9.12(d)). The light-dangerous and severe-dangerous zones are primarily found in the middle and high latitudes of the Northern Hemisphere, such as in Western North America, Southwest Russia, and Northeast China. The spatial distribution of ESI values from 2000 to 2015 can generally be captured by CMIP5 models, showing that the CMIP5-EM is suitable for predicting changes in ESI in the future.

The time series of the worldwide mean ESI over the upcoming periods is shown in Fig. 9.12(a). According to future forecasts, the ESI rises to ~1.8 by 2100 under the RCP8.5 scenario and falls to ~1.4 by 2100 in the RCP4.5 scenario from ~1.5 (~2020). These opposing patterns show that greenhouse gas emissions are the primary driver of ESI variance and that

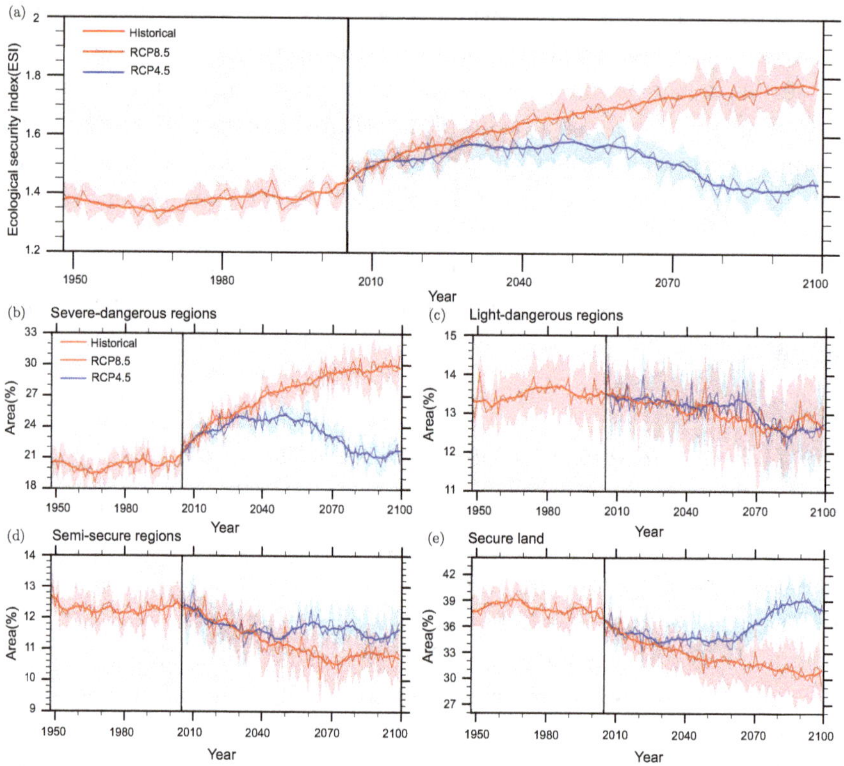

Fig. 9.12. Temporal variations in the global mean ESI and changes in the area coverage of different land surface types. Time series of globally averaged ESI values obtained from the CMIP5-EM (a). Area coverage (percentage) (b–e) is shown for severe-dangerous regions (b), light-dangerous regions (c), semi-secure regions (d), and secure regions (e). The thin red (blue) solid lines are the CMIP5-EM from the historical simulations and RCP8.5 (RCP4.5) projections. The shading denotes the 95% confidence intervals of the 18 models. Seven-year running means (thick colored lines) are shown to emphasize the Earth's health trends (Huang *et al*, 2020a).

limiting greenhouse gas emissions under RCP4.5 could reduce the severity of this disease. The time series of the four subtypes' area variations during the course of the past and the future are shown in Fig. 9.12(b–f). Under the RCP8.5 scenario, it is obvious that the future expansion of severely risky regions is important. Additionally, the RCP8.5 scenario predicts a decline in the area coverage of secure and semi-secure regions after 2005. However, in the RCP4.5 scenario, the overall amount of insecure land will not increase globally because carbon dioxide emissions will be constrained, reducing oxygen use. The area coverage of extremely risky places will first rise and then fall, whereas the area coverage of secure regions will initially drop and then rise. According to the RCP4.5 scenario, there will not be much of a difference between the four different land surface types' areas at the end of the century and those at its start. The areas of severe-dangerous regions and light-dangerous regions will somewhat shrink among them, while the areas of secure regions will slightly expand.

The area changes of insecure land from the RCP8.5 scenario as suggested by CMIP5-EM are shown in Fig. 9.13 together with the spatial distribution of the mean ESI in the years 2085–2100. By the end of the century, insecure land will cover 57.1% of the world's land area, mostly in Africa, India, Europe, and West and East Asia. On the other hand, the percentages of secure and semi-secure regions will be 27.1% and 13.1%, respectively, on the varied land surface types (60°S–60°N). Under the RCP8.5 scenario, the amount

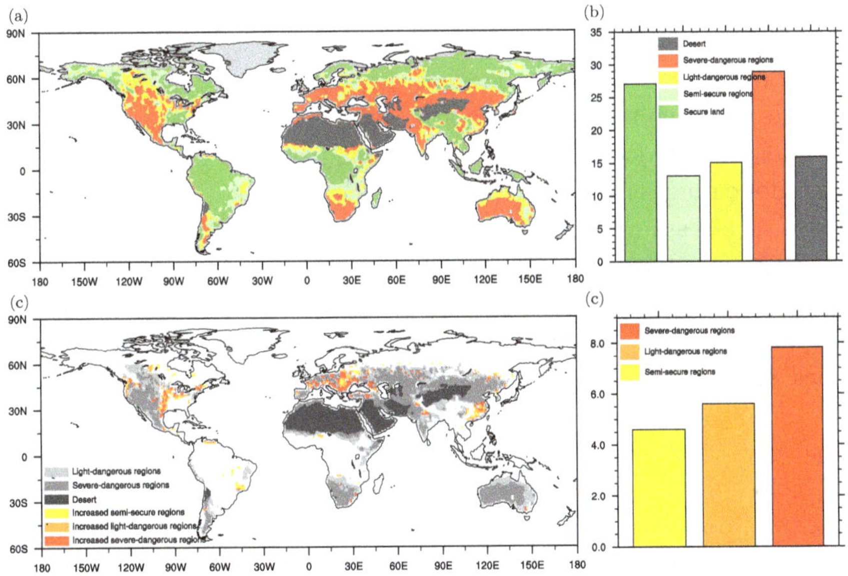

Fig. 9.13. Global distribution of ESI from CMIP5-EM and RCP8.5 scenarios for 2085–2100 and its changes in relation to observations from 2000 to 2015. Spatial distribution of projected ESI values for 2085–2100 (a). Area coverage (percentage) of different land surface types (60°S–60°N) based on projected ESI values from 2085 to 2100 (b). Projections of changes in subtypes from the CMIP5-EM and RCP8.5 scenarios are shown relative to the baseline period (observations, 2000–2015) for 2085–2100 (c). The gray shading denotes the baseline land surface types from 2000 to 2015. Changes include any transitions between adjacent and nonadjacent subtypes. For example, the "increased" category means that the indicated regions transitioned from more secure subtypes, i.e., the "increased semi-secure" category represents regions that changed from being secure to semi-secure. The increased area coverage (percentage) of different land surface types (60°S–60°N) (d) shown in (c) (Huang *et al*, 2020a).

of secure, semi-secure, and light-dangerous territory will be 5.6% less by ~2100 than it was at the start of this century (Fig. 9.10). Under the RCP8.5 scenario, the area of the world's severe-dangerous regions will grow from 23.3% to 28.9% of the total land area by ~2100. The increased areas of semi-secure, light-dangerous, and severe-dangerous regions are 4.6%, 5.6%, and 7.8%, respectively, when compared to those of insecure land from 2000 to 2015 and those from 2085 to 2100 under the RCP8.5 scenario. The net changes in secure land and severe-dangerous regions represent −4.1% and 5.6% of the world's land, respectively. The most important change (reaching up to 28.9% of the world's land area by 2100) is an increase in severely hazardous regions, which is mostly due to a notable decline in secure territory and a modest decline in semi-secure and light-dangerous regions.

9.4 Impacts of Soil Temperature on Desert Carbon Sink

The main method by which carbon leaves the soil and enters the atmosphere is through soil respiration (soil-to-atmosphere CO_2 flux) (Lou and Zhou, 2006). The projected yearly release of carbon from soil into the atmosphere over the globe ranges from 91 to 94 Pg C, with an additional 0.04 Pg C added each year as a result of this process (Xu and Shang, 2016; Bond-Lamberty, 2018). This procedure partially controls terrestrial ecosystems' carbon cycles (Hashimoto *et al.*, 2015). The atmospheric CO_2 content can be significantly impacted by even little changes in this process, which could provide a positive feedback effect in the context of climate change (Han *et al.*, 2018). Generally speaking, soil respiration is the sum of all biotic and abiotic processes in the soil and is very sensitive to changes in the environment. Due to the active biological processes present, especially the respiration of plant roots and soil microbes, soil respiration is extremely powerful in ecosystems with relatively high productivity. The contribution of abiotic processes to soil respiration is frequently minimal and easy to overlook. These processes include the dissolution/diffusion of CO_2, the adsorption/desorption of CO_2, the dissolution/precipitation of carbonate, the UV decomposition of organic carbon, and the chemical oxidation of carbon-containing compounds in soil. The importance of abiotic processes to soil respiration, however, cannot be disregarded in an extreme environment with relatively low productivity (such as saline soils, the Gobi Desert, and polar dry valleys) (Parsons *et al.*, 2004; Chen *et al.*, 2014). In some harsh conditions, the role of abiotic processes may even take precedence (Ball *et al.*, 2009). Additionally, in certain severe situations, negative CO_2 fluxes (i.e., CO_2 uptake by the soil) have been observed, indicating that these environments may be CO_2 sinks (Li *et al.*, 2015b).

About 1.8 Pg of carbon, or the missing carbon sink, is absent from the global carbon cycle each year, which has baffled scientists for a long time. The scientific community has recently shown a great deal of interest in the desert ecosystem, a typical widely dispersed severe habitat, as a result of the finding that it can store a lot of CO_2 to fill the gap in the carbon sink. Numerous studies have shown that abiotic processes primarily regulate soil respiration in deserts by quantitatively separating the contributions of biological and abiotic processes (Fa *et al.*, 2016). The main CO_2 emission processes in desert ecosystems include surface turbulence, air volume expansion, exudation of the CO_2 dissolved in soil water, desorption, carbonate precipitation, and breakdown of soil organic matter (Fa *et al.*, 2016). In addition, under the influence of soil moisture, possible chemical reactions may take place in saline and alkaline desert soils, promoting carbonate dissolution and soil CO_2 absorption (Xie *et al.*, 2009). Rapidly accelerating CO_2 absorption by the soil is possible with a decrease in soil temperature and an increase in soil moisture (Cuezva *et al.*, 2011; Fa *et al.*, 2014). Additionally, high-pH saline–alkali soil in deserts might facilitate CO_2 absorption in colder temperatures. Similar findings were made by Ma *et al.* (2013, 2014), who discovered that soil respiration in deserts was dominated by the inorganic CO_2 flux under the control of soil temperature and that changes in soil respiration were somewhat impacted by soil pH. It has been shown that the primary reason for carbon sequestration in deserts is the inorganic CO_2 flux in the soil at night. In conclusion, changes in the solubility of CO_2 governed by Henry's Law in soil–water films, pH-mediated CO_2 dissolution chemistry, surface adhesion of CO_2 to soil minerals (Schlesinger, 2017), and variations in the volume of gases caused by changes in pressure and temperature are the main abiotic soil carbon sequestration processes. We still do not know how much these abiotic processes actually contribute to the desert's carbon sink or how they interact with one another inside. Due to this significant knowledge vacuum, it is difficult to determine the desert ecosystem's role in the carbon cycle and the long-term trend of the desert as a carbon sink under climate change.

The Taklimakan Desert (TD), the second-largest shifting desert in the world, is distinguished by its remote location, dry climate, sparse vegetation, complex dune types, strong mobility of dunes, large areas of shifting sand, a thick shifting sand layer, and small sand particle size. It serves as an emblematic example of deserts throughout the world. Through field measurements and laboratory studies, the TD was utilized as a case study to quantitatively assess the contribution of each internal soil respiration activity within the shifting sand layer to the desert carbon sink and its primary driving mechanism. On the basis of this, the long-term trend of the carbon sink rate of shifting sand in TD under climate change was investigated, along with historical environmental monitoring data and Fifth Coupled Model Intercomparison Project (CMIP5) simulations. We must once again acknowledge the importance of deserts in the global carbon cycle in order to close the gap left by the lost carbon sink.

9.4.1 *Observational study on desert carbon sink*

(1) *Site description*

A homogeneous shifting sand layer primarily covers the TD's surface. The three monitoring sites' environmental parameters, particularly the soil temperature, were consistent from north to south. The desert's remote town of Tazhong was regarded as the research area most indicative of the TD. The Tazhong shifting sand (38°58′N, 83°39′E, 1099 m above sea level) is where this investigation was carried out. A continental, warm-tempered, arid desert climate prevails throughout the entire area. The intraannual distribution is extremely non-uniform, with precipitation concentrated in May–August. The annual mean precipitation is 25.9 mm. 3812.3 mm of potential evaporation occurs annually. The area has a wide diurnal temperature range and four distinct seasons. Maximum temperatures range from 40.0° to 46.0°, while minimum temperatures range from −20.0° to −32.6°. The average annual temperature is 12.1° (Yang *et al.*, 2017; C. Zhou *et al.*, 2019). Resources based on plants and animals are severely lacking in this area due to poor environmental conditions. In the TD, shifting sand occupies a total area of around 2.364×10^{11} m^2, or about 70% of the TD (Wang *et al.*, 2005). Silt and clay make up 1.5% of the shifting sand's composition, followed by fine and very fine sand (88.8%) and medium sand (9.7%) grains. Sand has an average particle size of 0.083–0.129 mm and a standard deviation that ranges from 0.03 to 0.98 mm. In comparison to other deserts, the surface sand in this area is finer (Huo *et al.*, 2011; Yang *et al.*, 2016a). The physicochemical characteristics of the shifting sand are shown in Table 9.1. The area suffers constant easterly winds, with an average of 11 days per year of gale-force weather, more than 157 days of floating dust and blowing sand, and an average of 16 days per year of sandstorms (Yang *et al.*, 2016b). The basic components of aeolian landforms are inter-dune corridors and linear, high,

Table 9.1. Description of the physical and chemical properties of the topsoil of the shifting sand in the hinterland of the Taklimakan Desert (TD) (mean ± standard deviation) (Yang *et al.*, 2020a).

Variable	Depth(cm)	
	0–5	5–10
Average gravimetric soil moisture content (%, kg kg^{-1})	0.0013 ± 0.0002	
Soil total salt (g kg^{-1})	3.45 ± 0.35	9.64 ± 1.10
Soil organic carbon (g kg^{-1})	0.56 ± 0.12	0.63 ± 0.03
Soil total nitrogen (g kg^{-1})	0.06 ± 0.01	0.09 ± 0.02
Ratio of soil organic carbon to soil total nitrogen	9.85 ± 0.83	7.66 ± 1.58
pH	7.33 ± 0.02	7.3 ± 0.04
Bacteria (organisms g^{-1} dry soil)	26,000	4,400
Fungi (organisms g^{-1} dry soil)	<100	<100
Actinomyces (organisms g^{-1} dry soil) Gu *et al.*(2000)	238	

and composite longitudinal dunes. The dunes are 40–50 m in height and are oriented in either a NE–SW or NNE–SSW orientation.

(2) *Disassembly experiment*

To facilitate analyses of the contributions of the various components of TD's shifting sand to soil respiration and to account for the fact that shifting sand has a simple structure, we hypothesized that soil respiration in shifting sand (R_s) was the sum of the contributions from sand (R_{sand}), moisture (R_{moisture}), salt/alkali ($R_{\text{salt/alkali}}$), and microbes (R_{microbes}) (Eq. 9.11). The present agreement is fully supported by the four processes involved in the soil respiration of the shifting sand (Parsons *et al.*, 2004; Fa *et al.*, 2014, 2016). Future research will take into account the intricate relationships between the numerous parts and variations in each process's ability to between the numerous parts and variations in each:

$$R_s = R_{\text{sand}} + R_{\text{moisture}} + R_{\text{salt/alkali}} + R_{\text{microbe}}. \tag{9.11}$$

From the shifting sand layer at the research site, we collected topsoil (0–10 cm) and separated it into four parts (Table 9.2). Sample 1 was the control and was not treated. To eliminate the contribution of soil microorganisms to soil respiration, Sample 2 was autoclaved. The autoclaving was done three times, for half an hour each, to make sure there was enough sterilization. As soon as the shifting sand had been sanitized, it was put in an oven to dry at 105°. Before being placed in a sterile UV chamber, the

Table 9.2. Methods of topsoil sample pretreatment (Yang *et al.*, 2020a).

Sample No.	Contribution of each component	Sample weight (kg)	Pretreatment
1	R_s	13.21	No treatment (control check)
2	$R_{\text{sand}} + R_{\text{water}} + R_{\text{solt/alkat}}$	13.22	Autoclaving → Drying → Add distilled water
3	$R_{\text{sand}} + R_{\text{water}}$	12.23	Remove soluble salts and alkalis → Autoclaving → Drying →Add distilled water
4	R_{sand}	12.33	Remove soluble salts and alkalis → Autoclaving → Drying

dried shifting sand was firmly sealed. Finally, a watering can was used to add 17.18 mL of distilled water to the sample after allowing it to stand in a sterile state for one day in order to match the soil's water content (0.0013 kg kg^{-1}). The distilled water was added while the shifting sand sample was evenly mixed. After two days, the sample was sealed and left to stand so that the CO_2 and water could come into balance. 30 L of distilled water was added to Sample 3 before it was put into plastic packaging. The samples were thoroughly mixed, then allowed to stand for 1 h before the top liquid layer was separated. To get rid of soluble salts and alkalis in the sample, this process was done three times. The sample was then subjected to the same three rounds of sterilization, followed by drying, adding distilled water (15.9 mL), and standing time, as with Sample 2. Sample 4 had the same treatment regimen as Sample 3, but distilled water was not added.

The four samples that were permitted to stand and equilibrate were put in plastic trays (diameter 40 cm, height 18 cm), as illustrated in Fig. 9.14. The four samples and their plastic trays were buried in the shifting sand 20 m from the observation tower, measuring 3 m in height, one day prior to soil respiration measurements. Sand levels inside and outside of the tray were maintained at the same levels. Four cylindrical soil collars (cross-sectional area of 371.8 cm^2 and a height of 10 cm) were embedded in the four samples up to a depth of 8 cm.

An automatic soil respiration measurement system (model LI-8100A fitted with a LI-8150 multiplexer, LI-COR, Lincoln, NE, USA) outfitted with four LI-8100–104 long-term monitoring chambers was used to track the daily dynamics of soil respiration continuously and synchronously in the four samples from October 10 to October 18, 2017. We calibrated the analyzer before taking the data to ensure accuracy and properly increased the time

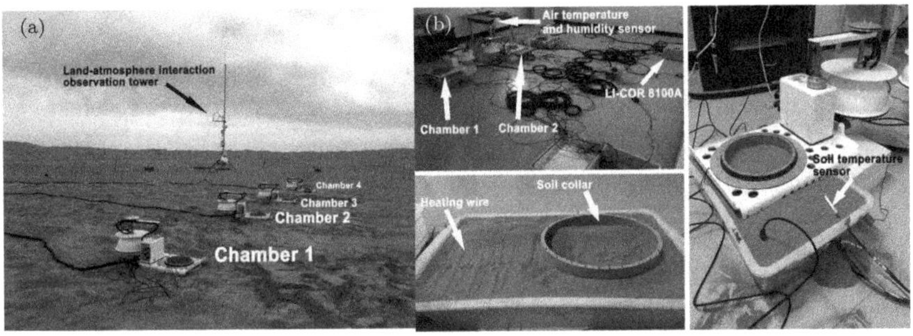

Fig. 9.14. Layout of the soil respiration experiment. Field monitoring of the contributions of various shifting sand components to soil respiration. Chambers 1–4 correspond to samples 1–4. The land–atmosphere interaction observation tower provided the historical soil temperature observation data used in this study (left). A temperature-controlled experimental setup to measure CO_2 flux in sand. Heating wires were placed at a spacing of 2 cm along the surface of the sand in one of the plastic boxes. Soil collars were embedded in the sand. Soil temperature sensors were installed in the sand at depths of 0 and 10 cm (right) (Yang *et al.*, 2020a).

for the entire testing process because soil respiration is low in deserts. Yang *et al.* (2017) provide more information about the equipment settings used for field measurements. To maintain the chamber's airtightness during the trial period, the seal ring was cleaned three times daily. Additionally, we carried out evaluation trials in the research region to boost our confidence in the application of the LI-8100A equipment to a desert environment. The soil collars' bottoms were entirely covered with plastic film, buried in the sand, and then totally sealed. The plastic layer prevented CO_2 from moving from the atmosphere to the shifting sand. The four chambers' respiration rates were all focused near zero with an accuracy of $\pm 0.02\ \mu\text{mol}\cdot \text{m}^{-2}\cdot s^{-1}$, and they did not change while the environment changed. As a result, the LI-8100A's measurement of soil respiration in shifting sand in this investigation was regarded as precise and dependable. Finally, we acquired CO_2 flux data for several components of TD's shifting sand via step-by-step disassembly.

(3) *Temperature-controlled experiment*

To avoid complex and uncontrolled climatic conditions in the field, a temperature-controlled experiment was carried out in the basement of the TD meteorological field experiment station (Fig. 9.14). The air in the basement was kept at a consistent 10° during the testing period. During the experiment, the air temperature did not vary by more than 0.79°C. The mean relative air humidity was 18.32%. The CO_2 level was kept constant at 405.8 ppm. Sand samples were placed in two plastic boxes according to the treatment procedure used for Sample 4. The plastic boxes contained integrated soil collars. In one of the plastic boxes, a heating wire was equally distributed at a 2-cm interval along the surface of the sand. By altering the voltage of the heating wire, the temperature of the sand could be managed (Fig. 9.14).

The second box was not heated and was only used as a control. The soil respiration in the sand was synchronously and continuously monitored from November 21–25, 2017, using an automatic soil respiration measurement system. The device settings matched those from the field monitoring experiment exactly. To determine the soil temperature at the appropriate depths ($\mathrm{T_{0cm}}$ and $\mathrm{T_{10cm}}$), soil temperature sensors (Model 109, Campbell Scientific, Logan, UT, USA) were inserted in the sand in the two plastic boxes at depths of 0 cm and 10 cm.

The synergistic effects of the soil temperature difference between 0 cm and 10 cm ($\mathrm{T_{0-10cm}}$) and the rate of soil temperature change at 10 cm ($\mathrm{\Delta T_{10cm}/\Delta t}$, °C· $\mathrm{min^{-1}}$; soil temperature difference at 10 cm every minute between two consecutive CO_2 flux measurements) on soil respiration were examined using four common nonlinear equations:

$$R = a + bT_{0-10\mathrm{cm}} + c(\Delta T_{10\mathrm{cm}}/\Delta t), \tag{9.12}$$

$$R = a + bT_{0-10\mathrm{cm}} + c(\Delta T_{10\mathrm{cm}}/\Delta t) + dT_{0-10\mathrm{cm}}(\Delta T_{10cm}/\Delta t), \tag{9.13}$$

$$R = a + bT_{0-10\mathrm{cm}} + c(\Delta T_{10\mathrm{cm}}/\Delta t) + d(T_{0-10\mathrm{cm}})^2 + e(\Delta T_{10\mathrm{cm}}/\Delta t)^2, \tag{9.14}$$

$$R = aT^b_{0-10\mathrm{cm}}(\Delta T_{10\mathrm{cm}}/\Delta t)^c, \tag{9.15}$$

in which R is soil respiration (such as R_{sand} or R_s), and a, b, c, d, e are fitting parameters.

(4) *Estimation of historical soil respiration in TD's shifting sand*

We chose the best-fitting equation for R_{sand} or R_s, and along with the hourly soil temperature observations in the study area at 0 and 10 cm between 2004 and 2017, we estimated the corresponding soil respiration, based on the regression equations.

(5) *Future trends in soil respiration in TD's shifting sand*

To extend the timescale over which the relationship was evaluated and to calculate future soil respiration in TD's shifting sand, we would like to use the relationship between soil temperature ($T_{0-10\mathrm{cm}}$ and $\Delta T_{10\mathrm{cm}}/\Delta t$) and soil respiration in conjunction with average soil temperature prediction data from various models in CMIP5. However, there were a number of obstacles to be overcome. From CMIP5, eight models that predicted data for monthly mean soil temperatures from 2006 to 2100 were chosen. Using monthly mean surface temperature data collected in the research area between 2006 and 2017, the simulated surface temperature data were independently confirmed. The three models with the best simulation results were identified.

These three models did not directly offer soil temperature at a depth of 10 cm and had varying soil depth levels. Soil temperature at a depth of 10 cm was determined using the regression relationship between the observed monthly mean soil temperatures at depths of 0 and 10 cm in the study area from 2006 to 2017. This relationship also included the average surface temperatures of the three chosen models. Between 2017 and 2100, the monthly average $\mathrm{T}_{0-10\mathrm{cm}}$ was measured. On a monthly timescale, the $\Delta T_{10\mathrm{cm}}/\Delta t$ parameter loses its original physical meaning, making it impossible to directly apply the previously established regression relationship between soil respiration, $T_{0-10\mathrm{cm}}$, and $\Delta T_{10\mathrm{cm}}/\Delta t$.

We reestablished a link between cumulative monthly soil respiration and the monthly average $T_{0-10\mathrm{cm}}$ using historical soil temperature data and the monthly values of cumulative soil respiration obtained in the previous section. At the monthly scale, there was a strong linear regression link between $T_{0-10\mathrm{cm}}$ and the soil respiration of TD's shifting sand:

$$R_s = 13641 \times T_{0-10\mathrm{cm}} - 18814 \quad R^2 = 0.928, \tag{9.16}$$

in which R_s (μmol m^{-2} month^{-1}) is soil respiration at the monthly scale and $T_{0-10\mathrm{cm}}$ is the soil temperature difference between 0 cm and 10 cm at the monthly scale. Finally, under the RCP4.5 and RCP8.5 scenarios, the annual total of soil respiration (i.e., carbon sink rate) was determined in TD's shifting sand from 2018 to 2100.

9.4.2 *Contribution of each component to soil respiration*

When compared to other ecosystems, the shifting sand's soil respiration is greatly reduced, with a negative CO_2 flux frequently present at night (Lou and Zhou, 2006; Yang *et al.*, 2017). The daily fluctuation varied from -0.026 to 0.101 μmol $\cdot$ m$^{-2}\cdot$ s^{-1}, and the average rate of soil respiration in the shifting sand was 28.7 $\pm 10^{-3}$ μmol$\cdot$ m$^{-2}\cdot$ s^{-1}. The contribution of the CO_2 fluxes of different components in TD's shifting sand to soil respiration was

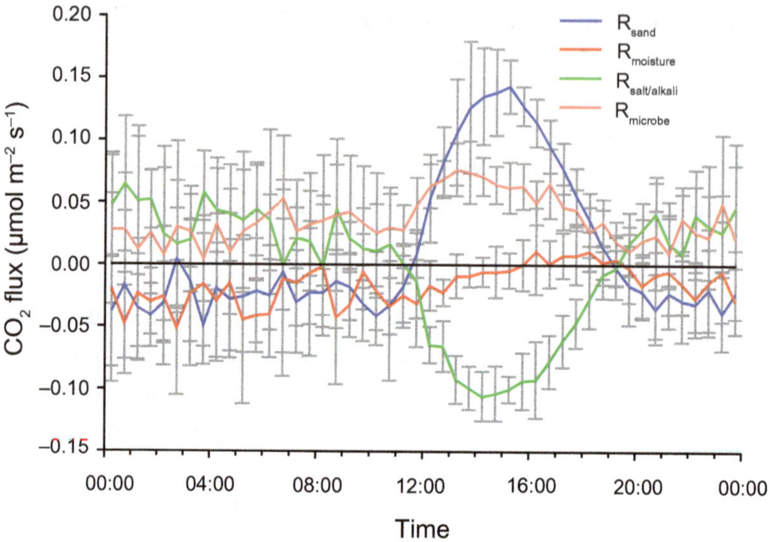

Fig. 9.15. The contribution of different components of shifting sand to the diurnal dynamics of soil respiration in the Taklimakan Desert (TD) shifting sand was disassembled. The diurnal contributions of the CO_2 flux of each shifting sand component are expressed as R_{sand}, R_{moisture}, $R_{\text{salt/alkali}}$, and R_{microbe}. The error bars represent the standard deviation of the repeated CO_2 flux observations over many days (Yang *et al.*, 2020a).

determined during the disassembly experiment and is depicted in Fig. 9.15. Sand and salts/alkalis made very large contributions, predominating the release and absorption of CO_2, respectively. In contrast to moving sand, the diurnal pattern of CO_2 flux for sand followed a unimodal curve. The sand encouraged the release of CO_2 into the atmosphere throughout the day, which peaked (0.143 μmol·m^{-2}) at 15:15 h (Beijing time, used hereafter). Sand encouraged stable CO_2 absorption from the atmosphere during the night, with a mean rate of -0.026 μmol·m^{-2}. The contribution of moisture to soil respiration was very limited, with a rate ranging from -0.051 to 0.011 μmol·m^{-2}· s^{-1}. Moisture promoted the release of CO_2 to the atmosphere between 15:45 and 19:15 h and promoted atmospheric CO_2 absorption at other times.

The impacts of soil temperature were the only factors studied in future tests because moisture had a negligible effect on soil respiration as a whole. Atmospheric CO_2 was strongly absorbed by salts and alkalis during the day, reaching a peak (–0.106 μmol·m^{-2}) at 14:15, but throughout the night, salts and alkalis encouraged a stable CO_2 release into the atmosphere. A minor percentage of the CO_2 released into the atmosphere was continuously caused by microbes, and these effects were stronger during the day than at night.

9.4.3 *Driving mechanisms of CO_2 flux in sand*

The temperature changes in the various strata for the whole observation period in the temperature-controlled experiments' sands without temperature regulation did not exceed 1°C. The associated CO_2 flux remained largely constant at zero. However, when the temperature was controlled, there was a large CO_2 absorption/release from the sand, and the varying rhythm of this absorption/release was generally consistent with variation in soil temperature (Fig. 9.16). The CO_2 flux increased as the soil temperature rose, as seen in the figure. The CO_2 flux changed from initially positive to negative values when the soil temperature dropped quickly. However, as soon as soil temperatures stabilized, the CO_2 flux intensity started to trend downward and eventually reached zero.

Based on the information above and considering that soil respiration indicates the overall performance of the soil surface layer (Lou and Zhou, 2006), we claim that the two main factors influencing the CO_2 flux in the sand (R_{sand}) are the difference in soil temperature between depths of 0 cm and 10 cm ($T_{0-10\text{cm}}$) and the rate of change in soil temperature at a depth of 10 cm ($\Delta T_{10\text{cm}}/\Delta t$). The $T_{0-10\text{cm}}$ parameter denotes the flow and intensity of soil heat, whereas $\Delta T_{10\text{cm}}/\Delta t$ denotes the soil's capacity to absorb and release heat (Ball *et al.*, 2009). $T_{0-10\text{cm}}$ and $\Delta T_{10\text{cm}}/\Delta t$, which showed strong linear and quadratic regression connections, respectively, were more

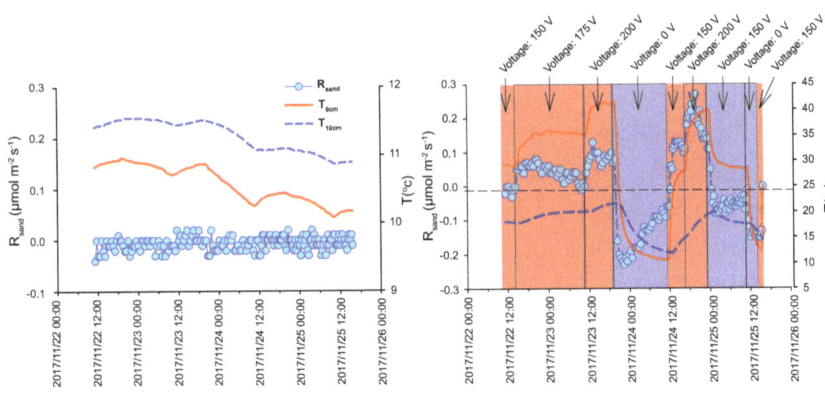

Fig. 9.16. Relationship between R_{sand} and soil temperatures at depths of 0 cm and 10 cm ($T_{0\text{cm}}$ and $T_{10\text{cm}}$, respectively) in a temperature-controlled experiment using a voltage adjustment. Left: Unregulated soil temperature. Right: Regulated soil temperature. Black vertical lines separate regions of different voltages. Red regions correspond to voltage increases compared to the voltage in the preceding region, and subsequent increases in soil temperature. Blue regions correspond to voltage decreases compared to the voltage of the preceding region, and subsequent decreases in soil temperature (Yang *et al.*, 2020a).

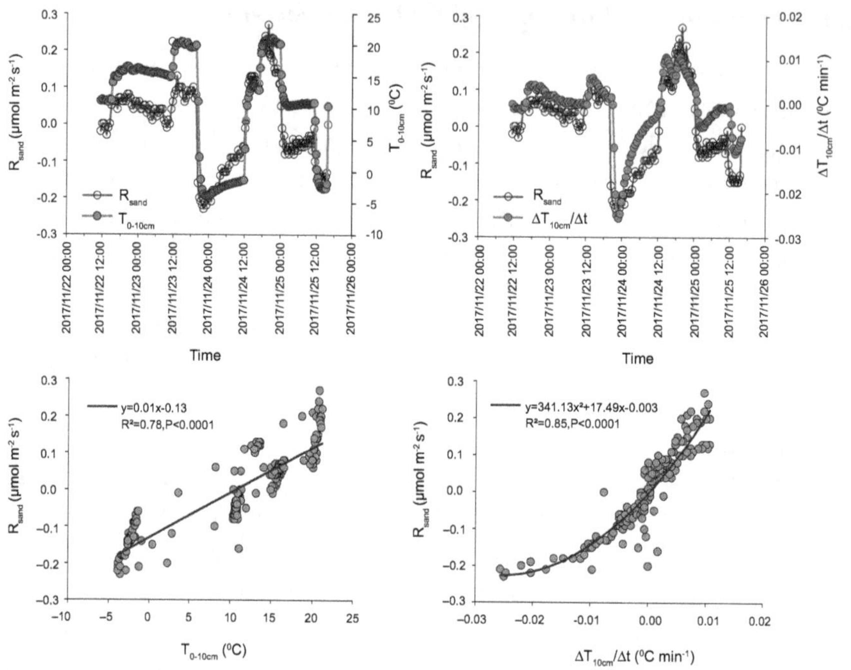

Fig. 9.17. Relationships between $T_{0-10\text{cm}}$, $\Delta T_{10\text{cm}}/\Delta t$, and R_{sand} in temperature-controlled experiments. Relationship between R_{sand} and $T_{0-10\text{cm}}$ (left); relationship between R_{sand} and $\Delta T_{10\text{cm}}/\Delta t$ (right) (Yang *et al*, 2020a).

consistent with R_{sand} than soil temperature (Fig. 9.17). It was more obvious that this uniformity might be improved once the soil temperature was constant. Additionally, investigations of the synergistic effects demonstrated that Eq. (9.14) could account for the complete response of R_{sand} to changes in $T_{0-10\text{cm}}$ and $\Delta T_{10\text{cm}}/\Delta t$ (Table 9.3). The calculated values of R_{sand} were in good agreement with the measured values using Eq. (9.14) and historical soil temperature data (Fig. 9.18).

9.4.4 *Effect of different layers' temperature on soil respiration*

We discovered through the disassembly tests that the main factor regulating the contribution of the various soil respiration components in the shifting sand was soil temperature. Due to this, we chose to conduct studies on the soil respiration monitoring data for shifting sand in the research region and the related soil temperatures of two layers ($T_{0\text{cm}}$ and $T_{10\text{cm}}$) on January 17–31, 2013, October 17–23, 2013, May 4–7, 2015, and July 16–23, 2019

Table 9.3. Synergistic effect analysis for R_{sand}/R_s with $T_{0-10\mathrm{cm}}$ and $\Delta T_{10\mathrm{cm}}/\Delta t$ (Yang *et al.*, 2020a).

	No.	Fitted equation	R^2
R_{sand}	2	$R_{\mathrm{sand}} = 0.007 + 0.002\ T_{0-10cm} + 2.958\ \Delta T_{10\mathrm{cm}}/\Delta t$	0.496**
	3	$R_{\mathrm{sand}} = -\ 0.010 + 0.002\ T_{0-10cm} + 1.542\ \Delta T_{10\mathrm{cm}}/\Delta t + 0.108\ T_{0-10cm}\Delta T_{10\mathrm{cm}}/\Delta \mathrm{t}$	0.614**
	4	$R_{\mathrm{sand}} = -0.031 - 0.002\ T_{0-10\mathrm{cm}} + 4.116\ \Delta T_{10\mathrm{cm}}/\Delta t + 0.0004\ T_{0-10\mathrm{cm}}^2 - 67.961\ (\Delta T_{10\mathrm{cm}}/\Delta t)^2$	0.748**
	5	Failed	/
R_{s}	2	$R_{\mathrm{s}} = -1.55 \times 10^{-4} + 0.008\ T_{0-10cm} - 2.445\ \Delta \mathrm{T}_{10\mathrm{cm}}/\Delta \mathrm{t}$	0.756**
	3	$R_{\mathrm{s}} = 0.007 + 0.008\ T_{0-10\mathrm{cm}} - 2.147\ \Delta T_{10\mathrm{cm}}/\Delta t - 0.048 \times T_{0-10\mathrm{cm}}\Delta \mathrm{T}_{10\mathrm{cm}}/\Delta \mathrm{t}$	0.768**
	4	$R_{\mathrm{s}} = -0.004 + 0.006\ T_{0-10m} - 0.194\ \Delta T_{10m}/\Delta t + 3.669 \times 10^{-5} T_{0-10\mathrm{cm}}^2 - 35.928\ (\Delta T_{10\mathrm{cm}}/\Delta t)^2$	0.792**
	5	Failed	/

Notes: R^2 is the goodness-of-fit of the regression equation; "*" and "**" indicate a significant correlation at p <0.05 and p <0.01 (two-tailed), respectively.

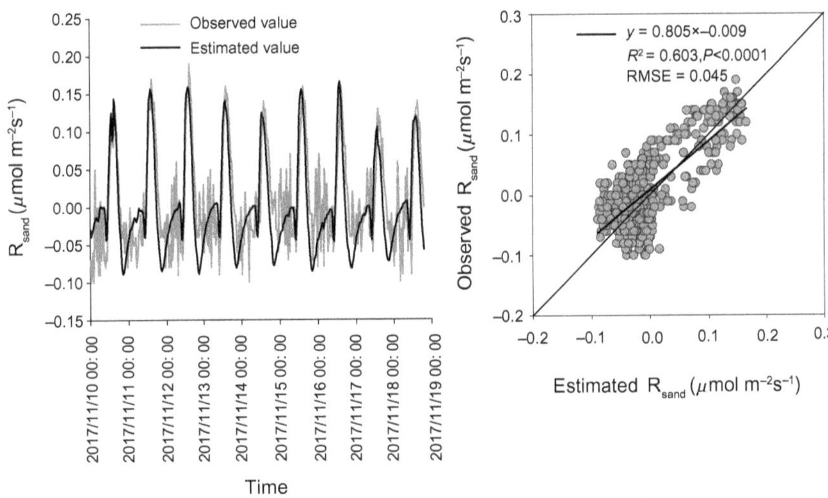

Fig. 9.18. Comparison of R_{sand} obtained from experimental data and estimated by Eq. (9.14) in sand. Time series of observed and estimated R_{sand} (left). Regression relationship between observed and estimated R_{sand} (right) (Yang *et al.*, 2020a).

(Fig. 9.19). With substantial linear regression connections ($R^2 = 0.756$ and 0.501, respectively; $p < 0.001$), the $T_{0-10\mathrm{cm}}$ and $\Delta T_{10\mathrm{cm}}/\Delta t$ were in agreement with R_s. Therefore, we examined the synergistic effects of $T_{0-10\mathrm{cm}}$ and $\Delta T_{10\mathrm{cm}}/\Delta t$ on R_s (Table 9.3). Except for a failed fitting analysis when

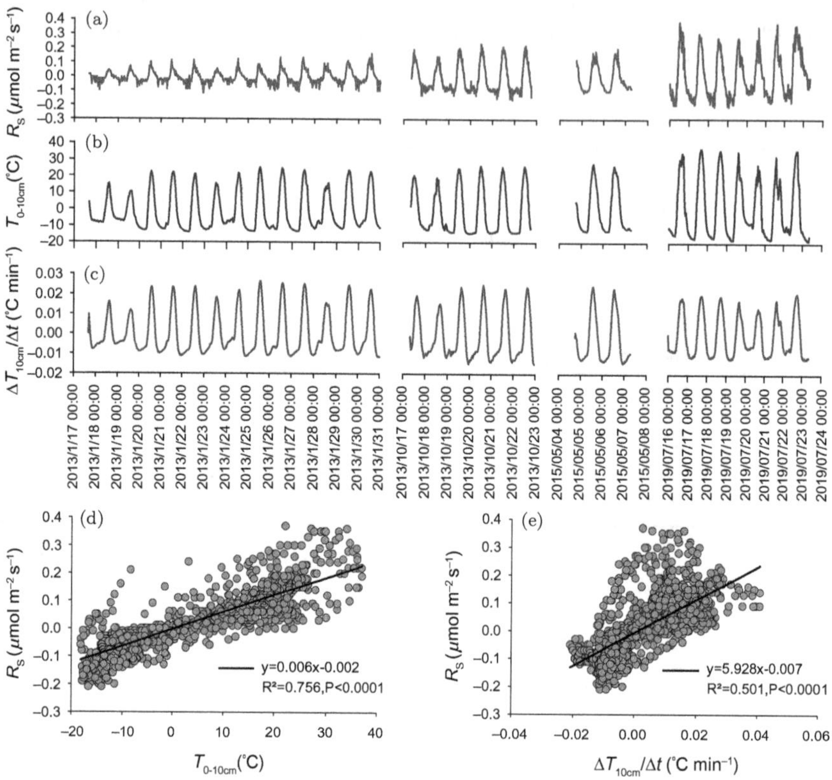

Fig. 9.19. Relationships among $T_{0-10\text{cm}}$, $\Delta T_{10\text{cm}}/\Delta t$, and R_s during January 17–31, 2013, October 17–23, 2013, May 4–7, 2015, and July 16–23, 2019. Time series of R_s (a), $T_{0-10\text{cm}}$ (b), and $\Delta T_{10\text{cm}}/\Delta t$ (c). Regression analyses between R_s and $\mathrm{T}_{0-10\text{cm}}$ (d). Regression analyses between R_s and $\Delta T_{10\text{cm}}/\Delta t$ (e) (Yang *et al.*, 2020a).

using Eq. (9.15), the other three equations explained the comprehensive responses of R_s to the variation in $T_{0-10\text{cm}}$ and $\Delta T_{10\text{cm}}/\Delta t$ ($R^2 > 0.756$) well and to a highly significant level ($p < 0.001$). This suggests that the main factors influencing soil respiration in shifting sand and the main factors driving the CO_2 flux in the sand were the synergistic impacts of $T_{0-10\text{cm}}$ and $\Delta T_{10\text{cm}}/\Delta t$. Other components, except the sand, also responded favorably to $T_{0-10\text{cm}}$ and $\Delta \mathrm{T}_{10\text{cm}}/\Delta t$. The predicted values of R_s were in good agreement with the observed values using Eq. (9.14) and the related observational data for soil temperature in various eras (Fig. 9.20).

The observational data from 2017 demonstrate the annual variability of the daily CO_2 exchange ($F_{C'-IRGASON}$) in the hinterland of the TD. From January to March and from October to December, most values are negative (i.e., the shifting sand absorbs atmospheric CO_2). In contrast, from April

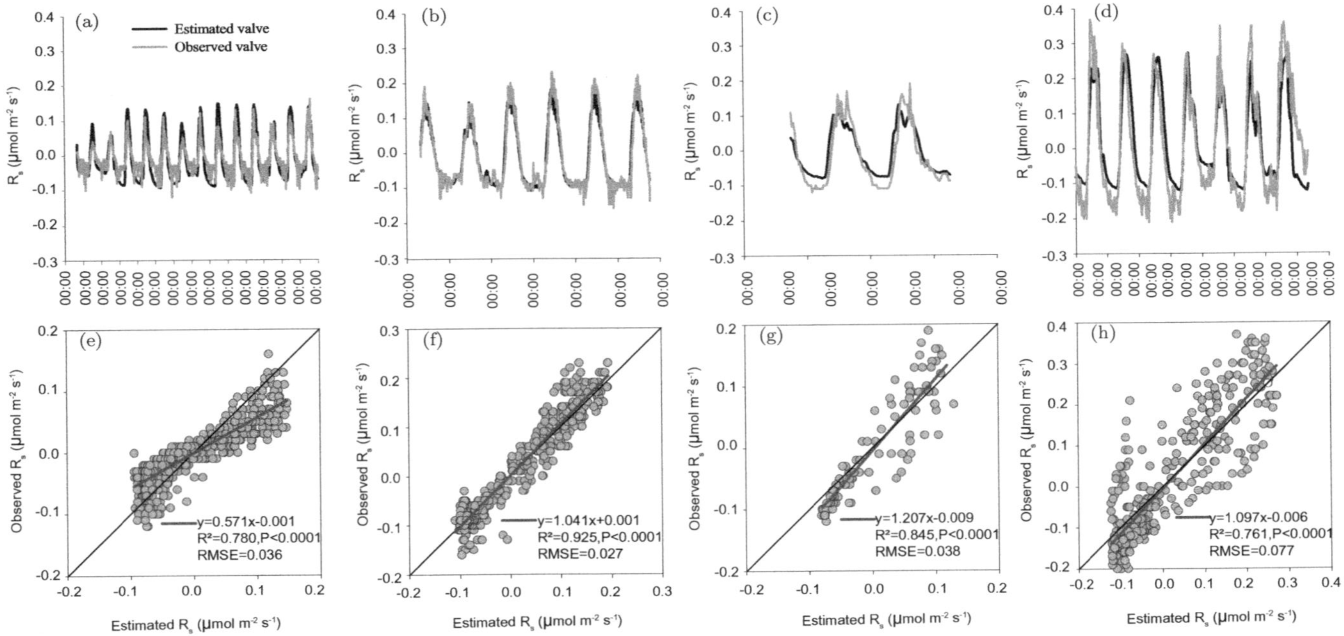

Fig. 9.20. Comparison of R_s obtained from experiments and estimated by Eq. (9.14) in shifting sand. Relationships between the observed and estimated R_s are shown for January 17–31, 2013 (a); October 17–23, 2013 (b); May 4–7, 2015 (c); July 16–23, 2019 (d); and (e–h). The regression relationships for the time periods are shown in (a–d) (Yang *et al.*, 2020a).

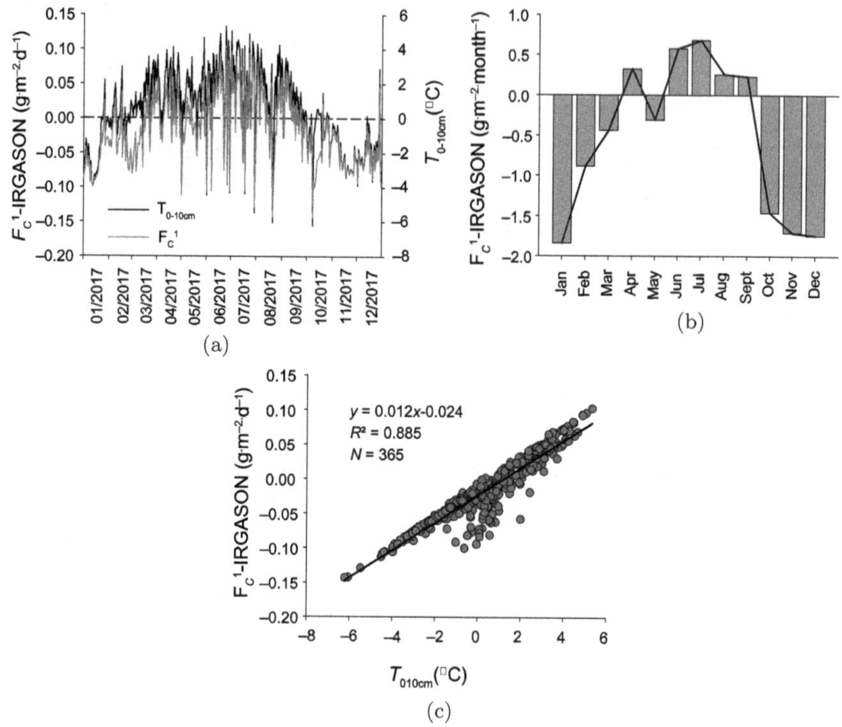

Fig. 9.21. High-frequency CO_2 flux values obtained by IRGASON. Time series of the corrected daily CO_2 exchange ($F'_{C-IRGASON}$) and the daily average soil temperature difference between 0 cm and 10 cm (T_{0-10cm}) in the hinterland of the TD in 2017 (a). Total monthly carbon sequestration of shifting sand in the hinterland of the TD (b). Regression relationship between $F'_{C-IRGASON}$ and T_{0-10cm} (c) (Yang *et al.*, 2022).

to September, most values are positive (i.e., the shifting sand releases CO_2) (Fig. 9.21). Additionally, there was a significant linear relationship between the daily CO_2 exchange and the difference in average soil temperature between 0 and 10 cm (T_{0-10cm}), which represents the direction and strength of the soil heat flow ($R^2 = 0.885$, $p < 0.01$). The conclusion was expanded upon by Gao *et al.* (2021) to include an examination of the effect of the air–soil temperature difference on desert carbon sequestration. In deserts, thermal convection, which is fueled by the temperature difference between the air and the soil, is crucial for CO_2 absorption and release. Additionally, we identified T_{0-10cm} at 2°C as a crucial threshold that regulated the changeover between shifting sand's carbon source and sink. When $T_{0-10cm} > 2$°C, the shifting sand in the desert released CO_2; when $T_{0-10cm} < 2$°C, it absorbs CO_2. Absorption is substantially stronger in the dynamic equilibrium of release and absorption over the year. According to this, the shifting sand in the TD's hinterland sequestered carbon at a rate of 6.31 g m^{-2} in 2017. The strongest absorption was in January (1.84 g m^{-2} $month^{-1}$), whereas the strongest release was in July (up to 1.84 g m^{-2} $month^{-1}$). This outcome is much lower than that of other deserts (Schlesinger, 2017).

9.4.5 *Carbon sink rate of shifting sand*

We discovered that the contribution of different shifting sand components to the total CO_2 flux in the shifting sand is strongly correlated with soil temperature in the hinterland of the TD through a combination of partitioning (dealkalization/desalination, sterilization, and drying) and temperature-controlled experiments. The two main parameters influencing CO_2 exchange in the shifting sand are the difference in soil temperature between depths of 0 and 10 cm and the rate of soil temperature change at a depth of 10 cm. Additionally, we measured the CO_2 fluxes of different components in TD's shifting sand to determine their contributions (Fig. 9.22(a)). Comparing shows that an unusually large portion of the overall CO_2 exchange is contributed by the expansion and contraction of soil air containing CO_2 brought on by heat fluctuations in shifting sand. In combination with salts/alkali chemistry, this long-hidden process predominates in the release and absorption of CO_2 in shifting sand, respectively (Fig. 9.22(b)).

We calculated the total annual CO_2 exchange in TD's shifting sand using the close connections between CO_2 flux and soil temperature, along with historical soil temperature data collected in the research region between 2004 and 2017 at intervals of 0 cm and 10 cm (Fig. 9.23). Between 2004 and 2017, the physical process of thermal expansion of soil air containing CO_2 in TD's shifting sand resulted in an average annual emission of 6.71×10^6 t of CO_2 to the atmosphere. Additionally, the remaining elements (salt/alkali + moisture + microbes) worked together to cause TD's shifting sand to absorb atmospheric CO_2 at an average rate of 8.31×10^6 t a^{-1} between 2004 and 2017. With a combined yearly carbon sink rate of 1.60×10^6 t a^{-1} over the years 2004–2017, these release and absorption processes caused the surface of TD's shifting sand to demonstrate a clear and stable carbon sink effect. Leaching causes the absorbed CO_2 to progressively reach the groundwater layer, where it eventually converges with groundwater flow in the saline water layer under the wide desert. This process is comparable to the ocean's inorganic carbon sinks (Ma *et al.*, 2014; Li *et al.*, 2015b).

In addition, although appearing lifeless and ignored for a long time, desert ecosystems have silently contributed much to the global carbon cycle.

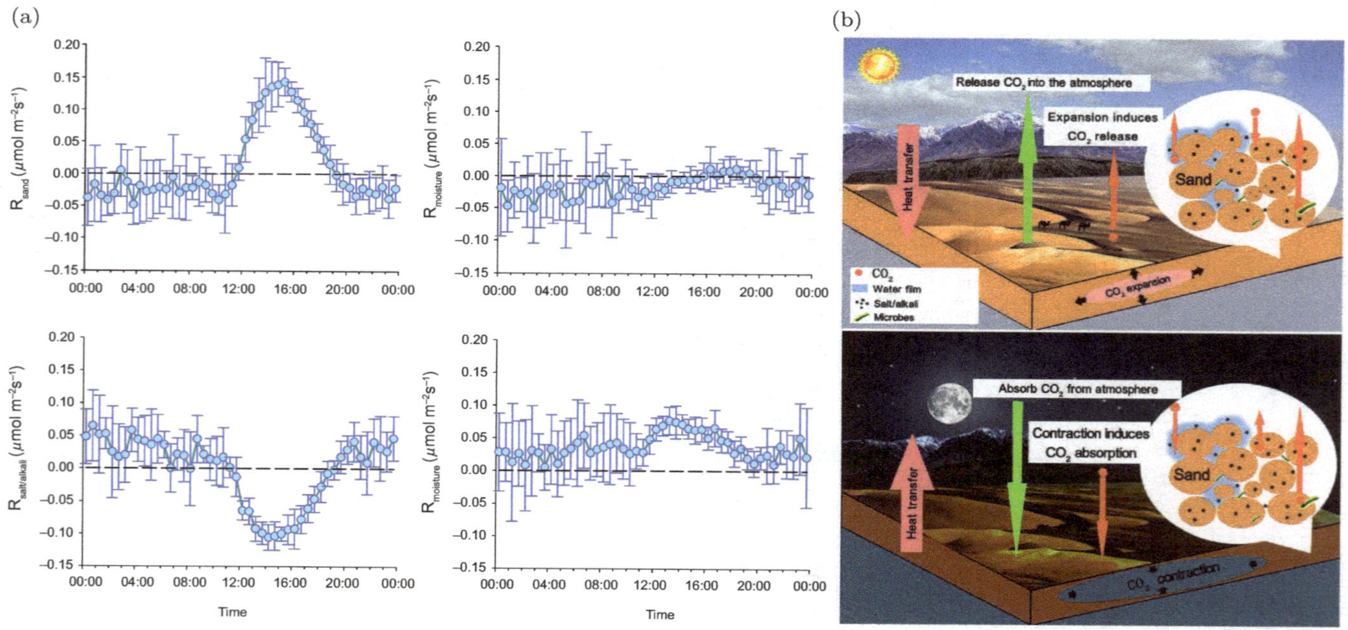

Fig. 9.22. Contributions of various shifting sand components to the daily dynamic of the total CO_2 flux in Tazhong shifting sand of the TD. The diurnal contributions of various shifting sand components are shown for R_{sand}, R_{moisture}, $R_{\text{salt/alkali}}$, and R_{microbe}. Error bars represent standard deviations (a). Overview of various processes of CO_2 exchange in TD's shifting sand (b). (Yang *et al.*, 2020b).

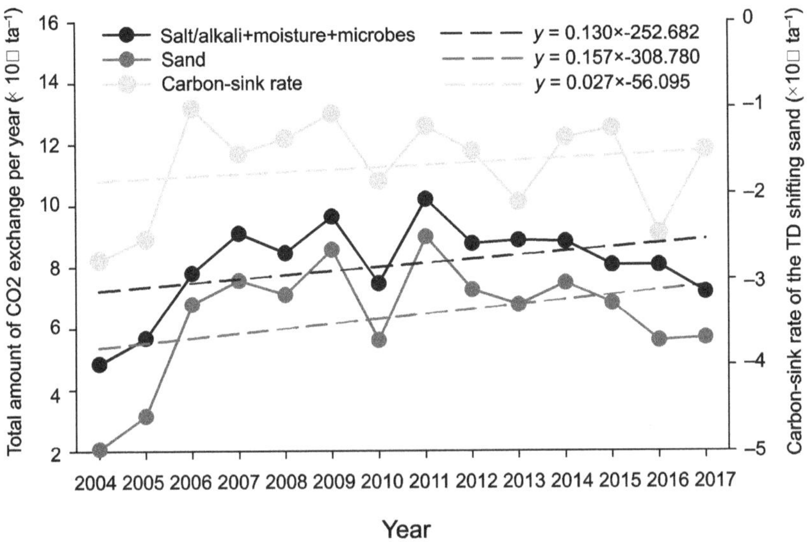

Fig. 9.23. Temporal variation in the total amount of CO_2 exchange per year for 2004–2017 in TD's shifting sand (Yang *et al.*, 2020b).

According to our research, the shifting sand in the TD region has an enormous potential carbon storage capacity. This is due to a combination of factors, including the TD's about 8 trillion cubic meters of groundwater and the region's widespread salinization from severe evaporation. Only a small portion of the world's deserts — roughly 0.747% — is represented by TD's shifting sand. The worldwide shifting desert may sequester about 2.125×10^8 t of CO_2 each year if all deserts on the planet are considered and the Taklimakan values are taken as an average uptake state. As a result, the role played by desert ecosystems in the global carbon cycle cannot be willfully disregarded. Even though this cannot entirely fix the issue of the "missing carbon sink," it can at least help to close the gap.

Increasing T_{0-10cm} (Fig. 9.24) in the future will more strongly stimulate the thermal expansion of soil air, pumping more CO_2 into the atmosphere, which will eventually cause the gradual weakening of the carbon sink rate of TD's shifting sand. This is in accordance with the relationship between T_{0-10cm} and the carbon sink rate of TD's shifting sand. As depicted in Fig. 9.24(b), under RCP4.5 and RCP8.5, the carbon sink rate of TD's shifting sand will decline at a rate of 0.43% and 1.20% per year, respectively. Under RCP8.5, it was predicted that the shifting sand surface in the TD would achieve a CO_2 absorption/release neutral state around 2100. The weakening of the carbon sink rate was more pronounced under the RCP8.5 scenario.

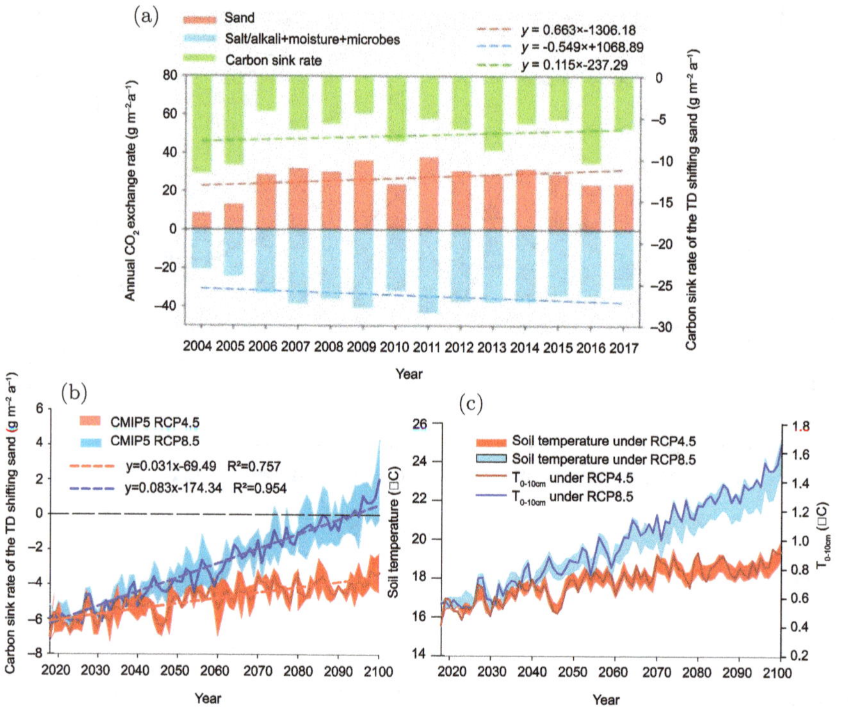

Fig. 9.24. Temporal variation of the annual CO_2 exchange rate of shifting sand in the Taklimakan Desert (TD). Temporal variation in the soil CO_2 exchange rate per year for 2004–2017 in TD's shifting sand (a). Under Representative Concentration Pathway (RCP) scenarios 4.5 and 8.5, temporal variation in the carbon sink rate for 2018–2100 in TD's shifting sand. Shading denotes the mean ± standard deviation of three models (MIROC5, MRI-CGCM3, and CMCC-CMS) (b). Temporal variation in soil temperature and $T_{0-10\text{cm}}$ under the RCP scenarios 4.5 and 8.5. The upper boundary of the shading denotes the soil temperature at 0 cm and the lower boundary denotes the soil temperature at 10 cm (c) (Yang *et al.*, 2020a).

9.5 Impacts of Desertification Control on Terrestrial Carbon Sink

Through a variety of processes, including the sequestration of atmospheric CO_2 during plant growth, the release of CO_2 during combustion and respiration, the release of methane and nitrous oxide during wild events, terrestrial ecosystems have the power to change the atmosphere's greenhouse gas concentrations. Typically, desertification causes a drop in productivity as well as a reduction in both above- and below-ground carbon stores. Overgrazing is one of the factors that cause a decline in both plant and SOC pools.

Desertification control can not only improve the regional ecological environment but also transforms desertified land into an important terrestrial carbon sink. This is the secret to strengthening and enhancing the ability of ecosystems in semi-arid regions to act as carbon sinks. An essential component of the terrestrial environment is the desert ecosystem. China's desert ecosystem covers 1.65 million square kilometers, or 17% of the country's total land area. It has a tremendous potential for carbon sequestration and a value of 718.743 billion for its carbon sequestration service. In the past, the function of preventing wind flow and controlling desertification through sand fixation received more attention, but a desert's benefits as a carbon sink were neglected for a very long time. The abiotic carbon sequestration capacity of deserts is not paid enough attention.

9.5.1 *Measures and experience of preventing and controlling desertification*

One of the countries in the world that is most severely impacted by desertification is China. The control of sandstorm sources in Beijing and Tianjin, the conversion of farmland to forests and grasslands, the construction of the Green Wall of China, and the comprehensive control of rocky desertification are just a few of the key national ecological projects that China has continued to carry out in recent years. As a result, the population loss caused by desert expansion in China has changed to desert decline caused by vegetation greening. The National Forestry and Grassland Administration conducted five desertification monitoring campaigns through 2014, identifying variations, disasters, and contributing factors. Since the reduction in 2004, China has maintained a "double reduction" in its desertification land area. Since China joined the United Nations Convention to Combat Desertification in 1996, the UNCCD and the international community have been very concerned about the issue and have acknowledged China's implementation. The relevant documents of the China–Africa and China–Arab Cooperation Forum and the BRICS leaders' declaration include information on cooperation to prevent and control desertification, and the United Nations Commission on Sustainable Development meeting noted that China is the world leader in this area.

China is crucial to the global effort to manage desertification, according to the secretariat of the United Nations Convention to Combat Desertification. Gansu Province is one of the provinces in China with the largest areas of land subject to desertification, active development, and significant injury. About 25 desertification land surveys have been conducted in Gansu Province, which has a thorough understanding of area, resources, impact mode, and intensity. The land types in Gansu Province that are subject to desertification are categorized on the basis of these surveys.

The following five experiences are based on years of preventing desertification:

(1) *Biological measures to combat desertification*

Biological measure is the principal method for combating desertification. The use of vegetation, such as trees, bushes, and herbs, can increase the amount of land covered and enhance the physical and chemical characteristics of the soil. Species selection and afforestation techniques are top priorities in terms of biological measures.

About 780 desert and semi-desert species were successfully introduced and domesticated in the Gansu Province, which also established 5 Desert Botanical Gardens and the Gansu Desert Plant Resource Information Sharing Platform; 150 adaptable species were chosen to control desertification. China's first and largest desert botanical garden is located in Minqin (1974). 626 desert and semi-desert species are collected there. In Gansu Province, breeding and afforestation techniques were figured out for 30 beneficial desert species. On desertified land in Gansu Province, windbreak and sand fixation forests were built, occupying 144,000 ha, or 1.2% of the entire sandy desertification land area.

Large shelterbelts have been built in Gansu since the 1950s. The paradigm of a short shelterbelt with a sparse network was chosen for extension up until the 1970s. For the renewal and transformation of the shelterbelt, anti-stem borer species were introduced at the start of 1990. In order to preserve 600,000 ha of farmland in the Hexi Corridor, Gansu built an 87,000 hm^2 farmland shelterbelts by 2010.

(2) *Mechanical measures to combat desertification*

The mechanical measure, which complements the biological measure and is not appropriate for large-scale extension due to its limited lifespan and high cost, includes technologies that release sand buildup and the installation of sand barriers. Effective ways to reduce and prevent risks connected with sand movement include mechanical stabilization and removing collected sand. One of the earliest and most basic techniques for managing shifting sands is the construction of sand-blocking walls (banks). Downwind of moving sand dunes and shifting sands, surrounding communities, mining or oil fields, and along communication or transportation facilities, walls or banks are put in place. One of the earliest techniques for stopping and settling shifting sand was the use of sand sedimentation ditches. Vertical sand barriers are another name for sand-blocking barriers. When there is turbulence, the barrier alters the wind flow and slows the wind movement. Before afforestation, barriers were put in place to stabilize shifting sand dunes and safeguard seedlings. In the province of Gansu, more than 25 different types of sand barriers were created or introduced. The four sand barriers — clay sand, straw checkerboard, cotton stalk, and plastic checkerboard — were widely used, particularly in arid or extremely arid regions with precipitation of less than 150 mm.

(3) *Enclosure measures to combat desertification*

The enclosure, one of the frequently employed methods, is meant to safeguard and promote the regeneration of desert vegetation. When degraded rangeland is enclosed, plant roots can start to receive enough moisture and mineral nourishment from the soil, and plant leaves can start to store energy and organic material through photosynthesis. This results in regular plant growth and reproduction, which boosts productivity and vegetation coverage, increases the vegetation's ability to act as a windbreak and stabilize sand, and prevents sand movement and the reactivation of permanent sand dunes. On the other hand, the vegetation's environment is altered by enclosure measures. Higher plant cover and a higher organic matter content in the soil can efficiently retain soil moisture and decrease soil erosion by lowering wind speeds, increasing the soil's capacity to hold water, and protecting it from wind and water erosion. In the desertified area, Gansu Province established 7 national nature reserves and 1 provincial nature reserve. The total area under protection spanned 2.18 million ha.

(4) *Ecological engineering and restoration*

The province of Gansu successfully managed desert encroachment through the implementation of numerous eco-restoration initiatives, including artificial measures. Between 1991 and 2001, the Desertification Prevention and Control Project was in operation. It included 168,200 ha of desertified land and 52,300 ha of sand fixation forests. By 2012, Gansu had established control over 50,000 km^2 of desertified land, 253,000 ha of sand fixation forest, more than 1200 km of significant windbreak and fixation shelterbelt, 470 blown sand passages accounting for 54% of all passages, and 200,000 ha of shifting sandy land.

(5) *Comprehensive measures to combat desertification*

An efficient and crucial method of preventing desertification in semi-arid regions is the use of comprehensive measures that combine biological and engineering methods. The integrated agricultural system and desertification control ideology developed in Gansu Province later became the foundational paradigm for desert management in China. In order to combat desertification, the province of Gansu developed more than 20 comprehensive techniques or models. These are shorthand for technologies to combat desertification. Since these expressions with Chinese characteristics are not the focus of this paragraph, deleting them makes it easier for readers to read. At the margin of the oasis in a typical desert location, there is a comprehensive management scheme that combines fixation, blocking, transportation, and containment. More than 99% less wind-blown silt was produced, and the wind speed at a height of 20 cm decreased 42.9–83.1%.

9.5.2 *Strengthen tree planting and promote forestry carbon sink capacity*

Reforestation and afforestation are two of the most economical ways to improve carbon storage and reduce the rate of increase in CO_2 concentration in terrestrial ecosystems. 1.83 t of CO_2 is typically absorbed and 1.62 t of O_2 is emitted for every 1m3 of forest growth. About 2/3 of the terrestrial environment comprises carbon that is fixed annually. Between 2021 and 2025, China will fully manage 6.7 million ha (100 million mu) of desertified land. To boost the sink capacity, there is a lot of potential for artificial afforestation in desertified areas. Semi-arid regions, where precipitation is infrequent, evaporation is strong, wind is strong, and sand is plentiful, are where the influence of natural environmental conditions on the effect of afforestation emission reduction is most apparent.

The CO_2 emission reduction assessment of artificial afforestation lacks objective and accurate data support. Therefore, the method for estimating large-scale CO_2 emission reductions by planting trees in arid areas is proposed. This is done by highlighting the importance of artificial afforestation for CO_2 emissions estimation research, basing it on CO_2 emission reduction calculation rules, combining it with the feature of planting areas in natural environments, and making it clear how artificial afforestation in arid areas contributes to global climate warming. It also offers theoretical foundations for desert forestry carbon sequestration management.

9.5.3 *Carbon sink technology of desert grassland*

The most efficient way to increase the ability of desert grasslands to sequester carbon is through optimal management, which includes fencing, reseeding, and the creation of artificial grasslands. China's desertification land area will increase from 2021 to 2025, and the protected area will reach 2 million ha (30 million mu). Our findings demonstrate that by restricting grazing operations, enclosure measures can successfully increase community composition, grassland productivity, and soil quality. However, a prolonged enclosure period, on the other hand, will encourage the invasion of annual weedy grasses, which is detrimental to the synthesis of soil organic carbon. In addition, litter accumulates in the soil surface, limiting the input of carbon from aboveground plants to the soil. Concerning the impact of conservation on carbon sequestration in desertification grassland, there is still a significant degree of doubt. It is therefore advised to carry out dynamic monitoring of carbon storage in desertification and containment reserves and investigate the technological mode of increasing sink capacity in desertification and containment reserves in accordance with the adaptive management law in order to provide a scientific foundation for enhancing the quality of containment and protection.

9.5.4 *Selecting suitable tree species to enhance plantations' carbon sink capacity*

To improve carbon storage in desert plantation ecosystems, tree species and afforestation pattern selection are important decision-making considerations. The growth of vegetation in semi-arid regions is restricted by local climate conditions and soil character. Plants which grow in semi-arid regions are tolerant to drought and reduce the consumption of moisture to maintain growth. The key adaptation traits of suitable plants include a developed root system, leaf deterioration, and high salt tolerance. Besides, the ideal carbon sequestration forest species have higher growth rate and wood density, and lower decomposition rates. The faster-growing species have a greater capacity to fix carbon than the slower-growing species, and mixed forests produce more biomass than pure forests.

At 6 and 10 years of stand age, *Caragana microphylla* and *Salix officinalis* reached their maximum rates of carbon fixation. At the age of 10 years, the artificial *C. ammodendrum* reached its highest carbon sequestration rate. Although artificial *C. ammodendrum* had a lower carbon sequestration capacity than the natural *C. ammodendri* shrub, it was still much better than *C. korshinskii, C. anthracis*, and *Seabuckthorn*. Although the exact mechanism is yet unknown, carbon sequestration is intimately related to biomass growth, soil carbon storage, and litter accumulation. How can the sparsely afforested areas fix more carbon? How can we create afforestation models that are good for sequestering carbon? Important theoretical issues like the ones mentioned must be resolved immediately in the future. In order to provide scientific and technological support for the planning and design of carbon sequestration forests in the forestry and grassland sectors, it is suggested that research be done on the carbon sequestration rate, process, and regulation mechanism of typical sand-fixing afforestation species in different climate zones.

9.5.5 *Rehabilitate degraded plantations to consolidate carbon storage*

An essential component of anthropogenic reduction of terrestrial ecosystem carbon release in the international deliberations of the United Nations Intergovernmental Panel on Climate Change (IPCC) is mitigation of forest degradation. In the Hexi Corridor of Gansu Province, artificial pickling forests had grown to a total area of 89,000 hectares by 2016. With a protected area of 43,500 hectares, Minqin County has the longest history of development and the most extensive level of afforestation. It is crucial for enhancing soil quality, maintaining biodiversity, boosting microclimate, and increasing the carbon sink. According to estimates, an artificial *Ammodendrum* ecosystem can fix 434 g of carbon per square meter each year. The Shiyang River Basin's artificial *Ammodensus* sand-fixing forest has 84 million tons of carbon storage capacity, and it has 195 million tons of carbon sequestration potential. The *Ammodendron* plantation's biomass and ecosystem carbon storage first increased and subsequently fell as the number of afforestation years increased. The primary driver behind the loss of the sand-fixing forest's ability to act as a carbon sink is the general degradation of artificial *Ammodendrum*. Shelterbelts like poplar and *zizyphus jujube* in the oasis in the arid desert region are also suffering greatly from neglect and poor management. "Reforestation, light management, and protection" are long-standing occurrences. It is recommended that forestry and grass departments focus more on caring for and managing deteriorated shelterbelts, investigating long-term mechanisms of forest management and protection, minimizing forest mortality, and strengthening the potential of plantation ecosystems in arid regions to store carbon.

9.5.6 *Explore sustainable mode of increasing carbon sink capacity*

The species of forest and grass with desert characteristics are resistant to barrenness, drought, and salinity, and have strong stress resistance and remarkable ecological and economic benefits. They are crucial germplasm sources for improving the land carbon sink for desertification. In addition to having high value as food and medicine, typical desert plants like *Cistanche deserticola, Cymphora, allium amurensis*, and *Amurensis* can fix atmospheric CO_2, which can achieve the unification of the development of the sand industry and increase carbon sequestration and sinks. The potential for artificial grassland to raise carbon sink capacity is enormous. Planting annual artificial grasslands is helpful for improving and restoring the land; however, perennial artificial grassland planting is better for transferring photosynthate to the soil.

Artificial alfalfa grasslands have a higher carbon density than natural ones, and as they expand year after year, they sequester more carbon. The sand industry and grass industry have not received enough attention, and the development mode is extensive and the scale is modest due to the long-term dominance of the conventional agricultural development model of "grain as the key link." The currently available grass and forest products are mostly primary goods with a short industrial chain. Arable land and desertification wasteland resources have been sitting inert for a very long period in desertification locations. It is recommended that relevant government departments incorporate the growth of industries with traits of the forest and grass into plans for rural revitalization, increase investment in R&D for products with those traits, innovate the way idle land is managed, and take advantage of the potential for carbon sequestration and income growth. The quality, efficiency, and exchange of the forest and grass industries in semi-arid regions can be increased by directing the flow of technology, talent, and capital to these industries.

9.6 Conclusions and Discussion

In this chapter, we created a desertification vulnerability map that shows the global region of the risk class for desertification while taking into account both global climate change and human activities. The creation of GDVI involved fusing the results of human activity, climate change, and vegetation conditions. Vegetation cover and the AI are significant components in desertification risk estimation (Jafari and Bakhshandehmehr, 2016; Lavado Contador *et al.*, 2009). ΔT is frequently used to assess recent climatic change (Dodd *et al.*, 2014). Pressure on the economic and environmental systems increases as PD rises. Economic growth and social growth are directly correlated with GDP and CO_2 emissions. Therefore, we combined the effects of PD, GDP, and CO_2 emissions using a weighted scale in order to evaluate the contribution of human activity to global desertification. To confirm the classification of the GDVI using the probability density function, the USDA's desertification vulnerability map (Fig. 9.3) was employed (Eswaran and Reich, 2003). The desertification vulnerability maps coincided in a majority of regions. By contrasting the GDVI with NPP and soil moisture, the capacity of the GDVI to evaluate global desertification vulnerability was also examined. Sparsely vegetated areas with low NPP levels and arid areas with little soil moisture are locations with a high GDVI value. NPP and soil moisture had a negative geographical and temporal correlation with GDVI. In general, GDVI's results on desertification vulnerability were fair and trustworthy.

The biggest danger of desertification, according to the GDVI distribution, is primarily seen in places near deserts or other arid terrain, particularly in Northern China, Northern Africa, Western America, India, and Mexico between 2000 and 2014 (Fig. 9.2). According to Chen and Tang (2005) and Feng *et al.* (2015), land use and management decisions in Northern China were primarily influenced by socioeconomic factors, which had an impact on desertification. This is in line with our findings, which demonstrate that the areas of Northern China are at a higher risk of desertification due to high levels of human activity and climatic aridity. We demonstrated that India experiences severe desertification. Both Arya *et al.* (2009) and Kundu *et al.* (2017) stressed how badly the land is degrading in India. G. Zhang *et al.* (2018) demonstrated increased desertification and degradation of grasslands in Central Asia between 2000 and 2014. Our findings also indicated that Central Asia faces a significant risk of desertification. In the literature, it has also been noted that there is a high danger of desertification and land degradation in Mexico and the Sahel (Becerril-Piña *et al.*, 2015). However, there are a few areas where our evaluation of desertification differs from that of earlier research. For instance, Tomasella *et al.* (2018) discovered that Northeastern Brazil is experiencing desertification; yet, our findings showed that this region is at a moderate risk of desertification due to high LAI and a low ΔT. Compared to earlier research, GDVI generally performs well in determining the danger of desertification.

The CMIP5 simulations under the RCP4.5 and the RCP8.5 conditions were used to project the GDVI's spatial and temporal fluctuation. When compared to observations, CMIP5 successfully replicated the spatial distribution of GDVI. RCP4.5 anticipates the implementation of strategies to reduce emissions (Thomson *et al.*, 2011). RCP8.5 is based on a scenario that assumes rapid population growth, moderate rates of technological advancement, and increases in energy efficiency. As a result, in the absence of climate change legislation, this scenario would result in high long-term energy consumption and greenhouse gas emissions (Riahi *et al.*, 2011). Human activity will be more intense under the RCP8.5 than under the RCP4.5 and climate change will be more severe as a result of different policies. Globally, the risk of desertification will rise as a result of warming, the expansion of arid and semi-arid regions, and intensified human activity. The coverage of low-risk desertification zones will significantly fall by the end of this century, from 47% to 24% under the RCP8.5 and to 35% under the RCP4.5. Under the RCP8.5, subtype alterations will be more significant than under the RCP4.5. Future impoverished people living in arid and semi-arid regions will experience more natural disasters as the high-risk desertification area grows, necessitating the creation of a worldwide action plan to lessen the risk of desertification. Although we use the multi-model ensemble mean of CMIP5 to lower the uncertainty of future predictions, it should be emphasized that uncertainties connected to climate change and human activities are predicted to increase over time. As a result, uncertainty associated with the projected GDVI will also increase.

It is observed that while GDVI can aid in the development of rehabilitation methods by combining the effects of climate change, human activity, and vegetation conditions to offer a picture of the risk state, it has the drawback of equally weighting each index. The causes and key elements contributing to the vulnerability of the world to desertification were not thoroughly examined. Ecological protection engineering and rehabilitation technology are anticipated to play an increasingly significant role in the control and mitigation of desertification as a result of technological advancements, improved knowledge of climate change, and anthropogenic contributions. It is challenging to quantify the contributions of land management and environmental protection to risk mitigation, particularly on a worldwide scale, because of how strongly they depend on the location in question. Therefore, engineering for environmental protection was not quantified in the HAI. Our findings might therefore overstate the probability of desertification in regions where environmental protection engineering has been put in place. Future research must carefully examine the application of the GDVI in a specific local area using extensive local data. Studies are also required to discover the major indicators and sub-indicators as well as the contributions of these components. To learn more about desertification on a global scale, biophysical and socioeconomic aspects such as soil quality, environmental protection engineering, and industrial structure should be considered for GDVI's future development.

The government intends to restart the Green Dam's rehabilitation by adding new ideas about sustainable development and climate change adaptation in order to prevent the consequences of desertification. In order to achieve strong sustainability, environmental sustainability has always been a crucial component of sustainable development. Creating a reasonable index can point policymakers in the proper direction for environmental sustainability. Of course, ecological security is a crucial sign as well. Ecological security has become a crucial policy concern on a global scale. The use of quantitative metrics to evaluate government performance on a variety of environmental protection and natural resource management issues is becoming more common. The ecological security index can make it simpler for us to discover issues, monitor trends, pinpoint best practices, and maximize the benefits from ecological civilization's constructions.

Since oxygen is necessary for all living things to survive, the amount of it directly affects the biomass of the planet. Reduced oxygen levels during historical eras have been linked to a number of historical mass extinctions. We have reached the threshold of one-third of the 9 planetary boundaries suggested by Rockström *et al.* (2009b) in the past 100 years due to frequent human intervention in the use of various resources on earth, including freshwater resources. Additionally, excessive use of fossil fuels has increased carbon dioxide emissions and human oxygen consumption to record levels, which has warmed the planet generally due to the greenhouse effect. The amount of oxygen produced by the earth's flora gradually reduced during the Anthropocene as a result of the growth of metropolitan areas and the acceleration of desertification. The world in which we live is in an unsustainable state as a result of rising oxygen use and falling oxygen generation. In order to assess past and future ecological security, an index of ecological security (ESI) was built, which includes biological, oxygen, carbon, thermal and hydrological cycles. Arid and semi-arid regions have experienced a catastrophic loss in ecological security over the past 60 years, which has spread to neighboring regions. It is anticipated that ecological security will respond strongly to human activity and global warming. Under a scenario with significant greenhouse gas emissions, the amount of insecure land will account for more than 57% of all land by about 2100.

Geoscientists can undoubtedly use the newly proposed ESI to reevaluate the vitality of land with various ecological statuses, particularly the emergence and spread of unstable land in the arid and semi-arid regions, and use the GDVI to identify desertification vulnerability. The systematic classification of land using ESI can warn us about both the overt threat from anthropogenic insecure land, which has largely gone unnoticed globally due to human activities motivated by avarice, and the concealed threat from climate crises caused by naturally insecure land. Extreme weather events and human disturbances, such as accelerating urbanization and population increase, have a growing impact on ecosystem stability as a result of climate change. The results show that safe land would become unsustainable and the risk of ecological security would occur when the external pressure is greater than the environmental capacity. According to the population forecasts made under SSP5 (RCP8.5), there will be 12.7 billion people on the planet by the year 2100, meaning that more people will be at risk from severe ecological security. Because it can further prevent unsecure land from joining altogether and slow the anticipated losses of biological species richness, identifying the beginning point of insecure land at an earlier stage and preventing its invasion are of particular relevance and a high priority. Therefore, more urgent measures must be taken to encourage oxygen generation and protect water resources, such as increasing the number of trees planted, reducing the growth of deserts and arid and semi-arid regions to prevent the spread of dust (Shugart *et al.*, 2003), and cultivating new green space among vast areas of unstable land. Moreover, we must prevent additional activities that use more oxygen and water, particularly by restricting the combustion of fossil fuels, banning garbage in the ocean and lakes, properly disposing of municipal and industrial waste, and dispersing industrial zones equally. More land will undoubtedly turn into desert and unstable land if we continue to take land resources for granted; this final stage is irreversible and could result in humanity losing its habitat permanently. Therefore, exchanging genuine land resources for quick economic development is foolish and lacks forethought. Without hesitation, we must adopt new policies to prevent the environment from getting any worse.

Numerous researchers have predicted that the severe environment of the desert, which is extensively scattered throughout the arid and semi-arid regions, may store a significant amount of CO_2. The second-largest shifting desert in the world, benefiting from the plentiful saline–alkali groundwater and considerable salinization, TD has a significant capacity to store carbon. The amount that each internal function contributes to the TD carbon sink and how these systems will adapt to climate change are both unknowns, though. For each part of the moving sand in the TD, we calculated the precise magnitude and process of the CO_2 fluxes in this chapter. We specifically considered the frequently overlooked processes of soil air expansion/contraction and the exaggerated impact of soil moisture on the CO_2 flux in a dry climate. Additionally, it was confirmed that saline/alkaline elements in the desert absorb CO_2. Finally, we discovered that from 2004 to 2017, TD's shifting sand had a yearly CO_2 uptake rate of 7.11 g m^{-2} a^{-1}. This makes it a stable carbon sink today. If all the world's shifting deserts are considered, it is impossible to deny how important desert ecosystems are to the global carbon cycle. These findings might contribute to filling in some of the gaps in the global carbon cycle's carbon sink. However, CMIP5 simulations show that the carbon sink contribution of TD's shifting sand will decline in the future due to overstimulation of the process of soil air expansion, which is driven by temperature and contributes to CO_2 release.

The contribution of the CO_2 fluxes of different components in TD's shifting sand to soil respiration was quantified based on the disassembly experiment. As far as we are aware, this is the first time that the unusual phenomenon of sand supporting a stable CO_2 exchange in the study of soil respiration throughout the TD region has ever been documented. We created upcoming temperature-controlled tests and offered ideas for the mechanisms impacting the diurnal pattern of sand CO_2 flux because there is now no good explanation for this stimulating action of sand. Through observational studies, we infer that the expansion/contraction of soil air containing CO_2 brought about a CO_2 release/absorption event in sand, which unintentionally aided in soil respiration. The rapid rise or fall in soil temperature stimulated the expansion or contraction of soil air, which led to the rapid release or absorption of CO_2. Additionally, the loose and porous nature of sand permits soil air to expand and contract in response to rapid temperature fluctuations. As a result, the most important factor influencing the CO_2 flux in the sand is soil temperature (Rodeghiero and Cescatti, 2005). The evaporation of deep soil water causes diurnal fluctuation at the same time as the movement of soil heat via shifting of sand. The moisture content of the extremely dry topsoil of shifting sand rises throughout the day due to an increase in the evaporation of deep soil water and the escape of water vapor from the surface.

This encourages salt and alkali to absorb ambient CO_2 and dissolve soil carbonates. In contrast, the carbonate reprecipitation at night encourages the release of CO_2 by salt and alkali (Schlesinger, 2017), which eventually results in salt and alkali levels that strongly encourage the absorption of atmospheric CO_2 in shifting sand during the day and release at night. The absorption and release of CO_2 in shifting sand were primarily influenced by the interaction between this process and the expansion and contraction of soil air. The extremely dry TD area has some slight variations in soil moisture, but these are not enough to counteract the drought's limiting effects on soil respiration; therefore, moisture only makes a very small contribution to soil respiration in shifting sand. The aforementioned scenario has changed how we think about moisture's function in the desert carbon cycle. There are signs of life in the desert, proving that it is not as lifeless as it first appeared, as evidenced by the limited contribution of bacteria to soil respiration, which revealed that microbial activity in the TD is very weak.

With an average annual CO_2 uptake rate of 7.11 g $m^{-2}\cdot a^{-1}$ between 2004 and 2017, TD's shifting sand has shown an amazing carbon sink impact as a result of the superimposition of numerous processes. This could partially explain the absence of the carbon sink in the global carbon cycle. Large volumes of CO_2 will be released into the atmosphere as a result of the decomposition of organic carbon being accelerated by global warming (Giardina *et al.*, 2014). The future thermal expansion of soil air will be substantially stimulated by increases in $T_{0-10\mathrm{cm}}$ owing to climate change, releasing more carbon dioxide from the sand into the atmosphere in desert ecosystems with extremely low organic carbon stocks. Eventually, this will cause the carbon sink rate in the TD moving sand to gradually decline. The rate of CO_2 uptake in TD's shifting sand will decline under RCP4.5 and RCP8.5 by 0.43% and 1.20% per year, respectively. According to RCP8.5, it was anticipated that the shifting sand surface in the TD will achieve a CO_2 absorption/release neutral state around 2100. This may have significant consequences on the welfare of society and would enable a positive feedback effect under climate change, generating accelerated heat in arid and semi-arid regions (Huang *et al.*, 2008a, 2016a, 2018). Desert ecosystems, which have historically received little attention, have always been significant carbon sinks in the background. The need for developing climate change countermeasures will nevertheless become more critical as deserts' capacity to act as carbon sinks gradually declines.

It goes without saying that if we do nothing to slow down climate change, by the end of the 21st century, more CO_2 will be released from the land into the atmosphere under the RCP8.5 scenario, and more areas will experience high desertification vulnerability based on GDVI, ESI projections, and the carbon sink capacity of deserts. Arid and semi-arid regions have a significant capability for penitential C sequestration, which allows them to absorb CO_2 produced by human activity. To stop the spread of desertification and restore degraded land, people have devised a corpus of acts, including biological and mechanical measures. Around the world, a wide range of initiatives and actions can aid in halting and reversing deterioration while also restoring soil C stocks and mitigating climate change. The physical, chemical, biological, and ecological features and activities of the soil, as well as its general health, are all significantly impacted by restoring the soil C stock. To increase SOC/SIC stocks, it is essential to implement solutions for sustainable land management in arid and semi-arid regions. Arid and semi-arid region forestation is a useful tactic to slow the rate of increase in atmospheric CO_2 concentration caused by human activity, which will help mitigate climate change and benefit the environment. However, because the reduced albedo of the forests might have significant warming impacts, the real climatic benefits of forested areas across the vast worldwide arid and semi-arid regions are questionable. However, there are other study areas for arid and semi-arid region afforestation that need to be looked at, including the choice of plant species, how to enhance nursery and planting practices, and the reactions and feedbacks of arid and semi-arid regions forests to climate change.

People should react positively and take good care of our earthly environment in order to protect our home. Raising people's awareness of the dangers of desertification will help prevent global desertification. Future efforts to control desertification will benefit from advancing research on the causes of desertification and its prevention, which will give policymakers more information on which to base their decisions. Although there have been some advancements in desertification studies recently, more research and understanding is still needed to increase efforts to mitigate desertification. International collaboration can be very successful in preserving the environment. The cooperation of nations in regional or global action plans is required as a result of the vast variations in the resources and capacities of different countries.

Environmental reductions are known to be fueled by human activity and climate change. Environmental change challenges to biodiversity, such as the effects of habitat fragmentation, overexploitation, invasive species, and pollution, may be much more serious than those we anticipate (Tilman *et al.*,

2017; Foden *et al.*, 2019; Newbold 2018). Additionally, over the past ten years, storm surges that were more frequent and intense, waves that overtook at least some of an island due to rising sea levels, and other factors all contributed to a major loss of habitat and perhaps even the direct extinction of some animals. The Bramble Cay melomys (Melomys rubicola) became the first modern animal to go extinct due to climate change in 2016 (Woinarski and Burbidge 2016). Even though human actions outweigh climate change as the primary cause of the loss of biodiversity worldwide now, this will likely alter in the ensuing decades. We put forth the idea of "land cancer" and create a method for identifying cancerous land by considering the Earth as a living entity. Earth cancer begins in arid and semi-arid regions where oxygen production is lowest and in metropolitan areas and heavily industrialized places where oxygen consumption is highest (Fig. 9.25). These areas are regarded as key contributors to the ongoing loss of biodiversity, and their impact on biodiversity is anticipated to grow in this century (Visconti *et al.*, 2016; Cardinale *et al.*, 2012).

Biodiversity is the basis for human survival on Earth, but the evidence is clear; we are destroying biodiversity at an unprecedented rate (Cardinale, *et al.*, 2012; Rodrigues, *et al.*, 2014). According to the International Union for the Conservation of Nature, one-fifth of terrestrial vertebrates worldwide are considered vulnerable, endangered, or critically endangered. Numerous studies have revealed that by 2050, agricultural growth will cause the habitats of 87.7% of these species to be lost, with 1,280 species estimated to lose ≥25% of their habitats (Williams *et al.*, 2021). By 2050, 27 European large mammal species are predicted to lose 25% of their habitats

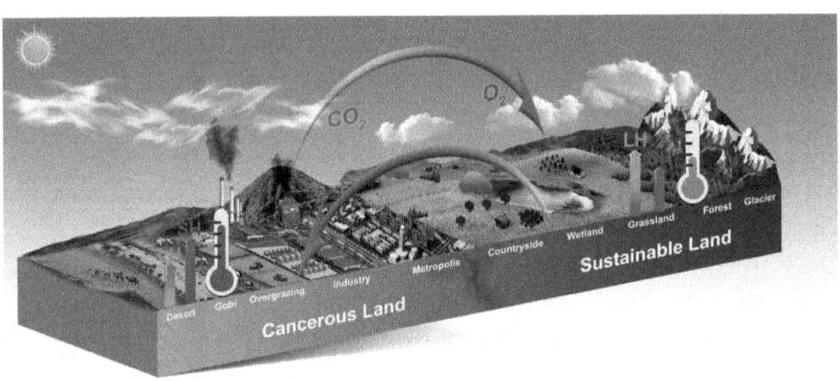

Fig. 9.25. Schematic diagram of the oxygen cycle between the land types of healthy land and cancerous land.

(Rondinini and Visconti, 2015) due to human habitat encroachment, and 440 mammalian carnivore and ungulate species are predicted to experience a decline in abundance by 18%–35% and an increase in extinction risk by 8%–23% by 2050 (Visconti *et al.*, 2016). Since 1998, the World Wide Fund for Nature has continuously produced the Earth Vitality Report to track the most recent status of the world's biodiversity in an effort to lessen the loss of species. The Living Planet Index (LPI), which is based on population trends of vertebrate species from all over the world, is a gauge of the state of the world's biological diversity. Thus far, thirteen issues have been released. The Earth is currently facing enormous challenges, as shown by the LPI, Ecological Footprint, Water Footprint, "Earth Boundary," and other indicators primarily utilized in the Earth Vitality Report. According to the LPI in 2020, the average number of mammals, birds, amphibians, reptiles, and fish declined by 68% (from −73% to −62%) between 1970 and 2016 (WWF *et al.*, 2020). Notably, different locations experience biodiversity loss at varying rates, with tropical areas experiencing the greatest losses. The LPI fell by 94% in the tropical subregion of North and South America, the largest decline in any region.

The main reasons include the expansion of grasslands, savannahs, woodlands, and wetlands; overgrazing, deforestation; climate change; and alien species invasion. This result suggests that to bring biodiversity back to its original level, we should not only avoid its loss but also foster its restoration. This task is not simple. However, if we can understand the interdependence of every component of the ecosystem and act swiftly based on this understanding, we will indeed bring the world back to life and make it healthier and more resilient. We are aware that our current way of life on Earth cannot continue and we must make changes. The most pressing danger to the health of our world comes from us. We have altered this planet so drastically that we have ushered in the Anthropocene, a new geological epoch. The global connections that all living things rely upon are breaking. However, if we move swiftly, we have the knowledge and the answers to restore the health of our world. The world's best problem-solvers have always been people.

Only urgent transformative change that addresses the core causes — the linked economic, sociocultural, demographic, political, institutional, and technical indirect drivers behind the direct drivers — can reverse recent decreases in biodiversity and usher in a sustainable global future. It is also possible to prevent unexpected outcomes by using what has been learned from prior attempts to promote sustainability. To support current efforts, policies to lessen the underlying dangers to biodiversity through proactive national collaboration, such as agricultural growth, will undoubtedly be needed. To stop and reverse the loss of nature by 2030 and to create a culture that values nature, there needs to be a global "new consensus on harmony between humans and nature." Ultimately, this path is our best assurance for human health, livelihood, and a bright future for our children.

Chapter 10

Conclusions and Perspectives

This book is an overview of recent progress in semi-arid climate change research and a systematic summary of the temporal and spatial characteristics of climate change over semi-arid regions. Modern observations have demonstrated that a massive increase in PET and a decrease in precipitation have resulted in a drying trend, which is expected to affect a substantial portion of arid and semi-arid regions (Zhao and Dai, 2015, 2017), leading to an approximately 10% increase in global arid and semi-arid regions (Feng and Fu, 2013). It was also indicated that the drying trend primarily emerged over East Asia, the Sahel in West Africa, Southern Africa, and Eastern Australia beginning in 1950 (Dai and Zhao, 2017; Huang *et al.*, 2016a). According to multi-model forecasts, drying is much more widespread than wetting in the tropics, subtropics, and mid-latitudes because of ubiquitous PET increases (Feng and Fu, 2013; Scheff and Frierson, 2015). Both observations and model simulations have indicated that during the 20th century, precipitation changes played a relatively important role in climate shifts. However, during the 21st century, the projected temperature changes have become increasingly important and will dominate the shifts in climate types as warming intensifies (Feng and Fu, 2013).

Recent studies have indicated that semi-arid regions in the Northern Hemisphere (NH) mid-latitudes contribute to continental warming to an extent of approximately 50% (see Chapter 2), which is particularly notable in Asia. Similarly, climate warming in the past 100 years is evident in Northeast, North, and Northwest China, which are majorly covered by semi-arid regions (IPCC, 2007). Li *et al.* (2015a) also showed that warming has become more serious in semi-arid regions in China, especially in Northeastern China (see Chapter 2). Global warming is non-uniform, and semi-arid regions have witnessed greater warming than others over the last century (see Chapter 3)

(Huang *et al.*, 2012; Ji *et al.*, 2014), especially during the cold season (November to March). Utilizing a recently developed dynamic adjustment method, Guan *et al.* (2015b) investigated the drivers of enhanced semi-arid warming (ESAW) in East Asia and suggested that the ESAW over East Asia is induced by a regional anthropogenic forcing closely related to local human activities (see Chapter 3). Additionally, Zhou *et al.* (2016) replicated the ESAW in historical simulations including both anthropogenic and natural forcings, but the ESAW disappeared when only the natural forcing was considered, revealing new potential footprints for anthropogenic warming. Nonetheless, the major reason for desert amplification is the increase in water vapor (Zhou *et al.*, 2016), which further indicates that water vapor in semi-arid regions is likely controlled by water vapor originating from the ocean or SST. However, in phase 5 of the Coupled Model Intercomparison Project (CMIP5) simulations, the SAT trend is underestimated in semi-arid areas and overestimated in humid areas, indicating that there are great uncertainties in the CMIP5 simulations about how SAT will respond to global warming (Huang *et al.*, 2016b). Furthermore, a warming climate may accelerate the ESAW, leading to significant societal and economic repercussions (see Chapter 3).

The temporal and spatial variations in the semi-arid climate may be attributed to both regional- and global-scale factors (Fig. 10.1) through

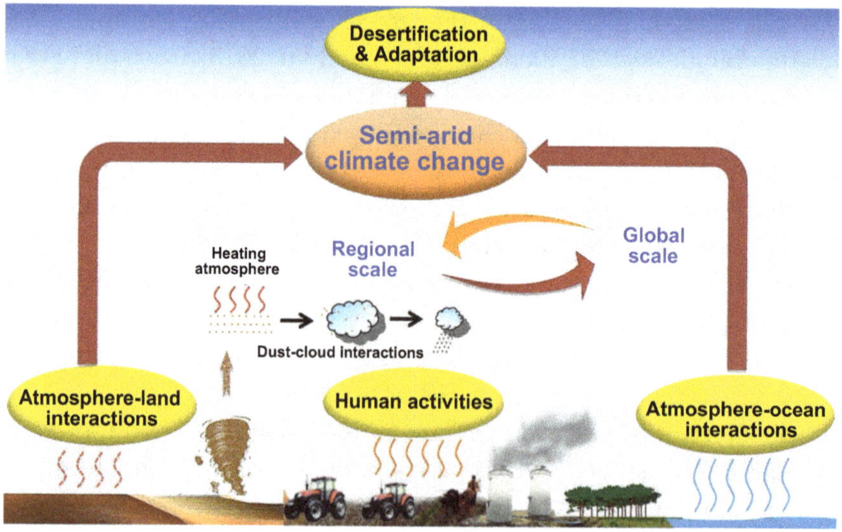

Fig. 10.1. Major drivers of semi-arid climate system (Huang *et al.*, 2017a).

physical processes and dynamic mechanisms, including dust–cloud–precipitation interactions, land–atmosphere interactions, ocean–atmosphere interactions, and human activities. In light of the local–cloud interactions, dust aerosols exert direct, indirect, and semidirect effects on the semi-arid climate (see Chapter 4). These aerosols can absorb solar radiation and subsequently expedite the evaporation of water droplets in lower-atmosphere water clouds, decreasing the cloud's liquid water path. Nevertheless, according to land–atmosphere interactions, the intensity of the response to climate change varies over different regions. In arid, semi-arid, and subhumid regions, drought-related changes to the land surface seem to give rise to a positive feedback loop that prolongs and intensifies droughts (see Chapter 5). The ocean–atmosphere interactions majorly determine the global decadal-to-interdecadal semi-arid climate changes. Additionally, the decadal-to-interdecadal variability in SSTs can alter the overlying atmospheric circulation and influence arid and semi-arid regions through teleconnection (see Chapters 6 and 8). Human beings have dramatically changed the global land surface and atmospheric composition through activities, such as deforestation, agriculture, and the burning of fossil fuels. The long-term change trends, particularly those of temperature and precipitation, will be predominated by human activities (see Chapter 7).

Mineral dust mainly originates from arid and semi-arid regions (Huang *et al.*, 2015) and has a significant impact on the regional climate by altering the energy balance and water cycles through aerosol–cloud–precipitation interactions (see Chapter 4). With increasing global temperature, the large-scale precipitation decrease leads to less wet deposition, thereby augmenting the amount of dust. Dust aerosols can alter the radiative energy of the Earth's surface atmosphere by directly attenuating solar radiation (Huang *et al.*, 2009) and indirectly modifying cloud macrophysical (areal coverage, structure, and altitude) and microphysical properties (droplet size and phase) in various ways (see Chapter 4). Due to greater surface albedo, the direct effects of dust aerosols over arid and semi-arid regions contribute to warming in the atmosphere, cooling at the surface, and warming or no change at the TOA. Asian dust aerosols absorb more solar radiative energy than those over the Sahara. Dust aerosols can evaporate clouds and thereby exaggerate global warming (Huang *et al.*, 2010b). In addition, dust can inhibit or enhance precipitation under certain conditions, which will influence the hydrological cycle. Whether dust speeds up (Huang *et al.*, 2014) or slows down (Miller *et al.*, 2004) desertification or whether it plays an indiscernible role in the expansion of arid and semi-arid regions (Zhao *et al.*, 2015) remains debated. Even though dust is normally associated with arid and semi-arid regions, significant dust events have also occurred at cold, high latitudes. Therefore, it is necessary to examine how high-latitude dust impacts the arid and semi-arid climate in the future.

Land–atmosphere interactions involve complex surface processes that exchange energy, water, and materials between the surface and the atmosphere. Large-scale atmospheric processes as well as long-term processes of land–atmosphere interactions are crucial for understanding climate change and its future implications (see Chapter 5). Different regions' responses to climate change vary in severity based on land–atmosphere interactions. Enhanced semi-arid region warming and decreased AI were identified and amplified by land–atmosphere interactions as a result of climate change responses to the energy cycle, water cycle, and carbon cycle. GPP is the most significant CO_2 flux of the carbon cycle in terrestrial ecosystems, influencing all carbon cycle variables. According to the study discussed in Chapter 5, in a future high-emission scenario, the total semi-arid regions' GPP will rise by 12%±3% by 2100 in comparison to the 2000–2014 baseline. This increase in GPP in semi-arid regions may not result in an increase in global GPP, as semi-arid regions will mostly encroach on historically productive ecosystems. Additionally, if historical semi-arid regions degrade faster than newly developed semi-arid regions, GPP per unit semi-arid region will decline. By reducing soil moisture, enhanced semi-arid regions' warming also exacerbates semi-arid region's expansion, potentially leading more energy to SH than to LH. Soil drying over East Asia is mainly caused by declining precipitation, which is almost doubly accelerated by rising temperatures. As soil moisture declines, future water shortages could be caused by the positive feedback between temperature and soil moisture. The most noticeable drying occurred over transitional zones between dry and wet climates. Beginning in the 1980s, there was obvious drying that started in the humid transitional zones and then spread to the dry transitional regions. Observations of land–atmosphere interactions are also essential. The characteristics of the land surface energy budget and seasonal and diurnal variations in moisture and heat flux over the Loess Plateau have been analyzed using observation data from the SACOL (see Chapter 5).

Global-scale interdecadal aridity changes are dominated by continent-scale atmospheric and oceanic oscillations as well as by pertinent global-scale factors, such as planetary-scale changes in atmospheric circulations (see Chapters 6 and 8), SSTs, and other external forcings (Guan *et al.*, 2015a). The aforementioned influences can be illustrated by two aspects.

First, Hadley cell expansion is essential in the global climate system, especially in subtropical arid and semi-arid regions. Second, ocean–atmosphere interactions regulate how these global-scale factors affect climate change in arid and semi-arid regions with varying regional performances. These interaction effects on the decadal to interdecadal variability in SST in different ocean basins can trigger overlying atmospheric circulation anomalies on the corresponding timescales and subsequently affect the climates over other regions via teleconnection (see Chapter 8). The compositions of different phases of the AMO, PDO, and ENSO (see Chapter 6) will enhance or offset changes in land–sea and north–south thermal contrasts, and as a result, they will affect the intensity of westerlies and planetary waves as well as the blocking frequency, accelerating or mitigating global warming and drying over a large scale (Huang *et al.*, 2017a).

Human activities exert a major influence on climate change over semi-arid regions, including enhanced warming, area expansion, extreme events, water shortages, desertification, biodiversity loss, and other forms of environmental degradation (see Chapter 7). Anthropogenic activities, such as increased GHG emissions, are strongly associated with enhanced warming, increased aridity, and area expansion in semi-arid regions, which may lead to desertification and pose a serious threat to the ecosystems and inhabitants of semi-arid regions, as well as exacerbate local tensions. Furthermore, recent research has revealed that many declining hotspots of terrestrial water storage (Famiglietti, 2004), a system that integrates surface waters, glaciers, canopy water, soil moisture, and groundwater, are concentrated in semi-arid regions around the world. The foremost reason for this phenomenon is that human water use grows with the increase in population and living standards. Large cities in semi-arid regions are also more likely to experience future severe hypoxia, which is directly linked to the decline in oxygen concentration and increased extremely calm weather due to global warming. Unrestricted increases in GHG emissions would result in a hypoxic environment in extremely calm weather, enhanced warming, more frequent heat waves, and water shortages expanding to additional areas near semi-arid regions. With the rapid increase in population, the adverse repercussions of human activities will affect more people around the world, particularly in semi-arid regions. The realization of a smart, sustainable, and healthy Earth is still a long way off (see Chapter 7). As an interdisciplinary subject, the study of climate change has grown rapidly and involves multidisciplinary fields. However, the development of climate change research is hindered by the lack of a theoretical system.

Due to complex climates, relatively high sensitivity to climate change, and the pressing need for regional development in semi-arid regions, the preceding chapters suggest that further investigation into semi-arid climate change is needed, although notable advances have been made in recent studies. Chapter 8 attempts to start with the construction of a theoretical system of semi-arid climate change and explore the dynamic framework of climate change. Semi-arid climate change is the coupling of long-term trends caused by external forcing and interdecadal oscillations regulated by ocean processes. It demonstrates an intensifying trend over a long time, as well as decadal oscillations and noteworthy regional differences in dry and wet changes, which is the result of the abnormal atmospheric circulation affected by the combination of different sea temperature phases. The dynamics of semi-arid climate systems were initially discussed in Chapter 8. In the future, we will construct a theoretical system of semi-arid climate change.

Desertification has attracted much attention as a phenomenon of land degradation in arid and semi-arid regions. Combating desertification expansion is crucial for ensuring ecological security and protecting the environment, and it also improves human welfare. One of the keys to preventing desertification is identifying important areas that are vulnerable to both climate change and human activities. According to the GDVI (see Chapter 9), which considers factors, such as climate, vegetation, and human activities, regions with a high risk of desertification are concentrated in North America, Central Asia, Australia, Northern China, and the edge of the Sahara Desert in Africa. Approximately 14% of the global land area is at high risk for desertification, whereas 42% of the global land area is at low risk. In addition to identifying regions with high desertification vulnerability, it is critical to improve the understanding of the mechanism and feedback during the desertification process. Based on the oxygen cycle, the ecological security assessment index and the notion of global land cancerization are proposed. Aside from losing its capacity to store groundwater and experiencing a decrease in precipitation, cancerous land is unable to absorb CO_2 through light and action to form O_2 vegetation. It will continue to spread to the surrounding land and endanger the surrounding ecological security. According to the results based on the terrestrial ecosystem security index, urbanized areas in humid zones and arid and semi-arid regions are where most global terrestrial ecosystem insecurity is located. Deserts, as a result of desertification, are primarily found in arid and semi-arid regions. Climate change also has an impact on desert ecosystems. These regions are considered to be major drivers of ongoing biodiversity loss, and the 21st century is anticipated to experience an even greater impact from them. Since 1998, the World Wide Fund For Nature (WWF) has continuously released the Earth Vitality Report to track the latest status of global biodiversity in an effort to reduce the loss of species (see Chapter 9).

The core of this book is semi-arid climate change; however, its contents cover all aspects of climate change. Although recent studies have made substantial progress in research on semi-arid climate change, it still needs to be further studied due to the complexity of the climates in these regions, their relative vulnerability to climate change, and the pressing need for regional development. Future research should focus on the following aspects:

(1) *Identifying the temporal and spatial characteristics of semi-arid climate change on different timescales*: Most researchers have focused on specific types of arid and semi-arid regions, while few have explored the temporal and spatial correlations of semi-arid climate change in different semi-arid regions around the world in a comprehensive way. In the future, multisource data, such as climate reconstruction data and modern observational data, should be combined in more studies to investigate the spatial variation and spatiotemporal correlation of semi-arid climate change as well as to create a complete picture of dry–wet oscillations on multiple timescales.

(2) *Detecting major factors in wetting/drying semi-arid regions*: Based on previous results, the time series and spatial distribution of semi-arid regions exhibit regional variability, and the eight typical global semi-arid areas are inconsistent. They also have oscillation changes, which reflect the reverse between wet and dry conditions. The causes of semi-arid regions' change are complicated, with one or more significant causes at specific timescales and spatial regions. Therefore, future work on detecting major factors affecting wetting/drying in semi-arid regions requires more investigation. Moreover, the transition zone between semi-arid and semi-humid regions is a key area for studying human activities.

(3) *Investigating the cycle of energy and hydrology*: Under enhanced warming in semi-arid regions, the energy and hydrology cycle becomes faster, and these changes have not been fully explored or considered. Both local and global circulations show increased energy and hydrology rates. This variability results from both natural and anthropogenic effects; while previous works have studied and discussed natural and anthropogenic effects, new findings show that natural oscillations have taken on new features, which have not been fully considered in previous research. Additionally, the anthropogenic

effect does not occur in only one way; it has direct, indirect, and semidirect effects. This aspect of the anthropogenic effect on climate change over semi-arid regions should receive more attention.

(4) *Studying land–atmosphere interactions over semi-arid regions by combining ground-based measurements and satellite observations*: High-quality datasets of ground-based measurements are required to correct satellite remote sensing retrievals and construct a time series of key variables, including the surface radiation budget, sensible heat flux, PET, surface temperature, and soil moisture. It is necessary to support specialized multi-scale data analysis experiments that investigate the interscale relationships between point measurements on the ground, such as skyscrapers that bridge the distance between the surface layer and airborne measurements, and observations retrieved from satellites at different scales and resolutions. Gridded datasets of key parameters for land–atmosphere interactions must be created by every data source. In addition, we should study the interaction between human activities and land–atmosphere systems, especially the cycling of energy, water, and materials in different arid and semi-arid regions. "Big data" can be used because of increasing ground-based observations, improved satellite retrieval products, climatic proxies, increasingly reasonable parameterization schemes for modeling, and the advancement of data assimilation techniques.

(5) *Examining the effects of ocean oscillations on extreme events*: The ocean has changed significantly over the last few decades, not only in terms of SST but also in terms of heat, oxygen, salinity, intensity, and other factors. These changes greatly affect both internal and external ocean activities, particularly the interaction between the atmosphere and the ocean. As the period and amplitude of typical ocean oscillations change, their single effect and combined effects on semi-arid regions also change. Except for the ENSO, PDO, AMO, and monsoon, the effect of the IOD has become more prominent and visible on the different timescales of climate change over semi-arid regions.

(6) *Understanding the dynamic mechanisms of semi-arid climate change*: Researchers should examine the dynamic mechanisms underlying the impacts of changes in atmospheric circulation on arid and semi-arid region climates by combining diagnostic analyses and numerical simulations. It is indispensable to explore how essential elements, such as clouds, radiation, precipitation, and evaporation, respond to changes in atmospheric circulation in arid and semi-arid regions and how the feedbacks function.

Further evaluations should be performed to determine the relative contributions of natural forcings (e.g., solar activities and volcanic eruptions) and human-induced perturbations (e.g., changes in GHG emissions, aerosols, and land use/land cover) to arid and semi-arid region climate change.

(7) *Quantifying feedback between socioeconomic and biogeochemical dynamics*: The carbon, nutrient, and water balances of vegetation and soil are either directly or indirectly influenced by the fluxes of various gases and particulate matter between the terrestrial biota and the atmosphere. To effectively implement adaptive management strategies, natural resource managers must have a thorough understanding of how changing climatic circumstances impact plant and animal phenology. To provide practical solutions and adaptation options for a variety of problems, such as land management, wildlife, and habitat conservation, quantification of feedback between land and atmosphere under the effect of social economy and natural climate change is needed.

(8) *Exploring the response of terrestrial aridity to aerosols*: Aerosol decreases relative to current levels are anticipated to result in an additional warming of the system in the 21st century, and could exacerbate the effects of GHG-induced warming on the climate. In addition, there are numerous ways in which various aerosol particles might affect the climate. For example, BC decreases the solar radiation that reaches the surface, resulting in a reduction in PET. However, BC-induced surface warming tends to increase PET. Therefore, more research is required to determine how various aerosol particles affect the semi-arid climate.

(9) *Significant impact of land use/cover change (LUCC) on the regional climate and environment*: LUCC impacts the environment by affecting the interchange of energy, momentum, and moisture exchange between the atmosphere and land surface. We need to focus on understanding the processes and mechanisms that control the interactions among LUCC, regional climate, and socio-ecosystems over semi-arid regions. In addition, it is still unclear how much drought is affected by humans and whether there will be any negative impact on society. Since it is unclear how human activities and drought interact with one another, drought management is ineffective. In this human-influenced era, we must reframe the concept of drought to reflect human contributions to drought mitigation and enhancement. Generally, semi-arid regions should be the focus of more attention because they are the most vulnerable to GHG-induced climate change and other human activities.

(10) *Improving the capability of climate models to simulate aridity*: The P/PET used in this book is only a measurement to assess aridity changes, which has some uncertainties. On the other hand, local land aridity changes are extremely reliant on the variables chosen. It is anticipated that a significant portion of global land areas will experience both increased vegetation growth and decreased soil moisture (Mankin *et al.*, 2017a, 2017b). Therefore, it is difficult to utilize one index to accurately capture all features of aridity worldwide. The latest climate models are inadequate for simulating and predicting precipitation, which is a barrier to studying variations and forecasting aridity. We should improve the precision of aridity forecasts.

(11) *Recognizing regional impacts and developing adaptive strategies*: For climate disaster management, a practical and quantitative index of the vulnerability of a nature–society system should be developed to predict the regional implications of climate hazards. In particular, it is vital to identify a catastrophic system threshold to provide an early warning for disasters, to act proactively, and finally to lessen their impacts. This process may be a central part of adaptive strategies, where China has set an excellent example.

(12) *Raising people's awareness of environmental protection and strengthening international cooperation*: To slow climate change, it is vital to absorb and fix the carbon dioxide production by human activities. Arid and semi-arid regions are widespread, and ecological projects, such as afforestation in the process of desertification control, can not only slow the expansion of desertification but also help to absorb carbon dioxide from the atmosphere and increase the capacity of terrestrial ecosystems to fix carbon. However, their potential contribution still needs to be further investigated. We can improve the assessment of the carbon sequestration capacity of different ecosystems in arid and semi-arid regions, and we may increase the terrestrial carbon sequestration capacity in the process of desertification control, contributing to the goal of climate change mitigation and sustainable development.

From 1978 to 2021, China implemented 14 successive conservation and restoration projects, including the "Green Great Wall" and "Converting Farmland to Forest," and proposed 8 ecological civilization concepts. The 14 projects can be categorized into restoration and ecological projects, which are intended to curb soil erosion and preserve biodiversity, particularly to safeguard arid and semi-arid region ecosystems and combat desertification. The 14 ecological conservation and restoration projects have altered land-use patterns through afforestation, forest reservation, aerial seeding, and desert renewal, which have exerted a significant impact on the ecosystems of semi-arid regions in China. Then, in 2020, China launched the Master Plan for Major Projects to Protect and Restore China's Major Ecosystems (2021–2035).

Due to large-scale ecological protection and restoration projects, China's arid and semi-arid areas are generally distinguished by an increase in grassland ecosystems and a decrease in desert ecosystems. The changing trends in woodland and grassland areas have coincided with the implementation of ecological restoration projects aimed at expanding vegetation coverage. Ecological protection and restoration projects initiated in 1999, such as "Converting Farmland to Forest," have substantially reduced soil erosion and land degradation in arid and semi-arid areas of China. This project has made substantial progress in land use and vegetation coverage by converting farmland with a slope of 25° and above into woodland, shrubland, or grassland.

As the transitional zone between arid and humid regions, semi-arid regions are ecological security barriers that prevent arid regions from expanding to desertification and ensure the development of human society. This book provides an overview of recent progress in semi-arid climate change research. Those results demonstrate that global climate governance requires concerted efforts from all countries. The COVID-19 pandemic, which seriously threatened human society for three years, reminds us of the significance of harmonious coexistence between human beings and nature, as life and health are closely linked to the natural environment. After COVID-19, people have become more aware of the significance of the relationship between human beings and nature. Climate change is a challenge for the whole of humanity. With the full implementation of the *Paris Agreement*, global climate governance is about to enter a new phase. Coping with global warming needs positive communication and coordination among major countries, and it requires global joint promotion to build a community of human destiny. The epidemic can be stopped by vaccines, but people cannot be immune to the global crisis caused by global warming. Only through the concerted efforts of all countries and scientific responses can we build a better world of cleanliness, beauty, common prosperity, and shared destiny.

Bibliography

Abrol, V. and Sharma, P. (2012). *Resource Management for Sustainable Agriculture*, (IntechOpen, London).

Adhikari, P. and Mejia, J. F. (2021). Influence of aerosols on clouds, precipitation and freezing level height over the foothills of the Himalayas during the Indian summer monsoon, *Clim. Dyn.*, 57, pp. 395–413.

AghaKouchak, A., Chiang, F., Huning, L., *et al.* (2020). Climate extremes and compound hazards in a warming world, *Annu. Rev. Earth Planet. Sci.*, 48, pp. 519–548.

Ahlström, A., Raupach, M. R., Schurgers, G., *et al.* (2015). The dominant role of semi-arid ecosystems in the trend and variability of the land CO2 sink, *Science*, 348, pp. 895–899.

Ajai, Arya, A. S., Dhinwa, P. S., *et al.* (2009). Desertification/land degradation status mapping of India, *Curr. Sci.*, 97, pp. 1478–1483.

Albrecht, B. A. (1989). Aerosols, cloud microphysics, and fractional cloudiness, *Science*, 245, pp. 1227–1230.

Alexander, L. V., Zhang, X., Peterson, T. C., *et al.* (2006). Global observed changes in daily climate extremes of temperature and precipitation, *J. Geophys. Res-Atmos.*, 111.

Alexander, M. (2013). Extratropical air–sea interaction, sea surface temperature variability, and the Pacific decadal oscillation, In: Sun, D. and Bryan, F. (eds.), *Climate Dynamics: Why Does Climate Vary?* (American Geophysical Union, Washington) pp. 123–148.

An, L., Wang, J., Huang, J., *et al.* (2021). Divergent causes of terrestrial water storage decline between drylands and humid regions globally, *Geophys. Res. Lett.*, 48.

An, Z., Kutzbach, J. E., Prell, W. L., *et al.* (2001). Evolution of Asian monsoons and phased uplift of the Himalaya-Tibetan plateau since Late Miocene times, *Nature*, 411, pp. 62–66.

Anderegg, W., Schwalm, C., Biondi, F., *et al.* (2015). Pervasive drought legacies in forest ecosystems and their implications for carbon cycle models, *Science*, 349, pp. 528–532.

Anderegg, W. R. L., Trugman, A. T., Bowling, D. R., *et al.* (2019). Plant functional traits and climate influence drought intensification and land-atmosphere feedbacks, *Proc. Natl. Acad. Sci. USA*, 116, pp. 14071–14076.

Andreae, M. O., Rosenfeld, D., Artaxo, P., *et al.* (2004). Smoking rain clouds over the Amazon. *Science*, 303, pp. 1337–1342.

Angeler, D. G., Allen, C. R., Birgé, H. E., *et al.* (2014). Assessing and managing freshwater ecosystems vulnerable to environmental change, *Ambio*, 43, pp. 113–129.

Armah, F. A., Odoi, J. O., Yengoh, G. T., *et al.* (2010). Food security and climate change in drought-sensitive savanna zones of Ghana, *Mitig. Adapt. Strateg. Glob. Change*, 16, pp. 291–306.

Arnone Iii, J. A., Verburg, J. P., Johnson, D., *et al.* (2008). Prolonged suppression of ecosystem carbon dioxide uptake after an anomalously warm year, *Nature*, 455, pp. 383–386.

Arya, A. S., Ajai, Dhinwa, P. S., Pathan, S. K., *et al.* (2009). Desertification/land degradation status mapping of India, *Current Science*, 97, pp. 1478–1483.

Ashok, K., Behera, S. K., Rao, S. A., *et al.* (2007). El Niño Modoki and its possible teleconnection, *J. Geophys. Res.*, 112.

Avila, F. B., Pitman, A. J., Donat, M. G., *et al.* (2012). Climate model simulated changes in temperature extremes due to land cover change, *J. Geophys. Res.*, 117.

Baldocchi, D. D., Vogel, C. A. and Hall, B. (1997). Seasonal variation of energy and water vapor exchange rates above and below a boreal jack pine forest canopy, *J. Geophys. Res.*, 102, pp. 28939–28951.

Ball, B. A., Virginia, R. A., Barrett, J. E., *et al.* (2009). Interactions between physical and biotic factors influence CO_2 flux in Antarctic dry valley soils, *Soil Biol. Biochem.*, 41, pp. 1510–1517.

Ballantyne, A., Smith, W., Anderegg, W., *et al.* (2017), Accelerating net terrestrial carbon uptake during the warming hiatus due to reduced respiration, *Nat. Clim. Change*, 7, pp. 148–152.

Barnett, J. and O'Neill, S. (2010). Maladaptation, *Global Environ. Change*, 20, pp. 211–213.

Barriopedro, D., García-Herrera, R., Lupo, A. R., *et al.* (2006). Climatology of Northern Hemisphere blocking, *J. Clim.*, 19, pp. 1042–1063.

Barsugli, J. J. and Battisti, D. S. (1998). The basic effects of atmosphere-ocean thermal coupling on midlatitude variability, *J. Atmos. Sci.*, 55, pp. 477–493.

Bauer, J., Herbst, M., Huisman, J. A., *et al.* (2008). Sensitivity of simulated soil heterotrophic respiration to temperature and moisture reduction functions, *Geoderma*, 145, pp. 17–27.

Becerril-Piña, R., Mastachi-Loza, C. A., González-Sosa, E., *et al.* (2015). Assessing desertification risk in the semi-arid highlands of central Mexico, *J. Arid Environ.*, 120, pp. 4–13.

Beer, C., Reichstein, M., Tomelleri, E., *et al.* (2010). Terrestrial gross carbon dioxide uptake: Global distribution and covariation with climate, *Science*, 329, pp. 834–838.

Benzi, R., Malguzzi P., Speranza, A., *et al.* (1986). The statistical properties of general atmospheric circulation: Observational evidence and a minimal theory of bimodality, *Q. J. R. Meteorol. Soc.*, 112, pp. 661–674.

Berdugo, M., Delgado-Baquerizo, M., Soliveres, S., *et al.* (2020). Global ecosystem thresholds driven by aridity, *Science*, 367, pp. 787–790.

Berg, A., Findell, K., Lintner, B., *et al.* (2016). Land–atmosphere feedbacks amplify aridity increase over land under global warming, *Nat. Clim. Change*, 6, pp. 869–874.

Berg, A., Sheffield, J. and Milly, P. C. D. (2017). Divergent surface and total soil moisture projections under global warming, *Geophys. Res. Lett.*, 44, pp. 236–244.

Bestelmeyer, B. T., Okin, G. S., Duniway, M. C., *et al.* (2015). Desertification, land use, and the transformation of global drylands, *Front. Ecol. Environ.*, 13, pp. 28–36.

Bi, J., Huang, J., Fu, Q., *et al.* (2013). Field measurement of clear-sky solar irradiance in Badain Jaran Desert of Northwestern China, *J. Quant. Spectrosc. Radiat. Transfer*, 122, pp. 194–207.

Bian, J., Li, D., Bai, Z., *et al.* (2020). Transport of Asian surface pollutants to the global stratosphere from the Tibetan Plateau region during the Asian summer monsoon, *Natl. Sci. Rev.*, 7, pp. 516–533.

Bierkens, M. F. P. and Wada, Y. (2019). Non-renewable groundwater use and groundwater depletion: A review, *Environ. Res. Lett.*, 14.

Bishop, J., Davis, R. E. and Sherman, J. T. (2002). Robotic observations of dust storm enhancement of carbon biomass in the North Pacific, *Science*, 298, pp. 817–821.

Bodas-Salcedo, A., Hill, P. G., Furtado, K., *et al.* (2016). Large contribution of supercooled liquid clouds to the solar radiation budget of the southern ocean, *J. Clim.*, 29, pp. 4213–4227.

Bond, T. C., Doherty, S. J., Fahey, D. W., *et al.* (2013). Bounding the role of black carbon in the climate system: A scientific assessment, *J. Geophys. Res. Atmos.*, 118, pp. 5380–5552.

Bond-Lamberty, B. (2018). New techniques and data for understanding the global soil respiration flux, *Earths Future*, 6, pp. 1176–1180.

Bopp, L., Resplandy, L., Orr, J. C., *et al.* (2013). Multiple stressors of ocean ecosystems in the 21st century: Projections with CMIP5 models, *Biogeosciences*, 10, pp. 6225–6245.

Branstator, G. (2002). Circumglobal teleconnections, the jet stream waveguide, and the North Atlantic oscillation, *J. Clim.*, 15, pp. 1893–1910.

Burba, G. G., Verma, S. B. and Kim, J. (1999). Surface energy fluxes of *Phragmites australis* in a prairie wetland, *Agric. For. Meteorol.*, 94, pp. 31–51.

Cai, W., Cowan, T. and Thatcher M. (2012). Rainfall reductions over Southern Hemisphere semi-arid regions: The role of subtropical dry zone expansion, *Sci. Rep.*, 2.

Cai, W., Wan, L., Jiang, Y., *et al.* (2015). Short-lived buildings in China: Impacts on water, energy, and carbon emissions, *Environ. Sci. Technol.*, 49, pp. 13921–13928.

Caminade, C. and Terray, L. (2010). Twentieth century Sahel rainfall variability as simulated by the ARPEGE AGCM, and future changes, *Clim. Dyn.*, 35, pp. 75–94.

Cao, H., Mitchell, J. F. B. and Lavery, J. R. (1992). Simulated diurnal range and variability of surface temperature in a global climate model for present and doubled CO_2 climates, *J. Clim.*, 5, pp. 920–943.

Cardinale, B. J., Duffy, J. E., Gonzalez, A., *et al.* (2012). Biodiversity loss and its impact on humanity, *Nature*, 486, pp. 59–67.

Ceballos, G., Ehrlich, P. R., Barnosky, A. D., *et al.* (2015). Accelerated modern human–induced species losses: Entering the sixth mass extinction, *Sci. Adv.*, 1.

Cesana, G., Chepfer, H., Winker, D., *et al.* (2016). Using in situ airborne measurements to evaluate three cloud phase products derived from CALIPSO, *J. Geophys. Res.-Atmos.*, 121, pp. 5788–5808.

Chang, L. L., Yuan, R., Gupta, H. V., *et al.* (2020). Why is the terrestrial water storage in dryland regions declining? A perspective based on gravity recovery and climate experiment satellite observations and Noah land surface model with multiparameterization schemes model simulations, *Water Resour. Res.*, 56.

Chapman, S., Watson, J. E. M., Salazar, A., *et al.* (2017). The impact of urbanization and climate change on urban temperatures: A systematic review, *Landscape Ecol.*, 32, pp. 1921–1935.

Charney, J. G. and Stern, M. (1962). On the stability of internal baroclinic jets in a rotating atmosphere, *J. Atmos. Sci.*, 19, pp. 159–172.

Charney, J. G. (1975). Dynamics of deserts and drought in the Sahel, *Q. J. R. Meteorol. Soc.*, 101, pp. 193–202.

Charney, J. G. and DeVore, J. G. (1979). Multiple flow equilibria in the atmosphere and blocking, *J. Atmos. Sci.*, 36, pp. 1205–1216.

Charney, J. G. and Straus, D. M. (1980). Form-drag instability, multiple equilibria and propagating planetary waves in baroclinic, orographically forced, planetary wave systems, *J. Atmos. Sci.*, 37, pp. 1157–1176.

Chen, C., Park, T., Wang, X., *et al.* (2019a). China and India lead in greening of the world through land-use management, *Nat Sustain*, 2, pp. 122–129.

Chen, F., Chen, J., Huang, W., *et al.* (2019b). Westerlies Asia and monsoonal Asia: Spatiotemporal differences in climate change and possible mechanisms on decadal to sub-orbital timescales, *Earth-Sci. Rev.*, 192, pp. 337–354.

Chen, G. and Huang, R. (2012). Excitation mechanisms of the teleconnection patterns affecting the July precipitation in northwest China, *J. Clim.*, 25, pp. 7834–7851.

Chen, H., Yu, B., Zhou, B., *et al.* (2020). Role of local atmospheric forcing and land–atmosphere interaction in recent land surface warming in the mid-latitudes over East Asia, *J. Clim.*, 33, pp. 2295–2309.

Chen, M., Xie, P., Janowiak, J. E., *et al.* (2002a). Global land precipitation: A 50-yr monthly analysis based on gauge observations, *J. Hydrometeorol.*, 3, pp. 249–266.

Chen, S., Jiang, N., Huang, J., *et al.* (2018). Quantifying contributions of natural and anthropogenic dust emission from different climatic regions, *Atmos. Environ.*, 191, pp. 94–104.

Chen, T. and Chen, X. (2007). Variation of diurnal temperature range in China in the past 50 years, *Plateau Meteorol.*, 26, pp. 150–157 (in Chinese).

Chen, W., Graf, H. F. and Takahashi, M. (2002b). Observed interannual oscillations of planetary wave forcing in the Northern Hemisphere winter, *Geophys. Res. Lett.*, 29, pp. 30-1–34-4.

Chen, W., Yang, S. and Huang, R. (2005). Relationship between stationary planetary wave activity and the East Asian winter monsoon. *J. Geophys. Res.*, 110.

Chen, X., Wang, W. F., Luo, G. P., *et al.* (2014). Can soil respiration estimate neglect the contribution of abiotic exchange, *J. Arid Land*, 6, pp. 129–135.

Chen, X. and Tung, K.-K. (2018). Global surface warming enhanced by weak Atlantic overturning circulation, *Nature*, 559, pp. 387–391.

Chen, Y. and Tang, H. (2005). Desertification in north China: Background, anthropogenic impacts and failures in combating it, *Land Degrad. Dev.*, 16, pp. 367–376.

Cheng, S., Guan, X., Huang, J., *et al.* (2015). Long-term trend and variability of soil moisture over East Asia, *J. Geophys. Res. Atmos.*, 120, pp. 8658–8670.

Cheng, S. and Huang, J. (2016). Enhanced soil moisture drying in transitional regions under a warming climate, *J. Geophys. Res. Atmos.*, 121, pp. 2542–2555.

Cheng, S., Huang, J., Ji., F., *et al.* (2017). Uncertainties of soil moisture in historical simulations and future projections, *J. Geophys. Res. Atmos.*, 122, pp. 2239–2253.

Cherlet, M., Hutchinson, C., Reynolds, J., *et al.* (2018). *World Atlas of Desertification*, (Publication Office of the European Union, Luxembourg).

Chin, M., Diehl, T., Ginoux, P., *et al.* (2007). Intercontinental transport of pollution and dust aerosols: Implications for regional air quality, *Atmos. Chem. Phys.*, 7, 5501–5517.

Choi, Y. S., Lindzen, R. S., Ho, C. H., *et al.* (2010). Space observations of cold-cloud phase change, *Proc. Natl. Acad. Sci. USA*, 107, pp. 11211–11216.

Chou, C., Neelin, J., Chen, C., *et al.* (2009). Evaluating the "rich-getricher" mechanism in tropical precipitation change under global warming, *J. Clim.*, 22, pp. 1982–2004.

Claquin, T., Schulz, M., Balkanski, Y., *et al.* (1998). Uncertainties in assessing radiative forcing by mineral dust, *Tellus, Ser. B*, 50, pp. 491–505.

Cohen, J. L., Furtado, J. C., Barlow, M., *et al.* (2012). Asymmetric seasonal temperature trends, *Geophys. Res. Lett.*, 39.

Cohen, J. L., Screen, J. A., Furtado, J. C., *et al.* (2014). Recent Arctic amplification and extreme mid-latitude weather, *Nat. Geosci.*, 7, pp. 627–637.

Cook, B. I., Smerdon, J. E., Seager, R., *et al.* (2014). Global warming and 21st century drying, *Clim. Dyn.*, 43, pp. 2607–2627.

Cook, B. I., Ault, T. R., Smerdon, J. E. (2015). Unprecedented 21st century drought risk in the American Southwest and Central Plains, *Sci. Adv.*, 1.

Cook, B. I., Cook, E. R., Smerdon, J. E., *et al.* (2016). North American megadroughts in the common era: Reconstructions and simulations, *Wiley Clim. Change*, 7, pp. 411–432.

Cook, J. and Highwood, E. J. (2004). Climate response to tropospheric absorbing aerosols in an intermediate general circulation model, *Q. J. R. Meteorol. Soc.*, 130, pp. 175–191.

Costa, M. J., Sohn, B. J., Levizzani, V., *et al.* (2006). Radiative forcing of Asian dust determined from the synergized GOME and GMS satellite data a case study, *J. Meteorol. Soc. Jpn.*, 84, pp. 85–95.

Cuezva, S., Fernandez-Cortes, A., Benavente, D., *et al.* (2011). Short-term CO_2(g) exchange between a shallow karstic cavity and the external atmosphere during summer: Role of the surface soil layer, *Atmos. Environ.*, 45, pp. 1418–1427.

Cui, J., Yang, H., Huntingford, C., *et al.* (2021). Vegetation response to rising CO_2 amplifies contrasts in water resources between global wet and dry land areas, *Geophys. Res. Lett.*, 48.

D'Andrea, F., Tibaldi, S., Blackburn, M., *et al.* (1998). Northern Hemisphere atmospheric blocking as simulated by 15 atmospheric general circulation models in the period 1979–1988, *Clim. Dyn.*, 14, pp. 385–407.

Dai, A., Fung, I. Y. and DelGenio, A. D. (1997). Surface observed global land precipitation variations during 1900–88, *J. Clim.*, 10, pp. 2943–2962.

Dai, A., Trenberth, K. E. and Qian, T. (2004). A global dataset of Palmer Drought Severity Index for 1870–2002: Relationship with soil moisture and effects of surface warming, *J. Hydrometeorol.*, 5, pp. 1117–1130.

Dai, A. (2011). Drought under global warming: A review, *WIRES Clim. Change*, 2, pp. 45–65.

Dai, A. (2013a). Increasing drought under global warming in observations and models, *Nat. Clim. Change*, 3, pp. 52–58.

Dai, A. (2013b). The influence of the inter-decadal Pacific oscillation on US precipitation during 1923–2010, *Clim. Dyn.*, 41, pp. 633–646.

Dai, A., Fyfe, J. C., Xie, S., *et al.* (2015). Decadal modulation of global surface temperature by internal climate variability, *Nat. Clim. Change*, 5, pp. 555–559.

Dai, A. and Zhao, T. (2017). Uncertainties in historical changes and future projections of drought, Part I: Estimates of historical drought changes, *Clim. Change*, 144, pp. 519–533.

Dai, A. and Song, M. (2020). Little influence of Arctic amplification on mid-latitude climate, *Nat. Clim. Change*, 10, pp. 231–237.

Darkoh, M. B. K. (1998). The nature, causes and consequences of desertification in the drylands of Africa, *Land Degrad. Dev.*, 9, pp. 1–20.

Das, R., Evan, A. T. and Lawrence, D. (2013). Contributions of long-distance dust transport to atmospheric P inputs in the Yucatan Peninsula, *Global Biogeochem. Cycles*, 27, pp. 167–175.

DeCruz, L., Demaeyer, J. and Vannitsem, S. (2016). The modular arbitrary-order ocean–atmosphere model: MAOOAM v1.0, *Geosci. Model Dev.*, 9, pp. 2793–2808.

DeMott, P. J., Sassen, K., Poellot, M. R., *et al.* (2003). African dust aerosols as atmospheric ice nuclei, *Geophys. Res. Lett.*, 30.

Diaz, R. J. and Rosenberg, R. (2008). Spreading dead zones and consequences for marine ecosystems, *Science*, 321, pp. 926–929.

Dickinson, R. E. (1995). Land-atmosphere interaction, *Rev. Geophys.*, 33, pp. 917–922.

Ding, L., Huang, J., Li, C., *et al.* (2022). Variations in terrestrial oxygen sources under climate change, *Chin. Sci. Bull.*, 65, pp. 1810–1823.

Ding, Q. and Wang, B. (2005). Circumglobal teleconnection in the northern hemisphere summer, *J. Clim.*, 18, pp. 3483–3505.

Dodd, E. M. A., Merchant, C. J., Rayner, N. A., *et al.* (2014). An investigation into the impact of using various techniques to estimate Arctic surface air temperature anomalies, *J. Clim.*, 28, pp. 1743–1763.

Dole, R. M. and Gordon, N. D. (1983). Persistent anomalies of the extratropical Northern Hemisphere wintertime circulation: Geographical distribution and regional persistence characteristics, *Mon. Wea. Rev.*, 111, pp. 1567–1586.

Donohue, R., McVicar, T. and Roderick, M. (2009). Climate-related trends in Australian vegetation cover as inferred from satellite observations, 1981–2006, *Globe Change Biol*, 15, pp. 1025–1039.

Eck, T. F., Holben, B. N., Sinyuk, A., *et al.* (2010). Climatological aspects of the optical properties of fine/coarse mode aerosol mixtures, *J. Geophys. Res.*, 115.

Egger, J. (1978). Dynamics of blocking highs, *J. Atmos. Sci.*, 35, pp. 1788–1801.

Egger, J. (1981). Stochastically driven large-scale circulations with multiple equilibria, *J. Atmos. Sci.*, 38, pp. 2606–2618.

Egger, J., Hoinka, K.-P. and Spengler, T. (2015). Aspects of potential vorticity fluxes: Climatology and impermeability, *J. Atmos. Sci.*, 72, pp. 3257–3267.

Enfield, D. B., Mestas-Nuñez, A. M. and Trimble, P. J. (2001). The Atlantic multidecadal oscillation and its relation to rainfall and river flows in the continental U.S., *Geophys. Res. Lett.*, 28, pp. 2077–2080.

Enomoto, T., Hoskins, B. J. and Matsuda, Y. (2003). The formation mechanism of the Bonin high in August, *Quart. J. Roy. Meteorol. Soc.* 129, pp. 157–178.

Ertel, H. (1942). Ein neuer hydrodynamischer Wirbelsatz, *Meteor. Z.*, 59, pp. 277–281.

Eswaran, H. and Reich, P. (2003). Global desertification vulnerability map. Accessible at https://upload.wikimedia.org/wikipedia/commons/6/68/Desertification_map.png.

Eswaran, H., Reich, P. and Vearasilp, T. (2009). Perspectives on desertification during the Anthropocene, Land degradation and rehabilitation: Dryland ecosystems. Papers presented at the Fourth International Conference on Land Degradation, Cartagena, Murcia, Spain, pp. 1–16.

Fa, K., Liu, J., Zhang, Y., *et al.* (2014). CO_2 absorption of sandy soil induced by rainfall pulses in a desert ecosystem, *Hydrol. Process.*, 29, pp. 2043–2051.

Fa, K., Zhang, Y., Wu, B., *et al.* (2016). Patterns and possible mechanisms of soil CO_2 uptake in sandy soil, *Sci. Total Environ.*, 544, pp. 587–594.

Falge, E., Baldocchi, D., Tenhunen, J., *et al.* (2002). Seasonality of ecosystem respiration and gross primary production as derived from FLUXNET measurements, *Agric. For. Meteorol.*, 113, pp. 53–74.

Famiglietti, J. S. (2004). Remote sensing of terrestrial water storage, soil moisture and surface waters, *Geophys. Monogr. Ser.*, 150, pp. 197–207.

Fan, J., Rosenfeld, D., Zhang, Y., *et al.* (2018). Substantial convection and precipitation enhancements by ultrafine aerosol particles, *Science*, 359, pp. 411–418.

Fan, Y., Dool and H. V. D. (2014). Climate prediction center global monthly soil moisture data set at 0.5 degree resolution for 1948 to present, *J. Geophys. Res. Atmos.*, 109.

FAO (Food and Agriculture Organization of the United Nations). (1977). *World Map of Desertification*, (Food and Agricultural Organization (FAO), Rome).

Favre, A. and Gershunov, A. (2006). Extra-tropical cyclonic/anticyclonic activity in North-Eastern Pacific and air temperature extremes in Western North America, *Clim. Dyn.*, 26, pp. 617–629.

Feng, Q., Ma, H., Jiang, X., *et al.* (2015). What has caused desertification in China, *Sci. Rep.*, 5.

Feng, S., Hu, Q. and Oglesby, R. (2011). Influence of Atlantic sea surface temperatures on persistent drought in North America, *Clim. Dyn.*, 37, pp. 569–586.

Feng, S. and Fu, Q. (2013). Expansion of global drylands under a warming climate. *Atmos. Chem. Phys.*, 13, pp. 10081–10094.

Feng, Y. J., Yang, Q. Q., Tong, X. H., *et al.* (2018). Evaluating land ecological security and examining its relationships with driving factors using GIS and generalized additive model, *Sci. Total Environ.*, 633, pp. 1469–1479.

Fensham, R., Fairfax, R. and Archer, S. (2005). Rainfall, land use and woody vegetation cover change in semi-arid Australian savanna, *J. Ecol.*, 93, pp. 596–606.

Ferguson, C. R., Wood, E. F. and Vinukollu, R. K. (2012). A global intercomparison of modeled and observed land–atmosphere coupling, *J. Hydrometeorol.*, 13, pp. 749–784.

Ficklin, D. L. and Novick, K. A. (2017). Historic and projected changes in vapor pressure deficit suggest a continental-scale drying of the United States atmosphere, *J. Geophys. Res. Atmos.* 122, pp. 2061–2079.

Fischer, E. M. and Knutti, R. (2014). Detection of spatially aggregated changes in temperature and precipitation extremes, *Geophys. Res. Lett.*, 41, pp. 547–554.

Foden, W. B., Young, B. E., Akcakaya, H. R., *et al.* (2019). Climate change vulnerability assessment of species, *Wires. Clim. Change*, 10.

Forzieri, G., Alkama, R., Miralles, D., *et al.* (2017). Satellites reveal contrasting responses of regional climate to the widespread greening of Earth, *Science*, 356, pp. 1180–1184.

Fouquart, Y., Bonnel, B., Brogniez, G., *et al.* (1987). Observations of Saharan aerosols: Results of ECLATS field experiment. Part II: Broadband radiative characteristics of the aerosols and vertical radiative flux divergence, *J. Clim. Appl. Meteorol.*, 26, pp. 38–52.

Francioni, M., D'Ottavio, P., Lai, R., *et al.* (2019). Seasonal soil respiration dynamics and carbon-stock variations in mountain permanent grasslands compared to arable lands, *Agriculture*, 9, pp. 1–12.

Fu, B. (2010). Trends and priority areas in ecosystem research of China, *Geogr. Res.*, 29, pp. 383–396. (In Chinese)

Fu, B. and Li, Y. (2016). Bidirectional coupling between the Earth and human systems is essential for modeling sustainability, *Natl. Sci. Rev.*, 3, pp. 397–398.

Fu, B., Tian, H., Tao, F., *et al.* (2017). The impact of global change on ecosystem services, *China Basic Sci.*, 19, pp. 14–18. (In Chinese)

Fu, C., Jiang, Z., Guan, Z., *et al.* (2008). *Regional Climate Studies of China*, 1st edn., (Springer Berlin, Heidelberg).

Fu, Q. and Feng, S. (2014). Responses of terrestrial aridity to global warming, *J. Geophys. Res.*, 119, pp. 7863–7875.

Fu, Q., Johanson, C. M., Wallace, J. M., *et al.* (2006). Enhanced mid-latitude tropospheric warming in satellite measurements, *Science*, 312, pp. 1179.

Fu, Q., Lin, L., Huang, J., *et al.* (2016). Changes in terrestrial aridity for the period 850–2080 from the community earth system model, *J. Geophys. Res.*, 121, pp. 2857–2873.

Gallagher, M. E., Masiello, C. A., Hockaday, W. C., *et al.* (2014). Controls on the oxidative ratio of net primary production in agricultural ecosystems, *Biogeochemistry*, 121, pp. 581–594.

Gallagher, M. E., Liljestrand, F. L., Hockaday, W. C., *et al.* (2017). Plant species, not climate, controls aboveground biomass O_2: CO_2 exchange ratios in deciduous and coniferous ecosystems, *J. Geophys. Res-Biogeo.*, 122, pp. 2314–2324.

Gan, Z., Guan, X., Kong, X., *et al.* (2019). The key role of Atlantic Multidecadal Oscillation in minimum temperature over North America during global warming slowdown, *Earth. Space. Sci.*, 6, pp. 387–397.

Gao, M., Yang, Y., Gong, D., *et al.* (2018). Footprints of Atlantic Multidecadal Oscillation in the low-frequency variation of extreme high temperature in the Northern Hemisphere, *J. Clim.*, 32, pp. 791–802.

Gao, Q., Kang, M., Xu, H., *et al.* (2010). Optimization of land use structure and spatial pattern for the semi-arid loess hilly–gully region in China, *Catena*, 81, pp. 196–202.

Gao, Y., Zhao, Z., Zhang, Y., *et al.* (2021). Response of abiotic soil CO_2 flux to the difference in air-soil temperature in a desert, *Sci. Total Environ.*, 785.

Garreaud, R., Lopez, P., Minvielle, M., *et al.* (2013). Large-scale control on the Patagonian climate, *J. Clim.*, 26, pp. 215–230.

Ge, J., Su, J., Ackerman, T. P., *et al.* (2010a). Dust aerosol optical properties retrieval and radiative forcing over northwestern China during the 2008 China-U.S. joint field experiment, *J. Geophys. Res.*, 115.

Ge, J., Su, J., Fu, Q., *et al.* (2010b). Dust aerosol forward scattering effects on ground-based aerosol optical depth retrievals, *J. Quant. Spectrosc. Radiat. Transfer*, 112, pp. 310–319.

Gentine, P., Massmann, A., Lintner, B. R., *et al.* (2019). Land–atmosphere interactions in the tropics — A review, *Hydrol. Earth Syst. Sci.*, 23, pp. 4171–4197.

Gershunov, A. (1998). ENSO influence on intraseasonal extreme rainfall and temperature frequencies in the contiguous United States: Implications for long-range predictability, *J. Clim.*, 11, pp. 3192–3203.

Ghosh, R., Müller, W. A. and Baehr, J. (2017). Impact of observed North Atlantic multidecadal variations to European summer climate: A linear baroclinic response to surface heating, *Clim. Dyn.*, 48, pp. 3547–3563.

Giardina, C. P., Litton, C. M., Crow, S. E., *et al.* (2014). Warming-related increases in soil CO2 efflux are explained by increased below-ground carbon flux, *Nat. Clim. Change*, 4, pp. 822–827.

Giddings, L., Soto, M., Rutherford, B., *et al.* (2005). Standardized precipitation index zones for Mexico, *Atmosfera*, 18, pp. 33–56.

Ginoux, P., Garbuzov, D. and Hsu, N. C. (2010). Identification of anthropogenic and natural dust sources using Moderate Resolution Imaging Spectroradiometer (MODIS). Deep blue level 2 data, *J. Geophys. Res.*, 115.

Ginoux, P., Chin, M., Tegen, I., *et al.* (2001). Sources and distributions of dust aerosols simulated with the GOCART model, *J. Geophys. Res.*, 106, pp. 20255–20273.

Ginzburg, A. S., Vinogradova, A. A., Fedorova, E. I., *et al.* (2014). Content of oxygen in the atmosphere over large cities and respiratory problems, *Izv. Atmos. Ocean. Phys.*, 50, pp. 782–792.

Giorgi, F., Coppola, E. and Raffaele, F. (2014). A consistent picture of the hydroclimatic response to global warming from multiple indices: Models and observations, *J. Geophys. Res.*, 119, pp. 11695–11708.

Glover, J. D., Reganold, J. R., Bell, W. L., *et al.* (2010). Increased food and ecosystem security via perennial grains, *Science*, 328, pp. 1638.

Gong, D. and Wang, S. (1999). Long-term variability of the Siberian high and the possible connection to global warming, *Acta. Geogr. Sin.*, 54, pp. 125–133.

Gong, D., Zhou, T. and Wang, S. (2001). Advance in the studies on North Atlantic Oscillation (NAO), *Adv. Earth Sci.*, 16, pp. 413–420 (in Chinese).

Greve, P., Orlowsky, B., Mueller, B., *et al.* (2014). Global assessment of trends in wetting and drying over land, *Nature Geosci.* 7, pp. 716–721.

Greve, P. and Seneviratne, S. I. (2015). Assessment of future changes in water availability and aridity, *Geophys. Res. Lett.*, 42, pp. 5493–5499.

Griffin-Nolan, R. J., Bushey, J. A., Carroll, C. J., *et al.* (2018). Trait selection and community weighting are key to understanding ecosystem responses to changing precipitation regimes, *Funct. Ecol.*, 32, pp. 1746–1756.

Groisman, P., Shugart, H., Kicklighter, D., *et al.* (2017). Northern Eurasia Future Initiative (NEFI): facing the challenges and pathways of global change in the twenty-first century, *Prog. Earth Planet. Sc.*, 4.

Gu, F., Wen, Q., Pan, B., *et al.* (2000). A preliminary study on soil micooriganisms of artificial vegetation in the center of Taklimakan Desert, *Chin. Biodivers.*, 8, pp. 297–303 (in Chinese).

Guan, X., Cao, C., Zeng, X., *et al.* (2022a). Evidence of decreasing diurnal temperature range in Eastern Northern Hemisphere, *Environ. Res. Commun.*, 4.

Guan, X., Gao, Z., Huang, J., *et al.* (2022b). Speeding extreme cold events under global warming, *Environ. Res. Lett.*, 17.

Guan, X., Huang, J., Guo, N., *et al.* (2009). Variability of soil moisture and its relationship with surface albedo and soil thermal parameters over the Loess Plateau, *Adv. Atmos. Sci.*, 26, pp. 692–700.

Guan, X., Huang, J., Guo, R., *et al.* (2015a). The role of dynamically induced variability in the recent warming trend slowdown over the Northern Hemisphere, *Sci. Rep.*, 5, pp. 12669.

Guan, X., Huang, J., Guo, R., *et al.* (2015b). Role of radiatively forced temperature changes in enhanced semi-arid warming in the cold season over East Asia, *Atmos. Chem. Phys.*, 15, pp. 13777–13786.

Guan, X., Huang, J., Zhang, Y., *et al.* (2016). The relationship between anthropogenic dust and population over global semi-arid regions, *Atmos. Chem. Phys.*, 16, pp. 5159–5169.

Guan, X., Huang, J. and Guo, R. (2017). Changes in aridity in response to the global warming hiatus, *J. Meteorol. Res.*, 31, pp. 117–125.

Guan, X., Ma, J., Huang, J., *et al.* (2019). Impact of oceans on climate change in drylands, *Sci. China Earth Sci.*, 62, pp. 891–908.

Guo, R., Guan, X., He, Y., *et al.* (2018). Different roles of dynamic and thermodynamic effects in enhanced semi-arid warming, *Int. J. Climatol.*, 38, pp. 13–22.

Guo, S., Guan, X., Gao, L., *et al.* (2022). Opposite Atlantic Multidecadal Oscillation effects on dry/wet changes over Central and East Asian drylands, *Atmos. Res.*, 271.

Guo, W., Wang, X., Sun, J., *et al.* (2016). Comparison of land-atmosphere interaction at different surface types in the mid- to lower reaches of the Yangtze River valley, *Atmos. Chem. Phys.*, 16, pp. 9875–9890.

Gutzler, D. S., Kann, D. and Thornbrugh, C. (2002). Modulation of ENSO-based long-lead outlooks of Southwestern U.S. winter precipitation by the pacific decadal oscillation, *Wea. Forecasting.*, 17, pp.1163–1172.

Hadgu, G., Tesfaye, K. and Mamo, G. (2015). Analysis of climate change in Northern Ethiopia: Implications for agricultural production, *Theor. Appl. Climatol.*, 121, pp. 733–747.

Hadley, O. L. and Kirchstetter, T. W. (2012). Black-carbon reduction of snow albedo, *Nat. Clim. Change*, 2, pp. 437–440.

Han, D., Huang, J., Ding, L., *et al.* (2021). Oxygen footprint: An indicator of the anthropogenic ecosystem changes, *Catena*, 206.

Han, Q., Luo, G., Li, C., *et al.* (2018). Response of carbon dynamics to climate change varied among different vegetation types in Central Asia, *Sustainability*, 10.

Harris, I., Jones, P. D., Osborn, T. J., *et al.* (2014). Updated high-resolution grids of monthly climatic observations-the CRU TS3.10 dataset, *Int. J. Climatol.*, 34, pp. 623–642.

Hartmann, D. L., Ockert-Bell, M. E. and Michelsen, M. L. (1992). The effect of cloud type on Earth's energy balance: global analysis, *J. Clim.*, 5, pp. 1281–1304.

Hashimoto, S., Carvalhais, N., Ito, A., *et al.* (2015). Global spatiotemporal distribution of soil respiration modeled using a global database, *Biogeosciences*, 12, pp. 4121–4132.

Hassan, R., Scholes, R. and Ash, N. (2005). Ecosystems and human well-being, current state and trends, In: Safriel, U. and Adeel, Z. (eds.), *Part III. "Dryland Systems,"* (Island Press, Washington) pp. 625–658.

Haverd, V., Ahlström, A., Smith, B., *et al.* (2017). Carbon cycle responses of semi-arid ecosystems to positive asymmetry in rainfall, *Glob. Change Biol.*, 23, pp. 793–800.

Haynes, J. M., Vonder Haar, T. H., L'Ecuyer, T., *et al.* (2013). Radiative heating characteristics of Earth's cloudy atmosphere from vertically resolved active sensors, *Geophys. Res. Lett.*, 40, pp. 624–630.

Haywood, J. M., Francis, P., Osborne, S., *et al.* (2003). Radiative properties and direct radiative effect of Saharan dust measured by the C-130 aircraft during SHADE: 1. Solar spectrum, *J. Geophys. Res.*, 108.

He, Y., Huang, J. and Ji, M. (2014). Impact of land–sea thermal contrast on interdecadal variation in circulation and blocking, *Clim. Dyn.*, 43, pp. 3267–3279.

He, Y., Huang, J., Li, D., *et al.* (2018). Comparison of the effect of land–sea thermal contrast on interdecadal variations in winter and summer blockings, *Clim. Dyn.*, 51, pp. 1275–1294.

Held, I. M. and Soden, B. J. (2006). Robust responses of the hydrological cycle to global warming, *J. Clim.*, 19, pp. 5686–5699.

Henley, B. J., Gergis, J., Karoly, D. J., *et al.* (2015). A tripole index for the inter-decadal Pacific oscillation, *Clim. Dyn.*, 45, pp. 3077–3090.

Herrmann, S. M., Anyamba, A. and Tucker, C. J. (2005). Recent trends in vegetation dynamics in the African Sahel and their relationship to climate, *Global Environ. Change*, 15, pp. 394–404.

Hess, M., Koepke, P. and Schult, I. (1998). Optical properties of Aerosols and clouds: The software package OPAC, *Bull. Am. Meteorol. Soc.*, 79, pp. 831–844.

Higginbottom, P. T. and Symeonakis, E. (2014). Assessing land degradation and desertification using vegetation index data: Current frameworks and future directions, *Remote Sens.*, 6, pp. 9552–9575.

Hirsch, A. L. and Coauthors, 2019: Amplification of Australian heatwaves via local land — atmosphere coupling, *J. Geophys. Res. Atmos.*, 124, pp. 13625–13647.

Hobbins, M., Wood, A., Struebel, D., *et al.* (2012). What drives the variability of evaporative demand across the conterminous United States, *J. Hydrometeorol.*, 13, pp. 1195–1214.

Hockaday, W. C., Masiello, C. A., Randerson, J. T., *et al.* (2009). Measurement of soil carbon oxidation state and oxidative ratio by ^{13}C nuclear magnetic resonance, *J. Geophys Res. Biogeo.*, 114.

Hoerling, M., Hurrell, J., Xu, T., *et al.* (2004). Twentieth century North Atlantic climate change. Part II: Understanding the effect of Indian Ocean warming, *Clim. Dyn.*, 23, pp. 391–405.

Hoerling, M., Hurrell, J., Eischeid, J., *et al.* (2006). Detection and attribution of twentieth-century northern and southern African rainfall change, *J. Clim.*, 19, pp. 3989–4008.

Holton, J. R., Curry, J. A. and Pyle, J. A. (2003) Encyclopedia of atmospheric sciences, In: Molteni, F. (eds.), *Weather Regimes and Multiple Eequilibria* (Academic Press, San Diego) pp. 2577–2586.

Holton, J. R. and Hakim, G. J. (2012). *An Introduction to Dynamic Meteorology*, 5th edn., (Academic Press, San Diego).

Hoover, D. L., Duniway, M. C. and Belnap, J. (2015). Pulse-drought atop press-drought: Unexpected plant responses and implications for dryland ecosystems, *Oecologia*, 179, pp. 1211–1221.

Hoskins, B.J. and Ambrizzi, T., (1993). Rossby-wave propagation on a realistic longitudinally varying flow, *J. Atmos. Sci.*, 50, pp. 1661–1671.

Hoskins, B. J. and Karoly, D. J. (1981). The steady linear response of a spherical atmosphere to thermal and orographic forcing, *J. Atmos. Sci.*, 38, pp. 1179–1196.

Hoskins, B. J., McIntyre, M. E. and Robertson, A. W. (1985). On the use and significance of isentropic potential vorticity maps, *Quart. J. Roy. Meteor. Soc.*, 111, pp. 877–946.

Hsu, N. C., Gautam, R., Sayer, A. M., *et al.* (2012). Global and regional trends of aerosol optical depth over land and ocean using SeaWiFS measurements from 1997 to 2010, *Atmos. Chem. Phys.*, 12, pp. 8037–8053.

Hu, M., Li, Z., Yuan, M., *et al.* (2019a). Spatial differentiation of ecological security and differentiated management of ecological conservation in the Pearl River Delta, *China. Ecol. Ind.*, 104, pp. 439–448.

Hu, Y. and Fu, Q. (2007). Observed poleward expansion of the Hadley circulation since 1979, *Atmos. Chem. Phys.*, 7, pp. 5229–5236.

Hu, Y., Winker, D., Vaughan, M., *et al.* (2009). CALIPSO/CALIOP cloud phase discrimination algorithm, *J. Atmos. Ocean. Technol.*, 26, pp. 2293–2309.

Hu, Z., Wang, L., Wang, Z., *et al.* (2015). Quantitative assessment of climate and human impacts on surface water resources in a typical semi-arid watershed in the middle reaches of the yellow river from 1985 to 2006, *Int. J. Climatol.*, 35, pp. 97–113.

Hu, Z., Zhao, C., Huang, J., *et al.* (2016). Trans-Pacific transport and evolution of aerosols: Evaluation of quasi-global WRF-Chem simulation with multiple observations, *Geosci. Model Dev.*, 9, pp. 1725–1746.

Hu, Z., Huang, J., Zhao, C., *et al.* (2019b). Modeling the contributions of Northern Hemisphere dust sources to dust outflow from East Asia, *Atmos. Environ.*, 202, pp. 234–243.

Hu, Z., Huang, J., Zhao, C., *et al.* (2020). Modeling dust sources, transport, and radiative effects at different altitudes over the Tibetan Plateau, *Atmos. Chem. Phys.*, 20, pp. 1507–1529.

Huang, H., Seager, R. and Kushnir, Y. (2005). The 1976/77 transition in precipitation over the Americas and the influence of tropical sea surface temperature, *Clim. Dyn.*, 24, pp. 721–740.

Huang, H., Thomas, G. E. and Grainger, R. G. (2010a). Relationship between wind speed and aerosol optical depth over remote ocean, *Atmos. Chem. Phys.*, 10, pp. 5943–5950.

Huang, J. and Yi, Y. (1991). Inversion of a nonlinear dynamic-model from the observation. *Sci. China Ser. B Chem. Life Sci. Earth Sci.*, 34, pp. 1246–1251.

Huang, J., Higuchi, K. and Shabbar, A. (1998). The relationship between the north Atlantic oscillation and El Nino-southern oscillation, *Geophys. Res. Lett.*, 25, pp. 2707–2710.

Huang, J., Ji, M., Higuchi, K., *et al.* (2006a). Temporal Structures of the North Atlantic Oscillation and Its Impact on the Regional Climate Variability, *Adv. Atmos. Sci.*, 23, pp. 23–32.

Huang, J., Lin, B., Minnis, P., *et al.* (2006b). Satellite-based assessment of possible dust aerosols semi-direct effect on cloud water path over East Asia, *Geophys. Res. Lett.*, 33.

Huang, J., Minnis, P., Lin, B., *et al.* (2006c). Possible influences of Asian dust aerosols on cloud properties and radiative forcing observed from MODIS and CERES, *Geophys. Res. Lett.*, 33.

Huang, J., Wang, Y., Wang, T., *et al.* (2006d). Dusty cloud radiative forcing derived from satellite data for middle latitude region of East Asia, *Prog. Nat. Sci.*, 10, pp. 1084–1089.

Huang, J., Ge, J. and Weng, F. (2007a). Detection of Asia dust storms using multisensor satellite measurements, *Remote Sens. Environ.*, 110, pp. 186–191.

Huang, J., Minnis, P., Yi, Y., *et al.* (2007b). Summer dust aerosols detected from CALIPSO over the Tibetan Plateau, *Geophys. Res. Lett.*, 34.

Huang, J., Zhang, W., Zuo, J. Q., *et al.* (2008a). An overview of the semi-arid climate and environment research observatory over the Loess Plateau, *Adv. Atmos. Sci.*, 25, pp. 906–921.

Huang, J., Minnis, P., Chen, B., *et al.* (2008b). Long-range transport and vertical structure of Asian dust from CALIPSO and surface measurements during PACDEX, *J. Geophys. Res.*, 113.

Huang, J., Fu, Q., Su, J., *et al.* (2009). Taklimakan dust aerosol radiative heating derived from CALIPSO observations using the Fu-Liou radiation model with CERES constraints, *Atmos. Chem. Phys.*, 9, pp. 4011–4021.

Huang, J., Minnis, P., Yan, H., *et al.* (2010b). Dust aerosol effect on semi-arid climate over Northwest China detected from A-Train satellite measurements, *Atmos. Chem. Phys.*, 10, pp. 6863–6872.

Huang, J., Fu, Q., Zhang, W., *et al.* (2011). Dust and black carbon in seasonal snow across northern China, *B. Am. Meteorol. Soc.*, 92, pp. 175–181.

Huang, J., Guan, X., Ji, F., *et al.* (2012). Enhanced cold-season warming in semi-arid regions, *Atmos. Chem. Phys.*, 12, pp. 5391–5398.

Huang, J., Ji, M., Liu, Y., *et al.* (2013). Review of climate change research in arid and semi-arid regions, *Adv. Clim. Change Res.*, 9, pp. 9–14 (in Chinese).

Huang, J., Wang, T., Wang, W., *et al.* (2014). Climate effects of dust aerosols over East Asian arid and semi-arid regions, *J. Geophys. Res. Atmos.*, 119, pp. 11398–11416.

Huang, J., Liu, J., Chen, B., *et al.* (2015). Detection of anthropogenic dust using CALIPSO lidar measurements, *Atmos. Chem. Phys.*, 15, pp. 11653–11665.

Huang, J., Ji, M., Xie, Y., *et al.* (2016a). Global semi-arid climate change over last 60 years, *Clim. Dyn.*, 46, pp. 1131–1150.

Huang, J., Yu, H., Guan, X., *et al.* (2016b). Accelerated dryland expansion under climate change, *Nat. Clim. Change*, 6, 166–171.

Huang, J., Li, Y., Fu, C., *et al.* (2017a). Dryland climate change: Recent progress and challenges, *Rev. Geophys.*, 55, pp. 719–778.

Huang, J., Xie, Y., Guan, X., *et al.* (2017b). The dynamics of the warming hiatus over the Northern Hemisphere, *Clim. Dyn.*, 48, pp. 429–446.

Huang, J., Yu, H., Dai, A., *et al.* (2017c). Drylands face potential threat under 2°C global warming target, *Nat. Clim. Change*, 7, pp. 417–422.

Huang, J., Huang, J., Liu, X. Y., *et al.* (2018). The global oxygen budget and its future projection, *Sci. Bull.*, 63, pp. 1180–1186.

Huang, J., Ma, J., Guan, X., *et al.* (2019). Progress in semi-arid climate change studies in china, *Adv. Atmos. Sci.*, 36, pp. 922–937.

Huang, J., Yu, H., Han, D., *et al.* (2020a). Declines in global ecological security under climate change, *Ecol. Indic.*, 117.

Huang, J., Zhang, G., Zhang, Y., *et al.* (2020b). Global desertification vulnerability to climate change and human activities, *Land Degrad. Dev.*, 31, pp. 1380–1391.

Huang, J., Liu, X., He, Y., *et al.* (2021). The oxygen cycle and a habitable earth, *Sci. China Ser D-Earth Sci.*, 64, pp. 511–528.

Huang, J., Zhou, X., Wu, G., *et al.* (2023). Global climate impacts of land-surface and atmospheric processes over the Tibetan Plateau, *Rev. Geophys.* (in press), doi: 10.1029/2022RG000771.

Hudgins, L. and Huang, J. (1996). Bivariate wavelet analysis of Asia monsoon and ENSO, *Adv. Atmos. Sci.*, 13, pp. 299–312.

Hugonnet, R., McNabb, R., Berthier, E., *et al.* (2021). Accelerated global glacier mass loss in the early twenty-first century, *Nature*, 592, pp. 726–731.

Humphrey, V., Gudmundsson, L. and Seneviratne, S. I. (2017). A global reconstruction of climate-driven subdecadal water storage variability, *Geophys. Res. Lett.*, 44, pp. 2300–2309.

Huneeus, N., Schulz, M., Balkanski, Y., *et al.* (2011). Global dust model intercomparison in AeroCom phase I, *Atmos. Chem. Phys.*, 11, pp. 7781–7816.

Huo, W., He, Q., Yang, X., *et al.* (2011). The research on grain size characteristic of desert in north of China, *Res. Soil Water Conserv.*, 18, pp. 6–11 (in Chinese).

Huss, M. (2013). Density assumptions for converting geodetic glacier volume change to mass change, *Cryosphere*, 7, pp. 877–887.

IPCC. (2007). Climate change 2007: The physical science basis. Contribution of Working Group I to the Fourth Assessment Report of the Intergovernmental Panel on Climate Change. (Cambridge University Press, Cambridge).

IPCC. (2013). Climate change 2013: The physical science basis. Working Group I Contribution to the Fifth Assessment Report of the Intergovernmental Panel On Climate Change. (Cambridge University Press, Cambridge).

IPCC. AR5 (2014). Climate change 2014: Mitigation of climate change. Contribution of Working Groups III to the Fifth Assessment Report of the Intergovernmental Panel on Climate Change. (Cambridge University Press, Cambridge).

IPCC. AR6. (2021). Climate change 2021: The physical science basis. Contribution of working group I to the Sixth Assessment Report of the Intergovernmental Panel on Climate Change. (Cambridge University Press, Cambridge).

Ishidoya, S., Sugawara, H., Terao, Y., *et al.* (2020). O_2: CO_2 exchange ratio for Net Turbulent Flux observed in an urban area of Tokyo, Japan, and its application to an evaluation of anthropogenic CO_2 emissions, *Atmos. Chem. Phys.*, 20, pp. 5293–5308.

Jafari, R. and Bakhshandehmehr, L. (2016). Quantitative mapping and assessment of environmentally sensitive areas to desertification in central Iran, *Land Degrad. Dev.*, 27, pp. 108–119.

Ji, F., Wu, Z., Huang, J., *et al.* (2014). Evolution of land surface air temperature trend, *Nat. Clim. Change*, 4, pp. 462–466.

Jiao, W., Wang, L., Smith, W. K., *et al.* (2021). Observed increasing water constraint on vegetation growth over the last three decades, *Nat. Commun.*, 12.

Jin, Q., Wei, J., Yang, Z., *et al.* (2015). Consistent response of Indian summer monsoon to Middle East dust in observations and simulations, *Atmos. Chem. Phys.*, 15, pp. 9897–9915.

Johansson, E., Devasthale, A., Ecuyer, T., *et al.* (2015). The vertical structure of cloud radiative heating over the Indian subcontinent during summer monsoon, *Atmos. Chem. Phys.*, 15, pp. 11557–11570.

Johnson, B. T. (2005). The semi-direct aerosol effect: Comparison of a single-column model with large eddy simulation for marine stratocumulus, *J. Clim.*, 18, pp. 119–130.

Jung, M., M. Reichstein, C. Schwalm, *et al.* (2017). Compensatory water effects link yearly global land CO_2 sink changes to temperature, *Nature*, 541, pp. 516–520.

Junkermann, W., Hacker, J., Lyons, T., *et al.* (2009). Land use change suppresses precipitation, *Atmos. Chem. Phys.*, 9, pp. 6531–6539.

Justice, C., Townshend, J., Vermote, E., *et al.* (2002). An overview of MODIS Land data processing and product status, *Remote Sens. Environ.*, 83, pp. 3–15.

Kaas, E. and Branstator, G. (1993). The relationship between a zonal index and blocking activity, *J. Atmos. Sci. USA.*, 50, pp. 3061–3077.

Källén, E. (1981). The nonlinear effects of orographic and momentum forcing in a low-order, barotropic model, *J. Atmos. Sci.*, 38, pp. 2150–2163.

Kane, R. P. (2009). Periodicities, ENSO effects and trends of some South African rainfall series: An update, *S. Afr. J. Sci.*, 105, pp. 199–207.

Kanniah, K. D., Beringer, J., Hutley, L. B., *et al.* (2009). Evaluation of collections 4 and 5 of the MODIS gross primary productivity product and algorithm improvement at a tropical savanna site in northern Australia, *Remote Sens. Environ.*, 113, pp. 1808–1822.

Kasoar, M., Voulgarakis, A., Lamarque, J. F., *et al.* (2016). Regional and global temperature response to anthropogenic SO_2 emissions from China in three climate models, *Atmos. Chem. Phys.*, 16, pp. 9785–9804.

Kassas, M. (1995). Desertification: A general review, *J. Arid Environ.*, 30, pp. 115–128.

Kawamoto, K. (2008). Effect of precipitation on water cloud properties over China, *Geophys. Res. Lett.*, 35.

Keeling, R. F. and Shertz, S.R. (1992). Seasonal and interannual variations in atmospheric oxygen and implications for the global carbon cycle, *Nature*, 358, pp. 723–727.

Keeling, R. F., Körtzinger, A. and Gruber, N. (2010). Ocean deoxygenation in a warming world, *Ann. Rev. Mar. Sci.*, 2, pp. 463–492.

Keeling, R. F. and Manning, A. C. (2014). *Treatise on Geochemistry*, 2nd edn., (Elsevier, Amsterdam). pp. 385–404.

Keenan, T. F., Prentice, I. C., Canadell, J. G., *et al.* (2016). Recent pause in the growth rate of atmospheric CO_2 due to enhanced terrestrial carbon uptake, *Nat. Commun.*, 7.

Kim, D., Chin, M., Bian, H., *et al.* (2013). The effect of the dynamic surface bareness on dust source function, emission, and distribution, *J. Geophys. Res.*, 118, pp. 871–886.

Kim, H. and Choi, M. (2015). Impact of soil moisture on dust outbreaks in East Asia: Using satellite and assimilation data, *Geophys. Res. Lett.*, 42, pp. 2789–2796.

Klein, C., Bliefernicht, J., Heinzeller, D., *et al.* (2017). Feedback of observed interannual vegetation change: A regional climate model analysis for the West African monsoon, *Clim. Dyn.*, 48, pp. 2837–2858.

Klyashtorin, L. B. and Lyubushin, A. A. (2007). *Cyclic Climate Changes and Fish Productivity.* (VNIRO Publishing, Moscow).

Knight, J. R., Folland, C. K. and Scaife, A. A. (2006). Climate impacts of the Atlantic multidecadal oscillation, *Geophys. Res. Lett.*, 33.

Knohl, A. and Cuntz, M. (2017). Tracing carbon fixation, *Nature Clim. Change*, 7, pp. 393–394.

Kool, D., Kustas, W. P., Ben-Gal, A., *et al.* (2016). Energy and evapotranspiration partitioning in a desert vineyard, *Agric. For. Meteorol.*, 218, pp. 277–287.

Köppen, W. (2011). The thermal zones of the earth according to the duration of hot, moderate and cold periods and to the impact of heat on the organic world (Translated from: Die Wärmezonen der Erde, nach der Dauer der heissen, gemässigten und kalten Zeit und nach der Wirkung der Wärme auf die organische Welt betrachtet, Meteorol Z 1884, 1, 215–226). *Meteorologische Zeitschrift*, 20, pp. 351–360.

Koren, I., Remer, L. A., Altaratz, O., *et al.* (2010). Aerosol-induced changes of convective cloud anvils produce strong climate warming, *Atmos. Chem. Phys.*, 10, pp. 5001–5010.

Koren, I., Altaratz, O., Remer, L. A., *et al.* (2012). Aerosol-induced intensification of rain from the tropics to the mid-latitudes, *Nat. Geosci.*, 5, pp. 118–122.

Kosaka, Y. and Xie, S. P. (2013). Recent global-warming hiatus tied to equatorial Pacific surface cooling, *Nature.*, 501, pp. 403–407.

Kottek, M., Grieser, J., Beck, C., *et al.* (2006). World map of the Koppen–Geiger climate classification updated, *Meteorol. Z.*, 15, pp. 259–263.

Kucharski, F., Molteni, F. and Bracco, A. (2005). Decadal interactions between the western tropical Pacific and the North Atlantic Oscillation, *Clim. Dyn.*, 26, pp. 79–91.

Kucharski, F., Zeng, N. and Kalnay, E. (2013). A further assessment of vegetation feedback on decadal Sahel rainfall variability, *Clim. Dyn.*, 40, pp. 1453–1466.

Kuebbeler, M., Lohmann, U., Hendricks, J., *et al.* (2013). Dust ice nuclei effects on cirrus clouds, *Atmos. Chem. Phys. Discuss.*, 13, pp. 9751–9799.

Kundu, A., Patel, N. R., Saha, S. K., *et al.* (2017). Desertification in western Rajasthan (India): An assessment using remote sensing derived rain-use efficiency and residual trend methods, *Nat. hazards*, 86, pp. 297–313.

Kushnir, Y., Robinson, W., Bladé, I., *et al.* (2002). Atmospheric GCM response to extratropical SST anomalies: Synthesis and evaluation, *J. Clim.*, 15, pp. 2233–2256.

Lal, A. (2003). Carbon sequestration in dryland ecosystems, *Environ. Manage.*, 33, pp. 528–544.

Latif, M. and Barnett, T. P. (1994). Causes of decadal climate variability over the North Pacific and North-America, *Science*, 266, pp. 634–637.

Lau, K. M. and Kim, K. M. (2010). Fingerprinting the impacts of aerosols on long — term trends of the Indian summer monsoon regional rainfall, *Geophys. Res. Lett.*, 37, pp. 127–137.

Lau, K. M., Kim, M. K. and Kim, K. M. (2006). Asian summer monsoon anomalies induced by aerosol direct forcing: The role of the Tibetan Plateau, *Clim. Dyn.*, 26, pp. 855–864.

Lau, W. K. M. and Kim, K. M. (2015). Robust Hadley circulation changes and increasing global dryness due to CO2 warming from CMIP-5 model projections, *Proc. Natl. Acad. Sci. USA*, 112, pp. 3630–3635.

Lavado Contador, J. F., Schnabel, S., Gutiérrez, A. G., *et al.* (2009). Mapping sensitivity to land degradation in Extremadura SW Spain, *Land Degrad. Dev.*, 20, pp. 129–144.

Lee, K. H., Li, Z., Wong, M. S., *et al.* (2007). Aerosol single scattering albedo estimated across China from a combination of ground and satellite measurements, *J. Geophys. Res.*, 112.

Lee, S. M. and Min, S. K. (2018). Heat stress changes over East Asia under 1.5°C and 2.0°C global warming targets, *J. Clim.*, 31, pp. 2819–2831.

Legras, B. and Ghil, M. (1985). Persistent anomalies, blocking and variations in atmospheric predictability, *J. Atmos. Sci.*, 42, pp. 433–471.

Lemaître, C., Flamant, C., Cuesta, J., *et al.* (2010). Radiative heating rates profiles associated with a springtime case of Bodélé and Sudan dust transport over West Africa, *Atmos. Chem. Phys.*, 10, pp. 8131–8150.

Levin, Z., Teller, A., Ganor, E., *et al.* (2005). On the interactions of mineral dust, sea-salt particles, and clouds: A measurement and modeling study from the Mediterranean Israeli Dust Experiment campaign, *J. Geophys. Res.*, 110.

Levy, H., Horowitz, L. W., Schwarzkopf, M. D., *et al.* (2013). The roles of aerosol direct and indirect effects in past and future climate change, *J. Geophys. Res. Atmos.*, 118, pp. 4521–4532.

Lewis, S. C. and Karoly, D. J. (2013). Evaluation of historical diurnal temperature range trends in CMIP5 models, *J. Clim.*, 26, pp. 9077–9089.

Li, C., Zhao, T. and Ying, K. (2015). Effects of anthropogenic aerosols on temperature changes in China during the twentieth century based on CMIP5 models, *Theor. Appl. Climatol.*, 125, pp. 529–540.

Li, C., Huang, J., Ding, L., *et al.* (2020). Increasing escape of oxygen from oceans under climate change, *Geophys. Res. Lett.*, 47, pp. 1–8.

Li, C., Huang, J., Ding, L., *et al.* (2021). Estimation of oceanic and land carbon sinks based on the most recent oxygen budget, *Earths Future*, 9.

Li, D., Zhou, T., Zou, L., *et al.* (2018a). Extreme high-temperature events over East Asia in 1.5°C and 2°C warmer futures: Analysis of NCAR CESM low-warming experiments, *Geophys. Res. Lett.*, 45, pp. 1541–1550.

Li, D., Zou, L. and Zhou, T. (2018b). Extreme climate event changes in China in the 1.5°C and 2°C warmer climates: Results from statistical and dynamical downscaling, *J. Geophys. Res.-Atmos.*, 123, pp. 10196–10211.

Li, D., He, Y., Huang, J., *et al.* (2018c). Multiple equilibria in a land-atmosphere coupled system, *J. Meteorol. Res.*, 32, pp. 950–973.

Li, H., Dai, A., Zhou, T., *et al.* (2010a). Responses of East Asian summer monsoon to historical SST and atmospheric forcing during 1950–2000, *Clim. Dyn.*, 34, pp. 501–514.

Li, J. and Wang, J. (2003). A modified zonal index and its physical sense, *Geophys. Res. Lett.*, 30.

Li, J., Lv, Q., Zhang, M., *et al.* (2017). Effects of atmospheric dynamics and aerosols on the fraction of supercooled water clouds, *Atmos. Chem. Phys.*, 17, pp. 1847–1863.

Li, J., Jian, B., Huang, J., *et al.* (2018d). Long-term variation of cloud droplet number concentrations from space-based Lidar, *Remote Sens. Environ.*, 213, pp. 144–161.

Li, M. and Luo, Z. (1983). Nonlinear mechanism of abrupt change of atmospheric circulation during June and October, *Scientia Sinica (Series B)*, 26, pp. 746–754.

Li, M. and Ma, M. (2018). Decadal changes in summer precipitation over arid northwest China and associated atmospheric circulations, *Int. J. Climatol.*, 38, pp. 4496–4508.

Li, S. and Ji, L. (2001). Persistent anomaly in Ural area in summer and its background circulation characteristics, *Acta. Meteorol. Sin.*, 59, pp. 280–293 (in Chinese).

Li, W., Migliavacca, M., Forkel, M., *et al.* (2022). Widespread increasing vegetation sensitivity to soil moisture, *Nat. commun.*, 13.

Li, Y. and Zhao, X. (2012). An empirical study of the impact of human activity on long-term temperature change in China: A perspective from energy consumption, *J. Geophys. Res.*, 117.

Li, Y., Huang, J., Ji, M., *et al.* (2015a). Dryland expansion in Northern China from 1948 to 2008, *Adv. Atmos. Sci.*, 32, pp. 870–876.

Li, Y., Wang, Y., Houghton, R. A., *et al.* (2015b). Hidden carbon sink beneath desert, *Geophys. Res. Lett.*, 42, pp. 5880–5887.

Li, Z., Chen, H., Cribb, M., *et al.* (2007a). Preface to special section on East Asian Studies of Tropospheric Aerosols: An International Regional Experiment (EAST-AIRE), *J. Geophys. Res.*, 112.

Li, Z., Xia, X., Cribb, M., *et al.* (2007b). Aerosol optical properties and their radiative effects in northern China, *J. Geophys. Res.*, 112.

Li, Z., Zhao, X., Kahn, R., *et al.* (2009). Uncertainties in satellite remote sensing of aerosols and impact on monitoring its long-term trend: A review and perspective, *Ann. Geophys.*, 27, pp. 2755–2770.

Li, Z., Lee, K.-H., Wang, Y., *et al.* (2010b). First observation-based estimates of cloud-free aerosol radiative forcing across China, *J. Geophys. Res.*, 115.

Li, Z., Li, C., Chen, H., *et al.* (2011). East Asian Studies of Tropospheric Aerosols and their Impact on Regional Climate (EAST-AIRC): An overview, *J. Geophys. Res.*, 116.

Li, Z., Lau, W. K.-M., Ramanathan, V., *et al.* (2016). Aerosol and monsoon climate interactions over Asia, *Rev. Geophys.*, 54, pp. 866–929.

Liao, H., Seinfeld, J. H., Adams, P. J., *et al.* (2004). Global radiative forcing of coupled tropospheric ozone and aerosols in a unified general circulation model, *J. Geophys. Res.*, 109.

Lin, L., Gettelman, A., Xu, Y., *et al.* (2016). Simulated responses of terrestrial aridity to black carbon and sulfate aerosols, *J. Geophys. Res. Atmos.*, 121, pp. 785–794.

Liu, C. and Tao, S. (1983). Northward jumping of subtropical highs and cusp catastrophe, *Scientia Sinica (Series B)*, 26, pp. 1065–1074.

Liu, D. and Chang, Q. (2015). Ecological security research progress in China, *Acta Ecol. Sin.*, 35, pp. 111–121.

Liu, H., Yin, Y., Tian, Y., *et al.* (2008a). Climatic and anthropogenic controls of topsoil features in the semi-arid East Asian steppe, *Geophys. Res. Lett.*, 35.

Liu, J., Zheng, Y., Li, Z., *et al.* (2011a). Transport, vertical structure and radiative properties of dust events in southeast China determined from ground and space sensors, *Atmos. Environ.*, 45, pp. 6469–6480.

Liu, J., Zheng, Y., Li, Z., *et al.* (2011b). Analysis of cloud condensation nuclei properties at a polluted site in southeastern China during the AMF-China Campaign, *J. Geophys. Res.*, 116.

Liu, J., Zheng, Y., Li, Z., *et al.* (2012). Seasonal variations of aerosol optical properties, vertical distribution and associated radiative effects in the Yangtze Delta region of China, *J. Geophys. Res.*, 117.

Liu, L., Li, Z., Yang, X., *et al.* (2016). The long-term trend in the diurnal temperature range over Asia and its natural and anthropogenic causes, *J. Geophys. Res. Atmos.*, 121, pp. 3519–3533.

Liu, L., Guo, J., Chen, W., *et al.* (2018). Large-scale pattern of the diurnal temperature range changes over East Asia and Australia in boreal winter: A perspective of atmospheric circulation, *J. Clim.*, 31, pp. 2715–2728.

Liu, L., Gudmundsson, L., Hauser, M. *et al.* (2020a). Soil moisture dominates dryness stress on ecosystem production globally, *Nat. Commun.*, 11.

Liu, S., Wu, Q., Ren, X. *et al.* (2017). Modeled Northern Hemisphere autumn and winter climate responses to realistic Tibetan Plateau and Mongolia snow anomalies, *J. Clim.*, 30, pp. 9435–9454.

Liu, X., Huang, J., Huang, J., *et al.* (2020b). Estimation of gridded atmospheric oxygen consumption from 1975 to 2018, *J. Meteorolog. Res.*, 34, pp. 1–13.

Liu, Y., Huang, J., Shi, G., *et al.* (2011c). Aerosol optical properties and radiative effect determined from sky-radiometer over Loess Plateau of northwest China, *Atmos. Chem. Phys.*, 11, pp. 11455–11463.

Liu, Y., Zhu, Q., Hua, S., *et al.* (2020c). Tibetan Plateau driven impact of Taklimakan dust on northern rainfall, *Atmos. Environ.*, 234.

Liu, Y., Huang, J., Wang, T., *et al.* (2022). Aerosol-cloud interactions over the Tibetan Plateau: An overview, *Earth-Sci. Rev.*, 234.

Liu, Z., Omar, A., Vaughan, M., *et al.* (2008b). CALIPSO lidar observations of the optical properties of Saharan dust: A case study of long-range transport, *J. Geophys. Res.*, 113.

Liu, Z., Kar, J., Zeng, S., *et al.* (2019). Discriminating between clouds and aerosols in the CALIOP version 4.1 data products, *Atmos. Meas. Tech.*, 12, pp. 703–734.

Livina, V. N., Vaz Martins, T. M. and Forbes, A. B. (2015). Tipping point analysis of atmospheric oxygen concentration. *Chaos*, 25.

Lohmann, U. and Feichter, J. (2001). Can the direct and semi-direct aerosol effect compete with the indirect effect on a global scale, *Geophys. Res. Lett.*, 28, pp. 159–161.

Lohmann, U. and Feichter, J. (2005). Global indirect aerosol effects: A review, *Atmos. Chem. Phys.*, 5, pp. 715–737.

Lou, Y. and Zhou, X. (2006). *Soil Respiration and the Environment*, (Academic Press, San Diego).

Lu, J. (2009). The dynamics of the Indian Ocean Sea surface temperature forcing of Sahel drought, *Clim. Dyn.*, 33, pp. 445–460.

Lu, R., Oh, J.-H. and Kim, B.-J. (2002). A teleconnection pattern in upper-level meridional wind over the North African and Eurasian Continent in Summer, *Tellus A: Dyn. Meteorol. Ocean.*, 54, pp. 44–55.

Lu, X., Wang, L. and McCabe, M. F. (2016). Elevated CO2 as a driver of global dryland greening, *Sci. Rep.*, 6.

Lueker, T. J., Keeling, R. F. and Dubey, M. K. (2001). The oxygen to carbon dioxide ratios observed in emissions from a wildfire in northern California, *Geophys. Res. Lett.*, 28, pp. 2413–2416.

Luo, D., Chen, X., Dai, A., *et al.* (2018). Changes in atmospheric blocking circulations linked with winter Arctic warming: A new perspective, *J. Clim.*, 31, pp. 7661–7678.

Luo, M., Liu, Y., Zhu, Q., *et al.* (2020). Role and mechanisms of black carbon affecting water vapor transport to Tibet, *Remote Sens.*, 12, pp. 231.

Luo, Z. (1987). Abrupt change of flow pattern in baroclinic atmosphere forced by joint effects of diabatic heating and orography, *Adv. Atmos. Sci.*, 4, pp. 137–144.

Lyon, B. (2004). The strength of El Niño and the spatial extent of tropical drought, *Geophys. Res. Lett.*, 31.

Ma, J., Wang, Z., Stevenson, B. A., *et al.* (2013). An inorganic CO_2 diffusion and dissolution process explains negative CO_2 fluxes in saline/alkaline soils, *Sci. Rep.*, 3.

Ma, J., Liu, R., Tang, L. S., *et al.* (2014). A downward CO_2 flux seems to have nowhere to go, *Biogeosciences*, 11, pp. 6251–6262.

Ma, K., Fu, B., Li, X., *et al.* (2004). The regional pattern for ecological security (RPES): The concept and theoretical basis, *Acta Ecologica Sinica*, 4, pp. 105–112 (in Chinese).

Ma, Z. and Fu, C. (2006). Some evidence of drying trend over northern China from 1951 to 2004, *Chinese Sci. Bull.*, 51, pp. 2913–2925.

Ma, Z. and Shao, L. (2006). Relationship between dry/wet variation and the Pacific decadal oscillation (PDO) in northern China during the last 100 years, *Chin. J. Atmos. Sci.*, 30, pp. 464–474.

Ma, Z. and Fu, C. (2007). Global aridification in the second half of the 20th century and its relationship to large-scale climate background, *Sci. China Ser. D*, 50, pp. 776–788.

MacDonald, A. M., Lark, R. M., Taylor, R. G., *et al.* (2021). Mapping groundwater recharge in Africa from ground observations and implications for water security, *Environ. Res. Lett.*, 16.

Maestre, F. T., Escolar, C., de Guevara, M. L., *et al.* (2013). Changes in biocrust cover drive carbon cycle responses to climate change in drylands, *Global Change Biol.*, 19, pp. 3835–3847.

Mahmood, R., Pieke, R. A. S., Hubbard, K. G., *et al.* (2014). Land cover changes and their biogeophysical effects on climate, *Int. J. Climatol.*, 34, pp. 929–953.

Mahowald, N. M., Kloster, S., Engelstaedter, S., *et al.* (2010) Observed 20th century desert dust variability: Impact on climate and biogeochemistry, *Atmos. Chem. Phys.*, 10, pp. 10875–10893.

Makowski, K., Wild, M. and Ohmura, A. (2008). Diurnal temperature range over Europe between 1950 and 2005, *Atmos. Chem. Phys.*, 8, pp. 7051–7084.

Mallet, M., Tulet, P. Serça, D., Solmon, F., *et al.* (2009). Impact of dust aerosols on the radiative budget, surface heat fluxes, heating rate profiles and convective activity over West Africa during March 2006, *Atmos. Chem. Phys.*, 9, pp. 7143–7160.

Mann, M. E., Steinman, B. A. and Miller, S. K. (2014). On forced temperature changes, internal variability and the AMO, *Geophys. Res. Lett.*, 41, pp. 3211–3219.

Mantua, N. J. and Hare, S. R. (2002). The Pacific decadal oscillation, *J. Oceanogr.*, 58, pp. 35–44.

Mantua, N. J., Hare, S. R., Zhang, Y., *et al.* (1997). A Pacific interdecadal climate oscillation with impacts on salmon production, *Bull. Am. Meteorol. Soc.*, 78, pp. 1069–1079.

Martin, D., McKenna, H. and Livina, V. (2017). The human physiological impact of global deoxygenation, *J. Physiol. Sci.*, 67, pp. 97–106.

Martínez-Valderrama, J., Ibáñez, J., Del Barrio, G., *et al.* (2016). Present and future of desertification in Spain: Implementation of a surveillance system to prevent land degradation, *Sci. Total Environ.*, 563–564, pp. 169–178.

Masiello, C. A., Gallagher, M. E., Randerson, J. T., *et al.* (2008). Evaluating two experimental approaches for measuring ecosystem carbon oxidation state and oxidative ratio, *J. Geophys. Res. Biogeo.*, 113.

McCabe, G. J., Palecki, M. A. and Betancourt, J. L. (2004). Pacific and Atlantic Ocean influences on multidecadal drought frequency in the United States, *Proc. Natl. Acad. Sci.*, 101, pp. 4136–4141.

McCabe, G. J., Ault, T. R., Cook, B. I., *et al.* (2012). Influences of the El Niño Southern Oscillation and the Pacific Decadal Oscillation on the timing of the North American spring, *Int. J. Climatol.*, 32, pp. 2301–2310.

McComiskey, A. and Feingold, G. (2008). Quantifying error in the radiative forcing of the first aerosol indirect effect, *Geophys. Res. Lett.*, 35.

McCormick, L. R., Levin, L. A. and Oesch, N. W. (2019). Vision is highly sensitive to oxygen availability in marine invertebrate larvae, *J. Exp. Biol.*, 222.

Meehl, G. A., Washington, W. M., Arblaster, J. M., *et al.* (2012). Climate system response to external forcings and climate change projections in CCSM4, *J. Clim.*, 25, pp. 3661–3683.

Miao, J. H. and Ding, M. F. (1985). Catastrophe theory of seasonal variation, *Scientia Sinica (Series B)*, 28, pp. 1079–1092.

Middleton, N. and Thomas, D. (1997). *World atlas of desertification*, 2nd edn., (Hodder Education Publishers, London).

Millennium Ecosystem Assessment (MEA). (2005). *Ecosystems and Human Well-being: Desertification Synthesis,* (World Resources Institute, Washington DC).

Miller, R. L., Tegen, I. and Perlwitz, J. (2004). Surface radiative forcing by soil dust aerosols and the hydrologic cycle, *J. Geophys. Res.*, 109.

Milly, P. C. D., Betancourt, J., Falkenmark, M., *et al.* (2008). Stationarity is dead: Whither water management, *Science*, 319, pp. 573–574.

Min, Q., Li, R., Lin, B., *et al.* (2009). Evidence of mineral dust altering cloud microphysics and precipitation, *Atmos. Chem. Phys.*, 9, pp. 3223–3231.

Minejima, C., Kubo, M., Tohjima, Y., *et al.* (2012). Analysis of $\Delta O2/\Delta CO2$ ratios for the pollution events observed at Hateruma Island, *Japan. Atmos. Chem. Phys.*, 12, pp. 2713–2723.

Minnis, P., Sun-Mack, S., Young, D. F., *et al.* (2011). CERES Edition-2 cloud property retrievals using TRMM VIRS and Terra and Aqua MODIS data, part I: Algorithms, *IEEE. T. Geosci. Remote.*, 49, pp. 4374–4400.

Misra, V. (2003). The influence of Pacific SST variability on the precipitation over southern Africa, *J. Clim.*, 16, pp. 2408–2418.

Molteni, F., King, M. P., Kucharski, F., *et al.* (2011). Planetary-scale variability in the northern winter and the impact of land–sea thermal contrast, *Clim. Dyn.*, 37, pp. 151–170.

Monin, A. S. (1986). *An Introduction to the Theory of Climate,* (D. Reidel Publishing Company, Dordrecht).

Mori, M., Watanabe, M., Shiogama, H., *et al.* (2014). Robust Arctic sea-ice influence on the frequent Eurasian cold winters in past decades, *Nat. Geosci.*, 7, pp. 869–873.

Morice, C. P., Kennedy, J. J., Rayner, N. A., *et al.* (2012). Quantifying uncertainties in global and regional temperature change using an ensemble of observational estimates: The HadCRUT4 dataset, *J. Geophys. Res.*, 117.

Müller, C., Abbasi, M. K., Kammann, C., *et al.* (2004). Soil respiratory quotient determined via barometric process separation combined with nitrogen-15 labeling, *Soil Sci. Soc. Am. J.*, 68, pp. 1610–1615.

Müller Schmied, H., Cáceres, D., Eisner, S., *et al.* (2021). The global water resources and use model WaterGAP v2.2d: Model description and evaluation, *Geosci. Model Dev.*, 14, pp. 1037–1079.

Namias, J. (1950). The index cycle and its role in the general circulation, *J. Meteor.*, 7, pp. 130–139.

Nazemosadat, M. J. and Ghasemi, A. R. (2004). Quantifying the ENSO-related shifts in the intensity and probability of drought and wet periods in Iran, *J. Clim.*, 17, pp. 4005–4018.

Newbold, T. (2018). Future effects of climate and land-use change on terrestrial vertebrate community diversity under different scenarios, *Proc. R. Soc. Lond. B.*, 285.

Newbold, T., Hudson, L. N., Hill, S. L. L., *et al.* (2015). Global effects of land use on local terrestrial biodiversity, *Nature*, 520, pp. 45–50.

Newman, M., Alexander, M. A., Ault T. R., *et al.* (2016). The Pacific Decadal Oscillation, Revisited, *J. Clim.*, 29, pp. 4399–4427.

Nicholson, N. E. (2002). What are the key components of climate as a driver of desertification, In: Reynolds, J. F. and Stafford Smith, D. M. (eds.), *Global Desertification: Do Humans Cause Deserts?* (Dahlem University Press, Berlin).

Nicholson, S. E. and Kim, J. (1997). The relationship of the El Niño Southern oscillation to African rainfall, *Int. J. Climatol.*, 17, pp. 117–135.

Nicholson, S. E. (2011). *Dryland Climatology*, 1st edn., (Cambridge University Press, Cambridge).

Nicholson, S. (2013). The West African Sahel: A review of recent studies on the rainfall regime and its interannual variability, *ISRN. Meteorol.*, 2013.

Notz, D. and Stroeve, J. (2016). Observed Arctic sea — ice loss directly follows anthropogenic CO2 emission, *Science*, 354, pp. 747–750.

O'Gorman, P. A., Allan, R. P., Byrne, M. P., *et al.* (2012). Energetic constraints on precipitation under climate change, *Surv. Geophys.*, 33, pp. 585–608.

Okin, G. S., Baker, A. R., Tegen, I., *et al.* (2011). Impacts of atmospheric nutrient deposition on marine productivity: Roles of nitrogen, phosphorus, and iron, *Global Biogeochem. Cycles*, 25.

Olson, J. M., Alagarswamy, G., Andresen, J. A., *et al.* (2008). Integrating diverse methods to understand climate-land interactions in East Africa, *Geoforum*, 39, pp. 898–911.

Osmond, B., Ananyev, G., Berry, J., *et al.* (2004). Changing the way we think about global climate change research: Scaling up in ecosystem experimental science, *Global Change Biol.*, 10, pp. 393–407.

Overland, J. E., Wood, K. R. and Wang M. (2011). Warm Arctic-cold continents: Climate impacts of the newly open Arctic Sea, *Polar Res.*, 30.

Pan, J., Ji, L. and Bueh, C. (2009). Intraseasonal climate characteristics of the summertime persistent anomalous circulation over Eurasian middle and high latitudes, *Chinese J. Atmos. Sci.*, 33, pp. 300–312.

Panagiotopoulos, F., Shahgedanova, M., Hannachi, A., *et al.* (2005). Observed trends and teleconnections of the Siberian high: A recently declining center of action, *J. Clim.*, 18, pp. 1411–1422.

Parsons, A. N., Barrett, J. E., Wall, D. H., *et al.* (2004). Soil carbon dioxide flux in Antarctic dry valley ecosystems, *Ecosystems*, 7, pp. 286–295.

Paustian, K., Andrén, O., Clarholm, M., *et al.* (1990). Carbon and nitrogen budgets of four agro-ecosystems with annual and perennial crops, with and without N fertilization, *J. Appl. Ecol.*, 27, pp. 60–84.

Pendergrass, A. G. and Hartmann, D. L. (2014). The atmospheric energy constraint on global-mean precipitation change, *J. Clim.*, 27, pp. 757–768.

Peng, J., Wang, A., Liu, Y., *et al.* (2015). Assessing the atmospheric oxygen balance in a region of rapid urbanization: A case study in the pearl river delta, china, *Sustainability*, 7, pp. 13055–13072.

Peng, S. *et al.* (2013). Asymmetric effects of daytime and night-time warming on Northern Hemisphere vegetation, *Nature*, 501, pp. 88–92.

Perlwitz, J. and Miller, R. L. (2010). Cloud cover increase with increasing aerosol absorptivity: A counterexample to the conventional semidirect aerosol effect, *J. Geophys. Res.*, 115.

Petsch, S.T. (2013). *The global oxygen cycle*, In: Holland, Heinrich D. and Turekian, Karl K. (eds.), *Treatise on Geochemistry*, 2nd edn., (Elsevier, Oxford, London) pp. 437–473.

Piao, S., Friedlingstein, P., Ciais, P., *et al.* (2007). Growing season extension and its impact on terrestrial carbon cycle in the Northern Hemisphere over the past 2 decades, *Global Biogeochem. Cycles*, 21.

Piao, S., Fang, J., Ciais, P., *et al.* (2009). The Carbon balance of terrestrial ecosystems in China, *Nature*, 458, pp. 1009–1013.

Piao, S., Ciais, P., Huang, Y., *et al.* (2010). The impacts of climate change on water resources and agriculture in China, *Nature*, 467, pp. 43–51.

Piao, S., Sitch, S., Ciais, P., *et al.* (2013). Evaluation of terrestrial carbon cycle models for their response to climate variability and to CO2 trends, *Global Change Biol.*, 19, pp. 2117–2132.

Piao, S., Wang, X., Park, T., *et al.* (2020). Characteristics, drivers and feedbacks of global greening, *Nat. Rev. Earth Environ.*, 1, pp. 14–27.

Pimm, S. L., Jenkins, C. N., Abell, R., *et al.* (2014). The biodiversity of species and their rates of extinction, distribution, and protection, *Science*, 344.

Plumb, R. A. and Hou, A. (1992). The response of a zonally symmetric atmosphere to subtropical thermal forcing: Threshold behavior, *J. Atmos. Sci.*, 49, pp. 1790–1799.

Ponce-Campos, G., M. Moran, Huete, A. *et al.* (2013). Ecosystem resilience despite large-scale altered hydroclimatic conditions, *Nature*, 494, pp. 349–352.

Poulsen, C. J., Tabor, C., White, J. D., (2015). Long-term climate forcing by atmospheric oxygen concentrations, *Science*, 348, pp. 1238–1241.

Poulter, B., Frank, D., Ciais, P., *et al.* (2014). Contribution of semi-arid ecosystems to interannual variability of the global carbon cycle, *Nature*, 509, pp. 600–603.

Power, S., Casey, T., Folland, C., *et al.* (1999). Inter-decadal modulation of the impact of ENSO on Australia, *Clim. Dyn.*, 15, pp. 319–324.

Pravalie, R., Savulescu, I., Patriche, C., *et al.* (2017). Spatial assessment of land degradation sensitive areas in southwestern Romania using modified MEDALUS method, *Catena*, 153, pp. 114–130.

Pu, W., Wang, X., Wei, H., *et al.* (2017). Properties of black carbon and other insoluble light-absorbing particles in seasonal snow of northwestern China, *The Cryosphere*, 11, pp. 1213–1233.

Qi, J., Chen, J., Wan, S., *et al.* (2012). Understanding the coupled natural and human systems in dryland East Asia, *Environ. Res. Lett.*, 7.

Qian, L., Cui, H. and Chang, J. (2006). Impacts of land use and cover change on land surface temperature in the Zhujiang Delta, *Pedosphere*, 16, pp. 681–689.

Qian, Y., Ruby Leung, L., Ghan, S. J., *et al.* (2003). Regional climate effects of aerosols over China: Modeling and observation, *Tellus, Ser. B*, 55, pp. 914–934.

Qian, Y., Gong, D., Fan, J., *et al.* (2009). Heavy pollution suppresses light rain in China: Observations and modeling, *J. Geophys. Res.*, 114.

Qian, Y., Flanner, M. G., Leung, L. R., *et al.* (2011). Sensitivity studies on the impacts of Tibetan Plateau snowpack pollution on the Asian hydrological cycle and monsoon climate, *Atmos. Chem. Phys.*, 11, pp. 1929–1948.

Qin, D., Ding, Y., Xiao, C., *et al.* (2018). Cryospheric science: Research framework and disciplinary system, *Natl. Sci. Rev.*, 5, pp. 255–268.

Qin, Z., Lin, Z., Chen, H., *et al.* (2011). The bias correction methods based on the EOF/SVD for short-term climate prediction and their applications, *Acta. Meteorol. Sin.*, 69, pp. 289–296.

Qu, X. and Huang, G. (2012). An enhanced influence of tropical Indian Ocean on the south Asia high after the late 1970s, *J. Clim.*, 25, pp. 6930–6941.

Ramanathan, V., Ramana, M. V., Roberts, G., *et al.* (2007). Warming trends in Asia amplified by brown cloud solar absorption, *Nature*, 448, pp. 575–578.

Rampone, S. and Valente, A. (2019). Assessment of desertification vulnerability using soft computing methods, *J. Amb. Intel. Hum. Comp.*, 10, pp. 701–707.

Reason, C. J. C. and Rouault, M. (2002). ENSO-like decadal variability and South African rainfall, *Geophys. Res. Lett.*, 29, pp. 16-1–16-4.

Reich, P. B., Sendall, K. M., Stefanski, A., *et al.* (2016). Boreal and temperate trees show strong acclimation of respiration to warming, *Nature*, 531, pp. 633–636.

Reichstein, M., Falge, E., Baldocchi, D., *et al.* (2005). On the separation of net ecosystem exchange into assimilation and ecosystem respiration: Review and improved algorithm, *Glob. Change Biol.*, 11, pp. 1424–1439.

Reichstein, M., Ciais, P., Papale, D., *et al.* (2007). Reduction of ecosystem productivity and respiration during the European summer 2003 climate anomaly: A joint flux tower, remote sensing and modelling analysis, *Global Change Biol.*, 13, pp. 634–651.

Reinhold, B. B. and Pierrehumbert, R. T. (1982). Dynamics of weather regimes: Quasi-stationary waves and blocking, *Mon. Wea. Rev.*, 110, pp. 1105–1145.

Reinhold, B. B. and Pierrehumbert, R. T. (1985). Corrections to "Dynamics of weather regimes: Quasi-stationary waves and blocking", *Mon. Weather Rev.*, 113, pp. 2055–2056.

Ren, H., Zhang, P., Chou, J., *et al.* (2006). Large-scale low-frequency rainfall regimes and their transition modes in summertime over China, *Chin. Sci. Bull.*, 51, pp. 1355–1367.

Ren, Y., Yu, H., Liu, C., *et al.* (2022). Attribution of dry and wet climatic changes over Central Asia, *J. Clim.*, 35, pp. 1399–1421.

Reynolds, J. F., Smith, D. M. S., Lambin, E. F., *et al.* (2007). Global desertification: Building a science for dryland development, *Science*, 316, pp. 847–851

Riahi, K., Rao, S., Krey, V., *et al.* (2011). RCP8.5 — A scenario of comparatively high greenhouse gas emissions, *Clim. Change*, 109, pp. 33.

Ricchiazzi, P., Yang, S., Gautier, C., *et al.* (1998). SBDART: A research and teaching software tool for plane — parallel Radiative transfer in the Earth's atmosphere, *Bull. Am. Meteorol. Soc.*, 79, pp. 2101–2114.

Rockström, J., Steffen, W. L., Noone, K., *et al.* (2009a). Planetary boundaries: Exploring the safe operating space for humanity, *Ecol. Soc.*, 14.

Rockström, J., Steffen, W. L., Noone, K., *et al.* (2009b). A safe operating space for humanity, *Nature*, 461, pp. 472–475.

Rodeghiero, M. and Cescatti, A. (2005). Main determinants of forest soil respiration along an elevation/temperature gradient in the Italian Alps, *Global Change Biol.*, 11, pp. 1024–1041.

Rodell, M., Houser, P. R., Jambor, U., *et al.* (2004). The global land data assimilation system, *Bull. Am. Meteorol. Soc.*, 85, pp. 381–394.

Rodell, M., Famiglietti, J. S., Wiese, D. N., *et al.* (2018). Emerging trends in global freshwater availability, *Nature*, 557, pp. 651–659.

Rodrigues, A. S. L., Brooks, T. M., Butchart, S. H. M., *et al.* (2014). Spatially Explicit Trends in the Global Conservation Status of Vertebrates, *PLoS ONE*, 9.

Rodwell, M. J., Rowell, D. P. and Folland, C. K. (1999). Oceanic forcing of the wintertime North Atlantic Oscillation and European climate, *Nature*, 398, pp. 320–323.

Rondinini, C. and Visconti, P. (2015). Scenarios of large mammal loss in Europe for the 21st century, *Conserv. Biol.*, 29, pp. 1028–1036.

Ropelewski, C. F. and Halpert, M. S., (1987). Global and regional scale precipitation patterns associated with the El Niño/Southern oscillation, *Mon. Weather. Rev.*, 115, pp. 1606–1626.

Rosenfeld, D., Lohmann, U., Raga, G. B., *et al.* (2008). Flood or drought: How do aerosol affect precipitation, *Science*, 321, pp. 1309–1313.

Rosenfeld, D., Yu, X., Liu, G., *et al.* (2011). Glaciation temperatures of convective clouds ingesting desert dust, air pollution and smoke from forest fires, *Geophys. Res. Lett.*, 38.

Rossby, C. G. (1939). Relation between variations in the intensity of the zonal circulation of the atmosphere and the displacements of the semi-permanent centers of action, *J. Marine Res.*, 2, pp. 38–55.

Rotenberg, E. and Yakir, D. (2010). Contribution of semi-arid forests to the climate system, *Science*, 327, pp. 451–454.

Running, S. W., Nemani, R. R., Heinsch, F. A., *et al.* (2004). A continuous satellite-derived measure of globalterrestrial primary production, *BioScience*, 54, pp. 547–560.

Ruprich-Robert, Y., Delworth, T., Msadek, R., *et al.* (2018). Impacts of the Atlantic multidecadal variability on North American summer climate and heat waves, *J. Clim.*, 31, pp. 3679–3700.

Sakaeda, N., Wood, R. and Rasch, P. J. (2011). Direct and semi-direct aerosol effects of southern African biomass burning aerosol, *J. Geophys. Res.*, 116.

Sanderson, B. M., Yu, Y., Tebaldi, C., *et al.* (2017). Community climate simulations to assess avoided impacts in 1.5°C and 2°C futures. *Earth System Dyn.*, 8, pp. 827–847.

Santanello Jr., J. A., Dirmeyer, P. A., Ferguson, C. R., *et al.* (2018). Land–atmosphere interactions: The LoCo perspective, *Bull. Am. Meteorol. Soc.*, 99, pp. 1253–1272.

Sarangi, C., Qian, Y., Rittger, K., *et al.* (2019). Impact of light-absorbing particles on snow albedo darkening and associated radiative forcing over high-mountain Asia: High-resolution WRF-Chem modeling and new satellite observations, *Atmos. Chem. Phys.*, 19, pp. 7105–7128.

Sargent, M., Barrera, Y., Nehrkorn, T., *et al.* (2018). Anthropogenic and biogenic CO2 fluxes in the Boston urban region, *Proc. Natl. Acad. Sci.*, 115, pp. 7491–7496.

Sassen, K. (2002). Indirect climate forcing over the Western US from Asian dust storms, *Geophys. Res. Lett.*, 29, pp. 103-1–103-4.

Scheff, J. and Frierson, D. M. W. (2015). Terrestrial aridity and its response to greenhouse warming across CMIP5 climate models, *J. Clim.*, 28, pp. 5583–5600.

Scheff, J. (2018). Drought indices, drought impacts, CO2, and warming: A historical and geologic perspective, *Curr. Clim. Change Rep.*, 4, pp. 202–209.

Schimel, D., Pavlick, R., Fisher, J., *et al.* (2015). Observing terrestrial ecosystems and the carbon cycle from space, *Glob. Change Biol.*, 21, pp. 1762–1776.

Schlesinger, W. H. (2017). An evaluation of abiotic carbon sinks in deserts, *Global Change Biol.*, 23, pp. 25–27.

Schlesinger, W. H., Reynolds, J. F., Cunningham, G. L., *et al.* (1990). Biological feedbacks in global desertification, *Science*, 247, pp. 1043–1048.

Schleussner, C. F., Lissner, T. K., Fischer, E. M., *et al.* (2016). Differential climate impacts for policy-relevant limits to global warming: The case of 1.5°C and 2°C, *Earth Syst. Dynam.*, 7, pp. 327–351

Schmidtko, S., Stramma, L. and Visbeck, M. (2017). Decline in global oceanic oxygen content during the past five decades, *Nature*, 542, pp. 335–339.

Schott, F. A., Xie, S. and McCreary, J. P. (2009). Indian Ocean circulation and climate variability, *Rev. Geophys.*, 47.

Schumacher, D. L., Keune, J., Dirmeyer, P., *et al.* (2022). Drought self-propagation in drylands due to land–atmosphere feedbacks, *Nat. Geosci.*, 15, pp. 262–268.

Screen, J. A. and Simmonds, I. (2010). The central role of diminishing sea ice in recent Arctic temperature amplification, *Nature*, 464, pp. 1334–1337.

Screen, J. A. and Simmonds, I. (2014). Amplified mid-latitude planetary waves favour particular regional weather extremes, *Nat. Clim. Change*, 4, pp. 704–709.

Seager, R., Naik, N. and Vecchi, G. A. (2010a). Thermodynamic and dynamic mechanisms for large-scale changes in the hydrological cycle in response to global warming, *J. Clim.*, 23, pp. 4651–4668.

Seager, R., Kushnir, Y., Herweijer, C., *et al.* (2005). Modeling of tropical forcing of persistent droughts and pluvial over Western North America: 1856–2000, *J. Clim.*, 18, pp. 4065–4088.

Seager, R., Naik, N., Baethgen, W., *et al.* (2010b). Tropical oceanic causes of interannual to multidecadal precipitation variability in Southeast South America over the past century, *J. Clim.*, 23, pp. 5517–5539.

Seinfeld, J. H., Carmichael, G. R., Arimoto, R., *et al.* (2004). ACE-ASIA: Regional climatic and atmospheric chemical effects of Asian dust and pollution, *Bull. Am. Meteorol. Soc.*, 85, pp. 367–380.

Seneviratne, S. I., Corti, T., Davin, E. L., *et al.* (2010). Investigating soil moisture–climate interactions in a changing climate: A review, *Earth Sci. Rev.*, 99, pp. 125–161.

Seneviratne, S. I., Donat, M., Mueller, B., *et al.* (2014). No pause in the increase of hot temperature extremes, *Nat. Clim. Change*, 4, pp. 161–163.

Seneviratne, S. I., Donat, M. G., Pitman, A. J., *et al.* (2016). Allowable CO2 emissions based on regional and impact-related climate targets, *Nature*, 529, pp. 477–483.

Severinghaus, J. P., Broecker, W. S., Dempster, W. F., *et al.* (1994). Oxygen loss in biosphere 2, *Eos Trans. AGU*, 75, pp. 33–37.

Severinghaus, J. P. (1995). *Studies of the Terrestrial O_2 and Carbon Cycles in Sand Dune Gases and in Biosphere*, 2nd edn., (Columbia University, New York).

Shabbar, A., Huang, J. and Higuchi, K. (2001). The relationship between the wintertime North Atlantic oscillation and blocking episodes in the north Atlantic, *Int. J. Climatol.*, 21, pp. 355–369.

Sheffield, J., Wood, E. F. and Roderick, M. L. (2012). Little change in global drought over the past 60 years, *Nature*, 491, pp. 435–438.

Sherwood, S. and Fu, Q. (2014). A drier future, *Science*, 343, pp. 737–739.

Shi, T., Pu, W., Zhou, Y., *et al.* (2020). Albedo of black carbon-contaminated snow across Northwestern China and the validation with model simulation, *J. Geophys. Res. Atmos.*, 125.

Shin, S. H., Chung, I. U. and Kim, H. J. (2012). Relationship between the expansion of drylands and the intensification of Hadley circulation during the late twentieth century, *Meteorol. Atmos. Phys.*, 118, pp. 117–128.

Shugart, H., Sedjo, R. and Sohngen, B. (2003). *Forests & Global Climate Change: Potential Impacts on U.S. Forest Resources*, (Pew Center on Global Climate Change), Arlington.

Shuman, J. K., Shugart, H. H. and O'halloran, T. L. (2011). Sensitivity of Siberian larch forests to climate change, *Global Change Bio.*, 17, pp. 2370–2384.

Sillmann, J., Kharin, V. V., Zhang, X., *et al.* (2013). Climate extremes indices in the CMIP5 multimodel ensemble: Part 2. Future climate projections, *J. Geophys. Res.-Atmos.*, 118, pp. 2473–2493.

Sjögersten, S. and Wookey, P. A. (2016). The impact of climate change on ecosystem carbon dynamics at the Scandinavian mountain birch forest-tundra heath ecotone, *Ambio*, 38, pp. 2–10.

Sjöström, M., Zhao, M., Archibald, S., *et al.* (2013). Evaluation of MODIS gross primary productivity for Africa using eddy covariance data, *Remote Sens. Environ.*, 131, pp. 275–286.

Smagorinsky, J. (1953). The dynamical influence of large-scale heat sources and sinks on the quasi-stationary mean motions of the atmosphere, *Q. J. R. Meteorol. Soc.*, 79, pp. 342–366.

Smith, R. B. (1979). The influence of mountains on the atmosphere, *Adv. Geophys.*, 21, pp. 87–230.

Smith, W. K., Reed, S. C., Cleveland, C. C., *et al.* (2015). Large divergence of satellite and earth system model estimates of global terrestrial CO2 fertilization, *Nat. Clim. Change*, 6, pp. 306–310.

Smoliak, B. V., Wallace, J. M., Lin, P., *et al.* (2015). Dynamical adjustment of the northern hemisphere surface air temperature filed: Methodology and application to observations, *J. Clim. Appl. Meteorol.*, 28, pp. 1613–1629.

Sohn, L. B. (1973). The Stockholm declaration on the human environment, *Harvard Int. Law J.*, 14, pp. 423–515.

Song, X., Hansen, M., Stehman, S., *et al.* (2018). Global land change from 1982 to 2016, *Nature*, 560, pp. 639–643.

Spinoni, J., Vogt, J., Naumann, G., *et al.* (2015). Towards identifying areas at climatological risk of desertification using the Köppen–Geiger classification and FAO aridity index, *Int. J. Climatol.*, 35, pp. 2210–2222.

Stefana, E., Marciano, F., Cocca, P., *et al.* (2015). Predictive models to assess Oxygen Deficiency Hazard (ODH): A systematic review, *Saf. Sci.*, 75, pp. 1–14.

Steinman, B. A., Mann, M. E. and Miller, S. K. (2015). Atlantic and Pacific multidecadal oscillations and Northern Hemisphere temperatures, *Science*, 347, pp. 988–991.

Stephens, G. L. and Ellis, T. D. (2008). Controls of global-mean precipitation increases in global warming GCM experiments, *J. Clim.*, 21, pp. 6141–6155.

Sterling, S. M., Ducharne, A. and Polcher, J. (2013). The impact of global land-cover change on the terrestrial water cycle, *Nat. Clim. Change*, 3, pp. 385–390.

Su, J., Huang J., Fu, Q., *et al.* (2008). Estimation of Asian dust aerosol effect on cloud radiation forcing using Fu-Liou radiative model and CERES measurements, *Atmos. Chem. Phys.*, 8, pp. 2763–2771.

Sui, Y., Lang, X. M. and Jiang, D. B. (2018). Projected signals in climate extremes over China associated with a 2°C global warming under two RCP scenarios, *Int. J. Climatol.*, 38, pp. E678–E697.

Sun, B., Li, H. X., and Zhou, B. T. (2019). Interdecadal variation of Indian Ocean basin mode and the impact on Asian summer climate, *Geophys. Res. Lett.*, 46, pp. 12388–12397.

Sun, C., Li, J., Ding, R., *et al.* (2017). Cold season Africa-Asia multidecadal teleconnection pattern and its relation to the Atlantic multidecadal variability, *Clim. Dyn.*, 48, pp. 3903–3918.

Sun, Z., Wang, X., Yamamoto, H., *et al.* (2018). An attempt to introduce atmospheric CO_2 concentration data to estimate the gross primary production by the terrestrial biosphere and analyze its effects, *Ecol. Indic.*, 84, pp. 218–234.

Sutera, A. (1986). Probability density distribution of large-scale atmospheric flow, *Adv. Geophys.*, 29, pp. 227–249.

Sutton, R. T. and Dong, B. (2012). Atlantic Ocean influence on a shift in European climate in the 1990s, *Nat. Geosci.*, 5, pp. 788–792.

Sutton, R. T. and Hodson, D. L. R. (2005). Atlantic Ocean forcing of North American and European summer climate, *Science*, 309, pp. 115–118.

Sutton, R. T., Dong, B. W. and Gregory, J. M. (2007). Land/sea warming ratio in response to climate change: IPCC AR4 model results and comparison with observations, *Geophys. Res. Lett.*, 34.

Svoboda, M., LeComte, D., Hayes, M., *et al.* (2002). The drought monitor, *Bull. Am. Meteorol. Soc.*, 83, pp.1181–1190.

Takemura, T., Okamoto, H., Maruyama, Y., *et al.* (2000). Global three-dimensional simulation of aerosol optical thickness distribution of various origins, *J. Geophys. Res.*, 105, pp. 17853–17873.

Takemura, T., Nozawa, T., Emori, S., *et al.* (2005). Simulation of climate response to aerosol direct and indirect effects with aerosol transport-radiation model, *J. Geophys. Res.*, 110.

Tang, X., Li, H., Liu, G., *et al.* (2015). Sensitivity of near real-time MODIS gross primary productivity in terrestrial forest based on eddy covariance measurements, *Chin. Geogra. Sci.*, 25, pp. 537–548.

Tao, W., Chen, J., Li, Z., *et al.* (2012). Impact of aerosols on convective clouds and precipitation, *Rev. Geophys.*, 50.

Tapley, B. D., Bettadpur, S., Ries, J. C., *et al.* (2004). GRACE measurements of mass variability in the Earth system, *Science*, 305, pp. 503–505.

Taylor, K. E., Stouffer, R. J. and Meehl, G. A. (2012). An overview of CMIP5 and the experiment design, *Bull. Am. Meteorol. Soc.*, 93, pp. 485–498.

Tegen, I. and Lacis, A. A. (1996). Modeling of particle size distribution and its influence on the radiative properties of mineral dust aerosol, *J. Geophys. Res.*, 101, pp. 19237–19244.

Teller, A. and Levin, Z. (2006). The effects of aerosols on precipitation and dimensions of subtropical clouds: A sensitivity study using a numerical cloud model, *Atmos. Chem. Phys.*, 6, pp. 67–80.

Thomson, A. M., Calvin, K. V., Smith, S. J., *et al.* (2011). RCP4.5: A pathway for stabilization of radiative forcing by 2100, *Clim. Change*, 109.

Thompson, D. W. J. and Wallace, J. M. (1998). The Arctic oscillation signature in the wintertime geopotential height and temperature fields, *J. Geophys. Res. Lett.*, 25, pp.1297–1300.

Tibaldi, S. and Molteni, F. (1990). On the operational predictability of blocking, *Tellus. A.*, 42, pp. 343–365.

Tilman, D., Clark, M., Williams, D. R., *et al.* (2017). Future threats to biodiversity and pathways to their prevention, *Nature*, 546, pp. 73–81.

Ting, M., Kushnir, Y., Seager, R., *et al.* (2009). Forced and internal twentieth-century SST trends in the North Atlantic, *J. Clim.*, 22, pp. 1469–1481.

Ting, M., Kushnir, Y., Seager, R., *et al.* (2011). Robust features of Atlantic multi-decadal variability and its climate impacts, *Geophys. Res. Lett.*, 38, pp. 351–365.

Tomasella, J., Silva Pinto Vieira, R. M., Barbosa, A. A., *et al.* (2018). Desertification trends in the Northeast of Brazil over the period 2000–2016, *Int. J. Appl. Earth Obs.*, 73, 197–206.

Tramontana, G., Jung, M., Schwalm, C., *et al.* (2016). Predicting carbon dioxide and energy fluxes across global FLUXNET sites with regression algorithms, *Biogeosciences*, 13, pp. 4291–4313.

Trenberth, K. E. and Hurrell, J. W. (1994). Decadal atmosphere–ocean variations in the Pacific, *Clim. Dyn.*, 9, pp. 303–319.

Trenberth, K. E., Stepaniak, D. P. and Caron, J. M. (2000). The global monsoon as seen through the divergent atmospheric circulation, *J. Clim.*, 13, pp. 3969–3993.

Trenberth, K. E., Branstator, G. W., Karoly, D., *et al.* (1998). Progress during TOGA in understanding and modeling global teleconnections associated with tropical sea surface temperatures, *J. Geophys. Res.*, 103, pp. 14291–14324.

Trenberth, K. E., Caron, J. M., Stepaniak, D. P., *et al.* (2002). Evolution of El Niño Southern Oscillation and global atmospheric surface temperatures, *J. Geophys. Res.*, 107, pp. AAC 5-11–AAC 5-17.

Trenberth, K. E., Dai, A., Schrier, G. V. D., *et al.* (2014a). Global warming and changes in drought, *Nat. Clim. Change*, 4, pp. 17–22.

Trenberth, K. E., Fasullo, J. T., Branstator, G., *et al.* (2014b), Seasonal aspects of the recent pause in surface warming, *Nat. Clim. Change*, 4, pp. 911–916.

Trenberth, K. E. (2015). Has there been a hiatus, *Science*, 349, pp. 691–692.

Trewartha, G. and Horn, L. (1980). *An Introduction to Climate*, 5th edn., (McGraw-Hill, New York).

Tung, K. K. and Zhou, J. S. (2013). Using data to attribute episodes of warming and cooling in instrumental records, *Proc. Natl. Acad. Sci.*, 110, pp. 2058–2063.

Turekian, K. K. and Holland, H. D. (2013). *Treatise on Geochemistry*, 2nd edn. Vol. 10.11. "The Global Oxygen Cycle". (Elsevier Science, Oxford), pp. 437–473.

Turner, D., Ritts, W., Cohen, W., *et al.* (2006). Evaluation of MODIS NPP and GPP products across multiple biomes, *Remote Sens. Environ.*, 102, pp. 282–292.

Twohy, C. H., Kreidenweis, S. M., Eidhammer, T., *et al.* (2009). Saharan dust particles nucleate droplets in eastern Atlantic clouds, *Geophys. Res. Lett.*, 36.

Twomey, S. (1977). The influence of pollution on the shortwave albedo of clouds, *J. Atmos. Sci.*, 34, pp. 1149–1152.

Ukkola, A. M., Prentice, I. C., Keenan, T. F., *et al.* (2015). Reduced streamflow in water-stressed climates consistent with CO2 effects on vegetation, *Nat. Clim. Change*, 6, pp. 75–78.

UNCCD. (1994). United Nations Convention to Combat Desertification in Countries Experiencing Serious Drought and/or Desertification, Particularly in Africa. A/AC.241/27, Paris.

Uno, I., Eguchi, K., Yumimoto, K., *et al.* (2009). Asian dust transported one full circuit around the globe, *Nat. Geosci.*, 2, pp. 557–560.

Valizadeh, J., Ziaei, S. M. and Mazloumzadeh, S. M. (2014). Assessing climate change impacts on wheat production (a case study), *J. Saudi Soc. Agric. Sci.*, 13, pp. 107–115.

Van der Werf, G. R., Randerson, J. T., Giglio, L., *et al.* (2017). Global fire emissions estimates during 1997–2016, *Earth Syst. Sci. Data*, 9, pp. 697–720.

Van Loon, A. F., Gleeson, T., Clark, J., *et al.* (2016). Drought in the anthropocene, *Nat. Geosci.*, 9, pp. 89–91.

Vannitsem, S., Demaeyer, J., Cruz, L. D., *et al.* (2015). Low-frequency variability and heat transport in a low-order nonlinear coupled ocean–atmosphere model, *Physica D*, 309, pp. 71–85.

Visconti, P., Bakkenes, M., Baisero, D., *et al.* (2016). Projecting global biodiversity indicators under future development scenarios, *Conserv. Lett.*, 9, pp. 5–13.

Vogel, M. M., Orth, R., Cheruy, F., *et al.* (2017). Regional amplification of projected changes in extreme temperatures strongly controlled by soil moisture-temperature feedbacks, *Geophys. Res. Lett.*, 44, pp. 1511–1519.

Wackernagel, M. and Rees, W. E. (1996). *Our Ecological Footprint: Reducing Human Impact on the Earth.* (New Society Publishers, British Columbia).

Walker, G. T. and Bliss, E. W. (1932). Quarterly journal of the royal meteorological society, *World. Weather. V.*, 4, pp. 53–84.

Wallace, J. M. and Gutzler, D. S. (1981). Teleconnections in the geopotential height field during the Northern Hemisphere winter, *Mon. Weather. Rev.*, 109, pp. 784–812.

Wallace, J. M., Zhang, Y. and Renwick, J. A. (1995). Dynamic contribution to hemispheric mean temperature trends, *Science*, 270, pp. 780–783.

Wallace, J. M., Zhang, Y. and Bajuk, L. (1996). Interpretation of interdecadal trends in Northern Hemisphere surface air temperature, *J. Clim.*, 9, pp. 249–259.

Wallace, J. M., Fu, Q., Smoliak, B. V., *et al.* (2012). Simulated versus observed patterns of warming over the extratropical Northern Hemisphere continents during the cold season, *Proc. Natl. Acad. Sci. USA*, 109, pp. 14337–14342.

Wang, B., and Fan, Z. (1999). Choice of South Asian summer monsoon indices, *Bull. Am. Meteorol. Soc.*, 80, pp. 629–638.

Wang, G., J. Huang, W. Guo., *et al.* (2010a). Observation analysis of land-atmosphere interactions over the Loess Plateau of northwest China, *J. Geophys. Res.*, 115.

Wang, J., Song, C., Reager, J. T., *et al.* (2018). Recent global decline in endorheic basin water storages, *Nat. Geosci.*, 11, pp. 926–932.

Wang, K. and Dickinson, R. E. (2013). Contribution of solar radiation to decadal temperature variability over land, *Proc. Natl. Acad. Sci.*, 110, pp. 14877–14882.

Wang, S., Huang. J., He, Y., *et al.* (2014). Combined effects of the Pacific Decadal Oscillation and El Niño-Southern Oscillation on Global Land Dry-Wet Changes, *Sci. Rep.*, 4.

Wang, K., Dickinson, R. E. and Liang, S. (2012). Global atmospheric evaporative demand over land from 1973 to 2008, *J. Clim.*, 25, pp. 8353–8361.

Wang, S., Meng, W., Jin, X., *et al.* (2015). Ecological security problems of the major key lakes in China, *Environ. Earth Sci.*, 74, pp. 3825–3837.

Wang, S., Zhang, Q., Yue, P., *et al.* (2020c). Effects of evapotranspiration and precipitation on dryness/wetness changes in China, *Theor. Appl. Climatol.*, 142, pp. 1027–1038.

Wang, S., He, Y., Hu, S., *et al.* (2022a). Enhanced warming in global dryland lakes and its drivers, *Remote Sens.-Basel.*, 14.

Wang, S., Huang, J., Huang, G., *et al.* (2022b). Enhanced impacts of Indian Ocean Sea surface temperature on the dry/wet variations over Northwest China, *J. Geophys. Res.*, 127.

Wang, T., Han, Y., Huang, J., *et al.* (2020a). Climatology of dust-forced radiative heating over the Tibetan Plateau and its surroundings, *J. Geophys. Res.*, 125.

Wang, T., Tang, J., Sun, M., *et al.* (2021). Identifying a transport mechanism of dust aerosols over South Asia to the Tibetan Plateau: A case study, *Sci. Total Environ.*, 758.

Wang, W., Huang, J., Minnis, P., *et al.* (2010b). Dusty cloud properties and radiative forcing over dust source and downwind regions derived from A-Train data during the Pacific Dust Experiment, *J. Geophys. Res.*, 115.

Wang, W., Huang, J., Zhou, T., *et al.* (2013). Estimation of radiative effect of a heavy dust storm over northwest China using Fu-Liou model and ground measurements, *J. Quant. Spectrosc. Radiat. Transfer*, 122, pp. 114–126.

Wang, X., Chen, F. and Dong, Z. (2006). The relative role of climatic and human factors in desertification in semiarid China, *Global Environ. Change*, 16, pp. 48–57.

Wang, X., Pu, W., Ren, Y., *et al.* (2017). Observations and model simulations of snow albedo reduction in seasonal snow due to insoluble light-absorbing particles during 2014 Chinese survey, *Atmos. Chem. Phys.*, 17, pp. 2279–2296.

Wang, X., Wei, H., Liu, J., *et al.* (2019a). Quantifying the light absorption and source attribution of insoluble light-absorbing particles on Tibetan Plateau glaciers between 2013 and 2015, *The Cryosphere*, 13, pp. 309–324.

Wang, X., Shi, T., Zhang, X., *et al.* (2020b). An overview of snow Albedo sensitivity to black carbon contamination and snow grain properties based on experimental datasets across the Northern Hemisphere, *Curr. Pollut. Rep.*, 6, pp. 368–379.

Wang, Y., Wang, J. and Yan, C. (2005). Data set of desert (sand) distribution in China with scale of 1:100000. Cold Arid Regions Sci. Data Center Lanzhou. https://doi.org/10.3972/westdc.006.2013.db.

Wang, Z., Letu, H., Shang, H., *et al.* (2019b). A supercooled water cloud detection algorithm using Himawari-8 satellite measurements, *J. Geophys. Res.*, 124, pp. 2724–2738.

Wardle, D. A. (2013). Ecology: Drivers of decoupling in drylands, *Nature*, 502, pp. 628–629.

Warren, A., Sud, Y. C. and Rozanov, B. (1996). The future of deserts, *J. Arid Environ.*, 32, pp. 75–89.

Wei, Y., Yu, H., Huang, J. *et al.* (2019) Drylands climate response to transient and stabilized 2°C and 1.5°C global warming targets. *Clim. Dyn.*, 53, pp. 2375–2389.

Wei, Y., Wu, J., Huang, J., *et al.* (2021a). Declining oxygen level as an emerging concern to global cities, *Envirson. Sci. Technol.*, 55, pp. 7808–7817.

Wei, Y., Yu, H., Huang, J., *et al.* (2021b). Improving China's summer precipitation prediction in 2020 by observational constrained bias correction, *Theor. Appl. Climatol.*, 145, pp. 1317–1331.

Wentz, F. J., Ricciardulli, L., Hilburn, K., *et al.* (2007). How much more rain will global warming bring, *Science*, 317, pp. 233–235.

Whittinghill, K. A., Currie, W. S., Zak, D. R., *et al.* (2012). Anthropogenic N deposition increases soil C storage by decreasing the extent of litter decay: Analysis of field observations with an ecosystem model, *Ecosystems*, 15, pp. 450–461.

Williams, D. R., Clark, M., Buchanan, G. M., *et al.* (2021). Proactive conservation to prevent habitat losses to agricultural expansion, *Nat. Sustain.*, 4, pp. 314–322.

Winslow, M., Shapiro, B. I., Thomas, R., *et al.* (2004). *Desertification, Drought, Poverty and Agriculture: Research Lessons and Opportunities.* (Joint publication of the International Center for Agricultural Research in the Dry Areas (ICARDA), the International Crops Research Institute for the Semi-Arid Tropics (ICRISAT), and the UNCCD Global Mechanism (GM), Aleppo, Syria; Patancheru, India; and Rome, Italy), p. 52.

Wishner, K. F., Seibel, B. A., Roman, C., *et al.* (2018). Ocean deoxygenation and zooplankton: Very small oxygen differences matter, *Sci. Adv.*, 4.

Woinarski, J. and Burbidge, A. A. (2016). Melomys rubicola. *The IUCN Red List of Threatened Species 2016*, e.T13132A195439637.

Woolway, R. I., Kraemer, B. M., Lenters, J. D., *et al.* (2020). Global lake responses to climate change, *Nat. Rev. Earth Env.*, 1, pp. 388–403.

Wu, G., Liu, Y., Zhang, Q., *et al.* (2007). The influence of mechanical and thermal forcing by the Tibetan Plateau on Asian climate, *J. Hydrometeorol.*, 8, pp. 770–789.

Wu, G., He, B., Liu, Y., *et al.* (2015). Location and variation of the summertime upper-troposphere temperature maximum over South Asia, *Clim. Dyn.*, 45, pp. 2757–2774.

Wu, M., Schurgers, G., Rummukainen, M., *et al.* (2016). Vegetation-climate feedbacks modulate rainfall patterns in Africa under future climate change, *Earth Syst. Dyn.*, 7, pp. 627–647.

Wu, Q. and Straus, D. M. (2004) AO, COWL, and observed climate trends, *J. Clim.*, 17, pp. 2139–2156.

Wu, Z., Huang, N., Wallace, J. M., *et al.* (2011). On the time-varying trend in global-mean surface temperature, *Clim. Dyn.*, 37, pp. 759–773.

WWF. (2020). The Living Planet Index database. Retrieved from http://www.livingplanetindex.org.

Xie, J., Li, Y., Zhai, C., *et al.* (2009). CO_2 absorption by alkaline soils and its implication to the global carbon cycle, *Environ. Geol.*, 56, pp. 953–961.

Xie, S., Annamalai, H., Schott, F. A., *et al.* (2002). Structure and mechanisms of south Indian Ocean climate variability, *J. Clim.*, 15, pp. 864–878.

Xie, Y., Huang, J. and Liu, Y. (2017). From accelerated warming to warming hiatus in China, *Int. J. Climatol.*, 37, pp. 1758–1773.

Xie, Y., Huang, J. and Ming, Y. (2019). Robust regional warming amplifications directly following the anthropogenic emission, *Earth's Future*, 7, pp. 363–369.

Xie, Y., Wu, G., Liu, Y., *et al.* (2020). Eurasian cooling linked with Arctic warming: Insights from PV dynamics, *J. Clim.*, 33, pp. 2627–2644.

Xie, Y., Wu, G., Liu, Y., *et al.* (2022a). A dynamic and thermodynamic coupling view of the linkages between Eurasian cooling and Arctic warming, *Clim. Dyn.*, 58, pp. 2725–2744.

Xie, Y., Nie, H. and He, Y. (2022b). Extratropical climate change during periods before and after an Arctic ice-free summer, *Earth's Future*, 10.

Xu, C., Ma, Y., Yang, K., *et al.* (2018). Tibetan Plateau impacts on global dust transport in the upper troposphere, *J. Clim.*, 31, pp. 4745–4756.

Xu, M. and Shang, H. (2016). Contribution of soil respiration to the global carbon equation, *J. Plant Physiol.*, 203, pp. 16–28.

Xu, P., Wang, L., Chen, W., (2019). The British–Baikal Corridor: A teleconnection pattern along the summertime polar front jet over Eurasia, *J. Clim.*, 32, pp. 877–896.

Xu, Y. (2014). *Climate Effects of Black Carbon and the Emission Reduction for Mitigating Climate Change*, (University of California, San Diego).

Xu, Y., Ramanathan, V. and Washington, W. M. (2016). Observed high-altitude warming and snow cover retreat over Tibet and the Himalayas enhanced by black carbon aerosols, *Atmos. Chem. Phys.*, 16, pp. 1303–1315.

Yang, F., Ali, M., Zheng, X., *et al.* (2017). Diurnal dynamics of soil respiration and the influencing factors for three land-cover types in the hinterland of the Taklimakan Desert China, *J. Arid Land*, 9, pp. 568–579.

Yang, F., Huang, J., He, Q., *et al.* (2020a). Impact of differences in soil temperature on the desert carbon sink, *Geoderma*, 379.

Yang, F., Huang, J., Zhou, C., *et al.* (2020b). Taklimakan desert carbon-sink decreases under climate change, *Sci. Bull.*, 65, pp. 431–433.

Yang, F., Huang, J., Zheng, X., *et al.* (2022). Evaluation of carbon sink in the Taklimakan Desert based on correction of abnormal negative CO_2 flux of IRGASON, *Sci. Total Environ.*, 838, 155988.

Yang, H., Shen, X., Yao, J., *et al.* (2020c). Portraying the impact of the Tibetan Plateau on global climate, *J. Clim.*, 33, pp. 3565–3583.

Yang, H. and Wen, Q. (2020). Investigating the role of the Tibetan Plateau in the formation of Atlantic meridional overturning circulation, *J. Clim.*, 33, pp. 3585–3601.

Yang, J., Liu, Q., Xie, S., *et al.* (2007). Impact of the Indian Ocean SST basin mode on the Asian summer monsoon, *Geophys. Res. Lett.*, 34.

Yang, J., Li, Y., Mu, J., *et al.* (2015). Analysis of relationship between sea surface temperature in tropical Indian Ocean and precipitation in East of Northwest China, *Plateau. Meteorology*, 34, pp. 690–699. (in Chinese).

Yang, X., Yang, F., Liu, X., *et al.* (2016a). Comparison of horizontal dust fluxes simulated with two dust emission schemes based on field experiments in Xinjiang China, *Theor. Appl. Climatol.*, 126, pp. 223–231.

Yang, X., Shen, S., Yang, F., *et al.* (2016b). Spatial and temporal variations of blowing dust events in the Taklimakan Desert, *Theor. Appl. Climatol.*, 125, pp. 669–677.

Yang, Z., Niu, G., Mitchell, K. E., *et al.* (2011). The community Noah land surface model with multiparameterization options (Noah-MP): 2. Evaluation over global river basins, *J. Geophys. Res.*, 116.

Yao, J., Liu, H., Huang, J., *et al.* (2020). Accelerated dryland expansion regulates future variability in dryland gross primary production, *Nat. Commun.*, 11.

Yeh, T. (1950). The circulation of the high troposphere over China in the winter of 1945–46, *Tellus*, 2, pp. 173–183.

Yeh, T.-C. and Chu, P. (1958). *Some Fundamental Problems of the General Circulation of the Atmosphere*, (Science Press, Beijing). (in Chinese).

Yin, D., Roderick, M. L., Leech, G., *et al.* (2014a). The contribution of reduction in evaporative cooling to higher surface air temperatures during drought, *Geophys. Res. Lett.*, 41, pp. 7891–7897.

Yin, K., Lu, D., Tian, Y., *et al.* (2014b). Evaluation of carbon and oxygen balances in urban ecosystems using land use/land cover and statistical data, *Sustainability*, 7, pp. 195–221.

Yin, Y. and Chen, L. (2007). The effects of heating by transported dust layers on cloud and precipitation: A numerical study, *Atmos. Chem. Phys.*, 7, pp. 3497–3505.

Yoden, S. (1983). Nonlinear interactions in a two-layer, quasi-geostrophic, low-order model with topography. Part I: Zonal flow-forced wave interactions, *J. Meteorol. Soc. Jpn.*, 61, pp. 1–38.

Yu, H., Remer, L. A., Chin, M., *et al.* (2008). A satellite-based assessment of transpacific transport of pollution aerosol, *J. Geophys. Res.*, 113.

Yu, H., Zhang, Q., Wei, Y., *et al.* (2022). Bias-corrections on aridity index simulations of climate models by observational constraints, *Int. J. Climatol.*, 42, pp. 889–907.

Yu, P., Han, D., Liu, S., *et al.* (2018). Soil quality assessment under different land uses in an alpine grassland, *Catena*, 171, pp. 280–287.

Yuan, T., Chen, S., Huang, J., *et al.* (2019). Influence of dynamic and thermal forcing on the meridional transport of Taklimakan Desert dust in spring and summer, *J. Clim.*, 32, pp. 749–767.

Yuan, T., Huang, J., Cao, J., *et al.* (2021). Indian dust-rain storm: Possible influences of dust ice nuclei on deep convective clouds, *Sci. Total Environ.*, 779.

Zeng, N., Neelin, J. D., Lau, K.-M., *et al.* (1999). Enhancement of interdecadal climate variability in the Sahel by vegetation interaction, *Science*, 286, pp. 1537–1540.

Zhai, P. and Zhou, Q. (1997). The change of northern hemisphere snow cover and its impact on summer rainfalls in China, *J. Appl. Meteor. Sci.*, 8, pp. 231–235 (in Chinese).

Zhang, D., Liu, D., Luo, T., *et al.* (2015). Aerosol impacts on cloud thermodynamic phase change over East Asia observed with CALIPSO and CloudSat measurements, *J. Geophys. Res.*, 120, pp. 1490–1501.

Zhang, G., Cai, M. and Hu, A. (2013). Energy consumption and the unexplained winter warming over Northern Asia and North America, *Nat. Clim. Change*, 3, pp. 466–470.

Zhang, G., Biradar, C. M., Xiao, X., *et al.* (2018a). Exacerbated grassland degradation and desertification in Central Asia during 2000–2014, *Ecol. Appl.*, 28, pp. 442–456.

Zhang, H., Ma, J. and Zheng, Y. (2010). Modeling study of the global distribution of radiative forcing by dust aerosol, *Acta Meteorol. Sin.*, 24, pp. 558–570.

Zhang, L. and Zhou, T. (2015). Drought over East Asia: A review, *J. Clim.*, 28, pp. 3375–3399.

Zhang, M., Yu, H., Huang, J., *et al.* (2019). Comparison of extreme temperature response to 0.5°C additional warming between dry and humid regions over East-central Asia, *Int. J. Climatol.*, 39, 3348–3364.

Zhang, Y., Wallace, J. M., and Battisti, D. S. (1997). ENSO-like interdecadal variability: 1900–93, *J. Clim.*, 10, pp. 1004–1020.

Zhang, Y., Guan, X., Yu, H., *et al.* (2017). Contributions of radiative factors to enhanced dryland warming over East Asia, *J. Geophys. Res.*, 122, pp. 7723–7736.

Zhang, Z., Sun, X. and Yang, X. (2018b). Understanding the interdecadal variability of East Asian summer monsoon precipitation: Joint influence of three Oceanic signals, *J. Clim.*, 31, pp. 5485–5506.

Zhao, C., Liu, X. and Leung, L. R. (2012). Impact of the desert dust on the summer monsoon system over Southwestern North America. *Atmos. Chem. Phys.*, 12, pp. 3717–3731.

Zhao, C., Chen, S., Ruby Leung, L., *et al.* (2013). Uncertainty in modeling dust mass balance and radiative forcing from size parameterization, *Atmos. Chem. Phys.*, 13, pp. 10733–10753.

Zhao, T. and Dai, A. (2017). Uncertainties in historical changes and future projections of drought, Part II: Model-simulated historical and future drought changes, *Clim. Change*, 144, pp. 535–548.

Zhao, Y. and Zhang, H. (2015). Impacts of SST warming in tropical Indian Ocean on CMIP5 model-projected summer rainfall changes over Central Asia, *Clim. Dyn.*, 46, pp. 3223–3238.

Zheng, F., Yuan, Y., Ding, Y., *et al.* (2021). The 2020/21 extremely cold winter in China influenced by the synergistic effect of La Niña and Warm Arctic, *Adv. Atmos. Sci.*, 39, pp. 546–552.

Zhou, C., Ali, M., Yang, F., *et al.* (2019a). Dust uplift potential in the Taklimakan Desert: An analysis based on different wind speed measurement intervals, *Theor. Appl. Climatol.*, 137, pp. 1449–1456.

Zhou, L., Dai, A., Dai, Y., *et al.* (2009). Spatial dependence of diurnal temperature range trends on precipitation from 1950 to 2004, *Clim. Dyn.*, 3, pp. 429–440.

Zhou, L., Chen, H., Hua, W., *et al.* (2016). Mechanisms for stronger warming over drier ecoregions observed since 1979, *Clim. Dyn.*, 47, pp. 2955–2974.

Zhou, S., Williams, A. P., Lintner, B. R., *et al.* (2021a). Soil moisture–atmosphere feedbacks mitigate declining water availability in drylands, *Nat. Clim. Change*, 11, pp. 38–44.

Zhou, T. and Zou, L. (2010). Understanding the predictability of East Asian summer monsoon from the reproduction of land–sea thermal contrast change in AMIP-type simulation, *J. Clim.*, 23, pp. 6009–6026.

Zhou, Y. and Wu, Z. (2016). Possible impacts of mega-El Niño/Southern Oscillation and Atlantic multidecadal oscillation on Eurasian heatwave frequency variability, *Q. J. R. Meteorol. Soc.*, 142, pp. 1647–1661.

Zhou, Y., Wen, H., Liu, J., *et al.* (2019b). The optical characteristics and sources of chromophoric dissolved organic matter (CDOM) in seasonal snow of northwestern China, *The Cryosphere*, 13, pp. 157–175.

Zhou, Y., West, C. P., Hettiyadura, A. P. S., *et al.* (2021b). Measurement report: Molecular composition, optical properties, and radiative effects of water-soluble organic carbon in snowpack samples from northern Xinjiang, China, *Atmos. Chem. Phys.*, 21, pp. 8531–8555.

Zhou, Y., West, C. P., Hettiyadura, A. P. S., *et al.* (2022). Molecular characterization of water-soluble brown carbon chromophores in snowpack from northern Xinjiang, *China. Environ. Sci. Technol.*, 56, pp. 4173–4186.

Zhu, K., Guan, X., Huang, J., *et al.* (2021). Precipitation over semi-arid regions of North Hemisphere affected by Atlantic Multidecadal Oscillation, *Atmos. Res.*, 262.

Zhu, Y. and Yang, X. (2003). Relationships between Pacific Decadal Oscillation (PDO) and climate variabilities in China, *Acta Meteorological Sinica.*, 61, pp. 641–654. (in Chinese).

Zhu, Z. and Zhu, B. (1982). Equilibrium states of ultra-long waves driven by non-adiabatic heating and blocking situation. *Scientia Sinica (Series B)*, 25, pp. 1201–1212.

Zhu, Z. (1985). Equilibrium states of planetary waves forced by topography and perturbation heating and blocking situation, *Adv. Atmos. Sci.*, 2, pp. 359–367.

Zhu, Z., Piao, S., Myneni, R. B., *et al.* (2016). Greening of the Earth and its drivers, *Nat. Clim. Change*, 6, pp. 791–795.

Index

F

G

H

I

J

K

L

M

N

O

P

R

www.ingramcontent.com/pod-product-compliance
Lightning Source LLC
LaVergne TN
LVHW081932200926
848484LV00009B/2406
9789811276170